I0131562

Multifunctional Inorganic Nanomaterials for Energy Applications

Multifunctional Inorganic Nanomaterials for Energy Applications provides deep insight into the role of multifunctional nanomaterials in the field of energy and power generation applications. It mainly focuses on the synthesis, fabrication, design, development, and optimization of novel functional inorganic nanomaterials for energy storage and saving devices. It also covers studies of inorganic electrode materials for supercapacitors, membranes for batteries and fuel cells, and materials for display systems and energy generation.

Features:

- Explores computational and experimental methods of preparing inorganic nanomaterials and their multifunctional applications.
- Includes synthesis and performance analysis of various functional nanomaterials for energy storage and energy saving applications.
- Reviews the current research directions and the latest developments in the field of energy materials.
- Discusses the importance of computational techniques in designing novel nanomaterials.
- Highlights the importance of multifunctional applications of nanomaterials in the energy sector.

This book is aimed at graduate students and researchers in materials science, electrical engineering, and nanomaterials.

Multifunctional Inorganic Nanomaterials for Energy Applications

Edited by
H.P. Nagaswarupa, Mika E.T. Sillanpää,
H.C. Ananda Murthy, and Ramachandra Naik

CRC Press
Taylor & Francis Group
Boca Raton London New York

CRC Press is an imprint of the
Taylor & Francis Group, an **informa** business

Designed cover image: Ramachandra Naik

First edition published 2024
by CRC Press
2385 NW Executive Center Drive, Suite 320, Boca Raton FL 33431

and by CRC Press
4 Park Square, Milton Park, Abingdon, Oxon, OX14 4RN

CRC Press is an imprint of Taylor & Francis Group, LLC

© 2024 selection and editorial matter, H.P. Nagaswarupa, Mika E.T. Sillanpää, H.C. Ananda Murthy and Ramachandra Naik; individual chapters, the contributors

Reasonable efforts have been made to publish reliable data and information, but the author and publisher cannot assume responsibility for the validity of all materials or the consequences of their use. The authors and publishers have attempted to trace the copyright holders of all material reproduced in this publication and apologize to copyright holders if permission to publish in this form has not been obtained. If any copyright material has not been acknowledged please write and let us know so we may rectify in any future reprint.

Except as permitted under U.S. Copyright Law, no part of this book may be reprinted, reproduced, transmitted, or utilized in any form by any electronic, mechanical, or other means, now known or hereafter invented, including photocopying, microfilming, and recording, or in any information storage or retrieval system, without written permission from the publishers.

For permission to photocopy or use material electronically from this work, access www.copyright.com or contact the Copyright Clearance Center, Inc. (CCC), 222 Rosewood Drive, Danvers, MA 01923, 978-750-8400. For works that are not available on CCC please contact mpkbookspermissions@tandf.co.uk

Trademark notice: Product or corporate names may be trademarks or registered trademarks and are used only for identification and explanation without intent to infringe.

ISBN: 9781032644189 (hbk)
ISBN: 9781032766072 (pbk)
ISBN: 9781003479239 (ebk)

DOI: 10.1201/9781003479239

Typeset in Times
by codeMantra

Contents

Preface

Welcome to *Multifunctional Inorganic Nanomaterials for Energy Applications*. In this comprehensive volume, we delve deep into the pivotal role of multifunctional nanomaterials in revolutionizing the landscape of energy and power generation applications.

The demand for sustainable energy solutions has never been more pressing, and the exploration of innovative materials is key to meeting this challenge. This book serves as a guiding light, focusing primarily on the synthesis, fabrication, design, development, and optimization of novel functional inorganic nanomaterials tailored specifically for energy storage and saving devices.

This book explores the realm of multifunctional inorganic nanomaterials for batteries, supercapacitors, and fuel cells. Through both computational and experimental methodologies, we unravel the synthesis techniques and performance analyses of these materials, shedding light on their promising applications in advancing energy storage technologies.

Further, we pivot to multifunctional inorganic nanomaterials for white light-emitting diodes (WLEDs) and optoelectronic devices. These materials hold immense potential in revolutionizing display systems and energy generation, and our exploration of their synthesis and performance is bound to inspire further innovation in this domain.

The importance of catalysis in energy conversion processes cannot be overstated, and this book delves into the multifunctional applications of inorganic nanomaterials in catalytic processes also. From enhancing efficiency to enabling new reaction pathways, these materials are at the forefront of catalytic innovation in the energy sector.

Finally, we delve into the design of novel inorganic nanomaterials using computational techniques. By leveraging the power of computational modeling and simulation, researchers can accelerate the discovery and optimization of materials with tailored properties, pushing the boundaries of what is achievable in energy applications.

This book is crafted for graduate students and researchers in materials science, electrical engineering, and nanomaterials, offering a comprehensive overview of the latest research directions and developments in the field of energy materials. We hope that the insights gleaned from these pages will inspire and empower future generations of innovators to continue pushing the boundaries of energy technology.

H.P. Nagaswarupa

Mika E.T. Sillanpää

H.C. Ananda Murthy

Ramachandra Naik

Contributors

Vinayak Adimule
Department of Chemistry
Angadi Institute of Technology and
    Management (AITM)
Belagavi, India

Javed Akhtar
Department of Chemistry
Gauhati University
Guwahati, India

Noora Al-Qahtani
Center for Advanced Materials (CAM)
Qatar University
Doha, Qatar

J. Anjaiah
Department of Physics
Geethanjali College of Engineering and
    Technology
Keesara, India

Arabinda Baruah
Department of Chemistry
Gauhati University
Guwahati, India

Praveen K. Bayannavar
Department of Studies in Chemistry
Karnatak University
Dharwad, India

Manoj Kumar Bharty
Departments of Chemistry, Institute of Science
Banaras Hindu University
Varanasi, India

K.N. Bhoomika
Department of Chemistry and Biochemistry,
    School of Sciences
Jain University
Bangalore, India

Rajender Boddula
Center for Advanced Materials (CAM)
Qatar University
Doha, Qatar

Soumya S. Bulla
Department of Studies in Physics
Davangere University
Davangere, India

Chetan Chavan
Department of Physics
V.V. Sangha's Kottureshwara Degree College
Kottur, India

H.A. Deepa
Department of Chemical Engineering
Dayananda Sagar College of Engineering
Bengaluru, India

H.M. Deepa
Department of Studies in Chemistry
Davangere University
Davangere, India

V.V. Deshmukh
Department of Studies in Chemistry
Shivagangotri-Davangere University
Davangere, India

S.V. Dhanyashree
Department of Studies in Chemistry
Davangere University
Davangere, India

K. Divyarani
Department of Studies and Research in
    Chemistry
Tumkur University
Tumakuru, India

B. Arifa Farzana
PG and Research Department of Chemistry
Jamal Mohamed College (Autonomous)
Tiruchirappalli, India

Vishwajit M. Gaikwad
Department of Physics
Amolakchand Mahavidyalaya
Yavatmal, India

M. Dilshad Begum Golgeri
Department of Life Sciences
Indian Academy Degree College (Autonomous)
Bangalore, India

Himangshu Prabal Goswami
Department of Chemistry
Gauhati University
Guwahati, India

H.V. Harini
Department of Studies in Chemistry
Davangere University
Davangere, India

J. Hemalatha
Department of Chemistry
Bharathiar University
Coimbatore, India

B.C. Indumukhi
Department of Studies in Chemistry
Shivagangotri-Davangere University
Davangere, India

S. Ishwarya
Department of Studies in Chemistry
Davangere University
Davangere, India

H.K. Jahnavi
Department of Studies in Chemistry
Davangere University
Davangere, India

Asha Kademane
Department of Life Science, School of Sciences
JC Road, Jain (Deemed-to-be) University
Bangalore, India

Anand S. Kakde
Department of Physics
Amolakchand Mahavidyalaya
Yavatmal, India

Trishna Kalita
Department of Chemistry
Gauhati University
Guwahati, India

Ravindra R. Kamble
Department of Studies in Chemistry
Karnatak University
Dharwad, India

M. Kanaka Durga
Department of Physics
Geethanjali College of Engineering and
    Technology
Keesara, India

Deepak R. Kasai
Department of Chemistry, Faculty of
    Engineering and Technology
Jain University
Bangalore, India

Ashlesha Kawale
Department of Chemistry
Guru Ghasidas Vishwavidyalaya Central
    University
Bilaspur, India

G. Krishnamurthy
Department of Studies in Chemistry
Bangalore University
Bangalore, India

Ajay B. Lad
Department of Physics
Amolakchand Mahavidyalaya
Yavatmal, India

G.M. Madhu
Department of Chemical Engineering
Ramaiah Institute of Technology
Bengaluru, India

S. Manjunatha
B.M.S. College of Engineering
Bangalore, India

Praveen Martis
Department of Chemistry
St. Aloysius College (Autonomous)
Mangalore, India

F.M. Mashood Ahamed
PG and Research Department of Chemistry
Jamal Mohamed College (Autonomous)
Tiruchirappalli, India

Soumya V. Menon
Department of Chemistry & Biochemistry,
    School of Sciences
Jain (Deemed-to-be) University
Bangalore, India

Jayanti Mishra
Department of Applied Sciences
New Horizon College of Engineering
Bangalore, India

A. Mushira Banu
PG and Research Department of Chemistry
Jamal Mohamed College (Autonomous)
Tiruchirappalli, Tamil Nadu, India

H.P. Nagaswarupa
Department of Studies in Chemistry
Davangere University
Davangere, India

G.H. Nagaveni
Department of Studies in Physics
Davangere University
Davangere, India

Ramachandra Naik
Department of Physics
New Horizon College of Engineering
Bangalore, India

S. Nandhabala
Department of Chemistry
Bharathiar University
Coimbatore, India

Yashwanth Narayan
Department of Life Science, School of Sciences
JC Road, Jain (Deemed-to-be) University
Bangalore, India

C.S. Naveen
Department of Physics, School of Engineering
Presidency University
Bengaluru, India

Manojna R. Nayak
Department of Studies in Chemistry
Karnatak University
Dharwad, India

G. Neeraja Rani
Department of Physics
Geethanjali College of Engineering and
    Technology
Keesara, India

K.M. Nikhileshwar
Department of Life Science, School of Sciences
JC Road, Jain (Deemed-to-be) University
Bangalore, India

G. Padma Priya
Department of Chemistry and Biochemistry
    School of Sciences
Jain University
Bangalore, India

L. Parashuram
Department of Chemistry
Nitte Meenakshi Institute of Technology
Bangalore, India

V.S. Patil
Department of Studies in Physics
Davangere University
Davangere, India

Nisha S. Pattanashetty
Department of Chemistry
G.M. Institute of Technology
Davangere, India.

Shivarudrappa Honnali Pattanashetty
Department of Chemistry
G.M. Institute of Technology
Davangere, India

Ramyakrishna Pothu
School of Physics and Electronics
College of Chemistry and Chemical Engineering
Hunan University
Changsha, China

I. Prabha
Department of Chemistry
Bharathiar University
Coimbatore, India

G.D. Prasanna
Department of Studies in Physics
Davangere University
Davangere, India

M.K. Prashanth
Department of Chemistry
BNM Institute of Technology
Bangalore, India

K. Preethi
Department of Chemistry
Bharathiar University
Coimbatore, India

Roshan Priyanshi
Department of Chemistry and Biochemistry,
    School of Sciences
Jain University
Bangalore, India

B. Pushpa
Department of Studies in Chemistry
Davangere University
Davangere, India

M. Pushpa
Department of Studies in Chemistry
Davangere University
Davangere, India

Ahmed Bahgat Radwan
Center for Advanced Materials (CAM)
Qatar University
Doha, Qatar

M.S. Raghu
Department of Chemistry
New Horizon College of Engineering
Bangalore, India

S. Rajendra Prasad
Department of Studies in Chemistry
Davangere University
Davangere, India

P. Raju
Department of Physics
Geethanjali College of Engineering and
    Technology
Keesara, India

R. Ravishankar
Department of Chemical Engineering
Dayananda Sagar College of Engineering
Bengaluru, India

Ashwini Rayar
Department of Studies in Physics
Davangere University
Davangere, India

Puli Mohith Vishnu Vardhan Reddy
Department of Chemistry and Biochemistry,
    School of Sciences
Jain University
Bangalore, India

Vinutha Reddy
Department of Chemistry and Biochemistry
    School of Sciences
Jain University
Bangalore, India

Kishor G. Rewatkar
Vidya Vikas Arts
    Commerce & Science College
Wardha, India

M. Rudresh
Department of Aeronautical Engineering
Dayananda Sagar College of Engineering
Bengaluru, India

Bullapura Matt Santhosh
Department of Chemistry
G. M. Institute of Technology
Davanagere, 577006
Karnataka, India

A.S. Santhosh Kumar
Department of Studies in Chemistry
Bangalore University
Bangalore, India

Manash Jyoti Sarmah
Department of Chemistry
Gauhati University
Guwahati, India

N.G.R.H.R. Senevirathne
Department of Chemistry & Biochemistry,
    School of Sciences
Jain (Deemed-to-be) University
Bangalore, India

C. Senthamil
Department of Chemistry
Bharathiar University
Coimbatore, India

H. Shanavaz
Department of Chemistry, Faculty of
    Engineering and Technology
Jain University
Bangalore, India

Nishant Shekhar
Department of Chemistry
Guru Ghasidas Vishwavidyalaya Central
    University
Bilaspur, India

S. Sreenivasa
Department of Studies and Research in
    Chemistry
Tumkur University
Tumakuru, India

Arti Srivastava
Department of Chemistry
Guru Ghasidas Vishwavidyalaya Central
    University
Bilaspur, India

G. Subbulakshmi
Department of Chemistry and Biochemistry,
    School of Sciences
Jain University
Bangalore, India

Vinayak Sunagar
Department of Studies in Chemistry
Davangere University
Davangere, India

Pravat K. Swain
Department of Chemistry
Dr. J N College
Balasore, India
Department of Chemistry
Berhampur Degree College
Balasore, India
Department of Basic Sciences and Humanities
Satyasai Engineering College (BPUT Rourkela)
Balasore, India

D.M. Tejashwini
Department of Studies in Chemistry
Shivagangotri-Davangere University
Davangere, India

Bitap Raj Thakuria
Department of Chemistry
Gauhati University
Guwahati, India

S. Thiyagaraj
PG Department of Physics and Electronics
    School of Sciences
Jain University
Bangalore, India

J.J. Umashankar
Department of Chemistry
Bharathiar University
Coimbatore, India

Basappa Chidananda Vasantha Kumar
Department of Chemistry
DRM Science College
Davangere, India

Jyothi Vaz
Department of Chemistry
St. Aloysius College (Autonomous)
Mangalore, India

C.J. Vijaykumar
Department of Studies in Physics
Davangere University
Davangere, India

R. Vijay Kumar
Department of Studies in Physics
Davangere University
Davangere, India

Anjana Vinod
Department of Chemistry
Nitte Meenakshi Institute of Technology
Bangalore, India

H.J. Amith Yadav
Department of Studies in Physics
Davangere University
Davangere, India

K. Yogesh Kumar
Department of Chemistry, Faculty of
 Engineering and Technology
Jain University
Bangalore, India

Editors

H.P. Nagaswarupa, PhD, is currently working as a Professor in the Department of Studies in Chemistry, Davangere University, India. He obtained his Master's degree in Chemistry from Bangalore University (2002) and Ph.D. degree from Bharathidasan University, Trichy (2012). He has been teaching chemistry at undergraduate and postgraduate levels for 20 years. He has guided six PhD and six MPhil students. He has published over 171 research papers in national/international journals and holds five patents. His papers are well cited with an h-index of 37. He has organized more than 50 workshops/conferences at national/international levels. He is also a member of various committees for the inspection of new and old institutions. His areas of interest include Nanoscience and Nanotechnology, Materials Science, and Corrosion Science.

Mika E.T. Sillanpää, PhD, is a Professor with over 20 years' experience of scholarly research, teaching, and supervision, and has engaged in synergistic collaboration with over 100 research partners from the world's leading laboratories in six continents. Currently, he is an Editor in *Inorganic Chemistry Communications* (Elsevier) and a Field Chief Editor in *Frontiers in Environmental Chemistry*. Having an h-index of 107, his publications have been cited over 55,000 times (Google Scholar). Two of his publications are among 0.5% top-cited publications in the history of the corresponding journal (*Journals Chemosphere and Science of the Total Environment*). Many of his scientific articles have been listed as the hottest papers in the journal or even the entire research field (chemical engineering and environmental science). Mika Sillanpää's research work centers on chemical treatment in environmental engineering and environmental monitoring and analysis. Sillanpää received his MSc (Eng.) and DSc (Eng.) degrees from Aalto University where he also completed an MBA degree in 2013. Since 2000, he has been a Full Professor/Adjunct Professor at the University of Oulu, the University of Eastern Finland, the LUT University, and the University of Eastern Finland. He has acted as a consultant in around ten companies ranging from global giants, such as Nokia, to SMEs. He has also served as a board member of numerous companies and has had R&D collaborating with over 80 companies in all sizes. He has successfully managed as a principal investigator substantial funding of over 30 million euros in highly competitive calls for research funding such as the EU Framework Programmes and Structural Funds, Business Finland, the Academy of Finland, Foundations, and industrial partners. He has supervised 60 PhDs and has been a reviewer in over 250 academic journals, many of which are top-ranked in their fields.

Mika Sillanpää has published more than 1,000 articles in peer-reviewed international journals, including; *Chemical Society Reviews, Advanced Materials, Environmental Science & Technology, Water Research, ACS Applied Materials and Interfaces, Applied Catalysis B: Environmental, Green Chemistry, Journal of Catalysis, Journal of Cleaner Production, Carbon, Journal of Hazardous Materials, Journal of Chromatography A, Environment International, Journal of Hazardous Materials, Bioresource Technology, Renewable Energy, Renewable and Sustainable Energy Reviews, Sustainable Energy & Fuels, Advances in Colloid and Interface Science, Electrochemistry Communications, Biosensors & Bioelectronics, Physical Chemistry Chemical Physics, Analytical Chemistry, Journal of Physical Chemistry C, Mass Spectrometry Reviews and Critical Reviews in Environmental Science & Technology, Desalination, Environmental Science: Nano, Separation and Purification Reviews, Environmental Pollution, Electrochimica Acta, Chemical Engineering Journal, Colloid Surfaces B – Biointerfaces, Langmuir, Trac-Trends in Analytical Chemistry, Ultrasonics Sonochemistry,* and *Coordination Chemistry Reviews* (all having impact factors >5).

H.C. Ananda Murthy (FRSC-UK) is currently working as Professor of Inorganic Chemistry in the Department of Applied Sciences, Papua New Guinea University of Technology, Lae, Morobe Province, 411, Papua New Guinea. He is also serving as an Adjunct Professor at Saveetha University, India. He has worked at various prestigious universities in India, Tanzania, and Ethiopia in the last 25 years. He has taught various chemistry courses to UG, PG, and PhD students of the universities.

Prof. Ananda has published more than 180 research/review articles in the journals of international repute (H-index = 27, i-10 index = 74). He has authored ten books, 22 book chapters, and six compendia. He has a total of five patents to his credit. He has carried out several projects and guided 16 UG, three PG, and three PhD students. At present, six PhD students are working under his supervision in the areas related to the synthesis of inorganic metal oxides–based nanomaterials for the biomedical and environmental applications.

He had successfully completed a project titled "Amphibious bicycle" sanctioned by National Innovation Foundation of India (NIF) in the year 2009. He had also worked as a researcher in association with Geological Survey of Tanzania, Dodoma, Tanzania. He had successfully completed a research project on green synthesis of silver and copper nanoparticles, granted by ASTU, Ministry of Science and Technology, Government of Ethiopia in the year 2021. He is currently working on three projects, one of which was sanctioned recently by Telecom Malaysia for the project on the development of electrochromic smart windows.

He is a member of ISTE, IRI, ECSI, CSI, NESA (India), CSE (Ethiopia), and ACS (USA) and a newly elected member of International Council on Materials Education, University of North Texas-USA.

Dr. H.C. Ananda Murthy has recently been elected as a Fellow of the Royal Society of Chemistry, London, UK (FRSC-UK) in the year 2022. He has been serving as a guest editor for *Journal of Nanomaterials*, *Evidence-Based Complementary & Alternative Medicine*, *Frontiers in Food Science and Technology*, and *Micro-machines*. He has received the best paper presentation award, Certificate of Excellence for community service, and Certificate of Excellence from Elsevier Researcher Academy.

He has recently received Excellent Researcher Award at Asia's Science, Technology, and Research Awards Congress held at Trichy, Tamil Nadu, India in July 2023.

His research interest mainly includes synthesis and applications of composite materials and nanomaterials for biomedical, sensor, and environmental applications.

Ramachandra Naik, PhD, is currently working as Senior Assistant Professor of Physics at New Horizon College of Engineering, Bangalore, India. He has published more than 61 research/review articles in the journals of international repute. He has authored six books and five book chapters. At present, two PhD students are working under his supervision in the areas related to the synthesis of multifunctional inorganic metal oxides–based nanomaterials for the energy applications. He has been serving as a guest editor for the special issue in *Results in Chemistry*.

Introduction

To achieve extremely efficient energy use, our society has been up against growingly difficult obstacles. Multifunctional nanomaterials have been gaining popularity in the area of energy applications. The electrical, thermal, mechanical, optical, and catalytic properties of materials are crucial for a variety of energy applications, including energy generation, conversion, storage, savings, and transmission. Thermoelectric, piezoelectric, triboelectric, photovoltaic, catalytic, and electrochromic properties of materials have significant impacts on a variety of energy applications at the nanoscale level. Due to their distinctive qualities, such as superior electrical and thermal conductivity, a large surface area, and chemical stability, inorganic nanoparticles are extremely competitive in the energy sector. The most recent findings and developments in multifunctional inorganic nanomaterials (NMs) for energy applications are compiled in this book from the standpoint of various energy applications. To increase their performances and the integration of individual functions of nanomaterials into a device, we also demonstrate the special functionalities of inorganic nanomaterials.

NMs have been generally considered and applied in different fields (e.g., drug conveyance, hardware, heat move, primary composites, and contamination anticipation) because of their fantastic properties, which are missing at the macroscale. In any case, customary examination strategies, including research facility studies and sub-atomic excitement, require a lot of chance to get the design and properties of NMs and track down new NMs with ideal capacities. Machine learning (ML), as an artificial intelligence method, can examine a lot of perplexing exploratory information with high effectiveness and decide the potential by overseeing rules. ML can help compound disclosure by separating synthetic information from informational indexes. The normal ML models incorporate Naïve Bayes (NB), support vector machine (SVM), decision tree (DT), random forest (RF), artificial neural network (ANN), logistic regression (LR), genetic algorithm (GA), k-nearest neighbor (KNN), and k-means models. Supervised, semisupervised, or unsupervised ML models (shallow or deep ML models) can be naturally improved by information preparing or past experience. Better than conventional density functional theory (DFT) investigation and molecular dynamics (MD) recreations as far as estimation speed and speculation capacity are concerned, the blend of NMs and ML will prompt incredible mechanical advancements in the fields of materials, science, and medication.

More than 10,000 papers per year associated to the structures, properties, adsorption properties, and catalytic behavior of NMs have been published in the Web of Science. Huge quantities of data are shaped every day, but big data are not sufficiently used for NMs. ML can facilitate the construction of quantitative structure–property relationship (QSPR) models, rapidly identifying promising materials for specific applications. With the rapid growth of ML related to NMs (from 14 papers in 2010 to 183 papers in 2019 based on the Web of Science), the number of low-cost or repeated studies has increased. In contrast, the knowledge gaps (e.g., accuracy and interpretability) remain large. To conserve scientific resources and promote the development of ML related to NMs, an investigation is urgently needed to summarize our accomplishments to date and identify what requires further attention in the future.

ML for investigation and design of NM structures: ML can not only support in the analysis of the structural defects of NMs, the design of NMs with ideal structures, and the detection of new NMs from structural databases but can also deliver vision into the design of protein structures in the field of biomimetic NMs. In the field of NMs, numerous structure descriptors have been developed; however, new controlling descriptors are urgently needed due to the limited applicability of the existing descriptors.

Understanding the relationships between structures and properties is critical to the theoretical research and application of NMs. Traditional trial-and-error experimentation is extensive and expensive. ML methods provide an accurate, fast, proven, and high-throughput alternative, promoting the discovery of new NMs with excellent electronic, thermal, and optical properties.

Research on optimization of multifunctional materials for energy applications has been carried out worldwide. Still, innovations on achieving materials optimization in terms of cost, energy saving and high efficiency simultaneously for energy applications are very much required.

1 Multifunctional Multiferroic Nanocomposites for Novel Device Applications

M. Kanaka Durga, G. Neeraja Rani, J. Anjaiah, and P. Raju

INTRODUCTION

The continuous quest for developing novel materials while designing new devices is a hot research area over the recent years. Various industrial sectors such as telecommunications, information technology, aerospace technology, surface mount devices, sensors, and EMI filters demand multifunctional characteristic features in a single material. This requirement has led to the development of new novel multiferroic nanocomposites with property interrelationships of matrix and filler phases [1–3].

Multiferroic nanocomposites exhibit a minimum of two ferroic orders, viz. ferroelectric, ferromagnetic, ferroelastic, or antiferromagnetic in a single material. More specifically, ferroelectric and ferromagnetic composites, owing to their excellent dielectric, magnetic properties and strong magnetoelectric coupling, fascinated many researchers in recent times [4–8]. Magnetic, dielectric, and electromagnetic responses of these multiferroics to external fields are the primary reason for these materials to be suitable for EMI filters or microwave absorbers in various sectors. This, in turn, depends on the filler concentration in the matrix, synthesis techniques and their operational conditions, etc. [7,8].

Though notable research has been done in this area over the past decades, it is still challenging to develop novel multifunctional materials fulfilling all requirements such as thin, lightweight, and broad bandwidth simultaneously, in addition to good absorption performance in microwave frequencies for shielding applications.

In the present context, the results of the work done on spinel $Mg_{0.48}Cu_{0.12}Zn_{0.4}Fe_2O_4$ (MCZ) ferrite and ferroelectric barium titanate phases have been considered. It has been proved that there exists excellent compatibility between two phases due to magnetoelectric coupling and the connectivity of the matrix and filler can be controlled by their relative compositions and synthesis methods [9].

This leads to the fact that material's dielectric and magnetic properties can be altered dramatically to tune to adjust the impedance matching, the resonant frequencies, quality factor and attenuation or amplification, etc. Elaborate work has been carried out on various ferrite–ferroelectric composites such as NCZ ferrite-$CaTiO_3$, NCZ ferrite-$BaTiO_3$, and NCMnZ ferrite-$BaTiO_3$ and their microwave absorption performance in X-band region [10–19].

It is concluded that MCZ ferrite possesses better electromagnetic properties than other ferrites due to high resistivity, low magnetostriction, high Curie temperature, and low cost [19].

In this chapter, we analyze the results of a series of MgCuZn ferrite and barium titanate with varying compositions, and study their dielectric, magnetic, and electromagnetic characteristics [7,8].

EXPERIMENTAL

The synthesis technique adopted in the current work was mechanical alloying, a physical method, for the preparation of MCZ ferrite +BT nanocomposites. The basic materials used for the synthesis

DOI: 10.1201/9781003479239-1

of nanopowders of ferrite and ferroelectric are more economical – both low cost and yield – for physical methods with those of chemical routes. Moreover, the other advantages are low-temperature processing and bulk production along with the simple preparation method. Various classes of high-energy ball milling devices are available based on their efficiency of grinding, capacity, operational conditions for heating, cooling, etc., which include SPEX Shaker Mills, Attritor Mills, and Planetary Ball Mills [20–22].

In the present context, a high-energy planetary ball mill, Retschco Planetary ball mill PM 100, was used to prepare MCZ powder and barium titanate nanopowders. The relative concentration of ZN, Cu, and Mg ions in $Mg_{0.48}Cu_{0.12}Zn_{0.4}Fe_2O_4$ was considered in the current work from the past [19].

Initially, $Mg_{0.48}Cu_{0.12}Zn_{0.4}Fe_2O_4$ (MCZ) and $BaTiO_3$ (BT) nanopowders were prepared separately by a mechanical alloying method, called high-energy planetary ball mill. The method of preparation of MCZ nanopowder is as follows: the starting materials Fe_2O_3, CuO, MgO, and ZnO, all with 99.8% purity from Aldrich, were used as received. The amounts of these powders were weighed and mixed in stoichiometric ratio and transferred to a hardened WC vial for milling along with 10, 12 mm WC balls. The ball mill operational parameters were set with a ball powder mass charge of 14:1 ratio and the speed of the mill as 400 rpm with 40-minute interval. Total grinding time was 20 h. The samples were taken for XRD, FTIR, and SEM characterizations at 5-h intervals of milling time for testing the ferrite phase formation.

The preparation of barium titanate nanopowder was carried out in a similar way, as follows: pure powders of BaO and TiO_2 were taken and mixed in stoichiometric ratio. The mixture was grounded in the ball mill. The samples were tested for perovskite phase formation using XRD, FTIR, and SEM characterization techniques at 5-h intervals. The operational settings were kept the same as for ferrite powder preparation.

The composites were prepared by mixing MCZ ferrite and BT nanopowders at different mol%, $(100-x)$ $Mg_{0.48}Cu_{0.12}Zn_{0.4}Fe_2O_4 + xBaTiO_3$ {where $x = [0$ (MCZ), 20 (MB1), 40 (MB2), 50 (MB3), 60 (MB4), 80 (MB5), 100 (BT)]}, and then milled for 40 h with the same operating conditions that were set as mentioned above. The nanocomposites of desired structure were achieved by carefully controlling the process parameters during milling of nanopowders.

The composite powders were made into pellets and toroids by a uniaxial pressing machine and then sintered for 2 h at 850°C in air for studying dielectric and magnetic characteristics. The concentration of constituent phase details and the sample names are given in Table 1.1 [7].

The complex permittivity and complex permeability studies of the prepared composite pellets and toroids were measured and analyzed using an LCR meter (Zentech 3305) and an impedance analyzer (Agilent 4291 B) at room temperature in two different steps: 100 kHz–1 MHz and 1 MHz–1.8 GHz. The EMI shielding properties were investigated by reflection and transmission methods using vector network analyzer (VNA, Agilent, 8722 ES) in X-band (8.2–12.4 GHz) frequency

TABLE 1.1

MCZBT Nanocomposite Compositions and Sample Names

S. No.	Sample Name	Composition
1	MCZ	$Mg_{0.48}Cu_{0.12}Zn_{0.4}Fe_2O_4$
2	MB1	80 mol% $Mg_{0.48}Cu_{0.12}Zn_{0.4}Fe_2O_4 + 20\%BaTiO_3$
3	MB2	60 mol% $Mg_{0.48}Cu_{0.12}Zn_{0.4}Fe_2O_4 + 40\%BaTiO_3$
4	MB3	50 mol% $Mg_{0.48}Cu_{0.12}Zn_{0.4}Fe_2O_4 + 50\%BaTiO_3$
5	MB4	40 mol% $Mg_{0.48}Cu_{0.12}Zn_{0.4}Fe_2O_4 + 60\%BaTiO_3$
6	MB5	20 mol% $Mg_{0.48}Cu_{0.12}Zn_{0.4}Fe_2O_4 + 80\%BaTiO_3$
7	BT	$BaTiO_3$

region. The powders were pressed into rectangular pellets of 11 mm×23 mm×3 mm measurements suitable to work as rectangular waveguide for EMI studies. The magnetic studies were performed by vibrating sample magnetometer (VSM: Lakeshore, Model 7404), and a ferroelectric test system was used to investigate ferroelectric hysteresis studies.

RESULTS AND DISCUSSION

XRD of sintered composites are shown in Figure 1.1. The composites display simultaneous existence of cubic spinel structure of ferrite phase and perovskite structure of ferroelectric BT phase with their respective characteristic peaks. All peaks are tallied with standard JCPDS reported values with no. 08-0234 for cubic spinel ferrite phase and card no. 50626 for perovskite BT phase. From Figure 1.1, it is noted that the height or intensity of major peaks of both the phases, namely, (311) for spinel and (110) for perovskite phase, depends on the relative concentrations of individual phase fractions.

The average lattice constant and theoretical density of each sample were calculated from XRD pattern. The bulk density was obtained for each sample using the Archimedes principle, and the average value of about 12 trials was calculated for each sample. From the theoretical and bulk

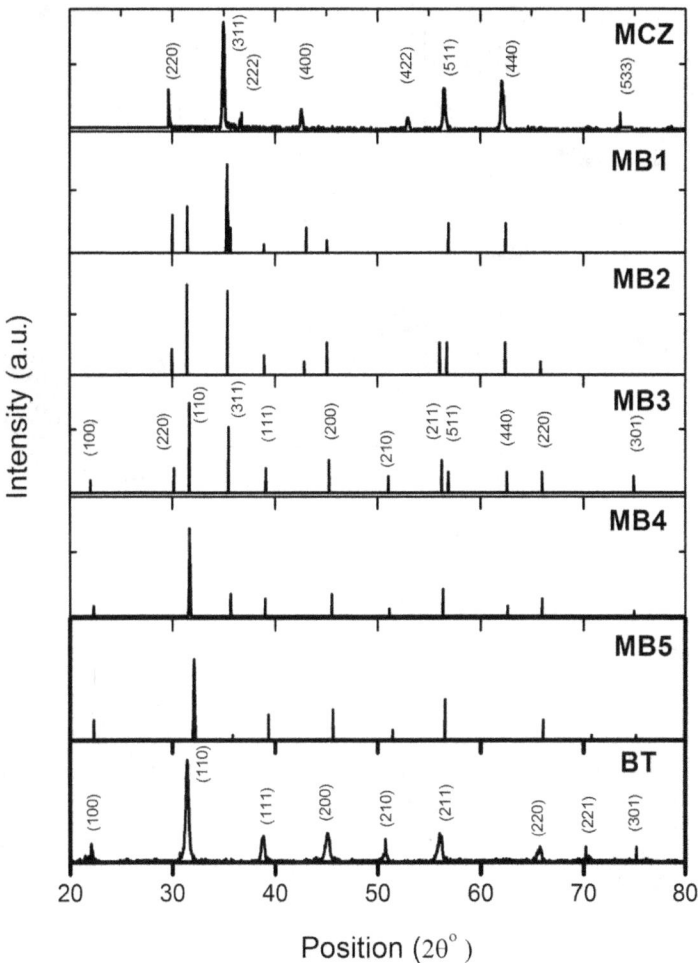

FIGURE 1.1 XRD of MCZBT nanocomposites.

density values, the porosity of the composite was estimated to be less than 10%, using the following formula:

$$\rho_{\text{porocity}} = \left(1 - \frac{d_{\text{bulk}}}{d_x}\right) \times 100$$

SEM and Edax of the nanocomposites with 20%, 50%, and 80% of BT phase, MB1, MB3, and MB5, are shown in Figure 1.2. The presence of ferrite and BT constituent elements in different proportions is confirmed from the energy dispersive X-ray spectroscopy and Edax figures. These pictures show that only the elements of ferrite and BT phases are present and no peak corresponding to other elements is seen in the composites. By using the line intercept method, the average grain size of the samples was determined to be in the range of 68–91 nm.

The frequency spectra of complex permittivity, both real and imaginary, have been analyzed in the range of 100 kHz–1.8 GHz. The real and imaginary permittivities were continued to be constant up to 1.18 GHz. This constant region was not shown in the plot for better clarity. On further increase

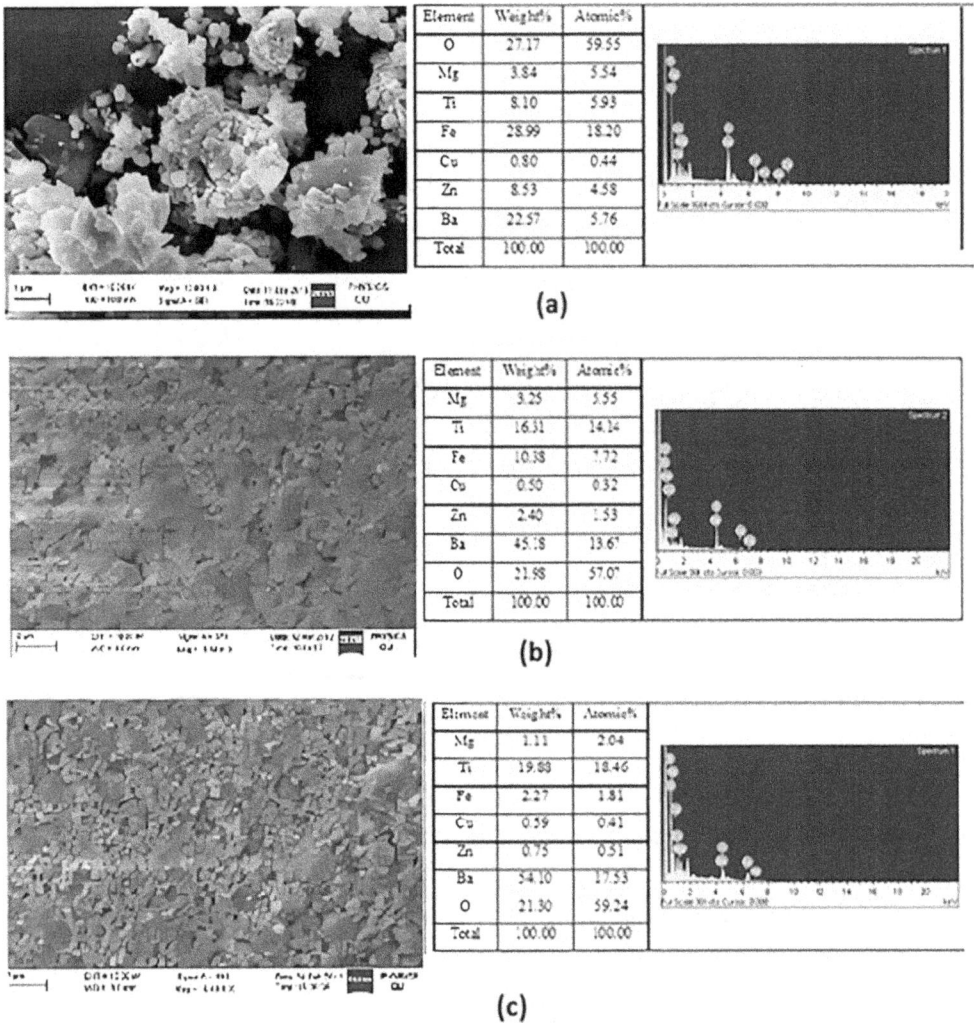

Element	Weight%	Atomic%
O	27.17	59.55
Mg	3.84	5.54
Ti	8.10	5.93
Fe	28.99	18.20
Cu	0.80	0.44
Zn	8.53	4.58
Ba	22.57	5.76
Total	100.00	100.00

(a)

Element	Weight%	Atomic%
Mg	3.25	5.55
Ti	16.31	14.14
Fe	10.38	7.72
Cu	0.50	0.32
Zn	2.40	1.53
Ba	45.18	13.67
O	21.98	57.07
Total	100.00	100.00

(b)

Element	Weight%	Atomic%
Mg	1.11	2.04
Ti	19.88	18.46
Fe	2.27	1.81
Cu	0.59	0.41
Zn	0.75	0.51
Ba	54.10	17.53
O	21.30	59.24
Total	100.00	100.00

(c)

FIGURE 1.2   SEM and Edax of (a) MB1, (b) MB3, and (c) MB5 samples.

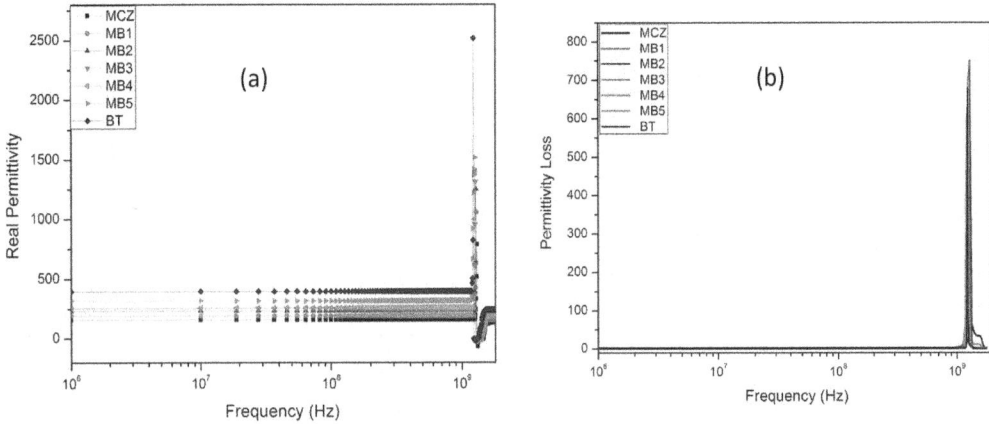

FIGURE 1.3 Frequency dependence of (a) real part and (b) imaginary part of permittivity of nanocomposites.

of frequency, ε' rises steeply and attains a maximum, which shows a resonance peak and then falls steeply to a very low value around 1.2 GHz. This feature is called relaxation, as can be noted from Figure 1.3(a). Figure 1.3(b) exhibits that ε'' increases and acquires a maximum at around the frequency of 1.25 GHz.

In general, imaginary part of permittivity, in the case of polycrystalline composites, results due to lag in polarization with respect to alternating applied electric field. This lag increases due to the presence of impurities and imperfections in the composite material, thereby increasing the dielectric loss. A major loss in the present composite samples comes from the electron hopping between Fe^{2+} and Fe^{3+} ions of ferrite magnetic phase. The resonance peak obtained in the ε versus frequency and ε'' versus frequency curves around 1.2 GHz in the present samples may be due to the fact that hopping frequency of electrons between Fe^{2+} and Fe^{3+} ions of ferrite phase is equal to the applied field frequency where the maximum electrical energy is transferred to the oscillating ions and power loss shoots up, thereby resulting in resonance. As the applied field frequency increases above the resonance frequency, the sample system passes through an anti-phase condition, where it opposes the vibrations of the charges and the polarization falls to a very low value, called relaxation. Hence, relaxation arises from the inertia of the system of charges and restoring forces acting on the charges which oppose the force due to the applied electric field. As the frequency is raised further, complete relaxation occurs and the polarization resumes a steady value in which the dielectric mechanism is no longer operative. This process gives the characteristic resonance shape to the permittivity curves of the present samples for ionic and electronic polarization mechanisms [23]. In the case of ferrite–ferroelectric nanocomposite materials, the electromagnetic interaction between ferroelectric and ferromagnetic phases influences the frequency and shape of the resonance peak. In the present nanocomposite materials, the electric field around ferrite grains will disturb and oppose the charged particles' motion in the ferroelectric phase. The equivalent damping for the charged particles' motion increases with the enhancement of magnetic phase, so the dielectric response changes from resonance to relaxation gradually. Since the frequency-dependent permittivity reflects the statistical average effect of nanoscale charged particles, the resonance peak turns flatter and shifts to a higher frequency with the rise of ferrite amount [29, 7]. The above explanation is taken from the work of Kanakadurga et al. [7].

The real component of complex permeability (μ') of the samples is shown in Figure 1.4(a). It was observed that the permeability (μ) for MB3 sample was around 462 at 100 kHz and continues to maintain constant up to 40 MHz, with an increase of frequency. On further increase of frequency, μ' is found to increase and shows a relaxation peak at around 150 MHz. A similar trend has also been seen for other samples as shown in Figure 1.4(a). In Figure 1.4 (μ''), for all samples, it is found to be

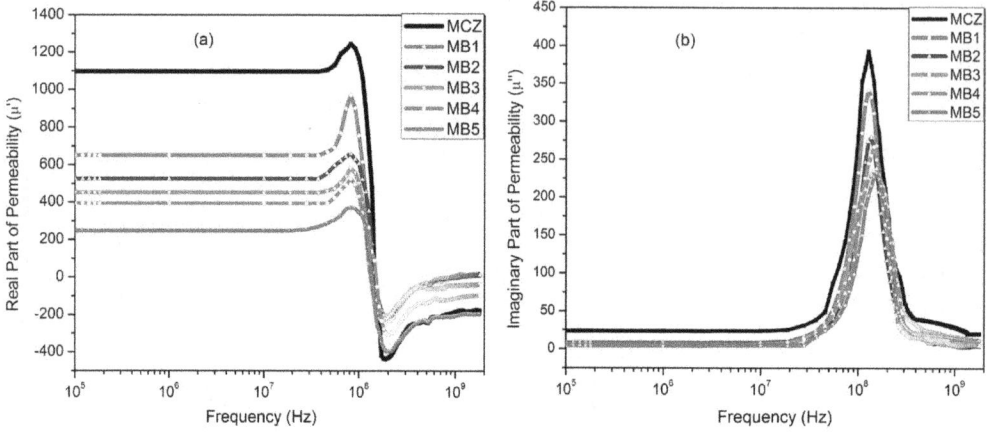

FIGURE 1.4   Frequency dependence of (a) real part and (b) imaginary part of permeability of nanocomposites.

constant on increase of frequency from 100 kHz to 40 MHz. Above 40 MHz, μ'' increases and shows a wide resonance near 150 MHz, where the real permeability rapidly decreases. This phenomenon is called domain wall relaxation for pure ferrites by Nakamura [24]. This behavior is similar to the earlier results of complex permeability of various composites reported by several investigators [23–25].

The electromagnetic properties of the composites have been investigated using vector network analyzer, and Figure 1.5 displays the reflection loss variation with frequency in X-band region. All the samples were made with an optimized thickness of 3 mm. The maximum absorption of microwaves can be interpreted as minimum reflection loss in Table 1.2, which gives the information about the minimum reflection loss, RL_{min}, bandwidth. The minimum reflection losses for MCZ and BT samples are found to be −14.84 dB at 9.37 GHz and −16.86 dB at 9.82 GHz, respectively. The composite samples exhibit the enhanced microwave-absorbing properties as can be seen from Figure 1.5 and Table 1.2. The bandwidth, the width of frequency in which the reflection loss is less than −10 dB, should be sufficiently broad for the material samples to be suitable for shielding applications. The current composites have the bandwidth above 0.3 GHz from 10.1 to 10.3 GHz, as given in Table 1.2.

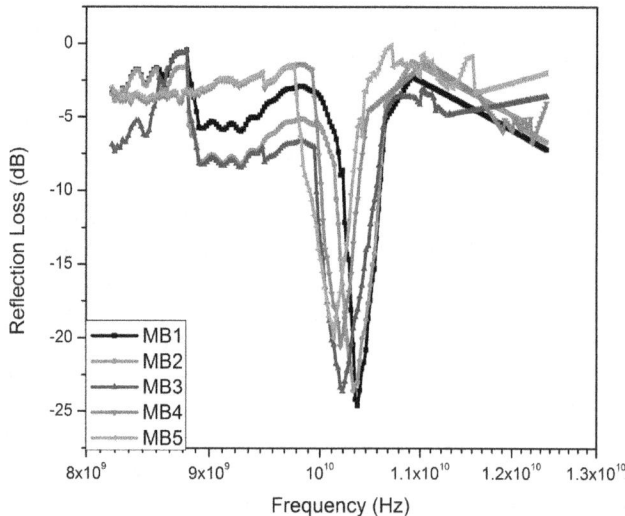

FIGURE 1.5   Frequency dependence of reflection loss of MCZBT nanocomposites.

TABLE 1.2

Minimum Reflection Loss of MCZBT Samples

S. No.	Sample Name	Minimum Reflection loss RL_{min} (dB)	Frequency at RL_{min} (GHz)	Bandwidth (GHz)	VSM Data M_s (emu/g)	H_c (Oe)	E_c kV/cm
1	MCZ	−14.84	9.37	0.4	21.18	11	−
2	MB1	−24.61	10.36	0.38	16.53	12	8.78
3	MB2	−23.59	10.32	0.42	12.33	14	6.16
4	MB3	−23.65	10.22	0.57	7.51	11	4.75
5	MB4	−20.56	10.19	0.34	3.79	10	2.56
6	MB5	−19.99	10.13	0.39	0.16	4	0.89
7	BT	−16.86	9.82	0.78	−	−	−

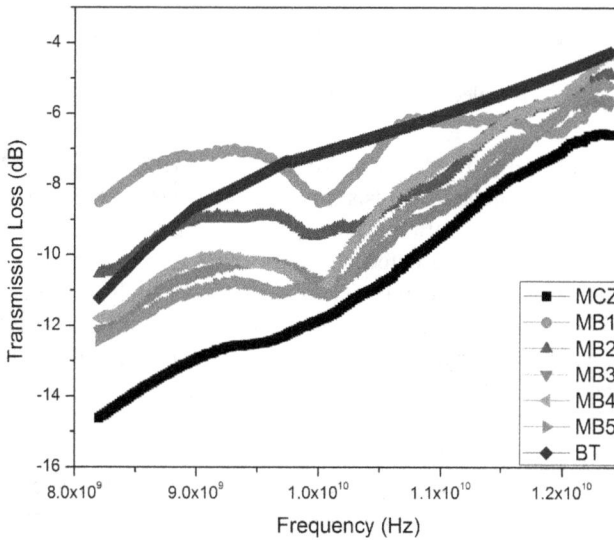

FIGURE 1.6 Frequency dependence of transmission loss for the composites.

Among all the samples, the maximum bandwidth was found to be 0.567 GHz for the 50% BT phase, i.e., for MB3 sample.

The nanocomposites of thickness 3 mm show the variation of transmission loss with frequency, as noted from Figure 1.6. The complete X-band range shows the transmission loss variation from −5 to −15 dB. From this figure, we can see that the transmission loss gradually decreases with frequency and attains a minimum near 10 GHz. From the reflection loss and transmission loss data, the effective absorption of microwaves can be determined. The incident electromagnetic wave energy can be divided into parts: reflected wave energy, transmitted wave energy, and absorption energy, when it is incident on the sample [26, 27]. Hence, the effective absorption of the wave can be found as follows: Absorption % = 100-reflected power-transmitted power. The composite with 80% of BT phase exhibits an effective absorption of more than 91% of incident wave, as compared to all other samples. At 10.13 GHz, the minimum reflection loss and transmission loss of MB5 are −20 and −11 dB. Other samples show an absorption of more than 80% of the incident wave. From these results, we can conclude that the newly synthesized materials are potential ones for microwave-absorbing shields in X-band region.

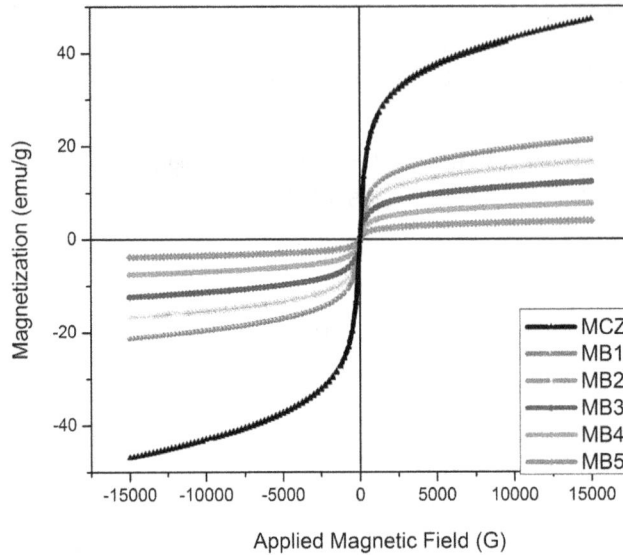

FIGURE 1.7   VSM of ferrite–ferroelectric composites.

Figure 1.7 displays magnetic hysteresis of MCZ, MB1, MB2, MB3, and MB4 at room temperature. The presence of magnetic hysteresis in the samples proves that the composites are magnetically ordered. The values of saturation magnetization (M_S) and coercivity (H_C) are obtained from the hysteresis loops and are presented in Table 1.2. It is evident clearly from the table that the coercivity (H_c) increases with the BT content up to 40% of BT (MB2), which reveals that the magnetization becomes weak because of the presence of nonmagnetic BT phase, leading to domain wall pinning, which, in turn, results in increase of coercivity. For $x > 40\%$, the coercivity decreases, which means that no domain wall pinning takes place. The magnetization saturation linearly decreases with BT phase. Our results are consistent with the Bruggeman theory [28, 29, 30]. According to this theory, given the properties of the individual components ψ_1, $m1$, and ψ_2, $m2$, predictions for effective dielectric and magnetic properties of the composite material that consists of the two component (MCZBT in the present case) phases, are best described by the equation $f\left\{\left(\psi_{1-}\psi_{\text{eff}}\right)/\left(\psi_1 + 2\psi_{\text{eff}}\right)\right\} + \left(1 - f\right)\left\{\left(\psi_2 - \psi_{\text{eff}}\right)/\left(\psi_2 + 2\psi_{\text{eff}}\right)\right\} = 0$, where f is the volume fraction, ψ_1 and ψ_2 are the frequency-dependent complex permittivity or permeability of the individual composite constituents, and ψ_{eff} is the effective complex permittivity or permeability of the composites [1]. In the current context, the above equation can be rewritten as $\left\{3\left(1 - f\right)/\left[2 + \psi_2/\psi_{\text{eff}}\right]\right\} + 3 f/2 = 1$, where ψ_1 is the permeability of one phase (BT in the present case), ψ_2 is the permeability of the second phase (MCZ), and ψ_{eff} is the effective complex permeability of composites. The ψ_1 value of BT is unity because of its inherent nonmagnetic nature. The above equation is deduced based on the assumptions that the ferrite phase is dominant and $\psi_{\text{eff}} \gg 1$. However, because BT predominates, the $\psi_1 : \psi_{\text{eff}}$ ratio cannot be neglected. From the above theory, it appears that with an increase of nonmagnetic volume fraction $\left(f\right)$, i.e., the increase of BT content, the effective permeability $\left(\psi_{\text{eff}}\right)$ of the composites decreases.

Figure 1.8 displays the ferroelectric hysteresis for the composite samples at room temperature. The figure shows that the composite samples possess ferroelectric hysteresis indicating that the samples are spontaneously polarized. The ferroelectric nature of the hysteresis loops gradually reduces with the decrease of ferroelectric amount, which is due to relatively lower electrical resistance of MCZ ferrite. The ferroelectric coercive field (EC) decreases with an increase of BT content, which implies that the composites can be easily polarized under the applied electric field.

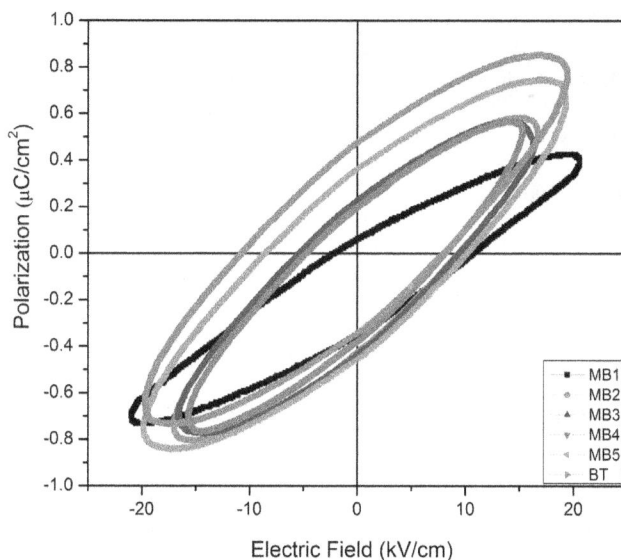

FIGURE 1.8   P-E loops of composite samples.

CONCLUSIONS

In the present chapter, development of multiferroic materials of magnetic and ferroelectric nanocomposites using physical methods has been discussed and their characteristic results have been analyzed. Dielectric, magnetic, and electromagnetic characterizations were investigated in frequency regions of 1 MHz–1.8 GHz and 8.2 GHz–12.4 GHz using impedance analyzer and vector network analyzer, respectively. From the results of the current studies, it is concluded that the composite with 80% of ferroelectric phase showed a microwave absorption of more than 91% of incident electromagnetic wave. All other samples also exhibit an effective absorption of more than 80% of incident wave. These results suggest that the current material compositions are potential ones for EMI shielding filter applications.

REFERENCES

[1] N.A. Spaldin, Multiferroics: Past, present, and future, *MRS Bull.*, 42 (2017), 385. https://doi:10.1557/mrs.2017.86

[2] I. Gruszka, A. Kania, E. Talik, M. Szubka, S. Miga, J. Klimontko, J. Suchanicz, Characterization of multiferroic $PbFe0.5Nb0.5O3$ and $PbFe0.5Ta0.5O3$ ceramics derived from citrate polymeric precursors, *J. Am. Ceram. Soc.*, 102(2019), 1296. https://doi.org/10.1111/jace.15998

[3] D. Bochenek, A. Chrobak, G. Ziółkowski, Electric and magnetic properties of the multiferroic composites made based on $Pb(Fe1/2Nb1/2)1–xMnxO3$ and the nickel-zinc ferrites, *Materials*, 16 (2023), 3785. https://doi.org/10.3390/ma16103785

[4] S. Goel, A. Garg, R.K. Gupta, A. Dubey, N.E. Prasad, S. Tyagi, Effect of neodymium doping on microwave absorption property of barium hexaferrite in X-band, *Mater. Res. Express.*, 7 (2020), 016109. https://doi:10.1088/2053-1591/ab6544

[5] M. Green, X. Chen, Recent progress of nanomaterials for microwave absorption, *J. Mater.*, 5 (2019), 503e541. https://doi.org/10.1016/j.jmat.2019.07.003

[6] I. Hajimiri, M.R. Khoshroo, Combination of perovskite and magnetic inverse spinel structures to improve microwave absorption properties, *Mater. Sci. Eng. B.*, 225 (2017), 75. https://doi.org/10.1016/j.mseb.2017.06.016

[7] M.K. Durga, P. Raju, Effect of $BaTiO3$ phase on frequency dispersion characteristics of $Mg0.48Cu0.12Zn0.4Fe2O4 + BaTiO3$ nanocomposites, *Mater. Sci. Eng. B.*, 272 (2021), 11534. https://doi.org/10.1016/j.mseb.2021.115340

[8] M. Kanakadurga, P. Raju, S.R. Murthy, Preparation and characterization of BaTiO3+MgCuZnFe2O4 nanocomposites, *J. Magn. Magn. Mater.*, 341 (2013), 112. https://doi.org/10.1016/j.jmmm.2013.04.037

[9] M. Liu, O. Obi, J. Lou, S. Stoute, J.Y. Huang, Z.U. Cai, K.S. Ziemer, N.X. Sun, Spin-spray deposited multiferroic composite Ni0.23Fe2.77O4/Pb(Zr,Ti)O3 with strong interface adhesion, *Appl. Phys. Lett.*, 92 (2008), 152504. https://doi.org/10.1063/1.2911743

[10] M.A. El Hiti, Dielectric behavior and ac electrical conductivity of Zn-substituted NiMg ferrites, *J. Magn. Magn. Mater.*, 164 (1996), 187. https://doi.org/10.1016/S0304-8853(96)00368-X

[11] M.A. Ahmed, M.A. El Hiti, M.M. Mosaad, S.M. Attia, Ac conductivity in CuCr ferrites, *J. Magn. Magn. Mater.*, 146 (1995), 84. https://doi.org/10.1016/0304-8853(94)01630-5

[12] K Sadhana, Development of nanocomposites for high frequency applications, PhD thesis, (2009), Osmania University, Hyderabad, India.

[13] W.W. Ling, H-W. Zhang, H.E. Ying, Y.X. Li, Y.Y. Wang, Magnetic and dielectric properties of low temperature fired ferrite/ceramic composite materials, *Prog. Nat. Sci.: Mater. Int.*, 21 (2011), 21. https://doi.org/10.1016/S1002-0071(12)60019-7

[14] Z.H. Wu, Y. Shi, Z.Z. Yong, Electric and magnetic properties of a new ferrite-ceramic composite material, *Chin. Phys. Lett.*, 19 (2002), 269. https://doi.org/10.1088/0256-307X/19/2/340

[15] P.R. Mandal, T.K. Nath, Investigation of magnetic and electrical properties of multiferroic CZFO-PZT nanocomposites, *AIP Conf. Proc.*, 1372 (2011), 98. https://doi.org/10.1063/1.3644425

[16] C.E. Ciomaga, A.M. Neagu, M.V. pop, M. Airimioaei, S. Tascu, G. Schileo, C. Galassi, L. Mitoseriu, Ferroelectric and dielectric properties of ferrite-ferroelectric ceramic composites, *J. Appl. Phys.*, 113 (2013), 074103. https://doi.org/10.1063/1.4792494

[17] A. Mandal, D. Ghosh, A. Malas, P. Pal, C.K. Das, Synthesis and microwave absorbing properties of Cu-Doped nickel zinc ferrite/Pb(Zr0.52Ti0.48)O3 Nanocomposites, *J. Eng.*, (2013), 391083. https://doi.org/10.1155/2013/391083

[18] A. Mandal, C.K. Das, Effect of BaTiO3 on the microwave absorbing properties of Co-doped Ni-Zn ferrite nanocomposites, *J. Appl. Polym. Sci.*, (2014), 39926. https://doi.org/10.1002/app.39926

[19] S.R. Murthy, Low temperature sintering of MgCuZn ferrite and its electrical and magnetic properties, *Bull. Mater. Sci.*, 24 (2001), 379. https://doi.org/10.1007/BF02708634

[20] R.W. Siegel, Nanophase materials: Synthesis, structure, and properties. In: Fujita, F.E. (eds) *Physics of New Materials*. Springer Series in Materials Science, vol. 27. (1998), Springer, Berlin, Heidelberg. https://doi.org/10.1007/978-3-642-46862-9_4

[21] R. Janot and D. Guerand, Ball-milling in liquid media: Applications to the preparation of anodic materials for lithium-ion batteries, *Prog. Mater. Sci.*, 50 (2005), 1. https://doi.org/10.1016/S0079-6425(03)00050-1

[22] Z. Yue, J. Chen, X. Qi, Z. Gui, L. Li, Preparation and electromagnetic properties of low-temperature sintered ferroelectric-ferrite composite ceramics, *J. Alloys Compd.*, 375 (2004), 243. https://doi.org/10.1016/j.jallcom.2003.11.156

[23] N. Rezlescu, E. Rezlescu, P.D. Popa, M.L. Craus, L. Rezlescu, Copper ions influence on the physical properties of a magnesium-zinc ferrite, *J. Magn. Magn. Mater.*, 182 (1998), 199. https://doi.org/10.1016/S0304-8853(97)00495-2

[24] T. Nakamura, Snoek's limit in high-frequency permeability of polycrystalline Ni-Zn, Mg-Zn, and Ni-Zn-Cu spinel ferrites, *J. Appl. Phys.*, 88 (2000), 348. https://doi.org/10.1063/1.373666

[25] R. Dosoudil, V. Olah, Complex permeability spectra of manganese-zinc ferrite and its composite, *J. Electr. Eng.*, 52 (2002), 24.

[26] K. Pubby, S.B. Narang, Ka band absorption properties of substituted nickel spinel ferrites: Comparison of open-circuit approach and short-circuit approach, *Ceram. Int.*, 45 (2019), 23673. https://doi.org/10.1016/j.ceramint.2019.08.081

[27] N.D. Patil, N.B. Velhal, V.R. Puri, Structural, electrical, and magnetic characteristics of Ni/Ti-modified BiFeO3 lead-free multiferroic material, *J. Mater. Sci. Mater. Electron.*, 28 (2017), 2. https://doi.org/10.1007/s10854-022-08329-z

[28] J.V. Mantese, A.L. Micheli, D.F. Dungan, R.G. Geyer, J. Baker-Jarvis, J Grosvenor, Applicability of effective medium theory to ferroelectric/ferrimagnetic composites with composition and frequency-dependent complex permittivities and permeabilities, *J. Appl. Phys.*, 79 (1996), 1655. https://doi.org/10.1063/1.361010

[29] J. Smith, H.P.J. Wijn, Ferrites-physical properties and technical applications, Wiley, Eindhoven, 1959, ark:/13960/t95751z41

[30] K. Sadhana, K. Praveena, S.R. Murthy, Magnetic properties of xNi0.53Cu0.12Zn0.35Fe1.88O4+ (1−x) BaTiO3 nanocomposites, *J. Magn. Magn. Mater.*, 322 (2010), 3729. https://doi.org/10.1016/j.jmmm.2010.06.007

2 Energy Storage in Metal Oxides and Composites

G. Krishnamurthy and A.S. Santhosh Kumar

INTRODUCTION

It is a matter of great concern that the conventional energy sources on our planet are getting exhausted day by day. As the energy consumption demand is constantly increasing, new alternative energy sources are being developed across the world. The highly threatening increase in pollution level and global warming invites our attention to the necessity of developing a clean energy portfolio. Special focus should be given on the development of alternative energy sources as well as their storage. The most well-known energy production and storage technologies are batteries, fuel cells and supercapacitors. In contrast to fuel cells and batteries, supercapacitors use a different energy generating technique. Even though these three systems have distinct energy storage and conversion processes, there exist some electrochemical similarities between them. The development of electrochemical energy storage (EES) techniques is largely attributed to the innovation and advance of electrode materials with tailored structure and high reactivity. To enhance the efficiency of materials, researchers are working on combining TMOs with other transition metals, metal oxides, carbon-based materials, etc. This can modify the surface area, pore characteristics, ion intercalation/deintercalation, conductivity, etc.

Among various energy storage systems, EES devices (such as batteries and supercapacitors) have been extensively studied and considered one of the most promising green energy storage systems for the sustainable energy supply in near future, due to their high efficiency, versatility and flexibility. An electrochemical cell consists of two electrodes (denoted as cathode/anode or positive/negative) separated by an ionically conductive, electronically insulating electrolyte. Batteries convert chemical energy into electrical energy through Faradaic charge transfer processes where (i) oxidation/reduction reactions occur within anode/cathode active materials and (ii) electrons are transported through an external circuit to maintain charge neutrality at each electrode. These reactions are irreversible in primary batteries (e.g. $Zn-MnO_2$ and $Li-MnO_2$) designed for single-use applications. On the other hand, secondary batteries (e.g. lead-acid, nickel-metal hydride, and Li-ion) leverage reversible redox processes and can be repeatedly charged/discharged, a requirement for many end-use applications (e.g. electric vehicles) [1].

Electrochemical capacitors, so-called double-layer capacitors, supercapacitors or ultracapacitors, are electrical power sources which utilize the capacitive properties at the interface between an electronic conductor (electrode) and liquid ionic conductor (electrolyte solution).

Supercapacitors store/deliver energy through non-Faradaic processes where ions are stored in the electrochemical double layer near the electrode surfaces. On the other hand, pseudocapacitive materials store energy through charge transfer reactions which may include (i) oxidation/reduction of the electrode surface and (ii) intercalation of ions into a host active material. Hybrid configurations utilising pseudocapacitive materials approach the specific energy of rechargeable batteries. Two important performance metrics of energy storage devices are specific energy (Wh/kg) and specific power (W/kg), which describe how much and how quickly energy can be stored/delivered, respectively. Analogous quantities normalised to system volume (i.e. energy/power densities with units of Wh/L and W/L) are also commonly used. Ragone plots (Figure 2.2) summarise these energy/power relationships and are useful to assess the viability of different energy storage platforms for

DOI: 10.1201/9781003479239-2

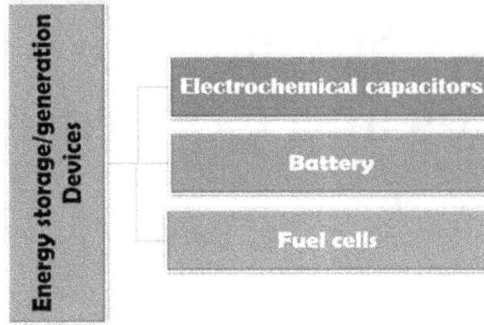

FIGURE 2.1   Electrochemical energy storer.

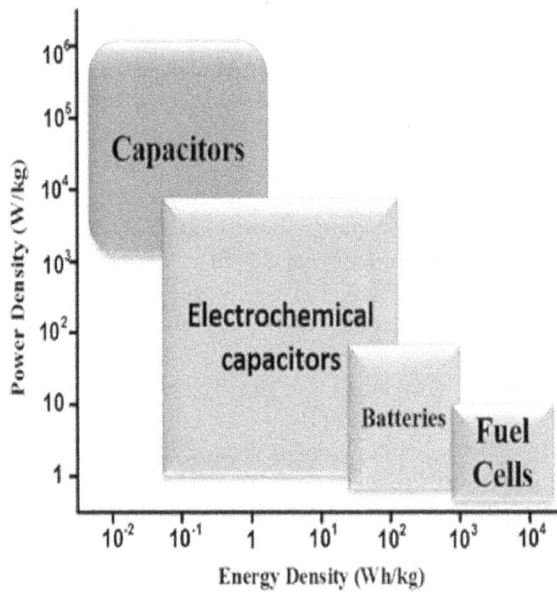

FIGURE 2.2   Ragone plot representing various EES.

a given application. Figure 2.1 shows a fundamental trade-off between a system's specific energy and power. For example, supercapacitors exhibit (i) high specific power due to rapid ion adsorption/desorption near electrode surfaces but (ii) low specific energy since charge storage only occurs within the electrochemical double layer. On the other hand, batteries store energy within the bulk structure of active materials, enabling high specific energy. The rate of energy storage/delivery in batteries is generally limited by solid-state diffusion or phase nucleation kinetics in the active material, resulting in lower specific power than supercapacitors. With these trends in mind, it should be emphasised that the energy/power characteristics of an electrochemical device are also highly dependent on design factors such as material selection, cell format, and electrode architecture [2,3].

The reason behind the overview of supercapacitors (SC) energy storage system is that SCs weigh less than that of battery with the same energy storage capacity, fast access to stored energy, charging very fast than battery, 106 times charge/discharge cycle, storage capacity independent of the number of charging discharging cycles, negligible environmental concerns, and energy density 10–100 times larger than that of traditional capacitors [4]. SCs can store substantially more energy than conventional capacitors because the charge separation takes place across a very small distance in the electrical double layer that constitutes the interface between an electrode and the adjacent electrolyte, and an increased amount of charge can be stored on the highly extended surface area electrode materials.

Electrochemical capacitors, also known as supercapacitors, exhibit high specific capacitance, high specific power, long life cycle and fast charge/discharge rate. Theoretical capacitance values of some TMOs are represented in Figure 2.3 as a bar diagram. Electrochemical capacitors charge and discharge more rapidly than batteries over longer cycles, but their practical applications remain limited owing to their significantly lower energy densities. Pseudocapacitors and hybrid capacitors have been developed to extend Ragone plots to higher energy density values, but they are also limited by insufficient breadth of options for electrode materials, which require materials storing alkali metal cations such as Li^+ and Na^+. Herein, we report a comprehensive and systematic review of emerging anion storage materials for performance- and functionality-oriented applications in electrochemical and battery-capacitor hybrid devices (Figure 2.4).

Supercapacitors are classified according to the energy storage mechanism as electrical double-layer capacitors (EDLCs), pseudocapacitors (PCs) and hybrid capacitors. In EDLCs, charge storage is based on reversible adsorption–desorption mechanism at the electrode–electrolyte

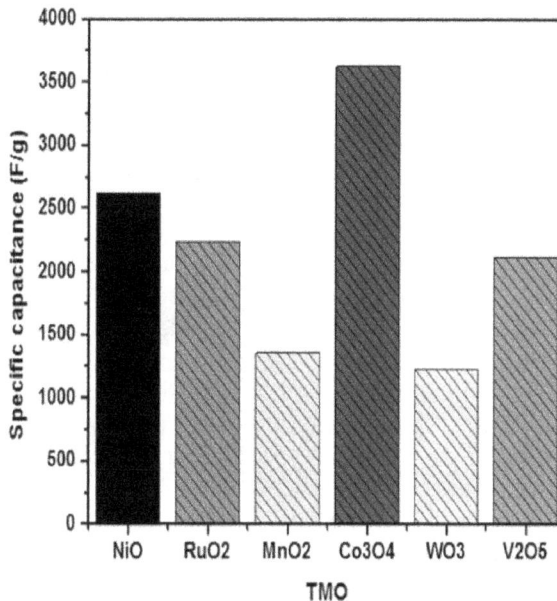

FIGURE 2.3 Theoretical specific capacitance of some TMOs.

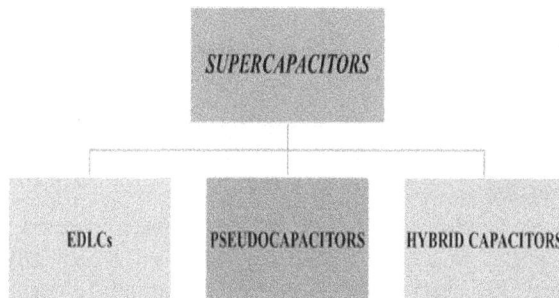

FIGURE 2.4 Types of capacitors.

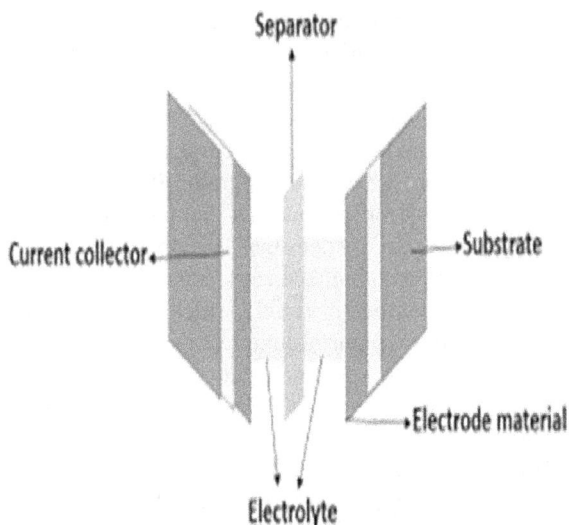

FIGURE 2.5   Schematic representation of supercapacitor model.

interface and not involving any Faradaic reaction. Electric double layers are formed with the accumulation of charge over the opposite electrodes. Here, the electrode's exposed surface area to the electrolyte determines the capacitance. In other words, pore size of the electrode material should match with the ion size of the electrolyte to avoid capacitance drop. On the other hand, in pseudocapacitors or redox capacitors, charge storage is based on rapid Faradaic reaction at the surface of electrode material. Generally, pseudocapacitors provide capacitance value higher than EDLCs. In hybrid systems, materials for EDLCs (capacitor-like power sources) and pseudocapacitors (energy sources that resemble batteries) are combined on a single electrode substrate [5]. Electrode is the key factor that determines the performance of any supercapacitor. Depending on the anode and cathode materials in SC device fabrication, it is classified into symmetric, asymmetric, and hybrid devices. If two electrodes of the SC are of the same material, it is called symmetric device and includes EDLC, pseudocapacitive, and hybrid-type material electrodes. Porosity of the electrode material should be well tuned according to the application where it is used. Small pores yield better surface area, which in turn enhances the specific capacitance and energy density. The better surface area of a porous material enables much more reactive sites and promotes the transfer of electrons and ions. However, this small pore increases the equivalent series resistance (ESR) and hence reduces the specific power. Therefore, less porous materials are preferred for applications where high peak current is demanded [6]. Schematic representation of supercapacitor device model is shown in Figure 2.5.

SYNTHESIS OF ELECTRODE MATERIALS

These modifications are aiming at improving the ion/electron transportation characteristic of metal oxides and obtaining high energy and power densities. It is well accepted that the EES systems involve electrochemical process with charge transfer and ion diffusion.

Unfortunately, most of the active electrode materials (e.g. NiO, MnO_2, Co_3O_4 and Fe_2O_3) are p-type semiconductors whose electric conductivity is too low to support fast electron transport required by high rates. Therefore, great efforts are dedicated to ameliorating the electrochemical activity and kinetic feature of metal oxides by designing composite materials with other highly conductive layers or scaffolds.

TABLE 2.1

Merits and Demerits of Different Solution-Based Methods

Methods	Applicable Objects	Merits	Demerits
Electrodeposition (ED)	Nanostructured films	Easy control in morphology, relatively fast	Unsuitable for large-scale production
Chemical precipitation (CP)	Powders, colloidal nanostructures	Fast deposition, large-scale production	Difficult in morphology control
Microwave synthesis (MS)	Powders, colloidal nanostructures	Much faster than CP, large-scale production	Difficult in morphology control
Sol-gel	Nanostructured film and powders	Large-scale production	Unwanted dense film without using template
Hydrothermal (HT) and Solvothermal synthesis (ST)	Nanostructured film and powders	Easy control in morphology, large-scale production	High temperature, slow growth
Chemical bath deposition (CBD)	Nanostructured film and powders	Easy control in morphology, large-scale production, faster than HT and ST	Limited to a few metal oxides

SYNTHESIS METHODS AND KEY CHARACTERISTICS OF METAL OXIDE NANOPARTICLES

The synthesis of nanoparticles can be achieved with either a "top-down" or "bottom-up" approach. In the top-down approach, bulk materials are broken down into nanoparticles (NPs) by size reduction (via various lithographic techniques, milling, grinding, laser ablation, sputtering, etc.). In the bottom-up approach, NPs are obtained by chemical, physical and biological techniques (plant material, microbes, biological products, etc.). Typically, chemical and physical synthesis routes (i.e., bottom-up approaches) are employed in the synthesis of MONPs and result in an efficient quantity of obtained nanoparticles, but they have the disadvantages of higher cost, presence of poisonous chemicals (e.g. absorbed on the NPs surface) leading to adverse effects when used in biomedical applications, the need to use of stabilisers, etc. [7–10]. Such synthesis routes include, but are not limited to (i) (chemical) precipitation, (ii) wet chemical synthesis, (iii) hydrothermal, (iv) solvothermal, (v) sol-gel, (vi) solid-state pyrolytic methods, (vii) thermal decomposition and (viii) microwave-assisted synthesis. Figure 2.6 provides an overview of the various synthesis routes used for manufacturing metal oxide nanoparticles.

TRANSITION METAL OXIDE AND COMPOSITES

Transition metal oxides (TMOs) are well studied by researchers in the energy storage field. Pseudocapacitive nature of these materials is due to the fast and reversible redox reactions at the surface of electrode material. Theoretical specific capacitance of some TMOs is shown in Figure 2.3. The idea of preparing a composite structure is always encouraged because it imposes a positive result on the overall electrochemical properties of the material. Under the combined effect of various materials/ions, many of the limitations of the TMOs can be surmount. Zheng et al. [13] fabricated a nanostructured electrode material by combining ZnO and NiO. Here ZnO acted as the electrode modifier, and hence, the specific capacitance was improved. Materials capable of exhibiting multiple oxidation states are of particular interest when it comes to electrode fabrication. NiO is one such candidate, whose multiple-valence states favour the fast redox reaction, which in turn enhances the specific capacitance.

FIGURE 2.6 Possible metal oxide nanoparticle synthesis methods ("bottom-up" approach).

TRANSITION METAL OXIDE AND CARBON-BASED COMPOSITE

Carbon-based materials are generally selected for various applications including supercapacitors, thanks to their high abundance, easy preparation and cost-effectiveness. The porous nature and excellent conductivity make carbon-based materials – such as graphene, carbon nanotubes and activated carbon – and various other carbon derivatives a widely accepted electrode materials for supercapacitor applications. As the synthesis techniques are approaching new heights, carbon-based electrodes are generated in various morphologies including nanofibers, nanoflowers, nanorods and nanotubes. MnO_2, V_2O_5, ZnO and RuO_2 are some of the promising electrode materials. The performance and structure of electrode material play crucial roles in determining the effective capacity of SC. The limitations of using MnO_2 as electrode material in SC, like poor conductivity, can be alleviated by combining with conductive materials such as carbon to make hybrid electrode material. Carbon-based materials, such as activated carbon, carbon nanotube and graphene, are attractive due to their pore structure, volume, specific surface area, and presence of functional groups [12]. Cai W. et al. proposed an effective method to synthesise N-doped carbon@MnO_2 3D core-shell composite, which shows excellent electrochemical performance. The composite shows high specific capacitance with excellent cyclic stability and high retention [15,16]. Carbon nanotube finds application as electrical double-layer capacitor with high specific surface area and conductivity. But the low specific capacitance limits its usage, thus combined with TMOs. Lei et al. demonstrated a facile method to synthesise MnO_2 nanosheets at graphenated CNTs by hydrothermal method.

It shows a high specific capacitance of 575.4 F/g at 0.5 mA/cm^2 and a considerably large energy density of about 51.2 Wh/kg [17]. Long et al. reported flexible SC electrode with delta MnO$_2$ nanosheets anchored on activated carbon cloth. Compared with all other carbon-based materials, graphene has attracted tremendous attention in high-performance energy storage systems because of its intriguing properties including large surface area, excellent conductivity and commendable thermal, optical, and mechanical properties. Basically, graphene is a single layer of atom constructed by sp^2-bonded carbon atoms arranged in a poly aromatic honeycomb crystal structure. Theoretical specific capacitance of graphene-based EDLC is 550 F/g. In literature, one can see that supercapacitors based on graphene exhibited a specific capacitance of 75 F/g with an energy density of 31.9 Wh/kg in ionic liquid electrolytes and 135 F/g specific capacitance in aqueous electrolyte. On the other hand, it showed a specific capacitance of 99 F/g in organic electrolytes [18]. But it is noted that the restacking of graphene sheets reduces its conductivity and results in poor specific capacitance. Restacking occurs due to the van der Waals interaction between the sheets, and it diminishes coulombic efficiency also. In order to improve the capacitive nature of graphene, they are usually made composites with other capacitive materials. Graphene–metal oxide composite seems to be a good combination since metal oxide hinders graphene from restacking. Metal oxides carry the role of a stabiliser, which prevents the accumulation of graphene sheets.

Combination of graphene and metal oxide compliments each other by eliminating the complications faced by these materials individually. Development of composites with pseudocapacitive materials possesses an advantage of generating capacitance from redox charge transfer in addition to the double-layer capacitance. In metal oxide graphene composite, graphene acts as a passage for charge transfer, whereas metal oxides provide pseudo-capacitance. ZnO is a versatile material possessing 3.37 eV bandgap, with an exciton binding energy of 60 meV, which makes it suitable for a wide range of applications including supercapacitors. Nevertheless, lower specific capacitance is exhibited by ZnO compared with other metal oxides, and it is superior with low cost, high abundance and less toxicity. Also, the electron donating nature of ZnO makes it a good partner for electron acceptor graphene to make efficient electrode materials. As reported by Dutta et al. [19], ZnO/rGO composite electrode can achieve a high specific capacitance of 1,012 F/g at a current density of 1 A/g with an outstanding power density of 3534.6 W/kg. Sreejesh et al. [20] further prepared ZnO/rGO nanocomposite by microwave-assisted technique to achieve a high capacitance up to 631 F/g and a long life cycle tested up to 2,000 cycles. Introduction of graphene can effectively transcend the poor electrical conductivity of NiO, another promising transition metal oxide candidate. The core-shell hybrid NiO/rGO electrode prepared by electrophoretic deposition method shows a capacitance of 940 F/g at a current density of 2 A/g [21]. Pore et al. reported the achievement of a specific capacitance of 727.1 F/g at 1 mA/cm^2 current density with a good cyclic stability of about 80.4% over 9,000 cycles for the hydrothermally prepared NiO/rGO electrodes [22].

ENERGY STORAGE OF METAL OXIDE/COMPOSITES

NiO and NiO-Based Composites

Nickel oxide is a classic example of transition metal materials with wide applications in electrochromic, catalysis, chemical sensor and EES. It can be used as an anode material for LIBs or a cathode material for supercapacitors. Pure NiO is an insulator with a bulk band gap of ~3.4 eV. It crystallises in the rock-salt structure (cubic crystal structure, JCPDS 4-0835) with a lattice constant of 0.417 nm and a high-spin antiferromagnetic spin structure at low temperatures. Lots of NiO nanostructures (such as nanowires, nanoflakes, nanobowls and nanospheres) have been prepared by different solution-based synthesis methods and applied as active materials for EES with enhanced electrochemical performance.

The name sol-gel derives from the fact that microparticles or molecules in a solution (sols) agglomerate and under controlled conditions eventually link together to form a coherent network (gel). There are two generic variations of the sol-gel technique. One is called the colloidal method, and the other is called the polymeric (or alkoxide) route. The differences between the two stem from the types of starting materials (precursors) that are used. Both routes involve suspending or dissolving the precursor(s) in a suitable liquid, usually water for the colloidal route and alcohol for the polymeric route. The precursor is then activated by the addition of an acid (such as hydrochloric acid) or a base (such as potassium hydroxide). The activated precursors react together to form a network. The network grows and ages with time and temperature until it is the size of the container. At this point, the viscosity of the liquid increases at an exponential rate until gelation occurs, that is, no more flow is observed. Many different industries could benefit from adopting sol-gel because of its versatility in fabricating a wide range of materials with different properties. Current examples are found in the construction, electronics, communications, automotive and EES.

To date, there are only a few reports about the sol-gel method for the synthesis of NiO and NiO-based composites for EES applications. Previously, our group used $NiCl_2$, HCl and formaldehyde to form $NiCl_2$/resorcinol-formaldehyde (RF) gel and finally obtained NiO hollow spheres after heat treatment [23]. This sol-gel process belongs to polymeric (or alkoxide) route. The as-prepared NiO hollow spheres present a high reversible LIBs capacity of 635 mAh/g at 200 mA/g. Kim et al. prepared NiO nanostructures with three distinct morphologies by a sol-gel method, and their morphology-dependent supercapacitor properties were exploited [24]. They used $Ni(NO_3)_2$, $Ni(CH_3COO)_2$, hexamethylene tetramine and ammonia as the starting materials and utilised different control reagents to fabricate three NiO nanostructures (flower-like, slice-like and particle-like NiO). Their sol-gel method is a typical colloidal route. These NiO samples can deliver a capacitance of 480 F/g at 0.5 A/g for supercapacitors. Meanwhile, Liu et al. reported a $NiO/NiCo_2O_4/Co_3O_4$ composite via a sol-gel method by using $Ni(NO_3)_2$, $CoCl_2$ and propylene oxide. The composite shows a high specific capacitance (1,717 F/g) and enhanced rate capability [25].

Microwave synthesis (MS) has become the method of choice for many chemists and biochemists for a multitude of reactions for one reason: it simply works better. Reactions that took hours, or even days, to complete can now be performed in minutes with better yields and cleaner chemistries. In many of the published examples, microwave heating has been shown to dramatically reduce reaction times, increase product yields and enhance product purities by reducing unwanted side reactions compared to conventional heating methods. The advantages of this enabling technology have, more recently, also been exploited in the context of multistep total synthesis of TMOs. MS is suitable for large-scale synthesis of powder products. According to the published papers about MS process of nanostructured NiO, notice the fact that the MS method can be regarded as a modified solution synthesis method built on the traditional chemical precipitation (CP) method. First, a precipitation is formed by simply mixing precipitants (ammonia or sodium hydroxide) and nickel sources. Then the precipitation is transferred into a microwave oven to undergo deep reaction to obtain NiO nanostructures such as nanospheres, nanoflakes, nanoplates and nanoparticles. These MS-NiO nanostructures show a capacity of 884 mAh/g and a pseudo-capacitance of (277–585 F/g).

COBALT OXIDES AND COBALT OXIDES–BASED COMPOSITES

Cobalt oxides (Co_3O_4 and CoO) have attracted great attention for EES due to their excellent electrochemical reactivity and redox reversibility since the pioneering works of Tarason's group and Conway et al. Co_3O_4 is a black antiferromagnetic solid with a band gap of ~2.0 V. As a mixed valence compound, its formula is sometimes written as $Co^{II}Co^{III}_2O_4$, which adopts the normal spinel structure, with Co^{2+} ions in tetrahedral interstices and Co^{3+} ions in the octahedral interstices of the cubic close-packed lattice of oxide anions. The spinel Co_3O_4 is a magnetic semiconductor with a lattice constant $a=0.808$ nm (JCPDS 42-1467). The other interesting cobalt oxide is CoO (cobalt monoxide), which has rock-salt structure (NaCl structure, JCPDS 09-0402) and consists of two interpenetrating fcc sublattices of Co^{2+} and O^{2-} with a lattice constant of 0.426 nm and a band gap of ~2.6 V. The CoO is not stable in the air and will be oxidised into the thermodynamically favoured form $-Co_3O_4$. In view of their high capacity (LIBs: the theoretical capacity of Co_3O_4 and CoO is 890 and 718 mAh/g, respectively) and capacitance (supercapacitors: a theoretical capacitance of >2,000 F/g), the cobalt oxides (Co_3O_4 and CoO) have been extensively studied as active materials for EES. Numerous nanostructured cobalt oxides (such as nanowires, nanoflakes, nanocage, nanorods, nanoparticles nanobowls and nanospheres) have been prepared by different solution-based synthesis methods and enhanced electrochemical performances have been demonstrated in these systems.

Among all the solution-based synthetic routes for the synthesis of cobalt oxides for EES, HT is the most popular solution synthesis method and the published HT papers account for ~40% in all the published solution-based literature. Combined with our works, we introduce one of the most fascinating HT methods for the preparation of cobalt oxides (Co_3O_4 and CoO) nanowire arrays and their composite materials. In our case, we used $Co(NO_3)_2$ (can also be $CoSO_4$ or $CoCl_2$), NH_4F and urea as the starting materials and the whole reactions were conducted at 110–120°C. The involved HT reactions can be illustrated as follows [26–28]:

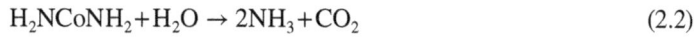

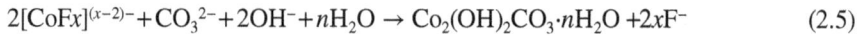

$$Co^{2+} + xF^- \rightarrow [CoFx]^{(x-2)-} \tag{2.1}$$

$$H_2NCoNH_2 + H_2O \rightarrow 2NH_3 + CO_2 \tag{2.2}$$

$$CO_2 + H_2O \rightarrow CO_3^{2-} + 2H^+ \tag{2.3}$$

$$NH_3 \cdot H_2O \rightarrow NH^{4+} + OH^- \tag{2.4}$$

$$2[CoFx]^{(x-2)-} + CO_3^{2-} + 2OH^- + nH_2O \rightarrow Co_2(OH)_2CO_3 \cdot nH_2O + 2xF^- \tag{2.5}$$

The Co_3O_4 and CoO nanowires can be formed after heat treatment under different atmospheres: Co_3O_4 in air and CoO in pure argon or nitrogen. If the argon is just normal, not high purity, the final product will also be Co_3O_4 nanowires. The different reactions in the annealing process are as follows:

During the annealing process for Co_3O_4 nanowires:

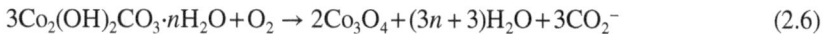

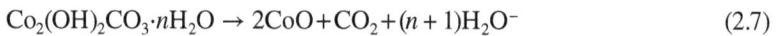

$$3Co_2(OH)_2CO_3 \cdot nH_2O + O_2 \rightarrow 2Co_3O_4 + (3n + 3)H_2O + 3CO_2^- \tag{2.6}$$

During annealing process for CoO nanowires

$$Co_2(OH)_2CO_3 \cdot nH_2O \rightarrow 2CoO + CO_2 + (n + 1)H_2O^- \tag{2.7}$$

The cobalt oxides (Co_3O_4 or CoO) nanowires show sharp tips and have an average diameter of ~80 nm, length up to around 5–25 μm. The length of nanowires could be easily controlled by the growth time. The as-prepared Co_3O_4 or CoO nanowires consist of numerous interconnected nanoparticles and present a rough appearance with a large quantity of mesoporous structure, which is ascribed to the successive release and loss of CO_2 and H_2O during the thermal decomposition of $Co_2(OH)_2CO_3$ precursor which is single crystalline in nature [11].

The above HT-cobalt oxides can be grown on almost all the substrates including ITO, FTO, glass slide, nickel foam, carbon cloth, Ti foil, nickel foil, copper foil, stainless steel and so on. In the HT solution, the powder products can be collected and are some interesting cobalt oxides (Co_3O_4 or CoO) nanostructures (such as balls, spheres and hexapods) assembled by nanowires. The NH_4F in the above HT method is acted as the growth promoter to improve the growth density of nanowires and prompt the formation of basic cobalt carbonate hydroxide precursor. Another problem to be noticed is that the reaction time also has great influence on the morphology of samples. When the time is less than 3 h, the products may not be nanowires, but nanoflakes. This special change is still under investigation. The as-prepared HT-Co_3O_4 and HT-CoO nanowire arrays are well characterised as electrode materials for LIBs and supercapacitors. For supercapacitors application, cobalt oxides (Co_3O_4 and CoO) undergo reversible redox reactions in the alkaline electrolyte. The reactions for Co_3O_4 in supercapacitors are illustrated in the following [14]:

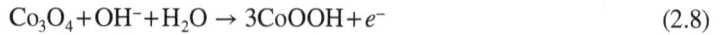

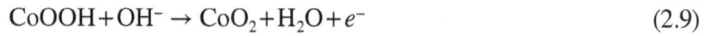

$$Co_3O_4 + OH^- + H_2O \rightarrow 3CoOOH + e^- \tag{2.8}$$

$$CoOOH + OH^- \rightarrow CoO_2 + H_2O + e^- \tag{2.9}$$

The electrochemical reactions for CoO are expressed in the following:

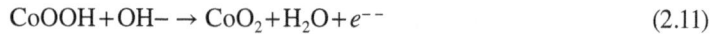

$$CoO + OH^- \rightarrow CoOOH + e^{--} \tag{2.10}$$

$$CoOOH + OH- \rightarrow CoO_2 + H_2O + e^{--} \tag{2.11}$$

These cobalt oxides (Co_3O_4 or CoO) nanowire arrays exhibit noticeable pseudocapacitive performances with a high capacitance of 754 F/g at 2 A/g and 610 F/g at 40 A/g as well as excellent cycling stability. Generally, two redox peaks will be found in the CV curves of cobalt oxides (Co_3O_4 or CoO) nanowire array, namely, $CoOOH$ and CoO_2, which will be formed at high oxidation potentials. $CoOOH$ has been detected and confirmed, but CoO_2, sometimes, is considered as a virtual state, or not stable.

The high capacity/capacitance and cycling stability in the core/shell systems are believed to benefit from the unique architecture including the porous cobalt oxides core nanowire, nanostructured shell and ordered array configuration.

(i) The highly porous system shortens the transportation/diffusion path for both electrons and ions, thus leading to faster kinetics and high-rate capability; (ii) high surface area of the mesoporous nanowire arrays favours the efficient contact between active materials and electrolytes, providing more active sites for electrochemical reactions; and (iii) the porous core/shell nanowire array architecture possesses favourable morphological stability, which helps to alleviate the structure damage caused by volume expansion during the cycling process. In short, these characteristics would lead to fast ion/electron transfer, sufficient contact between active materials and electrolyte, and enhanced flexibility, finally resulting in enhanced electrochemical performance. In addition, we found that the active material deposited on 3D porous substrates (such as nickel foam) shows better electrochemical performance than their counterparts on the flat substrate, due to the fact that the 3D porous substrates provide higher active surface area and inner space for fast ion/electron transfer and higher energy conversion efficiency. Notice that the cobalt oxides (Co_3O_4 or CoO) nanowires can also be prepared without NH_4F, but the obtained nanowires are much thinner and prone to be bending. The reactions of precursor are shown in the following

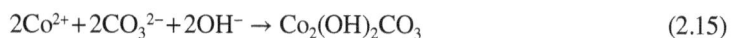

$$H_2NCoNH_2 + H_2O \rightarrow 2NH_3 + CO_2 \tag{2.12}$$

$$NH_3 \cdot H_2O \rightarrow NH_4 + OH^- \tag{2.13}$$

$$CO_2 + H_2O \rightarrow CO_3^{2-} + 2H^+ \tag{2.14}$$

$$2Co^{2+} + 2CO_3^{2-} + 2OH^- \rightarrow Co_2(OH)_2CO_3 \tag{2.15}$$

It is noteworthy that the chemical formula of $Co_2(OH)_2CO_3$ precursor is complex, and sometimes it will contain the anions in the cobalt source (e.g. Cl^-, SO_4^- and NO_3^{2-}). A lot of cobalt oxides nanostructures (e.g. nanorod, echinus-like, emongrass-like, chrysanthemumlike, nanosphere, nanowire, nanoflake, nanoplate, nanocube, nanonet) and composites (Co_3O_4/graphene and Co_3O_4/carbon) have been fabricated by the HT method only with urea and cobalt sources, and they showed a high

LIBs capacity of 700–100 mAh/g and delivered large specific capacitances (500–1,124 F/g) at high charge/discharge rate. Even so, researchers prefer the first HT method with NH_4F because the quality of the final cobalt oxides samples is much higher with better structural control and adhesion on the substrates and higher active materials load [29].

Moreover, the obtained porous HT-cobalt oxides (Co_3O_4 and CoO) nanowire arrays exhibit good EES performance with high capacitance/capacity. Additionally, they are good templates or scaffolds for constructing new high-performance core/shell nanoarrays or composites. Nowadays, in order to further improve the EES performance, great efforts are focusing on the modification of HT-cobalt oxides (Co_3O_4 and CoO) nanowires by introducing graphene, metal, carbon nanotube and conducting polymers to get composite materials with new or improved functions. The introduction of conductive backbone or coating can improve the conductivity of the bare nanowires leading to enhanced electrochemical properties. Solvothermal synthesis (ST) is the twin brother of HT. The literature of ST for the synthesis of cobalt oxides (Co_3O_4 and CoO) nanostructures is not as much as that of HT. Several cobalt oxides (Co_3O_4 and CoO) nanostructures (such as nanocapsules, nanosheets, nanospheres and nanoflower) have been obtained in the organic solvents including methanol, ethanol and ethylene glycol.

COPPER OXIDES (CUO) AND CUO-BASED COMPOSITES

Copper oxides are one of the most attractive classes of TMOs with wide applications in energy conversion devices (anode material for LIBs or cathode material for supercapacitors) and optoelectronic devices. Copper forms two oxides in accordance with its two valences: cuprous oxide (Cu_2O) and cupric oxide (CuO). Both are semiconductors with band gaps of 2.0 and 1.2 eV, respectively. Their band gaps make them good candidates for photovoltaic devices (e.g. solar cells and water splitting), catalysts, sensors and optoelectronic devices. Cu_2O crystallises in a simple cubic structure which can be considered as two sublattices, a face-centred cubic (fcc) sublattice of copper cations and a body-centred cubic (bcc) sublattice of oxygen anions (JCPDS 05-0667). The oxygen atoms occupy tetrahedral interstitial positions relative to the copper sublattice, so that oxygen is tetrahedrally coordinated by copper, whereas copper is linearly coordinated by two neighbouring oxygens [30].

Among all the solution-based synthetic routes for the synthesis of copper oxides for EES, CP is the most popular solution synthesis method and the published CP papers account for ~50% in all the published solution-based literature. Copper nitrate, copper chloride, copper sulphate and copper acetate are the most widely used copper sources. The adopted precipitants are common reagents such as ammonia and NaOH. In the following section, we introduce some typical CP methods for the synthesis of hierarchical copper oxides with various morphologies. Previously, we fabricated various CuO nanostructures via a facile CP method by using $CuSO_4$ as the copper source and ammonia as the precipitant. The involved CP reactions can be illustrated as follows:

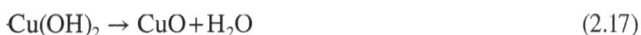

$$Cu^{2+}+2NH_3.H_2O \rightarrow Cu(OH)_2+2NH_4^+ \tag{2.16}$$

$$Cu(OH)_2 \rightarrow CuO+H_2O \tag{2.17}$$

The nanostructured CuO products are obtained after reacting for 12 h at 90°C. In our case, the pH value has a great influence on the morphology of the CuO products because there is a competing reaction in the solution expressed in the following [31]:

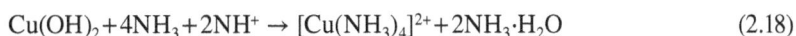

$$Cu(OH)_2+4NH_3+2NH^+ \rightarrow [Cu(NH_3)_4]^{2+}+2NH_3 \cdot H_2O \tag{2.18}$$

Obviously, $Cu(OH)_2$ could be dissolved due to the complex reaction for the existence of excess NH^{4+} and free NH_3 in the solution. There is no doubt that different pH values will lead to different concentrations of $[Cu(NH_3)_4]^{2+}$ and OH^- in the solutions, and finally resulting in the formation of CuO with various morphologies such as leaf, shuttle, flower, dandelion and caddice clew. These morphologies can be easily tailored by adjusting pH, indicating that pH is critical for the controllable growth of copper oxides in the CP method. It is observed that the LIBs performance of copper oxides is tightly associated with the morphology of samples. CuO samples with different morphologies show distinct electrochemical performances. Taking our work for example, compared to other CuO nanostructures, dandelion-like and caddice clew–like CuO prepared by our group exhibits reversible discharge capacities of 385 and 400 mAh/g at 0.1 C, 340 and 374 mAh/g at 0.5 C after 50 cycles, respectively. The higher discharge capacities and better cycling performances are attributed to their larger surface area and porosity, leading to better contact between CuO and electrolyte and shorter diffusion length of lithium ions. Inspired by the above works, we adopted a surfactant, cetyltrimethylammonium bromide (CTAB), to modify the surface morphology of CuO spheres via a similar CP method. Ordered nano-needle arrays can be formed on the surface of the CuO spheres by the aid of CTAB. Each CuO sphere is about 2 μm in diameter and possesses a large number of nanoneedles that are about 20–40 nm in width and more than 300 nm in length. The needle-like hierarchical structure can greatly increase the contact area between CuO and electrolyte, which provides more sites for Li^+ accommodation, shortens the diffusion length of Li^+ and enhances the reactivity of electrode reaction, especially at high rates. After 50 cycles, the reversible capacity of the prepared needle-like CuO can sustain a capacity of 441 mAh/g. Additionally, we used $N_2H_4 \cdot H_2O$ to replace ammonia to prepare Cu_2O/Cu core/shell nanosphere composites. The corresponding reactions are described in the following [32]:

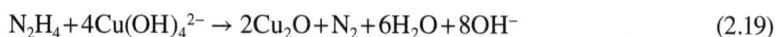

$$N_2H_4 + 4Cu(OH)_4^{2-} \rightarrow 2Cu_2O + N_2 + 6H_2O + 8OH^- \qquad (2.19)$$

$$N_2H_4 + Cu_2O \rightarrow 4Cu + N_2 + 2H_2O \qquad (2.20)$$

The core/shell Cu_2O/Cu exhibits weaker polarisation, better cycling life and higher coulombic efficiency than the pure octahedral Cu_2O due to the conductivity modification by Cu. Meanwhile, in order to further improve the LIBs performance of CuO, we prepared CuO/MWCNT nanocomposite by the CP method with CTAB The MWCNTs are incorporated into the leaf-like CuO nanoplates and build up a network to connect the CuO nanoleaves. The as-prepared CuO/ MWCNT exhibits superior reversible Li-ion storage, and the capacity maintains 627 mAh/g after 50 cycles. The improved capability is ascribed to the MWCNT conductive network in the composite. Graphene/CuO composites are also realised in the above CP method, and enhanced electrochemical performance is proved. Similar enhancement was reported by Ko et al., who prepared CP-mesoporous CuO/CNTs with a capacity of 600 mAh/g at 5C. When CuO is applied for supercapacitors, the following redox reactions may be involved in the change between Cu(I) and Cu(II) species [33,34]:

$$CuO + 1/2H_2O + e^- \rightarrow 1/2Cu_2O + OH^- \qquad (2.21)$$

$$Or\ CuO + H_2O + e^- \rightarrow CuOH + OH^- \qquad (2.22)$$

But we have to point out that the copper oxides are not good active materials candidate for supercapacitors due to its poor electrochemical redox reactivity and low discharge voltage in alkaline electrolyte, as compared to other metal oxides such as cobalt oxides and nickel oxides. On the other hand, high-performance copper oxides–based composites have attracted great attention, and great efforts are focusing on the modification of CP-copper oxides by introducing graphene, metal and carbon nanotube by CP method to further improve the EES performance. The introduction of conductive backbone or coating can improve the conductivity of the bare copper oxides leading to enhanced electrochemical properties.

Hydrothermal and Solvothermal Synthesis of CuO and CuO-Based Composites

HT and ST are facial solution-based ways to synthesise copper oxides. According to the published papers of HT/ST-copper oxides, one kind of copper sources is directly from the oxidisation of Cu foil substrate. The other kind of copper sources comes from copper salt, such as nitrate, copper sulphate and copper acetate. For the first kind of sources, we synthesised nanoflower-like CuO and CuO/Ni composite films directly on the copper foil in the mixed solution of NaOH and $(NH_4)_2S_2O_8$. The reactions occurred in this process can be summarised as follows [35,36]:

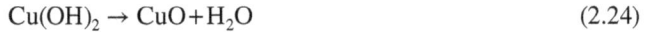

$$Cu+4OH^-+(NH_4)_2S_2O_8 \rightarrow Cu(OH)_2+2SO4^{2-}+2NH_3+2H_2O \quad (2.23)$$

$$Cu(OH)_2 \rightarrow CuO+H_2O \quad (2.24)$$

The copper substrates are used not only as a source of copper but also as a support for the copper compound films. Note the fact that pH plays a vital role during the formation of the CuO film. When the reaction is carried out in an acidic solution, no film is deposited on the copper surface. The copper will continuously be oxidised and dissolved, forming a blue solution. In pH range from 8 to 10, the copper foil loses its metallic lustre, but there is still no perceptible solid deposit on it. When the molar ratio of $[NaOH]/[(NH_4)_2S_2O_8]$ is >10 and the concentration of NaOH is >1 M, the above reaction leads to the deposition of a blue precursor film on the copper foil. At elevated temperature and high pH, $Cu(OH)_2$ can be easily converted into CuO. The formation of $Cu(OH)_2$ and CuO nanostructures on copper surfaces involves inorganic polymerisation (polycondensation) reactions under alkaline and oxidative conditions. Particle morphology may vary depending on synthetic conditions. Even ageing in aqueous solution may bring about significant dimensional, morphological, and structural changes. As a brief summary, the growth of fibre and scroll films of $Cu(OH)_2$ and copper oxides can be controlled by varying the concentration of NaOH and the reaction time if $[NaOH]/[(NH_4)_2S_2O_8]$ is in the range of 10–40. The obtained CuO/Ni composite film shows better LIBs performance than the pure CuO and delivers a better capacity retention (96.3%) than the pure CuO film (67.8%). In addition, Zhang et al. synthesised gear-like CuO film directly on Cu substrate by ammonia. The formation mechanism from Cu to porous CuO nanoarrays can be interpreted as follows:

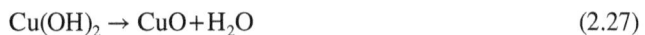

$$Cu^{2+}+4NH_3.H_2O \rightarrow Cu[NH_3]_4^{2+}+2OH^- \quad (2.25)$$

$$Cu[NH_3]_4^{2+}+2OH^- \rightarrow Cu(OH)_2+4NH_3 \quad (2.26)$$

$$Cu(OH)_2 \rightarrow CuO+H_2O \quad (2.27)$$

In the above reaction process, the HT parameters (such as the reaction time, concentration of ammonia solution and reaction temperature) have great influence on the reaction rate and morphology of the final CuO film. The as-prepared gear-like CuO delivers a capacitance of F/g at 1 A/g. But the discharge voltage is negative (−0.3 V), indicating that the CuO is not a suitable pseudocapacitive material due to the poor working voltage window.

As for the second HT-copper oxides, Gund et al. fabricated nanosheet clusters of caddice clew, yarn ball and cabbage slash–like microstructures of copper oxides in the presence of different surfactants. The possible chemical reactions in the solution can be supposed as follows

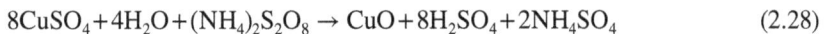

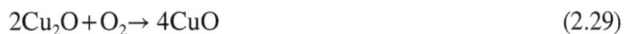

$$8CuSO_4+4H_2O+(NH_4)_2S_2O_8 \rightarrow CuO+8H_2SO_4+2NH_4SO_4 \quad (2.28)$$

$$2Cu_2O+O_2 \rightarrow 4CuO \quad (2.29)$$

These CuO microstructures show good surface properties like uniform surface morphology, good surface area and uniform pore size distribution. The obtained CuO samples present a high specific capacitance of 535 F/g. In addition, Gao et al. constructed complex CuO hollow micro/nanostructures with the use of $CuSO_4$, ammonia and tyrosine. These structures consist of CuO nanosheets, which self-organise into hollow micrometre-sized monoliths with a hierarchical architecture. The hierarchically hollow structures enhance mass transport in macroscales, promote the accessibility of the nanomaterials and greatly facilitate dispersion, transportation, separation, and recycling of nanomaterials. The synthesised hierarchical CuO hollow micro/nanostructures exhibit a high discharge/charge capacity of 560 mAh/g for LIBs.

In the following section, we introduce an important ST work of Lou's group for the synthesis of ST-copper oxides nanostructures including nanorods and nanosheets. They developed a facile ST method with $Cu(NO_3)_2$ source and 2-proponal or ethanol solvent, and the possible main reactions are expressed in the following:

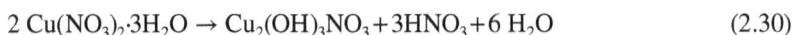

$$2\,Cu(NO_3)_2 \cdot 3H_2O \rightarrow Cu_2(OH)_3NO_3 + 3HNO_3 + 6\,H_2O \tag{2.30}$$

During the annealing process:

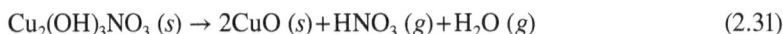

$$Cu_2(OH)_3NO_3\ (s) \rightarrow 2CuO\ (s) + HNO_3\ (g) + H_2O\ (g) \tag{2.31}$$

The morphology of $Cu_2(OH)_3NO_3$ nanostructures can be readily controlled by using different solvents, in which precursor supersaturation for nucleation and crystal growth is determined by solvent polarity and solubility. In the above method, notice that the precise control of precursor concentration and supersaturation plays a prime role in the synthesis of $Cu_2(OH)_3NO_3$ nanostructures and the degree of supersaturation directly determines the crystal growth mode. Generally both the concentration and supersaturation of the precursor for the formation of low-dimensional $Cu_2(OH)_3NO_3$ nanostructures must be kept at a relatively low level to avoid overwhelming isotropic growth or even homogeneous nucleation in the solution. The above chemistry reactions lead to the formation of unique copper oxides, nanorods (in 2-proponal) and nanosheets (in ethanol). The unique nanostructural copper oxide films show excellent electrochemical performance as demonstrated by high capacities of 450–650 mAhg^{-1} at 0.5–2^{0}C and almost 100% capacity retention over 100 cycles after the second cycle.

MnO₂ and MnO₂-Based Composites

Manganese oxides have several different forms and phases (such as MnO_2, Mn_2O_3, MnO and Mn_3O_4), and MnO_2 is the one of the most fascinating manganese oxides for EES applications due to its high capacity/capacitance, earth abundance, low cost and environmental friendliness. It is well known that MnO_2 can exist in different structural forms (e.g. α (JCPDS 44-0141), β (JCPDS 24-0735), γ (JCPDS 14-0644), δ (JCPDS 80-1098), ε (JCPDS 30-0820,), λ (JCPDS 44-0992) type), when the basic structural unit ([MnO_6] octahedron) is linked in different ways. The difference in the above polymorphs lies in the arrangement of the Mn4+ within the octahedral sites. Based on the different [MnO_6] links, MnO_2 can be divided into three categories: the chain-like tunnel structure such as α, β and γ and the sheet or layered structure such as δ-MnO_2, and the 3D structure such as λ-MnO_2. MnO_2 usually crystallises in the rutile crystal structure with three-coordinate oxide and octahedral metal centres. The band gap of MnO_2 is 0.26–0.7 eV depending on different polymorphs (β-MnO_2 is ~0.26 eV, and γ-MnO_2 is ~0.6 eV). MnO_2 has been extensively studied as an active material in EES systems since the last century.

The first kind of HT method of MnO_2 involves the chemical reactions between oxidant ($KMnO_4$) and reducing agents (such as C, Cu, ethanol and polyethylene glycol). This facile HT method fabricates self-supported Co_3O_4/MnO_2 core/shell nanowire arrays on Ti foil. In our case, we used amorphous carbon as the reducing agent and the reactions are expressed in the following:

$$4MnO^{4-}+3C+H_2O \rightarrow 4MnO_2+CO_3^{2-}+2HCO^{3-} \tag{2.32}$$

$$4MnO^{4-}+2H_2O \rightarrow 4MnO_2+4OH^-+3O_2 \tag{2.33}$$

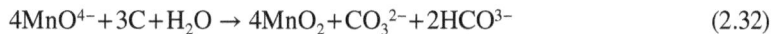

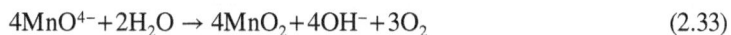

For the above reactions, when the $KMnO_4$ solution is mixed with the carbon source at room temperature before the hydrothermal processing, the nanocrystalline MnO_2 will be formed on the surface of the carbon source due to the slow redox process according to Equation (2.32). When the solution is further treated in the hydrothermal reaction, the nanoflaky MnO_2 grows from the preformed nanocrystalline due to the decomposition of $KMnO_4$ in water according to Equation (2.33), where MnO_2 nanoflakes will be formed. Except for the amorphous carbon, the carbon source can also be CNT, graphene, carbon cloth and so on.

The reaction in Equation (2.32) will promote the reaction in Equation (2.33) to take place. This main problem in this HT method is that it will consume the carbon source and it is hard to precisely control the morphology of MnO_2/carbon. In view of these characteristics, MnO_2 hollow nanostructures can be formed when the HT reaction occurs deeply and sacrifices the carbon template. The key point in this HT is to use a reducing agent to trigger the reaction. If the reducing agent is Cu or ethanol, the reactions in Equation (2.32) will be different as shown in the following [37]:

$$2MnO^{4-}+3Cu+4H_2O \rightarrow 2MnO_2+3Cu(OH)_2+2OH^- \tag{2.34}$$

$$MnO^{4-}+CH_3CH_2OH \rightarrow MnO_2+CH_3COOH+OH^-+0.5\,H_2 \tag{2.35}$$

The obtained HT-MnO_2 shows a nanoflake structure. In our experiment, the MnO_2 nanoflakes are well decorated on the surface of Co_3O_4 nanowire core forming core/shell nanowire arrays, which exhibit excellent supercapacitor performance with a high capacitance of 480 F/g at 2.67 A/g, good cycle life with 2.7% capacitance loss after 5,000 cycles and remarkable rate capability. It is reported that the pseudocapacitive (Faradic) reactions occurring on the surface and in the bulk of the electrode are the major charge storage mechanisms for manganese oxides. The surface Faradaic reaction involves the surface adsorption of electrolyte cations on the manganese oxide, illustrated as follows [38]. The bulk Faradaic reaction relies on the intercalation or deintercalation of electrolyte cations in the bulk of the manganese oxide:

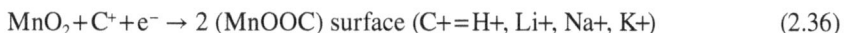

$$MnO_2+C^++e^- \rightarrow 2\,(MnOOC)\ surface\ (C+=H+,\,Li+,\,Na+,\,K+) \tag{2.36}$$

In our work, the redox process is mainly governed by the insertion and deinsertion of Li^+ and from the electrolyte into the porous nanostructured MnO_2 matrix. In addition, hierarchical TiO_2 nanobelts@MnO_2 ultrathin nanoflakes core/shell arrays were reported by the same HT method and showed a specific capacitance of 557 F/g as cathode for supercapacitors Some other MnO_2 nanostructures (such as nanosphere, hollow sphere and tubular nanostructure) have also been fabricated by this HT method and applied as electrode materials for EES. However, the above as-prepare pure MnO_2 nanostructures show a capacitance lower than 400 F/g due to the poor conductivity of MnO_2. To effectively utilise MnO_2 materials, MnO_2 composites with carbon (e.g. MnO_2/C nanosphere, MnO_2/CNT, MnO2/Graphene) are constructed and exhibit improved electrochemical performances than the pure counterparts. On the other hand, MnO_2 is a potential active material (either cathode or anode) for LIBs. The theoretical capacity of MnO_2 as LIBs anode is ~1,232 mAh/g, and the reactions are described in the following:

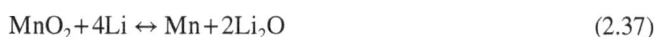

$$MnO_2+4Li \leftrightarrow Mn+2Li_2O \tag{2.37}$$

IRON OXIDES AND IRON OXIDES–BASED COMPOSITES

Iron is the fourth most common element in the earth's crust (6.3% by weight) and is usually oxidised into three kinds of oxides (Fe_2O_3, Fe_3O_4, and FeO), which are promising active materials for applications in photo-catalysis, EES and photo-electrochemical water splitting. Hematite Fe_2O_3 is the most thermodynamically stable form of iron oxide under ambient conditions, and as such, it is also the most common form of crystalline iron oxide. Fe_2O_3 has different polymorphs including α-Fe_2O_3 (rhombohedral structure, JCPDS 80-2377), β-Fe_2O_3 (orthorhombic structure, JCPDS 08-0093) and γ-Fe_2O_3 (cubic structure, JCPDS 89-5892) depending on the structure of $Fe(O)_6$ octahedra. The arrangement of cations produces pairs of $Fe(O)_6$ octahedra. Each octahedral shares edges with three neighbouring $Fe(O)_6$ octahedra in the same plane and one face with an octahedral in an adjacent plane.

To date, the HT+ST papers account for ~32% in all the published papers of solution synthesis of iron oxides for EES applications. $FeCl_3$, $FeSO_4$ and $Fe(NO_3)_3$ are the most widely used iron sources. The solvent includes water, ethanol and ethylene glycol. As one of strong acid weak alkali salts, iron salt is very easy to hydrolyse into ferric hydroxide precipitation in water at high temperature. Then, ferric hydroxide precipitation decomposes into ferric oxide and water. The reactions are as follows:

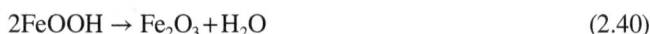

$$Fe^{3+}+6H_2O \rightarrow [Fe(H_2O)_6]^{3+} \tag{2.38}$$

$$[Fe(H_2O)_6]^{3+} \rightarrow FeOOH+3H^++4H_2O \tag{2.39}$$

$$2FeOOH \rightarrow Fe_2O_3+H_2O \tag{2.40}$$

The prepared porous Fe_3O_4/C core/shell nanorods by the above HT reactions demonstrated a high capacity of 762 mAh/g for LIBs application. Additionally, Zhang et al. reported an interesting carbon-coated Fe_3O_4 nanospindles by using $FeCl_3$ and NaH_2PO_4. The main reactions are similar to those above. But NaH_2PO_4 is important for controlling the morphology of the samples. The hydrolysis of NaH_2PO_4 produces OH^- as a precipitating agent. It controls the nucleation as well as growth of small iron oxide particles and finally forms Fe_3O_4 nanospindles. The obtained Fe_3O_4/C nanospindles show a high reversible capacity of 745 mAh/g. Meanwhile, the iron oxides and composites are also investigated as cathode materials for supercapacitors. The Fe_3O_4/reduced graphene oxide delivers a capacitance of 480 F/g. As for the supercapacitor electrode, the capacitance mechanism of ferric oxides in aqueous electrolytes is still not clear yet. It is noteworthy that the solvent has great influence on the morphology of the final products, especially in organic solvents. Therefore, solvothermal synthesis (ST) is developed for the synthesis of iron oxides and composites. Zeng et al. reported flower-like α-Fe_2O_3 by using ethanol as solvent and proposed that the tentative solvothermal precursor could be a composition of $FeOC_2H_5O$, which converts into iron oxides after annealing. [39].

In addition, they conducted a series of other investigations to further understand the effect of ethanol in the process. It is believed that ethanol may play important roles in at least three aspects: (i) acting as the solvent, (ii) acting as starting material to form the iron alkoxide precursor and (iii) providing water molecules in the system by etherification. They also found that the rate of reaction has a strong impact on the morphology of products.

As a kind of reducing agent, EG is always used to synthesise Fe_3O_4 in most cases. Our group synthesised hierarchical hollow-structured single-crystalline magnetite (Fe_3O_4) microspheres as LIBs anode. In our case, the solution is a mixture of ferric chloride, EG, polyvinylpyrrolidone (PVA) and NaAc, which is used as iron source, solvent and reducing agent, surfactant and precipitant, respectively. The possible reactions are shown in the following:

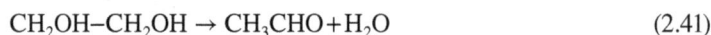

$$CH_2OH-CH_2OH \rightarrow CH_3CHO+H_2O \tag{2.41}$$

$$2CH_3CHO + Fe^{3+} \rightarrow Fe^{2+} + 2H^+ + CH_3COCOCH_3 \qquad (2.42)$$

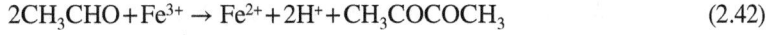

Meanwhile, NaAc hydrolyses to offer OH^- in the ST system:

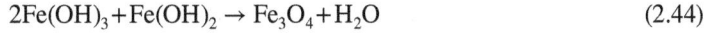

$$Fe^{3+} + 3OH^- \rightarrow Fe(OH)_3 \qquad (2.43)$$

$$2Fe(OH)_3 + Fe(OH)_2 \rightarrow Fe_3O_4 + H_2O \qquad (2.44)$$

The as-prepared hierarchical hollow-structured Fe_3O_4 shows a high specific capacity and excellent cycle performance (851 mAh/g at 1 C and 750 mAh/g at 3 C up to 50 cycles). However, the poor intrinsic electrical conductivity of iron oxides hinders its practical applications for EES. There are different approaches to overcome these problems. One solution is to fabricate iron oxide nanostructures. Other strategies are to confine the iron oxide nanostructures with carbon to form a composite with the carbon host matrix. Fe_3O_4/C core/shell structures are synthesised to improve the conductivity of the integrate electrode.

VANADIUM OXIDES AND VANADIUM OXIDES–BASED COMPOSITES

The hydrothermal and solvothermal synthesis methods are popular with materials researchers and have been widely used for the fabrication of vanadium oxides and composites. Generally, there are two common types of HT routes for the synthesis of V_2O_5 for EES. One HT method is to use the metavanadate ($NaVO_3$ or NH_4VO_3) and HCl as the starting materials and utilise the hydrolysis of ammonium metavanadate in the acid solution to form vanadium oxides after heat treatment. The other HT method is to utilise V_2O_5 sols precursor resulting from the reactions between commercial V_2O_5 powder and H_2O_2, and finally obtain V_2O_5 nanostructures after hydrothermal synthesis and annealing. Carbon sources (such as graphene, carbon nanotube and sucrose) and V_2O_5 are usually used as the starting materials to prepare HT-VO_2 and their composites [41]:

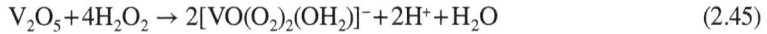

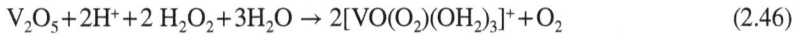

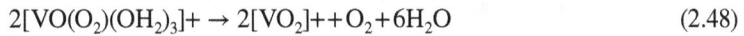

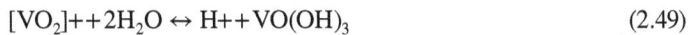

$$V_2O_5 + 4H_2O_2 \rightarrow 2[VO(O_2)_2(OH_2)]^- + 2H^+ + H_2O \qquad (2.45)$$

$$V_2O_5 + 2H^+ + 2 H_2O_2 + 3H_2O \rightarrow 2[VO(O_2)(OH_2)_3]^+ + O_2 \qquad (2.46)$$

$$2[VO(O_2)_2(OH_2)]^- + 4H^+ + 2H_2O \rightarrow 2[VO(O_2)(OH_2)_3]^+ + O_2 \qquad (2.47)$$

$$2[VO(O_2)(OH_2)_3]+ \rightarrow 2[VO_2]+ + O_2 + 6H_2O \qquad (2.48)$$

$$[VO_2]+ + 2H_2O \leftrightarrow H+ + VO(OH)_3 \qquad (2.49)$$

During the annealing process:

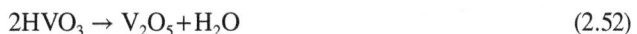

$$2VO(OH)_3 \rightarrow V_2O_5 + 3H_2O \qquad (2.50)$$

In addition, the V_2O_5 nanobelt shows a high reversible specific capacity of 163 mAh/g at 6 C. All the above HT vanadium oxides are power materials, but not films.

The other HT of V_2O_5 is associated with the hydrolysis of metavanadate. For example:

$$2NH_4VO_3 + H_2O \rightarrow NH_3 \cdot H_2O + HVO_3 \qquad (2.51)$$

During the annealing process:

$$2HVO_3 \rightarrow V_2O_5 + H_2O \qquad (2.52)$$

VO_2/graphene nanoribbon and nanobelts reported by Yan's group and Yang et al. exhibit a capacity of 160–300 mAh/g. Nanobelt and nanoribbon are very thin and favourable for fast ion/electron transfer. The introduction of graphene improves the conductivity of composite leading to enhanced properties and cycling stability. In parallel to HT, solvothermal synthesis (ST) of vanadium oxides is also extensively studied to construct vanadium oxides nanostructures including rod-like, nanosphere, urchin-like and hollow microflower. The vanadium sources of ST are different from those of HT. Most of the vanadium sources are vanadium-based organic solvents such as vanadium oxytripropoxide and vanadium isopropoxide. Usually, the solvents are ethylene glycol, ethanol or N-methylpyrrolidone. The specific reactions are still not fully clear and need further investigation. The ST-vanadium oxides nanostructures could deliver a higher specific capacitance of 537 F/g at a current density of 1 A/g in neutral aqueous electrolytes. The ST-V_2O_5 hollow microflowers exhibit a remarkable reversible capacity of 211 mAh/g and good cycling stabilities and excellent rate capabilities [40,43].

Tungsten Oxides and Tungsten Oxides–Based Composites

Tungsten oxide (WO_3), also known as tungsten trioxide, has attracted extensive attention because of its distinctive physical and chemical properties, making it suitable for applications in electrochromic devices, photo-catalysis, gas sensing and EES. To date, most of the nanostructured WO_3 for EES have been fabricated by the HT method, which is classified into two types according to different mechanism and tungsten sources: Na_2WO_4 and metal W. For the first one, the HT-WO_3 uses Na_2WO_4 and HC as the starting materials and the WO_3 precipitation formed in a concentrated acid solution containing tungsten ions. The reactions are described in the following [42,43]:

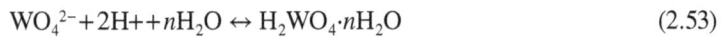

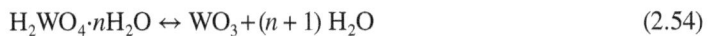

$$WO_4^{2-}+2H++nH_2O \leftrightarrow H_2WO_4{\cdot}nH_2O \qquad (2.53)$$

$$H_2WO_4{\cdot}nH_2O \leftrightarrow WO_3+(n+1)\,H_2O \qquad (2.54)$$

Different WO_3 nanostructures and composites such as nanoflake, nanowire, nanoplate, nanoflowers WO_3/SnO_2 and WO_3/graphene have been prepared by the HT method and applied as active materials for EES. It is noteworthy that WO_3 nanowires can be formed with the help of $(NH_4)_2SO_4$ and each nanowire is a hexagonal single crystal and their long axes are oriented toward the [0001] direction. In the presence of an appropriate amount of ammonium sulphate in the solution, WO_3 primary particles aggregate along the [0001] direction of the h-WO_3 unit cell via self-assembly, because sulphate ions preferentially adsorb on the faces parallel to the c-axis of the WO_3 nanocrystal and thus 1D single crystal nanowires are finally formed. These WO_3-based nanostructures show quite good electrochemical performances. The ordered WO_3 nanowire arrays on carbon cloth exhibit a high specific capacitance of 521 F/g at 1 A/g. The WO_3/graphene nanocomposite electrode delivers a reversible lithium storage capacity of 656 mAh/g after 100 cycle at 100 mA/g and an enhanced cyclability compared with the bare WO_3 nanowires electrode. The obtained capacitance and capacity are comparable to those of other metal oxides. The other HT can be called sol-based HT method, which first prepares WO_3 sols by reacting metal W with H_2O_2. Then the WO_3 sols are treated in HT condition to form WO_3 nanomaterials. The possible reaction is very complex, as shown here

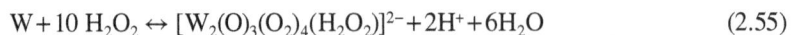

$$W+10\,H_2O_2 \leftrightarrow [W_2(O)_3(O_2)_4(H_2O_2)]^{2-}+2H^++6H_2O \qquad (2.55)$$

The hexagonal WO_3 nanostructures are fabricated by the sol-based HT and show a discharge capacity of 215 mAh/g. The first HT method is much easier than the second one, which is tedious and difficult for the preparation of sols.

CONCLUSIONS

EES devices and systems are playing an important role in developing a secure and green energy future of human society. The performance of EES systems, both lithium ion batteries and supercapacitors, is mainly determined by the electrode materials. Among the electrode materials studied so far, metal oxides–based electrodes can provide high energy density with the Faradic reaction, and thus have drawn much attention for next-generation, high-performance EES devices. In the past decades, the nanoscale metal oxides and their composites have elicited much interest due to their distinctive structural features and intriguing properties. Generally, compared to bulk materials, nanostructured metal oxides can provide many merits, including enlarged electrode/electrolyte contact area for reactions, short ion diffusion distance and better charge transfer kinetics.

REFERENCES

[1] S. Kulandaivalu, M. Z. Hussein, A. M. Jaafar, M. A. Abdah, N. H. Azman and Y. Sulaiman, A simple strategy to prepare a layer-by-layer assembled composite of Ni-Co LDHs on polypyrrole/rGO for a high specific capacitance supercapacitor. *RSC Adv.* 2019; **9**(69): 40478–40486.

[2] R. Liang et al. Transition metal oxide electrode materials for supercapacitors: A review of recent developments. *Nanomaterials* 2021; **11**: 1248. DOI: 10.3390/nano11051248

[3] D. Choi, G. E. Blomgren and P. N. Kumta, Fast and reversible surface redox reaction in nanocrystalline vanadium nitride supercapacitors. *Adv. Mater.* 2006; **18**: 1178–1182.

[4] A. S. Arico, P. Bruce, B. Scrosati and J. M. Tarascon, and W. Van Schalkwijk, Nanostructured materials for advanced energy conversion and storage devices. *Nat. Mater.* 2005; **4**: 366–377.

[5] J. Hassoun and B. Scrosati, A high-performance polymer tin sulfur lithium ion battery. *Angew. Chem. Int. Ed.* 2010; **49**: 2371–2374.

[6] H. Mirzaei and M. Darroudi, Zinc oxide nanoparticles: Biological synthesis and biomedical applications. *Ceram. Int.* 2017; **43**: 907–914.

[7] A. Khorsand Zak, R. R. Razali, W. H. Abd Majid and M. Darroudi, Synthesis and characterization of a narrow size distribution of zinc oxide nanoparticles. *Int. J. Nanomed.* 2011; **6**: 1399.

[8] M. Hudlikar, S. Joglekar, M. Dhaygude and K. Kodam, Latex-mediated synthesis of ZnS nanoparticles: Green synthesis approach. *J. Nanopart. Res.* 2012; **14**: 865.

[9] G. Singhal, R. Bhavesh, K. Kasariya, A. R. Sharma and R. P. Singh, Biosynthesis of silver nanoparticles using Ocimum sanctum (Tulsi) leaf extract and screening its antimicrobial activity. *J. Nanopart. Res.* 2011; **13**: 2981–2988.

[10] A. Ray, A. Roy, S. Saha and S. Das, Transition metal oxide-based Nanomaterials for energy storage application. In: *Science, Technology and Advanced Application of Supercapacitors.* London: Intechopen. 2018 [Online]. DOI: 10.5772/intechopen.80298

[11] J. Noh, C. M. Yoon, Y. K. Kim and J. Jang, High performance asymmetric supercapacitor twisted from carbon fiber/MnO2 and carbon fiber/MoO3. *Carbon* 2017; **116**: 470–478.

[12] J. H. Zheng, R. M. Zhang, X. G. Wang and P. F. Yu, Synthesizing a flower-like NiO and ZnO composite for supercapacitor applications. *Res. Chem. Intermed.* 2018; **44**(9): 5569–5582.

[13] B. Varshney, M. J. Siddiqui, A. H. Anwer, M. Z. Khan, F. Ahmed, A. Aljaafari et al. Synthesis of mesoporous SnO2/NiO nanocomposite using modified sol-gel method and its electrochemical performance as electrode material for supercapacitors. *Sci. Rep.* 2020; **10**(1): 13.

[14] Y. Q. Du, P. Xiao, J. Yuan and J. W. Chen, Research Progress of graphene-based materials on flexible supercapacitors. *Coatings* 2020; **10**: 892.

[15] W. Cai, R. Kumar Kankala, M. Xiao, N. Zhang, and X. Zhang, Three-dimensional hollow N-doped ZIF-8-derived carbon@ MnO2 composites for supercapacitors. *Appl. Surf. Sci.* 2020: 146921. DOI: 10.1016/j.apsusc.2020.146921

[16] L. Ye, C. Wang and Y. Le et al. Construction of MnO2 nanosheets @graphenated carbon nanotube networks core-shell heterostructure on 316L stainless steel as binder-free supercapacitor electrodes. *Int. J. Hydrogen Energy.* 2020; **45**(53): 28930–28939. DOI: 10.1016/j.ijhydene.2019.09.070

[17] K.J. Gilmore, G. G. Wallace, D. Li, B. H. Chen and M. B. Mu, Mechanically strong, electrically conductive, and biocompatible graphene paper. DOI: 10.1002/adma.200800757

[18] A. Dutta, S. Mishra, S. K. Saha, S. Sarkar, A. Guchhait and A. J. Akhtar, Boosting the Supercapacitive performance of ZnO by 3-dimensional conductive wrapping with graphene sheet. *J. Inorg. Organomet. Polym. Mater.* 2022; **32**(1): 180–190.

[19] M. Sreejesh, S. Dhanush, F. Rossignol and H. S. Nagaraja, Microwave assisted synthesis of rGO/ZnO composites for non-enzymatic glucose sensing and supercapacitor applications. *Ceram. Int.* 2017; **43**: 4895–4903.

[20] J. Yus, Y. Bravo, A. J. Sanchez-Herencia, B. Ferrari and Z. Gonzalez, Electrophoretic deposition of RGO-NiO coreshell nanostructures driven by heterocoagulation method with high electrochemical performance. *Electrochim. Acta.* 2019; **308**: 363–372.

[21] O. C. Pore, A. V. Fulari, V. G. Parale, H. H. Park, R. V. Shejwal, V. J. Fulari et al. Facile hydrothermal synthesis of NiO/rGO nanocomposite electrodes for supercapacitor and nonenzymatic glucose biosensing application. *J. Porous Mater.* 2022: 1–1.

[22] M. A. A. M. Abdah, N. H. N. Azman, S. Kulandaivalu et al. Review of the use of transition-metal-oxide and conducting polymer-based fibres for high-performance supercapacitors. *Mater. Des.* 2019; **186**: 108199. DOI: 10.1016/j.matdes.2019.108199

[23] X. H. Huang, J. P. Tu, C. Q. Zhang and F. Zhou, Hollow microspheres of NiO as anode materials for lithium-ion batteries. *Electrochim. Acta* 2010; **55**: 8981–8985.

[24] S. I. Kim, J. S. Lee, H. J. Ahn, H. K. Song and J. H. Jang, Facile route to an efficient NiO supercapacitor with a three-dimensional nanonetwork morphology. *ACS Appl. Mater. Interfaces* 2013; **5**: 1596–1603.

[25] M. C. Liu, L. B. Kong, C. Lu, X. M. Li, Y. C. Luo and L. Kang, NiCo$_2$S$_4$ on yeast-templated porous hollow carbon spheres for supercapacitors. *ACS Appl. Mater. Interfaces* 2012; **4**: 4631–4636.

[26] J. P. Liu, J. Jiang, C. W. Cheng, H. X. Li, J. X. Zhang, H. Gong and H. J. Fan, Co$_3$O$_4$ Nanowire@MnO$_2$ ultrathin nanosheet core/shell arrays: a new class of high-performance pseudocapacitive materials. *Adv. Mater.* 2011; **23**: 2076–2081.

[27] X. H. Xia, J. P. Tu, Y. Q. Zhang, J. Chen, X. L. Wang, C. D. Gu, C. Guan, J. S. Luo and H. J. Fan, Hollow nickel nanocorn arrays as three-dimensional and conductive support for metal oxides to boost supercapacitive performance. *Chem. Mater.* 2012; **24**: 3793–3799.

[28] C. Guan, J. P. Liu, C. W. Cheng, H. X. Li, X. L. Li, W. W. Zhou, H. Zhang and H. J. Fan, Hybrid structure of cobalt monoxidenanowire @ nickel hydroxidenitrate nanoflake aligned on nickel foam for high-rate supercapacitor. *Energy Environ. Sci.* 2011; **4**: 4496–4499.

[29] J. Chen, X. H. Xia, J. P. Tu, Q. Q. Xiong, Y. X. Yu, X. L. Wang and C. D. Gu, Three dimensional hierarchical pompon-like Co3O4 porous spheres for high-performance lithium-ion batteries. *J. Mater. Chem.* 2012; **22**: 15056–15061.

[30] G. Filipic and U. Cvelbar, Copper oxide nanowires: a review of growth. *Nanotechnology* 2012: 23.

[31] J. Xiang, J. Tu, L. Zhang, Y. Zhou, X. Wang and S. Shi, Self-assembled synthesis of hierarchical nanostructured CuO with various morphologies and their application as anodes for lithium ion batteries. *J. Power Sources* 2010; **195**: 313–319.

[32] B. Heng, C. Qing, D. Sun, B. Wang, H. Wang and Y. Tang, Rapid synthesis of CuO nanoribbons and nanoflowers from the same reaction system, and a comparison of their supercapacitor performance. *RSC Adv.* 2013; **3**: 15719–15726.

[33] J. Xiang, J. Tu, X. Huang and Y. Yang, Surface treatment of copper (II) oxide nanoparticles using citric acid and ascorbic acid as biocompatible molecules and their utilization for the preparation of poly(vinyl chloride) novel nanocomposite films. *J. Solid State Electr.* 2008; **12**: 941–945.

[34] J. Xiang, J. Tu, Y. Yuan, X. Wang, X. Huang and Z. Zeng, Nanostructured CuO/Co$_2$O$_4$@ nitrogen doped MWCNT hybrid composite electrode for high-performance supercapacitors. *Electrochim. Acta.* 2009; **54**: 1160–1165.

[35] L. Yu, Y. Jin, L. Li, J. Ma, G. Wang, B. Geng and X. Zhang, Facile synthesis of hierarchical CuO nanorod arrays on carbon nanofibers for high-performance supercapacitors. *Cryst. Eng. Comm.* 2013; **15**: 7657–7662.

[36] G. S. Gund, D. P. Dubal, D. S. Dhawale, S. S. Shinde and C. D. Lokhande, Porous CuO nanosheet clusters prepared by a surfactant assisted hydrothermal method for high performance supercapacitors. *RSC Adv.* 2013; **3**: 24099–24107.

[37] J. L. Liu, L. Z. Fan and X. H. Qu, Facile synthesis of ordered porous Si@C nanorods as anode materials for Li-ion batteries. *Electrochim. Acta* 2012.

[38] S. W. Bian, Y. P. Zhao and C. Y. Xian, MnO2-based nanostructures for high-performance supercapacitors. *Mater. Lett.* 2013; **111**: 75–77.

[39] S. Zeng, K. Tang, T. Li, Z. Liang, D. Wang, Y. Wang, Y. Qi and W. Zhou, Controlled Growth of Large-Area, Uniform, Vertically Aligned Arrays of α-Fe$_2$O$_3$ Nanobelts and Nanowires. *J. Phys. Chem. C* 2008; **112**: 4836–4843.

[40] A. Q. Pan, H. B. Wu, L. Zhang and X. W. Lou, Uniform V$_2$O$_5$nanosheet-assembled hollow microflowers with excellent lithium storage properties. *Energy Environ. Sci.* 2013; **6**: 1476.

[41] S. Yang, Y. Gong, Z. Liu, L. Zhan, D. P. Hashim, L. Ma, R. Vajtai and P. M. Ajayan, Two-dimensional nanosheets as building blocks to construct three-dimensional structures for lithium storage. *Nano Lett.* 2013; **13**: 1596–1601.

[42] X. X. Li, G. Y. Zhang, F. Y. Cheng, B. Guo and J. Chen, Compact TiO_2@SnO_2@C heterostructured particles as anode materials for sodium-ion batteries with improved volumetric capacity. *J. Electrochem. Soc.* 2006; 153: H133–H137.

[43] C. Y. Kim, M. Lee, S. H. Huh and E. K. Kim, Sol–gel based materials for biomedical applications. *J. Sol-Gel Sci. Technol.* 2010; **53**: 176–183.

3 Inorganic Nanomaterials
A Review of Synthesis, Properties and Recent Breakthroughs in Battery and Fuel Cell Technology

M. Pushpa, B. Pushpa, Soumya S. Bulla, and Chetan Chavan

INTRODUCTION

In the contemporary period, there exists a pressing want for power-generating methods that prioritise environmental preservation. Consequently, this necessity has spurred substantial scholarly investigation into energy conversion technologies, including batteries and fuel cells [1]. Considerable efforts have been directed toward the development of economically viable batteries and fuel cells, with the aim of reducing reliance on oil and mitigating environmental damage. The majority of technologies are closely linked to energy consumption, and when technical advancements occur, there is an inevitable and concurrent rise in energy use. Additionally, in the twenty-first century, oil has emerged as the predominant source of energy. The use of these particular fuels for energy production is linked to significant releases of carbon, sulphur and nitrogen oxides, along with byproducts resulting from incomplete combustion, into the Earth's atmosphere. The escalating levels of contamination of the environment and the diminishing availability of natural gas and oil reserves have prompted a pressing need to explore sustainable and environmentally friendly sources of energy. Energy consumption, on the other hand, has a contrasting pattern. Wind power producers also face similar challenges [2]. This necessitates the use of energy storage technologies. Among the many options, batteries and fuel cells (FC)–driven systems are widely recognised as among the most viable alternatives. The yearly energy variations cannot be effectively mitigated by the installation of batteries due to their high self-discharge rate. Nevertheless, they are ideal for the purpose of short-term energy storage. Another noteworthy concern is associated with the notion of electrical power sources for handheld gadgets and cordless electric tools. Currently, there is a significant emphasis on inorganic nanoparticles (NPs) for batteries and fuel cell technology due to their high capacity and reliable performance. This study primarily focuses on exploring the applications of these materials in the context of batteries and fuel cells. Inorganic NPs possess distinctive characteristics, including exceptional electrical and thermal conductive properties, substantial surface area and chemical stability. Consequently, these materials assume a pivotal function in the advancement of new energy applications, facilitating improved performance [3,4]. In order to extend the distinctive use and employment of various nanomaterials (NMs), NPs are synthesised in discrete manner to obtain numerous forms such as clusters, bulk materials and powder NPs. The methods for the synthesis of NMs are mentioned in Figure 3.1.

Inherent to their composition, inorganic NPs possess exceptional physicochemical properties, including catalysts, optical and thermal energy, electrical and magnetic attributes. These distinctive qualities enable them to provide significant capabilities, including but not limited to medication transport, diagnostics, imaging and treatment. NMs often have a surface layer made up of metal ions, polymers and tiny molecules. The exterior layer of inorganic NMs with material that differs

DOI: 10.1201/9781003479239-3

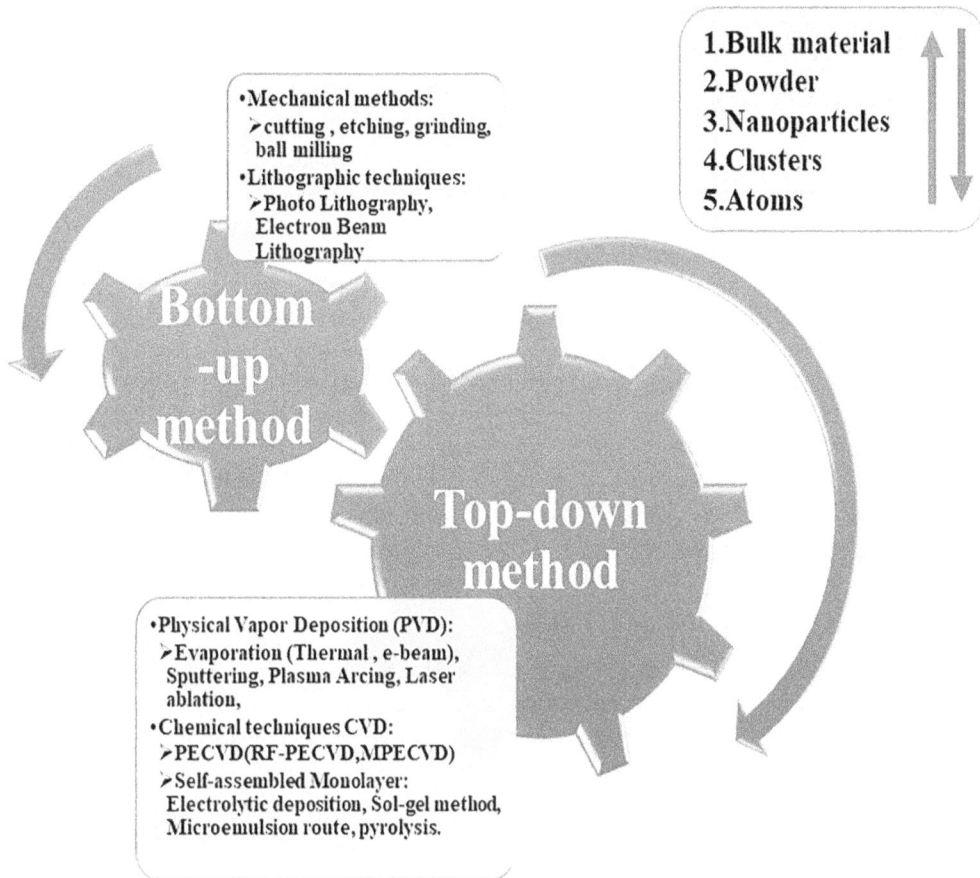

1. Bulk material
2. Powder
3. Nanoparticles
4. Clusters
5. Atoms

• Mechanical methods:
  ➤ cutting, etching, grinding, ball milling
• Lithographic techniques:
  ➤ Photo Lithography, Electron Beam Lithography

Bottom-up method

Top-down method

• Physical Vapor Deposition (PVD):
  ➤ Evaporation (Thermal, e-beam), Sputtering, Plasma Arcing, Laser ablation,
• Chemical techniques CVD:
  ➤ PECVD (RF-PECVD, MPECVD)
  ➤ Self-assembled Monolayer: Electrolytic deposition, Sol-gel method, Microemulsion route, pyrolysis.

FIGURE 3.1 Synthesis process of nanomaterials.

chemically from the core is referred to as the shell. Particular features of NMs depend on the composition of the core, which is a crucial component of NMs. There are two distinct approaches to the nanoscale: the top-down approach and the bottom-up approach. Reducing the size of the structure to the nanoscale is what the top-down technique entails. On the other hand, the bottom-up approach entails the construction of bigger nanostructures by the assembly of smaller molecular and atomic particles. Different top-down methods (ball milling, thermal evaporation, laser ablation and sputtering) are used to synthesise NMs. Along with these methods there are some other different ways from which the NMs are being created such as chemical vapour deposition, thermal decomposition, hydrothermal synthesis, solvothermal synthesis, templating, combustion, microwave synthesis, gas phase method and traditional sol-gel method [5]. These are categorised as bottom-down approach methods. The synthesised distinct NMs have been employed for abundant zones to study their applications or implementations. This paper provides an information about the recent breakthroughs and about the forthcoming challenges in use of these NMs in the energy storage section. Before studying the applications of NMs, we should know about the basic terms of NMs such as detailed descriptions on definition, classification, types, synthetic routes and their particular utilisation in energy storage like batteries and fuel cells. Reportedly, Nanotechnology and Nanoscience constitute where the adventure originates.

NANOMATERIALS

The significance of evolution in the fields of nanotechnology and nanoscience has been underscored by recent technological progress. Nanotechnology is a multidisciplinary academic domain that spans several scientific disciplines, including physics, chemistry, materials science and engineering. Nanotechnology encompasses the manipulation of materials and devices by the modification of their size and form at the nanoscale. This field involves the processes of growth and development, and the synthesis process, analysis and application of such materials and technologies. The use of nanotechnology is seeing a notable rise across several domains of research and technology. The distinction between nanoscience and nanotechnology lies in their respective focuses and applications. Nanoscience primarily aims to elucidate the organisation and fundamental characteristics of atoms at the nanoscale. By contrast, nanotechnology pertains to the manipulation and control of matter at the atomic level, enabling the development of novel NMs endowed with diverse properties [6–8]. The "nanomaterials" are the fundamental and important components of nanotechnology. Materials with a size of less than 100 nm, at least in one dimension, are referred to as nanomaterials. This indicates that they are much smaller than microscale. The size of NMs is typically 10^{-9} m, or one billionth of a metre. Nanotechnology involves several grouping or classes of NMs that are classified in accordance with their size and shapes.

CLASSIFICATION OF NANOMATERIALS

In spite of the fact that this technology was used for a lot of work, there is still room for creating unique NMs in a variety of domains for human advancement. Physical, chemical and mechanical processes have advanced significantly over time. The manipulation of shape and dimension at the level of the nanoscale results in the emergence of unique characteristics and capabilities in NMs. Based on their size at the nanoscale level precisely at 100 nm, the objects under investigation are divided into different categories: zero-dimensional (0-D), one-dimensional (1-D), two-dimensional (2-D), three-dimensional (3-D) and bulk NMs. In general, the organic, inorganic and carbon-based NPs are categorised as having better qualities than bigger sizes of the corresponding materials [9–11]. Based on their nanoscale size, NMs are classified into the following types as shown in Figure 3.2.

As mentioned earlier, the classification of NMs can also be categorised as organic, inorganic and carbon-based, which is done on the foundation of the presence of carbon and other metals in NMs. These types of NMs are briefly discussed as follows.

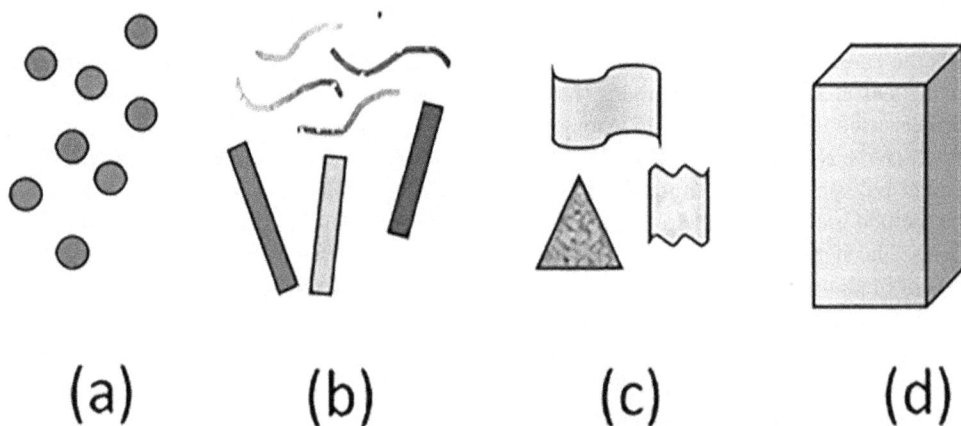

(a)   (b)   (c)   (d)

FIGURE 3.2 Classification of nanomaterials: (a) zero-dimensional, (b) one-dimensional, (c) two-dimensional and (d) three-dimensional.

Organic nanomaterials: The majority of organic-based NMs come from non-carbon chemicals so that carbon compounds are the lone exception. Natural or artificial organic compounds serve as templates for organic NPs. Only organic components, excluding carbon molecules, are employed to create organic-based NMs. Dendrimers, cyclodextrin, liposomes and micelles represent a selection of exemplars [12].

Inorganic nanomaterials: The primary distinction between organic and inorganic NPs is that the former are composed of molecules with a carbon base, whereas the latter are composed of molecules without a carbon base. Elemental metals, metal oxides and metal salts are only a few examples of the wide variety of chemicals that make up inorganic NPs.

Carbon-based nanomaterials: The family of carbon allotropes, known as carbon NMs, is large. Research in nanoscience and nanotechnology has been sparked by this substance due to the extraordinary electrical, mechanical and chemical characteristics of carbon-based NMs. Carbon nanotubes, carbon nanofibers, graphene nanosheets, graphene quantum dots, graphene nanoribbons and graphene NPs are examples of carbon NMs.

All the types of NMs exhibit different kinds of properties to enhance their applications in most of the fields.

PROPERTIES OF NANOMATERIALS

Different from those of bulk materials, NMs exhibit unique features. The majority of nanostructure materials are naturally crystalline, and they possess special qualities that result in notable increases in their mechanical properties. Many researchers have recently intensified their efforts to learn more about the physical characteristics of metallic NPs. NMs display various characteristics, including physical, chemical, mechanical, electrical, magnetic and optical characteristics. The consequences of size dependence are increasingly prominent at the nanoscale, area of contact, molecular effects of magnetism, superior mechanical characteristics, high thermal conductivity and electrical conductivity, exemplary backing for catalysts and bacterial resistance. The NMs with these unique properties are synthesised by employing various methods or synthetic routes.

SYNTHESIS OF NANOMATERIALS

Three alternative methods can be used to create NPs. They are biological, physical and chemical methods. The emerging discipline of nanotechnology exhibits the capacity to supplant conventional micron technologies and provide advantageous material characteristics that are contingent upon size. A variety of innovative materials must be affordable and have well-controlled size, forms, porosity, crystalline phases and structures. Top-down methods for the creation of NMs include the mechanical/ball milling, electrospinning, lithography, sputtering, arc discharge method and laser ablation. Bottom-down approach methods include solvothermal/hydrothermal, soft and hard templating methods, chemical vapour deposition (CVD) and sol-gel method [13]. The classification of synthesis techniques of NMs is portrayed in Figure 3.3.

As shown in Figure 3.3, some of the techniques are discussed briefly as follows.

TOP-DOWN APPROACH

Mechanical milling: A very practical and promising method of creating powdered NPs is mechanical milling. These days, industry frequently uses magnetic oxide particles produced directly chemically. It is common knowledge that ball milling is a technique for combining, blending, shaping and reducing particle sizes.

Laser ablation: A laser beam is focused on a sample surface during laser ablation (LA), which removes material from the irradiated area. Many technical applications have been thought of and applied with laser ablation-producing NMs, depositing thin metallic and dielectric layers, creating

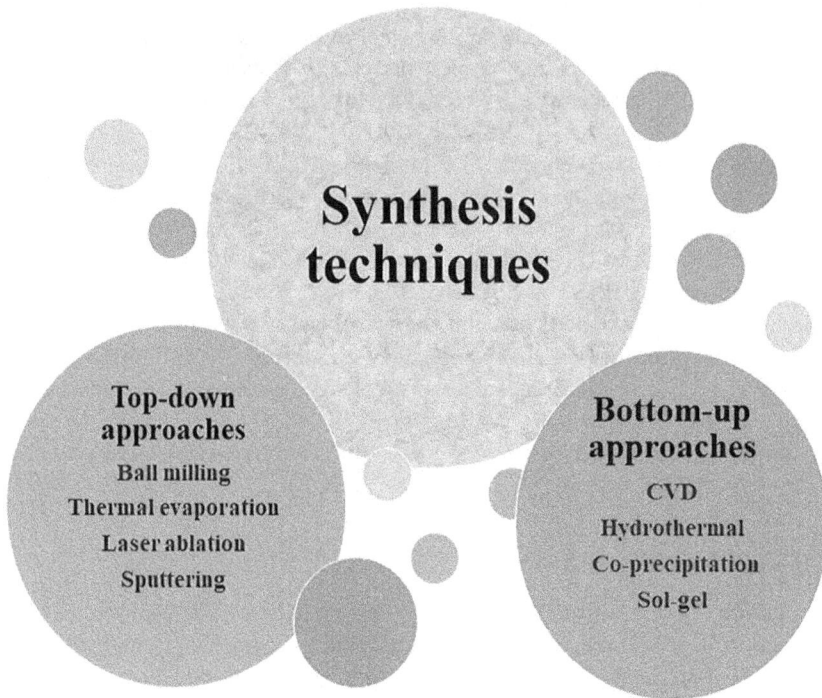

FIGURE 3.3   Methods for the synthesis of nanomaterials.

superconductors, welding and bonding metal parts on a regular basis, and micromachining MEMS systems.

Sputtering: Sputter deposition is the typical definition of a sputtering method of coating that uses high vacuum and is part of the PVD process family. Sputtering is also used in surface physics to prepare high-purity surfaces by cleaning them, as well as to determine the chemical makeup of surfaces.

Lithography: For creating nanoarchitectures, lithography uses a concentrated electron or light beam. To create the necessary structure and shape, print the needed structure or shape onto the material utilised to create large-area periodic nanostructures with control over material, size and period. It is considered a low cost-effective technique.

Similarly, the other variety used for the synthesis of NMs is bottom-down approach.

BOTTOM-DOWN APPROACH

Chemical vapour deposition (CVD): One of the most often utilised bottom-up techniques for the creation of thin films and NPs is chemical vapour deposition (CVD). It is more challenging to deposit chemical vapour than physical vapour. It involves the reaction or dissolution of one or more gaseous adsorption species.

Sol-gel method: For the creation of different nanostructures, the sol-gel procedure is a more chemical (wet chemical) method, most notably metal oxide NPs. It is used in businesses that produce specialised garments, building insulation and surface coating. Compared to other techniques, the sol-gel process is more widely used and has more industrial applications.

Hydrothermal/solvothermal: The term "solvothermal synthesis" refers to a method used in the production of a diverse array of materials, such as metals, semiconductors, ceramics and polymers. A solvent is used in the procedure at a moderate to high pressure and high temperature that encourages the precursors' contact with one another during synthesis. Water should be the solvent. The process is also known as "hydrothermal synthesis."

The classification of NMs involves two sets of categories; that is, on the basis of size, they are grouped as 0-D, 1-D, 2-D and 3-D NMs. Along with this, there is an another class of NMs which is characterised on the basis of the elements (like carbon, metals and organic moieties) that form NMs and on the composition of NMs.

TYPES OF NANOMATERIALS

The field of nanotechnology is related to a lot of new material terminology. A brief summary of a few of the numerous varieties of NMs are provided below. Interesting physical, chemical and biological properties are provided by NMs. Inorganic-, carbon-, organic- and composite-based NMs are the four different types of NMs that can be distinguished [14]. Different metal and metal oxide NPs are typically included in inorganic-based NMs. Silver (Ag), gold (Au), aluminium (Al), cadmium (Cd), copper (Cu), iron (Fe), zinc (Zn) and lead (Pb) NPs are a few examples of metal-based inorganic NMs. Similarly, the examples of inorganic NMs based on metal oxides can include silica (SiO_2), iron oxide (Fe_3O_4), magnesium aluminium oxide ($MgAl_2O_4$), titanium dioxide (TiO_2), cerium oxide (CeO_2) and zinc oxide (ZnO) [15]. The types of inorganic NMs can also categorised as carbon-based NMs which include the examples like the single-walled carbon nanotube, multi-walled carbon nanotube, carbon fibre, activated carbon, and carbon black, graphene and fullerene. The composite-based NMs are another form of inorganic NMs which are made by two or more inorganic NMs that are having different physical and chemical properties. The simple classification of inorganic NMs is picturised in Figure 3.4.

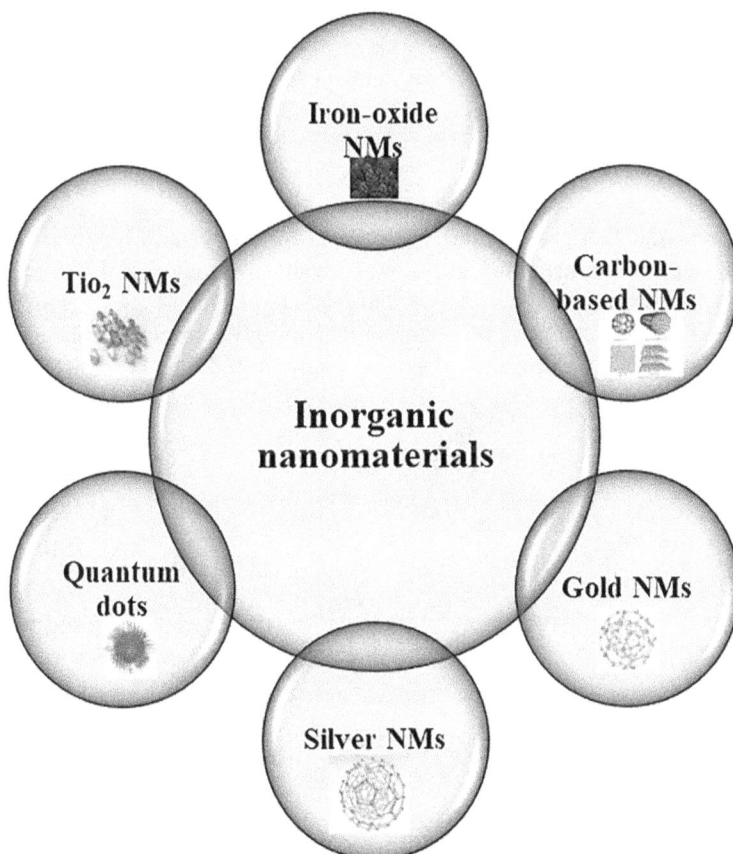

FIGURE 3.4 Simple representation of types of inorganic nanomaterials.

In recent times, there has been a significant surge in interest around inorganic NMs possessing two dimensions, often referred to as 2D-LINs. This heightened emphasis may be attributed to the unique physical, chemical and technological characteristics that are contingent upon the thickness of these materials. In recent times, there have been successful advancements in the synthesis of two-dimensional layered inorganic NMs, including MoS_2, WS_2 and SnS_2. The use of inorganic NMs is rising quickly because of their essential categories and significant themes. Inorganic NM's distinctive carbon structure gives them outstanding performance in the electrical, optoelectronic, mechanical and chemical fields. By changing the composition, structure and morphology of inorganic NPs, the properties of those materials can be further modified. In spite of all these properties and types of NMs that are synthesised by numerous techniques, NMs extend their unique applications and are employed for many uses including the automobiles, engineering, sports, textiles, food packaging and conservation, and pharmaceutical industries. This book chapter is mainly focusing on the energy storage applications such as batteries and fuel cell.

APPLICATIONS OF NANOMATERIALS

To prevent environmental contamination, which is a major source of fossil fuels, an economical and environmentally acceptable energy storage technology is crucial. Super capacitors, fuel cells and rechargeable batteries are potential energy storage technologies. The general or overall applications of NMs are outlined in Figure 3.5.

It is used to create structures made of coal, silicon, inorganic minerals, metals and semiconductors that don't function in humid environments [16]. As mentioned in the above figure, electronics, energy biomedicine, environment, food, textile, cosmetics, catalysis, agriculture, etc. are all the most common applications of NMs. As mentioned earlier, this study emphasises the discussion about the applications of inorganic NMs in the field of energy storage such as batteries and fuel cell applications [17].

BATTERIES

Currently, batteries and fuel cells are crucial power sources because they will still be utilised in numerous consumer, industrial and military applications throughout the twenty-first century. Leading the way in research for a sustainable future is the creation of energy conversion and storage technologies. Battery technology is becoming increasingly important as the electronics industry develops, allowing utility electricity to be used freely for lighting, portable electronic devices,

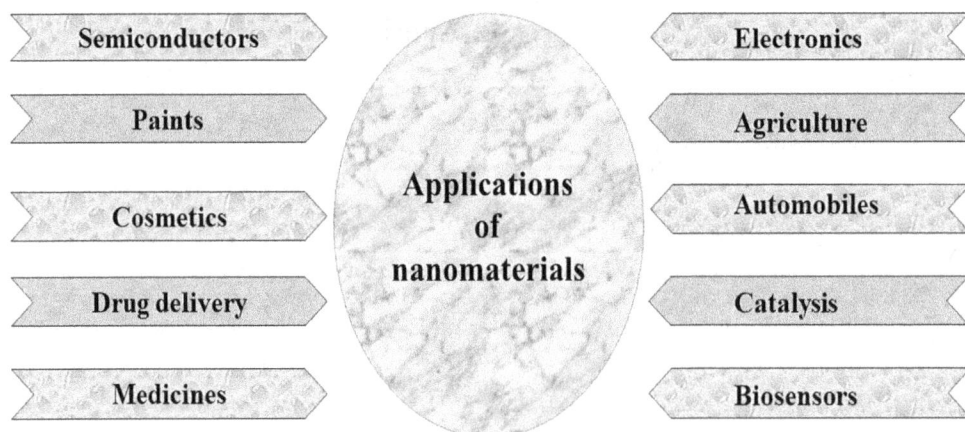

Semiconductors		Electronics
Paints	Applications of nanomaterials	Agriculture
Cosmetics		Automobiles
Drug delivery		Catalysis
Medicines		Biosensors

FIGURE 3.5 General applications of nanomaterials.

cameras, watches, calculators, memory backup and a wide range of other applications. Compared to alternative power sources, batteries provide a number of benefits. Typically, they are self-contained, effective, convenient and reliable, and they may be quickly customised to meet user needs. In addition to batteries, fuel cells are a highly effective and environmentally friendly kind of power generation [18].

NMs based on graphene have created a possibility for the next-generation energy storage device due to their distinct characteristics, especially for sodium-ion battery (SIB) and lithium-sulphur battery. A new industry that is highly competitive on a worldwide scale is sustainable development, the essential truth about unsustainable energy use, environmental deterioration and in particular, the growing desire to enhance living conditions. Unfortunately, lithium-ion batteries continue to have a low energy density. Thus, the development of battery life as a whole is significantly constrained, and competition is reduced. In order to realise the great vision, various new-concepts of batteries have emerged. High energy density without the requirement for the costly metals is present in alternative variants. Vanadium flow batteries, Li–S and sodium-ion batteries, Li-O batteries, thermal batteries, sodium-sulphur batteries, calcium-ion batteries (CIBs), magnesium-ion batteries, high-energy supercapacitors, and others are used as an inorganic NMs for energy storage applications like batteries, and the literature of the same has been depicted in Table 3.1.

FUEL CELLS

Fuel cells are becoming one of the most promising sources of energy for the future due to their many benefits and uses. Despite numerous obstacles to commercialisation, nanostructured materials have produced fresh innovations and solutions to these problems. Fuel cells are extremely effective and environmentally beneficial devices. Experience the electrochemical reactions for producing electricity [29]. Because of the fact that they are regarded as renewable energy sources. Unlike the traditional coal combustion system, they do not emit harmful pollutants including carbon dioxide (CO_2), carbon monoxide (CO), nitrogen dioxide (NO_2) and sulphur dioxide (SO_2). Numerous fuel cell types have been investigated such as proton exchange membrane fuel cell (PEMFC), phosphoric acid fuel cell (PAFC), molten carbonate fuel cell (MCFC), solid oxide fuel cell (SOFC) and microbial fuel cell (MFC). Fuel cells and batteries are fundamentally comparable in terms of their componentry. Electrochemical systems, including fuel cells and water splitting devices, are widely recognised as highly efficient and environmentally sustainable technologies for energy conversion and storage. Development of tightly connected inorganic/nanocarbon hybrid materials recently made to enhance electrocatalytic activities and metal-nitrogen complexes, hydroxides, sulphides

TABLE 3.1
A Detailed Overview of the Recent Progress in Inorganic Nanomaterials for Battery

Sl. No.	Type of Nanoparticle	Particle Size	Shape of NPs	Application	References
1.	TiO_2/α-Al_2O_3 nanocomposite	9–11 nm	Cubic-like	Heterogeneous catalysis & solar cells	[19]
2.	Au/Ag NPs	3–20 nm	Spherical	Biomedical	[20]
3.	Lithium-ion NPs	5–20 μm	Spherical	Energy storage	[21]
4.	Graphene oxide	151.8±12.7 nm	Rectangular	Energy storage	[22]
5.	Potassium-ion batteries (PIBs)	138 pm	Tetrameric	Batteries	[23]
6.	Bivalent zinc-ion batteries (ZIBs)	30–150 nm	Isometric	Energy storage	[24]
7.	Sodium-ion batteries (SIBs)	1.02 Å	BCC	Energy storage	[25]
8.	Graphite	25–60 nm	Tetrahedral cubic	Electrochemical storage	[26]
9.	Lithium-sulphur	500 μm	Cubic	Electrochemical storage	[27]
10.	$NiFe_2O_4$	58–37.5 nm	Rectangular	Electrochemical storage	[28]

and inorganic metal oxides all exhibit stability. NMs (platinum, ruthenium, silver and zirconium NPs) help low-temperature fuel cells operate quickly. Additionally, fuel cell components are made from carbon NMs such as carbon nanotubes, graphene and fullerenes. An electrochemical device known as a fuel cell transforms the chemical energy contained in fuels directly into useful electrical energy. Fuel cells function similar to batteries, except they don't lose power or require recharge. The proton-conducting polymer membrane used in polymer electrolyte membrane (PEM) fuel cells, also known as proton exchange membrane fuel cells, serves as the electrolyte. An electrical current generated by a fuel cell is intended to be used for work outside of the cell. There are some disadvantages regarding the fuel cells such as costly to produce because of the high price of catalysts (platinum). The availability of hydrogen is not wide, and also, it is an expensive gas to utilise. And there is a lack of infrastructure to manage the hydrogen distribution. It is very expensive to manufacture the cells. Some of the metals used for the application of fuel cell are Pt, Pd, Co, Cu, Fe, and Ti. Along with these, their bimetallic and trimetallic alloys are also preferred for the fuel cell applications. Moreover, some of the carbon-based NMs like MWCNTs and SWCNTs are probably applicable for fuel cells. The narrated literature has been illustrated in Table 3.2.

CONCLUSION

There is a long history of unintentional human use of NMs. The field of nanotechnology has made significant advancements. NMs are anything with a size between one and one hundred nanometres. Top-down strategies are one of the key methods for the synthesis of NMs. Bottom-up strategies are used in the second strategy. Numerous unusual characteristics of NMs have been demonstrated. NMs differ from their bulk counterparts in several ways. NMs possess several notable attributes, including a large surface area, magnetism, quantum phenomena, antibacterial properties, and high thermal and electrical conductivities. NMs with regulated morphologies and dimensions at the nanoscale level produce desired effects. Using nanotechnology, certain products have already been released for sale. One of the primary developments in contemporary energy technology is the development of alternative energy sources. The most significant of these are fuel cells and lithium-ion batteries. The crucial factor that will influence how they will evolve in the future is the rise in energy and power density.

TABLE 3.2

A Detailed Overview of the Recent Progress in Inorganic Nanomaterials for Fuel Cells

Sl. No.	Type of Nanoparticle	Particle Size	Shape of NPs	Application	References
1.	Pd and Pd-based catalyst	3.7 nm	FCC	Fuel Cell	[30]
2.	Pt and Pt-based Catalyst	0.5–2 nm	FCC	Fuel Cell	[31]
3.	Ti-Fe bimetallic alloys	~2–6 nm	Hexagonal	Energy Storage	[32]
4.	Co-Cu bimetallic alloys	0.087–0.079 nm	Cluster	Energy Storage	[33]
5.	Phosphoric acid fuel cell (PAFC)	$10-200\,kW\,560\,mW/cm^2$	Matrix	Energy Storage	[34]
6.	Proton exchange membrane fuel cell (PEMFC)	<1 to kW ~100	Hierarchical	Fuel Cell	[35]
7.	PdCoAu-trimetallic alloys	5–11 nm	Dogbone-Shaped	Energy Storage	[36]
8.	PdCoMo-trimetallic alloys	5–11 nm	Dogbone-Shaped	Fuel Cell	[37]
9.	MWCNT-supported Pt	10–20 nm	Multilayered	Fuel Cell	[38]
10.	SWCNT-supported PtRu	<2 nm	Single-Layered	Fuel Cell	[39]

REFERENCES

[1] Srinivasan S, Mosdale R, Stevens P, Yang C. Fuel cells: reaching the era of clean and efficient power generation in the twenty-first century. *Annu Rev Energy Environ* 1999; 24: 281–328. https://doi.org/10.1146/annurev.energy.24.1.281

[2] Dincer I. Energy and environmental impacts: present and future perspectives. *Energy Sources* 1998; 20: 427–53. https://doi.org/10.1080/00908319808970070

[3] Sazali N, Wan Salleh WN, Jamaludin AS, Mhd Razali MN. New perspectives on fuel cell technology: A brief review. *Membranes (Basel)* 2020; 10: 99. https://doi.org/10.3390/membranes10050099

[4] Wang H, Dai H. Strongly coupled inorganic-nano-carbon hybrid materials for energy storage. *Chem Soc Rev* 2013; 42: 3088–113. https://doi.org/10.1021/ja3089923

[5] Biswas A, Bayer IS, Biris AS, Wang T, Dervishi E, Faupel F. Advances in top-down and bottom-up surface nanofabrication: Techniques, applications & future prospects. *Adv Colloid Interface Sci* 2012; 170: 2–27. https://doi.org/10.1016/j.cis.2011.11.001

[6] Nasrollahzadeh M, Sajadi SM, Sajjadi M, Issaabadi Z. An introduction to nanotechnology. *Interface Sci Technol* 2019; 28: 1–27. https://doi.org/10.1016/B978-0-12-813586-0.00001-8

[7] Покропивнψ ς, Lohmus R, Hussainova I, Pokropivny A, Vlassov S. *Introduction to Nanomaterials and Nanotechnology.* Tartu University Press, Ukraine; 2007. ISBN: 978-9949-11-741-3

[8] Kelsall RW, Hamley IW, Geoghegan M. *Nanoscale Science and Technology 2005.* John Wiley, Chichester, First published: 25 February 2005, https://doi.org/10.1002/0470020873

[9] Glezer AM. Structural classification of nanomaterials. *Russ. Metall. (Met.)* 2011; 2011: 263–9. DOI 10.1088/2053-1591/ab3175

[10] Khan FA. Nanomaterials: types, classifications, and sources. *Appl Nanomater Hum Health* 2020: 1–13. https://doi.org/10.33263/BRIAC131.041

[11] Sannino D. Types and classification of nanomaterials. *Nanotechnol Trends Future Appl* 2021: 15–38. https://doi.org/10.1007/978-981-15-9437-3_2

[12] Chavan C, Bhajantri RF, Cyriac V, Bulla S, Ravikumar HB, Raghavendra M et al. Exploration of free volume behavior and ionic conductivity of PVA: x (x= 0, Y2O3, ZrO2, YSZ) ion-oxide conducting polymer ceramic composites. *J Non Cryst Solids* 2022; 590: 121696. https://doi.org/10.1016/j.jnoncrysol.2022.121696

[13] Khan FA. Synthesis of nanomaterials: methods & technology. *Appl Nanomater Hum Health* 2020: 15–21. https://doi.org/10.1007/978-981-15-4802-4_2

[14] Guo Z, Tan L. *Fundamentals and Applications of Nanomaterials.* Artech House, Norwood, MA, 2009.

[15] Bratovcic A. Different applications of nanomaterials and their impact on the environment. *SSRG Int J Mater Sci Eng* 2019; 5: 1–7. https://doi.org/10.5772/intechopen.91362

[16] Ali AS. Application of nanomaterials in environmental improvement. *Nanotechnol Environ* 2020. https://doi.org/10.5772/intechopen.91438

[17] Bhajantri RF, Chavan C, Cyriac V, Bulla SS. Investigation on the structural and ion transport properties of magnesium salt doped HPMC-PVA based polymer blend for energy storage applications. *J Non Cryst Solids* 2023; 609: 122276. https://doi.org/10.1016/j.jnoncrysol.2023.122276

[18] Chavan C, Bhajantri RF, Cyriac V, Ismayil, Bulla SS, Sakthipandi K. Investigations on anomalous behavior of ionic conductivity in NaPF6 salt loaded hydroxyethyl cellulose biodegradable polymer electrolyte for energy storage applications. *Polym Adv Technol* 2023; 34: 1698–715. https://doi.org/10.1002/pat.6004

[19] Luna M, Delgado JJ, Romero I, Montini T, Gil MLA, Martínez-López J et al. Photocatalytic TiO2 nanosheets-SiO2 coatings on concrete and limestone: an enhancement of de-polluting and self-cleaning properties by nanoparticle design. *Constr Build Mater* 2022; 338: 127349. https://doi.org/10.1016/j.conbuildmat.2022.127349

[20] Vahl A, Lupan O, Santos-Carballal D, Postica V, Hansen S, Cavers H et al. Surface functionalization of ZnO: Ag columnar thin films with AgAu and AgPt bimetallic alloy nanoparticles as an efficient pathway for highly sensitive gas discrimination and early hazard detection in batteries. *J Mater Chem A* 2020; 8: 16246–64. https://www.doi.org/10.17035/d.2019.0081500252

[21] Sandí G, Carrado KA, Joachin H, Lu W, Prakash J. Polymer nanocomposites for lithium battery applications. *J Power Sources* 2003; 119: 492–6. https://doi.org/10.1016/S0378-7753(03)00272-6

[22] Gupta RK, Alahmed ZA, Yakuphanoglu F. Graphene oxide based low cost battery. *Mater Lett* 2013; 112: 75–7. https://doi.org/10.1016/j.matlet.2013.09.011

[23] Verma R, Didwal PN, Ki H-S, Cao G, Park C-J. SnP3/carbon nanocomposite as an anode material for potassium-ion batteries. *ACS Appl Mater Interfaces* 2019; 11: 26976–84. https://doi.org/10.1021/acsami.9b08088

[24] Huang A, Zhou W, Wang A, Chen M, Tian Q, Chen J. Molten salt synthesis of α-MnO2/Mn2O3 nanocomposite as a high-performance cathode material for aqueous zinc-ion batteries. *J Energy Chem* 2021; 54: 475–81.https://doi.org/10.1016/j.jechem.2020.06.041

[25] Liang Y, Lai W, Miao Z, Chou S. Nanocomposite materials for the sodium-ion battery: a review. *Small* 2018; 14: 1702514. https://doi.org/10.1002/smll.201702514

[26] Morishita T, Hirabayashi T, Okuni T, Ota N, Inagaki M. Preparation of carbon-coated Sn powders and their loading onto graphite flakes for lithium ion secondary battery. *J Power Sources* 2006; 160:638–44. https://doi.org/10.1016/j.jpowsour.2006.01.087

[27] Manthiram A, Fu Y, Su Y-S. Challenges and prospects of lithium-sulfur batteries. *Acc Chem Res* 2013; 46: 1125–34. https://doi.org/10.1021/ar300179v

[28] Mujahid M, Khan RU, Mumtaz M, Soomro SA, Ullah S. NiFe2O4 nanoparticles/MWCNTs nanohybrid as anode material for lithium-ion battery. *Ceram Int* 2019; 45: 8486–93. https://doi.org/10.1016/j.ceramint.2019.01.160

[29] Chavan C, Bhajantri RF, Bulla S, Ravikumar HB, Raghavendra M, Sakthipandi K et al. Ion dynamics and positron annihilation studies on polymer ceramic composite electrolyte system (PVA/NaClO4/Y2O3): application in electrochemical devices. *Ceram Int* 2022; 48:1 7864–84. https://doi.org/10.1016/j.ceramint.2022.03.058

[30] Yin Z, Lin L, Ma D. Construction of Pd-based nanocatalysts for fuel cells: opportunities and challenges. *Catal Sci Technol* 2014; 4: 4116–28. https://doi.org/10.1039/C4CY00760C

[31] Ren X, Lv Q, Liu L, Liu B, Wang Y, Liu A et al. Current progress of Pt and Pt-based electrocatalysts used for fuel cells. *Sustain Energy Fuels* 2020; 4: 15–30 https://doi.org/: 10.1039/C9SE00460B

[32] Schmauss TA, Mogni L, Barnett SA. Phase Formation during Reduction-Oxidation of Ni-Substituted Sr (Ti, Fe) O3-δ Solid Oxide Fuel Electrodes. *Electrochemical Society Meeting Abstracts* 2021; 239: 1134. https://doi.org/10.1149/MA2021-01371134mtgabs

[33] Demirci UB. Theoretical means for searching bimetallic alloys as anode electrocatalysts for direct liquid-feed fuel cells. *J Power Sources* 2007; 173: 11–8. https://doi.org/10.1016/j.jpowsour.2007.04.069

[34] Sammes N, Bove R, Stahl K. Phosphoric acid fuel cells: Fundamentals and applications. *Curr Opin Solid State Mater Sci* 2004; 8: 372–8. https://doi.org/10.1016/j.cossms.2005.01.001

[35] Peighambardoust SJ, Rowshanzamir S, Amjadi M. Review of the proton exchange membranes for fuel cell applications. *Int J Hydrogen Energy* 2010; 35: 9349–84. https://doi.org/10.1016/j.ijhydene.2010.05.017

[36] Kivrak H, Atbas D, Alal O, Çögenli MS, Bayrakceken A, Mert SO et al. A complementary study on novel PdAuCo catalysts: synthesis, characterization, direct formic acid fuel cell application, and exergy analysis. *Int J Hydrogen Energy* 2018; 43: 21886–98. https://doi.org/10.1016/j.ijhydene.2018.09.135

[37] Mohanraju K, Cindrella L. Impact of alloying and lattice strain on ORR activity of Pt and Pd based ternary alloys with Fe and Co for proton exchange membrane fuel cell applications. *RSC Adv* 2014; 4: 11939–47. https://doi.org/10.1039/C3RA47021K

[38] Guo D-J, Qiu X-P, Zhu W-T, Chen L-Q. Synthesis of sulfated ZrO2/MWCNT composites as new supports of Pt catalysts for direct methanol fuel cell application. *Appl Catal B* 2009; 89: 597–601. https://doi.org/10.1016/j.apcatb.2009.01.025

[39] Santasalo-Aarnio A, Borghei M, Anoshkin I V, Nasibulin AG, Kauppinen EI, Ruiz V et al. Durability of different carbon nanomaterial supports with PtRu catalyst in a direct methanol fuel cell. *Int J Hydrogen Energy* 2012; 37: 3415–24. https://doi.org/10.1016/j.ijhydene.2011.11.009

4 Green Synthesis of Silver-Based Nanomaterials and Nanocomposites, and Sustainable Breakthrough in the Battery Application
A Review

C.J. Vijaykumar, Soumya S. Bulla,
Chetan Chavan, and B. Pushpa

INTRODUCTION

Due to a number of factors, bulk energy storage has recently attracted significant attention. The concerned scientific community is looking into the potential for increased use of renewable energy sources due to the depleting and unsustainable nature of conventional energy sources [1]. As an outcome, the most recent research literatures address the obstacles to the dispatchability of renewable energy sources that impede their maximal utilization literatures [2,3]. In view of this, one of the most significant possibilities for maximizing the effectiveness of unconventional energy sources is bulk energy storage, specifically battery energy storage [4,5]. In order to improve the functionality and effectiveness of battery energy storage systems, metallic nanoparticles (NPs) are being investigated as a potential option.

Particles in the nanometer range (i.e., 1–100 nm) have been termed as nanoparticles (NPs) or nanomaterials. They are distinct from larger constituents in that they have peculiar physical, chemical, and biologic features. With the progress of nanotechnology, several novel nanomaterials with unique attributes are emerging, opening up a wide range of applications and research prospects [6]. Owing to their structural configuration, number of dimensions, pore dimensions, origin, and their ability to cause toxicity, nanomaterials may be categorized into five primary categories. There are numerous ways of synthesizing NPs. However, most of them are extremely costly, highly energy-intensive, or harmful toward humans as well as the environment. In green synthesis, natural antioxidants are used in place of traditional reducing agents. These "green synthesis" methods are based on green chemistry ideas. In green synthesis, microorganisms or biological products are employed, such as fruit, leaf, and seed extracts. This process was non-toxic, cost-effective, one step, and environmentally friendly. All products that are hazardous to the environment or human health are replaced with more effective and environmentally friendly alternatives [7].

Recent years have witnessed an extensive examination of the physical, chemical, and biological characteristics of silver NPs (AgNPs), which vary in size, shape, function, crystallinity, and topologies. Greater numbers of individuals are using silver NPs (AgNPs) and silver nanocomposites. Solar energy absorption, storage batteries, optical receptors, catalysis, bio-labeling, and food preparation are a few instances of the applications. The AgNPs may be employed for a variety of purposes and are biocompatible. These purposes include cancer treatment and diagnosis, as well as medical care, drug delivery, and cosmetics. Overconsumption of energy causes environmental problems and energy shortages. To gain access to alternate energy sources, sustainable energy development is necessary. Nevertheless, effective means for converting and storing energy involve silver-based NPs (AgNPs) and silver-based nanocomposites developed through a more environmentally friendly method.

Researchers are increasingly interested in developing a sustainable and cost-effective "green method" for synthesizing AgNPs. This is driven by the potential applications of AgNPs in battery storage, the discipline of electrochemistry, actuators, optical devices, devices for storing energy, catalysis, and other fields. The aim is to address the limitations associated with existing chemical and physical techniques of synthesis. The synthetic methods that are utilized to create AgNPs and their applications in various domains, notably those that relate to energy, are examined in the current work. The classification of NPs and their methods of production, particularly "green synthesis" of silver NPs, are covered in the first section of this chapter. The use of AgNPs in numerous fields is then described, with a focus on battery storage.

CLASSIFICATION OF NANOPARTICLES

Based on their structural configuration, number of dimensions, pore dimensions, origin, and potential toxicity, nanomaterials can be generally divided into five primary categories as outlined in Figure 4.1 [8].

CLASSIFICATION OF NANOMATERIALS BASED ON SOURCE

Based on the source, nanomaterials can be classified into two groups: natural and synthetic NPs.

NATURAL NANOMATERIALS

The natural environment encompasses a diverse array of nanomaterials that occur organically. These include infectious agents, protein fragments, minerals such as clay, liquid colloidal substances like

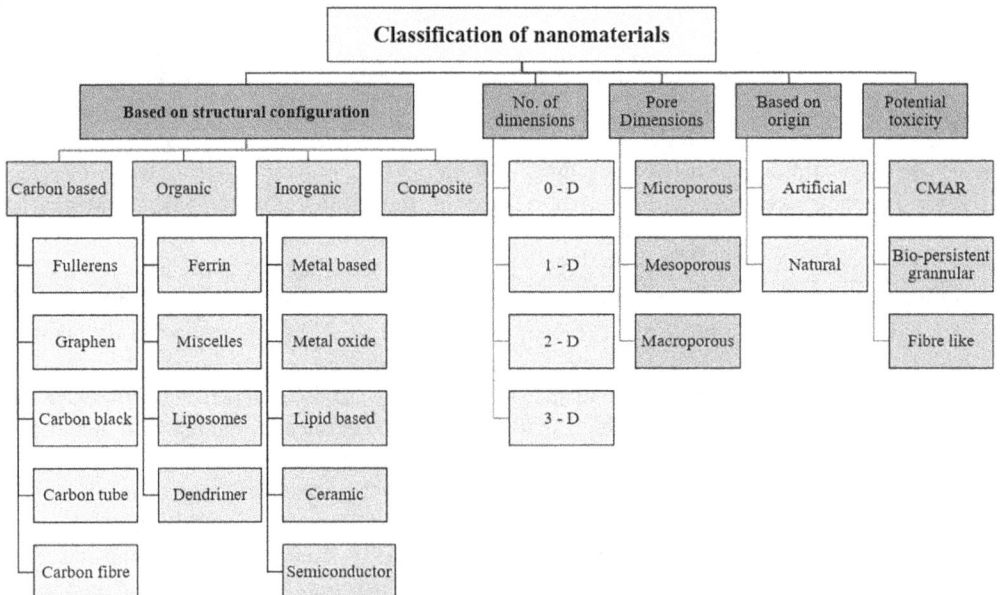

FIGURE 4.1   General classification of nanomaterials.

milk and plasma, aerosol mist, gelatinous substances, minerals materials like oysters, coral reefs, and cartilage, insect feathers and gemstones, cobwebs, lotus foliage, geckos feet, ash from volcanic eruptions, and ocean spray.

Synthetic Nanomaterials

Quantum dots (QDs), a type of semiconductor NP, and carbon nanotubes are two examples of man-made nanomaterials that are created intentionally utilizing exact mechanical and manufacturing techniques. Depending on their structural makeup, nanomaterials can be classified as composites, dendrimers, or metal-based materials.

Structure-/Composition-Based Classification of Nanomaterials

NPs can be generically categorized into four types based on their structural makeup: organic/dendrimers, inorganic, carbon-based, and composite.

Organic Nanomaterials

At the nanoscale, organic molecules undergo a process of transformation, resulting in the formation of organic nanomaterials. Ferritin, liposomes, dendrimers, micelles, and organic NPs or polymers represent a selection of exemplars. Nanocapsule amphipathic molecules are examples of innocuous sustainable NPs characterized by their hollow interiors, which exhibit sensitivity to heat, electromagnetic (EM) rays, and light. The surface of a dendrimer is adorned with several chain terminations that possess the capability to execute certain chemical reactions. Dendrimers are used in a variety of different types of systems, including those for molecular identification, sensing, light harvesting, and opto-electrochemical reactions. In addition, the use of three-dimensional dendrimers in drug delivery holds promise due to their inherent structural features, including the presence of internal cavities capable of accommodating supplementary molecules.

Inorganic Nanomaterials

NPs that don't contain carbon atoms are referred to as inorganic NPs. Inorganic NPs are often categorized as those made of nanomaterials with a metal or metal oxide foundation.

   a. Metal-based nanoparticles

      There are two ways for developing metal-based NPs: destructively or constructively. Metal materials that are often utilized in NP production include aluminum (Al), titanium (Ti), cadmium (Cd), platinum (Pt), cobalt (Co), cerium (Ce), copper (Cu), thallium (Tl), gold (Au), nickel (Ni), iron (Fe), silica (Si), lead (Pb), calcium (Ca), silver (Ag), bismuth (Bi), and zinc (Zn). Metal NPs offer exceptional ultraviolet-visible sensitivity and electrical, catalytic, thermal, and antibacterial capabilities because of their quantum effects and high surface-to-volume ratio. Due to their exceptional optical characteristics, metal NPs are utilized in a wide range of scientific domains.

   b. Metal oxide nanoparticles

      Metal oxide NPs are made of negative oxygen ions and positive metallic ions, also referred to as metal oxide nanomaterials. Metal oxide NPs like tin oxide (SnO_2), silicon dioxide (SiO_2), titanium oxide (TiO_2), zinc oxide (ZnO), tungsten oxide (WO_3), and aluminum oxide (Al_2O_3) are examples of those that are routinely produced using synthetic means. When contrasted to their metal analogs, these NPs have exceptional characteristics.

   c. Semiconductor nanomaterials

      Nanomaterials for semiconductors have characteristics similar to those of insulators and metals. They are divided into three categories:

I.   Concentrated magnetic semiconductor nanomaterials: These have intrinsic magnetic order and can adopt the shape of an antiferromagnetic binary compound, such as EuTe.

II. Non-magnetic semiconductor nanomaterials: Electron charge in semiconductors has been extremely successful in non-magnetic semiconductors that do not include magnetic ions and are utilized for networking as well as data processing, but they are not employed for mass information storage in disposable technological advances.

III. Diluted magnetic semiconductor nanomaterials: By introducing certain magnetic dopants to the hosts matrix, this results in the random replacement of certain portions of the diamagnetic parent cations with magnetic cations, and the semiconducting substances are made magnetic. These materials combine the magnetic properties of conventional and magnetic semiconductors with semiconducting qualities that they retain.

d. Ceramic nanomaterials

Ceramic NPs are inorganic solid materials that are synthesized by subjecting carbide compounds, carbonate elements, oxides, and phosphate minerals to precise climate-controlled heating and cooling processes. Drug delivery systems that specifically target tumors, glaucoma, and various bacterial infections can be made with ceramic NPs. Furthermore, because of its usage in processes like photocatalysis, photodegradation of dyes, imaging applications, and catalysis, nanomaterials are also attracting a lot of research interest.

e. Lipid-based nanomaterials

The predominant morphology observed in NPs composed of lipids is spherical, with dimensions typically ranging from 10 to 100 nm. It consists of a matrix composed of soluble lipophilic molecules around a solid core composed of lipids. Lipid-based NPs find use in the biomedical field as a means of delivering medications and as a therapeutic approach for cancer treatment involving the release of RNA.

CARBON-BASED NANOMATERIALS

As seen in the diagram, there are five fundamental categories of carbon-based nanomaterials, including nanotubes made of carbon, graphene, fullerenes, carbon nanofiber, and carbon black. Fullerenes, referred to as Bucky balls, are carbon nanomaterials characterized by their spherical and ellipsoidal structures.

Fullerenes are a class of carbon-based structures characterized by their spherical shape, composed of a varying number of carbon atoms ranging from 28 to 1,500. These structures have diameters that may reach up to 8.2 nm in single-layered fullerenes, while multilayered fullerenes have diameters ranging from 4 to 36 nm. Graphene has a two-dimensional (2D) planar surface composed of carbon atoms arranged in a hexagonal honeycomb lattice, with a sheet thickness of around 1 nm. The term "nanotubes" is used to describe cylindrical lattices. Graphene nanostructures are used in the production of carbon nanofibers, whereas an unstructured variant of carbon with varying widths between 20 and 70 nm is often referred to as carbon black. Nanotubes are produced by shaping hollow cylinders, which may have diameters as small as 0.7 nm for a single-layered nanotube and 100 nm for an intricate carbon nanotube. The lengths of these nanotubes range from a few micrometers to several millimeters. Nanomaterials composed of carbon have superior durability compared to steel, and possess thermal conductivity along the longitudinal axis of the tube, but lacking heat conductivity across it.

COMPOSITES NANOMATERIALS

Composite nanomaterials are a class of material made by the nanoscale fusion of two or more components. Dispersed nanoscale particles that improve the material's characteristics are often added to a matrix material that offers structural support. The dispersed particles are often NPs of a different material, whereas the matrix material might be made of a polymer, metal, or ceramic.

Nanomaterials are now used in a variety of items, from vehicle parts to packaging materials, to enhance their mechanical, thermal, and flame-retardant qualities.

CLASSIFICATION OF NANOMATERIALS BASED ON THE NUMBER OF DIMENSIONS

According to their size dimensions, nanomaterials can be divided into four categories: 0D, 1D, 2D, and 3D.

- Zero-dimensional (0-D) nanomaterials

    0-D nanomaterials refer to a class of materials that either exhibit a lack of dimensions, with a size greater than 10 nm, or possess each of the three dimensions (x, y, and z) within the nanoscale range. Some examples of zero-dimensional nanomaterials are quantum dots (QDs), fullerenes, and NPs. Various shapes and forms may be seen in these entities, which may exhibit metallic or ceramic properties. Additionally, they are translucent and crystalline in nature, and have single-crystalline or polycrystalline structures.

- One-dimensional (1-D) nanomaterials

    One-dimensional nanomaterials are characterized by having a single dimension (x-axis) at the nanoscale, whereas the remaining dimensions (y and z axes) exhibit much larger scales. These NPs may exhibit unique properties such as high aspect ratios, exceptional mechanical durability, and enhanced conductivity along the entire length. Carbon nanotubes and carbon nanowires are prominent well-recognized examples of 1-D nanomaterials.

- Two-dimensional (2-D) nanomaterials

    Nanoscale-thick ultrathin films or sheets with substantially greater lateral dimensions are known as two-dimensional nanomaterials. Due to their atomic-scale thickness, these materials have distinctive electrical, optical, and mechanical capabilities. The most well-known example of a two-dimensional nanomaterial is graphene, which is a single sheet of carbon atoms organized in a 2-D honeycomb lattice.

- Three-dimensional (3-D) nanomaterials

    Materials that do not possess limitations in terms of size or dimensions inside the nanoscale are often referred to as 3D nanostructures or bulk materials. The bulk material consists of discrete blocks that are within the nanometer scale, often ranging from 1 to 100 nm. On the other hand, 3D nanomaterials possess three dimensions that exceed 100 nm, since all of their dimensions are beyond the nanometer range or are larger. The structural components of 0D, 1D, and 2D are situated in close proximity to each other, resulting in the formation of interfaces within multi-nano layers, dispersion of NPs, and aggregation of nanowires and nanotubes. Three-dimensional nanomaterials include a variety of structures, such as colloids, free NPs exhibiting diverse morphological characteristics, and thin films characterized by atomic-scale porosity.

CLASSIFICATION BASED ON PORE DIMENSIONS

Based on its length and diameter, a nanomaterial can be distinguished between three categories: microporous materials, mesoporous materials, and macroporous materials. In the meaning intended, their size, distribution, and interacting characteristics of the molecules are revealed by the pore diameter. If the intruding molecules are lower than the size of the pore, the diffusion process will include fewer molecule–wall contacts along with greater molecule–molecule interaction. They are beneficial for applications that depend on this characteristic for adsorption and diffusion.

a. Microporous

    Materials that possess holes with a width of less than 2 nm are often known as microporous materials. These substances may consist only of minuscule molecules, such as gases or linear molecules. They exhibit strong interaction qualities and sluggish diffusion kinetics. Na-Y and genuine minerals from clay are two instances of microporous materials. They are utilized in membrane filters, gas storage materials, and gas purification systems.

b. Mesoporous

Pores in mesoporous materials have a diameter that allows them to accommodate certain big molecules that are greater than 2 nm but lesser than 50 nm. Mesoporous materials can be utilized as nanoreactors for polymerization or as adsorbing systems for liquids or vapors. A few instances are MCM-41, MCM-48, SBA-15, and carbon mesoporous materials.

c. Macroporous

Porous materials that can hold massive molecules, including polyaromatic compounds or tiny biological particles, are known as macroporous materials. These pores have a diameter of greater than 50 nm. Examples of macroporous materials are titania, silica, and zirconia. These substances are mostly employed as sensing materials, scaffolds for grafting functional groups like catalytic centers, and matrices to store functional molecules.

Toxicity-Based Nanomaterial Classification

Nanomaterials are categorized into one of three distinct groups according to the degree of toxicity that they might potentially exhibit. These include CMAR NPs (carcinogenic, mutagenic, asthmagenic, reproductive toxin), persistent granular NPs, and fiber-like NPs.

Fibrous Nanoparticles

Fiber-like NPs appear to be extremely hard, biodegradable nanotubes of carbon, fiber-like oxides of metals, and carbon nanotubes but fail to possess their asbestos-like properties. The work environment exposure standards for durable bio-nanotubes of carbon and rigid nanomaterials vary from 10^4 to 10^5 fibers/m^3.

Bio-Persistent Granular Nanoparticles

The suggested exposure thresholds for bio-persistent amorphous NPs are 2×10^7 particles/m^3, which are similar to the exposure limits for metallic elements such as Fe, Au, Ag, Co, La, Pb, and its oxides. An insoluble NPs concentration of 0.3 mg/m^3 has been proposed as a recommended threshold in the absence of an established occupational exposure limit.

CMAR Nanoparticles

Examples of CMAR NPs include nickel (Ni), cadmium (Cd) containing QDs, chromium VI (Cr-VI), beryllium (Be), arsenic (As), and zinc chromate (ZnCrO$_4$). Accordingly, suggested exposure limits are from 2×10^7 to 4×10^7 particles/m^3. 0.003 mg/m^3 has been suggested for non-work exposure-limited hydrophobic NPs. 1.5 mg/m^3 is the suggested value for aqueous NPs having no functional exposure limit.

METHODS OF PREPARATION OF NPS

There are several techniques for producing NPs, and these techniques can be divided into two categories as shown in Figure 4.2.

Top-Down Methods

The top-down technique, also known as the destructive approach, involves the fragmentation of macroscopic materials into smaller entities, leading to the formation of nanomaterials. A few examples of the top-down technique in nanofabrication include lithography, ball milling

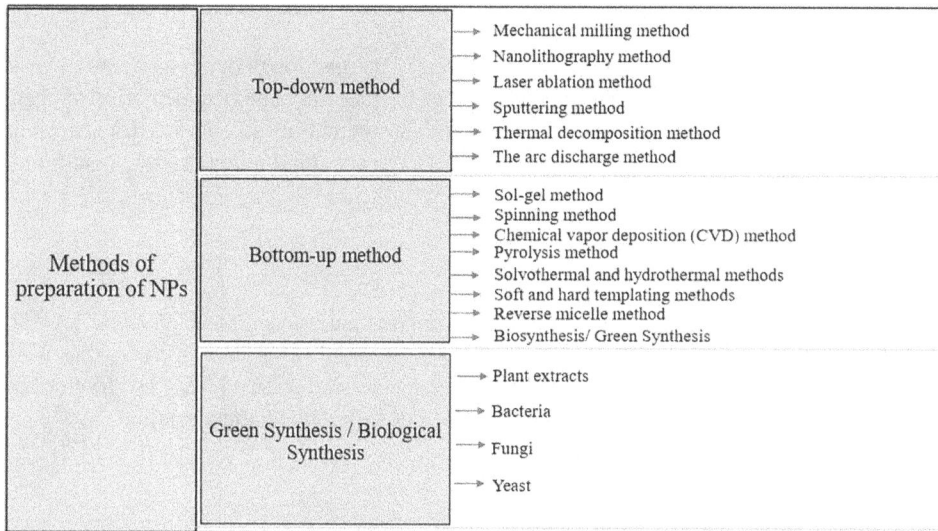

FIGURE 4.2 Methods of preparation of nanoparticles.

or mechanical grinding [9], sputtering, ablation with lasers, electron bursting, arc discharge, and thermal disintegration.

MECHANICAL MILLING TECHNIQUE

Mechanical milling is a predominant, well-recognized top-down approach for the fabrication of diverse NPs. The use of this technique is seen in the fabrication of a diverse range of nanocomposite materials, encompassing resilient to wear spray coatings, nanoalloys comprising aluminum, nickel, magnesium, and copper, as well as alloys fortified with oxide and carbide.

NANOLITHOGRAPHY TECHNIQUE

The process involves the use of a material that is sensitive to light to produce a desired form or structure via the act of printing, followed by the selective removal of a specific region to achieve the intended shape or structure. Lithography uses a focused electron or light beam to create nano-architectures in a practical way. The capacity of nanolithography to create a cluster with the desired shape and size from a single NP is one of its key benefits. The need for sophisticated equipment and the resulting expenses are drawbacks.

LASER ABLATION TECHNIQUE

NPs are generated using laser ablation synthesis, wherein a high-intensity laser beam is used to impact the target material. In a laser ablation technique, the process of metal atom evaporation occurs, followed by immediate solvation of these atoms by surfactant molecules. This solvation process leads to the formation of NPs inside the solution.

SPUTTERING METHOD

The process of sputtering refers to the phenomenon that occurs when ejected particles collide with ions to deposit NPs. Sputtering is a procedure that takes place after depositing a thin coating of NPs and before undergoing the annealing step.

THERMAL DECOMPOSITION METHOD

The disintegration was initiated by the presence of heat. The reaction under consideration is classified as an endothermic process. The use of heat induces the breaking and fragmentation of chemical bonds, resulting in the formation of smaller bonds. The formation of NPs occurs via a chemical process that takes place when the metal is subjected to specified temperatures, resulting in its breakdown.

THE ARC-DISCHARGE METHOD

This approach has the capability to generate a wide range of nanostructured materials. Several carbon-based products are produced such as the natural compound fuller carbon nanohorns (CNHs), carbon nanotubes, few-layer graphene, and amorphous sphere-like carbon NPs. The aforementioned procedure has significant importance in the production of fullerene nanomaterials.

BOTTOM-UP METHOD

Atoms, clusters, and NPs all function as building blocks in the bottom-up, or constructive, process. Some instances of bottom-up strategies are biological synthesis, sol-gel, spinning, pyrolysis, and chemical vapor deposition (CVD).

SOL-GEL METHOD

The implementation of an appropriate solution of chemicals as a precursor constitutes the procedural step. The sol-gel process often employs metal oxide and chloride precursors. Metal oxides and chlorides are often used as sol-gel precursors in many applications.

SPINNING METHOD

The spinning disc reactor (SDR) is a device used for the synthesis of NPs via the process of coalescing spinning particles. The apparatus consists of a chamber or reactor containing a revolving disc, which allows for the control of physical parameters such as temperature. Several factors, such as the disc surface, fluid/precursor ratio, disc rotational speed, liquid flow rate, and feed location, have an impact on it. The technique of spin disc processing was employed for the production of magnetic NPs.

CHEMICAL VAPOR DEPOSITION (CVD) METHOD

A substrate is subjected to CVD in order to deposit a thin layer of gaseous reactants onto its surface. A chemical reaction happens when a hot substrate and a mixed gas come into contact. A thin coating of the product created by this reaction is recovered and utilized again. The need for specialized equipment and the hazardous nature of the gaseous by-products are drawbacks of CVD.

PYROLYSIS METHOD

Pyrolysis is the predominant process used in companies for the large-scale production of NPs. Pyrolysis has the advantages of being an easy, economically viable, and uninterrupted procedure, resulting in a substantial output.

SOLVOTHERMAL AND HYDROTHERMAL METHODS

The aforementioned procedure employs an aqueous hydrothermal method to facilitate a heterogeneous reaction, leading to the formation of nanostructured materials. Hydrothermal and solvothermal methods are often used in closed systems. Hydrothermal and solvothermal methods are advantageous in the fabrication of several nano-geometries of materials, including nanowires, nanosheets, nanorods, and nanospheres.

SOFT AND HARD TEMPLATING METHODS

Nanoporous materials are manufactured by the use of both hard and soft template methods. The soft template method is a conventional and uncomplicated approach used in the production of nano-structured materials. The methodology used in this study involves utilization of a diverse range of pliable templates, such as block copolymers, flexible organic compounds, and anionic, cationic, and non-ionized surfactants, in order to fabricate nanoporous materials.

REVERSE MICELLE TECHNIQUE

The reverse micelle method may also be used to produce nanomaterials in the required shapes and sizes. Reversible micelles naturally develop in the situation of a water-in-oil emulsion, with the heads with hydrophilic characteristics pointing toward a water-containing core.

BIOSYNTHESIS/GREEN SYNTHESIS

Biosynthesis is an environmentally benign and green way of producing non-toxic and biological NPs. Green NP production is well suited and utilized due to its multiple benefits over conventional physical and chemical approaches. Microbes (bacteria, algae, and fungus), bio-building blocks, and other plant components are used in biosynthesis.

BIOSYNTHESIS OR GREEN SYNTHESIS

Biological synthesis refers to the process of generating NPs by the use of plant extracts and micro-organisms, such as bacteria, yeast, and fungus.

Phytonanotechnology has revealed a new area for the eco-friendly, straightforward, and economically advantageous synthesis of NPs. Scalability, biocompatibility, and the capacity to synthesize NPs using water as a universal solvent are benefits of phytonanotechnology. Plants are used in the field of phytonanotechnology for the purpose of synthesizing NPs. Various components of plants, such as the roots, fruit, stems, seeds, and leaves, are employed in the production of NPs. The precise mechanisms by which plants are used in the synthesis of NPs remain uncertain. The dependence of NP production on many organic substances, including organic acids, proteins, vitamins, and secondary metabolites such as flavonoids, alkaloids, terpenoids, polysaccharides, and heterocyclic compounds, has been empirically shown.

Microorganisms are widely recognized as nanofactories with significant potential as ecologically friendly and cost-efficient agents for NP production, circumventing the need for harsh and dangerous chemicals while using little energy. Microorganisms have the capacity to accumulate and mitigate the presence of harmful metallic elements as a result of the abundance of reduction enzymes inside their systems. The reductase enzymes play a crucial role in the conversion of metallic salts into NPs. In recent years, a multitude of bacteria, yeast, and fungus have been identified. Proteins, reduction cofactors, metal resistant genetic material, enzymatic agents, and organic materials play crucial roles as capping and terminating agents in the process of synthesizing NPs.

   a. Plant extracts

      The NPs can be reduced and stabilized using various plants. Numerous researchers have employed biological methods to synthesize metal or metal oxide NPs utilizing various plant parts, including leaves, stems, roots, and fruit. Proteins, coenzymes, and carbohydrates are just a few of the biomolecules found in plants that help turn metal salts into NPs.

b. Bacteria

Bacteria are capable of reducing metal ions, which makes them useful in the creation of NPs. Metal and other new NPs are created using an extensive variety of bacterial species.

c. Fungi

A very effective method with clearly defined morphology is employing fungi to biologically synthesize metal or metal oxide NPs. Fungi serve as a biological agent for the creation of NPs because they contain an intracellular enzyme. Bacteria are incapable of producing a comparable number of NPs as fungi produce.

d. Yeast

Unicellular microscopic organisms constitute yeast. There are more than 1,500 yeast species known. Numerous scientists have reported employing yeast in the synthesis of NPs. Yeast is used to create many NPs or nanomaterials.

COMPARISON OF BIOLOGICAL, CHEMICAL, AND PHYSICAL APPROACHES

There is a continuous development of novel techniques for the synthesis of NPs. The aforementioned strategies may be categorized into three primary classifications: biological methodologies, physical approaches, and chemical protocols. The biological technique is the ideal way for creating NPs since it is uncomplicated, non-toxic, and economical. In the creation of NPs, capping and reducing agents are crucial. The chemical and physical methods for synthesizing NPs employ dangerous and extremely poisonous substances, which are the reason for environmental damage. In chemical and physical procedures, the reducing or capping agent is expensive. Chemicals created through biological processes and used in biological processes are safe for the environment. Therefore, a biological technique is the most preferred way to create NPs.

ADVANTAGES OF GREEN OR BIOLOGICAL SYNTHESIZE

Several advantages of green synthesis are presented in Figure 4.3.

SILVER NANOPARTICLES AND COMPOSITES: THEIR GREEN SYNTHESIS AND APPLICATIONS

SILVER NANOPARTICLES

Recent years have seen a detailed investigation of the physical, chemical, and biological characteristics of silver NPs (AgNPs), which vary in size, shape, function, crystallinity, and topologies. The capacity of silver to exert excellent potential for application in a wide range of disciplines has led to the description of silver as "dynamic."

Researchers have recently paid a great deal of attention to silver NPs (AgNPs) because of their amazing resistance to a variety of bacteria as well as the emergence of antibiotic medication resistance. AgNPs are useful in a variety of disciplines, including biomedicine, medication delivery, water treatment, and agriculture [10] due to their outstanding properties. Due to their high conductivity, AgNPs are used in inks, adhesives, electronic devices, pastes, and other materials.

Physical-chemical methods include mechanical milling, nanolithography, laser ablation, thermal decomposition, sputtering, arc discharge, sol-gel, spinning, CVD, and pyrolysis that have been used to create AgNPs. These techniques produce good results, but they have drawbacks such the use of

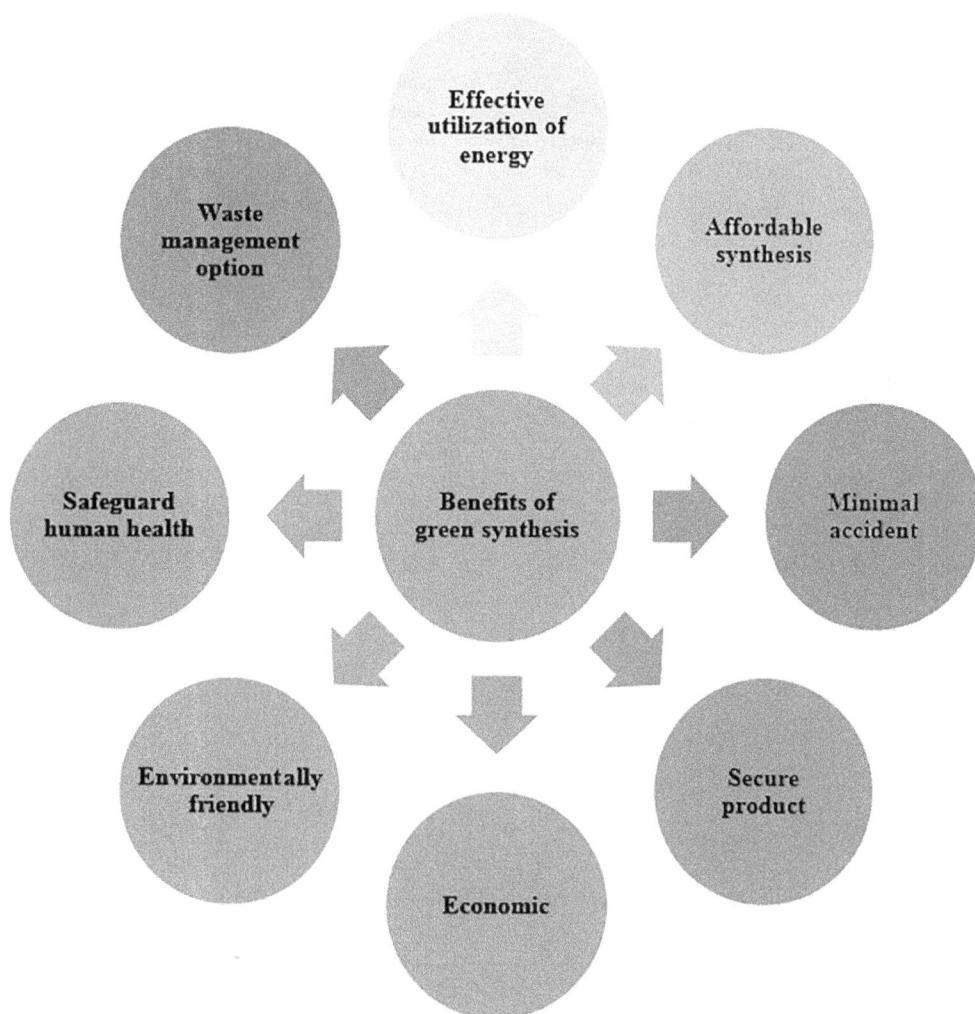

FIGURE 4.3 Benefits of green synthesis.

hazardous chemicals, high operating costs, and energy requirements. A safe, cost-effective, and energy-efficient novel alternative for the synthesis of AgNP employing green synthesis (also known as biological synthesis) with plant extracts and microorganisms like bacteria and fungi as reducing and capping agents is quickly emerging in light because of the disadvantages of conventional physio-chemical approaches.

The combination of green chemistry and nanotechnology will increase the variety of metallic NPs that are compatible with biology and cytology. Silver metal ion solution and a reducing biological agent are the main requirements for the environmentally friendly synthesis of AgNPs. Most of the time, reducing agents or other cell components serve as stabilizing and capping agents, overcoming the need to add external capping and stabilizing agents. Recently, a lot of researchers have been concentrating on environmentally friendly AgNP synthesis. The majority of these studies concentrated on various plant sources for its synthesis and various applications. A basic overview of the subject is provided in the following table.

Author/year	Green reducing agent	Application	Reference
Alharbi et al. (2022)	Medicinal plant extracts	Therapeutic agents against bacteria and fungi	[11]
Khane et al. (2022)	Aqueous *Citrus limon* zest extract	Antioxidant and antimicrobial activities	[12]
Hawar et al. (2022)	*Alhagi graecorum* leaf extract	Cytotoxicity and antifungal activities	[13]
Naseer et al. (2022)	*Allium cepa* extract	Antimicrobial activity	[7]
C Singh et al. (2023)	Root extract of *Premna integrifolia*	Cytotoxic and antibacterial activities	[14]
Widatalla et al. (2022)	Green tea leaf extract	Antimicrobial activity	[15]
Yassin et al. (2022)	Leaf extract of *Origanum majorana*	Bioactivity against multidrug-resistant bacterial strains	[16]
R. Pungle et al. (2022)	*Tridax Procumbens* plant extract	Antimicrobial and anticancer activities	[17]
Khanal et al. (2022)	Root extract of *Rubus ellipticus*	Antioxidant and antimicrobial activities	[18]
Wulandar et al. (2022)	*Kepok* banana peel extract (*Musa paradisiaca* L.)	Non-alcoholic hand sanitizer	[19]
Alarjani et al. (2022)	*Pisum sativum* pod extract	Multidrug-resistant foodborne pathogens	[20]
Baran et al. (2023)	*Allium cepa* L. Peel Extract	Antioxidant, antipathogenic, and anticholinesterase activities	[21]
Hatipoğlu et al. (2023)	*Raphanus sativus* leaf aqueous extract	Toxicological/microbiological activities	[22]
Takcı et al. (2023)	*Salvia officinalis* aqueous extract	Antibacterial activity	[23]
Chirumamilla et al. (2023)	Leaf extract of *Solanum khasianum*	Biological applications	[24]
Chakravarty et al. (2023)	Fruits extracts of *Syzygium cumini*	Biological applications	[25]
Goh et al. (2023)	Banana peel extract	Banana preservation	[26]
Nam et al. (2023)	*Andrographis paniculata* extract	Cytotoxicity assessment and hydrogen peroxide electro-sensing	[27]
Ahmad et al. (2023)	*Piper cubeba* ethanolic extract	Enzyme inhibitory activities	[28]
P sharma et al. (2023)	Extract of *Murraya koenigii* (*Mk*) leaf	Sensor, catalyst, and antibacterial agent	[29]
Borah et al. (2023)	*Cyanobacterium Nostoc* carneum	Photocatalytic, antibacterial, and anticoagulative activities	[30]

Silver Nanocomposites

A "composite material" is defined as a combination of substances with remarkably dissimilar macroscopic physical and chemical properties [31]. Composite materials may be engineered to possess certain thermal expansion coefficients, exhibit superior strength and stiffness at elevated temperatures, demonstrate resistance to corrosion, display increased conductivity, and exhibit the ability to withstand severe temperature extremes [32]. The matrix, often serving as the ongoing stage in composite substances, encompasses the other phase(s) known as "reinforcement" that are embedded inside it.

A variety of composite and nanocomposite substances have been used in the manufacturing process, including a broad selection of matrices including polymer compounds, carbon, metals, and ceramics, along with reinforcements such as particulates, fibers, and layered materials. Matrix materials possess a diverse range of attributes, one of which is the capacity of successfully binding the dispersed phase. The substance effectively preserves its dispersed state in the suitable position and alignment, while safeguarding it from chemical reactivity [33].

Based on matrix material and reinforcements, composites can be broadly classified as follows.

CERAMIC MATRIX NANOCOMPOSITES

Ceramic matrix composites mostly consist of ceramic materials, which constitute the primary portion of their overall volume. The chemical compound known as ceramic is classified within the same group as oxides, nitrites, borides, and other related substances. In a majority of cases, the metallic material is used as the second constituent in ceramic matrix nanocomposites. In theory, the metallic and ceramic constituents are intricately intermixed to attain a homogeneous dispersion and are mutually incorporated to generate the desired nanoscale characteristics. Ceramic makes up the majority of the volume of ceramic matrix composites. As a result, nanocomposites are created, which exhibit enhanced optical, electrical, and magnetic capabilities. In addition to optical, thermal, conductive, and electrical features, these materials also exhibit exceptional corrosion resistance and other protective qualities [34].

METAL–MATRIX NANOCOMPOSITES

Metal-reinforced nanocomposites are used as a matrix to create composite materials. These composites fall into two categories: continuous reinforced materials and materials with discontinuous reinforcement. Metal and carbon nanotubes incorporated into a matrix to generate composites (CNT-MMC) are the most essential type of nanocomposites. This newly developing trend in materials is being explored to optimize the excellent tensile strength and conductivity of electricity of CNTS.

POLYMER NANOCOMPOSITES

The study and use of nanoscience in the fields connected to matrices made up of polymer-NPs is known as polymer nanoscience. These composite materials are constructed of a polymer or copolymer matrix that has NPs or nanofiller particles scattered within. These reinforcements could take on a variety of forms (such as platelets, fibers, or spheroids), but at least one of them needs to have dimensions between 1 and 50 nm. These polymeric nanocomposites are establishing the way for a brand-new class of macromolecular materials with low densities and numerous potential applications [35].

POLYMER/SILVER NANOCOMPOSITES

The physicochemical characteristics of NPs created with the use of nanotechnology differ significantly from those of bulk materials [36]. Silver NPs (also known as nanosilver or AgNPs) stand out among them due to their distinct physical, chemical, and biological characteristics when compared to gold and platinum, which are their adversaries [6]. Silver NPs have gained significant attention in the polymer composites sector because of their exceptional properties, including dielectric property [37], high electrical and thermal efficiency, surface-enhanced scattering by Raman, chemically stable enzyme activity, and nonlinear optoelectronic behavior. These characteristics of polymer/silver nanocomposites enable their use in applications such as catalysis, energy storage and transport, and optical sensing. Additionally, silver's antibacterial properties have boosted its use in many different kinds of commodities, such as clothing, footwear, paints, wound plasters, appliances, cosmetics, and plastics.

A brief summary of selected silver/polymer nanocomposites and their use in diverse fields is provided in the following table:

Author/Year	Silver Composite	Application	References
Thinh et al. (2023)	Silver-doped graphene oxide nanocomposite	Optical sensing, adsorption, catalysis, antibacterial activity	[38]
Zhijian Sun et al. (2022)	Silver-epoxy nanocomposites	Packaging applications	[39]
Atta et al. (2021)	Polyaniline/silver oxide/silver nanocomposite	Electrode for a super capacitor	[40]
Hasanin et al. (2021)	Tertiary silver nanocomposite consisting starch, oxidized cellulose, and ethyl cellulose	Antibacterial, antifungal, and antiviral activities	[41]
Attia et al. (2021)	Chitosan-silver nanocomposites	Antiparasitic activity	[42]
Rehan et al. (2020)	Polyacrylonitrile/silver nanocomposite films	Antimicrobial activity, catalytic activity, electrical conductivity, UV protection	[43]
Miao Li et al. (2022)	Boron nitride/silver nanocomposite	Oral denture bases	[44]
Salem et al. (2022)	Silver nanocomposite based on carboxymethyl cellulose	Antibacterial, antifungal, and anticancer activities	[45]
Kumar Krishnan et al. (2021)	Chitosan/silver nanocomposites	Synergistic antibacterial action	[46]
Gupta et al. (2010)	Polyaniline–silver nanocomposite	Optical and electrical transport	[47]

SILVER NANOPARTICLES AND COMPOSITES: THE BATTERY APPLICATION

The research and utilization of clean and renewable energies have received a lot of attention recently as an outcome of fossil fuel usage and the ever-growing needs of humans. As result, numerous efforts were concentrated on developing and creating novel materials for batteries and other energy storage and production systems [48]. A battery is an electrochemical device that uses electrodes to undergo oxidation and reduction reactions to store energy. Energy is both stored and released throughout the charging and discharging processes. Good storage capacity, extended life, good performance, and economical battery production are all crucial attributes [49]. Anode (negative electrode), cathode (positive electrode), and electrolyte separator (which works as an ion-conducting channel between cathode and anode) are the three primary components of a standard battery cell. The electrolyte serves only as an ion-conductive medium and is not engaged in any battery reactions [4,50]. The schematic diagram of battery cell is shown in Figure 4.4.

The first association between batteries and nanotechnology was made in 1992. The interaction is greatly reliant on the anode, cathode, and electrolyte designs used in batteries. Additionally, it has improved the characteristics of batteries by increasing conductivity and decreasing side reactions that could lead to battery deterioration [49].

The benefits of employing nanomaterials in batteries include the following:

1. Nanomaterials have a higher specific area based on their dimension, which leads to shorter electron transfer pathways and a faster electron transfer rate.
2. The chemical reactions result in clear volume changes in the electrodes. The management of this shrinkage through the use of nanoscale materials has led to an improvement in battery efficiency and a longer life cycle.
3. When utilizing NPs, the increased contact surface also boosts the electrolytes and the electrode interacting. It enhances the pace at which ions move through the electrolyte, improving the battery's efficiency [51]. Liquid electrolytes (LEs) and solid electrolytes (SEs) are the two main categories used to classify electrolytes. LEs often exhibit strong ionic conductivity and diffusion coefficient, and are nonaqueous liquids. However, LEs have several drawbacks as well, including as flammability, explosion risk, and dendrite development [52]. So solid polymer electrolytes (SPEs) have been recommended as one of the finest alternatives for achieving safe battery technology [51].

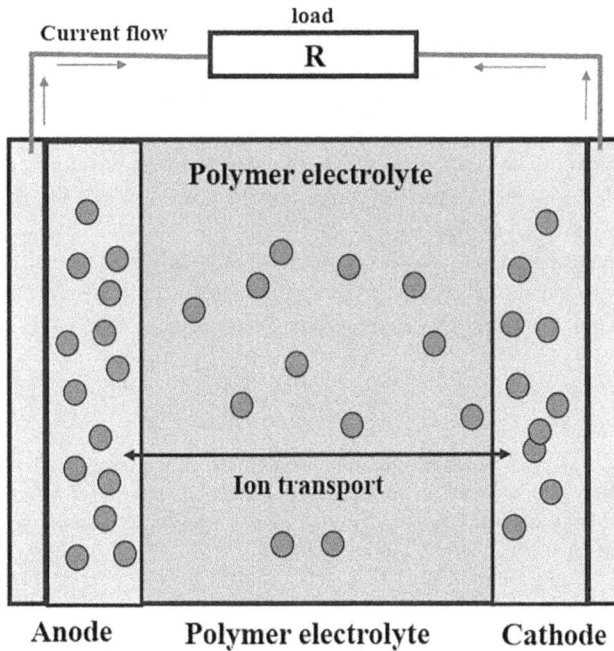

FIGURE 4.4 Schematic diagram of battery cell.

SPEs can stop the growth of dendrites and have superior mechanical qualities to LEs. The combination of polymer electrolytes with metal electrodes for high energy density in battery manufacturing is another benefit. However, the fundamental issue with polymer-based electrolytes relates to their poor mechanical and ionic conductivity qualities. Recently, interest has grown in using NPs as a polymer electrolyte additive to solve these problems. In general, NPs can improve mechanical qualities because of their innate bulk and surface characteristics, and they can also improve ionic conductivity because of their conducting nature. To date, a variety of NPs, including metal-based NPs (such as silver, gold, iron, and aluminum NPs), carbon-based NPs (such as graphene oxide (GO) NPs and carbon nanotubes), ceramic-based NPs (such as zinc oxide NPs, titanium oxide NPs, and silica), and polymer-based NPs, have been employed to enhance electrolytes' capacity for storing energy devices like batteries. Metallic NPs like silver NPs have demonstrated greater ionic conductivity than all other NPs used in polymer matrix. In a polymer matrix, it has been observed that AgNPs increase surface contacts and form transverse linkages. Additionally, the amorphous component enhances the electrolyte's free mobile ions [51]. As a result, silver NPs improve the electrochemical stability and conductivity of polymer electrolytes. Below is a list of some of the key studies on silver-based polymer electrolytes.

Author/Year	Silver Polymer Electrolyte	Ionic Conductivity	References
Verma et al. (2015)	PEO-based silver polymer electrolyte	2.2×10^{-6} S/cm	[53]
Aziz et al. (2019)	Chitosan: silver polymer electrolyte	2.3×10^{-6} S/cm	[54]
E. Correa et al. (2017)	Polycaprolactone: silver nitrate Polymer electrolyte	9.02×10^{-3} S/m	[55]
G. Hirankumar et al. (2005)	Polyvinyl alcohol: silver polymer electrolyte	1.48×10^{-5} S/m	[56]
Aziz et al. (2010)	Chitosan: silver triflate polymer electrolyte	4.25×10^{-8} S/m	[57]
Vipin Cyriac et al. (2022)	PVA/NaBr with AgNPs polymer electrolyte	1.22×10^{-4} S/cm	[58]
Hadi et al. (2020)	$CS:AgNO_3:Al_2O_3$ polymer electrolyte	3.7×10^{-4} S/cm	[59]
Suthanthiraraj et al. (2011)	$(PEO–AgCF_3 SO_3)/MgO$-based polymer electrolyte	2×10^{-6} S/cm	[60]

CONCLUSION

Metal NP 'green' synthesis is a growing and important area of study. A variety of natural extracts such as yeast, fungus, bacteria, and extracts of plants have all been successfully used as bio-ingredients in the production of products. Extracts of plants have been found to perform well among them. This current chapter is developed to include insights on the environmentally friendly production of AgNPs and their utilization in battery applications. Based on the literature that is currently accessible, the green synthesis of AgNPs and applications in batteries have undergone a thorough literature review. Given the explanation above, it has been widely accepted that using green chemistry is a suitable way to create AgNPs and use them in battery applications. This allows for a more sustainable choice of materials and processes, which is a crucial requirement.

REFERENCES

[1] Sahana Nayak, A. A. Kittur, and Shravankumar Nayak. "Green Synthesis of Silver-Zirconia Composite Using Chitosan Biopolymer Binder for Fabrication of Electrode Materials in Supercapattery Application for Sustainable Energy Storage." *Current Research in Green and Sustainable Chemistry* (2022) 5: 100292. https://doi.org/10.1016/j.crgsc.2022.100292

[2] Ziyu Zhang, Tao Ding, Quan Zhou, Yuge Sun, Ming Qu, Ziyu Zeng, Yuntao Ju, Li Li, Kang Wang, and Fangde Chi. "A Review of Technologies and Applications on Versatile Energy Storage Systems." *Renewable and Sustainable Energy Reviews* (2021) 148: 111263. https://doi.org/10.1016/j.rser.2021.111263

[3] Bo Yang, Jun Pan, Yishun Zhu, Xurui Huang, Chutong Wang, Chuangxin Guo, and Yizong Guo. "A Review of Energy Storage System Study." Pp. 2858–63 In 2020 IEEE 4th Conference on Energy Internet and Energy System Integration (EI2). IEEE (2020). https://doi.org/10.1109/EI250167.2020.9347112

[4] Rajashekhar F. Bhajantri, Chetan Chavan, Vipin Cyriac, and Soumya S. Bulla. "Investigation on the Structural and Ion Transport Properties of Magnesium Salt Doped HPMC-PVA Based Polymer Blend for Energy Storage Applications." *Journal of Non-Crystalline Solids* (2023) 609: 122276. https://doi.org/10.1016/j.jnoncrysol.2023.122276

[5] Chetan Chavan, R. F. Bhajantri, Soumya Bulla, H. B. Ravikumar, M. Raghavendra, K. Sakthipandi, K. Yogesh Kumar, and B. P. Prasanna. "Ion Dynamics and Positron Annihilation Studies on Polymer Ceramic Composite Electrolyte System (PVA/NaClO4/Y2O3): Application in Electrochemical Devices." *Ceramics International* (2022) 48(12): 17864–84. https://doi.org/10.1016/j.ceramint.2022.03.058

[6] Virender K Sharma, Ria A. Yngard, and Yekaterina Lin. "Silver Nanoparticles: Green Synthesis and Their Antimicrobial Activities." *Advances in Colloid and Interface Science* (2009) 145(1–2): 83–96. https://doi.org/10.1016/j.cis.2008.09.002

[7] Naseer Ahmed, Munawar Iqbal, Salman Ali, Arif Nazir, Mazhar Abbas, and Naveed Ahmad. "Green Synthesis of Silver Nanoparticles Using Allium Cepa Extract and Their Antimicrobial Activity Evaluation." *Chemistry International* (2022) 8: 89–94. https://doi.org/10.5281/zenodo.6862470

[8] Bawoke Mekuye and Birhanu Abera. "Nanomaterials: An Overview of Synthesis, Classification, Characterization, and Applications." *Nano Select.* (2023) https://doi.org/10.1002/nano.202300038

[9] Chetan Chavan, Rajashekhar Fakeerappa Bhajantri, Soumya Siddalingappa Bulla, and Kathiresan Sakthipandi. "Indigenously Designed and Fabricated Mechanical Milling Set-up to Synthesis Nanoparticles: A Cost-Effective Method." (2021) https://nopr.niscpr.res.in/handle/123456789/5813

[10] Sista Kameswara Srikar, Deen Dayal Giri, Dan Bahadur Pal, Pradeep Kumar Mishra, and Siddh Nath Upadhyay. "Green Synthesis of Silver Nanoparticles: A Review." *Green and Sustainable Chemistry* (2016) 6(1): 34–56. https://dx. doi.org/10.4236/gsc.2016.61004

[11] Njud S. Alharbi, Nehad S. Alsubhi, and Afnan I. Felimban. "Green Synthesis of Silver Nanoparticles Using Medicinal Plants: Characterization and Application." *Journal of Radiation Research and Applied Sciences* (2022) 15(3): 109–24. https://doi.org/10.1016/j.jrras.2022.06.012

[12] Yasmina Khane, Khedidja Benouis, Salim Albukhaty, Ghassan M. Sulaiman, Mosleh M. Abomughaid, Amer Al Ali, Djaber Aouf, Fares Fenniche, Sofiane Khane, and Wahiba Chaibi. "Green Synthesis of Silver Nanoparticles Using Aqueous Citrus Limon Zest Extract: Characterization and Evaluation of Their Antioxidant and Antimicrobial Properties." *Nanomaterials* (2022.) 12(12): 2013. https://doi.org/10.3390/nano12122013

[13] Sumaiya N Hawar, Hanady S. Al-Shmgani, Zainb A. Al-Kubaisi, Ghassan M. Sulaiman, Yaser H. Dewir, and Jesamine J. Rikisahedew. "Green Synthesis of Silver Nanoparticles from Alhagi Graecorum Leaf Extract and Evaluation of Their Cytotoxicity and Antifungal Activity." *Journal of Nanomaterials* 2022: 1–8. https://doi.org/10.1155/2022/1058119

[14] Chandrashekhar Singh, Sumit Kumar Anand, Richa Upadhyay, Nidhi Pandey, Pradeep Kumar, Deepjyoti Singh, Punit Tiwari, Rajesh Saini, Kavindra Nath Tiwari, and Sunil Kumar Mishra. "Green Synthesis of Silver Nanoparticles by Root Extract of Premna Integrifolia L. and Evaluation of Its Cytotoxic and Antibacterial Activity." *Materials Chemistry and Physics* (2023) 297: 127413. https://doi.org/10.1016/j.matchemphys.2023.127413

[15] Hiba Abbas Widatalla, Layla Fathi Yassin, Ayat Ahmed Alrasheid, Shimaa Abdel Rahman Ahmed, Marvit Osman Widdatallah, Sahar Hussein Eltilib, and Alaa Abdulmoneim Mohamed. "Green Synthesis of Silver Nanoparticles Using Green Tea Leaf Extract, Characterization and Evaluation of Antimicrobial Activity." *Nanoscale Advances* (2022) 4(3): 911–15. https://doi.org/10.1039/D1NA00509J

[16] Mohamed Taha Yassin, Ashraf Abdel-Fattah Mostafa, Abdulaziz Abdulrahman Al-Askar, and Fatimah O. Al-Otibi. "Facile Green Synthesis of Silver Nanoparticles Using Aqueous Leaf Extract of Origanum Majorana with Potential Bioactivity against Multidrug Resistant Bacterial Strains." *Crystals* (2022) 12(5): 603. https://doi.org/10.3390/cryst12050603

[17] Rohini Pungle, Shivraj Hariram Nile, Nilesh Makwana, Ragini Singh, Rana P. Singh, and Arun S. Kharat. "Green Synthesis of Silver Nanoparticles Using the Tridax Procumbens Plant Extract and Screening of Its Antimicrobial and Anticancer Activities." *Oxidative Medicine and Cellular Longevity* 2022. https://doi.org/10.1155/2022/9671594

[18] Lekha Nath Khanal, Khaga Raj Sharma, Hari Paudyal, Kshama Parajuli, Bipeen Dahal, G. C. Ganga, Yuba Raj Pokharel, and Surya Kant Kalauni. "Green Synthesis of Silver Nanoparticles from Root Extracts of Rubus Ellipticus Sm. and Comparison of Antioxidant and Antibacterial Activity." *Journal of Nanomaterials* 2022: 1–11. https://doi.org/10.1155/2022/1832587

[19] Ika O Wulandari, Baiq E. Pebriatin, Vita Valiana, Saprizal Hadisaputra, Agus D. Ananto, and Akhmad Sabarudin. "Green Synthesis of Silver Nanoparticles Coated by Water Soluble Chitosan and Its Potency as Non-Alcoholic Hand Sanitizer Formulation." *Materials* (2022) 15(13): 4641. https://doi.org/10.3390/ma15134641

[20] Khaloud Mohammed Alarjani, Dina Huessien, Rabab Ahmed Rasheed, and M. Kalaiyarasi. "Green Synthesis of Silver Nanoparticles by Pisum Sativum L. (Pea) Pod against Multidrug Resistant Foodborne Pathogens." *Journal of King Saud University-Science* (2022) 34(3): 101897. https://doi.org/10.1016/j.jksus.2022.101897

[21] Mehmet Fırat Baran, Cumali Keskin, Ayşe Baran, Abdulkerim Hatipoğlu, Mahmut Yildiztekin, Selçuk Küçükaydin, Kadri Kurt, Hülya Hoşgören, Md Moklesur Rahman Sarker, and Albert Sufianov. "Green Synthesis of Silver Nanoparticles from Allium Cepa L. Peel Extract, Their Antioxidant, Antipathogenic, and Anticholinesterase Activity." *Molecules* (2023) 28(5): 2310. https://doi.org/10.3390/molecules28052310

[22] Abdulkerim Hatipoğlu, Ayşe Baran, Cumali Keskin, Mehmet Fırat Baran, Aziz Eftekhari, Sabina Omarova, Dawid Janas, Rovshan Khalilov, Mehmet Tevfik Adican, and Sevgi İrtegün Kandemir. "Green Synthesis of Silver Nanoparticles Based on the Raphanus Sativus Leaf Aqueous Extract and Their Toxicological/Microbiological Activities." *Environmental Science and Pollution Research* (2023): 1–13. https://doi.org/10.1007/s11356-023-26499-z

[23] Deniz Kadir Takcı, Melis Sumengen Ozdenefe, and Sema Genc. "Green Synthesis of Silver Nanoparticles with an Antibacterial Activity Using Salvia Officinalis Aqueous Extract." *Journal of Crystal Growth* (2023) 614: 127239. https://doi.org/10.1016/j.jcrysgro.2023.127239

[24] Pavani Chirumamilla, Sunitha Bai Dharavath, and Shasthree Taduri. "Eco-Friendly Green Synthesis of Silver Nanoparticles from Leaf Extract of Solanum Khasianum: Optical Properties and Biological Applications." *Applied Biochemistry and Biotechnology* (2023) 195(1): 353–68. https://doi.org/10.1007/s12010-022-04156-4

[25] Archana Chakravarty, Iftkhar Ahmad, Preeti Singh, Mehraj Ud Din Sheikh, Gulshitab Aalam, Suresh Sagadevan, and Saiqa Ikram. "Green Synthesis of Silver Nanoparticles Using Fruits Extracts of Syzygium Cumini and Their Bioactivity." *Chemical Physics Letters* (2022) 795: 139493. https://doi.org/10.1016/j.cplett.2022.139493

[26] Hui Tung Goh, Choon Yoong Cheok, and Swee Pin Yeap. "Green Synthesis of Silver Nanoparticles Using Banana Peel Extract and Application on Banana Preservation." *Food Frontiers* (2023) 4(1): 283–88. https://doi.org/10.1002/fft2.206

[27] Quach Thi Thanh Huong, Nguyen Thanh Hoai Nam, Bui Thanh Duy, Hoang An, Nguyen Duy Hai, Hoang Thuy Kim Ngan, Lam Thanh Ngan, Tran Le Hoai Nhi, Dang Thi Yen Linh, and Tran Nhat Khanh. "Structurally Natural Chitosan Films Decorated with Andrographis Paniculata Extract and Selenium Nanoparticles: Properties and Strawberry Preservation." *Food Bioscience* (2023) 53: 102647. https://doi.org/10.1016/j.fbio.2023.102647

[28] Khalil Ahmad, Hafiz Muhammad Asif, Taimoor Afzal, Mohsin Abbas Khan, Muhammad Younus, Umair Khurshid, Maryem Safdar, Sohaib Saifulah, Bashir Ahmad, and Abubakar Sufyan. "Green Synthesis and Characterization of Silver Nanoparticles through the Piper Cubeba Ethanolic Extract and Their Enzyme Inhibitory Activities." *Frontiers in Chemistry* (2023) 11: 1065986. https://doi.org/10.3389/fchem.2023.1065986

[29] Partha Pratim Sarma, Kailash Barman, and Pranjal K. Baruah. "Green Synthesis of Silver Nanoparticles Using Murraya Koenigii Leaf Extract with Efficient Catalytic, Antimicrobial, and Sensing Properties towards Heavy Metal Ions." *Inorganic Chemistry Communications* (2023) 152: 110676. https://doi.org/10.1016/j.inoche.2023.110676

[30] Debasish Borah, Neeharika Das, Pampi Sarmah, Kheyali Ghosh, Madhurya Chandel, Jayashree Rout, Piyush Pandey, Narendra Nath Ghosh, and Chira R. Bhattacharjee. "A Facile Green Synthesis Route to Silver Nanoparticles Using Cyanobacterium Nostoc Carneum and Its Photocatalytic, Antibacterial and Anticoagulative Activity." *Materials Today Communications* (2023) 34: 105110. https://doi.org/10.1016/j.mtcomm.2022.105110

[31] Oluwatimilehin O Fadiran, Natalie Girouard, and J. Carson Meredith. "Pollen Fillers for Reinforcing and Strengthening of Epoxy Composites." *Emergent Materials* (2018) 1: 95–103. https://doi.org/10.1007/s42247-018-0009-x

[32] Soumya S Bulla, R. F. Bhajantri, Chetan Chavan, and K. Sakthipandi. "Synthesis and Characterization of Polythiophene/Zinc Oxide Nanocomposites for Chemiresistor Organic Vapor-Sensing Application." *Journal of Polymer Research* (2021) 28(7): 251. https://doi.org/10.1007/s10965-021-02618-7

[33] Kishor Kumar Sadasivuni, Sunita Rattan, Sadiya Waseem, Snehal Kargirwar Brahme, Subhash B. Kondawar, S. Ghosh, A. P. Das, Pritam Kisore Chakraborty, Jaideep Adhikari, and Prosenjit Saha. "Silver Nanoparticles and Its Polymer Nanocomposites-Synthesis, Optimization, Biomedical Usage, and Its Various Applications." *Polymer Nanocomposites in Biomedical Engineering* (2019): 331–73. https://doi.org/10.1007/978-3-030-04741-2_13

[34] Anton Popelka, P. Noorunnisa Khanam, and Mariam Ali AlMaadeed. "Surface Modification of Polyethylene/Graphene Composite Using Corona Discharge." *Journal of Physics D: Applied Physics* (2018) 51(10): 105302. https://doi.org/10.1088/1361-6463/aaa9d6

[35] Sehrish Habib, Eman Fayyed, Rana Abdul Shakoor, Ramazan Kahraman, and Aboubakr Abdullah. "Improved Self-Healing Performance of Polymeric Nanocomposites Reinforced with Talc Nanoparticles (TNPs) and Urea-Formaldehyde Microcapsules (UFMCs)." *Arabian Journal of Chemistry* (2021) 14(2): 102926. https://doi.org/10.1016/j.arabjc.2020.102926

[36] Deepalekshmi Ponnamma, Alper Erturk, Hemalatha Parangusan, Kalim Deshmukh, M. Basheer Ahamed, and Mariam Al Ali Al-Maadeed. "Stretchable Quaternary Phasic PVDF-HFP Nanocomposite Films Containing Graphene-Titania-SrTiO 3 for Mechanical Energy Harvesting." *Emergent Materials* (2018)1: 55–65. https://doi.org/10.1007/s42247-018-0007-z

[37] Soumya S Bulla, R. F. Bhajantri, Chetan Chavan, and K. Sakthipandi. "Biosynthesized Silver Nanoparticles Encapsulated in a Poly (Vinyl Alcohol) Matrix: Dielectric and Structural Properties." *ChemistrySelect* (2022) 7(47): e202201771. https://doi.org/10.1002/slct.202201771

[38] Doan Ba Thinh, Nguyen Minh Dat, Nguyen Ngoc Kim Tuyen, Le Tan Tai, Nguyen Duy Hai, Ninh Thi Tinh, Le Minh Huong, Tran Do Dat, Quach Thi Thanh Huong, and Nguyen Thanh Hoai Nam. "A Review of Silver-dopped Graphene Oxide Nanocomposite: Synthesis and Multifunctional Applications." *Vietnam Journal of Chemistry* (2022) 60(5): 553–70. https://doi.org/10.1002/vjch.202200034

[39] Zhijian Sun, Jiaxiong Li, Michael Yu, Mohanalingam Kathaperumal, and Ching-Ping Wong. "A Review of the Thermal Conductivity of Silver-Epoxy Nanocomposites as Encapsulation Material for Packaging Applications." *Chemical Engineering Journal* (2022) 446: 137319. https://doi.org/10.1016/j.cej.2022.137319

[40] A Atta, M. M. Abdelhamied, Doaa Essam, Mohamed Shaban, Alhulw H. Alshammari, and Mohamed Rabia. "Structural and Physical Properties of Polyaniline/Silver Oxide/Silver Nanocomposite Electrode for Supercapacitor Applications." *International Journal of Energy Research* (2022) 46(5): 6702–10. https://doi.org/10.1002/er.7608

[41] Mohamed Hasanin, Mostafa A. Elbahnasawy, Amr M. Shehabeldine, and Amr H. Hashem. "Ecofriendly Preparation of Silver Nanoparticles-Based Nanocomposite Stabilized by Polysaccharides with Antibacterial, Antifungal and Antiviral Activities." *BioMetals* (2021) 34: 1313–28. https://doi.org/10.1007/s10534-021-00344-7

[42] Marwa M. Attia, Nahed Yehia, Mohamed Mohamed Soliman, Mustafa Shukry, Mohamed T. El-Saadony, and Heba M. Salem. "Evaluation of the Antiparasitic Activity of the Chitosan-Silver Nanocomposites in the Treatment of Experimentally Infested Pigeons with Pseudolynchia Canariensis." *Saudi Journal of Biological Sciences* (2022) 29(3): 1644–52. https://doi.org/10.1016/j.sjbs.2021.10.067

[43] Mohamed Rehan, Amr A. Nada, Tawfik A. Khattab, Nayera A. M. Abdelwahed, and Amira Adel Abou El-Kheir. "Development of Multifunctional Polyacrylonitrile/Silver Nanocomposite Films: Antimicrobial Activity, Catalytic Activity, Electrical Conductivity, UV Protection and SERS-Active Sensor." *Journal of Materials Research and Technology* (2020) 9(4): 9380–94. https://doi.org/10.1016/j.jmrt.2020.05.079

[44] Miao Li, Sifan Wang, Ruizhi Li, Yuting Wang, Xinyue Fan, Wanru Gong, and Yu Ma. "The Mechanical and Antibacterial Properties of Boron Nitride/Silver Nanocomposite Enhanced Polymethyl Methacrylate Resin for Application in Oral Denture Bases." *Biomimetics* (2022) 7(3): 138. https://doi.org/10.3390/biomimetics7030138

[45] Salem S. Salem, Amr H. Hashem, Al-Aliaa M. Sallam, Ahmed S. Doghish, Abdulaziz A. Al-Askar, Amr A. Arishi, and Amr M. Shehabeldine. "Synthesis of Silver Nanocomposite Based on Carboxymethyl Cellulose: Antibacterial, Antifungal and Anticancer Activities." *Polymers* (2022) 14(16): 3352. https://doi.org/10.3390/polym14163352

[46] Siva Kumar-Krishnan, Evgen Prokhorov, Monserrat Hernández-Iturriaga, Josué D. Mota-Morales, Milton Vázquez-Lepe, Yuri Kovalenko, Isaac C. Sanchez, and Gabriel Luna-Bárcenas."Chitosan/Silver Nanocomposites: Synergistic Antibacterial Action of Silver Nanoparticles and Silver Ions." *European Polymer Journal* (2015) 67: 242–51. https://doi.org/10.1016/j.eurpolymj.2015.03.066

[47] K Gupta, P. C. Jana, and A. K. Meikap. "Optical and Electrical Transport Properties of Polyaniline-Silver Nanocomposite." *Synthetic Metals* (2010) 160(13–14): 1566–73. https://doi.org/10.1016/j.synthmet.2010.05.026

[48] Mahtab Hamrahjoo, Saeed Hadad, Elham Dehghani, Mehdi Salami-Kalajahi, and Hossein Roghani-Mamaqani. "Poly (Poly (Ethylene Glycol) Methyl Ether Methacrylate-Co-Acrylonitrile) Gel Polymer Electrolytes for High Performance Lithium-Ion Batteries: Comparing Controlled and Conventional Radical Polymerization." *European Polymer Journal* (2022) 173: 111276. https://doi.org/10.1016/j.eurpolymj.2022.111276

[49] Ashish Bhatnagar, Manoj Tripathi, Shalu, and Abhimanyu Prajapati. "Nanotechnology for Batteries." Pp. 29–48 In Nanotechnology for Electronic Applications. Springer. (2022) https://doi.org/10.1007/978-981-16-6022-1_2

[50] Chetan Chavan, Rajashekhar F. Bhajantri, Vipin Cyriac, Soumya S. Bulla, and K. Sakthipandi. "Investigations on Anomalous Behavior of Ionic Conductivity in NaPF6 Salt Loaded Hydroxyethyl Cellulose Biodegradable Polymer Electrolyte for Energy Storage Applications." *Polymers for Advanced Technologies.* (2023) https://doi.org/10.1002/pat.6004

[51] Amirhossein Enayati-Gerdroodbar, Svetlana N. Eliseeva, and Mehdi Salami-Kalajahi. "A Review on the Effect of Nanoparticles/Matrix Interactions on the Battery Performance of Composite Polymer Electrolytes." *Journal of Energy Storage* (2023) 68: 107836. https://doi.org/10.1016/j.est.2023.107836

[52] Ziqi Guo, Shuoqing Zhao, Tiexin Li, Dawei Su, Shaojun Guo, and Guoxiu Wang. "Recent Advances in Rechargeable Magnesium-based Batteries for High-efficiency Energy Storage." *Advanced Energy Materials* (2020) 10(21): 1903591. https://doi.org/10.1002/aenm.201903591

[53] Mohan L Verma, and Homendra D. Sahu. "Ionic Conductivity and Dielectric Behavior of PEO-Based Silver Ion Conducting Nanocomposite Polymer Electrolytes." *Ionics* (2015) 21(12): 3223–31. https://doi.org/10.1007/s11581-015-1517-9

[54] Shujahadeen B Aziz., M. A. Brza, Pshko A. Mohamed, M. F. Z. Kadir, M. H. Hamsan, Rebar T. Abdulwahid, and H. J. Woo. "Increase of Metallic Silver Nanoparticles in Chitosan: AgNt Based Polymer Electrolytes Incorporated with Alumina Filler." *Results in Physics* (2019) 13: 102326. https://doi.org/10.1016/j.rinp.2019.102326

[55] E Correa, M. E. Moncada, and V. H. Zapata. "Electrical Characterization of an Ionic Conductivity Polymer Electrolyte Based on Polycaprolactone and Silver Nitrate for Medical Applications." *Materials Letters* (2017) 205: 155–57. https://doi.org/10.1016/j.matlet.2017.06.046

[56] G Hirankumar, S. Selvasekarapandian, M. S. Bhuvaneswari, R. Baskaran, and M. Vijayakumar. "Ag+ Ion Transport Studies in a Polyvinyl Alcohol-Based Polymer Electrolyte System." *J Solid State Electrochemistry* (2006) 10: 193–97. https://doi.org/10.1007/s10008-004-0612-z

[57] Shujahadeen B. Aziz, Zul Hazrin Zainal Abidin, and Abdul Kariem Arof. "Effect of Silver Nanoparticles on the DC Conductivity in Chitosan-Silver Triflate Polymer Electrolyte." *Physica B: Condensed Matter* (2022) 405(21): 4429–33. https://doi.org/10.1016/j.physb.2010.08.008

[58] Vipin Cyriac, Shilpa Molakalu Padre, Ismayil, Gurumurthy Sangam Chandrashekar, Chetan Chavan, Rajashekhar Fakeerappa Bhajantri, and Mudiyaru Subrahmanya Murari. "Tuning the Ionic Conductivity of Flexible Polyvinyl Alcohol/Sodium Bromide Polymer Electrolyte Films by Incorporating Silver Nanoparticles for Energy Storage Device Applications." *Journal of Applied Polymer Science* (2022) 139(28): e52525. https://doi.org/10.1002/app.52525

[59] Jihad M. Hadi, Shujahadeen B. Aziz, Muaffaq M. Nofal, Sarkawt A. Hussein, Muhamad H. Hafiz, Mohamad A. Brza, Rebar T. Abdulwahid, Mohd F. Z. Kadir, and Haw J. Woo. "Electrical, Dielectric Property and Electrochemical Performances of Plasticized Silver Ion-Conducting Chitosan-Based Polymer Nanocomposites." *Membranes* (2020) 10(7): 151. https://doi.org/10.3390/membranes10070151

[60] S. Austin Suthanthiraraj, and M. Kumara Vadivel. "Electrical and Structural Properties of Poly (Ethylene Oxide)/Silver Triflate Polymer Electrolyte System Dispersed with MgO Nanofillers." *Ionics* (2012) 18: 385–94. https://doi.org/10.1007/s11581-011-0637-0

5 Electrochemical Studies of Hybrid Nanovanadates and Nanotungstates

S. Ishwarya, Vinayak Sunagar, H.V. Harini,
H.P. Nagaswarupa, and Ramachandra Naik

INTRODUCTION

As a result of the massive consumption of fossil fuels by the transportation and industrial sectors, which contributed to global warming, energy storage has become one of the most important problems in the twenty-first century. For our society and country to develop sustainably, many energy systems have been devised [1,2]. Appropriate to their prospective and technical uses in electrical and magnetic devices, nanomaterials such as metallic and ceramic ones have captured the scientific community's interest in recent years. The effectiveness of physical and chemical characteristics can be greatly increased by nanoparticles since they typically contain low density and huge exterior area per unit volume. Because of their quantum scale, physical and chemical performances of nanomaterials differ noticeably from that of bulk materials, prompting their usage in the chemical, mechanical, and information technology sectors. Several sectors of matter science and nanotechnology are increasingly emphasizing the significance of producing and characterizing inorganic objects among nano-sized proportions along with their morphological distinctiveness [3]. Vanadates and tungstates have piqued the curiosity of many people due to their remarkable catalytic and electrochemical properties, which include the possibility of being employed in a variety of industries such as magnetic property applications, humidity sensors, gas sensors, and lithium batteries [4–6]. The best performance of lithium-ion batteries (LIBs) has been demonstrated by transition metals vanadates and tungstates (Mg, Ti, Co, Ag, Mn, Fe). Because of their exceptional ion storage potential, LIBs are a leading example of renewable energy storage and are widely used in contemporary society [7,8]. It is possible to think of vanadates as a derivative of vanadium oxide that has hybridized with other metal ions or clusters [9]. Wolframite-type zinc tungstate nanorods have been created in order to fulfill the rising energy demand and environmental awareness. These nanorods have the potential to produce reversibly maintained high capacity of more than 420 mAh/g subsequent to 150 cycles [10]. Metal vanadates such as $BiVO_4$, $Na_2V_6O_{16}.1.63H_2O$, $NH_4V_4O_{10}$, and $CeVO_4$ have recently been developed. Because of their adaptable layered architectures and the fact that vanadates may exist in a variety of valence states, vanadium-based cathodes are becoming more popular. The electrochemical super capacitors are regarded as an emerging technique to supply multiple devices with a constant power supply [11]. It can meet these needs since it excels in the realm of quick charging and discharging processes with high power densities and low maintenance costs [12]. The tungstates based on transition metals have received exacting focus recently for electrochemical applications. Tungstates enhance the overall electrochemical presentation of the material by increasing its conductivity. In the earlier period, transition metal tungstates are investigated for utilizing in supercapacitor. The main issues with the preponderance of these materials are achieving high specific capacity. Although cycling has been revealed to progress capacitive presentation, a lot of attempts have been made to create the primary capacitive performance and cycle strength of tungstate-based material by synthesizing composite materials with conducting carbon support and mixed metal

DOI: 10.1201/9781003479239-5

oxide, among other things [13]. Based on the electrode materials used, performance of supercapacitor can be tailored. Electrical double-layer capacitors (EDLCs) and pseudocapacitors are two different types of supercapacitors that store electric charge.

Due to its strong catalytic and electrochemical properties, tungstate has attracted a great deal of attention. These qualities have been waged extensively in a multiplicity of sectors, including the use of magnetic properties, humidity sensors, and gas sensors. Many applications employ copper tungstate, an eminent n-type semiconductor among a band space of 2.25 eV. CuWO4 thin film gas sensor created by PLD technique with remarkable response to NO and WO3@CuWO4-based gas sensor with enhanced efficacy for CO gas detection at ambient temperature [6]. Similar to the potential use of m-BiVO$_4$ in gas sensing, we have looked into its electrochemical behavior. It exhibits a high sensitivity to formaldehyde and ethanol gases [14]. According to the grain size effect, nanostructures with a variety of forms and morphologies have been investigated for improved gas-sensing capabilities [15,16]. Optical sensing is widely used in the field of analytical sensing and optical imaging because of its high sensitivity, technical simplicity, and quick response time [17–19]. The sensing targets range from cations to anions, as well as small biologically active molecules to tumor micro-environment-related parameters, which include polarity, temperature, and viscosity. Research on fluorescence sensors has advanced significantly in recent years. The growth of this field has resulted in significant fluorescence sensing systems, which offer a dependable fluorescent response for the investigation of biological or environmental targeted analytes. Small molecule-based sensors are sometimes used, which enables the extremely selective quantitative measurement of a specific analyte [20]. The multi-valence, wide band gap, thermal stability, layered structure, and chemical stability of vanadium-and tungstate-based materials make them promising candidates for future development [21]. Rare earth elements readily form various coordination numbers with the majority of elements and valence state compounds [22]. Due to their special 4f orbital electrons, high spin, and large atomic magnetic moment, rare earth vanadate–based gas sensors would exhibit good gas sensitivity and thermal stability [23]. Vanadate-and tungstate-based gas sensors have been reported [6,24].

DIFFERENT SYNTHESIS METHODS OF NANOVANADATE AND NANOTUNGSTATE NANOPARTICLES

The different methods for producing nanoparticles can be categorized as top-down or bottom-up.

BOTTOM-UP METHOD

Ina bottom-up or constructive approach, materials are formed from atoms through clusters and nanoparticles. The most widely used bottom-up processes for creating nanoparticles include sol-gel, spinning, chemical vapor deposition (CVD), pyrolysis, and biosynthesis [25].

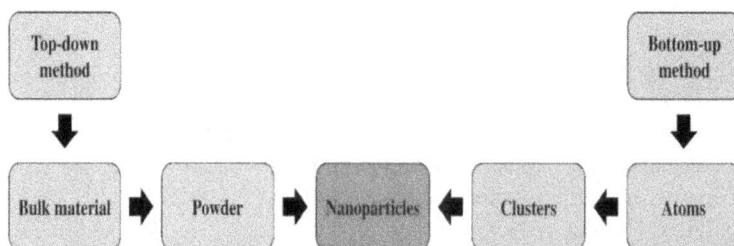

FIGURE 5.1   Synthesis methods of nanovanadates and nanotungstates.

Top-Down Method

The top-down or destructive procedure involves breakdown of a huge substance into smaller parts. Some of the most well-liked methods for producing nanoparticles consist of mechanical milling, thermal breakdown, nanolithography, laser ablation, sputtering, and nanolithography [25].

The production of nanoparticles has greatly benefited from the use of natural plant extracts such as roots, leaves, stems, seeds, fruits, flowers, and petals. Synthesis is made possible by the high photochemical content of extracts, which acts as a reducing, capping, and stabilizing agent and eliminates the need for external organic solvents like surfactants, reducing agents, and stabilizers. The chemical approach uses toxic chemicals that are not only harmful to the environment, but are also dangerous for the persons conducting the process. The most significant reaction that took place was oxidation or reduction. Sol-gel, combustion, microwave, co-precipitation, hydrothermal, and sonication techniques are examples of chemical approaches. Additional capping and stabilizing chemicals are required for physical and chemical techniques [26].

TABLE 5.1
Synthesis of Vanadates and Tungstates, Methods and Calcinated Temperature

Vanadate NPs	Synthesis Method	Calcinated Temp. (°C)	References
$BiVO_4$	Hydrothermal	333k for 6 hours	[27]
$Zn_3(VO_4)_2$	Impregnation	500 for 1 hour	[28]
$Zn_2V_2O_7$	Hydrothermal	450 for 5 hours	[29]
AlV_3O_9	Hydrothermal	500 for 2 hours	[30]
$FeVO_4$	Hydrothermal	60 for 10 hours	[31]
$Zn_3V_2O_8$	Hydrothermal	400 for 1 hour	[32]
$Ni_3V_2O_8$	Hydrothermal	450 for 3 hours	[33]
$CuVO_4$	Hydrothermal	–	[34]
$CeVO_4$	Hydrothermal	180 for 24 hours	[35]
$GdVO_4$	Hydrothermal	650 for 3 hours	[36]
$CuVO_4$	Hydrothermal	180 for 24 hours	[37]
$Ag_2V_4O_{11}$	Hydrothermal	Room temp.	[38]
$NH_4V_4O_{10}$	Hydrothermal	120 for 24 hours	[39]
$Zn_3(VO_4)_2$	Co-precipitation	600 for 6 hours	[40]
$FeVO_4$	Hydrothermal	600 for 3 hours	[41]
$K_{0.25}V_2O_5$	Sol-gel	550 for 4 hours	[42]
$BiVO_4$	Hydrothermal	200 for 4 hours	[43]
NaV_3O_8 & $Zn_3V_2O_8$	Sonication	80 for 12 hours	[44]
Li_3VO_4	Hydrothermal	500 for 5 hours	[45]
$BiVO_4$	Hydrothermal	500 for 1 hour	[46]
$AgVO_4$	Solvothermal	230 for 6 hours	[47]
CaV_4O_9	Hydrothermal	550 for 3 hours	[48]
$Zn_3V_2O_8$	Hydrothermal	80 for 24 hours	[49]
$Ca_{10}V_6O_{25}$	Hydrothermal	60 for 12 hours	[50]
KV_5O_{13}	Hydrothermal	60 for 12 hours	[51]
$Ni_3V_2O_8$	Co-precipitation	500 for 3 hours	[52]
$NaVO_3$	Co-precipitation	250 for 5 hours	[53]
$Co_3V_2O_8$	Co-precipitation	450 for 3 hours	[54]
$K_2V_8O_{21}$ and $K_{0.25}V_2O_5$	Hydrothermal	520 and 550 for 3 and 4 hours, respectively	[55]

(Continued)

TABLE 5.1 (*Continued*)

Synthesis of Vanadates and Tungstates, Methods and Calcinated Temperature

Vanadate NPs	Synthesis Method	Calcinated Temp. (°C)	References
MoV_2O_8	Ultra-sonication method	550 for 2 hours	[56]
FeV_xO_y 1-D nanostructures	Solvothermal	450 for 2 hours	[57]
$Zn_3(OH)_2(V_2O_7)$	Microwave	100 for 6 hours	[58]
$LiZnVO_4$	Sol-gel Combustion route	600	[59]
Ce-doped $NdVO_4$	Hydrothermal method	200 for 24 hours	[60]
NaV_3O_8:Zn	Hydrothermal	80 for 10 hours	[61]
Li_3VO_4:Ca	Solid-state reaction	80 for 12 hours	[62]
Pt:$FeVO_4$	Hydrothermal	500 for 2 hours	[63]
La-doped $BiVO_4$	Co-precipitation	550 for 4 hours	[64]
$Ba_2YV_3O_{11}$:Eu^{3+}	Combustion	500 for 4 hours	[65]
$LiSrVO_4$:Eu^{3+}	Combustion method	300 for 3 hours	[66]
$Ca_2NaMg_2V_3O_{12}$:Eu^{+3} and $Ca_2NaMg_2V_3O_{12}$:Sm^{+3}	Combustion method	800	[67]
$NiO/Ni_3V_2O_8$	Hydrothermal	80 for 12 hours	[68]
$Cu_2V_2O_7$/S-rGO	Sonication	Room temp.	[69]
$BiVO_4$/CuO	Sol-gel	500	[70]
$CeVO_4$/CNT	Silicone oil-bath method.	400	[71]
GO@$LaVO_4$	Hydrothermal	80 for 12 hours	[72]
P-CVO–N@GO	Hydrothermal	60 for 48 hours	[73]
PEDOT:PSS intercalated (($NH_4)_2V_6O_{16}$)	Sonication	–	[74]
$BiVO_4$/FTO	Solvothermal	450 for 2 hours	[75]
$LaVO_4$@BN	Hydrothermal	400 for 2 hours	[76]
$Ni_3V_2O_8$/NG hybrid & Fe_2VO_4/NG hybrid	Hydrothermal	60 for 10 hours	[77]
GO-Mo$BiVO_4$	Co-precipitation	500 for 2 hours	[78]
Ag-NPs@$AgVO_3$-NRs	Hydrothermal	50 for 2 hours	[79]
$Mn_2V_2O_7$-CoV_2O_6	Hydrothermal	450 for 4 hours	[80]

	Tungstates		
Tungstate NPs	Synthesis method	Calcinated temp.(°C)	Reference
$FeWO_4$	Hydrothermal	–	[81]
$FeWO_4$	Hydrothermal	700°C	[82]
$FeO4W$	Hydrothermal	–	[83]
$CeWO_4$	Precipitation	–	[84]
TiO_2/Bi_2WO_6	Solvothermal	–	[85]
$NiWO_4$	Sonochemical	550°C	[86]
$NiWO_4$	Hydrothermal	–	[87]
$NiWO_4$	Hydrothermal	700°C	[88]
$NiWO_4$	Hydrothermal	400°C	[89]
$La_2(WO_4)_3$	Sonochemical	800°C	[90]
Ag_2WO_4	Co- Precipitation	–	[91]
$CoWO_4$	Hydrothermal	–	[92,93]
$CoWO_4$	Co-precipitation	450°C	[94]
$CoWO_4$	Sonochemical	–	[95]
Yb_2WO_6	Co-precipitation	600°C	[96]
$CuWO_4$	Microwave irradiation	600°C	[97]
$CuWO_4$	Hydrothermal	500°C	[98]
Co-$NiWO_4$	Hydrothermal	–	[99]

(Continued)

TABLE 5.1 (*Continued*)
Synthesis of Vanadates and Tungstates, Methods and Calcinated Temperature

Vanadate NPs	Synthesis Method	Calcinated Temp. (°C)	References
$Li_2Ni(WO_4)_2$	Combustion	650°C	[100]
$Ag@MoS_2/WO_3$	Photo reduction	–	[101]
$Ni-CoWO_4$	Hydrothermal	400°C	[102]
$Eu_2(WO_4)_3$	Hydrothermal	–	[103]
$ZnWO_4$	Ultrasonic-assisted co-precipitation	500°C	[104]
$ZnWO_4$	Hydrothermal	550	[10]
$CoNiWO_4$	Hydrothermal	–	[105]
$Li_2Co(WO_4)_2$@Ag composite	Auto-combustion	500°C	[106]
$PANI-CaWO_4$	Precipitation	750°C	[107]
$Sm_2(WO_4)_3$	Ultra-sonication	800°C	[108]
$Sm_2(WO_4)_3$	Sonochemical	700°C	[108]
$Tb_2(WO_4)_3$	Sonochemical	800°C	[109]
$Pr_2(WO_4)_3$	Precipitation	600°C	[110]
Bi_2WO_6	Hydrothermal	–	[110]

CHARACTERIZATION

The phase purity and crystallinity of the synthesized samples were examined using the X-ray diffraction (XRD) analysis. When using a high-energy electron beam to scan the material in a raster scan pattern, an electron microscope is said to be using scanning electron microscopy (SEM) [28]. To determine the size of particles, transmission electron microscopy (TEM) is utilized. By using Raman spectroscopy, the band structures were assessed. UV-VIS spectroscopy was used to examine the optical characteristics and bond energy gap. All electrochemical research, including sensors, cyclic voltammetry (CV), AC impedance, and galvanostatic charge–discharge (GCD), was conducted by the CH-instrument. Information about functional groups is revealed by the FTIR.

X-RAY DIFFRACTION (XRD) ANALYSIS

To check the crystallinity of samples, vanadium-based nanoparticles underwent XRD analysis. The hydrothermal technique used to manufacture hyperbranched $BiVO_4$ in Table 5.1 will display the XRD pattern. With the JCPDS number 83-1700, it displays the monoclinic structure with the cell constants $a = 5.200$, $b = 5.097$, and $c = 11.74$ A0. Impurities were not found [27]. Using XRD spectroscopy, the purity and oxidation state of $Zn_3V_2O_8$ nanoplatelets were identified. It displayed the entire $Zn_3V_2O_8$ nanoparticle survey spectrum. It demonstrates the presence of the state's Zn_{2p}, O_{1S}, V_{2P}, and C1S. $Zn_{2p3/2}$ and $Zn_{2p1/2}$ are represented by the two peaks on the Zn_{2p} spectra at 1021.58 and 1044.68 eV, respectively. The JCPDS No. for all diffraction peaks was 87-0417 [30].The hydrothermal process was used to create $GdVO_4$. By using XRD, it was discovered that several distinct peaks with two values were found with JCPDS card no. 17-0260. It demonstrates that the materials were synthesized with only Gd, V, and O components and without any other diffraction peaks with impurities [36]. By using the co-precipitation approach, $NaVO_3$ was created. Phase purity and crystal structure were found. It displays an orthorhombic structure with cell constants of 5.36, 14.15, and 3.65 A°, respectively, for a, b, and c. JCPDS No. 32-1198 was used [52]. They are within a well-crystallized state, as shown by the unique sharp and well-defined peaks. There were no distinct peaks for various impurities, suggesting that the synthesized compound is pure [90].

FOURIER TRANSFORM INFRARED SPECTROSCOPY (FTIR)

The chemical composition and functional elements of the synthesized tungstate nanoparticles are explored using the Fourier transform infrared spectroscopy (FTIR) technology [81]. The FTIR spectra of sodium vanadates show the O–V–O asymmetric vibration band at around $560\,cm^{-1}$, the V–=–O stretching vibration bands at the $950–880\,cm^{-1}$ area. While the bands at 755 and $540\,cm^{-1}$ are attributed to asymmetric stretching vibrations of V–O–V bonds, the bands at $1,012–953\,cm^{-1}$ are assigned to V–O stretching vibrations that are somewhat shifted to higher wave number compared with those of $NaVO_3$. The sodium vanadate ($NaVO_3$) from the reagent bottle and the FTIR data from the $NaVO_3$ nanobelts are in good agreement [53]. On a Spectrum-2000 FTIR spectrophotometer (PerkinElmer Corp., USA), FTIR spectra of the $CeVO_4$ NRs mixed with KBr were captured in the $400–4,000\,cm^{-1}$ range. Using a monochromatic Al K1,2 source (1,486.60 eV) and a PHI Quantum 2000 Scanning ESCA Microprobe (PHI. Corp., USA) equipment, the surface characteristics of the materials were analyzed. All binding energies were calculated using the surface adventitious carbon's C1s peak at 284.8 eV [35].The out-of-plane V–O–V vibration could be seen at $537\,cm^{-1}$, and the V–O stretching bond could be found at $1,004\,cm^{-1}$ in the infrared spectroscopy (IR) spectrum. It is possible that the addition of ammonium metavanadate created the absorption bands at $1400\,cm^{-1}$, which indicates the presence of NH^{4+}. The IR spectrum also shows the presence of water in addition to the vanadium and oxide states. Adsorbed water and crystal water, respectively, exhibit stretching vibrations, which are represented by the absorption bands at 3,417 and $3,191\,cm^{-1}$. Water molecules are also indicated by the symmetric band at $1,621\,cm^{-1}$, which is caused by H_2O vibrations [80].

SCANNING ELECTRON MICROSCOPY (SEM)

Electron microscopy examined the morphologies and microstructures of the produced vanadate and tungstate nanoparticles [82]. The as-prepared KVO is made of nanofibers, which are typically roughly several micrometers in length and dozens of nanometers in width, according to the pictures from SEM and TEM. The solid-state diffusion path of the Zn^{2+} ions is undoubtedly shortened by the assembly of KVO nanobelts from ultrathin nanobelts with a thickness of less than $10\,nm$ [51]. The accumulation of irregular and varied particles is shown in the SEM image. The dimension of a few ten nanometers shows that the particle size was greatly decreased by the ABR technique. Ca-doped samples produced by the same process display a comparable morphology. But the surface was dramatically changed by the presence of Ca^{2+} ions [62].

TRANSMISSION ELECTRON MICROSCOPY (TEM)

TEM surveys the morphology of vanadate- and tungstate-based nanomaterials. It indicates the production of densely packed yet finely scattered nanoparticles [111]. Clear lattice fringes with d-spacing of 0.297 and 0.253 nm, indexed to the interplanar spacing of (120) and (221) planes of triclinic $FeVO_4$, are seen in the HR-TEM image of the chosen region of the single rod figure. By contrast, the Pt-decorated $FeVO_4$ nanorods exhibit the (120) triclinic $FeVO_4$ lattice as well as extra Pt lattice fringes with a d-spacing of 0.225 nm, which corresponds to the d-spacing of (111) plane of Pt [63,112]. TEM images were used to evaluate the surface morphology and crystallite size of $Ba_2Y_{0.95}Eu_{0.05}V_3O_{11}$ phosphor. The shape of nanocrystals appears to be spherical with some modest aggregation. Sizes of the particles seemed to range from 35 to 95 nm. Therefore, it can be said that the particle size determined by TEM examination is in good accordance with the results we derived from Scherer's equation, thereby validating the nano-dimensional profile of $Ba_2Y_{0.95}Eu_{0.05}V_3O_{11}$ crystals [65].

ULTRAVIOLET–VISIBLE DIFFUSE REFLECTANCE SPECTROSCOPY (UV-DRS)

UV-visible diffuse reflectance spectroscopy is used to inquire the optical distinctiveness of the synthesized vanadate and tungstate nanocomposites. Using Tauc's relation: $(hv) = A (hv-Eg)n$, the bandgap energies were computed by the data acquired from UV-DRS [91].The bandgap energy of 2.43 eV, which was determined using diffuse reflectance, was one of the other properties of the $BiVO_4$ photo anode films that were obtained [46].A useful tool for describing optical absorption properties, which are well known to be one of the key determinants of photocatalytic activity, is UV-visible diffuse reflectance spectroscopy (DRS). It shows the undoped, Nd^{+3}-, and Er^{+3}-doped YVO_4 diffuse reflectance spectra in the UV-VIS region between 200 and 800 nm. The formula for the optical absorption close to the band edge is $\alpha(hv)^n = A(hv-Eg)$, where α, h, v, E_g, and A are respectively, the absorption coefficient, Planck constant, light frequency, bandgap energy, and a constant [113].

ENERGY-DISPERSIVE X-RAY (EDAX) SPECTROSCOPY

The homogeneous distribution of the Mg, V, and O elements is shown in images of the single MgVO nanobelt taken using energy-dispersive X-ray spectroscopy (EDAX) [114].The zinc, vanadium, carbon, and oxygen elements were amply supported by the EDX spectrum that was exhibited. The results of the ZVC composite further demonstrated the product's purity by showing no additional atoms or elements. ZVC has an atomic ratio of 2.41:39.11:14.27:44.21, and its weight percentage is displayed in Ref. [115].

THERMOGRAVIMETRY AND RAMAN STUDIES

A Renishaw micro-Raman spectrometer (RE-04) connected to a solid-state laser with a diode pumped at 514 nm was used to evidence the Raman spectra of nanomaterials. TGA examines the thermal venture of the materials [105].

RAMAN SPECTROSCOPY

The accelerated electron-hole migration across the valence and conduction bands in $GdVO_4$ crystals may have been influenced by the production of symmetrical and asymmetrical vibrations in the Raman spectrum of $GdVO_4$ [116–118]. The chemical structure, polymorphism, crystallinity, and interactions between molecules were studied using the Raman scattering spectrum. For (i) $Cu_2V_2O_7$, (ii) S-rGO, and (iii) $Cu_2V_2O_7$/S-rGO composites, a specific peak is dominated under a Raman shift of 500–2,000 cm^{-1}, especially for (i) $Cu_2V_2O_7$. The vibration bands are shown by the Raman peaks at 631, 816, 843, 872, 909, and 949 cm^{-1}. However, the Raman bands that are found between 631 and 909 cm^{-1} are attributed to the symmetric $v_1(VO_3)$ vibration at 949 cm^{-1}, the antisymmetric stretching $v_2(VO_3)$ vibration at 909 cm^{-1}, and the $V_3(V-O-V)$ antisymmetric vibration at 872 cm^{-1}. The bands for $Cu_2V_2O_7$ occurred at 631 cm^{-1} for $V_2(VO_3)$ bending vibration and 816 and 843 cm^{-1} for $V_4(VO_3)$ bending vibration, respectively [69].

ENERGY APPLICATIONS OF VANADATES AND TUNGSTATES

APPLICATIONS FOR BATTERIES

Due to their greater reversible capacity, higher Li+ diffusion coefficients, and improved rate capability compared to typical bulk materials, several nanostructured tungstate materials have generated a lot of attention as anodes for Li-ion batteries. Additionally, by employing conversion reaction processes, it has been lightened that the particle size influence on lithium response for nanophase transition metal oxides ($MxOy$:M=Co, Cu, Ni, Fe) can produce significantly greater capacities than graphite. In these materials, lithium is reversibly extracted from Li_2O with concurrent oxidation/reduction of the transition metal ($MxOy + ne- + nLi+ = xM0 + LinOy$) [82].

Cyclic voltammetry, one way to determine a material's capacitive nature is to look at its electric double-layer capacitance or pseudo-capacitance. In a setup using three electrodes, Ag/AgCl, platinum wire, and $NiO/Ni_3V_2O_8$ coated on Ni-foam, respectively, served as the reference, counter, and working electrodes. By using CV, electrochemical characteristics can be assessed. At various scan speeds of 10, 20, 30, 50, 80, and 100 mV/s, the CV study was put to the test. The active electrode ($NiO/Ni_3V_2O_8$) has a pseudo-capacitive character, which is demonstrated by the well-shaped CV curve, which also displays oxidation and reduction peaks in the potential window of 0–0.6 V. The detected redox peaks reveal a shift as the applied potential scan rate increased from 10 to 100 mV/s due to an increase in internal diffusion resistance within the pseudo-active materials. Straight spikes could be seen for all products that result from the electron/ion transfer mechanism at low frequencies of applied AC voltage and exhibit capacitive identity with variable slopes directly according to its conductivity [68]. Different concentrations of NLT are applied to the produced material to test the effects of concentration on the $Cu_2V_2O_7$/S-rGO/SPCE electrochemical performance toward NLT. After extensive testing with a range of NLT concentrations from 30 to 180 M at 0.1 M of PB at pH (7.0) with a scan rate of 50 mV/s, it is clear from the concentration effect that as the NLT concentration gradually increased, the reduction peak potential current also increased. This is because the ionic strength of the electrolyte increased along with the formation and interaction of the reduction products with the surface area of the prepared material [69].

Applications for Supercapacitor

An effectual technique for calculating the supercapacitive efficiency of an electrode that is exposed to regulated conditions is GCD investigation. Three electrodes are utilized to carry out the technique. The IR reductions of CWO-NPs were tested for 50 segments in a GCD experiment for the 400 mV window of potential. The current density used for GCD measurements was 10 mA/g. GCD curves of CWO-NPs show continuous reversible cycles through charge–discharge pathway for 50 segments. Surprisingly, relatively tiny IR dips were detected throughout the discharge procedure. An indication of low surface resistance and loss of energy is a minimal IR decrease of CWO-NPs. The outstanding efficiency of the particular capacitance of CWO-NPs is revealed in the discharge time of 1,100 seconds for 50 segments at 10 mA/g that is calculated for these devices [119].

At a current density of 100 mAg1, the GCD curves of the pure $CeVO_4$ and $CeVO_4$/CNT HCNS electrodes were measured. Three different sloping voltage curves may be seen for the pure $CeVO_4$ electrode during the initial discharge. The continuous reduction of V^{3+} to V^{2+} and V^{2+} to V^+, respectively, may be attributed to the first and second long sloping curves between 1.37 and 1.61 V (Region-I) and 0.9 and 1.02 V (Region-II), while the oxidation of V^{2+} to V^{3+} and V^+ to V^{2+} during initial charge is represented by the broad sloping curve between 1.4 and 2.1 V. The production of SEI film is associated with the third discharge sloping curve between 0.72 and 0.83 V. With the obtained peaks in CV, the initial charge–discharge slope curves fit well. The pure $CeVO_4$ and $CeVO_4$/CNT HCNS electrodes were found to have initial charge–discharge capacity values of 659/425 and 433/245 mAh/g, respectively. Initial Columbic efficiency measurements were made, yielding values of 64% for pure $CeVO_4$ and 56% for the $CeVO_4$/CNT HCNS electrode [71]. According to the charge–discharge curves of PICs at various voltage windows, when the voltage window rises to 4 V (current density, 0.2 Ag1), the discharge time is the longest. Additionally, the equivalent energy density was greatly increased, going from 57.4 to 118.2 Wh/kg. The PIC's optimal working voltage range is therefore set to 0.5–4 V. A high energy density of 150.8 Wh/kg can be reached at a power density of 112.5 W/kg in the galvanostatic charge/discharge profiles of the P-CVO-N@GO//AC PIC at various current densities ranging from 0.05 to 10 A/g At a high-power density of 22,500 W/kg, the device nevertheless managed to achieve an energy density of 78.2 Wh/kg [73].

APPLICATIONS FOR SENSORS

By monitoring photocurrent at a specific potential that changed with glucose concentration, the sensing capability of the BiVO$_4$-based PEC sensor was assessed. The photocurrent shows a tidily increasing glucose content in the 5–35 mM range. The fitting curve shows that there is a strong linear association between photocurrent and glucose levels, with a coefficient of 0.991.

Tungstate-based nanoparticles increased electrical conductivity which is utilized for recognizing gases such as NO$_2$ and NH$_3$. Because of the charge transfer from nanoparticles to NO$_2$ when the gas molecules bind them together, pores of nanoparticles expand, resulting in excellent gas sensors [25]. The MoSe$_2$/CuWO$_4$ film sensor has great reproducibility and responsiveness. The alteration in capacitance may be seen on a smartphone by attaching the humidity sensor to a portable device at the same time as flexible humidity testing tests were conducted. The MoSe$_2$/CuWO$_4$ humidity sensor appears to be useful for detecting approaching fingers, counting water droplets, and human breath [6].

By observing the photocurrent of glucose oxidation while turning the light on and off 100 times, the sensor's stability was assessed. As shown, there is almost no photocurrent degradation, indicating that BiVO$_4$ is stable enough to oxidize glucose. Five independent BiVO$_4$ photo anodes were made under identical conditions in order to evaluate the reproducibility of the PEC sensor, and the sensitivity findings revealed a relative standard deviation (RSD) of 7.64% between them, showing an appropriate reproducibility [75]. The catalytic activity of electrochemical sensors was significantly influenced by the pH of the supporting electrolyte. In 0.05 M PBS at various pH levels, the electrochemical signals of LaV/F-BN-modified electrode toward 200 M FZD were captured. The electrochemical signal increased when the electrolyte pH was between 3.0 and 7.0, and it decreased between pH 9.0 and 11.0. It should be highlighted that a high reduction peak current at pH 7.0 was attained, which is more suited for actual use in biological systems. For pH values v/s reduction potential (Epc) potential, a linear regression plot was created with an R^2 of 0.99. For further investigation, 0.05 M PBS with a pH of 7.0 is used. In 0.05 M PBS (pH 7.0), the electrochemical performance of LaV/F-BN-modified RDGCE is assessed at 1,200 rpm. An amperometric approach was used to estimate the linear range and LOD of FZD for the LaV/F-BN/RDGCE. In the picture, the typical amperometric signal of the LaV/F-BN/RDGCE to consecutive addition of FZD in the range of 0.015–639 M was shown. Each addition of FZD triggers a progressive response from LaV/FBN/RDGCE that resembles a staircase [76]. Biological samples and cationic chemicals interfered with the DPV responses in GO-MoBiVO$_4$. By adding 300 mL of TCP and 0.1 M of N$_2$ saturating PBS (pH 5) along with biomolecules and cation compounds such glucose, 4 nitro phenol (4NP), copper (Cu^{2+}), and zinc (Zn^{2+}), this response was ascertained. The modified electrode that was suggested demonstrated great selectivity and successfully counteracted the impact of common interference compounds [78].

CONCLUSION

The process of direct calcination, which is based on electrochemical, sol-gel, solution combustion, and co-precipitation methods, is used to create tungstate nanoparticles. The researchers were able to analyze the characteristics of the recently developed nanoparticles using FT-IR, XRD, TEM, and SEM. In the current work, we investigated how tungstate nanoparticles enhance conductivity, offer adequate energetic sites and storage space, and generate ultra-fast charge transfer channels that enhance the kinetics of electrochemical events. Vanadates and tungstates are versatile materials that have been widely explored for various energy applications, such as batteries, supercapacitors, and sensors. They exhibit multiple oxidation states, layered structures, and synergistic effects between transition metals and vanadium, which enable them to store and deliver charge efficiently and reversibly. Vanadates and tungstates can be fabricated into various nanostructures with different dimensions and morphologies, which can enhance their specific surface area, electroactive sites, and ion diffusion pathways. Metal tungstates are electrode materials that have excellent cyclic performance, high rate capability, and specific capacitance. Nanosensors have distinct physical properties that allow them to give sensitivity instructions of magnitude superior than traditional devices while also providing performance benefits such as quick response and mobility.

Moreover, vanadates can be hybridized with other materials, such as carbonaceous materials, metal oxides, and polymers, to improve their conductivity, stability, and flexibility. Vanadates have shown promising performance in aqueous zinc-ion batteries, pseudo-supercapacitors, and electrochemical sensors, among others. However, there are still some challenges and limitations that need to be addressed, such as low electronic conductivity, structural degradation, poor cycling life, and environmental toxicity. Therefore, further research and development are needed to optimize the synthesis methods, design strategies, and device configurations of vanadates for energy applications.

FUTURE OUTCOMES

Applications of energy from vanadates and tungstates will keep generating more interest in research and development in the future. The development of novel synthesis methods and design approaches to yield a variety of vanadate nanostructures and hybrids with adjustable form, composition, and properties, exploring novel vanadate types and tungstates that have been modified physically, redoxically, and by adding other transition metals or other materials. To improve the electrochemical stability and performance of vanadates, research is being done on their interfacial phenomena and reaction processes in different electrolytes and circumstances, expanding the numerous potential uses of vanadates by including more energy-related technologies such as fuel cells, thermoelectric power, and photocatalysis, developing sustainable and ecologically acceptable processes for the production and use of vanadate energy, as well as evaluating the applications' potential economic benefits and environmental effects.

REFERENCES

[1] Li, Cheng, and Limin Qi. "Colloidal-crystal-assisted patterning of crystalline materials." *Advanced Materials* 22, no. 13 (2010): 1494–1497.

[2] Sedighizadeh, Mostafa, Masoud Esmaili, and S. Mohammadreza Mousavi-Taghiabadi. "Optimal joint energy and reserve scheduling considering frequency dynamics, compressed air energy storage, and wind turbines in an electrical power system." *Journal of Energy Storage* 23 (2019): 220–233.

[3] Gul, Fiza, Muhammad Athar, and Muhammad Asim Farid. "Nanocomposites of transition metals tungstate for potential applications in magnetic and microwave devices." *Journal of Electroceramics* 40 (2018): 300–305.

[4] Pei, Lizhai, Nan Lin, Tian Wei, Handing Liu, and Haiyun Yu. "Formation of copper vanadate nanobelts and their electrochemical behaviors for the determination of ascorbic acid." *Journal of Materials Chemistry A* 3, no. 6 (2015): 2690–2700.

[5] Sajid, Muhammad Munir, Sadaf Bashir Khan, Naveed Akthar Shad, Nasir Amin, and Zhengjun Zhang. "Visible light assisted photocatalytic degradation of crystal violet dye and electrochemical detection of ascorbic acid using a $BiVO_4/FeVO_4$ heterojunction composite." *RSC Advances* 8, no. 42 (2018): 23489–23498.

[6] Zhang, Dongzhi, Mengyu Wang, Wenyuan Zhang, and Qi Li. "Flexible humidity sensing and portable applications based on $MoSe_2$ nanoflowers/copper tungstate nanoparticles." *Sensors and Actuators B: Chemical* 304 (2020): 127234.

[7] Goriparti, Subrahmanyam, Ermanno Miele, Francesco De Angelis, Enzo Di Fabrizio, Remo Proietti Zaccaria, and Claudio Capiglia. "Review on recent progress of nanostructured anode materials for Li-ion batteries." *Journal of Power Sources* 257 (2014): 421–443.

[8] Armand, Michel, and J-M. Tarascon. "Building better batteries." *Nature* 451, no. 7179 (2008): 652–657.

[9] Xu, Xiaoming, Fangyu Xiong, Jiashen Meng, Xuanpeng Wang, Chaojiang Niu, Qinyou An, and Liqiang Mai. "Vanadium-based nanomaterials: a promising family for emerging metal-ion batteries." *Advanced Functional Materials* 30, no. 10 (2020): 1904398. https://doi.org/10.1002/adfm.201904398

[10] Yang, Lijuan, Xin He, ChunjuLv, Lidong Jiang, Bojian Wang, and Kangying Shu. "One-step preparation and characterization of zinc tungstate-carbon nanoparticles with application to lithium-ion batteries." *Instrumentation Science & Technology* 44, no. 6 (2016): 603–613. https://doi.org/10.1080/10739149.2016.1184160

[11] Guardia, Laura, Loreto Suárez, NausikaQuerejeta, Viliam Vretenár, Peter Kotrusz, Viera Skákalová, and Teresa A. Centeno. "Biomass waste-carbon/reduced graphene oxide composite electrodes for enhanced supercapacitors." *Electrochimica Acta* 298 (2019): 910–917. https://doi.org/10.1016/j.electacta.2018.12.160.

[12] Cheng, Yan, Yifu Zhang, Hanmei Jiang, Xueying Dong, Changgong Meng, and Zongkui Kou. "Coupled cobalt silicate nanobelt-on-nanobelt hierarchy structure with reduced graphene oxide for enhanced supercapacitive performance." *Journal of Power Sources* 448 (2020): 227407. https://doi.org/10.1016/j.jpowsour.2019.227407

[13] Mallick, Sourav, Amit Mondal, and C. Retna Raj. "Rationally designed mesoporous carbon-supported Ni-NiWO4@ NiS nanostructure for the fabrication of hybrid supercapacitor of long-term cycling stability." *Journal of Power Sources* 477 (2020): 229038. https://doi.org/10.1016/j.jpowsour.2020.229038

[14] Saravanakumar, Balakrishnan, Chandran Radhakrishnan, Murugan Ramasamy, Rajendran Kaliaperumal, Allen J. Britten, and Martin Mkandawire. "Copper oxide/mesoporous carbon nanocomposite synthesis, morphology and electrochemical properties for gel polymer-based asymmetric supercapacitors." *Journal of Electroanalytical Chemistry* 852 (2019): 113504. https://doi.org/10.1016/j.jelechem.2019.113504

[15] Vinothkumar, Venkatachalam, Gajapaneni Venkata Prasad, Shen-Ming Chen, Arumugam Sangili, Seung-Joo Jang, Hong Chul Lim, and Tae Hyun Kim. "One-step synthesis of calcium-doped copper oxide nanoparticles as an efficient bifunctional electrocatalyst for sensor and supercapacitor applications." *Journal of Energy Storage* 59 (2023): 106415. https://doi.org/10.1016/j.est.2022.106415

[16] Zhao, Yu, Yi Xie, Xi Zhu, Si Yan, and Sunxi Wang. "Surfactant-free synthesis of hyperbranched monoclinic bismuth vanadate and its applications in photocatalysis, gas sensing, and lithium-ion batteries." *Chemistry-A European Journal* 14, no. 5 (2008): 1601–1606. https://doi.org/10.1002/chem.200701053

[17] Mirzaei, Ali, Jae-Hun Kim, Hyoun Woo Kim, and Sang Sub Kim. "How shell thickness can affect the gas sensing properties of nanostructured materials: Survey of literature." *Sensors and Actuators B: Chemical* 258 (2018): 270–294. https://doi.org/10.1016/j.snb.2017.11.066

[18] Yuan, Hongye, Saif Abdulla Ali AlateeqiAljneibi, Jiaren Yuan, Yuxiang Wang, Hui Liu, Jie Fang, Chunhua Tang et al. "ZnO nanosheets abundant in oxygen vacancies derived from metal-organic frameworks for ppb-level gas sensing." *Advanced Materials* 31, no. 11 (2019): 1807161. https://doi.org/10.1002/adma.201807161

[19] Ashton, Trent D., Katrina A. Jolliffe, and Frederick M. Pfeffer. "Luminescent probes for the bioimaging of small anionic species in vitro and in vivo." *Chemical Society Reviews* 44, no. 14 (2015): 4547–4595. https://doi.org/10.1039/C4CS00372A

[20] Lin, Vivian S., Wei Chen, Ming Xian, and Christopher J. Chang. "Chemical probes for molecular imaging and detection of hydrogen sulfide and reactive sulfur species in biological systems." *Chemical Society Reviews* 44, no. 14 (2015): 4596–4618. https://doi.org/10.1039/C4CS00298A

[21] Yang, Zhigang, Jianfang Cao, Yanxia He, Jung Ho Yang, Taeyoung Kim, Xiaojun Peng, and Jong Seung Kim. "Macro-/micro-environment-sensitive chemosensing and biological imaging." *Chemical Society Reviews* 43, no. 13 (2014): 4563–4601. https://doi.org/10.1039/C4CS00051J

[22] Yu, Minghao, Yan Zeng, Yi Han, Xinyu Cheng, Wenxia Zhao, Chaolun Liang, Yexiang Tong, Haolin Tang, and Xihong Lu. "Valence-optimized vanadium oxide supercapacitor electrodes exhibit ultrahigh capacitance and super-long cyclic durability of 100 000 cycles." *Advanced Functional Materials* 25, no. 23 (2015): 3534–3540. https://doi.org/10.1002/adfm.201501342

[23] Dehnicke, Kurt, and Andreas Greiner. "Unusual Complex Chemistry of Rare-Earth Elements: Large Ionic Radii-Small Coordination Numbers." *AngewandteChemie International Edition* 42, no. 12 (2003): 1340–1354. https://doi.org/10.1002/anie.200390346

[24] Yan, Zheng-Guang, and Chun-Hua Yan. "Controlled synthesis of rare earth nanostructures." *Journal of Materials Chemistry* 18, no. 42 (2008): 5046–5059. https://doi.org/10.1039/B810586C

[25] Ealia, S. A. M., and M. P. Saravanakumar. (2017, November). IOP conference series: materials science and engineering, IOP Publishing. (Vol. 263, No. 3, p. 032019). https://doi.org/10.1088/1757-899X/263/3/032019

[26] Chen, Limiao. "Hydrothermal synthesis and ethanol sensing properties of CeVO4 and CeVO4-CeO2 powders." *Materials Letters* 60, no. 15 (2006): 1859–1862. https://doi.org/10.1016/j.matlet.2005.12.037

[27] Gómez-López, Paulette, Alain Puente-Santiago, Andrés Castro-Beltrán, Luis Adriano Santos do Nascimento, Alina M. Balu, Rafael Luque, and Clemente G. Alvarado-Beltrán. "Nanomaterials and catalysis for green chemistry." *Current Opinion in Green and Sustainable Chemistry* 24 (2020): 48–55. https://doi.org/10.1016/j.cogsc.2020.03.001

[28] Zhao, Yu, Yi Xie, Xi Zhu, Si Yan, and Sunxi Wang. "Surfactant-free synthesis of hyperbranched monoclinic bismuth vanadate and its applications in photocatalysis, gas sensing, and lithium-ion batteries." *Chemistry-A European Journal* 14, no. 5 (2008): 1601–1606. https://doi.org/10.1002/chem.200701053

[29] Zhang, Le-Xi, Gui-Nian Li, Yan-Yan Yin, Yue Xing, Heng Xu, Jing-Jing Chen, and Li-Jian Bie. "Zn3 (VO4) 2-decoration induced acetone sensing improvement of defective ZnO nanosheet spheres." *Sensors and Actuators B: Chemical* 325 (2020): 128805. https://doi.org/10.1016/j.snb.2020.128805

[30] Zhang, Qi, Qiqi Shi, Hongdong Li, Zhenyu Xiao, Kun-Peng Wang, Lingbo Zong, and Lei Wang. "Hydrothermal synthesis and electrochemical properties of 3D Zn2V2O7 microsphere for alkaline rechargeable battery." *Journal of Power Sources* 439 (2019): 227087. https://doi.org/10.1016/j.jpowsour.2019.227087

[31] Yan, Yan, Hao Xu, Wei Guo, Qingli Huang, Mingbo Zheng, Huan Pang, and Huaiguo Xue. "Facile synthesis of amorphous aluminum vanadate hierarchical microspheres for supercapacitors." *Inorganic Chemistry Frontiers* 3, no. 6 (2016): 791–797. https://doi.org/10.1039/C6QI00089D

[32] Xu, Wangwang, Lei Zhang, Kangning Zhao, Xiuxuan Sun, and Qinglin Wu. "Layered ferric vanadate nanosheets as a high-rate NH4+ storage electrode." *Electrochimica Acta* 360 (2020): 137008. https://doi.org/10.1016/j.electacta.2020.137008

[33] Vijayakumar, Subbukalai, Seong-Hun Lee, and Kwang-Sun Ryu. "Synthesis of Zn 3 V 2 O 8 nanoplatelets for lithium-ion battery and supercapacitor applications." *RSC Advances* 5, no. 111 (2015): 91822–91828. https://doi.org/10.1039/C5RA13904J

[34] Ramavathu, Lakshmana Naik, Seshagiri Rao Harapanahalli, Nagaraja Pernapati, and Bala Narsaiah Tumma. "Synthesis and characterization of Nickel Metavanadate (Ni3V2O8)- application as photocatalyst and supercapacitor" *International Journal of Nano Dimension* 12, no. 4 (2021): 411–421.

[35] Pei, L. Z., T. Wei, N. Lin, and Z. Y. Cai. "The electrochemical detection of tartaric acid using Cu vanadate nanorods modified electrode." *International Journal of Nano and Biomaterials* 6, no. 1 (2015): 41–51. https://doi.org/10.1504/IJNBM.2015.073157

[36] Hou, Jimin, Huihan Huang, Zhizhong Han, and Haibo Pan. "The role of oxygen adsorption and gas sensing mechanism for cerium vanadate (CeVO 4) nanorods." *RSC Advances* 6, no. 18 (2016): 14552–14558. https://doi.org/10.1039/C5RA20049K

[37] He, Aijiang, Li Feng, Lixiu Liu, Junlin Peng, Yuning Chen, Xuhao Li, Wencong Lu, and Junyang Liu. "Design of novel egg-shaped GdVO 4 photocatalyst: a unique platform for the photocatalyst and supercapacitors applications." *Journal of Materials Science: Materials in Electronics* 31 (2020): 13131–13140. https://doi.org/10.1007/s10854-020-03864-z

[38] Han, Gui-hong, Shu-zhen Yang, Yan-fang Huang, Yang Jing, Wen-cui Chai, Rui Zhang, and De-liang Chen. "Hydrothermal synthesis and electrochemical sensing properties of copper vanadate nanocrystals with controlled morphologies." *Transactions of Nonferrous Metals Society of China* 27, no. 5 (2017): 1105–1116. https://doi.org/10.1016/S1003-6326(17)60129-8

[39] Fu, Haitao, Xiaohong Yang, Xuchuan Jiang, and Aibing Yu. "Silver vanadate nanobelts: A highly sensitive material towards organic amines." *Sensors and Actuators B: Chemical* 203 (2014): 705–711. https://doi.org/10.1016/j.snb.2014.07.057

[40] Fang, Dong, Yunhe Cao, Ruina Liu, Weilin Xu, Suqin Liu, Zhiping Luo, Chaowei Liang, Xiaoqing Liu, and Chuanxi Xiong. "Novel hierarchical three-dimensional ammonium vanadate nanowires electrodes for lithium ion battery." *Applied Surface Science* 360 (2016): 658–665. https://doi.org/10.1016/j.apsusc.2015.11.038

[41] Arasi, S. Ezhil, P. Devendran, R. Ranjithkumar, S. Arunpandiyan, and A. Arivarasan. "Electrochemical property analysis of zinc vanadate nanostructure for efficient supercapacitors." *Materials Science in Semiconductor Processing* 106 (2020): 104785. https://doi.org/10.1016/j.mssp.2019.104785

[42] Kesavan, Ganesh, Praveen Kumar Gopi, Shen-Ming Chen, and Venkatachalam Vinothkumar. "Iron vanadate nanoparticles supported on boron nitride nanocomposite: electrochemical detection of antipsychotic drug chlorpromazine." *Journal of Electroanalytical Chemistry* 882 (2021): 114982 https://doi.org/10.1016/j.jelechem.2021.114982

[43] Fang, Guozhao, Jiang Zhou, Yang Hu, XinXin Cao, Yan Tang, and Shuquan Liang. "Facile synthesis of potassium vanadate cathode material with superior cycling stability for lithium ion batteries." *Journal of Power Sources* 275 (2015): 694–701. https://doi.org/10.1016/j.jpowsour.2014.11.052

[44] Liu, Ying, Xiaocui Xu, Churong Ma, Feng Zhao, and Kai Chen. "Morphology effect of bismuth vana-date on electrochemical sensing for the detection of paracetamol." *Nanomaterials* 12, no. 7 (2022): 1173. https://doi.org/10.3390/nano12071173

[45] Xie, Zhiqiang, Jianwei Lai, Xiuping Zhu, and Ying Wang. "Green synthesis of vanadate nanobelts at room temperature for superior aqueous rechargeable zinc-ion batteries." *ACS Applied Energy Materials* 1, no. 11 (2018): 6401–6408. https://doi.org/10.1021/acsaem.8b01378

[46] Ni, Shibing, XiaohuLv, Jianjun Ma, Xuelin Yang, and Lulu Zhang. "Electrochemical characteristics of lithium vanadate, Li3VO4 as a new sort of anode material for Li-ion batteries." *Journal of Power Sources* 248 (2014): 122–129. https://doi.org/10.1016/j.jpowsour.2013.09.050

[47] Ribeiro, Francisco Wirley Paulino, Fernando Cruz Moraes, Ernesto Chaves Pereira, Frank Marken, and Lucia Helena Mascaro. "New application for the BiVO4 photoanode: a photoelectroanalytical sensor for nitrite." *Electrochemistry Communications* 61 (2015): 1–4. https://doi.org/10.1016/j.elecom.2015.09.022

[48] Rostamzadeh, Taha, Shiva Adireddy, and John B. Wiley. "Formation of scrolled silver vanadate nano-peapods by both capture and insertion strategies." *Chemistry of Materials* 27, no. 10 (2015): 3694–3699. https://doi.org/10.1021/acs.chemmater.5b01161

[49] Xu, Xiaoming, Chaojiang Niu, Manyi Duan, Xuanpeng Wang, Lei Huang, Junhui Wang, Liting Pu et al. "Alkaline earth metal vanadates as sodium-ion battery anodes." *Nature Communications* 8, no. 1 (2017): 460. https://doi.org/10.1038/s41467-017-00211-5

[50] Suganya, B., J. Chandrasekaran, S. Maruthamuthu, B. Saravanakumar, and E. Vijayakumar. "Hydrothermally synthesized zinc vanadate rods for electrochemical supercapacitance analysis in various aqueous electro-lytes." *Journal of Inorganic and Organometallic Polymers and Materials* 30 (2020): 4510–4519. https://doi.org/10.1007/s10904-020-01581-y

[51] Pei, Lizhai, Yinqiang Pei, Yikang Xie, Chuangang Fan, Diankai Li, and Qianfeng Zhang. "Formation process of calcium vanadate nanorods and their electrochemical sensing properties." *Journal of Materials Research* 27, no. 18 (2012): 2391–2400. https://doi.org/10.1557/jmr.2012.254

[52] Qiu, Nan, Zhaoming Yang, Rui Xue, Yuan Wang, Yingming Zhu, and Wei Liu. "Toward a high-perfor-mance aqueous zinc ion battery: Potassium vanadate nanobelts and carbon enhanced zinc foil." *Nano Letters* 21, no. 7 (2021): 2738–2744. https://doi.org/10.1021/acs.nanolett.0c04539

[53] Sambandam, Balaji, Vaiyapuri Soundharrajan, Jinju Song, Sungjin Kim, Jeonggeun Jo, Duong Tung Pham, Seokhun Kim et al. "Ni3V2O8 nanoparticles as an excellent anode material for high-energy lith-ium-ion batteries." *Journal of Electroanalytical Chemistry* 810 (2018): 34–40. https://doi.org/10.1016/j.jelechem.2017.12.083

[54] Reddy, Ch V. Subba, In-Hyeong Yeo, and Sun-il Mho. "Synthesis of sodium vanadate nanosized mate-rials for electrochemical applications." *Journal of Physics and Chemistry of Solids* 69, no. 5-6 (2008): 1261–1264. https://doi.org/10.1016/j.jpcs.2007.10.072

[55] Soundharrajan, Vaiyapuri, Balaji Sambandam, Jinju Song, Sungjin Kim, Jeonggeun Jo, Pham Tung Duong, Seokhun Kim, Vinod Mathew, and Jaekook Kim. "Facile green synthesis of a Co3V2O8 nanoparticle electrode for high energy lithium-ion battery applications." *Journal of Colloid and Interface Science* 501 (2017): 133–141. https://doi.org/10.1016/j.jcis.2017.04.048

[56] Tang, Boya, Guozhao Fang, Jiang Zhou, Liangbing Wang, Yongpeng Lei, Chao Wang, Tianquan Lin, Yan Tang, and Shuquan Liang. "Potassium vanadates with stable structure and fast ion diffusion chan-nel as cathode for rechargeable aqueous zinc-ion batteries." *Nanoenergy* 51 (2018): 579–587. https://doi.org/10.1016/j.nanoen.2018.07.014

[57] Shahid, Muhammad, Jingling Liu, Zahid Ali, Imran Shakir, and Muhammad Farooq Warsi. "Structural and electrochemical properties of single crystalline MoV2O8 nanowires for energy storage devices." *Journal of Power Sources* 230 (2013): 277–281. https://doi.org/10.1016/j.jpowsour.2012.12.033

[58] Huang, Lei, Liyi Shi, Xin Zhao, Jing Xu, Hongrui Li, Jianping Zhang, and Dengsong Zhang. "Hydrothermal growth and characterization of length tunable porous iron vanadate one-dimensional nanostructures." *CrystEngComm* 16, no. 23 (2014): 5128–5133. https://doi.org/10.1039/C3CE42608D

[59] Pathak, Nimai, Santosh K. Gupta, Angelina Prince, R. M. Kadam, and V. Natarajan. "EPR investiga-tion on synthesis of Lithium zinc vanadate using sol-gel-combustion route and its optical properties." *Journal of Molecular Structure* 1056 (2014): 121–126. https://doi.org/10.1016/j.molstruc.2013.10.024

[60] Ying, Meihui, Jimin Hou, Wenqiang Xie, Yuanjie Xu, Shuifa Shen, Haibo Pan, and Min Du. "Synthesis, semiconductor characteristics and gas-sensing selectivity for cerium-doped neodymium vanadate nanorods." *Sensors and Actuators B: Chemical* 260 (2018): 125–133. https://doi.org/10.1016/j.snb.2017.12.192

[61] Ying, Meihui, Jimin Hou, Wenqiang Xie, Yuanjie Xu, Shuifa Shen, Haibo Pan, and Min Du. "Synthesis, semiconductor characteristics and gas-sensing selectivity for cerium-doped neodymium vanadate nanorods." *Sensors and Actuators B: Chemical* 260 (2018): 125–133. https://doi.org/10.1016/j.snb.2017.12.192

[62] Wan, Fang, Linlin Zhang, Xi Dai, Xinyu Wang, Zhiqiang Niu, and Jun Chen. "Aqueous rechargeable zinc/sodium vanadate batteries with enhanced performance from simultaneous insertion of dual carriers." *Nature Communications* 9, no. 1 (2018): 1656. https://doi.org/10.1038/s41467-018-04060-8

[63] Tran Huu, Ha, Ngoc Hung Vu, Hyunwoo Ha, Joonhee Moon, Hyun You Kim, and Won Bin Im. "Sub-micro droplet reactors for green synthesis of Li3VO4 anode materials in lithium ion batteries." *Nature Communications* 12, no. 1 (2021): 3081. https://doi.org/10.1038/s41467-021-23366-8

[64] Kaneti, Yusuf Valentino, Minsu Liu, Xiao Zhang, Yanru Bu, Yuan Yuan, Xuchuan Jiang, and Aibing Yu. "Synthesis of platinum-decorated iron vanadate nanorods with excellent sensing performance toward n-butylamine." *Sensors and Actuators B: Chemical* 236 (2016): 173–183. https://doi.org/10.1016/j.snb.2016.05.142

[65] Golmojdeh, Hosein, and Mohamad Ali Zanjanchi. "Ethanol gas sensor based on pure and La-doped bismuth vanadate." *Journal of Electronic Materials* 43 (2014): 528–534. https://doi.org/10.1007/s11664-013-2921-4

[66] Dalal, Jyoti, Avni Khatkar, Mandeep Dalal, Sangeeta Chahar, Priya Phogat, V. B. Taxak, and S. P. Khatkar. "Ba2YV3O11: Eu3+− Density functional and experimental analysis of crystal, electronic and optical properties." *Journal of Alloys and Compounds* 821 (2020): 153471. https://doi.org/10.1016/j.jallcom.2019.153471

[67] Zhou, Xianju, Lingni Chen, Sha Jiang, Guotao Xiang, Li Li, Xiao Tang, Xiaobing Luo, and Yu Pang. "Eu3+ activated LiSrVO4 phosphors: Emission color tuning and potential application in temperature sensing." *Dyes and Pigments* 151 (2018): 219–226. https://doi.org/10.1016/j.dyepig.2017.12.059

[68] Zhou, Huitao, Ning Guo, Xiang Lü, Yu Ding, Lu Wang, Ruizhuo Ouyang, and Baiqi Shao. "Ratiometric and colorimetric fluorescence temperature sensing properties of trivalent europium or samarium doped self-activated vanadate dual emitting phosphors." *Journal of Luminescence* 217 (2020): 116758. https://doi.org/10.1016/j.jlumin.2019.116758

[69] Vishnukumar, P., B. Saravanakumar, G. Ravi, V. Ganesh, Ramesh K. Guduru, and R. Yuvakkumar. "Synthesis and characterization of NiO/Ni3V2O8 nanocomposite for supercapacitor applications." *Materials Letters* 219 (2018): 114–118. https://doi.org/10.1016/j.matlet.2018.02.084

[70] Sharma, Tata Sanjay Kanna, and Kuo-Yuan Hwa. "Rational design and preparation of copper vanadate anchored on sulfur doped reduced graphene oxide nanocomposite for electrochemical sensing of antiandrogen drug nilutamide using flexible electrodes." *Journal of Hazardous Materials* 410 (2021): 124659. https://doi.org/10.1016/j.jhazmat.2020.124659

[71] do Prado, Thiago M., Carolina C. Badaró, Rafaela G. Machado, Pedro S. Fadini, Orlando Fatibello-Filho, and Fernando C. Moraes. "Using bismuth vanadate/copper oxide nanocomposite as photoelectrochemical sensor for naproxen determination in sewage." *Electroanalysis* 32, no. 9 (2020): 1930–1937. https://doi.org/10.1002/elan.202000055

[72] Narsimulu, D., A.K. Kakarla, and J.S. Yu. Cerium vanadate/carbon nanotube hybrid composite nanostructures as a high-performance anode material for lithium-ion batteries. *Journal of Energy Chemistry*, 58 (2021): 25–32. https://doi.org/10.1016/j.jechem.2020.09.028

[73] Maheshwaran, Selvarasu, Muthumariappan Akilarasan, Tse-Wei Chen, Shen-Ming Chen, Elayappan Tamilalagan, Ting-Yu Jiang, Eman A. Alabdullkarem, and Mustafa Soylak. "Electrocatalytic evaluation of graphene oxide warped tetragonal t-lanthanum vanadate (GO@ LaVO4) nanocomposites for the voltammetric detection of antifungal and antiprotozoal drug (clioquinol)." *Microchimica Acta* 188 (2021): 1–9. https://doi.org/10.1007/s00604-021-04758-5

[74] Liang, Huanyu, Yongcheng Zhang, Shujin Hao, Luhan Cao, Yanhong Li, Qiang Li, Dong Chen, Xia Wang, Xiangxin Guo, and Hongsen Li. "Fast potassium storage in porous CoV2O6 nanosphere@ graphene oxide towards high-performance potassium-ion capacitors." *Energy Storage Materials* 40 (2021): 250–258. https://doi.org/10.1016/j.ensm.2021.05.013

[75] Lee, Se Hun, Jae Hoon Bang, Jichang Kim, Changyong Park, Myung Sik Choi, Ali Mirzaei, Seung Soon Im, Heejoon Ahn, and Hyoun Woo Kim. "Sonochemical synthesis of PEDOT: PSS intercalated ammonium vanadate nanofiber composite for room-temperature NH3 sensing." *Sensors and Actuators B: Chemical* 327 (2021): 128924. https://doi.org/10.1016/j.snb.2020.128924

[76] He, Lihua, Zhe Yang, Chunli Gong, Hai Liu, Fei Zhong, Fuqiang Hu, Yaoyao Zhang, Guangjin Wang, and Bingqing Zhang. "The dual-function of photoelectrochemical glucose oxidation for sensor application and solar-to-electricity production." *Journal of Electroanalytical Chemistry* 882 (2021): 114912. https://doi.org/10.1016/j.jelechem.2020.114912

[77] Kokulnathan, Thangavelu, Ghzzai Almutairi, Shen-Ming Chen, Tse-Wei Chen, Faheem Ahmed, Nishat Arshi, and Bandar AlOtaibi. "Construction of lanthanum vanadate/functionalized boron nitride nanocomposite: The electrochemical sensor for monitoring of furazolidone." *ACS Sustainable Chemistry & Engineering* 9, no. 7 (2021): 2784–2794. https://doi.org/10.1021/acssuschemeng.0c08340

[78] Guo, Meng, Jayaraman Balamurugan, Nam Hoon Kim, and Joong Hee Lee. "High-energy solid-state asymmetric supercapacitor based on nickel vanadium oxide/NG and iron vanadium oxide/NG electrodes." *Applied Catalysis B: Environmental* 239 (2018): 290–299. https://doi.org/10.1016/j.apcatb.2018.08.026

[79] Gopi, Praveen Kumar, Chandan Hunsur Ravikumar, Shen-Ming Chen, Tse-Wei Chen, Mohammad Ajmal Ali, Fahad MA Al-Hemaid, Mohammad Suliman El-Shikh, and A. K. Alnakhli. "Tailoring of bismuth vanadate impregnated on molybdenum/graphene oxide sheets for sensitive detection of environmental pollutants 2, 4, 6 trichlorophenol." *Ecotoxicology and Environmental Safety* 211 (2021): 111934. https://doi.org/10.1016/j.ecoenv.2021.111934

[80] Barveen, Nazar Riswana, Tzyy-Jiann Wang, and Yu-Hsu Chang. "Photochemical decoration of silver nanoparticles on silver vanadate nanorods as an efficient SERS probe for ultrasensitive detection of chloramphenicol residue in real samples." *Chemosphere* 275 (2021): 130115. https://doi.org/10.1016/j.chemosphere.2021.130115

[81] Jadhav, Sagar, Pratiksha D. Donolikar, Nilesh R. Chodankar, Tukaram D. Dongale, Deepak P. Dubal, and Deepak R. Patil. "Nano-dimensional iron tungstate for super high energy density symmetric supercapacitor with redox electrolyte." *Journal of Solid State Electrochemistry* 23 (2019): 3459–3465. https://doi.org/10.1007/s10008-019-04427-x

[82] Shim, Hyun-Woo, In-Sun Cho, Kug Sun Hong, Won Il Cho, and Dong-Wan Kim. "Li electroactivity of iron (II) tungstate nanorods." *Nanotechnology* 21, no. 46 (2010): 465602. https://doi.org/10.1088/0957-4484/21/46/465602

[83] Kumar, Ponnaiah Sathish, Periakaruppan Prakash, Alagar Srinivasan, and Chelladurai Karuppiah. "A new highly powered supercapacitor electrode of advantageously united ferrous tungstate and functionalized multiwalled carbon nanotubes." *Journal of Power Sources* 482 (2021): 228892. https://doi.org/10.1016/j.jpowsour.2020.228892

[84] Naderi, Hamid, Hossein Sobati, Ali Sobhani-Nasab, Mehdi Rahimi-Nasrabadi, Mohammad Eghbali-Arani, Mohammad Reza Ganjali, and Hermann Ehrlich. "Synthesis and supercapacitor application of cerium tungstate nanostructure." *Chemistry Select* 4, no. 10 (2019): 2862–2867. https://doi.org/10.1002/slct.201803753

[85] Xu, Hui, Hongyuan Shang, Qingyun Liu, Cheng Wang, Junwei Di, Chunyan Chen, Liujun Jin, and Yukou Du. "Dual mode electrochemical-photoelectrochemical sensing platform for hydrogen sulfide detection based on the inhibition effect of titanium dioxide/bismuth tungstate/silver heterojunction." *Journal of Colloid and Interface Science* 581 (2021): 323–333. https://doi.org/10.1016/j.jcis.2020.07.120

[86] Kothandan, Vivekanandan Alangadu, Sivakumar Mani, Shen-ming Chen, and Shih-Hsun Chen. "Ultrasonic-assisted synthesis of nickel tungstate nanoparticles on poly (3, 4-ethylene dioxythiophene): poly (4-styrene sulfonate) for the effective electrochemical detection of caffeic acid." *Materials Today Communications* 26 (2021): 101833. https://doi.org/10.1016/j.mtcomm.2020.101833

[87] Ikram, Muhammad, Yasir Javed, Naveed Akhtar Shad, Muhammad Munir Sajid, Muhammad Irfan, Anam Munawar, Tousif Hussain, Muhammad Imran, and Dilshad Hussain. "Facile hydrothermal synthesis of nickel tungstate (NiWO4) nanostructures with pronounced supercapacitor and electrochemical sensing activities." *Journal of Alloys and Compounds* 878 (2021): 160314. https://doi.org/10.1016/j.jallcom.2021.160314

[88] Packiaraj, R., P. Devendran, K. S. Venkatesh, and N. Nallamuthu. "Investigation on the structural, morphological and electrochemical properties of nickel tungstate for energy storage application." *Inorganic Chemistry Communications* 126 (2021): 108490. https://doi.org/10.1016/j.inoche.2021.108490

[89] Sundaresan, Periyasamy, Periyasami Gnanaprakasam, Shen-Ming Chen, Ramalinga Viswanathan Mangalaraja, Wu Lei, and Qingli Hao. "Simple sonochemical synthesis of lanthanum tungstate (La2 (WO4) 3) nanoparticles as an enhanced electrocatalyst for the selective electrochemical determination of anti-scald-inhibitor diphenylamine." *Ultrasonics Sonochemistry* 58 (2019): 104647. https://doi.org/10.1016/j.ultsonch.2019.104647

[90] Manickavasagan, Abinaya, Rajakumaran Ramachandran, Shen-Ming Chen, and Muthuraj Velluchamy. "Ultrasonic assisted fabrication of silver tungstate encrusted polypyrrole nanocomposite for effective photocatalytic and electrocatalytic applications." *Ultrasonics Sonochemistry* 64 (2020): 104913. https://doi.org/10.1016/j.ultsonch.2019.104913

[91] Patil, Swati J., Nilesh R. Chodankar, Yun Suk Huh, Young-Kyu Han, and Dong Weon Lee. "Bottom-up Approach for Designing Cobalt Tungstate Nanospheres through Sulfur Amendment for High-Performance Hybrid Supercapacitors." *ChemSusChem* 14, no. 6 (2021): 1602–1611. https://doi.org/10.1002/cssc.202002968

[92] Sohouli, Esmail, Kourosh Adib, Bozorgmehr Maddah, and Mostafa Najafi. "Manganese dioxide/cobalt tungstate/nitrogen-doped carbon nano-onions nanocomposite as new supercapacitor electrode." *Ceramics International* 48, no. 1 (2022): 295–303. https://doi.org/10.1016/j.ceramint.2021.09.104

[93] Naderi, Hamid Reza, Ali Sobhani-Nasab, Mehdi Rahimi-Nasrabadi, and Mohammad Reza Ganjali. "Decoration of nitrogen-doped reduced graphene oxide with cobalt tungstate nanoparticles for use in high-performance supercapacitors." *Applied Surface Science* 423 (2017): 1025–1034. https://doi.org/10.1016/j.apsusc.2017.06.239

[94] Sundaresan, Periyasamy, Alagumalai Krishnapandi, and Shen-Ming Chen. "Design and investigation of ytterbium tungstate nanoparticles: An efficient catalyst for the sensitive and selective electrochemical detection of antipsychotic drug chlorpromazine." *Journal of the Taiwan Institute of Chemical Engineers* 96 (2019): 509–519. https://doi.org/10.1016/j.jtice.2018.10.021

[95] Kumar, R. Dhilip, and S. Karuppuchamy. "Microwave-assisted synthesis of copper tungstate nanopowder for supercapacitor applications." *Ceramics International* 40, no. 8 (2014): 12397–12402. https://doi.org/10.1016/j.ceramint.2014.04.090

[96] Wei, Chao, Ying Huang, Xin Zhang, Xuefang Chen, and Jing Yan. "Soft-template hydrothermal synthesis of nanostructured copper (II) tungstate cubes for electrochemical charge storage application." *Electrochimica Acta* 220 (2016): 156–163. https://doi.org/10.1016/j.electacta.2016.10.056

[97] Huang, Biao, Huayu Wang, Shunfei Liang, Huizhen Qin, Yang Li, Ziyang Luo, Chenglan Zhao, Li Xie, and Lingyun Chen. "Two-dimensional porous cobalt-nickel tungstate thin sheets for high performance supercapattery." *Energy Storage Materials* 32 (2020): 105–114. https://doi.org/10.1016/j.ensm.2020.07.014

[98] Mahieddine, Abdelkadir, Leila Adnane-Amara, and Noureddine Gabouze. "The effect of alkaline electrolytes and silver nanoparticles on the electrochemical performance of the dilithium nickel bis (tungstate) as electrode materials for high-performance asymmetric supercapacitor." *Journal of Alloys and Compounds* 882 (2021): 160754. https://doi.org/10.1016/j.jallcom.2021.160754

[99] Poudel, Milan Babu, Hem Prakash Karki, and Han Joo Kim. "Silver nanoparticles decorated molybdenum sulfide/tungstate oxide nanorods as high performance supercapacitor electrode." *Journal of Energy Storage* 32 (2020): 101693. https://doi.org/10.1016/j.est.2020.101693

[100] Prabhu, S., C. Balaji, M. Navaneethan, M. Selvaraj, N. Anandhan, D. Sivaganesh, S. Saravanakumar, Periyasamy Sivakumar, and R. Ramesh. "Investigation on mesoporous bimetallic tungstate nanostructure for high-performance solid-state supercapattery." *Journal of Alloys and Compounds* 875 (2021): 160066. https://doi.org/10.1016/j.jallcom.2021.160066

[101] Rahimi-Nasrabadi, Mehdi, Vafa Pourmohamadian, Meisam Sadeghpour Karimi, Hamid Reza Naderi, Mohammad Ali Karimi, Khadijeh Didehban, and Mohammad Reza Ganjali. "Assessment of supercapacitive performance of europium tungstate nanoparticles prepared via hydrothermal method." *Journal of Materials Science: Materials in Electronics* 28 (2017): 12391–12398. https://doi.org/10.1007/s10854-017-7059-3

[102] Rajakumaran, Ramachandran, Alagumalai Krishnapandi, Shen-Ming Chen, Karuppaiah Balamurugan, Fu Mao Chang, and Subramanian Sakthinathan. "Electrochemical investigation of zinc tungstate nanoparticles; a robust sensor platform for the selective detection of furazolidone in biological samples." *Microchemical Journal* 160 (2021): 105750. https://doi.org/10.1016/j.microc.2020.105750

[103] Rajpurohit, Anuja S., Ninad S. Punde, Chaitali R. Rawool, and Ashwini K. Srivastava. "Fabrication of high energy density symmetric supercapacitor based on cobalt-nickel bimetallic tungstate nanoparticles decorated phosphorus-sulphur co-doped graphene nanosheets with extended voltage." *Chemical Engineering Journal* 371 (2019): 679–692. https://doi.org/10.1016/j.cej.2019.04.100

[104] Mahieddine, Abdelkadir, Leila Adnane-Amara, Noureddine Gabouze, Ahmed Addad, Abir Swaidan, and Rabah Boukherroub. "Self-combustion synthesis of dilithium cobalt bis (tungstate) decorated with silver nanoparticles for high performance hybrid supercapacitors." *Chemical Engineering Journal* 426 (2021): 131252. https://doi.org/10.1016/j.cej.2021.131252

[105] Zafar, K., M. Wasim, B. Fatima, D. Hussain, R. Mehmood, andd M. Najam-ul-Haq. "Quantification of tramadol and serotonin by cobalt nickel tungstate in real biological samples to evaluate the effect of analgesic drugs on neurotransmitters." *Scientific Reports* 13(1) (2023): 10239. https://doi.org/10.1038/s41598-023-37053-9

[106] Sobhani-Nasab, Ali, Hamid Naderi, Mehdi Rahimi-Nasrabadi, and Mohammad Reza Ganjali. "Evaluation of supercapacitive behavior of samarium tungstate nanoparticles synthesized via sono-chemical method." *Journal of Materials Science: Materials in Electronics* 28 (2017): 8588–8595. https://doi.org/10.1007/s10854-017-6582-6

[107] Sobhani-Nasab, Ali, Mehdi Rahimi-Nasrabadi, Hamid Reza Naderi, Vafa Pourmohamadian, Farhad Ahmadi, Mohammad Reza Ganjali, and Hermann Ehrlich. "Sonochemical synthesis of terbium tungstate for developing high power supercapacitors with enhanced energy densities." *Ultrasonics Sonochemistry* 45 (2018): 189–196. https://doi.org/10.1016/j.ultsonch.2018.03.011

[108] Sundaresan, P., A. Yamuna, and S. M. Chen. "Sonochemical synthesis of samarium tungstate nanoparticles for the electrochemical detection of nilutamide." *Ultrasonics Sonochemistry* 67 (2020): 105146. https://doi.org/10.1016/j.ultsonch.2020.105146

[109] Zhang, Jin, Xingzhong Yuan, Longbo Jiang, Zhibin Wu, Xiaohong Chen, Hou Wang, Hui Wang, and Guangming Zeng. "Highly efficient photocatalysis toward tetracycline of nitrogen doped carbon quantum dots sensitized bismuth tungstate based on interfacial charge transfer." *Journal of Colloid and Interface Science* 511 (2018): 296–306. https://doi.org/10.1016/j.jcis.2017.09.083

[110] Shad, Naveed A., Sadia Z. Bajwa, Nasir Amin, Ayesha Taj, Sadaf Hameed, Yaqoob Khan, Zhifei Dai, Chuanbao Cao, and Waheed S. Khan. "Solution growth of 1D zinc tungstate (ZnWO4) nanowires; design, morphology, and electrochemical sensor fabrication for selective detection of chloramphenicol." *Journal of Hazardous Materials* 367 (2019): 205–214. https://doi.org/10.1016/j.jhazmat.2018.12.072

[111] Peng, Zhuo, Qiulong Wei, Shuangshuang Tan, Pan He, Wen Luo, Qinyou An, and Liqiang Mai. "Novel layered iron vanadate cathode for high-capacity aqueous rechargeable zinc batteries." *Chemical Communications* 54, no. 32 (2018): 4041–4044. https://doi.org/10.1039/C8CC00987B

[112] Zhang, Yongguang, Yan Zhao, Aishuak Konarov, Denise Gosselink, Zhi Li, Mahmoudreza Ghaznavi, and P. Chen. "One-pot approach to synthesize PPy@ S core-shell nanocomposite cathode for Li/S batteries." *Journal of Nanoparticle Research* 15 (2013): 1–7. https://doi.org/10.1007/s11051-013-2007-5

[113] Chandrashekar, C. K., P. Madhusudan, H. P. Shivaraju, C. P. Sajan, B. Basavalingu, S. Ananda, and K. Byrappa. "Synthesis of rare earth-doped yttrium vanadate polyscale crystals and their enhanced photocatalytic degradation of aqueous dye solution." *International Journal of Environmental Science and Technology* 15 (2018): 427–440. https://doi.org/10.1007/s13762-017-1401-4

[114] Zhang, L. F., J. Tang, S. Y. Liu, O. W. Peng, R. Shi, B. N. Chandrashekar, ... and C. Cheng. "A laser irradiation synthesis of strongly-coupled VOx-reduced graphene oxide composites as enhanced performance supercapacitor electrodes." *Materials Today Energy*, 5 (2017): 222–229. https://doi.org/10.1016/j.mtener.2017.07.004

[115] R. Marnadu, Abdullah M. Al-Enizi, and Mohd Ubaidullah. "Design of zinc vanadate (Zn3V2O8)/ nitrogen doped multiwall carbon nanotubes (N-MWCNT) towards supercapacitor electrode applications." *Journal of Electroanalytical Chemistry* 881 (2021): 114936. https://doi.org/10.1016/j.jelechem.2020.114936

[116] Yang, Man, Guozhi Ma, Hongli Yang, Zhan Xiaoqiang, Weiyou Yang, and Huilin Hou. "Advanced strategies for promoting the photocatalytic performance of FeVO4 based photocatalysts: A review of recent progress." *Journal of Alloys and Compounds* (2023): 168995. https://doi.org/10.1016/j.jallcom.2023.168995

[117] Li, Guangzhi, Zhenling Wang, Min Yu, Zewei Quan, and Jun Lin. "Fabrication and optical properties of core-shell structured spherical SiO2@ GdVO4: Eu3+ phosphors via sol-gel process." *Journal of Solid State Chemistry* 179, no. 8 (2006): 2698–2706. https://doi.org/10.1016/j.jssc.2006.05.019

[118] Reza Naderi, H., A. Sobhani-Nasab, E. Sohouli, K. Adib, and E. Naghian. (2020). *Analytical and Bioanalytical Electrochemistry* 12(2): 263–276.

[119] Ahmed, Jahangeer, Tansir Ahamad, Norah Alhokbany, Basheer M. Almaswari, Tokeer Ahmad, Afzal Hussain, Eida Salman Saad Al-Farraj, and Saad M. Alshehri. "Molten salts derived copper tungstate nanoparticles as bifunctional electro-catalysts for electrolysis of water and supercapacitor applications." *ChemElectroChem* 5, no. 24 (2018): 3938–3945. https://doi.org/10.1002/celc.201801196

6 Nanohybrid Titanates and Phosphates for Energy Applications

H.M. Deepa, B.C. Indumukhi, H.V. Harini,
H.P. Nagaswarupa, and Ramachandra Naik

INTRODUCTION

One of the biggest tasks for the twenty-first century was energy storage. In the modern society due to ecological concerns, we have to find low-cost and environmentally friendly energy conversion and energy storage systems whose performance depends on the property of the material. Nanostructural titanate materials have received great attention in the recent years because of their mechanical, electric, and optical properties. Nanostructural materials' importance has increased in electrochemical energy storage. We have to escalate the advantages and disadvantages of the titanates for energy conservation and storage, as well as their synthesis and properties. Lithium-ion batteries are one of the successful materials in electrochemistry. Their science and technology have been extensively reported for new generations of rechargeable lithium batteries, not only for applications in consumer electronics but especially for clean energy storage and use in hybrid electric vehicles. Further invention of new materials and new pathways is essential. Among the new pathways is the one of nanomaterial titanates used for lithium-ion batteries. Supercapacitors are used for supporting voltage systems in equipment and electric vehicles during increased loads. These are located in the area between the batteries and dielectric capacitors. Supercapacitors have high power and long life cycle but are very expensive [1]. Because of their outstanding performances, metal phosphates being acidic solids have a wide range of applications such as batteries, fuel cell, electrodes, and supercapacitors [2]. Due to the development of industries, a lot of industrial waste dump into the environment. Some of these wastes cause severe effects to the systems of human body such as immune system dysfunction and reproductive system disorder, which cause changes in behaviors of humans. So, research for the development of sensors to detect even a minute concentration of any chemical is of prime importance. Electrochemical detection is more advantageous than the conventional techniques in modifying the surface. Electrochemical sensors are the fast-growing chemical sensors. Amperometric sensor has the potential to be applied between reference and working electrode to cause oxidation or reduction of the species [3]. Electrochemical water splitting employed for the production of hydrogen transforms electrical energy into chemical energy (fuel). The capacity of energy transformation depends on the overpotentials arises due to oxygen growth near anode and the hydrogen growth near cathode. The efficiency of electrocatalysts is enhanced by employing noble metals and their oxides for HER (hydrogen evolution reaction) and OER (oxygen evolution reaction). The transition bimetallic phosphates act as active catalysts during the reaction. The corrosion and aggregation of electrodes are avoided by shelling them with active compounds and conducting carbon, which improves electrochemical water splitting [4]. Phosphates are employed as photocatalysts, electrocatalysts, and heterogenous catalysts. Rechargeable metal air batteries with varying electronic density in active sites showed a good anti-poisoning capacity and an electrocatalytic activity [5]. Transition metals (M=Fe, Ni, Co) enhance their catalytic activity.

DOI: 10.1201/9781003479239-6

Sun et al. synthesized Co-based phosphate catalysts at neutral pH, which releases 3D OER. Surface of the electrode covered by phosphates improves the contact between electrodes and electrolytes [6].

METHOD FOR THE SYNTHESIS OF NANOTITANATES AND NANOPHOSPHATES

Nanophosphates are synthesized by a number of methods depending on the nature of the material. The main two types are top-down and bottom-up methods. In the top-down method, bigger-size materials are reduced to nanoparticles, whereas in the bottom-up method, nanoparticles are formed from elementary size [7] (Figure 6.1).

CHEMICAL METHOD

Chemical methods involve the use of chemicals that are not eco-friendly and hazardous for the people handling the process due to their toxic properties. Oxidation/reduction is the most important reaction occurred. Chemical methods include sol-gel, combustion, microwave, co-precipitation, hydrothermal, and sonication methods. Physical and chemical methods require additional capping and stabilizing agents [8]. Chemical synthesis of titanates by different methods are given in Table 6.1 (Figures 6.2–6.5).

GREEN APPROACH

Green approach has gained importance in recent years due to its significant merits such as simplicity, biocompatibility, and safety. This approach has replaced the physiochemical methods by focusing on improving environmental sustainability and reducing toxicity. The approach integrates to synthesize titanate nanoparticles from different parts of plants and microorganisms containing useful biomedical applications [9]. Accordingly, green synthesis is found to be more reliable and

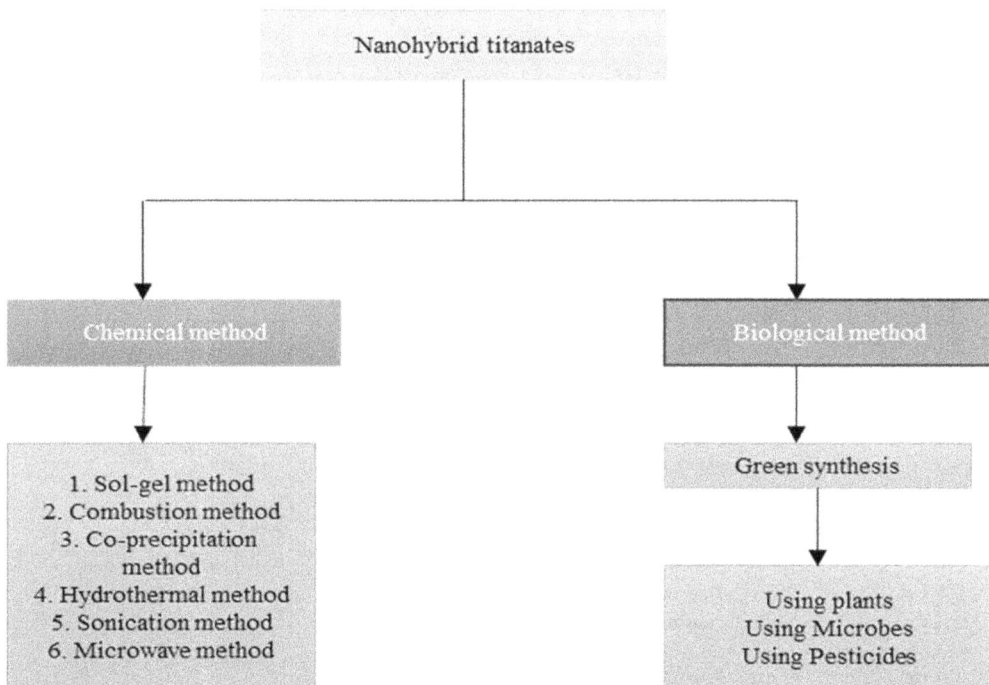

FIGURE 6.1 Synthesis methods of nanotitanates.

TABLE 6.1
Different Synthesis Methods of Titanate and Phosphate Nanoparticles

Metals	Methods of Preparation	Application	References
TiO_2	Hydrothermal method	Battery	[12]
$ZnTiO_3$	Hydrothermal method	Solar cell	[13]
$BaTiO_3$	Solid-state method	Energy storage	[14]
$SrTiO_3$	Solid-state sintering method	High energy storage	[15]
$NaTiO_3$	Hydrothermal method	Energy storage and photo electrode	[16]
$CuTiO_3$	Hydrothermal method	Gas sensing	[17]
$NiTiO_3$	Sol-gel method	Semiconductors	[18]
$Y_2Ti_2O_7$	Co-precipitation method	Optical devices	[19]
$PbTiO_3$	Sol-gel method	Dielectric properties	[20]
$Bi_4Ti_3O_{12}$	Hydrothermal method	Cool pigments	[21]
KTi_8O_{16}	Combustion method	Energy application	[22]
$Li_4Ti_5O_{12}$	Sol-gel method	Li-ion battery	[23]
$FeTiO_3$	Hydrothermal method	Gas sensing	[24]
$Na_2Ti_2O_4(OH)_2$	Hydrothermal method	Supercapacitor	[25]
$Li_4Ti_5O_{12}$	Solid-state method	Li-ion batteries	[26]
$LiLnTiO_3$	Sol-gel synthesis	Anisotropic conductivity	[27]
$CaCu_3Ti_4O_{12}$	Combustion method	Sensors	[28]
$NaBiTiO_3$	Solid-state method	Energy-stored ceramics	[29]
$BaZrTiO_3$	Sol-gel method	High-energy capacitors	[30]
$PbZrTiO_3$	Chemical method	Electric properties	[31]
$CaPbSrTiO_3$	Co-precipitation method	Ferroelectric property	[32]
$BiNaKTiO_3$	Solid-state method	Energy harvesting	[33]
Pd/Pt: TiO_3	Hydrothermal method	Sensor	[33]
Mg: $SrTiO_3$	Hydrothermal method	Dielectric property	[34]
Ln: $SrTiO_3$	Sol-gel method	Supercapacitor	[35]
Eu: $MgTiO_3$	Combustion method	LEDs	[36]
Sm: Zn_2TiO_4	Combustion method	Sensors	[37]
Co: $MgTiO_3$	Solid-state method	Dielectric resonator antenna	[38]
Fe: $BaTiO_3$	Solid-state method	Study of ferromagnetic property	[39]
Y_2O_5:$BaTiO_3$	Solid-state method	Multilayer capacitor	[40]
Fe: $SrTiO_3$	Sol-gel method	Fuel cell	[41]
Ni: $SrTiO_3$	Sol-gel method	Electric properties	[42]
Sr: $ZnTiO_3$	Solid-state method	Gas sensors	[43]
V: $NaTiO_3$	Sol-gel method	Sodium-ion battery	[44]
Ca: $Y_2Ti_2O_7$	Solid-state method	Dielectric property	[45]
Y: $SrTiO_3$	Auto-combustion method	SOFC	[46]
Bi: $BaSrTiO_3$	Sol-gel method	Dielectric property	[47]
Mn/Y: $BaSrTiO_3$	Sol-gel method	Energy storage	[48]
Fe/Ni: $BaSrTiO_3$	Sol-gel method		[49]
Ln: $BiNaTiO_3$	Solid-state method	Energy storage	[50]
Eu: Ba_3MoTiO_8	Combustion method	LED	[51]
Ni: $LnSrTiO_3$	Sol-gel method	Fuel cell	[52]
Sr: $ZnMnTiO_3$	Solid-state method	Dielectric studies	[53]
Sr: $CaCu_3Ti_4O_{12}$	Solid-state method	Dielectric studies	[54]
W: $BaSrTiO_3$	Solid-state method	Energy storage Solar cell	[52]
Sr: $ZnMnTiO_5$	Solid-state method	Ceramics	[55]

(Continued)

TABLE 6.1 (*Continued*)
Different Synthesis Methods of Titanate and Phosphate Nanoparticles

Metals	Methods of Preparation	Application	References
Mn: $BaZrTiO_3$	Solid-state method	Energy storage for ceramics	[56]
Gd: Nb-$BaTiO_3$	Solid-state method	Ceramics	[57]
$LiSr_2(PO_4)_3$	Hydrothermal method	Solid electrolytes and sensors	[58]
$LiTi_2(PO_4)_3$	Hydrothermal method, Pechini's method	Electrode material in energy storage devices	[59]
Ni-P	Hydrothermal	Supercapacitor	[60]
Ni–Mo, Co–Mo, Co–Ni PO_4 / nickel foam	Hydrothermal method	Electrocatalyst for OER and HER	[61]
Ni–Mn $(PO_4)_2$ / graphene foam	Hydrothermal method	Supercapacitors	[62]
Mn, Fe, Co, Ni–P	Solution combustion method	Catalyst for oxygen reduction, OER reaction	[63]
Ni–Fe PO_4	Hydrothermal method	Good conductors with large electrochemical surface area	[64]
Mn–Co PO_4	Hydrothermal method	Electrocatalyst for OER and HER	[65]
Al/Ag_3PO_4	Co-precipitation method	Catalyst	[66]
Cr–Zr PO_4	Sol-gel method	Catalysts for oxidation of hydrocarbons	[67]

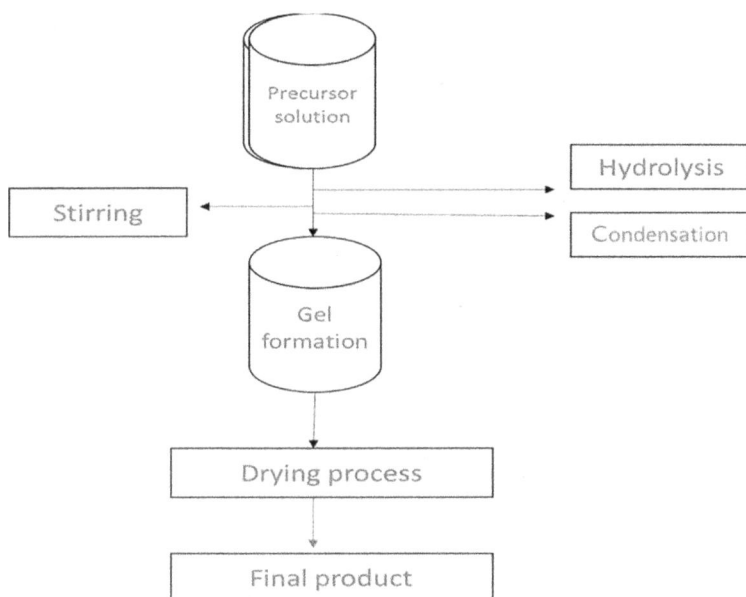

FIGURE 6.2 Sol-gel method.

less time consuming for the synthesis of various titanate nanoparticles in large scale without any impurities [10]. Extracts of roots, leaves, stems, seeds, fruits, flowers, and petals of plants have played a great role in the biosynthesis of nanoparticles. These extracts due to their rich phytochemical contents act as reducing, capping, and stabilizing agents and avoid utilization of external organic solvents (surfactants, reducing agents, and stabilizers) [11] (Figures 6.6).

FIGURE 6.3   Sonication method.

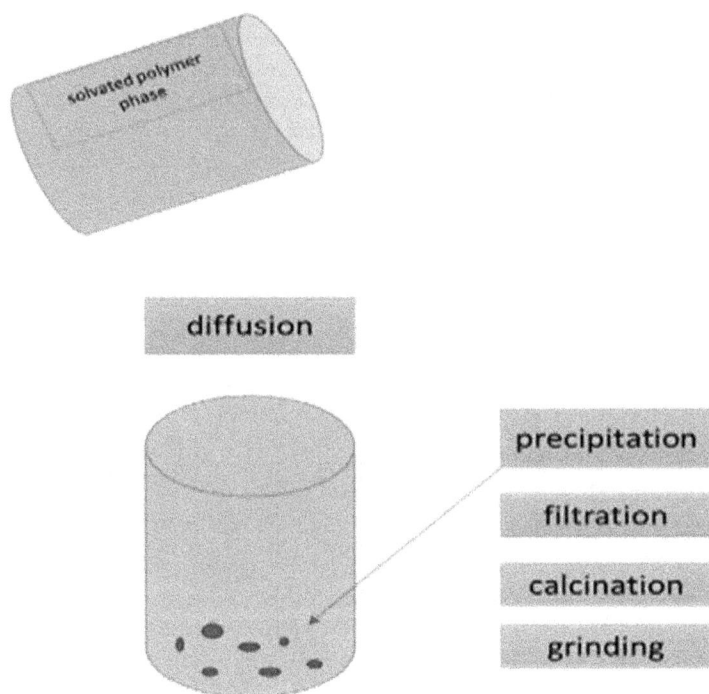

FIGURE 6.4   Co-precipitation method.

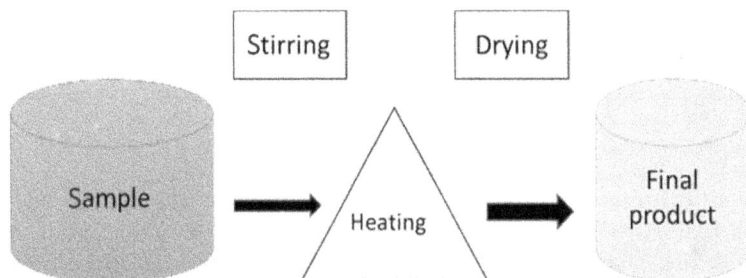

FIGURE 6.5   Hydrothermal method.

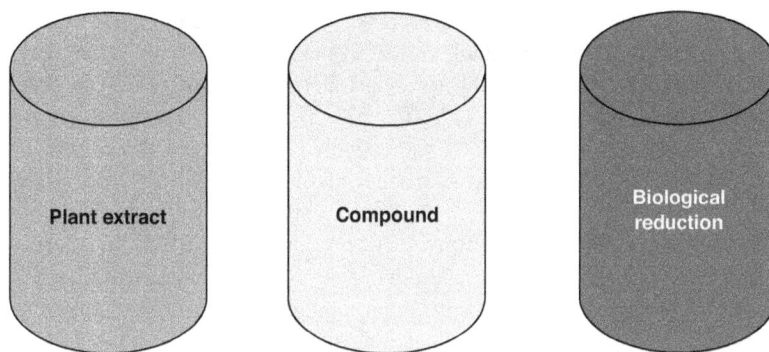

FIGURE 6.6 Green synthesis method.

CHARACTERIZATION

Physical and chemical properties of the designed nanohybrid titanates and phosphates such as particle size, surface morphology, composition, porosity, purity, and molecular weight are characterized by several spectroscopic instruments, such as powder X-ray diffraction (PXRD) to determine structure of crystalline material, Fourier transform infrared spectroscopy (FTIR) to identify molecular compounds, scanning electron microscopy (SEM) for high-resolution image of surface material, transmission electron microscopy (TEM) to observe the features of very small specimen, energy dispersive X-ray analysis (EDAX) to acquire the elemental composition of samples, UV-DRS spectral studies to measure sample surface and interior followed by other spectroscopic techniques.

X-RAY DIFFRACTION

XRD is used to determine the atomic and molecular crystal. A beam of X-rays strikes a crystal and scatters into many different directions. By measuring the angle and intensity of these diffracted beams, it gives the 3D picture of electron. This provides the information about the mean position of the atom.

Powder X-ray diffraction (PXRD) is a common characterization technique for nanoscale materials. XRD provides important information about various microscopic and spectroscopic methods, such as phase identification, sample purity, crystallite size, and, in some cases, morphology. As a bulk technique, the information it provides can be correlated with microscopy data to test if microscopic observations on a small number of particles are representative of the majority of the sample [68]. By treating with different techniques in different temperatures, the crystalline size and morphology can be changed. For samarium titanates, the calcination temperature was 1,100°C with a diffraction peak 2θ of 30.35°. The crystalline size was 140 nm, and it was thermally stable [69]. Barium calcium titanates were calcinated at 1,500°C for 5 h to reduce the orthorhombic pecks. Here, tetrahedral pecks are observed more than the orthorhombic pecks [70]. Tan Guo synthesized nickel iron phosphate by a one-step hydrothermal method which showed a strong peak at 30°, which confirms $Ni_3Fe_4(PO_4)_6$ along with nickel hydroxide dihydrate, hydroxy nickelate, and hydroxy ferrate in small quantities. By employing X-ray photoelectron spectroscopy, the elements present in the surface composition were observed, which provides a high-resolution spectrum having two sets of spin-orbit doublet of nickel in nickel iron phosphate at 855.76, 873.37, 856.88, and 874.85 eV, representing Ni^{2+} and Ni^{3+} [71]. In barium strontium titanates XRD shows the 2θ at 46°C at 25 splitting with tetragonal structure [72]. In aluminum-doped lithium titanates, the particle size was 140 nm, which is much smaller than that of lithium titanate [73]. D. Arumugam et al. synthesized magnesium-doped lithium iron phosphate by the sol-gel method. The XRD analysis reveals the structure of orthorhombic-olivine for lithium iron phosphate. Intensity of the peak increased by increase in calcination temperature also enhances the nature of crystallinity and purity of a compound. It also shows the shifting of XRD curve toward left side of $2\ominus$ axis with original compound [74]. Shuting Wen synthesized iron-nickel-phosphide by the hydrothermal method. Here the tetragonal phase of nickel phosphate shows planes at 310, 112, 202, 321, 330, 420, 312 and the apex at $2\ominus=32.61°$, $38.26°$, $40.54°$, $41.56°$, $44.38°$, $46.84°$, $48.86°$, respectively [75]. Zhixiao Qin et al. synthesized hexagonal nickel cobalt phosphate by the hydrothermal method with JCPDS No. 71-2336. It did not show any extra peaks [76,77]. Abdulmajid, A. et al. synthesized sodium-nickel-phosphate composite by the precipitation method. Scanning electron microscopy (SEM) was used to investigate the morphology of the pristine and composite materials with a diameter of 49 nm [78]. Xueying Song et al. synthesized nickel phosphate by different methods which showed nanotubes of different lengths. The diameter of synthesized nanotubes may vary from 8 to 10 nm, having large porosity that enhances the rate of diffusion and catalytic reactive sites [79]. Nickel cobalt phosphate shows a hexagonal plate-like structure with a diameter of 20 nm [80]. The elemental analysis of nanophosphates is carried out by EDX. For example, nickel cobalt phosphate contains Ni, Co, P, and O atoms. It also confirms the metal phosphate shell formation carried out either by solid-state reaction or by air oxidation [81]. Iron-doped cobalt copper phosphate shows a flower-like structure.

SCANNING ELECTRON MICROSCOPY

SEM is used for characterization of nanoparticles. SEM produces an image of a material by scanning it with a beam of electron. The contact of electron produces a different signal which contains the data about the surface structure and configuration of the sample. In addition to surface structure and configuration information, SEM can detect and analyze surface fracture, provide information about microstructure, examine surface contamination, reveal spatial variation in chemical composition, provide qualitative chemical analysis, and identify crystalline structure. SEM can observe bulk samples of several centimeters in size and is consequently more flexible in its application than TEM [82]. For all the samples, SEM shows the similar morphology, but the main difference is crystallite cluster. The crystallite diameter was 60 nm in the cases [24]. Barium strontium titanate was synthesized by a solid-state method with a grain size of 0.3–3 nm. Here, the grain size was increased to the maximum [72]. Magnesium-doped strontium titanates synthesized by the hydrothermal method show a spherical structure with smooth and regular shape, and the average size of the particle was 1.8 μm [34]. SEM images of strontium-doped zinc titanates show an average particle size of 1 μm [43]. Bismuth-doped barium strontium titanates show an average particle size of 35 nm [47].

TRANSMISSION ELECTRON MICROSCOPY

In TEM, an energetic beam of electron is used to provide topographical, morphological, compositional, and crystallographic information about the sample. TEM finds applications in cancer research, material science, nanotechnology, and semiconductor research. The lattice structure of a material can be determined by TEM. The shape and size of the sample can be determined from the image [66]. In TEM images of samarium titanates, a rod-like shape was found with an average width of 5–15 nm and a length of 20–50 nm [69]. TEM images of iron and copper titanate particles show a regular sphere shape with a diameter of 100–500 nm [17]. Dielectric properties of magnesium-doped strontium titanates were studied in the frequency range from 1,000 to 10,000 Hz at 25°C [34]. The particles of bismuth-doped barium strontium titanates are arranged in a uniform size and shape. To determine the average particle size, we must measure the diameter of each particle [47].

UV-VISIBLE SPECTROSCOPY

Nanoparticles have optical properties. They show delicate size, shape, concentration, and collection state, and their refractive index was nearer the nanoparticle surface, which makes UV-visible spectroscopy a valuable device for identifying, characterizing, and studying materials. Nanoparticles made from certain metals, such as gold and silver, strongly interact with definite wavelengths of light, and the unique optical properties of these materials are the foundation for the field of plasmonic. Zinc titanates were calcinated to 500°C, and the band gap of the sample was 3.71 ev. As we increase the temperature, the band gap also increases to 3.85 ev, which clearly shows that the value of band gap depends on the oxygen vacancy [13]. TEM images of sodium titanates confirm the different crystal structure in high-resolution image. The samples M_2 and M_5 show different structures. M_2 exists in a nanotube and nanoparticle structure with a diameter of 10 nm, with interplanar distance 1.88 A° and 3.49 A°. M_5 condition nanotube becomes nanorods [16]. Samarium titanates were calcinated under 600 and 1,100°C for 3 h. The band gap was calculated by using UV data and Tauc's equation [17]. UV-visible spectra of iron and nickel co-doped barium strontium titanates were observed at 231.85 and 228.64, respectively. In comparison, the bad gap increases with doping of Fe, and the band gap decreases with doping of nickel. The band gap energy was found to be 4.97 and 4.67, respectively [49].

RAMAN SPECTROSCOPY

Raman spectroscopy is a device used to recognize the phase transition of different nanoparticles and other nano design materials. It is used to determine which area of nanomaterials is amorphous and crystalline, detect any defect in the nanomaterial, identify the size of different nanomaterials, determine if the nanomaterial is similar or not, and determine the shape of the nanomaterial. Zinc titanates were heated at 500°C. Here, we cannot observe many active sites. This shows low crystallinity, as we increase the temperature, and we observe the active site [13]. Raman spectra of sodium titanates were observed at the range between 100 and 1,600 cm^{-1}. The sample was with the same structure with series pecks of 910, 667, 450, and 280 cm^{-1}. The first one is considered as Ti–O stretching, and the last three are related to a different Ti–O–Ti stretching mode. Here, we observe 96% of sodium titanates and 4% of analyte. The presence of peak of Raman spectra was 1,080 cm^{-1} [16].

TITANATES AND PHOSPHATES FOR ENERGY APPLICATION

BATTERY

Titanates attract attention toward rechargeable lithium batteries with a mesoporous structure and a good ion exchange ratio, which results in the good charge and discharge capacity, good kinetic, and robustness and safety. Titanate nanotube anodes in Li battery demonstrate an initial 282 mAh/g and discharge capacity at 0.24 mAh/g. Here the counter- and reference electrodes were made of lithium metal. The cyclic voltammetry shows an extensive cathode/anode peck at 1.69 and 2.08 V relative to a lithium electrode at a scan rate of 0.05 mV/s. Electron microscopy studies show morphology of titanate nanotubes that remains unchanged after 50 charge/discharge cycles [12]. Lithium titanates have the reversible discharge/charge curves of the LTO and LGC-10% electrodes at 0.5C and 1C, ant their charge capacity was 163 and 201 mAh/g at 0.5C, respectively. This increase of reversible capacity can be attributed to graphene oxide [69]. Zero-strain material lithium titanate as anode gives a good life cycle and rate capability to lithium batteries, and it also improves safety features. Lithium titanates are used in nanostructure forms. When doping with suitable anode or cathode, it improves the electric conductivity and power capability of lithium titanates. Carbon coating was another good method to improve the electrical conductivity of the material, and it helps to control the particle size of lithium titanates [26].

In twenty-first century, transition metal phosphates (TMP) were employed successfully for energy storage and efficient power generation. Phosphates act as cathode materials in lithium-ion batteries [83]. In at anode of lithium-ion batteries under discharging process, lithium atoms ionize to release electrons and the ions move toward cathode through an electrolyte to combine with their own electrons. Being smaller in size, lithium ions have the capability to pass through even micro-permeable separator present between cathode and anode [58]. G. X. Wang et al. synthesized Mg-doped LiFePO$_4$ (lithium iron phosphates) through supervalent ion technique to enhance the electronic conductivity of phosphates. LiFePO$_4$ doped with Mg enhances its magnitude of electronic conductivity property by 10^4 (i.e., from 10^{-8} to 10^{-4} Scm^{-1}) [59]. Sun et al. synthesized Li$_3$V$_2$(PO$_4$)$_3$ by a solid-state method, having the highest ionic conductivity possess 3D walk-way for the diffusion of lithium compared to LiFePO$_4$. 3 moles of Li+ ions are extracted and simultaneously inserted when Li$_3$V$_2$(PO$_4$)$_3$ charged at 4.8V. Doping enhances the electrochemical performance of a battery. Generally, the doping elements used are Fe^{2+}, Mg^{2+}, Zr^{4+}, Ge^{4+}, and Ce^{3+}. Li$_3$V$_2$(PO$_4$)$_3$ showed excellent performance of high-rate discharge after doping with Mg^{2+} ion. Doping of anions also increases electronic conductivity. Doping of both cation and anion enhances different properties of battery. Cation affects the size of a synthesized material, whereas anion decreases the charge transfer resistance by increasing conductance [60]. Synthesized titanium phosphate and lithium titanium phosphate by the hydrothermal method for the study of active electrode material in energy storage devices [61].

SUPERCAPACITOR

Nowadays, capacitor attracts much attention due to their power density based on ultrafast charge–discharge capability. Its limitation for the energy storage application is low discharge energy densities. In the recent years, electronic power system and electronic capacitor with high energy density can reduce the volume, weight, cost, and energy storage device. Their easy breakdown strength and easy discharge capacitors still limit their application for voltage equipment and electronic devices [14]. The CV curve of sodium titanates in 1 M KCl solution shows the peak potential at 0.15 and 0.55 V at a scan rate of 100 mV/s. Here, oxidation and reduction of sodium titanates show the energy storage under pseudo-capacitance mode. During charging, oxidation state of titanates changes from +3 to +4, and in discharging, the reverse takes place. In redox capacitance, sodium titanates obtained high capacitance at lower galvanostatic current during the charging–discharging process. The charging–discharging curves are independent of current density. This indicates that sodium titanates can be used to develop electrode material for supercapacitor [25]. The charge/discharge curve of lanthanum-doped strontium titanates at different current densities in 1 M KOH was taken. The CV was taken at different scan rates. Cyclic voltammograms were cycled between −0.3 and 0.3 v. Here, capacitance mainly depends on the redox reaction because its shape of CV is different from the shape of electric double-layer capacitors. Charge–discharge feature shows good pseudo-capacitive behaviors and redox reaction. Here, we observed that charging curves are glass symmetric to their discharge counterparts in whole potential range because redox reaction occurs on the surface of lanthanum strontium titanates. Lanthanum-doped strontium titanate shows the longest charging time and the highest capacitance, which is in good agreement with CV result. Here, the specific capacitance of lanthanum-doped strontium titanate samples calculated from the GCD curve using a current 1 mA is 306.74 F/g. The specific capacitance decreases with the increase in scan rate [35].

Supercapacitors, also known as ultracapacitors, designed for the storage of electrical energy are under vast development in recent years. The parameters of energy storage devices are energy and power density. Depending on the energy storage, potential supercapacitors are classified into electrochemical double-layer capacitor, pseudo-capacitor, and hybrid capacitor. Electrochemical double-layer capacitor was designed as the two-plate capacitor, in which charge accumulator placed between an electrode and an electrolyte. During charging of a supercapacitor, the movement of electrons takes place from cathode to anode through an external circuit. At the same time, cations move toward a cathode (negative electrode), and anions move toward anode (positive electrode). Phosphates are generally employed in supercapacitors, due to their special features like more power density, fast recharge capacity, long life cycle, good conductivity, more capacitance, easy and safety operating condition, lightweight, thin, and eco-friendliness. Metal phosphates secure their place in microchips as well as in portable roll-up electronic devices [61].

Phosphates are non-toxic and cheap with higher conductivity. They perform as a good electrode material in supercapacitor than metal oxides, hydroxides, and sulfides. They show outstanding chemical stability due to the presence of strong covalent bond between phosphorous and oxygen atoms along with good electronic and magnetic properties. Iftikhar Hussain et al. synthesized a binder-free zinc–cobalt–gallium phosphate in which zinc and cobalt enhances electronic conductivity with sufficient oxidation states in redox reactions [62] (Figures 6.7 and 6.8).

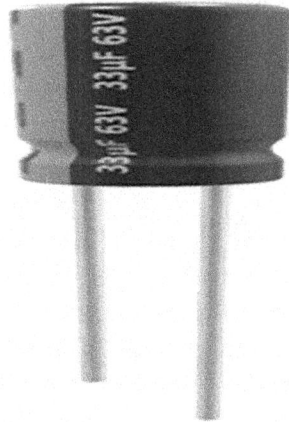

FIGURE 6.7   Supercapacitor.

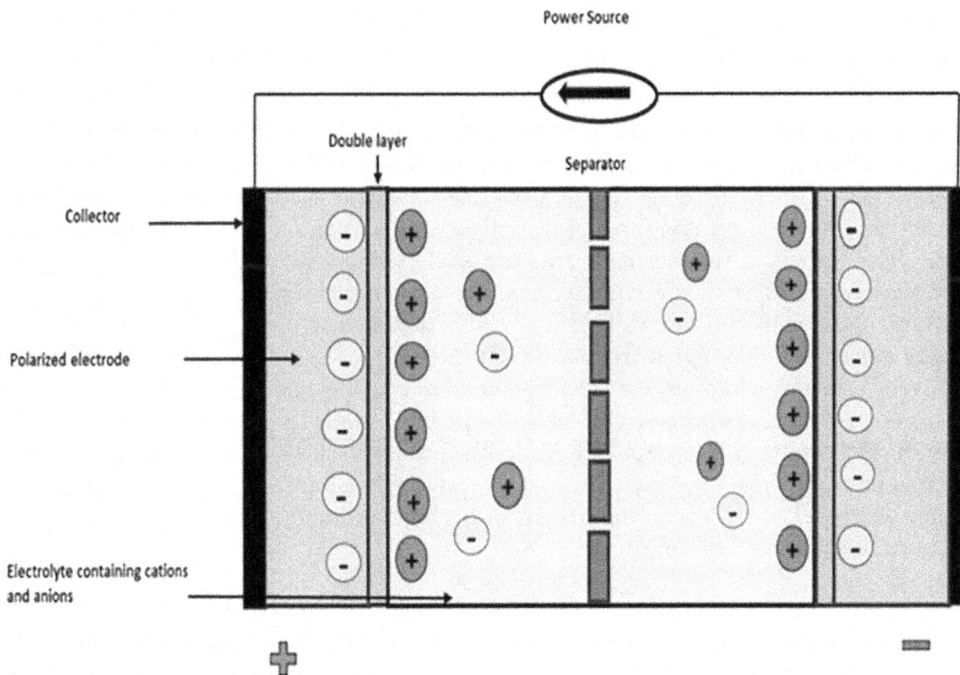

FIGURE 6.8   Construction of supercapacitor.

SENSOR

Normally gas sensors become sensitive only at high temperature. Various parameters attract the gas sensitivity such as thickness, crystal size, porosity, nature and amount of dopant, catalyst, and surface oxide [67]. Copper titanate has a large surface area, and it was responsive on ethanol and showed n-type semiconductor. The depletion layer forms near the surface of the particle by observing the O^- which controls the resistance of gas-sensing material at low temperatures. The sensitivity, selectivity, and long-lasting stability of these sensors were also good, indicating that they have potential applications in trace ethanol detection [17]. Pd and Pt are active for oxidation because the heat of absorption of oxygen on Pd and Pt is satisfactorily low to allow relatively low activation energy of oxidation and fast rate of reaction. It shows the maximum sensor response at 250°C, and at this temperature, the rate of reaction was fast, resulting in the large change in the temperature and the circuit voltage. Another reason for good sensor activity was better adsorption of hydrogen on the titanate nanotube surface, which facilitates the oxidation of hydrogen reaction by the Pd and Pt catalysts. This results in increase of Pd and Pt sensor activity because of fast reaction between absorption of H_2 and O_2 [33].

Sensors are considered detection devices that can sense the required data and convert the data into electrical signals [84]. Sensors are classified as active, passive, analog, digital, thermo-optic, thermoelectric, biological, radioactive, electrochemical, soil moisture, temperature, IR, heart-beat, light and ultrasonic sensors, etc. depending on their applications [85]. Using these sensors in our daily life becomes routine for our safety as well as healthy lifestyle. Nowadays living in pollution-free environment becomes a challenge to every human beings. To get pollution-free environment, we have to adopt some of the sensors containing nanophosphates, which have the capacity to sense the toxic, harmful chemicals and heavy metals (such as Cd^{2+}, Pb^{2+}, Hg^{2+}, and As^{3+}). Harmful radiations are already present and also enter into the environment. Sensors contain two working parts, namely, receptors and transducers. With large specificity, receptors increase the sensitivity of detecting the elements, whereas transducers are physical or chemical sensing components working on the optical, thermal, electrochemical, and piezoelectric principles [86,87].

CONCLUSION

Titanate nanoparticles are successfully synthesized by chemical methods. The physical-chemical properties of the designed nanoparticle were confirmed by XRD, SEM, TEM, and FTIR. These were utilized as catalysts for the degradation of toxic dyes under UV or visible or sunlight. Morphology, particle size, and specific surface area of the series of photocatalyst can influence the degradation activity. The efficient active sites on the surface of the nanoparticles reveal that they can act as effective catalysts. Undergoing reaction system, titanate nanocrystals can be recovered and reused for several cycles until they lose their catalytic activity. Nanophosphates synthesized by different methods with desired morphology are considered as highly reactive electrocatalyst and are employed for the redox reaction of some of the harmful as well as toxic organic compounds. Phosphates also act as very good electrode materials for the generation and storage of energy sources such as HER and OER. The doping of rare earth elements to the metal phosphate enhances the energy storage property of supercapacitors and batteries. Metal phosphates are eco-friendly and cost-effective, and are employed in sensor application due to their speed, high specificity, and high sensitivity. But they need more development in this field.

REFERENCES

[1] Arico, Antonino Salvatore, Peter Bruce, Bruno Scrosati, Jean-Marie Tarascon, and Walter Van Schalkwijk. "Nanostructured materials for advanced energy conversion and storage devices." *Nature Materials* 4, no. 5 (2005): 366–377.

[2] Colodrero, Rosario MP, Pascual Olivera-Pastor, Aurelio Cabeza, and Montse Bazaga-García. "Properties and applications of metal phosphates and pyrophosphates as proton conductors." *Materials* 15, no. 4 (2022): 1292. https://doi.org/10.3390/ma15041529

[3] Girish, K. M., S. C. Prashantha, H. Nagabhushana, C. R. Ravikumar, H. P. Nagaswarupa, Ramachandra Naik, H. B. Premakumar, and B. Umesh. "Multi-functional Zn2TiO4: Sm3+ nanopowders: excellent performance as an electrochemical sensor and an UV photocatalyst." *Journal of Science: Advanced Materials and Devices* 3, no. 2 (2018): 151-160.151-160. https://doi.org/10.1016/j.jsamd.2018.02.001

[4] Zhao, Dandan, Qi Shao, Ying Zhang, and Xiaoqing Huang. "N-Doped carbon shelled bimetallic phosphates for efficient electrochemical overall water splitting." *Nanoscale* 10, no. 48 (2018): 22787–22791. https://doi.org/10.1039/C8NR07756H

[5] Zhang, Jiachen, Li Wu, Lin Xu, Dongmei Sun, Hanjun Sun, and Yawen Tang. "Recent advances in phosphorus containing noble metal electrocatalysts for direct liquid fuel cells." *Nanoscale* 13, no. 38 (2021): 16052–16069. https://doi.org/10.1039/D1NR04218A

[6] Zhang, Yu, Tingting Qu, Feifei Bi, Panpan Hao, Muhong Li, Shanyong Chen, Xiangke Guo, Mingjiang Xie, and Xuefeng Guo. "Trimetallic (Co/Ni/Cu) hydroxyphosphate nanosheet array as efficient and durable electrocatalyst for oxygen evolution reaction." *ACS Sustainable Chemistry & Engineering* 6, no. 12 (2018):16859–16866. https://doi.org/10.1021/acssuschemeng.8b04180

[7] Baig, Nadeem, Irshad Kammakakam, and Wail Falath. "Nanomaterials: A review of synthesis methods, properties, recent progress, and challenges." *Materials Advances* 2, no. 6 (2021): 1821–1871. https://doi.org/10.1039/D0MA00807A

[8. Li, Jianlin, Qingliu Wu, and Ji Wu. "Synthesis of Nanoparticles via Solvothermal and Hydrothermal Methods 12." In Aliofkhazraei (ed.), *Handbook of Nanoparticles*, Springer International Publishing, Switzerland (2016). https://doi.org/10.1007/978-3-319-15338-4_17

[9] Hutchison, James E. "Greener nanoscience: a proactive approach to advancing applications and reducing implications of nanotechnology." *ACS Nano* 2, no. 3 (2008): 395–402. https://doi.org/10.1021/nn800131j

[10] Gómez-López, Paulette, Alain Puente-Santiago, Andrés Castro-Beltrán, Luis Adriano Santos do Nascimento, Alina M. Balu, Rafael Luque, and Clemente G. Alvarado-Beltrán. "Nanomaterials and catalysis for green chemistry." *Current Opinion in Green and Sustainable Chemistry* 24 (2020): 48–55. https://doi.org/10.1016/j.cogsc.2020.03.001

[11] Bavykin, Dmitry V., Jens M. Friedrich, and Frank C. Walsh. "Protonated titanates and TiO2 nanostructured materials: synthesis, properties, and applications." *Advanced Materials* 18, no. 21 (2006): 2807–2824. https://doi.org/10.1002/adma.200502696

[12] Gonzales, Leandro Lemos, Marlon da Silva Hartwig, Rafael Uarth Fassbender, Eduardo Ceretta Moreira, Marcelo Barbalho Pereira, Pedro Lovato Gomes Jardim, Cristiane Wienke Raubach, Mário Lucio Moreira, and Sérgio da Silva Cava. "Properties of zinc titanates synthesized by microwave assisted hydrothermal method." *Heliyon* 7, no. 3 (2021). https://doi.org/10.1016/j.heliyon.2021.e06521

[13] Li, Wen-Bo, Di Zhou, Li-Xia Pang, Ran Xu, and Huan-Huan Guo. "Novel barium titanate based capacitors with high energy density and fast discharge performance." *Journal of Materials Chemistry A* 5, no. 37 (2017): 19607–19612. https://doi.org/10.1039/C7TA05392D

[14] Haibo, Yang, Yan Fei, Lin Ying, and Wang Tong. "Novel strontium titanate-based lead-free ceramics for high-energy storage applications." (2017).

[15] Amy, Lucia, Sofia Favre, Daniel L. Gau, and Ricardo Faccio. "The effect of morphology on the optical and electrical properties of sodium titanate nanostructures." *Applied Surface Science* 555 (2021): 149610. https://doi.org/10.1016/j.apsusc.2021.149610

[16] Geioushy, R. A., O. A. Fouad, M. M. Rashad, D. A. Rayan, and A. T. Kandil. "Facile synthesis of nanosized samarium titanate (Sm 2 Ti 2 O 7) powders: structural, composition, thermal stability, optical and magnetic properties." *New Journal of Chemistry* 45, no. 17 (2021): 7799–7807. https://doi.org/10.1039/D1NJ01156A

[17] Ruiz-Preciado, M. A., A. Kassiba, Alain Gibaud, and A. Morales-Acevedo. "Comparison of nickel titanate (NiTiO3) powders synthesized by sol-gel and solid state reaction." *Materials Science in Semiconductor Processing* 37 (2015): 171–178. https://doi.org/10.1016/j.mssp.2015.02.063

[18] Wang, Zhengjuan, Xiaojun Wang, Guohong Zhou, Jianjun Xie, and Shiwei Wang. "Highly transparent yttrium titanate (Y2Ti2O7) ceramics from co-precipitated powders." *Journal of the European Ceramic Society* 39, no. 10 (2019): 3229–3234. https://doi.org/10.1016/j.jeurceramsoc.2019.04.018

[19] Lee, Jeon-Kook, Chong Hee Kim, and Hyung-Jin Jung. "Capacitance voltage characteristics of sol-gel derived lead titanate thin films on a silicon substrate." *Journal of Materials Science: Materials in Electronics* 2 (1991): 58–62.

[20] Meenakshi, Paraman, and Muthiah Selvaraj. "Bismuth titanate as an infrared reflective pigment for cool roof coating." *Solar Energy Materials and Solar Cells* 174 (2018): 530–537.https://doi.org/10.1016/j.solmat.2017.09.048

[21] He, Chunyong, Tao Bo, Yubin Ke, Bao-tian Wang, Juzhou Tao, and Pei Kang Shen. "Black potassium titanate nanobelts: Ultrafast and durable aqueous redox electrolyte energy storage." *Journal of Power Sources* 483 (2021): 229140. https://doi.org/10.1016/j.jpowsour.2020.229140

[22] Xie, Zhengwei, Xiang Li, Wen Li, Mianzhong Chen, and Meizhen Qu. "Graphene oxide/lithium titanate composite with binder-free as high capacity anode material for lithium-ion batteries." *Journal of Power Sources* 273 (2015): 754–760. https://doi.org/10.1016/j.jpowsour.2014.09.154

[23] Jin, Jiawei, Ye Zhang, Guochen Li, Zengyong Chu, and Gongyi Li. "Synthesis and enhanced gas sensing properties of iron titanate and copper titanate nanomaterials." *Materials Chemistry and Physics* 249 (2020): 123016. https://doi.org/10.1016/j.matchemphys.2020.123016

[24] Abd Aziz, Radhiyah, Izan IzwanMisnon, Kwok Feng Chong, Mashitah M. Yusoff, and Rajan Jose. "Layered sodium titanate nanostructures as a new electrode for high energy density supercapacitors." *Electrochimica Acta* 113 (2013): 141–148. https://doi.org/10.1016/j.electacta.2013.09.128

[25] Sandhya, C. P., Bibin John, and C. Gouri. "Lithium titanate as anode material for lithium-ion cells: a review." *Ionics* 20 (2014): 601–620.

[26] Stramare, S., V. Thangadurai, and W. Weppner. "Lithium lanthanum titanates: a review." *Chemistry of Materials* 15, no. 21 (2003): 3974–3990. https://doi.org/10.1021/cm0300516

[27] Matos, M., and L. Walmsley. "Cation-oxygen interaction and oxygen stability in CaCu3Ti4O12 and CdCu3Ti4O12 lattices." *Journal of Physics: Condensed Matter* 18, no. 5 (2006): 1793. https://doi.org/10.1088/0953-8984/18/5/030

[28] Bian, Shuaishuai, Zhenxing Yue, Yunzhou Shi, Jie Zhang, and Wei Feng. "Ultrahigh energy storage density and charge-discharge performance in novel sodium bismuth titanate-based ceramics." *Journal of the American Ceramic Society* 104, no. 2 (2021): 936–947. https://doi.org/10.1111/jace.17486

[29] Puli, Venkata Sreenivas, Dhiren K. Pradhan, Brian C. Riggs, Douglas B. Chrisey, and Ram S. Katiyar. "Investigations on structure, ferroelectric, piezoelectric and energy storage properties of barium calcium titanate (BCT) ceramics." *Journal of Alloys and Compounds* 584 (2014): 369–373. https://doi.org/10.1016/j.jallcom.2013.09.108

[30] Ioachim, A., M. I. Toacsan, M. G. Banciu, L. Nedelcu, F. Vasiliu, H. V. Alexandru, C. Berbecaru, and G. Stoica. "Barium strontium titanate-based perovskite materials for microwave applications." *Progress in Solid State Chemistry* 35, no. 2–4 (2007): 513–520. https://doi.org/10.1016/j.progsolidstchem.2007.01.017

[31] Sharma, Annu, Nandhini J. Usharani, and S. S. Bhattacharya. "Dielectric and ferroelectric properties of multicomponent equiatomic calcium lead strontium titanate (Ca0. 33Pb0. 33Sr0. 33) TiO3." *Open Ceramics* 6 (2021): 100130. https://doi.org/10.1016/j.oceram.2021.100130

[32] Camargo, Javier, Santiago Osinaga, Mariano Febbo, Sebastián P. Machado, Fernando Rubio-Marcos, Leandro Ramajo, and Miriam Castro. "Piezoelectric and structural properties of bismuth sodium potassium titanate lead-free ceramics for energy harvesting." *Journal of Materials Science: Materials in Electronics* 32, no. 14 (2021): 19117–19125.

[33] Han, Chi-Hwan, Dae-Woong Hong, Il-Jin Kim, Jihye Gwak, Sang-Do Han, and Krishan C. Singh. "Synthesis of Pd or Pt/titanate nanotube and its application to catalytic type hydrogen gas sensor." *Sensors and Actuators B: Chemical* 128, no. 1 (2007): 320–325. https://doi.org/10.1016/j.snb.2007.06.025

[34] Deshmukh, V. V., H. P. Nagaswarupa, C. R. Ravikumar, and N. Raghvendra. "Cyclic voltammetry performance of La-doped SrTiO3 electrodes for supercapacitor applications." In Proceedings of the eight DAE-BRNS interdisciplinary symposium on materials chemistry. 2021.

[35] Thammanna, B. M., K. Viswanathan, H. P. Nagaswarupa, and K. R. Vishnumahesh. "Novel MgTiO3: Eu3+ nanophosphor its photometric analysis for multifunctional applications." *Materials Today: Proceedings* 4, no. 11 (2017): 12306–12313. https://doi.org/10.1016/j.matpr.2017.09.164

[36] Girish, K. M., S. C. Prashantha, H. Nagabhushana, C. R. Ravikumar, H. P. Nagaswarupa, Ramachandra Naik, H. B. Premakumar, and B. Umesh. "Multi-functional Zn2TiO4: Sm3+ nanopowders: excellent performance as an electrochemical sensor and an UV photocatalyst." *Journal of Science: Advanced Materials and Devices* 3, no. 2 (2018): 151–160. https://doi.org/10.1016/j.jsamd.2018.02.001

[37] Tripathi, P., B. Sahu, S. P. Singh, O. Parkash, and D. Kumar. "Preparation and characterization of liquid phase (55B2O3-45Bi2O3) sintered cobalt doped magnesium titanate for wideband stacked rectangular dielectric resonator antenna (RDRA)." *Ceramics International* 41, no. 2 (2015): 2908–2916. https://doi.org/10.1016/j.ceramint.2014.10.116

[38] Singh, Devendra, Anju Dixit, and P. S. Dobal. "Ferroelectricity and ferromagnetism in Fe-doped barium titanate ceramics." *Ferroelectrics* 573, no. 1 (2021): 63–75. https://doi.org/10.1080/00150193.2021.1890464

[39] Ravanamma, R., K. Muralidhara Reddy, K. Venkata Krishnaiah, and N. Ravi. "Structure and morphology of yttrium doped barium titanate ceramics for multi-layer capacitor applications." *Materials Today: Proceedings* 46 (2021): 259–262. https://doi.org/10.1016/j.matpr.2020.07.646

[40] Shah, MAK Yousaf, Yuzheng Lu, Naveed Mushtaq, Sajid Rauf, Muhammad Yousaf, Muhammad Imran Asghar, Peter D. Lund, and Bin Zhu. "Demonstrating the potential of iron-doped strontium titanate electrolyte with high-performance for low temperature ceramic fuel cells." *Renewable Energy* 196 (2022): 901–911. https://doi.org/10.1016/j.renene.2022.06.154

[41] Sohib, Ahmad, Slamet Priyono, Wahyu Bambang Widayatno, Achmad Subhan, Sherly Novia Sari, Agus SukartoWismogroho, ChairulHudaya, and Bambang Prihandoko. "Electrochemical performance of low concentration Al doped-lithium titanate anode synthesized via sol-gel for lithium ion capacitor applications." *Journal of Energy Storage* 29 (2020): 101480. https://doi.org/10.1016/j.est.2020.101480

[42] Chang, Yee-Shin, Yen-Hwei Chang, In-Gann Chen, Guo-Ju Chen, Yin-Lai Chai, Sean Wu, and Te-Hua Fang. "The structure and properties of zinc titanate doped with strontium." *Journal of Alloys and Compounds* 354, no. 1-2 (2003): 303–309. https://doi.org/10.1016/S0925-8388(02)01362-2

[43] Chandel, Sakshee, Jaekook Kim, and Alok Kumar Rai. "Effect of vanadium doping on the electrochemical performances of sodium titanate anode for sodium ion battery application." *Dalton Transactions* 51, no. 31 (2022): 11797–11805. https://doi.org/10.1039/D2DT01626E

[44] Wen, Qinlong, Wancheng Zhou, Yiding Wang, Fa Luo, Dongmei Zhu, Zhibin Huang, and Yuchang Qing. "Effects of calcium doping on yttrium titanate for microwave absorbing applications." *Journal of Alloys and Compounds* 741 (2018): 700–706. https://doi.org/10.1016/j.jallcom.2018.01.124

[45] Singh, Saurabh, Priyanka A. Jha, Sabrina Presto, Massimo Viviani, A. S. K. Sinha, Salil Varma, and Prabhakar Singh. "Structural and electrical conduction behaviour of yttrium doped strontium titanate: anode material for SOFC application." *Journal of Alloys and Compounds* 748 (2018): 637–644.https://doi.org/10.1016/j.jallcom.2018.03.170

[46] Attar, Abbas Sadeghzadeh, Ehsan Salehi Sichani, and Shahriar Sharafi. "Structural and dielectric properties of Bi-doped barium strontium titanate nanopowders synthesized by sol-gel method." *Journal of materials research and technology* 6, no. 2 (2017): 108–115. https://doi.org/10.1016/j.jmrt.2016.05.001

[47] Liu, Wenlong, Yingjie Lei, Wei Feng, Dengjun He, Yunfeng Zhang, Jing Li, Jiaxuan Liao, and Lingzhao Zhang. "Comprehensive dielectric performance of alternately doped BST multilayer films coated with strontium titanate thin layers." *Journal of Materials Research and Technology* 13 (2021): 385–396. https://doi.org/10.1016/j.jmrt.2021.04.085

[48] Shaban, Naghi, and Mahmood Bahar. "Synthesis and characterization of Fe and Ni Co-doped Ba0. 6Sr0. 4TiO3 prepared by Sol-Gel technique." *J. Theor. Comput. Sci* 4, no. 2 (2017). https://doi.org/10.4172/2376-130X.1000157

[49] Yang, Haibo, Fei Yan, Ying Lin, and Tong Wang. "Enhanced energy-storage properties of lanthanum-doped Bi0. 5Na0. 5TiO3-based lead-free ceramics." *Energy Technology* 6, no.2 (2018): 357–365. https://doi.org/10.1002/ente.201700504

[50] Rohilla, Pooja, and A. S. Rao. "Synthesis optimisation and efficiency enhancement in Eu3+ doped barium molybdenum titanate phosphors for w-LED applications." *Materials Research Bulletin* 150 (2022): 111753.https://doi.org/10.1016/j.materresbull.2022.111753

[51] Cavazzani, Jonathan, Enrico Squizzato, Elena Brusamarello, and Antonella Glisenti. "Exsolution in Ni-doped lanthanum strontium titanate: a perovskite-based material for anode application in ammonia-fed Solid Oxide Fuel Cell." *International Journal of Hydrogen Energy* 47, no. 29 (2022): 13921–13932. https://doi.org/10.1016/j.ijhydene.2022.02.133

[52] Maddaiah, M., A. Guru Sampath Kumar, L. Obulapathi, T. Sofi Sarmash, K. Chandra Babu Naidu, D. Jhansi Rani, and T. Subba Rao. "Synthesis and characterization of strontium doped zinc manganese titanate ceramics." *Digest Journal of Nano materials and Biostructures* 10 (2015): 155–159.

[53] Amhil, S., S. Ben Moumen, A. Bourial, and L. Essaleh. "Evidence of large hopping polaron conduction process in strontium doped calcium copper titanate ceramics." *Physica B: Condensed Matter* 556 (2019): 36–41. https://doi.org/10.1016/j.physb.2018.12.032

[54] Kavitha, V., P. Mahalingam, M. Jeyanthinath, and N. Sethupathi. "Optical and structural properties of tungsten-doped barium strontium titanate." *Materials Today: Proceedings* 23 (2020): 12–15.https://doi.org/10.1016/j.matpr.2019.05.351

[55] Sangwan, Kanta Maan, Neetu Ahlawat, R. S. Kundu, Suman Rani, Sunita Rani, Navneet Ahlawat, and Sevi Murugavel. "Improved dielectric and ferroelectric properties of Mn doped barium zirconium titanate (BZT) ceramics for energy storage applications." *Journal of Physics and Chemistry of Solids* 117 (2018): 158–166. https://doi.org/10.1016/j.jpcs.2018.01.051

[56] Batoo, Khalid Mujasam, Ritesh Verma, Ankush Chauhan, Rajesh Kumar, Muhammad Hadi, Omar M. Aldossary, and Yarub Al-Douri. "Improved room temperature dielectric properties of Gd3+ and Nb5+ co-doped Barium Titanate ceramics." *Journal of Alloys and Compounds* 883 (2021): 160836. https://doi.org/10.1016/j.jallcom.2021.160836

[57] Holder, Cameron F., and Raymond E. Schaak. "Tutorial on powder X-ray diffraction for characterizing nanoscale materials." *Acs Nano* 13, no. 7 (2019): 7359–7365.

[58] Sun, Shuting, Ruhong Li, Deying Mu, Zeyu Lin, Yuanpeng Ji, Hua Huo, Changsong Dai, and Fei Ding. "Magnesium/chloride co-doping of lithium vanadium phosphate cathodes for enhanced stable lifetime in lithium-ion batteries." *New Journal of Chemistry* 42, no. 16 (2018): 13667–13673. https://doi.org/10.1039/C8NJ02165A

[59] Vaidyanath, Y. N., Ashamanjari, K. G., Mahesh, K. V., Mylarappa, M., Ramu, M. B., Prashantha, S. C., ... & Siddeswara, D. M. K. Journal homepage 5(7), 917–925. July-2017. www.journalijar.com

[60] Zhao, Junhong, Shaomei Wang, Zhen Run, Guangqin Zhang, Weimin Du, and Huan Pang. "Hydrothermal synthesis of nickel phosphate nanorods for high-performance flexible asymmetric all-solid-state supercapacitors." *Particle & Particle Systems Characterization* 32, no. 9 (2015): 880–885. https://doi.org/10.1002/ppsc.201500005

[61] Pandey, Manish, and Mishra, Guarav. "Types of sensor and their applications, advantages, and disadvantages." In *Emerging Technologies in Data Mining and Information Security: Proceedings of IEMIS 2018*, 3 (2019): 791–804. Springer Singapore.

[62] Mirghni, Abdulmajid A., Kabir O. Oyedotun, Okikiola Olaniyan, Badr A. Mahmoud, Ndeye Fatou Sylla, and Ncholu Manyala. "Electrochemical analysis of Na-Ni bimetallic phosphate electrodes for supercapacitor applications." *RSC Advances* 9, no. 43 (2019): 25012–25021. https://doi.org/10.1039/C9RA04487F

[63] Zhan, Yi, Meihua Lu, Shiliu Yang, Chaohe Xu, Zhaolin Liu, and Jim Yang Lee. "Activity of transition-metal (manganese, iron, cobalt, and nickel) phosphates for oxygen electrocatalysis in alkaline solution." *ChemCatChem* 8, no. 2 (2016): 372–379. https://doi.org/10.1002/cctc.201500952

[64] Guo, Tan, Lijing Zhang, Shan Yun, Jiadong Zhang, Litao Kang, Yanxing Li, Huaju Li, and Aibin Huang. "One-step synthesis of bimetallic Ni-Fe phosphates and their highly electrocatalytic performance for water oxidation." *Materials Research Bulletin* 114 (2019): 80–84. https://doi.org/10.1016/j.materresbull.2019.02.026

[65] Chinnadurai, Deviprasath, Aravindha Raja Selvaraj, Rajmohan Rajendiran, G. Rajendra Kumar, Hee-Je Kim, K. K. Viswanathan, and Kandasamy Prabakar. "Inhibition of redox behaviors in hierarchically structured manganese cobalt phosphate supercapacitor performance by surface trivalent cations." *ACS Omega* 3, no. 2 (2018): 1718–1725. https://doi.org/10.1021/acsomega.7b01762

[66] El Hallaoui, Achraf, Tourya Ghailane, Soukaina Chehab, Youssef Merroun, Rachida Ghailane, Said Boukhris, Taoufik Guédira, and Abdelaziz Souizi. "Synthesis of new bimetallic phosphate (Al/Ag3PO4) and study for its catalytic performance in the synthesis of 1, 2-dihydro-l-phenyl-3H-naphth [1, 2-e]-[1, 3] oxazin-3-one derivatives." *Mediterranean Journal of Chemistry* 11, no.3 (2021):215–228. https://doi.org/10.13171/mjc02108091579elhallaoui

[67] Sun, Lei, Deyu Kong, Fang Wang, Wei Luo, Yanqiu Chen, Zhouzhou, and Junhua Liu. "Amorphous Porous Chromium-Zirconium Bimetallic Phosphate: Synthesis, Characterization and Application in Liquid Phase Oxidation of Hydrocarbons by Different Oxygen Sources." *ChemistrySelect* 5, no. 4 (2020): 1552–1559. https://doi.org/10.1002/slct.201904073

[68] Murty, B. S., P. Shankar, Baldev Raj, B. B. Rath, James Murday, B. S. Murty, P. Shankar, Baldev Raj, B. B. Rath, and James Murday. "Tools to characterize nanomaterials." *Textbook of Nanoscience and Nanotechnology* (2013): 149–175. https://doi.org/10.1007/978-3-642-28030-6_5

[69] Jin, Jiawei, Ye Zhang, Guochen Li, Zengyong Chu, and Gongyi Li. "Synthesis and enhanced gas sensing properties of iron titanate and copper titanate nanomaterials." *Materials Chemistry and Physics* 249 (2020): 123016. https://doi.org/10.1016/j.matchemphys.2020.123016

[70] Puli, Venkata Sreenivas, Ashok Kumar, Douglas B. Chrisey, M. Tomozawa, J. F. Scott, and Ram S. Katiyar. "Barium zirconate-titanate/barium calcium-titanate ceramics via sol-gel process: novel high-energy-density capacitors." *Journal of Physics D: Applied Physics* 44, no. 39 (2011): 395403. https://doi.org/10.1088/0022-3727/44/39/395403

[71] Wen, Shuting, Guangliang Chen, Wei Chen, Mengchao Li, Bo Ouyang, Xingquan Wang, Dongliang Chen et al. "Nb-doped layered FeNi phosphide nanosheets for highly efficient overall water splitting under high current densities." *Journal of Materials Chemistry A* 9, no. 15 (2021): 9918–9926. https://doi.org/10.1039/D1TA00372K

[72] Kotova, N. M., K. A. Vorotilov, D. S. Seregin, and A. S. Sigov. "Role of precursors in the formation of lead zirconate titanate thin films." *Inorganic Materials* 50 (2014): 612–616. https://doi.org/10.1134/S0020168514060107

[73] Mizera, Adrian, and Ewa Drożdż. "Studies on structural, redox and electrical properties of Ni-doped strontium titanate materials." *Ceramics International* 46, no. 15 (2020): 24635–24641. https://doi.org/10.1016/j.ceramint.2020.06.252

[74] Guo, Tan, Lijing Zhang, Shan Yun, Jiadong Zhang, Litao Kang, Yanxing Li, Huaju Li, and Aibin Huang. "One-step synthesis of bimetallic Ni-Fe phosphates and their highly electrocatalytic performance for water oxidation." *Materials Research Bulletin* 114 (2019): 80–84. https://doi.org/10.1016/j.materresbull.2019.02.026

[75] Arumugam, D., G. Paruthimal Kalaignan, and P. Manisankar. "Synthesis and electrochemical characterizations of nano-crystalline LiFePO4 and Mg-doped LiFePO4 cathode materials for rechargeable lithium-ion batteries." *Journal of Solid State Electrochemistry* 13 (2009): 301–307. https://doi.org/10.1007/s10008-008-0533-3

[76] Wen, Shuting, Guangliang Chen, Wei Chen, Mengchao Li, Bo Ouyang, Xingquan Wang, Dongliang Chen et al. "Nb-doped layered FeNi phosphide nanosheets for highly efficient overall water splitting under high current densities." *Journal of Materials Chemistry A* 9, no. 15 (2021): 9918–9926. https://doi.org/10.1039/D1TA00372K

[77] Qin, Zhixiao, Yubin Chen, Zhenxiong Huang, Jinzhan Su, and Liejin Guo. "A bifunctional NiCoP-based core/shell cocatalyst to promote separate photocatalytic hydrogen and oxygen generation over graphitic carbon nitride." *Journal of Materials Chemistry A* 5, no. 36 (2017): 19025–19035. https://doi.org/10.1039/C7TA04434H

[78] Mirghni, Abdulmajid A., Kabir O. Oyedotun, Okikiola Olaniyan, Badr A. Mahmoud, Ndeye Fatou Sylla, and Ncholu Manyala. "Electrochemical analysis of Na-Ni bimetallic phosphate electrodes for supercapacitor applications." *RSC advances* 9, no. 43 (2019): 25012–25021. https://doi.org/10.1039/C9RA04487F

[79] Song, Xueying, Li Gao, Yamin Li, Wei Chen, Liqun Mao, and Jing-He Yang. "Nickel phosphate-based materials with excellent durability for urea electro-oxidation." *Electrochimica Acta* 251 (2017): 284–292. https://doi.org/10.1016/j.electacta.2017.08.117

[80] Petersen, Hilke, Niklas Stegmann, Michael Fischer, Bodo Zibrowius, Ivan Radev, Wladimir Philippi, Wolfgang Schmidt, and Claudia Weidenthaler. "Crystal structures of two titanium phosphate-based proton conductors: Ab initio structure solution and materials properties." *Inorganic Chemistry* 61, no. 5 (2021): 2379–2390. https://doi.org/10.1021/acs.inorgchem.1c02613

[81] Mondal, Anjon Kumar, Dawei Su, Shuangqiang Chen, Katja Kretschmer, Xiuqiang Xie, Hyo-Jun Ahn, and Guoxiu Wang. "A Microwave Synthesis of Mesoporous NiCo2O4 nanosheets as electrode materials for lithium-ion batteries and supercapacitors." *ChemPhysChem* 16, no. 1 (2015): 169–175. https://doi.org/10.1002/cphc.201402654

[82] Moseley, P. T., and B. C. Tofield (Eds.). *Solid-State Gas Sensors*, pp. 12–31. Bristol/Philadelphia: Hilger, 1987.

[83] Wang, G. X., S. L. Bewlay, K. Konstantinov, H. K. Liu, S. X. Dou, and J-H. Ahn. "Physical and electrochemical properties of doped lithium iron phosphate electrodes." *Electrochimica Acta* 50, no.2–3 (2004): 443–447. https://doi.org/10.1016/j.electacta.2004.04.047

[84] Su, Shao, Wenhe Wu, Jimin Gao, Jianxin Lu, and Chunhai Fan. "Nanomaterials-based sensors for applications in environmental monitoring." *Journal of Materials Chemistry* 22, no. 35 (2012): 18101–18110. https://doi.org/10.1039/C2JM33284A

[85] Xu, Kebin, Yuki Kitazumi, Kenji Kano, and Osamu Shirai. "Phosphate ion sensor using a cobalt phosphate coated cobalt electrode." *Electrochimica Acta* 282 (2018): 242–246. https://doi.org/10.1016/j.electacta.2018.06.021

[86] Wen, Shuting, Guangliang Chen, Wei Chen, Mengchao Li, Bo Ouyang, Xingquan Wang, Dongliang Chen et al. "Nb-doped layered FeNi phosphide nanosheets for highly efficient overall water splitting under high current densities." *Journal of Materials Chemistry A* 9, no. 15 (2021): 9918–9926. https://doi.org/10.1039/D1TA00372K

[87] Vaidyanath, Y. N., K. G. Ashamanjari, M. Mylarappa, Bhargava Ramu, K. R. Vishnu Mahesh, S. Prashanth Chandra, H. P. Nagaswarupa, and N. Raghavendra. "Comparative study of different immobilization of strontium inlisr2 (po4) 3 crystal through hydrothermal process." *IOSR Journal of Applied Physics* 9, no. 4 (2017): 13–19.

7 Design and Development of Novel Nanoferrite and Metal Oxide Composites for Energy Applications
A Review

S.V. Dhanyashree, H.K. Jahnavi, H.V. Harini,
S. Rajendra Prasad, H.P. Nagaswarupa,
and Ramachandra Naik

INTRODUCTION

NANOCOMPOSITES

Nowadays, nanocomposites gained a lot of interest. Nanocomposite material properties are influenced by both their morphology and interfacial characteristics as well as those of their individual parents [1]. Materials with a solid structure, known as nanocomposites, have a gap between the phases that is minimally produced by a dimension of nanoscale size and generally take the form of an inorganic matrix placed in an organic phase, or vice versa. Individual features such as mechanical, chemical, electrical, optical, and catalytic ones can be found in nanocomposite materials [2]. According to their matrix, there are three different forms of nanocomposites: metal matrix nanocomposites (MMNC), polymer matrix nanocomposites (PMNC), and ceramic matrix nanocomposites (CMNC) [3]. Composites or hybrid oxides of metal are created when several metal oxides combine. Metal oxide composites are essential in many industrial applications, including environmental cleanup, biomedical applications, sensors, and catalysis [4]. The automotive and aerospace sectors frequently use composite materials based on metal oxide. Composite materials combine ceramics, metals, and polymers for high performance and superior properties in advanced technology [5]. The primary focus of supercapacitors (SCs) and batteries is strong stability, extraordinary power density, and good electrochemical reversibility [6].

NANOFERRITES

Ferrites normally occur from iron oxides with a backbone structure of Fe_2O_3/Fe_3O_4 and have ferromagnetic properties. They can be transformed by doping different transition metal oxides [7,8]. Nanoferrites are extremely abundant, inexpensive, and highly biocompatible materials with distinctive magnetic and electrical properties. The synthesis of diverse processes may result in differences in particle size, surface areas, and application regions. Ferrites are effective materials for addressing environmental concerns as well as sustainable energy conversion and storage [9].

DOI: 10.1201/9781003479239-7

METAL OXIDE NANOCOMPOSITES AND NANOFERRITES FOR BATTERY

Batteries are regarded as a crucial strategy for maximizing energy consumption efficiency. For battery applications, ideal electrodes should be affordable and environmentally friendly, possess substantial energy and power densities, and survive for a long time [10]. The high energy density of lithium batteries makes them interesting for applications like electrification of vehicles, but concerns about safety, long-term stability (>20 years), and cost make these materials less ideal for larger batteries [11]. There has been a sharp increase in demand for effective energy storage technologies, such as batteries, solar cells, and fuel cells, as a result of recent worries over energy and environmental challenges [12].

Batteries can be classified into five categories:

1. Lithium-ion batteries
2. Lithium-sulphur batteries
3. Sodium-ion batteries
4. Dye-sensitised solar cells
5. Fuel cells (Figure 7.1)

LITHIUM-ION BATTERIES

Due to its superior stability, longer cycle life, and cheaper price, lithium-ion batteries are widely employed in microelectronics, energy storage, and power systems [13]. The widespread usage of lithium-ion batteries is due to their high electrochemical performance, which is greatly reliant on

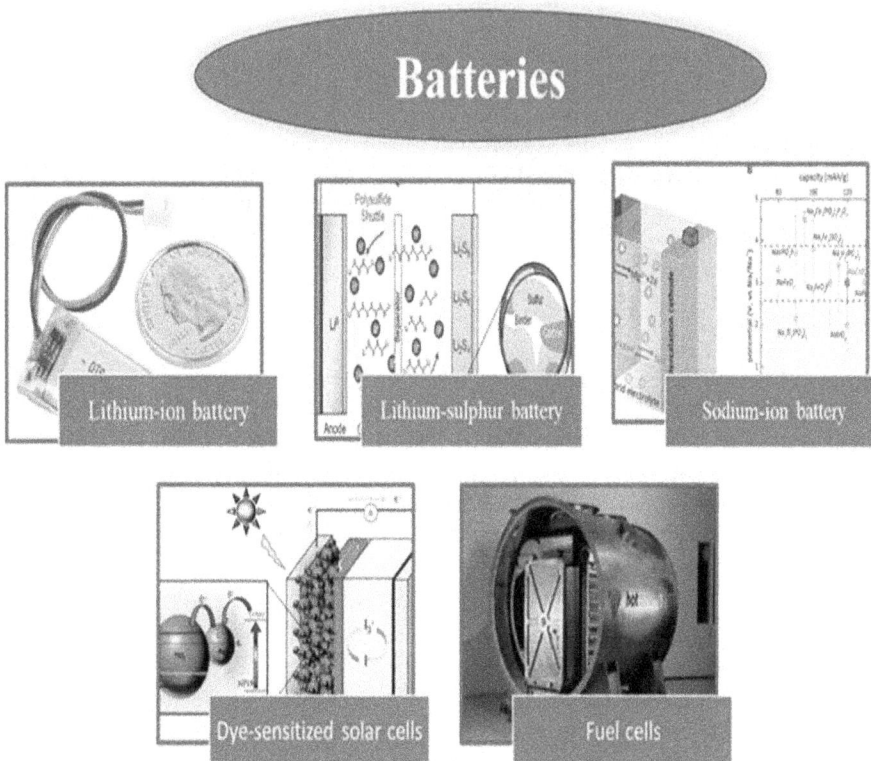

FIGURE 7.1 Schematic representation of batteries.

the cathode and anode materials, which are the primary lithium-ion battery components [14]. The charge/discharge efficiency has been limited by the poor lithium-ion transmission productivity, and the durability and stability of lithium-ion batteries are significantly impacted by the huge volume effect. For lithium-ion batteries, several innovative anode materials have been created to increase battery capacity and energy density [15].

LITHIUM-SULPHUR BATTERIES

Lithium-sulphur batteries provide a high potential and energy efficiency, and are environmentally protected and cost-effective [16]. However, the progressive electrochemical reaction develops soluble polysulfides, which could dissolve into the electrolytes and result in poor electrical conductivity, space shuttle effect and permanent capacity decline [17]. Theoretically, lithium-sulphur batteries with a sulphur composite cathode, lithium metal as the anode, and an organic liquid electrolyte might have capacities of 1,675 mAh/g and 2,567 Whk/g [18]. Due to the low cost, nontoxicity, and extremely high hypothetical energy density of sulphur cathode material, lithium-sulphur batteries are one of the most viable possibilities for future devices that store energy [19].

SODIUM-ION BATTERIES

NiO/graphene composites also show promising performance as sodium-ion batteries, and they're desirable low-cost alternatives to lithium-ion batteries, anode materials [20]. Due to the reduced cost and infinite sodium supplies, using Na+ instead of Li+ might result in the production of a reusable electrolytic battery that stores energy in an efficient manner [21].

DYE-SENSITISED SOLAR CELLS

As the next generation of solar cells, dye-sensitised solar cells which are capable of converting solar energy into electricity have received a lot of interest. Due to their exceptional energy conversion efficiency, straightforward production procedure, and inexpensive fabrication cost, these solar cells represent an alternative to traditional silicon-based photovoltaic systems [22]. Improving the creation of I^- from I^{3-} in the presence of a counterelectrode, which is heavily dependent on the characteristics of the counterelectrode that is used to catalyse the I^-/I^{3-} redox process, is one of the challenges [23] (Figure 7.1).

FUEL CELLS

Fuel cells, a promising environmentally friendly and extremely effective alternative energy source, have been the subject of extensive research in recent years [24]. Fuel cell technologies have gained interest from all around the world recently because of their high efficiency and low emissions. Polymer electrolyte membranes are used in the construction of fuel cells as proton conductors and electrolytic catalyst for electrochemical operations at low temperatures [25].

SUPERCAPACITOR

Due to their higher power efficiency and higher energy density, supercapacitors serve as a bridge between batteries and dielectric capacitors. For use in supercapacitors, electrode materials must be very porous and conduct electricity [26]. The fundamental function of a supercapacitor is to store energy by dispersing ions with charge in an electrolyte on its electrode surfaces [27].

Based on the energy storage method, ECCs are divided into electrical double-layer capacitors (EDLC), pseudocapacitors (PSCs), and hybrid capacitors (HCs) (Figure 7.2):

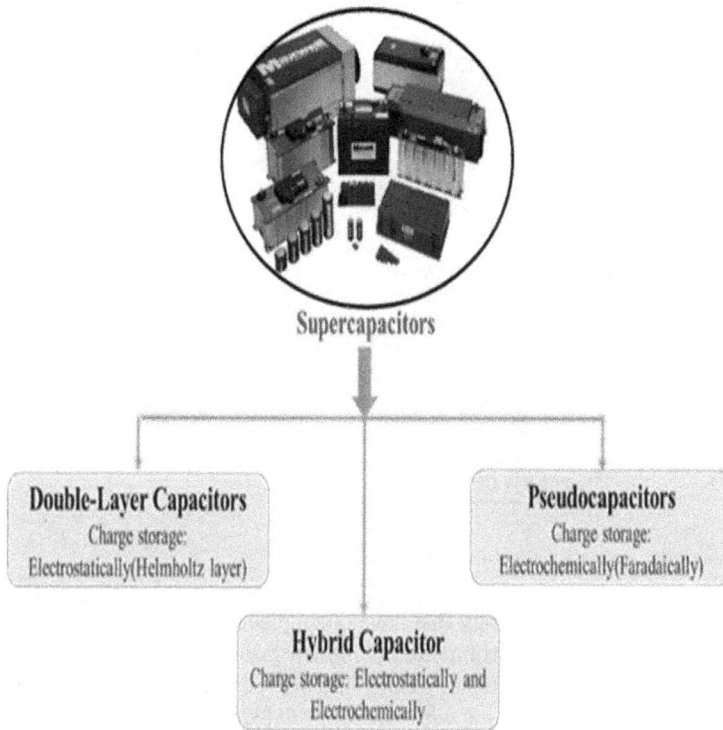

FIGURE 7.2 Schematic representation of supercapacitors.

a. Double-layer capacitors
b. Pseudocapacitors
c. Hybrid capacitors

SENSORS

Sensors are analytical tools that can detect any change or response and can operate in response to information from the outside world [28]. Electrochemistry-based biomarker sensors are one of the most popular biomarker detection technologies. They are applied in many detection domains, such as space exploration, environmental protection, and clinical diagnosis [29] (Figure 7.3).

Further sensors can be classified into three categories:

1. Biosensors
2. Gas sensors
3. Chemical sensors (Figure 7.3)

Biosensors

Among the various biosensor kinds, glucose sensors have drawn a lot of interest since diabetes is a global health issue that results in loss and mortality worldwide. Additionally, the food sector, bio-processing, and sustainable as well as renewable fuel cells may all benefit from the use of glucose control [30].

Gas Sensors

Gas-sensing devices are now widely utilised in a variety of fields, including interior environments, industries, aircraft, and detectors for different dangerous residence gases and vapours. This is due

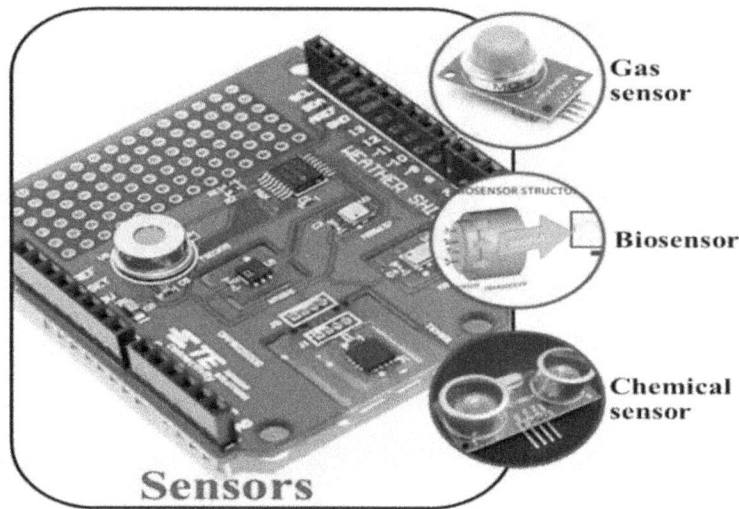

FIGURE 7.3   Schematic representation of sensors.

to the significant advances in technology that have taken place in recent years [31]. Gas sensors are crucial for mining, industry, and environmental monitoring due to their redox reactions at gas interfaces. Zhang et al. found that n-type metal oxides like SnO_2 can enhance gas-sensing performance by forming effective p-n heterojunctions with NiO or rGO [32].

Chemical Sensors

Chemical data are converted into a useful output signal by a chemical sensor. It frequently converts physical, chemical, or chemical interaction characteristics (such as concentration or overall composition of a particular species) into an output signal. The receptor and the transducer are the two fundamental components that make up chemical sensors. The receptor component of the sensor is where chemical information is converted into a type of energy that the transducer can measure [33]

SYNTHESIS OF METAL OXIDE COMPOSITES BY CHEMICAL METHODS

Due to the presence of some harmful chemicals absorbed on the surface, chemical synthesis techniques have been associated with a number of adverse side effects [34]. In terms of shape, purity, stability, and surface area, the synthesis methods play a significant part in the production of high-quality ferrite materials. There are numerous methodologies for generating nanoparticles. Therefore, these topics are addressed in detail.

In terms of shape, purity, stability, and surface area, the synthesis methods play a significant role in the production of high-quality ferrite materials. For the synthesising of ferrite materials, a variety of synthesis techniques are used, each of which has both advantages and drawbacks.

The two important methods for the synthesis of nano ferrites structures are as follows (Figure 7.4):

1. Chemical method
2. Green method

Chemical Methods

Combustion Method

Solution combustion synthesis is a cost-effective method for producing simple and complex oxides, particularly in ceramics, due to its simplicity and efficiency [35,36]. The synthesis of ferrite

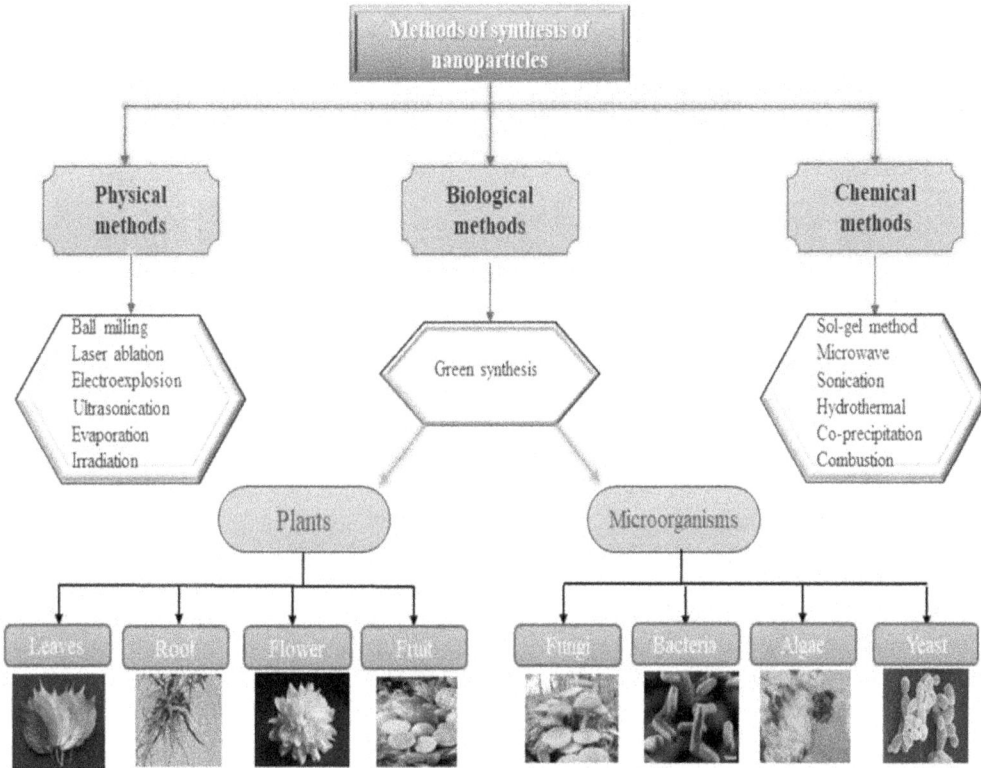

FIGURE 7.4 Schematic representation of synthesis of metal oxide composites.

nanomaterials can be carried out easily and inexpensively by using chemical methods. The same method is used to create ZnO–TiO$_2$, which has applications as a gas sensor, where it has been discovered that ZnO nanoplatelets include distributed TiO$_2$ nanoparticles. CNTs, Fe–Ni, and TiO$_2$ are all produced via chemical vapour deposition as Fe–Ni-doped TiO$_2$. MWCNT/ZnO was also created [37].

Sol-Gel Method

A flexible way for creating a large range of materials is offered by the sol-gel method. It allows for the effective application of several techniques and the ability to construct a wide range of nano- and microstructures [38]. For the synthesis of nanoferrites and their composites, the sol-gel process is well known and widely utilised. In general, metal alkoxide solutions are used in the sol-gel synthesis process. A variety of transition metal oxide–based nanocomposites, including CNTs/TiO$_2$, WO$_3$/TiO$_2$, ZnO/reduced grapheme oxide, ZnO–SiO$_2$, CdO–ZnO, CoTiO$_3$/CoFe$_2$O$_4$, and ZnO–SnO$_2$, are produced using the simple sol-gel synthesis method [39].

Sonication Method

Sonication is one of the most effective methods for reducing aggregation and improving the dispersion of nanostructures in the cement matrix. Indirect sonication or a bath sonicator using the cement matrix was shown to be the optimum method for improving the dispersion of clay nanoparticles [40]. Sonication method is the fastest and safest approach. It is the technique of transforming sound energy into physical vibrations that can be utilised to break down big particles in solution into smaller size nanoparticles. Sonication can be carried out using either an ultrasonic bath or an ultrasonic probe (sonicator) [41].

Co-precipitation Method

A solvent displacement technique is another name for the wet chemical process [42]. This procedure is a key tool for producing nanomaterials. Precipitates in solution happen when a substance's concentration reaches an excess. The precipitation will grow as a result of diffusion taking place on its surface, which will lead to the creation of nanoparticles [43]. Over the years, various combinations of transition metal oxides, including $CuWO_4$/ NiO, NiO/ZnO, CeO/.CuO/ ZnO, CeO_2/ZnO/ $ZnAl_2O_4$, ZnO/Ag and more have been observed [44].

Microwave Method

Microwave-assisted technologies are an interesting, ecologically harmless way for producing nanomaterials and nanocomposites [45]. The microwave-aided approach was developed only in the 1950s, but it has achieved popularity in the last two decades. The greatest advantage of microwave method over other conventional methods is quick reaction time and the ability to heat up the materials evenly. This technique allows for the manufacture of widely dispersed, small-sized particles. When fabricating Ag/ZnO/graphene for photocatalytic activity, Ag/ZnO–TiO_2 [46] and Mn_3O_4–Fe_2O_3/Fe_3O_4/rGO nanocomposites are used for supercapacitors and many more synthesised by using this method [47].

Hydrothermal Method

Nanocomposites based on transition metal oxides are manufactured by a hydrothermal technique. Hydrothermal synthesis is the most recent method for separating compounds from highly heated solutions at high vapour pressures [48]. A variety of TMONCs were synthesised via simple hydrothermal methods including carbon nanotube/cubic ferrite, Fe_2O_3–ZnO, MnO_2–CNT, TiO_2/reduced ferrite, and CuO–ZnO [49].

Green Method

The creation of "green" chemicals and nanomaterials has attracted considerable interest in the field of materials science as a reliable, long-lasting, and ecologically responsible method [50]. Bacteria, algae, and fungi make up the majority of microorganisms employed in green synthesis. As an alternative, extracts from different plant parts such as leaves, plants, seeds, fruits, and peelings have been employed to create different nanoparticles since the phytochemicals in these materials work as a preserving and reducing agent [51].

Green nanotechnology is easy to use, affordable, and environmentally beneficial. It has been receiving a lot of attention. Many of the green nanotechnology produced nanoparticles have been utilised successfully in a variety of fields. This synthesis can be carried out with algae, fungi, bacteria, plants, etc. Due to the presence of phytochemicals in their extract, which serve as a stabilising and reducing agent, several plant parts, including leaves, fruits, roots, stems, and seeds, have been employed for the production of different nanoparticles. The main benefits of producing ferrite nanoparticles using plant extracts are cost efficiency, ecological friendliness, and nontoxicity; thus, it may be utilised as an useful and inexpensive material which serve as substitute for the large-scale manufacturing of metal nanoparticles.

Using Plant Extracts

Due to the lack of hazardous chemicals used in the processes using plant materials, plants are great alternatives for producing nanoparticles. Plants are described as the environment's low-cost, low-maintenance production facilities [52]. To create nanoparticles, plant extracts are simply mixed with a solution of metallic salts at room temperature. It takes only a few minutes to complete the response. Several plants, including neem, *Aloe vera*, tea, *Cinnamomum camphora*, geranium leaf, and lemon grass, are used to create nanoparticles [53]. Numerous plant extract nanocomposites have been synthesised including AgZnO nanocomposite made from *Trigonella foenum-graecum* leaf extract, Fe_2O_3–Ag TMONCs made from *Psidium guajava* leaf extract, ZnO–NiO nanocomposite made from *Azadirachta indica* (neem) leaves extract, and ZnO/CuO nanocomposite made from cacao seed bark extract and more [54].

Using Microorganisms

Microorganisms are important nanofactories with a lot of potential for being both ecologically and economically beneficial, avoiding harmful, poisonous chemicals, and having a low energy need for physiochemical biosynthesis [55]. Investigations on the microbial synthesis of metal nanoparticles are ongoing. These microbes include mushrooms, yeast, and bacteria. Use of microorganisms is risky due to the pathogenic issue. As a result, it's critical to create nanoparticles utilising environmentally friendly methods [56] (Tables 7.1–7.3).

State of the Art

TABLE 7.1
Synthesis of Metal Oxide Composites by Chemical and Green Methods

Metal Oxide Composite	Synthesis Methods	Application	References
$LiMn_2O_4$	Sol-combustion	Lithium battery	[57]
Y_2O_3–ZrO_2/YSZ	Sol-combustion	SOFC electrolyte	[58]
ZnO-Activated carbon	Co-precipitation	Supercapacitor	[59]
Mullite-SiC	Sol-gel	Coating for carbon materials	[60]
ZnO–CdO	Sol-gel	Gas sensing	[61]
$TiO2$–Fe_2O_3	Sol-gel	Dye decolourisation and supercapacitors	[62]
Al_2O_3–TiO_2	Hydrothermal	Dye-sensitised solar cells	[63]
$CoFe_2O_4$/Graphite	Hydrothermal	Electrochemical supercapacitor	[64]
MnO_2–CNT	Hydrothermal	Supercapacitor	[65]
CdO–ZnO	Microwave	Gas sensing	[66]
CeO_2/SeO_2/ZrO_2	Co-precipitation	Fuel cells	[67]
ZnO–SiO_2	Sol-gel	Optoelectronic devices and as sensors	[68]
Mn_3O_4–Fe_2O_3/Fe_3O_4@rGO	Microwave	Electrochemical performance	[69]
ZnO-reduced graphene oxide	Green method	Supercapacitor and photocatalysis	[70]
Zinc iron oxide ($ZnFe_2O_4$)	Biological method	Electrochemical applications	[71]

TABLE 7.2
Chemical Methods for Various Nanoferrites with Particle Size

Ferrites	Synthesis	Particle Size (nm)	References
$SrFe_{12}O_{19}$	Co-precipitation	98–150	[72]
$NiFe_2O_4$	Co-precipitation	>50	[73]
$Mg_{0.5}Ni_{0.5}Fe_2O_4$	Co-precipitation	290–340	[74]
$CoCr_xFe_{2-x}O_4$	Co-precipitation	15–23	[75]
$NiFe_2O_4$	Co-precipitation	15	[76]
$NiFe_2O_4$	Sol-gel	16.7	[76]
$MnFe_2O_4$	Co-precipitation	72.5	[77]
$MnFe_2O_4$	Sol-gel	113.4	[77]
$MnFe_2O_4$	Co-precipitation	36	[78]

(Continued)

TABLE 7.2 (*Continued*)

Chemical Methods for Various Nanoferrites with Particle Size

Ferrites	Synthesis	Particle Size (nm)	References
$MnFe_2O_4$	Sol-gel	45	[78]
$MnFe_2O_4$	Hydrothermal	16	[78]
$Co_{0.3}Zr_{0.7}Fe_2O_4$	Sol-gel	19	[79]
$ZnFe_2O_4$	Sol-gel	35	[80]
$Ni_{0.8}Zn_{0.2}Fe_2O_4$	Sol-gel	73	[80]
$Co_{0.8}Zn_{0.2}Fe_2O_4$	Sol-gel	74	[80]
$MgFe_2O_4$	Co-precipitation	12.1	[81]
$MgFe_2O_4$	Sol-gel	9.5	[81]
$CoFe_2O_4$	Sol-gel	18	[82]
$CoFe_2O_4$	Co-precipitation	9.2	[83]
$CoFe_2O_4$	Sol-gel	30.5	[84]
$CoFe_2O_4$	Co-precipitation	35.1	[84]
$Co_{1-x}Ca_xFe_2O_4$	Combustion	7.97–37.73	[85]
$Li_{0.5}Fe_{2.5}O_4$	Sol-gel	8–30	[86]
$CoFe_{2-x}Sm_xO_4$	Sol-gel	8–86	[87]
$Co_{1-x}Ni_xFe_2O_4$	Sol-gel	20–25	[88]
$Mn_{1-x}Zn_xFe_2O_4$	Sol-gel	500	[89]
$Zn_{1-x}Co_xFe_2O_4$	Combustion	30–40	[90]
$CdFe_2O_4$	Combustion	12–27	[91]
$MgFe_2O_4$	Combustion	12–25	[92]
$BaNi_2Fe_{16}O_{27}$	Combustion	10–70	[93]
$Mn_{1-x}Zn_xFe_2O_4$	Hydrothermal	15	[94]
$NiFe_2O_4$	Hydrothermal	9	[95]
$Co_{1-x}Zn_xFe_2O_4$	Hydrothermal	51–70	[96]
$Ni_xCo_{1-x}Fe_2O_4$	Microwave	15	[97]
$Zn_{1-x}Cr_xLa_{0.1}Fe_{1.9}O_4$	Sonication	35–51	[98]
$Ni_{0.6}Zn_{0.4}Fe_2O_4$	Microwave	63	[99]
$CoTb_xFe_{2-x}O_4$	Sonication	11–14	[100]
$NiFe_2O_4$	Sonication	9–17	[101]
$Mg_{1-y}Mn_yAg_xFe_{2-x}O_4$	Sol-gel	51–65	[102]
$Mn_{0.5}Zn_{0.5}Fe_2O_4$	Co-precipitation	11.38	[103]
$Co_yZn_yMn_{1-2y}Fe_2O_4$	Hydrothermal	14.68–8.22	[104]
$Ni_{1-x}Co_xFe_2O_4$	Co-precipitation	16–40	[105]
$CoFe_2O_4$	Co-precipitation	15–23	[106]
$Mn_{0.4}Zn_{0.6}In_{0.5}Fe_{1.5}O_4$	Co-precipitation	35–49	[107]
$ZnCr_xFe_{2x}-O_4$	Co-precipitation	22–31	[108]
$Ni_{0.2}Cu_{0.2}Zn_{0.6}Fe_{2-x}Al_xO_4$	Co-precipitation	11–21	[109]
$Mg_{0.9}Mn_{0.1}Fe_2O_4$	Hydrothermal	5.8–2.6	[110]
$Mn_{1-x}Cu_xLa_{0.1}Fe_{1.9}O_4$	Sonication	37–60	[111]
$NiLa_xFe_{2-x}O_4$	Sonication	40–54	[112]
$CoTb_xFe_{2-x}O_4$	Sonication	11–14	[113]
$CuFe_2O_4$	Combustion	14	[114]
$ZnFe_2O_4$	Combustion	8	[114]
$NiFe_2O_4$	Combustion	11	[114]
$MnFe_2O_4$	Sonication	16–24	[115]
$NiFe_2O_4$	Sonication	9–17	[116]
$CuFe_2O_4$	Sonication	35–40	[117]

TABLE 7.3
Green Methods for Nanoferrites Using Different Green Reagents with Particle Size

Ferrites	Green Reagent	Synthesis	Particle Size (nm)	Ref. No.
$CoFe_2O_4$	Flower and leaves extracts of *Hibiscus*	Co-precipitation	23	[118]
$CoFe_2O_4$	Honey	Combustion	24–41	[119]
$CoFe_2O_4$	Aqueous extracts of ginger root	Self-combustion	100	[120]
$CoFe_2O_4$	Aqueous extracts of cardamom seeds	Self-combustion	100	[120]
$CoFe_2O_4$	*Aloe vera*	Sol-gel	50–65	[121]
$CoFe_2O_4$	Aqueous extracts of sesame	Self-combustion	3	[122]
$CoFe_2O_4$	*Crataegus microphylla* fruit extract	Sonication	70–80	[123]
$CoFe_{1.9}Sm_{0.1}O_4$	Lemon juice	Sol-gel auto-combustion	10–22	[124]
$CoCr_xFe_{2-x}O_4$	Honey	Sol-gel auto-combustion	12.2–27.9	[125]
$Ni_{1-x}Zn_xFe_2O_4$	Bovine gelatine	Self-combustion	5	[126]
$NiFe_2O_4$	*Aloe vera*	Hydrothermal	6–28	[127]
$Mn_{0.5}Zn_{0.5}Fe_2O_4$	Squash plant (*Cucurbita pepo* L.)	Microwave	10–12	[128]
$Zn_xCo_{1-x}Fe_2O_4$	Curd	Combustion	~12–21	[129]
$NiFe_2O_4$	*Monsonia burkeana* (M. Burkeana)	Hydrothermal	12–42	[130]
$CuFe_2O_4$	Neem leaves	Solution combustion	250	[131]
$MgFe_2O_4$	Lemon juice	Combustion	10.1–32.7	[132]
$Ni_{0.4}Ag_x$ $Co_{0.6-x}Fe_2O_4$	*Aloe vera*	Combustion	200–300	[133]
$Co_xZn_{(0.90-x)}Al_{0.10}Fe_2O_4$	Lemon juice	Sol-gel	20–26	[134]
$CoFe_2O_4$	Grape peel and grape pulp	Sol-gel	~5–~25	[135]
$MgFe_2O_4$	Albumen	Auto-combustion	24–56	[136]

ANALYSIS OF METAL OXIDE NANOCOMPOSITES USING DIFFERENT TECHNIQUES

Once ferrite nanoparticles and metal oxide composites have been developed, it is essential to char-acterise them in order to determine their performance and intended application. The many charac-teristics of FNPs are examined through various kinds of characterisation approaches. Some of the important and essential techniques are X-ray diffraction (XRD), UV-visible spectroscopy (UV), scanning electron microscopy (SEM), transmission electron microscopy (TEM), and Fourier trans-form infrared spectroscopy (FTIR) (Figure 7.5).

FIGURE 7.5 Synthesis of metal oxide composites by different techniques.

X-Ray Diffraction (XRD)

The X-ray diffraction study examines the crystalline quality and phase analysis of synthesised transition metal oxide-based nanocomposites and it is a crucial and well-liked technique for material characterisation [137]. The XRD results of their sample demonstrated that we had synthesised a graphene oxide/Fe_3O_4 nanocomposite [138]. An XRD data analysis further supported the fabrication success of iron–manganese metal oxide nanocomposite [139]. XRD is a very reliable analytical method used largely to determine the phase of crystalline materials and cell dimensions. Powder XRD allows for the detection of crystallite phases in a material, as well as the assessment of phase composition, grain size, strain, and structural flaws. In this method, the sample is illuminated by an X-ray beam with a wavelength of $\lambda = 1.5406$ Å at an angle. With the rotation of sample and detector, the reflected X-ray's intensity is measured. Using the Debye–Scherrer equation, the crystallite size can be determined (Figure 7.5).

Similar to this, other researchers created nanocomposites based on transition metal oxides and investigated the XRD results for a variety of materials, including ZnO, $NiFe_2O_4$, and ZnO–CuO [140]. TEM and XRD analyses are used to determine the sizes and structures of the nanoparticles in nanocomposites. The fundamental and essential characterisation method for nanomaterials is XRD [141].

UV-Visible Spectroscopy

UV-vis spectroscopy is thought to be the most important and popular spectrophotometric technique for the analysis of a wide variety of substances. The foundation of this technique is the measurement of the electromagnetic radiation's (EMR) interaction with a certain wavelength of matter [142]. A beam that travels directly from the light source to the detector with no contact with the sample is referred as reference beam. As a result of the sample beam's interactions with the sample, the sample is subjected to continuously varying UV light. When the wavelength emitted coincides to the energy level that stimulates an electron to a higher molecular orbital, energy is absorbed. The detector records the intensity ratio of the sample and reference beams. The wavelength is determined by the sample's highest absorption level.

When the light beam passes through the solution, some of it is absorbed and the remaining part is transmitted. Transmittance is defined as the ratio of light entering the sample to light exiting the sample at a particular wavelength. Absorbance is defined as the negative logarithm of transmittance:

$$A = \varepsilon c l$$

where A is the absorbance, ε is the absorptivity, c is the concentration, and l is the path length.

Fabrication of CdO–MgO–Fe_2O_3 also reported UV-visible spectra for the same. We observed a wide absorption band in UV-visible range of 340–500 nm. That nanocomposite showed red shift as compared to individual CdO, MgO, and Fe_2O_3 nanoparticles. He had also calculated band gap which was 1.76 eV for CdO–MgO–Fe_2O_3 nanocomposite [143].

Fourier Transform Infrared Spectroscopy (FTIR)

The FTIR method is easy, rapid, reliable, risk-free, and affordable. In order to better understand various substance chemical processes, vibrational spectroscopy of the Fourier transform infrared (FTIR) is used [144].

The FTIR spectrum reveals 505 cm^{-1} peak for hexagonal ZnO and the 861 cm^{-1} peak for cubic CdO. When we studied CuO–ZnO composites and examined at their FTIR spectra, we saw that all of the 0.4% CuO–ZnO, 2% CuO–ZnO, 10% CuO–ZnO, and 50% CuO–ZnO displayed a band for metal–oxygen (M–O) stretching vibrations in the 400–600 cm^{-1} range. When compared to standard spectrophotometers, Fourier transform spectrophotometers deliver the IR spectrum faster.

The FTIR spectra may be used to qualitatively evaluate the presence of functional groups or any other impurity in the synthesised nanoferrites. Moreover, it identifies the inorganic and polymeric materials by scanning the samples with infrared light. Changes in the distinctive pattern of absorption bands simply indicate a material composition change [145]. Changes in the distinctive pattern of absorption bands simply indicate a material composition change. FTIR may be used to identify and characterise unknown compounds, discover impurities in a substance, determine additives, and observe breakdown and oxidation. Infrared radiation of around $10,000–100\,cm^{-1}$ is passed through the sample, with some of the radiation absorbed and some flowing through. The sample converts absorbed radiation into vibrational or rotational energy. The resulting signal at the detector is a spectrum ranging from $4,000$ to $400\,cm^{-1}$, representing the samples' molecular fingerprint. Each molecule has a distinct fingerprint, which makes FTIR a powerful tool for chemical identification [109–111].

SCANNING ELECTRON MICROSCOPY (SEM)

Scanning electron microscopes are able to observe the sample surface by recognising the extra electrons that the material emits after interacting with the impacting electron beam [146]. The SEM image of the materials demonstrates the production of $CuO/TiO_2/ZnO$ nanocomposites [147]. ZnO/CuO nanocomposite was created and then examined using an SEM to produce a picture that reveals irregularly shaped semi-spheroidal nanoparticles [148]. An essential technology for characterisation of materials at the nanoscale is SEM. The basic tenet of SEM is that energetic electrons interact with condensed materials, mostly solids, to create a range of quantifiable signals [149]. SEM is a type of electron microscope that forms images of samples by scanning the surface with a focused beam of electrons [150]. With high-energy electron beams, SEM examines the sample's surface. In SEM, when the electron beam strikes the specimen surface, it interacts with the surface. When the incident electron beam hits the specimen, it emits X-rays as well as three different types of electrons, namely, backscattered (or primary) electrons, secondary electrons, and Auger electrons.

TRANSMISSION ELECTRON MICROSCOPY (TEM)

A high-resolution determining technique called transmission electron microscopy (TEM) records the movement of an electron beam through a material [151]. We found that all of the CuO–ZnO composites we tested and examined in their FTIR spectra had a band for metal–oxygen (M–O) stretching vibrations in the $400–600\,cm^{-1}$ region [152]. TEM is a type of microscopy in which an image is created by passing an electron beam through a specimen. TEM is used to obtain structural and morphological data on nanoparticles in the micron to nanometer range. TEM pictures provide details regarding the size, shape, arrangement, and defects in a few nanometer areas. When an electron beam passes through a sample, it interacts with sample, and the transmitted electrons are employed to create the image by magnifying and focussing them with an objective lens.

NiO/TiO_2 with 1%, 2%, 5%, and 10% of NiO were synthesised. We examined TEM pictures and discovered that the amount of nickel oxide decreased as it increased. Additionally, the nanocomposite's size range was 4–32 nm. ZnO/CuO nanocomposite synthesised and analysed under TEM for morphology and size study of his synthesised TiO_2-Ag nanocomposite. TEM confirmed agglomeration of spheroidal particles (10 nm) connected with the rod-shaped particles (50 nm) [153].

ENERGY-DISPERSIVE X-RAY ANALYSIS (EDAX)

EDAX is useful for determining the elemental composition of various nanomaterials as well as synthetic nanocomposites based on transition metal oxides. In order to investigate nanoparticles using SEM, the energy-dispersive X-ray analysis technique is applied [154]. Oxygen, silver, and iron were all present in the synthesised Fe_2O_3–Ag nanocomposite, according to the EDX analysis (weight percentages is 51.12, 23.25, and 25.63, respectively). In the synthesised Ag–ZnO nanocomposite, EDX revealed that the weight percentage of Ag was 5.60 and the weight of ZnO was 86.02 [155]. Using EDX, the elemental makeup of the synthesised Au/TiO$_2$ nanocomposite is determined. In Au/TiO$_2$ nanocomposite, EDX revealed the presence of Au (weight percentage is 9.53), Ti (weight percentage is 32.66), and O (weight percentage is 57.81) components [156].

Brunauer–Emmett–Teller (BET) Surface Area

The BET is useful for surface area observations and pore properties analysis. The study focuses on the ZnO/CuO nanocomposite. We observed images with mesopores and calculated specific surface area for ZnO/CuO (5%) and ZnO/CuO (10%) were 36.0 and 26.0 m^2/g, respectively. Also we calculated pore volumes for ZnO/CuO (5%) and ZnO/CuO (10%) were 0.0939 and 0.0984 cm^3/g, respectively [157].

ENERGY APPLICATIONS FOR BATTERY, SUPERCAPACITOR AND SENSOR

Ferrite nanoparticles and metal oxide composites are broadly used in each and every field due to their magnetic, electrical, optical and chemical properties. Their applications range from medical to modern industries. They are used in biomedical, wastewater treatment, catalysis, and information technology [26].

Energy storage devices like batteries and supercapacitors, transformer cores, sensors, microwave applications, and medicinal applications are utilized for various purposes [112,113]. These rechargeable energy storage devices such as batteries and supercapacitors are attractive for applications in portable electronics and electric vehicles. Such innovations will aid in supplying the rising need for ecologically friendly energy sources [114]. Among these energy storage devices, supercapacitor technology has attracted tremendous attention to be used in high-power application because of their higher power density and longer cycling life [115]. And also, nanoferrites are useful in diverse sensor applications in various fields. Researchers have developed nanoferrite materials for various sensor applications using various techniques [116,117].

Battery

Rechargeable electrochemical batteries have attracted the most popularity among energy storage technologies because of their high energy and power densities, low price, environmental friendliness, and long-term stability. Batteries are devices that transform electrical energy into chemical energy and store it for later release gradually. Because this chemical reaction in rechargeable batteries is reversible, the battery may be used repeatedly. Many technological fields, including transportation, portable gadgets, medical equipment, power tools, and the storage of electricity generated by intermittent renewable sources, depend on batteries [158]. Half cells, where alkali metal functions as both the reference and counterelectrode, are frequently set up for electrochemical measurements in the study of rechargeable alkali (Li, Na, K) ion batteries because it is thought that the theoretical value of alkali ion/alkali metal does not change with the passage of current [159]. Batteries involve complex situations like solid–liquid interface, ion diffusion, and multiple electrode reactions. Researchers typically use specific values from cyclic voltammetry (CV) curves to understand electrode reactions. Nernst equations are still used for qualitative and half-quantitative analyses of real batteries [160]. Battery research involves measuring impedance under low sinusoidal voltage over a wide frequency range, understanding electrochemical cell characteristics in an electrical circuit and the subsequent processes in electrode reactions [161]. Rechargeable electrochemical batteries gain popularity for high energy density, low price, environmental friendliness, and long-term stability [162]. It is generally known that the majority of the battery performance is controlled by the electrode materials, whose physical/chemical characteristics have a significant impact on the final electrochemical properties [115,116].

SUPERCAPACITORS

Supercapacitors enhance power density and energy density, connecting dielectric capacitors and batteries. Carbon-based nanoparticles, metal oxide nanoparticles, and metal oxide composites are attractive due to their specific capacitance and longer cycle life. The supercapacitors, also known as electrochemical capacitors (ECCs) and ultracapacitors (UCs), are the most desirable options among the ECSDs because they are affordable, have a high power density, have a long cycle life, need little charging time, and have a reasonable gravimetric/volumetric energy density. Metal oxide composites have been used for ultracapacitor and supercapacitor applications by a number of researchers. For instance, Guangyu et al. reported Co_3O_4/graphene nanocomposite with applications in supercapacitor devices. He discovered that the Co_3O_4/graphene nanocomposite had a maximum specific capacitance of 430 F/g at 1 A/g and 215 F/g at 0.4 A/g. Two-electrode systems provide less specific capacitance than three-electrode systems. He was successful in getting Co_3O_4/GO to have high cycle stability in both three-electrode and two-electrode systems [163]. For application in supercapacitors, D.V. Leontyeva et al. synthesised a carbon-supported NiO (NiO/C) nanocomposite and investigated its electrochemical behaviour. With an increase in scan rate from 5 to 40 mV/s, they computed specific capacitance from the CV curve to range from 1,100 to 777 F/g. Additionally, the galvanostatic charge–discharge curve was used to determine 970 F/g at 0.5 A/g mass-normalised current. He carried out a thousand cycles and discovered that (with 840 F/g) the cycle life was great [164]. Numerous other researchers have also looked into the use of transition metal oxide–based nanocomposites for supercapacitor applications. For example, Sivalingam Ramesh et al. reported on the use of $MWCNT/GO/NiCo_2O_4$ nanocomposites in supercapacitors [165], and Ian Y.Y. Bu et al. studied the use of ZnO/reduced graphene oxide in supercapacitors [166].

SENSORS

In the vast area of sensors, selectivity and sensitivity are crucial and difficult to achieve. Doping, surface modification, and other techniques have all been utilised to enhance sensor performance. Because of their increased surface area, surface activity, the addition of meso-porosity, etc., TMONC sensors have been shown to have greater selectivity and sensitivity than isolated transition metal oxide nanoparticles [167]. Nanoferrites are now attracting and valuable materials for sensors in many different kinds of industries and also in the wide range of sectors today [125]. Sensors based on ferrite NPs are very sensitive, with low detection limits and a high signal-to-noise ratio [126]. The use of new sensors is also based on the optical absorption of nanoferrites [127]. One of the most common applications for sensors is to detect variations in humidity [128]. Different gas sensor technologies, such as semiconductor gas sensors, catalytic gas sensors, electrochemical gas sensors, optical gas sensors, and acoustic gas sensors, have been employed for gas detection for many centuries. Biomolecules can be accurately detected and identified in experiments through the use of nanosensors such as giant magnetoresistance GMR sensor. PPy/ZnO and PPy/SnO₂ nanocomposite sensors for ammonia detection were developed by Mehrnaz Joulazadeh et al. According to them, PPy/ZnO nanocomposite made of transition metal oxides responded to ammonia the most 34% compared to PPy/SnO_2 nanocomposite (25%) and only PPy nanocomposite (15%). He proposed that the presence of Zn^{2+} cations in PPy/ZnO may have increased its conductivity, explaining why it responded more strongly than PPy/SnO_2 [168]. In reaction to formaldehyde and ethanol, Meenakshi Dutt et al. employed synthesised Li_2O-doped Fe_2O_3 and SnO_2–Fe_2O_3 nanocomposites as sensors in the working temperature range of 25–200°C. Li_2O-doped Fe_2O_3 sensor has a reported response of 6.08 and 5.82 for HCHO and C_2H^2OH, respectively, at 25°C, while SnO_2–Fe_2O_3 has a response of 6.44 and 4.93 for HCHO and C_2H_2OH, respectively, at 150°C [169]. As a case study, Bose Dinesh et al. constructed rGO–Co_3O_4 for efficient serotonin sensing in the presence of dopamine and the antioxidant ascorbic acid [170], and R. Sivasubramanian et al. synthesised copper (I) oxide reduced oxide of graphene nanocomposite material for dopamine sensing [171]. A number of other investigators also examined the sensing the ability of their synthesised transition metal oxide–based nanocomposites. Every sensor's performance is dependent on a number of factors, including its sensitivity, selectivity, detection limit, reaction time, and recovery time [123,129]. Lithium ferrites ($Li_{0.5}Fe_{2.5}O_4$) are excellent materials for sensing applications [130]. In the areas of engineering, medicine, and ecology, electrochemical sensor–based biosensors have discovered a number of fascinating uses [131] (Table 7.4).

In Table 7.4 P. Ramesh Kumar et al. successfully synthesised the nanocrystalline nickel ferrite particles using a sol-gel combustion technique. As an outstanding high rate (for example, 20 C rate) electrode, a nanosized $NiFe_2O_4$-based conversion anode with an alginate binder is used. More than 98% of the initial charge storage capacity of a nickel ferrite electrode is recovered across a significant number of cycles. Galvanostatic studies on such electrodes have shown that $NiFe_2O_4$ nanoparticles can produce 740 mAh/g capacity at a current rate of 1 C while retaining a high capacity. The remarkable electrochemical performance, stability, and coulombic efficiency will be possible to develop a ternary metal oxide–based electrode as an anode for next-generation lithium-ion batteries [172].

H. Kennaz et al. synthesised $CoFe_2O_4$ nanoparticles using a hydrothermal method. By measuring CV in a 6 M KOH electrolyte, the electrochemical performances were evaluated utilising a three-electrode configuration. On a Ni foam current collector, CV measurements have been carried out using three different loading weights of the produced $CoFe_2O_4$ as an active material. These ferrite nanoparticles exhibit high specific capacitance values, with 315 F/g at a scan rate of 5 mV/s, which make it a good candidate for supercapacitor application [182].

For the first time, Chu Xiangfeng et al. synthesised $NiFe_2O_4$ nanorods and nanocubes hydrothermally without the use of a surfactant and investigated their gas-sensing properties. At 175°C, the

TABLE 7.4
Energy Applications of Nanoferrite

Name of the Ferrite	Method of Synthesis	Applications	Specific Capacitance (F/g)	Ref. No.
$NiFe_2O_4$	Co-precipitation	In lithium nickel iron oxide cathodes for lithium-ion microbatteries	–	[34]
$Li_{0.5}Fe_{2.5}O_4$	Sol-gel	For cathode in rechargeable lithium batteries	–	[43]
$NiFe_2O_4$	Sol-gel	Anode for Li-ion battery	–	[172]
$MgFe_2O_4$	Citrate combustion	Anode for lithium battery	–	[173]
$Li_{1.1}Co_{0.3}Fe_{2.1}O_4$	Citrate auto-combustion	Anode for lithium battery	–	[174]
$NiFe_2O_4$	Combustion	Supercapacitors	454	[175]
$Co_xZn_{0.04-x}Fe_2O_4$	Co-precipitation	Supercapacitors	377	[176]
$NiFe_2O_4$	Hydrothermal	Supercapacitors	~120	[177]
$Cd_xCo_{1-x}Fe_2O_4$	Co-precipitation	Supercapacitors	395	[178]
$Cu_{1-x}Ni_xMn_{1.0}Fe_{1.0}O_4$	Sol-gel	Supercapacitors	975	[179]
$CoFe_2O_4/SiO_2$	Sol-gel	Supercapacitors	316.14	[180]
Zn-doped $MgFe_2O_4$	Sol-gel	Supercapacitors	484.6	[181]
$CoFe_2O_4$	Hydrothermal	Supercapacitors	315	[182]
$Mg_{0.5}Ni_{0.5}Fe_2O_4$	Co-precipitation	Sensors	–	[31]
$CoCr_xFe_{2-x}O_4$	Co-precipitation	Gas sensors	–	[32]
$Co_{1-x}Ni_xFe_2O_4$	Sol-gel	Data storage, sensors	–	[45]
$Mn_{1-x}Zn_xFe_2O_4$	Sol-gel	Gas sensors	–	[46]
$CdFe_2O_4$	Combustion	Gas and electrochemical sensing	–	[48]
$MgFe_2O_4$	Combustion	Sensors	–	[49]
$NiFe_2O_4$	Hydrothermal	Gas sensors	–	[183]

$NiFe_2O_4$ nanorod–based sensor was sensitive and selective to low concentrations of triethylamine, and it was particularly capable of detecting 1 ppm triethylamine. The $NiFe_2O_4$ nanocube-based sensor observed the opposite behaviour of the anomalous conductivity rise in a reducing gas atmosphere, with a response of 0.033 at 500 ppm triethylamine [183].

POSSIBLE OUTCOMES

Many practical approaches, including the sol-gel method, hydrothermal techniques, combustion method, microwave method, sonication method, co-precipitation method, and green methods, have been studied for the synthesis of various transition metal oxide–based composites. Here, several synthetic methods have been used to synthesise and modify transition metal oxide–based nanocomposites. For structural and morphological research, we have also evaluated several issued characterisation methods of synthesised metal oxide–based nanocomposites, such as XRD, EDAX, SEM, TEM, and FTIR. We have examined a special use for oxides of metal nanocomposites based on their various topologies. For these transition metal oxide–based composites to be used in practical applications, morphologies and surface characteristics are crucial. Transition metal oxide–based composites offer a wide range of reported possible real-world uses, including sensors, photocatalytic degradation of harmful dyes, medicinal applications (including drug transport, antimicrobials, biosensors, and anticancer), energy generation through solar cells and H_2 synthesis, and energy storage from supercapacitors.

REFERENCES

[1] Okpala, C. C. (2013). Nanocomposites-an overview. *International Journal of Engineering Research and Development*, 8(11), 17–23.
[2] Rahimi, M. R., & Mosleh, S. (2021). *Intensification of Sorption Processes: Active and Passive Mechanisms*. Elsevier, 219–221.
[3] Al-Mutairi, N. H., Mehdi, A. H., & Kadhim, B. J. (2022). Nanocomposites materials definitions, types and some of their applications: A review. *European Journal of Research Development and Sustainability*, 3(2), 102–108.
[4] Delekar, S. D. (Ed.). (2022). *Advances in Metal Oxides and Their Composites for Emerging Applications*. Elsevier, 57–96.
[5] Eray, S. (2020). Application of metal oxides in composites. In *Metal Oxide Powder Technologies*, Yarub Al-Douri (Ed.) (pp. 101–119). Elsevier.
[6] Raccichini, R., Varzi, A., Passerini, S., & Scrosati, B. (2015). The role of graphene for electrochemical energy storage. *Nature Materials*, 14(3), 271–279.
[7] Mohapatra, M., & Anand, S. (2010). Synthesis and applications of nano-structured iron oxides/hydroxides-a review. *International Journal of Engineering, Science and Technology*, 2(8), 127–146.
[8] Masunga, N., Mmelesi, O. K., Kefeni, K. K., & Mamba, B. B. (2019). Recent advances in copper ferrite nanoparticles and nanocomposites synthesis, magnetic properties and application in water treatment. *Journal of Environmental Chemical Engineering*, 7(3), 103179.
[9] Malaie, K., & Ganjali, M. R. (2021). Spinel nano-ferrites for aqueous supercapacitors; linking abundant resources and low-cost processes for sustainable energy storage. *Journal of Energy Storage*, 33, 102097.
[10] Koo, B., Xiong, H., Slater, M. D., Prakapenka, V. B., Balasubramanian, M., Podsiadlo, P., … Shevchenko, E. V. (2012). Hollow iron oxide nanoparticles for application in lithium ion batteries. *Nano Letters*, 12(5), 2429–2435.
[11] Trinh, N. D., Crosnier, O., Schougaard, S. B., & Brousse, T. (2011). Synthesis, characterization and electrochemical studies of active materials for sodium ion batteries. *ECS Transactions*, 35(32), 91.
[12] Palacin, M. R. (2009). Recent advances in rechargeable battery materials: A chemist's perspective. *Chemical Society Reviews*, 38(9), 2565–2575.
[13] Chen, Y., Kang, Y., Zhao, Y., Wang, L., Liu, J., Li, Y., … Li, B. (2021). A review of lithium-ion battery safety concerns: The issues, strategies, and testing standards. *Journal of Energy Chemistry*, 59, 83–99.
[14] Wang, H., Wang, L., Lin, J., Yang, J., Wu, F., Li, L., & Chen, R. (2021). Structural and electrochemical characteristics of hierarchical $Li_4Ti_5O_{12}$ as high-rate anode material for lithium-ion batteries. *Electrochimica Acta*, 368, 137470.

[15] Wang, F., Wang, B., Li, J., Wang, B., Zhou, Y., Wang, D., ... Dou, S. (2021). Prelithiation: A crucial strategy for boosting the practical application of next-generation lithium ion battery. *ACS Nano*, 15(2), 2197–2218.

[16] Zhang, Z., Luo, D., Li, G., Gao, R., Li, M., Li, S., ... Chen, Z. (2020). Tantalum-based electrocatalyst for polysulfide catalysis and retention for high-performance lithium-sulfur batteries. *Matter*, 3(3), 920–934.

[17] Yin, Y. X., Xin, S., Guo, Y. G., & Wan, L. J. (2013). Lithium-sulfur batteries: Electrochemistry, materials, and prospects. *Angewandte Chemie International Edition*, 52(50), 13186–13200.

[18] Cheong, K. Y., Impellizzeri, G., & Fraga, M. A. (Eds.). (2018). *Emerging Materials for Energy Conversion and Storage*. Elsevier.

[19] Zhou, L., Utetiwabo, W., Chen, R., & Yang, W. (2019). Layer by layer assemble of colloid nanomaterial and functional multilayer films for energy storage and conversion. In *Comprehensive Nanoscience and Nanotechnology*, David L. Andrews, Robert H. Lipson and Thomas Nann (Eds.) (vol. 2, pp. 255–278). Elsevier.

[20] Wang, X., Li, G., Li, J., Zhang, Y., Wook, A., Yu, A., & Chen, Z. (2016). Structural and chemical synergistic encapsulation of polysulfides enables ultralong-life lithium-sulfur batteries. *Energy & Environmental Science*, 9(8), 2533–2538.

[21] He, Y., Li, A., Dong, C., Li, C., & Xu, L. (2017). Mesoporous tin-based oxide nanospheres/reduced graphene composites as advanced anodes for lithium-ion half/full cells and sodium-ion batteries. *Chemistry-A European Journal*, 23(55), 13724–13733.

[22] Ren, H., Shao, H., Zhang, L., Guo, D., Jin, Q., Yu, R., ... Wang, D. (2015). A new graphdiyne nanosheet/Pt nanoparticle-based counter electrode material with enhanced catalytic activity for dye-sensitized solar cells. *Advanced Energy Materials*, 5(12), 1500296.

[23] Ranganathan, P., Sasikumar, R., Chen, S. M., Rwei, S. P., & Sireesha, P. (2017). Enhanced photovoltaic performance of dye-sensitized solar cells based on nickel oxide supported on nitrogen-doped graphene nanocomposite as a photoanode. *Journal of Colloid and Interface Science*, 504, 570–578.

[24] Choi, S. H., Ko, Y. N., Lee, J. K., & Kang, Y. C. (2014). Rapid continuous synthesis of spherical reduced graphene ball-nickel oxide composite for lithium ion batteries. *Scientific Reports*, 4(1), 5786.

[25] Wang, Y., Diaz, D. F. R., Chen, K. S., Wang, Z., & Adroher, X. C. (2020). Materials, technological status, and fundamentals of PEM fuel cells-a review. *Materials Today*, 32, 178–203.

[26] Yadav, S., Rani, N., & Saini, K. (2022, February). A review on transition metal oxides based nanocomposites, their synthesis techniques, different morphologies and potential applications. In *IOP Conference Series: Materials Science and Engineering* (Vol. 1225, No. 1, p. 012004). IOP Publishing.

[27] Şahİn, M. E., Blaabjerg, F., & Sangwongwanİch, A. (2021). Modelling of supercapacitors based on simplified equivalent circuit. *CPSS Transactions on Power Electronics and Applications*, 6(1), 31–39.

[28] Szunerits, S., & Boukherroub, R. (2018). Graphene-based biosensors. *Interface Focus*, 8(3), 20160132.

[29] Liu, R., Ye, X., & Cui, T. (2020). Recent progress of biomarker detection sensors. *Research* (Vol. 2020).

[30] Labib, M., Sargent, E. H., & Kelley, S. O. (2016). Electrochemical methods for the analysis of clinically relevant biomolecules. *Chemical Reviews*, 116(16), 9001–9090.

[31] Majhi, S. M., Mirzaei, A., Kim, H. W., Kim, S. S., & Kim, T. W. (2021). Recent advances in energy-saving chemiresistive gas sensors: A review. *Nano Energy*, 79, 105369.

[32] Hou, Y. L., Chen, J. Z., Qin, T., Guan, H. B., Wang, S. G., Zeng, R., & Zhao, D. L. (2023). Constructing hierarchical SnS_2 hollow micron cages anchored on S-doped graphene as anodes for superior performance alkali-ion batteries. *Electrochimica Acta*, 439, 141590.

[33] Eddaif, L., & Shaban, A. (2023). Fundamentals of sensor technology. In *Advanced Sensor Technology*, Ahmed Barhoum and Zeynep Altintas (Eds.) (pp. 17–49). Elsevier.

[34] Hasan, S. (2015). A review on nanoparticles: Their synthesis and types. *Research Journal of Recent Sciences*, 2277, 2502.

[35] Carlos, E., Martins, R., Fortunato, E., & Branquinho, R. (2020). Solution combustion synthesis: Towards a sustainable approach for metal oxides. *Chemistry-A European Journal*, 26(42), 9099–9125.

[36] Parauha, Y. R., Sahu, V., & Dhoble, S. J. (2021). Prospective of combustion method for preparation of nanomaterials: A challenge. *Materials Science and Engineering: B*, 267, 115054.

[37] Yadav, S., Rani, N., & Saini, K. (2022, February). A review on transition metal oxides based nanocomposites, their synthesis techniques, different morphologies and potential applications. In *IOP Conference Series: Materials Science and Engineering* (Vol. 1225, No. 1, p. 012004).

[38] Owens, G. J., Singh, R. K., Foroutan, F., Alqaysi, M., Han, C. M., Mahapatra, C., & Knowles, J. C. (2016). Sol-gel based materials for biomedical applications. *Progress in Materials Science*, 77, 1–79.

[39] Hamrouni, A., Moussa, N., Parrino, F., Di Paola, A., Houas, A., & Palmisano, L. (2014). Sol-gel synthesis and photocatalytic activity of ZnO-SnO$_2$ nanocomposites. *Journal of Molecular Catalysis A: Chemical*, 390, 133–141.

[40] Chung, D. D. L. (2017). Carbon fibers, nanofibers, and nanotubes. *Carbon Composites*, 2, 12–47.

[41] Altammar, K. A. (2023). A review on nanoparticles: Characteristics, synthesis, applications, and challenges. *Frontiers in Microbiology*, 14, 1155622.

[42] Nam, N. H., & Luong, N. H. (2019). Nanoparticles: Synthesis and applications. In *Materials for Biomedical Engineering,* Valentina Grumezescu and Alexandru Mihai Grumezescu (Eds.) (pp. 211–240). Elsevier.

[43] Kumar, A., Kuang, Y., Liang, Z., & Sun, X. (2020). Microwave chemistry, recent advancements, and eco-friendly microwave-assisted synthesis of nanoarchitectures and their applications: A review. *Materials Today Nano*, 11, 100076.

[44] Sedighi, F., Esmaeili-Zare, M., Sobhani-Nasab, A., & Behpour, M. (2018). Synthesis and characterization of CuWO4 nanoparticle and CuWO$_4$/NiO nanocomposite using co-precipitation method; application in photodegradation of organic dye in water. *Journal of Materials Science: Materials in Electronics*, 29, 13737–13745.

[45] Dahiya, M. S., Tomer, V. K., & Duhan, S. (2018). Metal-ferrite nanocomposites for targeted drug delivery. In *Applications of Nanocomposite Materials in Drug Delivery*, Inamuddin, Abdullah M. Asiri and Ali Mohammad (Eds.) (pp. 737–760). Woodhead publishing.

[46] Li, L., Zhang, X., Zhang, W., Wang, L., Chen, X., & Gao, Y. (2014). Microwave-assisted synthesis of nanocomposite Ag/ZnO-TiO$_2$ and photocatalytic degradation Rhodamine B with different modes. *Colloids and Surfaces A: Physicochemical and Engineering Aspects*, 457, 134–141.

[47] Kumar, R., Youssry, S. M., Ya, K. Z., Tan, W. K., Kawamura, G., & Matsuda, A. (2020). Microwave-assisted synthesis of Mn$_3$O$_4$-Fe$_2$O$_3$/Fe$_3$O$_4$@ rGO ternary hybrids and electrochemical performance for supercapacitor electrode. *Diamond and Related Materials*, 101, 107622.

[48] da Silva, A. K., Ricci, T. G., de Toffoli, A. L., Maciel, E. V. S., Nazario, C. E. D., & Lanças, F. M. (2020). The role of magnetic nanomaterials in miniaturized sample preparation techniques. In *Handbook on Miniaturization in Analytical Chemistry*, Chaudhery Mustansar Hussain (Ed.) (pp. 77–98). Elsevier.

[49] Zhang, G. Y., Sun, Y. Q., Gao, D. Z., & Xu, Y. Y. (2010). Quasi-cube ZnFe$_2$O$_4$ nanocrystals: Hydrothermal synthesis and photocatalytic activity with TiO$_2$ (Degussa P25) as nanocomposite. *Materials Research Bulletin*, 45(7), 755–760.

[50] Arsiya, F., Sayadi, M. H., & Sobhani, S. (2017). Green synthesis of palladium nanoparticles using Chlorella vulgaris. *Materials Letters*, 186, 113–115.

[51] Jadoun, S., Arif, R., Jangid, N. K., & Meena, R. K. (2021). Green synthesis of nanoparticles using plant extracts: A review. *Environmental Chemistry Letters*, 19, 355–374.

[52] Devi, H. S., Boda, M. A., Shah, M. A., Parveen, S., & Wani, A. H. (2019). Green synthesis of iron oxide nanoparticles using Platanus orientalis leaf extract for antifungal activity. *Green Processing and Synthesis*, 8(1), 38–45.

[53] Singh, P., Kim, Y. J., Zhang, D., & Yang, D. C. (2016). Biological synthesis of nanoparticles from plants and microorganisms. *Trends in Biotechnology*, 34(7), 588–599.

[54] Elemike, E. E., Onwudiwe, D. C., & Singh, M. (2020). Eco-friendly synthesis of copper oxide, zinc oxide and copper oxide-zinc oxide nanocomposites, and their anticancer applications. *Journal of Inorganic and Organometallic Polymers and Materials*, 30, 400–409.

[55] Yadav, S., Rani, N., & Saini, K. (2022, February). A review on transition metal oxides based nanocomposites, their synthesis techniques, different morphologies and potential applications. In *IOP Conference Series: Materials Science and Engineering* (Vol. 1225, No. 1, p. 012004) IOP Publishing.

[56] Vithiya, K., & Sen, S. (2011). Biosynthesis of nanoparticles. *International Journal of Pharmaceutical Sciences and Research*, 2(11), 2781.

[57] Park, H. B., Kim, J., & Lee, C. W. (2001). Synthesis of LiMn$_2$O$_4$ powder by auto-ignited combustion of poly (acrylic acid)-metal nitrate precursor. *Journal of Power Sources*, 92(1–2), 124–130.

[58] Patil, K. C., Aruna, S. T., & Mimani, T. (2002). Combustion synthesis: an update. *Current Opinion in Solid State and Materials Science*, 6(6), 507–512.

[59] Ghaedi, M., Ghayedi, M., Kokhdan, S. N., Sahraei, R., & Daneshfar, A. (2013). Palladium, silver, and zinc oxide nanoparticles loaded on activated carbon as adsorbent for removal of bromophenol red from aqueous solution. *Journal of Industrial and Engineering Chemistry*, 19(4), 1209–1217.

[60] Warrier, K. G. K., & Anilkumar, G. M. (2001). Densification of mullite-SiC nanocomposite sol-gel precursors by pressureless sintering. *Materials Chemistry and Physics*, 67(1–3), 263–266.

[61] Mosquera, E., del Pozo, I., & Morel, M. (2013). Structure and red shift of optical band gap in CdO-ZnO nanocomposite synthesized by the sol gel method. *Journal of Solid State Chemistry*, 206, 265–271.

[62] Bulin, C., Li, B., Zhang, Y., & Zhang, B. (2020). Removal performance and mechanism of nano α-Fe_2O_3/ graphene oxide on aqueous Cr (VI). *Journal of Physics and Chemistry of Solids*, 147, 109659.

[63] Zou, J. F., Chen, H. F., & Yan, T. (2012). Synthesis and adsorption properties of nano Al_2O_3/TiO_2 by hydrothermal method. In *Advanced Materials Research* (Vol. 580, pp. 509–512) Trans Tech Publications.

[64] Gan, L., Shang, S., Yuen, C. W. M., Jiang, S. X., & Hu, E. (2015). Hydrothermal synthesis of magnetic $CoFe_2O_4$/graphene nanocomposites with improved photocatalytic activity. *Applied Surface Science*, 351, 140–147.

[65] Teng, F., Santhanagopalan, S., Wang, Y., & Meng, D. D. (2010). In-situ hydrothermal synthesis of three-dimensional MnO_2-CNT nanocomposites and their electrochemical properties. *Journal of Alloys and Compounds*, 499(2), 259–264.

[66] Karthik, K., Dhanuskodi, S., Gobinath, C., & Sivaramakrishnan, S. (2015). Microwave-assisted synthesis of CdO-ZnO nanocomposite and its antibacterial activity against human pathogens. *Spectrochimica Acta Part A: Molecular and Biomolecular Spectroscopy*, 139, 7–12.

[67] Liu, M., He, C., Wang, J., Wang, W. G., & Wang, Z. (2010). Investigation of $(CeO_2)x(Sc_2O_3)(0.1_{1-x})(ZrO_2)$ $0.89_{(x=0.01-0.10)}$ electrolyte materials for intermediate-temperature solid oxide fuel cell. *Journal of Alloys and Compounds*, 502(2), 319–323.

[68] Grigorie, A. C., Muntean, C., Vlase, T., Locovei, C., & Stefanescu, M. (2017). ZnO-SiO_2 based nanocomposites prepared by a modified sol-gel method. *Materials Chemistry and Physics*, 186, 399–406.

[69] Kumar, R., Youssry, S. M., Ya, K. Z., Tan, W. K., Kawamura, G., & Matsuda, A. (2020). Microwave-assisted synthesis of Mn_3O_4–Fe_2O_3/Fe_3O_4@ rGO ternary hybrids and electrochemical performance for supercapacitor electrode. *Diamond and Related Materials*, 101, 107622.

[70] Zhang, L., Du, G., Zhou, B., & Wang, L. (2014). Green synthesis of flower-like ZnO decorated reduced graphene oxide composites. *Ceramics International*, 40(1), 1241–1244.

[71] Matinise, N., Kaviyarasu, K., Mongwaketsi, N., Khamlich, S., Kotsedi, L., Mayedwa, N., & Maaza, M. (2018). Green synthesis of novel zinc iron oxide ($ZnFe_2O_4$) nanocomposite via Moringa Oleifera natural extract for electrochemical applications. *Applied Surface Science*, 446, 66–73.

[72] Hessien, M. M., Rashad, M. M., & El-Barawy, K. (2008). Controlling the composition and magnetic properties of strontium hexaferrite synthesized by co-precipitation method. *Journal of Magnetism and Magnetic Materials*, 320(3–4), 336–343.

[73] Dehghan, R, Ebrahimi, S. S., & Badiei, A. (2008). Investigation of the effective parameters on the synthesis of Ni-ferrite nanocrystalline powders by coprecipitation method. *Journal of Non-Crystalline Solids*, 354(47–51), 5186–5188.

[74] Hankare, P. P., Jadhav, S. D., Sankpal, U. B., Chavan, S. S., Waghmare, K. J., & Chougule, B. K. (2009). Synthesis, characterization and effect of sintering temperature on magnetic properties of MgNi ferrite prepared by co-precipitation method, *Journal of Alloys and Compounds*, 475(1–2), 926–929.

[75] Vadivel, M., Babu, R. R., Sethuraman, K., Ramamurthi, K., & Arivanandhan, M. Synthesis, structural, dielectric, magnetic and optical properties of Cr substituted $CoFe_2O_4$ nanoparticles by co-precipitation method, *Journal of Magnetism and Magnetic Materials*, 2014, 362, 122–129.

[76] Jacob, B. P., Kumar, A., Pant, R. P., Singh, S., & Mohammed, E. M. (2011). Influence of preparation method on structural and magnetic properties of nickel ferrite nanoparticles. *Bulletin of Materials Science*, 34, 1345–1350.

[77] Ahmed, M. A., Okasha, N., & El-Dek, S. I. (2008). Preparation and characterization of nanometric Mn ferrite via different methods. *Nanotechnology*, 19(6), 065603.

[78] Kafshgari, L. A., Ghorbani, M., & Azizi, A. (2018). Synthesis and characterization of manganese ferrite nanostructure by co-precipitation, sol-gel, and hydrothermal methods. *Particulate Science and Technology*, 37(7), 904–910.

[79] Anand, S., Amaliya, A. P., Janifer, M. A., & Pauline, S. (2017). Structural, morphological and dielectric studies of zirconium substituted CoFe2O4 nanoparticles. *Modern Electronic Materials*, 3(4), 168–173.

[80] Zālīte, I., Heidemane, G., Krūmiņa, A., Rašmane, D., & Maiorov, M. (2017). ZnFe2O4 containing nanoparticles: Synthesis and magnetic properties. *Material Science & Applied Chemistry*, 34(1), 38.

[81] Thankachan, S., Xavier, S., Jacob, B., & Mohammed, E. M. (2013). A comparative study of structural, electrical and magnetic properties of magnesium ferrite nanoparticles synthesised by sol-gel and co-precipitation techniques. *Journal of experimental Nanoscience*, 8(3), 347–357.

[82] Avazpour, L., Toroghinejad, M. R., & Shokrollahi, H. (2015). Synthesis of single-phase cobalt ferrite nanoparticles via a novel EDTA/EG precursor-based route and their magnetic properties. *Journal of Alloys and Compounds*, 637, 497–503.

[83] Kim, Y. I., Kim, D., & Lee, C. S. (2003). Synthesis and characterization of $CoFe_2O_4$ magnetic nanoparticles prepared by temperature-controlled coprecipitation method. *Physica B: Condensed Matter*, 337(1–4), 42–51.

[84] Kurian, M., Thankachan, S., Nair, D. S., EK, A., Babu, A., Thomas, A., & Krishna KT, B. (2015). Structural, magnetic, and acidic properties of cobalt ferrite nanoparticles synthesised by wet chemical methods. *Journal of Advanced Ceramics*, 4, 199–205.

[85] Ateia, E. E., & Farag, M. (2019). Synthesis of cobalt/calcium nanoferrites with controllable physical properties. *Applied Physics A*, 125(5), 324.

[86] Verma, S., & Joy, P. A. (2008). Low temperature synthesis of nanocrystalline lithium ferrite by a modified citrate gel precursor method. *Materials Research Bulletin*, 43(12), 3447–3456.

[87] Rashad, M. M., Mohamed, R. M., & El-Shall, H. (2008). Magnetic properties of nanocrystalline Sm-substituted $CoFe_2O_4$ synthesized by citrate precursor method. *Journal of Materials Processing Technology*, 198(1–3), 139–146.

[88] Hankare, P. P., Sanadi, K. R., Garadkar, K. M., Patil, D. R., & Mulla, I. S. (2013). Synthesis and characterization of nickel substituted cobalt ferrite nanoparticles by sol-gel auto-combustion method. *Journal of Alloys and Compounds*, 553, 383–388.

[89] Martins, M. L., Florentino, A. O., Cavalheiro, A. A., Silva, R. I., Dos Santos, D. I., & Saeki, M. J. (2014). Mechanisms of phase formation along the synthesis of Mn-Zn ferrites by the polymeric precursor method. *Ceramics International*, 40(10), 16023–16031.

[90] Manikandan, A., Kennedy, L. J., Bououdina, M., & Vijaya, J. J. (2014). Synthesis, optical and magnetic properties of pure and Co-doped $ZnFe_2O_4$ nanoparticles by microwave combustion method. *Journal of Magnetism and Magnetic Materials*, 349, 249–258.

[91] Kaur, H., Singh, J., & Randhawa, B. S. (2014). Essence of superparamagnetism in cadmium ferrite induced by various organic fuels via novel solution combustion method. *Ceramics International*, 40(8), 12235–12243.

[92] Kaur, N., & Kaur, M. (2014). Comparative studies on impact of synthesis methods on structural and magnetic properties of magnesium ferrite nanoparticles. *Processing and Application of Ceramics*, 8(3), 137–143.

[93] Sharma, R., Agarwala, R. C., & Agarwala, V. (2007). A study on the heat-treatments of nanocrystalline nickel substituted BaW hexaferrite produced by low combustion synthesis method. *Journal of Magnetism and Magnetic Materials*, 312(1), 117–125.

[94] Zhuang, L., Zhang, W., Zhao, Y., Li, D., Wu, W., & Shen, H. (2012). Temperature sensitive ferrofluid composed of $Mn_{1-x}Zn_xFe_2O_4$ nanoparticles prepared by a modified hydrothermal process. *Powder Technology*, 217, 46–49.

[95] Babu Naidu, K. C., & Madhuri, W. (2017). Hydrothermal synthesis of $NiFe_2O_4$ nano-particles: Structural, morphological, optical, electrical and magnetic properties. *Bulletin of Materials Science*, 40, 417–425.

[96] Nakamura, T., & Hatakeyama, K. I. (2000). Complex permeability of polycrystalline hexagonal ferrites. *IEEE Transactions on Magnetics*, 36(5), 3415–3417.

[97] Chen, B. Y., Chen, D., Kang, Z. T., & Zhang, Y. Z. (2015). Preparation and microwave absorption properties of Ni-Co nanoferrites. *Journal of Alloys and Compounds*, 618, 222–226.

[98] Lenin, N., Karthik, A., Srither, S. R., Sridharpanday, M., Surendhiran, S., & Balasubramanian, M. (2021). Synthesis, structural and microwave absorption properties of Cr-doped zinc lanthanum nanoferrites $Zn_1\text{-}xCr_xLa_{0.1}Fe_{1.9}O_4$ (x = 0.09, 0.18, 0.27 and 0.36). *Ceramics International*, 47(24), 34891–34898.

[99] Lee, M. Y., Choi, Y. J., Woo, H. J., Kim, S. W., & Lee, B. W. (2022). Microwave absorption properties of $Ni_{0.6}Zn_{0.4}Fe_2O_4$ composites with nonmagnetic nanoferrite percentages. *Ceramics International*, 48(14), 20187–20193.

[100] Sadaqat, A., Almessiere, M., Slimani, Y., Guner, S., Sertkol, M., Albetran, H., … Ercan, I. (2019). Structural, optical and magnetic properties of Tb3+ substituted Co nanoferrites prepared via sonochemical approach. *Ceramics International*, 45(17), 22538–22546.

[101] Amulya, M. S., Nagaswarupa, H. P., Kumar, M. A., Ravikumar, C. R., Prashantha, S. C., & Kusuma, K. B. (2020). Sonochemical synthesis of $NiFe_2O_4$ nanoparticles: Characterization and their photocatalytic and electrochemical applications. *Applied Surface Science Advances*, 1, 100023.

[102] Jasrotia, R., Kumar, G., Batoo, K. M., Adil, S. F., Khan, M., Sharma, R., … Singh, V. P. (2019). Synthesis and characterization of Mg-Ag-Mn nano-ferrites for electromagnet applications. *Physica B: Condensed Matter*, 569, 1–7.

[103] Thakur, P., Sharma, R., Sharma, V., & Sharma, P. (2017). Structural and optical properties of $Mn_{0.5}Zn_{0.5}Fe_2O_4$ nano ferrites: effect of sintering temperature. *Materials Chemistry and Physics*, 193, 285–289.

[104] Asiri, S., Sertkol, M., Guner, S., Gungunes, H., Batoo, K. M., Saleh, T. A., … Baykal, A. (2018). Hydrothermal synthesis of $Co_yZn_yMn_{1-2y}Fe_2O_4$ nanoferrites: magneto-optical investigation. *Ceramics International*, 44(5), 5751–5759.

[105] Maqsood, A., & Khan, K. (2011). Structural and microwave absorption properties of $Ni_{(1-x)}Co_{(x)}Fe_2O_4$ (0.0≤ x≤ 0.5) nanoferrites synthesized via co-precipitation route. *Journal of Alloys and Compounds*, 509(7), 3393–3397.

[106] Albalah, M. A., Alsabah, Y. A., & Mustafa, D. E. (2020). Characteristics of co-precipitation synthesized cobalt nanoferrites and their potential in industrial wastewater treatment. *SN Applied Sciences*, 2, 1–9.

[107] Mathur, P., Thakur, A., & Singh, M. (2008). Low temperature synthesis of $Mn_{0.4}Zn_{0.6}In_{0.5}Fe_{1.5}O_4$ nanoferrite for high-frequency applications. *Journal of Physics and Chemistry of Solids*, 69(1), 187–192.

[108] Lassoued, A., & Li, J. F. (2022). Structure and optical, magnetic and photocatalytic properties of Cr3+ substituted zinc nano-ferrites. *Journal of Molecular Structure*, 1262, 133021.

[109] Shinde, B. L., Mandle, U. M., Pachpinde, A. M., & Lohar, K. S. (2022). Synthesis and characterization of Al3+ substituted Ni-Cu-Zn nano ferrites. *Journal of Thermal Analysis and Calorimetry*, 147(4), 2947–2956.

[110] Tsay, C. Y., Chiu, Y. C., & Lei, C. M. (2018). Hydrothermally synthesized Mg-based spinel nanoferrites: Phase formation and study on magnetic features and microwave characteristics. *Materials*, 11(11), 2274.

[111] Palaniappan, P., Lenin, N., & Uvarani, R. (2022). Copper substitution effect on the structural, electrical, and magnetic properties of manganese and lanthanum ($Mn_{1-x}Cu_xLa_{0.1}Fe_{1.9}O_4$) nanoferrites. *Journal of Alloys and Compounds*, 925, 166717.

[112] Lenin, N., Kanna, R. R., Sakthipandi, K., & Kumar, A. S. (2018). Structural, electrical and magnetic properties of $NiLa_xFe_{2-x}O_4$ nanoferrites. *Materials Chemistry and Physics*, 212, 385–393.

[113] Sadaqat, A., Almessiere, M., Slimani, Y., Guner, S., Sertkol, M., Albetran, H., … Ercan, I. (2019). Structural, optical and magnetic properties of Tb3+ substituted Co nanoferrites prepared via sono-chemical approach. *Ceramics International*, 45(17), 22538–22546.

[114] Shetty, K., Renuka, L., Nagaswarupa, H. P., Nagabhushana, H., Anantharaju, K. S., Rangappa, D., … Ashwini, K. (2017). A comparative study on $CuFe_2O_4$, $ZnFe_2O_4$ and $NiFe_2O_4$: morphology, impedance and photocatalytic studies. *Materials Today: Proceedings*, 4(11), 11806–11815.

[115] Amulya, M. S., Nagaswarupa, H. P., Kumar, M. A., Ravikumar, C. R., & Kusuma, K. B. (2021). Sonochemical synthesis of $MnFe_2O_4$ nanoparticles and their electrochemical and photocatalytic properties. *Journal of Physics and Chemistry of Solids*, 148, 109661.

[116] Amulya, M. S., Nagaswarupa, H. P., Kumar, M. A., Ravikumar, C. R., Prashantha, S. C., & Kusuma, K. B. (2020). Sonochemical synthesis of $NiFe_2O_4$ nanoparticles: Characterization and their photocatalytic and electrochemical applications. *Applied Surface Science Advances*, 1, 100023.

[117] Amulya, M. S., Nagaswarupa, H. P., Kumar, M. A., Ravikumar, C. R., Kusuma, K. B., & Prashantha, S. C. (2021). Evaluation of bifunctional applications of $CuFe_2O_4$ nanoparticles synthesized by a sono-chemical method. *Journal of Physics and Chemistry of Solids*, 148, 109756.

[118] Kushwaha, P., & Chauhan, P. (2021). Facile green synthesis of $CoFe_2O_4$ nanoparticles using hibiscus extract and their application in humidity sensing properties. *Inorganic and Nano-Metal Chemistry*, 1–8.

[119] Satheeshkumar, M. K., Kumar, E. R., Srinivas, C., Suriyanarayanan, N., Deepty, M., Prajapat, C. L., … Sastry, D. L. (2019). Study of structural, morphological and magnetic properties of Ag substituted cobalt ferrite nanoparticles prepared by honey assisted combustion method and evaluation of their antibacterial activity. *Journal of Magnetism and Magnetic Materials*, 469, 691–697.

[120] Tamboli, Q. Y., Patange, S. M., Mohanta, Y. K., Sharma, R., & Zakde, K. R. (2023). Green synthesis of cobalt ferrite nanoparticles: An emerging material for environmental and biomedical applications. *Journal of Nanomaterials*, 2023, 1–15.

[121] Routray, K. L., Saha, S., & Behera, D. (2019). Green synthesis approach for nano sized $CoFe_2O_4$ through aloe vera mediated sol-gel auto combustion method for high frequency devices. *Materials Chemistry and Physics*, 224, 29–35.

[122] Gingasu, D., Mindru, I., Mocioiu, O. C., Preda, S., Stanica, N., Patron, L., … Chifiriuc, M. C. (2016). Synthesis of nanocrystalline cobalt ferrite through soft chemistry methods: A green chemistry approach using sesame seed extract. *Materials Chemistry and Physics*, 182, 219–230.

[123] Naghizadeh, A., Mohammadi-Aghdam, S., & Mortazavi-Derazkola, S. (2020). Novel $CoFe_2O_4@$ $ZnO-CeO_2$ ternary nanocomposite: Sonochemical green synthesis using Crataegus microphylla extract, characterization and their application in catalytic and antibacterial activities. *Bioorganic Chemistry*, 103, 104194.

[124] Ravindra, K. N., Krishna, K. S., Akhilesh, K. P., & Vaishali, B. (2020, June). Green synthesis, characterization and magnetic properties of Sm doped cobalt ferrite nanoparticles. In *AIP Conference Proceedings* (Vol. 2244, No. 1, p. 070026). AIP Publishing LLC.

[125] Tiwari, P., Kane, S. N., Deshpande, U. P., Tatarchuk, T., Mazaleyrat, F., & Rachiy, B. (2020). Cr content-dependent modification of structural, magnetic properties and bandgap in green synthesized Co-Cr nano-ferrites. *Molecular Crystals and Liquid Crystals*, 699(1), 39–50.

[126] Taha, T. A., Elrabaie, S., & Attia, M. T. (2018). Green synthesis, structural, magnetic, and dielectric characterization of $NiZnFe_2O_4/C$ nanocomposite. *Journal of Materials Science: Materials in Electronics*, 29, 18493–18501.

[127] Hermosa, G. C., Liao, C. S., Wu, H. S., Wang, S. F., Liu, T. Y., Jeng, K. S., ... Sun, A. C. A. (2022). Green synthesis of magnetic ferrites (Fe 3 O 4, $CoFe_2O_4$, and $NiFe_2O_4$) stabilized by aloe vera extract for cancer hyperthermia activities. *IEEE Transactions on Magnetics*, 58(8), 1–7.

[128] Shebl, A., Hassan, A., Salama, D., Abd El-Aziz, M. E., & Abd Elwahed, M. (2019). Green synthesis of manganese zinc ferrite nanoparticles and their application as nanofertilizers for Cucurbita pepo L. *Beilstein Archives*, 2019(1), 45.

[129] Naik, M. M., Naik, H. B., Nagaraju, G., Vinuth, M., Vinu, K., & Viswanath, R. (2019). Green synthesis of zinc doped cobalt ferrite nanoparticles: Structural, optical, photocatalytic and antibacterial studies. *Nano-Structures & Nano-Objects*, 19, 100322.

[130] Makofane, A., Maake, P. J., Mathipa, M. M., Matinise, N., Cummings, F. R., Motaung, D. E., & Hintsho-Mbita, N. C. (2022). Green synthesis of $NiFe_2O_4$ nanoparticles for the degradation of Methylene Blue, sulfisoxazole and bacterial strains. *Inorganic Chemistry Communications*, 139, 109348.

[131] Reddy, B. C., Manjunatha, S., Manjunatha, H. C., Vidya, Y. S., Sridhar, K. N., Seenappa, L., ... Pasha, U. M. (2022). NeemLeaves mediated green synthesis of copper ferrite decorated reduced graphene oxide nanocomposite for photoluminescence, gamma/X-ray radiation shielding, antimicrobial and anti-cancer properties. *Solid State Sciences*, 134, 107029.

[132] Shunmuga Priya, R., Ranjith Kumar, E., Balamurugan, A., & Srinivas, C. (2021). Green synthesized $MgFe_2O_4$ ferrites nanoparticles for biomedical applications. *Applied Physics A*, 127(7), 538.

[133] Dhanda, N., Thakur, P., Sun, A. C. A., & Thakur, A. (2023). Structural, optical and magnetic properties along with antifungal activity of Ag-doped Ni-Co nanoferrites synthesized by eco-friendly route. *Journal of Magnetism and Magnetic Materials*, 572, 170598.

[134] Rahman, M. A., Islam, M. T., Singh, M. J., Hossain, I., Rmili, H., & Samsuzzaman, M. (2022). Magnetic, dielectric and structural properties of $Co_xZn_{(0.90-x)}Al0.10Fe_2O_4$ synthesized by sol-gel method with application as flexible microwave substrates for microstrip patch antenna. *Journal of Materials Research and Technology*, 16, 934–943.

[135] Tatarchuk, T., Danyliuk, N., Shyichuk, A., Kotsyubynsky, V., Lapchuk, I., & Mandzyuk, V. (2021). Green synthesis of cobalt ferrite using grape extract: The impact of cation distribution and inversion degree on the catalytic activity in the decomposition of hydrogen peroxide. *Emergent Materials*, 1–15.

[136] Udhaya, P. A., Meena, M., & Queen, M. A. J. (2019). Green synthesis of $MgFe_2O_4$ nanoparticles using albumen as fuel and their physicochemical properties. *Int. J. Sci. Res. Phys. Appl. Sci*, 7, 71–74.

[137] Pu, S., Xue, S., Yang, Z., Hou, Y., Zhu, R., & Chu, W. (2018). In situ co-precipitation preparation of a superparamagnetic graphene oxide/Fe 3 O 4 nanocomposite as an adsorbent for wastewater purification: synthesis, characterization, kinetics, and isotherm studies. *Environmental Science and Pollution Research*, 25, 17310–17320.

[138] Eslami, H., Ehrampoush, M. H., Esmaeili, A., Ebrahimi, A. A., Ghaneian, M. T., Falahzadeh, H., & Salmani, M. H. (2019). Synthesis of mesoporous Fe-Mn bimetal oxide nanocomposite by aeration co-precipitation method: physicochemical, structural, and optical properties. *Materials Chemistry and Physics*, 224, 65–72.

[139] Egizbek, K., Kozlovskiy, A. L., Ludzik, K., Zdorovets, M. V., Korolkov, I. V., Marciniak, B., & Kontek, R. (2020). Stability and cytotoxicity study of $NiFe_2O_4$ nanocomposites synthesized by co-precipitation and subsequent thermal annealing. *Ceramics International*, 46(10), 16548–16555.

[140] Akash, M. S. H., & Rehman, K. (2020). *Essentials of Pharmaceutical Analysis* (pp. 29–56). Springer.

[141] Fadlelmoula, A., Pinho, D., Carvalho, V. H., Catarino, S. O., & Minas, G. (2022). Fourier transform infrared (FTIR) spectroscopy to analyse human blood over the last 20 years: A review towards lab-on-a-chip devices. *Micromachines*, 13(2), 187.

[142] Rahman, A., Sabeeh, H., Zulfiqar, S., Agboola, P. O., Shakir, I., & Warsi, M. F. (2020). Structural, optical and photocatalytic studies of trimetallic oxides nanostructures prepared via wet chemical approach. *Synthetic Metals*, 259, 116228.

[143] Vladár, A. E., & Hodoroaba, V. D. (2020). Characterization of nanoparticles by scanning electron microscopy. In *Characterization of Nanoparticles* (pp. 7–27). Elsevier.

[144] Karthik, K., Dhanuskodi, S., Gobinath, C., & Sivaramakrishnan, S. (2015). Microwave-assisted synthesis of CdO-ZnO nanocomposite and its antibacterial activity against human pathogens. *Spectrochimica Acta Part A: Molecular and Biomolecular Spectroscopy*, 139, 7–12.

[145] Mansournia, M., & Ghaderi, L. (2017). CuO@ ZnO core-shell nanocomposites: novel hydrothermal synthesis and enhancement in photocatalytic property. *Journal of Alloys and Compounds*, 691, 171–177.

[146] Liang, J., Xiao, X., Chou, T. M., & Libera, M. (2021). Analytical cryo-scanning electron microscopy of hydrated polymers and microgels. *Accounts of Chemical Research*, 54(10), 2386–2396.

[147] Taufik, A., Albert, A., & Saleh, R. (2017). Sol-gel synthesis of ternary $CuO/TiO_2/ZnO$ nanocomposites for enhanced photocatalytic performance under UV and visible light irradiation. *Journal of Photochemistry and Photobiology A: Chemistry*, 344, 149–162.

[148] Lavin, A., Sivasamy, R., Mosquera, E., & Morel, M. J. (2019). High proportion ZnO/CuO nanocomposites: Synthesis, structural and optical properties, and their photocatalytic behavior. *Surfaces and Interfaces*, 17, 100367.

[149] de Haan, K., Ballard, Z. S., Rivenson, Y., Wu, Y., & Ozcan, A. (2019). Resolution enhancement in scanning electron microscopy using deep learning. *Scientific Reports*, 9(1), 1–7.

[150] Sharma, I., Sharma, M. V., & Sharma, P. (2022). A review on synthesis and characterization of MnZn ferrite nanoparticles via citrate precursor method. *International Journal of Multidisciplinary Educational Research*, 6(6), 11.

[151] Hamrouni, A., Moussa, N., Parrino, F., Di Paola, A., Houas, A., & Palmisano, L. (2014). Sol-gel synthesis and photocatalytic activity of $ZnO-SnO_2$ nanocomposites. *Journal of Molecular Catalysis A: Chemical*, 390, 133–141.

[152] Khan, M. S. I., Oh, S. W., & Kim, Y. J. (2020). Power of scanning electron microscopy and energy dispersive X-ray analysis in rapid microbial detection and identification at the single cell level. *Scientific Reports*, 10(1), 1–10.

[153] Biswal, S. K., Panigrahi, G. K., & Sahoo, S. K. (2020). Green synthesis of Fe_2O_3-Ag nanocomposite using Psidium guajava leaf extract: An eco-friendly and recyclable adsorbent for remediation of Cr (VI) from aqueous media. *Biophysical Chemistry*, 263, 106392.

[154] Noohpisheh, Z., Amiri, H., Farhadi, S., & Mohammadi-Gholami, A. (2020). Green synthesis of Ag-ZnO nanocomposites using Trigonella foenum-graecum leaf extract and their antibacterial, antifungal, antioxidant and photocatalytic properties. *Spectrochimica Acta Part A: Molecular and Biomolecular Spectroscopy*, 240, 118595.

[155] Yulizar, Y., Apriandanu, D. O. B., & Wibowo, A. P. (2019). Plant extract mediated synthesis of Au/TiO_2 nanocomposite and its photocatalytic activity under sodium light irradiation. *Composites Communications*, 16, 50–56.

[156] Saadatkhah, N., Carillo Garcia, A., Ackermann, S., Leclerc, P., Latifi, M., Samih, S., & Chaouki, J. (2020). Experimental methods in chemical engineering: Thermogravimetric analysis-TGA. *The Canadian Journal of Chemical Engineering*, 98(1), 34–43.

[157] Wong, K., & Dia, S. (2017). Nanotechnology in batteries. *Journal of Energy Resources Technology*, 139(1), 014001.

[158] Voiry, D., Chhowalla, M., Gogotsi, Y., Kotov, N. A., Li, Y., Penner, R. M., ... Weiss, P. S. (2018). Best practices for reporting electrocatalytic performance of nanomaterials. *ACS Nano*, 12(10), 9635–9638.

[159] Augustyn, V., Come, J., Lowe, M. A., Kim, J. W., Taberna, P. L., Tolbert, S. H., ... & Dunn, B. (2013). High-rate electrochemical energy storage through Li+ intercalation pseudocapacitance. *Nature Materials*, 12(6), 518–522.

[160] Stoller, M. D., & Ruoff, R. S. (2010). Best practice methods for determining an electrode material's performance for ultracapacitors. *Energy & Environmental Science*, 3(9), 1294–1301.

[161] Liu, S., Xia, X., Zhong, Y., Deng, S., Yao, Z., Zhang, L., ... Tu, J. (2018). 3D TiC/C core/shell nanowire skeleton for dendrite-free and long-life lithium metal anode. *Advanced Energy Materials*, 8(8), 1702322.

[162] Trinh, N. D., Crosnier, O., Schougaard, S. B., & Brousse, T. (2011). Synthesis, characterization and electrochemical studies of active materials for sodium ion batteries. *ECS Transactions*, 35(32), 91.

[163] He, G., Li, J., Chen, H., Shi, J., Sun, X., Chen, S., & Wang, X. (2012). Hydrothermal preparation of Co_3O_4@ graphene nanocomposite for supercapacitor with enhanced capacitive performance. *Materials Letters*, 82, 61–63.

[164] Leontyeva, D. V., Leontyev, I. N., Avramenko, M. V., Yuzyuk, Y. I., Kukushkina, Y. A., & Smirnova, N. V. (2013). Electrochemical dispergation as a simple and effective technique toward preparation of NiO based nanocomposite for supercapacitor application. *Electrochimica Acta*, 114, 356–362.

[165] Ramesh, S., Vikraman, D., Kim, H. S., Kim, H. S., & Kim, J. H. (2018). Electrochemical performance of MWCNT/GO/NiCo2O4 decorated hybrid nanocomposite for supercapacitor electrode materials. *Journal of Alloys and Compounds*, 765, 369–379.

[166] Bu, I. Y., & Huang, R. (2015). One-pot synthesis of ZnO/reduced graphene oxide nanocomposite for supercapacitor applications. *Materials Science in Semiconductor Processing*, 31, 131–138.

[167] Yadav, S., Rani, N., & Saini, K. (2022, February). A review on transition metal oxides based nanocomposites, their synthesis techniques, different morphologies and potential applications. In *IOP Conference Series: Materials Science and Engineering* (Vol. 1225, No. 1, p. 012004). IOP Publishing.

[168] Joulazadeh, M., & Navarchian, A. H. (2015). Ammonia detection of one-dimensional nano-structured polypyrrole/metal oxide nanocomposites sensors. *Synthetic Metals*, 210, 404–411.

[169] Dutt, M., Ratan, A., Tomar, M., Gupta, V., & Singh, V. (2020). Mesoporous metal oxide-α-Fe_2O_3 nanocomposites for sensing formaldehyde and ethanol at room temperature. *Journal of Physics and Chemistry of Solids*, 145, 109536.

[170] Sivasubramanian, R., & Biji, P. (2016). Preparation of copper (I) oxide nanohexagon decorated reduced graphene oxide nanocomposite and its application in electrochemical sensing of dopamine. *Materials Science and Engineering: B*, 210, 10–18.

[171] Dinesh, B., Veeramani, V., Chen, S. M., & Saraswathi, R. (2017). In situ electrochemical synthesis of reduced graphene oxide-cobalt oxide nanocomposite modified electrode for selective sensing of depression biomarker in the presence of ascorbic acid and dopamine. *Journal of Electroanalytical Chemistry*, 786, 169–176.

[172] Kumar, P. R., & Mitra, S. (2013). Nickel ferrite as a stable, high capacity and high rate anode for Li-ion battery applications. *RSC Advances*, 3(47), 25058–25064.

[173] Narsimulu, D., Rao, B. N., Venkateswarlu, M., Srinadhu, E. S., & Satyanarayana, N. (2016). Electrical and electrochemical studies of nanocrystalline mesoporous $MgFe_2O_4$ as anode material for lithium battery applications. *Ceramics International*, 42(15), 16789–16797.

[174] Ateia, E. E., Ateia, M. A., Fayed, M. G., El-Hout, S. I., Mohamed, S. G., & Arman, M. M. (2022). Synthesis of nanocubic lithium cobalt ferrite toward high-performance lithium-ion battery. *Applied Physics A*, 128(6), 483.

[175] Venkatachalam, V., & Jayavel, R. (2015, June). Novel synthesis of Ni-ferrite ($NiFe_2O_4$) electrode material for supercapacitor applications. In *AIP Conference Proceedings* (Vol. 1665, No. 1, p. 140016). AIP Publishing LLC.

[176] Rani, B. J., Ravi, G., Yuvakkumar, R., Ganesh, V., Ravichandran, S., Thambidurai, M., … Sakunthala, A. (2018). Pure and cobalt-substituted zinc-ferrite magnetic ceramics for supercapacitor applications. *Applied Physics A*, 124, 1–12.

[177] Kumar, R., Kumar, R., Sahoo, P. K., Singh, M., & Soam, A. (2022). Synthesis of nickel ferrite for supercapacitor application. *Materials Today: Proceedings*, 67, 1001–1004.

[178] Rajeevgandhi, C., & Sivagurunathan, P. (2022). Excellent performance of electrical and supercapacitor application of cadmium cobalt ferrite nanoparticles synthesized by chemical co-precipitation technique. *Journal of Materials Science: Materials in Electronics*, 33(21), 16791–16804.

[179] Agale, P., Salve, V., Patil, K., Mardikar, S., Uke, S., Patange, S., & More, P. (2023). Synthesis, characterization, and supercapacitor applications of Ni-doped $CuMnFeO_4$ nano Ferrite. *Ceramics International*, 49(16), 27003–27014

[180] Racik, K. M., Anand, S., Muniyappan, S., Nandhini, S., Rameshkumar, S., Mani, D., … Ramasamy, P. (2022). Preparation of $CoFe_2O_4/SiO_2$ nanocomposite as potential electrode materials for supercapacitors. *Inorganic Chemistry Communications*, 146, 110036.

[181] Uke, S. J., Mardikar, S. P., Bambole, D. R., Kumar, Y., & Chaudhari, G. N. (2020). Sol-gel citrate synthesized Zn doped $MgFe_2O_4$ nanocrystals: A promising supercapacitor electrode material. *Materials Science for Energy Technologies*, 3, 446–455.

[182] Kennaz, H., Harat, A., Guellati, O., Manyala, N., & Guerioune, M. (2016). Synthesis of cobalt ferrite nanoparticles by hydrothermal method for supercapacitors application. *Materials Science, Engineering*.

[183] Xiangfeng, C., Dongli, J., & Chenmou, Z. (2007). The preparation and gas-sensing properties of $NiFe_2O_4$ nanocubes and nanorods. *Sensors and Actuators B: Chemical*, 123(2), 793–797.

8 Battery and Fuel Cell Applications
Nanohybrid Aluminates, Ferrites, and Vanadates

H.V. Harini, D.M. Tejashwini, S. Ishwarya, V.V. Deshmukh, H.P. Nagaswarupa, and Ramachandra Naik

INTRODUCTION

The fundamental component of nanoscience and nanotechnology is nanomaterials. The definition mentioned previously emphasizes size a lot. Analysis of the distinctive and varied physical/chemical properties of nanomaterials/nanostructures at the nanoscale dimension (100 nm) than their bulk counterparts led to the discovery of numerous applications. This makes it possible to categorize nanomaterials into various surface-to-volume advances, such as nanosheets, nanofilaments, nanowires, nanotubes, and quantum dots. High surface to volume ratio, which was the primary phenomenon for exceptional physicochemical characteristics including mechanical, thermal, chemical, magnetic, structural, and electrical capabilities, is periodically the cause of these properties [1,2].

Secondary batteries are a type of energy storage technology that is widely used. Lithium-ion batteries have been used extensively in electric automobiles, mobile gadgets, and grid energy storage since its commercialization in 1991 [3]. The instance of lithium-ion battery electrode design, the choice of anode features, such as rapid electrode kinetics, cheap material cost and safety, and a very low redox potential vs. Li/ Li$^+$ compared to cathode materials, is critical. The anode materials used in the early stages of lithium battery research were low potential insertion compounds such as Li_2WO_2 [4]. Later, these metal oxides were supplanted by low-density, carbon-based materials, which are today and have been enjoying broad commercial success for more than two decades [5]. Despite the advancements in carbon-based anodes, it is commonly acknowledged that new anode materials are necessary due to current energy needs. Carbon anodes have certain drawbacks. For instance, carbon materials may suffer from solvent co-intercalation, leading to significant interlayer expansion, which exfoliates when reactions have a high rate of charge–discharge. Carbon-based substances have inherent limitations in terms of gravimetric and volumetric capabilities, which inhibit the development of high-energy-density lithium batteries [6].

Metaloxide semiconductors, which have crucial uses in lithium-ion batteries, have received a lot of attention during the past few decades [7]. Due to their high energy density, lithium-ion batteries have been widely employed in portable gadgets and investigated for use in electric vehicles and large-scale energy storage systems [8,9]. Lithium has become a significant issue recently. Because they are expensive, scarce, and take a long time to produce, Li and some of the other transition metals now utilized in Li-ion batteries could eventually become a significant issue and lead to a scarcity [10]. Due to its unique features, metal aluminates, ferrites, and vanadates are potential innovative materials that are employed in many applications, including batteries, supercapacitors,

DOI: 10.1201/9781003479239-8

and sensors [11]. The wide band gap, thermal stability, layered structure, chemical stability, and multivalence of metal aluminates, ferrites, and vanadium-based materials make them promising candidates for future development [12].

Due to their unique characteristics, fuel cells are poised to bring about a profound shift in the realm of electricity. By definition, a fuel cell is an electrochemical device that converts chemical energy from fuel into electrical energy without burning the fuel. As a result, the chemical energy associated with the electrochemical reaction of the fuel and oxidant in a fuel cell system is immediately converted into water, electricity, and heat. Fuel cells often employ fuels like H_2, methanol, and ethanol. In conclusion, the reactions carried out in a fuel cell can be described as follows: Electrons released as hydrogen in the anode electrode change into a hydrogen ion. The electrical current is created by these electrons as they travel through the foreign circuit concerning the cathode [13]. The membrane electrode assembly (MEA), which is composed of an electrocatalyst and a membrane, is the most significant component, or, to put it another way, the fuel cell's core. This section will handle the electrochemical reaction that produces electrical current. Conduction of generated protons from the anode to the cathode is the function of the membrane between the electrodes [14]. Due to its high specific energy at low operating temperatures and capacity to provide eco-friendly energy conversion with a high efficiency, a direct methanol fuel cell (DMFC) is typically regarded as a promising candidate for future energy generation technologies with a variety of applications. Therefore, direct methanol fuel cell (DMFC) development is the main focus of study [15,16]. Due to many reasons, including its potential use in high-performance luminescent devices, catalysts magnets, and other functional materials, rare earth compounds have drawn a lot of attention. This is due to the optoelectronic and chemical characteristics derived from the 4f shell of rare earth ions. Due to their electron transitions within the 4f shell, compounds of the lanthanum vanadate in particular have numerous applications [17–19] (Figure 8.1).

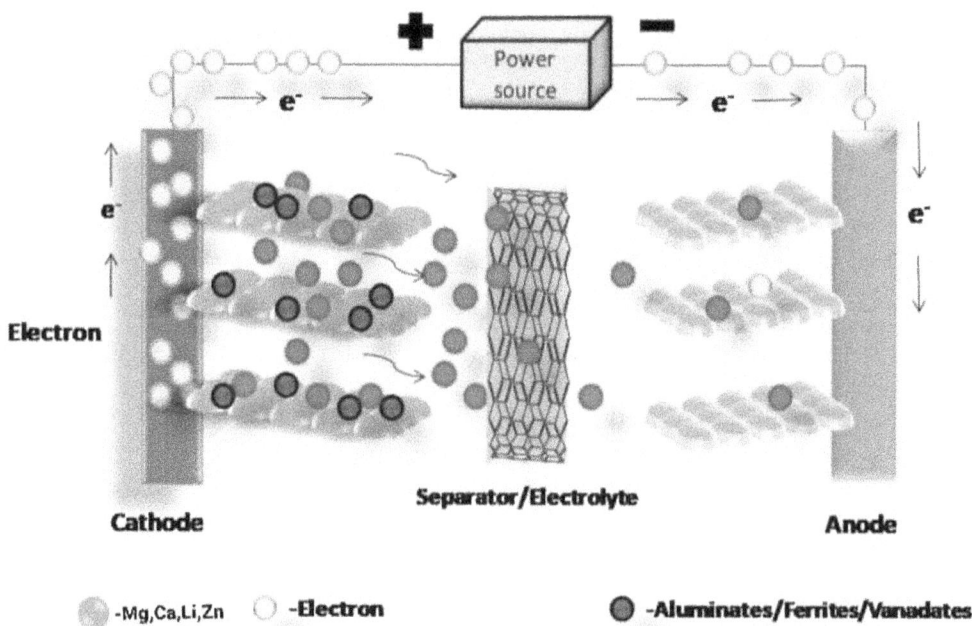

FIGURE 8.1   Schematic representation of anode and cathode materials in batteries.

NANOALUMINATES (ANODIC AND CATHODIC)

Finding positive electrode materials that can reversibly intercalate or store high-valent ions without substantial structural modifications or expansion of volume is one of the most difficult tasks. This is critical for preserving the battery's stability and cycle life. Sulfides, oxides, phosphates, and organic compounds are among the interesting choices for positive electrode materials because they have high theoretical capacities and open architectures that can house massive ions [20]. Locating electrolytes that can enable high ionic conductivity and the stability of high-valent ions is another difficult task. In order to avoid undesirable side effects or deterioration, the electrolytes must be compatible with both the positive and negative electrodes. Ionic liquids, solid-state substances, organic solvents, and molten salts are a few potential electrolyte choices [21]. The optimization of battery design and architecture to provide high energy, power, safety, and cost-effectiveness is a third problem. This entails weighing the trade-offs between various performance measures and choosing the ideal mixture of components and specifications for each battery system. The thickness, porosity, shape, loading, and current collector of the electrodes are just a few of the elements that have an impact on the design of the battery [22].

Various surveys have scrutinized the attempts of metal aluminum or hybrid aluminum (doped and composites) in the anode electrode due to their abundance, affordability, and recyclability. It has greater stability in the atmosphere compared to lithium-based materials, lessens the possible safety threat while also making handling easier in the atmosphere surroundings. Additionally, Al batteries use an ionic liquid electrolyte that is non-volatile and inert, thereby decreasing the safety concern, and also offer a lot of potential for sustainable energy applications [23]. However, as we indicated, there are still numerous obstacles to overcome, such as finding adequate positive electrodes and electrolytes capable of supporting high aluminum-ion mobility and reversibility [24]. Aluminum anode possesses four times higher volumetric capacity than lithium, i.e., 8.04 Ah/cm^3, which is also having a higher magnitude capacity than graphite anode used in lithium-ion batteries (0.84 Ah/cm^3) [25].

Here we discuss about different cathodes for aluminum anode electrode.

V$_2$O$_5$

One of the candidates is vanadium pentoxide (V$_2$O$_5$) for a cathode electrode, which is a layered material that can undergo structural changes and redox reactions upon intercalation of various ions. Al3+ ions are reversibly stored in the framework of V$_2$O$_5$ nanowires, which serve as the cathode in renewable aluminum batteries. Al3+ ions can intercalate and de-intercalate into V$_2$O$_5$ nanowires via a phase transition reaction, which results in redox of V$_2$O$_5$ and structural changes on the nanowires' surface. They show that in comparison to the traditional AlCl4- intercalation method, the reversible Al3+ storage mechanism produces a battery with a large capacity, good rate performance, and extended cycle life. Based on experimental evidence and theoretical calculations, it has a phase diagram and a potential reaction mechanism for the Al3+/V$_2$O$_5$ system. It can claim that research offers fresh opportunities for producing high-performance, environmentally friendly rechargeable aluminum batteries due to its efficient reversible redox reaction and galvanostatic cycles with high capacity [26,27].

MNO$_2$

Another cathode material that functions with an aluminum battery is manganese dioxide. This was based on the qualities of an ionic liquid called 1-ethyl-3-methylimidazolium chloride (EMIMCl) and a supporting substance called MnO$_2$. These two properties together have produced a material that can store charge by undergoing a redox reaction with water that is both common and inexpensive. In comparison to traditional lithium-ion batteries, which have an energy density of about 200 Wh/kg, Al ion battery can attain a high energy density of roughly 500 Wh/kg. The battery can also be recharged more than 1,000 times without suffering any material degradation or significant capacity loss. Under various circumstances, including high temperature, overcharge, short circuit, and mechanical abuse, the battery is also secure and stable [28]. Utilizing an in situ electrochemical transformation reaction with aluminum anode and spinel layer cathode materials, AlxMnO$_2$nH$_2$O was produced. They electrochemically oxidized the Mn$_3$O$_4$ in an electrolyte of 5 M Al(CF$_3$SO$_3$)$_3$-H$_2$O. This battery can attain a record-breaking 481 Wh/kg of energy density and a high specific capacity of 467 mAh/g. Additionally, they demonstrate that their battery can be recharged more than 1,000 times without suffering any material degradation or capacity loss. This technology is still in its infancy and that further research is required to fully understand the general electrochemical mechanism. Gravimetric capacity and other metrics, however, suggest that this cell shape is promising [29].

MXenes, a class of two-dimensional transition metal carbides and nitrides with high conductivity, sizable surface areas, and variable interlayer spacing, are some additional cathode materials for aluminum-ion batteries. For aluminum-ion batteries, MXenes can attain high capacities of up to 400 mAh/g [30]. Polymers can be used to create intercalation- or conversion-type cathodes for aluminum-ion batteries, such as polyacrylonitrile (PAN), polyaniline (PANI), and polyimide (PI). High flexibility, low cost, and environmental friendliness can all be provided by polymers [31]. MOFs have high porosity, a wide surface area, and changeable pore sizes, making them suitable as host materials for aluminum ions [32]. Prussian blue analogues (PBAs) are a set of coordination compounds with cyano groups and metallic ions arranged in a cubic arrangement. Due to their open lattice structure, PBAs can show a rapid and reversible insertion and extract of aluminum ions [33].

FERRITES FOR BATTERIES AND FUEL CELL (ANODIC AND CATHODIC MATERIALS)

Among all of the documented metal and mixed metal oxides, inexpensive iron oxides have piqued the interest of anode developers because they are environmentally acceptable, abundant, and can accommodate more than six lithium ions during deep discharge [34–36,1–3]. Excellent electrochemical properties exhibited by porous structures of different types of ferrites such as Fe$_3$O$_4$ [37,4], Fe$_2$O$_3$ [38,5], and hollow Fe$_2$O$_3$ spheres [39,6], these materials can act as a buffer in change of anodes during the conversion reaction and the repeated Li$^+$ insertion/extraction, reduces the pulverization problem and improving cycling performance. Pure Fe$_2$O$_3$ reacts with divalent metal oxides, and stable MFe$_2$O$_4$ is formed, essentially, depending on the preference energy, ionic radius of the metal ions, and possibility of various valency states, such ternary oxide spinel's crystallized into normal, partially inverse or inverse forms.

Due to its low cost, stable metal/oxide reversibility, high mechanical strength at the oxygen potential of fuel, and selective catalysis enabling fuel conversion, Fe, which is the most common transition metal in the earth's crust, has been used in the creation of solid oxide fuel cell. Increasing fuel adsorption/dissociation and electronic/ionic conduction by infiltration or exsolution is a common way to improve the anode's performance [40,41]. The most common cations used for the synthesis of a stable oxide anode in a reducing atmosphere are Cr3+, Ti4+, Nb5+, W6+, and Mo6+ in perovskite-type ferrite, whereas ferrites with Ni2+, Cu2+, and Co2+/3+ substitution tend to destabilize the structure and induce the formation of layered perovskite during fuel cell operation. Several decades ago, materials with an iron composition have been chosen to produce important solid oxide fuel cell. Because of the stability and "flexibility" of the Fe-O link, various highly active elements may be effectively doped into FeO6 octahedra. Due to the numerous states of valence and fluctuating coordination numbers of Fe in iron-based oxides, it also offers changeable conductivity. Numerous industries are designing novel iron-based materials as a result of the intriguing interaction of Fe-O [42].

Ferrite materials with diverse morphologies have been created and used as anode materials in lithium-ion batteries with improved characteristics, including macroporous particles [43], flakes

[44], hollow nanospheres [45], and films [46]. Because of their surface shape, surface area, and pore size distribution, spinel ferrites and related nanocomposites are extensively researched for battery applications. Many transitional metal oxide nanostructures like $MnFe_2O_4$ synthesized through Solvothermal process obtaining microsphere morphology used as anodic material for battery exhibiting as superior theoretical capacity [47]. Sol-gel-assisted synthesis of $NiFe_2O_4$ nanoparticle with the theoretical capacity of 915 mAh/g which exhibits better performance than the graphite anode with the 375 mAh/g exhibiting superior capacity and high-rate anodic material for batteries [4]. An oxidation peak and a reduction peak at 1.9 and 0.5 V, respectively, were seen in a $NiFe_2O_4/$ Li (PVDF) sample. The strength of the peaks varies with cycle numbers and indicates reduced stability against Li. The cathodic reactions of Fe_2O_3 to Fe and NiO to Ni are responsible for the two decreasing peaks located at 0.83 and 1.6 V against Li/Li+, respectively. The decrease in peak intensity indicates that electrode stability is a major problem following the main cycle. The oxidation of iron and nickel to produce Fe_3 and Ni_2 causes a primary broad peak to appear during the first anodic cycle, which shifts to a higher potential during the subsequent cycle, about 1.9 V against Li/ Li+ [45]. Due to their short diffusion route lengths, $NiFe_2O_4$ nanoparticles also offer good charge transfer capabilities. In consideration of this, we came to the conclusion that the $NiFe_2O_4$ electrode's increased performance was inextricably linked to the stability of the interfaces, which provided steady values of capacity at greater cycling. The research clarifies the $NiFe_2O_4$ electrode's discharge capability when used with various binders, including PVDF, CMC, and alginate. When comparing the $NiFe_2O_4$ electrode's discharge capacity at a 0.1 C (91.5 mA/g) rate to the number of $NiFe_2O_4$/Li (PVDF) cycles, the first cycle's value was 1,440 mAh/g, while the 50th cycle's value was 180 mAh/g. Around 85% of the discharge capacity is lost as a result of inadequate interfacial stability [48].

$CoFe_2O_4$ flower-like microspheres were created using a hydrothermal technique aided by ascorbic acid, certainly exciting to comprehend that nanostructures self-assemble and useful in synthesized performing materials. When used $CoFe_2O_4$ microspheres as the anode material for lithium-ion battery, it significantly improved electrochemical performance. This improvement is mostly due to the adequate contact between the active material and the electrolyte, the short Li+ diffusion length, the wide surface area, and the effective absorption of the active material's expansion in volume [49]. Additionally, fast electrode charge transfer kinetics, cost-effectiveness, material safety, and a weak redox potential versus Li/Li+ compared to cathode substances are crucial anode features in the manufacturing of lithium-ion battery electrodes. Low potential supplement substances like metal oxides were employed as anode materials in the first stages of lithium-ion battery development [4]. The synthesized spinel ferrite along with their nanocomposite with the improved asset is employed as an anodic material in batteries. But their charge capacity and rate performance decrease rapidly below 700 mAh/g and must be upgraded sooner to a viable lithium-ion battery more significantly at high charge/discharge rates. The previously mentioned constraints might be removed by utilizing at least two efficient methods to enhance their cycle performance: through hybrid or composite materials.

The theoretical capacity of the graphite anode utilized in lithium-ion battery is 372 mAh/g. To get over this issue, several researchers are working on developing additional anode materials using silicon, oxysalt, and transition metal oxide [50–54]. Zinc ferrite ($ZnFe_2O_4$) is a potential anode material for lithium-ion battery due to its affordability, environmental friendliness, lack of toxicity, and superior structural stability [55]. Li-ion and $ZnFe_2O_4$ react readily, producing a theoretical specific capacitance of 1,000.5 mAh/g [56,57]. $ZnFe_2O_4$/graphene oxide composites of diameters around 10 nm were created by using a two-step process. These composites had great reversible capacity, strong cycle stability, and remarkable rate capability. A large quantity of $ZnFe_2O_4$/graphene oxide composite may also be produced using this technique of production, which makes it useful in different scenarios [58]. The ionic conductivity of lithium-ion battery increases because of the rapid electron transit between the $ZnFe_2O_4$ nanoparticle and graphene oxide, which lowers internal resistance and increases specific capacitance at higher current densities. The composite material made of $ZnFe_2O_4$ and graphene oxide offers the greatest charge/discharge capacities of 1,181/1,744 mAh/g [59].

NANOVANADATE (ANODIC AND CATHODIC MATERIALS)

As cathode materials for rechargeable batteries produced by monovalent and divalent charge carriers (Li, Na, K, and Zn ions), vanadium vanadate including oxides has been thoroughly explored due to their diverse structures and exceptional electrochemical behaviors created by the multiple valence states of vanadium [60–63]. $K_2V_3O_8$ is a new anode material for KIBs that Lu et al. introduced recently [64], probably activated by a conversion process. Due to their enormous theoretical capacity (500/1000 mAh/g) and innovative redox reaction, transition metal oxides (TMOS) have emerged as intriguing anode choices for Li-ion batteries [65]. Numerous types of TMOS, including Fe_3O_4, Fe_2O_3, and Co_3O_4 as anodes for Li-ion batteries, have already been the subject of in-depth research [66–68], and its electrochemical performance is remarkable. As one form of particular TMOS, vanadium oxides (V_xO_y) commonly show layered or channel-like structure, enduring an intercalation/deintercalation reaction process for lithium ions storage and generally serving as cathode materials for Li-ion batteries [69–71]. The addition of metal elements (Fe, Cu, Zn, etc.) to vanadium oxides has led to the development of numerous novel ternary compounds, which, due to their unusual structures, display alluring electrochemical properties. Vanadium element–based TMOS (TVOS) of several types, including ZnV_2O_4, $FeVO_4$, ZnV_2O_6, CuV_2O_6, $Cu_2V_2O_7$, $Cu_5V_2O_{10}$, and $Ag_2V_4O_{11}$, have been made, and their electrochemical performance has been examined [72–74]. These materials exhibit vastly different electrochemical properties.

IMPORTANCE

Aluminum is a promising anode material for batteries because it has a high theoretical capacity of 2,980 mAh/g, which means it can store a lot of charge per unit mass. Aluminum ions have a high-valent ions that are being researched for secondary post-lithium batteries. The advantage of increasing the number of electrons involved in the electrochemical process, which can lead to higher capacity values. However, they also face similar challenges, such as finding suitable materials that can accommodate their large size and high charge.

Some of the cathode materials that have been explored for aluminum-ion batteries are vanadates, manganese oxides, and titanates. These materials have high theoretical capacities and multiple oxidation states that can accommodate the insertion and extraction of aluminum ions. However, these materials also have some drawbacks, such as low electronic conductivity, poor structural stability, and large volume change during cycling. Therefore, various strategies have been employed to improve the performance of these cathode materials, such as nanostructuring, doping, coating, and compositing.

Spinel ferrites and related nanocomposites are being studied extensively for battery applications and fuel cell applications owing to their surface shape, surface area, and pore size distribution. The most frequently used materials for studies of magnetism have been metal ferrites, which have also demonstrated enormous potential for numerous important technological applications, including electronic devices, medical diagnostics, and drug delivery. They are also excellent dielectric materials and energy storage materials. Large surface areas of nanomaterial are advantageous for absorbing and storing a significant amount of Li ions. Spinel ferrites based on transition metals and their composites are now being investigated for application throughout an array of industries, including the electrical industry and magnetic recording. Spinel ferrite offers a tripled density and a doubled capacity, making it an ideal material for battery devices [75]. In lithium-ion batteries, materials based on iron have received an abundance of curiosity as electrode materials due to their affordability and nontoxicity. Iron oxide in spinel forms, such as Fe_3O_4, has previously been intensively studied as a potential lithium insertion electrode.

Future research may concentrate on exploring in situ growth of different alloy nanoparticles, which may be employed as high-performance catalysts for mixed gases as fuels. Nano-sized metallic particles can be exsolved from iron-based oxides to increase the electrochemical activity.

With all factors considered, iron-based materials have demonstrated significant promise for use in solid fuel cell and are connected to intriguing areas, allowing for an increase in the commercialization of these energy conversion technologies in the near future [76].

Devices that use electrochemistry to generate electricity, such as fuel cells and metal-air batteries, do so without harming the environment. Earth-abundant transition metal oxides (TM oxides) are desirable substitutes for the pricey noble metals employed in these green energy technologies [77]. Vanadium oxides are one of the TMOs that have attracted interest for over four decades because of their abundant natural supplies, rich vanadium valences, and numerous structural options. By doping or structural modulation, the families are extremely modulable. Vanadium oxides are a branch of mixed transition metal oxides (MTMOs) with increased reversible capacity and electronic/ionic conductivity. "Vanadates" are the variants of vanadium oxides hybridized (or high-ratio doped) with other metal ions or clusters (i.e. NH4) [78].

Lowering the temperature at which it works for fuel cells (FCs), gained a lot of attention globally. The enhancement of electrolyte materials that function at lower temperatures has received a great deal of attention in the literature. It has been discovered that the family of MeVOX (Me-dopant metal, V-vanadium, Ox-oxide) materials has particular characteristics as an oxide ion conductor (electrolyte) at low operation temperature [79].

FUTURE OUTCOMES

The development of aluminates, ferrite, vanadium and their application in battery technology is anticipated to transform clean energy that has revolutionary future. Batteries will aid in the transition to a cleaner, more sustainable energy future since they are highly effective, energy-dense, stable, and have extended lifespan. Highly recommended aluminates, ferrite, and vanadium with superior high-rate discharge performance, cyclicity, and electrode density may be created as anode materials for lithium-ion battery devices. In light of these facts, we predict that the recently created ferrite-based, aluminate-based, and vanadium-based anode electrode materials with mesoporous characteristics and narrow pore size distribution will be desirable choices for creating lithium-ion batteries in the future. Energy shortages and environmental pollution are major issues for people today. Massive attempts have been made to replace fossil fuels with alternative energy sources, such as its synonym clean fuel, in order to solve these difficulties. Due to their unique characteristics, fuel cells are poised to bring about a profound shift in the realm of electricity.

REFERENCES

[1] Chen, Guanying, Indrajit Roy, Chunhui Yang, and Paras N. Prasad. "Nanochemistry and nanomedicine for nanoparticle-based diagnostics and therapy." *Chemical Reviews* 116, no. 5 (2016): 2826–2885.

[2] Yang, Xuan, Miaoxin Yang, Bo Pang, Madeline Vara, and Younan Xia. "Gold nanomaterials at work in biomedicine." *Chemical Reviews* 115, no. 19 (2015): 10410–10488.

[3] Li, Yaqi, Jia Guo, Kjeld Pedersen, Leonid Gurevich, and Daniel-Ioan Stroe. "Investigation of multi-step fast charging protocol and aging mechanism for commercial NMC/graphite lithium-ion batteries." *Journal of Energy Chemistry* 80 (2023): 237–246.

[4] Kumar, P. Ramesh, and Sagar Mitra. "Nickel ferrite as a stable, high capacity and high rate anode for Li-ion battery applications." *RSC Advances* 3, no. 47 (2013): 25058–25064.

[5] Yoshio, Masaki, Takaaki Tsumura, and Nikolay Dimov. "Silicon/graphite composites as an anode material for lithium ion batteries." *Journal of Power Sources* 163, no. 1 (2006): 215–218.

[6] Alcántara, R., M. Jaraba, P. Lavela, and J. L. Tirado. "NiCo$_2$O$_4$ spinel: First report on a transition metal oxide for the negative electrode of sodium-ion batteries." *Chemistry of Materials* 14, no. 7 (2002): 2847–2848.

[7] Wang, Zhiyu, Zichen Wang, Wenting Liu, Wei Xiao, and Xiong Wen David Lou. "Amorphous CoSnO 3@ C nanoboxes with superior lithium storage capability." *Energy & Environmental Science* 6, no. 1 (2013): 87–91.

[8] Armand, Michel, and J-M. Tarascon. "Building better batteries." *Nature* 451, no. 7179 (2008): 652–657.

[9] Dunn, Bruce, Haresh Kamath, and Jean-Marie Tarascon. "Electrical energy storage for the grid: A battery of choices." *Science* 334, no. 6058 (2011): 928–935.

[10] Vikström, Hanna, Simon Davidsson, and Mikael Höök. "Lithium availability and future production outlooks." *Applied Energy* 110 (2013): 252–266.

[11] Yu, Minghao, Yan Zeng, Yi Han, Xinyu Cheng, Wenxia Zhao, Chaolun Liang, Yexiang Tong, Haolin Tang, and Xihong Lu. "Valence-optimized vanadium oxide supercapacitor electrodes exhibit ultrahigh capacitance and super-long cyclic durability of 100 000 cycles." *Advanced Functional Materials* 25, no. 23 (2015): 3534–3540.

[12] Peighambardoust, S. Jamai, Soosan Rowshanzamir, and Mehdi Amjadi. "Review of the proton exchange membranes for fuel cell applications." *International Journal of Hydrogen Energy* 35, no. 17 (2010): 9349–9384.

[13] Ahmad, M. I. et al. "Proton conductivity and characterization of novel composite membranes for medium-temperature fuel cells." *Desalination* 193, no. 1–3 (2006): 387–397.

[14] Bianchini, C. and P. K. Shen. *Chemical Reviews*, 2009, 109, 4183–4206.

[15] Antolini, E. and E. R. Gonzalez. *Journal of Power Sources*, 195 (2010) 3431–3450.

[16] Yin, Z., M. Chi, Q. Zhu, D. Ma, J. Sune and X. Bao. *Journal of Materials Chemistry A*, 1 (2013) 9157–9163.

[17] (a) Xu, A. W., Y. P. Fang, L. P. You and H. Q. Liu. *Journal of the American Chemical Society*, 125(6) (2003) 1494–1495. (b) M. Oshikiri, J. Ye and M. Boero. *Journal of Physical Chemistry C* 118 (2014) 8331–8341.

[18] (a) Kaczmarek, A. M. and R. V. Deun. *Chemical Society Reviews*, 42 (2013), 8835–8848; (b) Neeraj, S., N. Kijima and A. K. Cheetham. *Chemical Physics Letters*, 387 (2004), 2–6; (c) Tomaszewicz, E., S. M. Kaczmarek and H. Fuks. *Journal of Rare Earths*, 27 (2009) 569–573.

[19] Xu, Z., B. Feng, Y. Gao, Q. Zhao, D. Sun, X. Gao, K. Li, F. Ding and Y. Sun. *CrystEngComm*, 14 (2012) 5530–5538.

[20] Verma, Jaya, and Deepak Kumar. "Metal-ion batteries for electric vehicles: Current state of the technology, issues and future perspectives." *Nanoscale Advances* 3, no. 12 (2021): 3384–3394.

[21] Canepa, Pieremanuele, Gopalakrishnan Sai Gautam, Daniel C. Hannah, Rahul Malik, Miao Liu, Kevin G. Gallagher, Kristin A. Persson, and Gerbrand Ceder. "Odyssey of multivalent cathode materials: Open questions and future challenges." *Chemical Reviews* 117, no. 5 (2017): 4287–4341.

[22] Li, Bin, and Jun Liu. "Progress and directions in low-cost redox-flow batteries for large-scale energy storage." *National Science Review* 4, no. 1 (2017): 91–105.

[23] Xie, Jian, Wangqiao Chen, Guankui Long, Weibo Gao, Zhichuan J. Xu, Ming Liu, and Qichun Zhang. "Boosting the performance of organic cathodes through structure tuning." *Journal of Materials Chemistry A* 6, no. 27 (2018): 12985–12991.

[24] Leisegang, Tilmann, Falk Meutzner, Matthias Zschornak, Wolfram Münchgesang, Robert Schmid, Tina Nestler, Roman A. Eremin, Artem A. Kabanov, Vladislav A. Blatov, and Dirk C. Meyer. "The aluminum-ion battery: A sustainable and seminal concept?" Frontiers in Chemistry 7 (2019): 268.

[25] Reed, Luke. "Aluminum ion batteries: Electrolytes and cathodes." PhD diss., UC Merced, 2015.

[26] González, J. R., F. Nacimiento, M. Cabello, R. Alcántara, P. Lavela, and J. L. Tirado. "Reversible intercalation of aluminium into vanadium pentoxide xerogel for aqueous rechargeable batteries." *RSC Advances* 6, no. 67 (2016): 62157–62164.

[27] Song, Ming, Hua Tan, Dongliang Chao, and Hong Jin Fan. "Recent advances in Zn-ion batteries." *Advanced Functional Materials* 28, no. 41 (2018): 1802564.

[28] Zhao, Qing, Michael J. Zachman, Wajdi I. Al Sadat, Jingxu Zheng, Lena F. Kourkoutis, and Lynden Archer. "Solid electrolyte interphases for high-energy aqueous aluminum electrochemical cells." *Science Advances* 4, no. 11 (2018): eaau8131.

[29] Wu, Chuan, Sichen Gu, Qinghua Zhang, Ying Bai, Matthew Li, Yifei Yuan, Huali Wang et al. "Electrochemically activated spinel manganese oxide for rechargeable aqueous aluminum battery." *Nature Communications* 10, no. 1 (2019): 73.

[30] Li, Changgang, Xudong Zhang, and Wen He. "Design and modification of cathode materials for high energy density aluminum-ion batteries: A review." *Journal of Materials Science: Materials in Electronics* 29 (2018): 14353–14370.

[31] Zafar, Zahid Ali, Sumair Imtiaz, Rameez Razaq, Shengnan Ji, Taizhong Huang, Zhaoliang Zhang, Yunhui Huang, and James A. Anderson. "Cathode materials for rechargeable aluminum batteries: Current status and progress." *Journal of Materials Chemistry A* 5, no. 12 (2017): 5646–5660.

[32] Hu, Yuxiang, Hongjiao Huang, Deshuang Yu, Xinyi Wang, Linlin Li, Han Hu, Xiaobo Zhu, Shengjie Peng, and Lianzhou Wang. "All-Climate Aluminum-Ion Batteries Based on Binder-Free MOF-Derived FeS 2@ C/CNT Cathode." Nano-Micro Letters 13 (2021): 1–12.

[33] Yang, Yujie, Jianbin Zhou, Linlin Wang, Zheng Jiao, Meiyi Xiao, Qiu-an Huang, Minmin Liu, Qinsi Shao, Xueliang Sun, and Jiujun Zhang. "Prussian blue and its analogues as cathode materials for Na-, K-, Mg-, Ca-, Zn-and Al-ion batteries." *Nano Energy* 99 (2022): 107424.

[34] Veluri, P. S., and S. Mitra. "Enhanced high rate performance of α-Fe 2 O 3 nanotubes with alginate binder as a conversion anode." *RSC Advances* 3, no. 35 (2013): 15132–15138.

[35] Xu, Xiaodong, Ruiguo Cao, Sookyung Jeong, and Jaephil Cho. "Spindle-like mesoporous α-Fe$_2$O$_3$ anode material prepared from MOF template for high-rate lithium batteries." *Nano Letters* 12, no. 9 (2012): 4988–4991.

[36] Mitra, Sagar, Philippe Poizot, Alexandre Finke, and J-M. Tarascon. "Growth and electrochemical characterization versus lithium of Fe$_3$O$_4$ electrodes made by electrodeposition." *Advanced Functional Materials* 16, no. 17 (2006): 2281–2287.

[37] Xiong, Q. Q., J. P. Tu, Y. Lu, J. Chen, Y. X. Yu, Y. Q. Qiao, X. L. Wang, and C. D. Gu. "Synthesis of hierarchical hollow-structured single-crystalline magnetite (Fe$_3$O$_4$) microspheres: The highly powerful storage versus lithium as an anode for lithium ion batteries." *The Journal of Physical Chemistry C* 116, no. 10 (2012): 6495–6502.

[38] Yao, Xiayin, Changlin Tang, Guoxia Yuan, Ping Cui, Xiaoxiong Xu, and Zhaoping Liu. "Porous hematite (α-Fe$_2$O$_3$) nanorods as an anode material with enhanced rate capability in lithium-ion batteries." *Electrochemistry Communications* 13, no. 12 (2011): 1439–1442.

[39] Wang, Bao, Jun Song Chen, Hao Bin Wu, Zhiyu Wang, and Xiong Wen Lou. "Quasiemulsion-templated formation of α-Fe$_2$O$_3$ hollow spheres with enhanced lithium storage properties." *Journal of the American Chemical Society* 133, no. 43 (2011): 17146–17148.

[40] Hua, Bin, Meng Li, Yi-Fei Sun, Jian-Hui Li, and Jing-Li Luo. "Enhancing perovskite electrocatalysis of solid oxide cells through controlled exsolution of nanoparticles." *ChemSusChem* 10, no. 17 (2017): 3333–3341.

[41] Gu, Xiang-Kui, Samji Samira, and Eranda Nikolla. "Oxygen sponges for electrocatalysis: Oxygen reduction/evolution on nonstoichiometric, mixed metal oxides." *Chemistry of Materials* 30, no. 9 (2018): 2860–2872.

[42] Ni, Chengsheng, Jun Zhou, Ziye Zhang, Shuangbin Li, Jiupai Ni, Kai Wu, and John TS Irvine. "Iron-based electrode materials for solid oxide fuel cells and electrolysers." *Energy & Environmental Science* 14, no. 12 (2021): 6287–6319.

[43] Wang, Ying, Dawei Su, Alison Ung, Jung-ho Ahn, and Guoxiu Wang. "Hollow CoFe$_2$O$_4$ nanospheres as a high capacity anode material for lithium ion batteries." *Nanotechnology* 23, no. 5 (2012): 055402.

[44] Chu, Yan-Qiu, Zheng-Wen Fu, and Qi-Zong Qin. "Cobalt ferrite thin films as anode material for lithium ion batteries." *Electrochimica Acta* 49, no. 27 (2004): 4915–4921.

[45] Lavela, P., and J. L. Tirado. "CoFe$_2$O$_4$ and NiFe$_2$O$_4$ synthesized by sol-gel procedures for their use as anode materials for Li ion batteries." *Journal of Power Sources* 172, no. 1 (2007): 379–387.

[46] Li, Jingfa, Jiazhao Wang, Xin Liang, Zhijia Zhang, Huakun Liu, Yitai Qian, and Shenglin Xiong. "Hollow MnCo$_2$O$_4$ submicrospheres with multilevel interiors: From mesoporous spheres to yolk-in-double-shell structures." *ACS Applied Materials & Interfaces* 6, no. 1 (2014): 24–30.

[47] Zhang, Zailei, Yanhong Wang, Qiangqiang Tan, Ziyi Zhong, and Fabing Su. "Facile solvothermal synthesis of mesoporous manganese ferrite (MnFe$_2$O$_4$) microspheres as anode materials for lithium-ion batteries." *Journal of Colloid and Interface Science* 398 (2013): 185–192.

[48] Vidal-Abarca, C., P. Lavela, and J. L. Tirado. "The origin of capacity fading in NiFe$_2$O$_4$ conversion electrodes for lithium ion batteries unfolded by 57Fe Mossbauer spectroscopy." *The Journal of Physical Chemistry C* 114, no. 29 (2010): 12828–12832.

[49] Xiong, Q. Q., J. P. Tu, S. J. Shi, X. Y. Liu, X. L. Wang, and C. D. Gu. "Ascorbic acid-assisted synthesis of cobalt ferrite (CoFe$_2$O$_4$) hierarchical flower-like microspheres with enhanced lithium storage properties." *Journal of Power Sources* 256 (2014): 153–159.

[50] Jiang, Jian, Yuanyuan Li, Jinping Liu, Xintang Huang, Changzhou Yuan, and Xiong Wen Lou. "Recent advances in metal oxide-based electrode architecture design for electrochemical energy storage." *Advanced Materials* 24, no. 38 (2012): 5166–5180.

[51] Reddy, Mogalahalli V., G. V. Subba Rao, and B. V. R. Chowdari. "Metal oxides and oxysalts as anode materials for Li ion batteries." *Chemical Reviews* 113, no. 7 (2013): 5364–5457.

[52] Fan, Yu, Qing Zhang, Qizhen Xiao, Xinghui Wang, and Kai Huang. "High performance lithium ion battery anodes based on carbon nanotube-silicon core-shell nanowires with controlled morphology." *Carbon* 59 (2013): 264–269.

[53] Shendkar, Janardhan H., Vijaykumar V. Jadhav, Pritamkumar V. Shinde, Rajaram S. Mane, and Colm O'Dwyer. "Hybrid composite polyaniline-nickel hydroxide electrode materials for supercapacitor applications." *Heliyon* 4, no. 9 (2018).

[54] Li, Ying, Bingkun Guo, Liwen Ji, Zhan Lin, Guanjie Xu, Yinzheng Liang, Shu Zhang et al. "Structure control and performance improvement of carbon nanofibers containing a dispersion of silicon nanoparticles for energy storage." *Carbon* 51 (2013): 185–194.

[55] Sharma, Yogesh, N. Sharma, GV Subba Rao, and B. V. R. Chowdari. "Li-storage and cyclability of urea combustion derived $ZnFe_2O_4$ as anode for Li-ion batteries." *Electrochimica Acta* 53, no. 5 (2008): 2380–2385.

[56] Xu, Huayun, Xianglan Chen, Liang Chen, Li'E. Li, Liqiang Xu, Jian Yang, and Yitai Qian. "A comparative study of nanoparticles and nanospheres $ZnFe_2O_4$ as anode material for lithium ion batteries." *International Journal of Electrochemical Science* 7, no. 9 (2012): 7976–7983.

[57] Guo, Xianwei, Xia Lu, Xiangpeng Fang, Ya Mao, Zhaoxiang Wang, Liquan Chen, Xiaoxue Xu, Hong Yang, and Yinong Liu. "Lithium storage in hollow spherical $ZnFe_2O_4$ as anode materials for lithium ion batteries." *Electrochemistry Communications* 12, no. 6 (2010): 847–850.

[58] Yao, Xiayin, Junhua Kong, Dan Zhou, Chenyang Zhao, Rui Zhou, and Xuehong Lu. "Mesoporous zinc ferrite/graphene composites: Towards ultra-fast and stable anode for lithium-ion batteries." *Carbon* 79 (2014): 493–499.

[59] He, Yang, Ling Huang, Jin-Shu Cai, Xiao-Mei Zheng, and Shi-Gang Sun. "Structure and electrochemical performance of nanostructured Fe_3O_4/carbon nanotube composites as anodes for lithium ion batteries." *Electrochimica Acta* 55, no. 3 (2010): 1140–1144.

[60] Z. Yu, B. Wang, X. Liao, K. Zhao, Z. Yang, F. Xia, C. Sun, Z. Wang, C. Fan, J. Zhang, Y. Wnag. *Advanced Energy Materials* (2020) 2000907.

[61] M. Song, H. Tan, D. Chao, H. J. Fan. *Advanced Functional Materials* 28 (2018) 1802564.

[62] V. Verma, S. Kumar, W. Manalastas Jr, R. Satish, M. Srinivasan. *Advanced Sustainable Systems* 3 (2019) 1800111.

[63] K. Zhao, C. Wang, Y. Yu, M. Yan, Q. Wei, P. He, Y. Dong, Z. Zhang, X. Wang, L. Mai. *Advanced Materials Interfaces* 5 (16) (2018) 1800848.

[64] W. Xu, C. Sun, K. Zhao, X. Cheng, S. Rawal, Y. Xu, Y. Wang. *Energy Storage Materials* 16 (2019) 527–534.

[65] P. Poizot, S. Laruelle, S. Grugeon, L. Dupont, J. M. Tarascon. *Nature* 407 (2000) 496e499.

[66] G. M. Zhou, D. W. Wang, F. Li, L. L. Zhang, N. Li, Z. S. Wu, L. Wen, G. Q. Lu, H. M. Cheng. *Chemistry of Materials* 22 (2010) 5306e5313.

[67] B. Wang, J. S. Chen, H. B. Wu, Z. Y. Wang, X. W. Lou. *Journal of the American Chemical Society* 133 (2011) 17146e17148.

[68] X. W. Lou, D. Deng, J. Y. Lee, J. Feng, L. A. Archer. *Advanced Materials* 20 (2008) 258e 262.

[69] N. A. Chernova, M. Roppolo, A. C. Dillon, M. S. Whittingham. *Journal of Materials Chemistry* 19 (2009) 2526e2552.

[70] L. Q. Mai, X. Xu, L. Xu, C. H. Han, Y. Z. Luo. *Journal of Materials Research* 26 (2011) 2175e2185.

[71] H. M. Liu, Y. G. Wang, K. X. Wang, E. Hosono, H. S. Zhou. *Journal of Materials Chemistry* 19 (2009) 2835e2840.

[72] L. F. Xiao, Y. Q. Zhao, J. Yin, L. Z. Zhang. *Chemistry – A European Journal* 15 (2009) 9442e9450.

[73] D. H. Sim, X. H. Rui, J. Chen, H. T. Tan, T. M. Lim, R. Yazami, H. H. Hng, Q. Y. Yan. *RSC Advances* 2 (2012) 3630e3633.

[74] H. W. Liu, D. G. Tang. *Materials Chemistry and Physics* 114 (2009) 656e659.

[75] Mu, C., Mao, J., Guo, J., Guo, Q., Li, Z., Qin, W., … & Qiao, S. Z. (2020). Rational design of spinel cobalt vanadate oxide Co_2VO_4 for superior electrocatalysis. *Advanced Materials*, 32(10), 1907168.

[76] Qiao, H., Luo, L., Chen, K., Fei, Y., Cui, R., & Wei, Q. (2015). Electrospun synthesis and lithium storage properties of magnesium ferrite nanofibers. *Electrochimica Acta*, *160*, 43–49.

[77] Ruan, W., Yue, X., Ni, J., Xie, D., & Ni, C. (2021). α-PbO_2-type niobate as efficient cathode materials for steam and CO2 electrolysis. *Journal of Power Sources*, *483*, 229234.

[78] Xia, D., Gao, H., Li, M., Gong, F., & Li, M. (2021). Transition metal vanadates electrodes in lithium-ion batteries: A holistic review. *Energy Storage Materials*, *35*, 169–191.

[79] Khaerudini, D. S., Guan, G., Zhang, P., Hao, X., & Abudula, A. (2014). Prospects of oxide ionic conductivity bismuth vanadate-based solid electrolytes. *Reviews in Chemical Engineering*, 30(6), 539–551.

9 The Role of Inorganic Nanomaterials in Energy Storage Devices

*A. Mushira Banu, B. Arifa Farzana,
and F.M. Mashood Ahamed*

FUEL CELLS

INTRODUCTION

Numerous energy-related technologies, including solar photovoltaic systems, fuel cells, solar thermal systems, lithium-ion batteries, and lighting, have contributed to the conversion and storage of renewable energy. Additionally, a new generation of solar collectors based on nanotechnology is known as nanofluid-based solar collectors. It might boost collector efficiency by as much as 30%. Fuel cells have a wide range of uses, including producing electricity for transportation, commercial, industrial, and residential structures, as well as long-term energy storage for the grid in reversible systems. Compared to traditional combustion-based technologies, which are now employed in many power plants and automobiles, fuel cells provide a number of advantages. Compared to combustion engines, fuel cells run more efficiently and are capable of directly converting the chemical energy in fuel into electrical energy at efficiencies of more than 60%. Nanostructured materials have brought a fresh innovation and discovering to overcome them despite numerous obstacles to commercialization. The use of nanomaterials in the catalyst, electrolyte/membrane, and electrodes in fuel cells can overcome numerous barriers to commercialization, including high costs of materials and fuel cross-over. The unique characteristics of nanomaterials, such as their high surface area and special size effect, can significantly improve both overall efficiency and cell performance (Figure 9.1).

DEVELOPMENTS

Fuel cells are extremely effective and green devices that generate energy through electrochemical reactions. They are regarded as green energy sources since they do not release harmful pollutants like carbon dioxide (CO_2), carbon monoxide (CO), nitrogen dioxides (NO_2), and sulfur dioxides (SO_2), which are produced by conventional coal combustion systems.

Each type of fuel cell has a varied amount of efficiency, which affects how much power is produced. PEMFC and DMFC are typically employed in low-power applications (1 W–100 kW) due to their low operating temperatures. On the other hand, SOFC and MCFC have a better chance of being used in large-scale power plants and high-power applications (100 kW–5 MW). Low-temperature operation has several advantages, including faster system startup and longer gadget life. High-temperature fuel cells, SOFCs, and MCFCs can reform a range of fuels to produce hydrogen due to their fuel flexibility. The high operating temperature, on the other hand, promotes material degradation and eventually limits the lifespan of the materials. Furthermore, a lengthy startup period is required, and they are limited to stationary applications.

For the past few years, hydrogen fuel has been seen as a new energy currency. Hydrogen fuel cells (Figure 9.2) are an essential use of hydrogen energy because they are highly efficient in converting

DOI: 10.1201/9781003479239-9

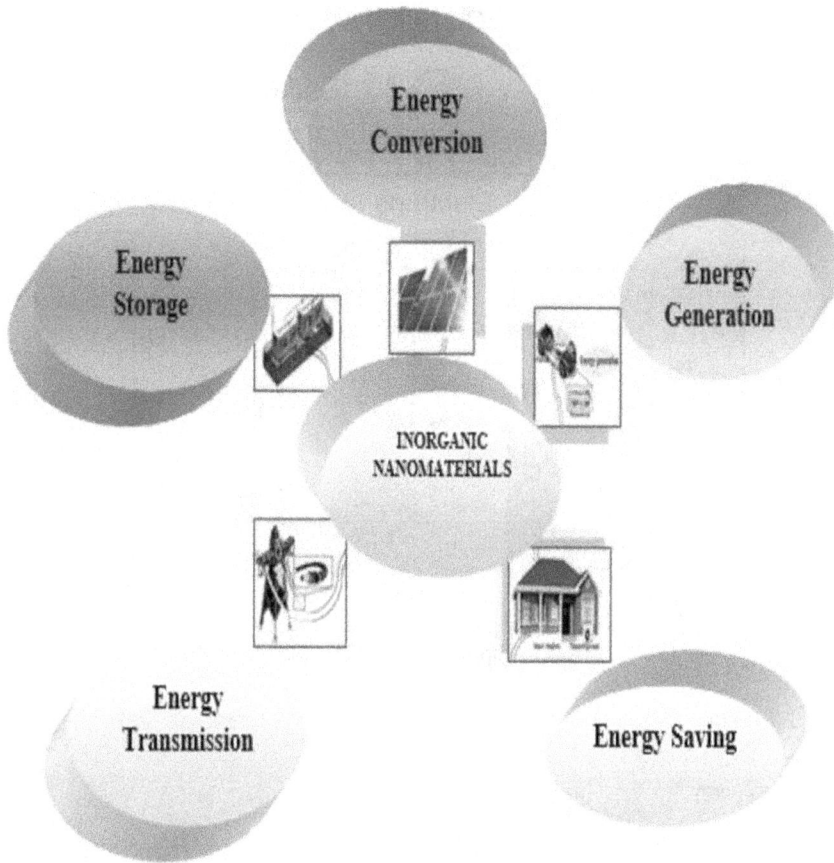

FIGURE 9.1   The role of inorganic nanomaterials in energy storage devices.

chemical energy into electric energy, which may be used in many forms of transportation. However, we needed extremely high-quality and reliable supply pressure vessels in order to store this new gasoline.

The following succinctly describes the qualities that need to be improved:

1. Lightweight complicated hydrides must be used to increase the system's low storage capacity [1,2]. Two distinct multifunctional hydrides can be effective agents when doped into nonporous framework materials, which is an interesting finding [3].
2. By adjusting the catalytic characteristics and changing the doping methods, they may achieve the purpose. Catalyzed and nano-confined samples together are possible candidates for utilization in fuel cells.

Throughout the last few decades, numerous researchers have investigated and analyzed the creation, use, and importance of nanomaterials. Nanoparticles of a diameter of a few nanometers can change the surfaces, behaviors, and electrochemical characteristics of materials. This modification might enhance the system's functionality and effectiveness.

NANOSTRUCTURED MATERIALS AS FUEL CELLS

Fuel cells are effectively utilizing nanostructured materials to increase the conversion of hydrogen energy to electricity. One of the most promising methods for utilizing a variety of energy sources

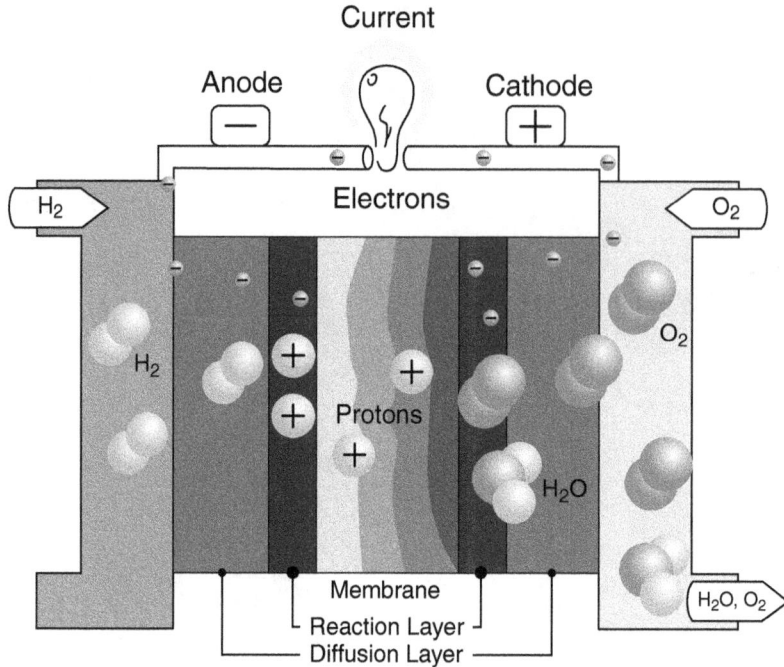

FIGURE 9.2   Fuel cells.

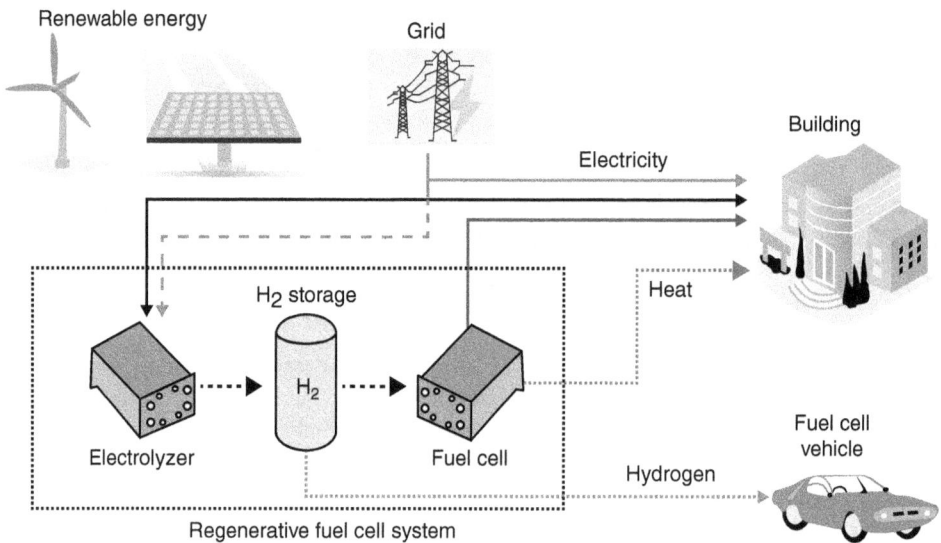

FIGURE 9.3   Different energy applications: energy generation, storage, conversion, and saving up on nano-materials substances [12].

(Figure 9.3), as well as for ensuring the sustainability of both the energy system and the environment, is fuel cell technology [4]. In a fuel cell, hydrogen and oxygen combine to make water, which produces heat and electricity. This happens in an eco-friendly manner with no damaging carbon dioxide (CO_2) emissions, as shown in Figure 9.2.

The fuel cell has a lot of benefits, but it also has certain disadvantages, such as high cost, operability, and durability difficulties. These issues may be resolved via nanotechnology. The fuel cell membrane divides hydrogen into protons and electrons. Carbon nanotubes (CNTs), catalysts, electrodes, and polymer electrolyte membranes are all examples of practical applications of nanomaterials. Another significant constraint on fuel cells is the large amounts of hydrogen that can be stored inside nanomaterials like carbon nanotubes (CNTs) and carbon nanofibers. The eco-friendliest source of energy, carbon nanotube fuel cells, are now used to store hydrogen. The layered graphene tubular structure of carbon nanotubes makes them electrically conductive.

Hydrogen is frequently utilized in space spacecraft because it has the highest energy-to-weight ratio of any fuel [5]. Due to their distinct structural characteristics, CNTs appear to be an effective and appealing catalyst support in hydrogen production. CNT surfaces are typically changed to form functional groups for specific purposes. Nanotube diameter affects the hydrogenation of single-walled carbon nanotubes (SWCNTs) using atomic hydrogen as the hydrogenation agent, with diameter values around 2.0 nm being the most efficient, according to Nikitin et al. [6]. As a result, hydrogenation (resulting in stable nanotube–hydrogen complexes) was almost complete. This leads to a 7% increase in hydrogen storage capacity. The power density of a platinum-catalyzed hydrogen cell supported by single-walled carbon nanotubes (SWCNTs) was investigated [7]. They concluded that CFE/SWCNT/Pt electrodes had a nearly 20% higher maximum power density than that of CFE/CB/Pt electrodes. Additionally, the hydrogen storage capacity of SWNTs and MWNTs is less than 1 wt% at room temperature, but by lowering the adsorption temperature or changing the CNTs, the capacity may be greatly enhanced between 4 and 8 wt% [8]. Statistics on hydrogen storage capacity in CNT differ from one another because of their unique characteristics. Additionally, it was established that the impregnation of Ni and Pd/Ni nanoparticles on the Vulcan XC-72R carbon black electrode catalyzed the oxidation of methanol in a hydrogen cell [9]. Pd/Ni/C electrocatalysts are more stable than Pd/C and Ni/C electrocatalysts, according to their investigation into the methanol oxidation reaction at Ni/C and Pd/Ni/C electrocatalysts in (0.2 M MeOH, 0.5 M KOH) solution. Pd/Ni/C is therefore an excellent and less expensive electrocatalyst for methanol oxidation that can be applied to fuel cells. According to Khater et al.'s investigation into the impact of bifunctional manganese oxide-silver nanocomposites anchored on graphitic mesoporous carbon on oxygen reduction and cathodic biofilm growth in microbial fuel cells for long-term operation, MnOx-Ag/GMC nanocomposites demonstrated high antibacterial activity in MFCs, inhibiting biofilm growth on the cathode [10]. Also shown are extremely effective, affordable, and environmentally acceptable electrocatalysts for the oxygen and hydrogen evolution processes needed for inexpensive water splitting. A simple adsorption-growth technique was used to anchor FeS_2 nanoparticles to the surface of MXene (FeS_2@MXene) [11].

Pt-based electrocatalysts are typically synthesized using a variety of colloidal preparation techniques, either unsupported or supported on high-surface-area carbon [13]. One recent development in the realm of CO-tolerant catalysts is the fabrication of new nanostructures by spontaneous deposition of Pt sub-monolayers on carbon-supported Ru nanoparticles [14,15]. There is still much to learn about the surface chemistry of electrocatalyst nanoparticles and the effects of strong metal-support interactions on the dispersion and electrical characteristics of platinum sites. An alternative to employing carbon blacks is to create porous silicon catalyst support structures with a 5 m pore diameter and a thickness of around 500 m. Due to their large surface areas, these shapes are desirable for small PEM fuel cells [16]. In DMFCs [16], a uniformly distributed, finely dispersed catalyst layer is coated on the silicon pore walls, where it interacts with the ionomer to produce a three-phase reaction zone that is highly efficient and capable of high power generation. Organic transition metal complexes, such as iron or cobalt organic macrocycles from the families of phenylporphyrins [17] and nanocrystalline transition metal chalcogenides [18], are being investigated as a substitute for platinum for oxygen reduction due to their high selectivity toward the ORR and tolerance to methanol cross-over in DMFCs. In order to produce a nanostructured product with a distinctive atomic arrangement, the metal-organic macrocycle is supported on carbon with a wide surface area and heated to high temperatures (between 500°C and 800°C). Even though their electrocatalytic activity is less than that of Pt, these materials display adequate performance [17].

As with batteries, a vital part of the fuel cell assembly is the electrolyte. Due to its outstanding conductivity and electrochemical stability [19], H_2 or methanol/air fuel cells use perfluorosulphonic polymer electrolyte membranes as its electrolyte. Unfortunately, they have a number of shortcomings, including membrane dehydration and methanol cross-over. The latter significantly impairs fuel cell functioning beyond 100°C, a need for the proper oxidation of small organic molecules that involves the development of heavily adsorbed chemical intermediates like CO-like species [13,20,21]. Alternative membranes based on block co-polymer ion-channel-forming materials or poly(arylene ether sulfone) [22], sulfonated poly(ether ketone) [23], or both, have also been proposed [24–26]. Many relationships between membrane nanostructure and transport characteristics as conductivity, diffusion, permeation, and electro-osmotic drag have been demonstrated [27]. It's interesting to note that sulfonated poly(ether ketone) has greater sulfonic group separation and fewer connected hydrophilic channels than Nafi on, which minimizes water permeation and electro-osmotic drag while maintaining high protonic conductivity [27]. Furthermore, by combining polymeric N-bases and polymeric sulfonic acids, nano-separated acid–base polymer blends have shown increased thermal and mechanical stability [23]. Over the past ten years, significant efforts have been made to build composite membranes. Ionomeric membranes modified by dispersing insoluble acids, oxides, zirconium phosphate, and other materials within their polymeric matrix are instances of this [24]. Ionomers and inorganic solid acids with high proton conductivity incorporated in porous non-proton-conducting polymers are more examples. By producing multilayer, insoluble Zr phosphonates in ionomeric membranes, Alberti and Casciola [24] have produced a nanocomposite that is superior to Zr phosphates and equal to Nafi, with the aim of reducing the drawbacks of electrolytes. These substances, including $Zr(O_3P\text{-}OH)(O_3P\text{-}C_6H_4\text{-}SO_3H)$, have conductivities that are significantly greater than those of perfluorosulphonic membranes. Nanoceramic fillers have been included into the polymer electrolyte network. By using tiny amounts of SiO_2 and Pt/TiO_2 (7 nm) nanoparticles to keep the water generated electrochemically inside the membrane, Stonehart, Watanabe, and colleagues [28] were able to alleviate the humidification restrictions in PEMFCs. It has been shown that the operating temperature of similarly modified membranes filled with nano-crystalline ceramic oxide is 150°C [29]. More than the theory that the inorganic filler induces structural changes in the polymer matrix, the presence of acidic functional groups on the surface of nanoparticle fillers appears to favor the water-retention mechanism [24–26]. The filler's impact on the transport properties does not appear to be considerable at this time [30]. However, the "vehicular mechanism" of proton conduction that occurs at lower temperatures [31] should be encouraged by the composite membrane's greater capacity to hold water at high temperatures (130–150°C) and low humidity levels [32]. The three-phase reaction zone is extended by the composite catalyst layer created in the PEMFCs by the Pt/C catalyst's close interaction with the electrolyte ionomer. Similar to the composite cathode strategy used in lithium-ion batteries, where the electrode is made up of two interpenetrating networks for electron and ion conduction, the benefit of this method is an improvement of the interfacial region between catalyst particles and ionomer [19,33]. It has been shown that the Pt content can be decreased to much less than 0.5 mg/cm^2 without affecting the cell's function or lifespan [19,33]. Durable multi-level MEAs are being created by the 3M Corporation employing automated assembly techniques and high-speed precision coating technologies [34]. A platinum-coated nano-whisker-based thin-film catalyst with nanostructures is a component of the MEA. The method makes use of single-crystalline whiskers of an organic pigment material that are highly oriented and have high aspect ratios. The application of the catalysts is made possible by this support, which also facilitates production and processing. The electro-catalytic activity that were so obtained, though, are comparable to those of catalyst-ionomer inks. Nanomaterials are more common, and their use is not just for low-temperature fuel cells. IT-SOFCs, or intermediate-temperature solid oxide fuel cells, are increasingly using nanostructured electro-ceramic materials. While nanoscale particles can serve as a starting point for temperatures beyond 1,000°C needed to build SOFCs [35], they are typically used to create micro-structured parts with unique ion-conduction and electro-catalytic properties.

Nanosized YSZ (8% Y_2O_3-ZrO_2) and ceria-based (CGO, SDC, YDC) powders enable a reduction in the firing temperature during the membrane-forming step in the cell manufacturing operation because their sintering properties are different from those of polycrystalline powders [36]. Furthermore, charge transfer reactions at the electrode–electrolyte interface are promoted by nanocrystalline ceria, which exhibits mixed electronic–ionic conduction capabilities [36]. Point flaws are the source of ionic charge carriers in electro-ceramic materials. Due to the substantially larger area of the interface and grain boundaries, the density of mobile faults in the space-charge region increases in nanostructured systems. The electrochemical behavior is completely different from that of bulk polycrystalline materials as a result [35,38]. These 'trivial' size effects should be separated from 'real' size effects, according to Maier [37], which take place when the particle size is less than four times the Debye length and where the local characteristics are changed in terms of ionic and electronic charge-carrier transport. Fascinatingly, Maier used the same space-charge model that was previously discussed to explain the observed increase in ionic conductivity of dry or hybrid Li-based polymer systems loaded with nano-inorganic fillers (SiO_2, Al_2O_3, TiO_2). Despite the importance of nanostructured materials, real fuel cell performance is still influenced by scale-up, stack housing design, gas manifold, and sealing. Despite advances in the direct electrochemical oxidation of alcohol and hydrocarbon fuels, hydrogen still serves as the primary fuel for most fuel cells [13,39,40]. Many studies are currently being conducted on alanates for hydrogen storage, carbon nanotubes, nanomagnesium-based hydrides, metal-hydride/carbon nanocomposites, and nanomagnesium-based hydrides [41,42]. However, the US Department of Energy has set the minimum hydrogen storage level for automotive applications at 6.5% [47], and current reliable hydrogen storage capacities in single-walled carbon nanotubes appear to be on par with or even lower than those of metal hydrides [45–47]. Early research revealed that [43,44] K- and Li-doped carbon nanotubes had hydrogen adsorption capabilities of 14–20 wt%. Mg-hydrides have been modified with transition metals in this field recently [48,49], and boron-nitride nanostructures have been studied [50]. Magnesium hydride, or MgH_2, is commonly changed by high-energy ball milling with alloying elements including [48] Ni, Cu, Ti, Nb, and Al to generate nanoparticles in the range of 20–30 nm that, after 20 hours of milling, have hydrogen storage capacities of roughly 6–11 wt% [48,49]. The partial substitution of Al for Mg in Mg_2Ni hydrogen storage systems appears to increase the life-cycle properties in addition to the hydrogen sorption kinetics of MgH_2 because an Al_2O_3 coating is present on the alloy surface [48]. Further advancements were made as a result of research into MgH_2-V nanocomposites 100 and the inclusion of very small amounts of nanoparticulate transition metal oxides (Nb_2O_5, WO_3, Cr_2O_3) to Mg-based hydrides. Additionally, the adsorption and desorption of transition metal oxide promoters appear to be greatly improved by the catalytic activity of substances like Cr_2O_3, reactive mechanical grinding, and nanosized particles [50].

REMARKS

Fuel cells have undergone extensive investigation because of the strains of the energy demand and the environmental issues brought on by traditional combustion-based power producers. Due to its high energy efficiency and low carbon and NOx emissions, fuel cells have become a promising alternative, clean, and sustainable power generating technology. The results demonstrate the huge potential of nanomaterials to enhance fuel cell performance. Whether formed from bulk to nanoscale particles (top-down) or from groups of atoms or molecules assembled to create nanoparticles, having nanometer range particles can significantly affect the properties of fuel cell components (electrocatalyst, PEM, electrodes, electrolyte), as well as their durability and stability.

SUPERCAPACITORS

INTRODUCTION

A supercapacitor (SC), commonly referred to as an ultracapacitor, is a high-capacity capacitor with a capacitance value much higher than regular capacitors but less voltage constraints. It fills the hole

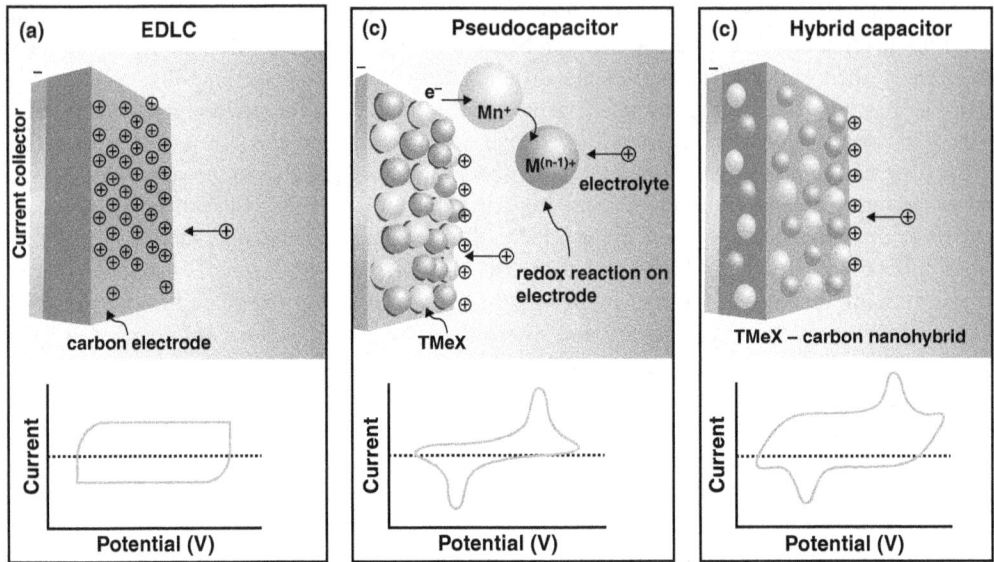

FIGURE 9.4 Different kinds of supercapacitors.

left by electrolytic capacitors and rechargeable batteries. It typically has ten to one hundred times the energy storage capacity per unit volume or mass of electrolytic capacitors, can absorb and disperse charge much more quickly than batteries, and can survive a much larger number of charge and discharge cycles than rechargeable batteries [51–52]

Electrochemical double-layer capacitors (EDLC), pseudo-capacitors (PC), and hybrid supercapacitors (HSC) are the three different types of supercapacitors (Figure 9.4). Due to their superior high power and cyclability, EDLCs electrostatically store energy at the electrode/electrolyte border. By storing energy generated by surface redox reactions on the electrode, PCs make ideal energy storage devices. The advantages of both EDLCs and PCs are combined in HSCs. The performance of hybrids is affected by the type of interface and its accessibility between the double-layer and pseudocapacitive components. The EDLC/PC interface will be covered in detail in this chapter using concrete examples. Graphene, turbostratic carbons with different morphologies, including carbon nanofibers (CNFs), activated carbons (ACs), carbon nanotubes (CNTs), and templated carbons (TCs) are a few of the materials employed in EDLC [53–54]. ACs, CNTs, graphene, turbostratic carbons with various morphologies, including CNFs, and TCs are among the materials used in EDLC. Transition metal oxides, nitrides, sulfides, and carbides that are redox active will make up the majority of the components in the PC.

To evaluate the performance of a hybrid electrode material for supercapacitors, several electrochemical characterization approaches are applied. The three techniques that are most frequently employed are cyclic voltammetry (CV), galvanostatic charging and discharging (GCD), and electrochemical impedance spectroscopy (EIS). CV provides a variety of data, including double-layer capacitance and redox reaction signatures, and evaluates the current response to a linear potential sweep between two sites [55]. GCD shows capacitance, power density, and energy density in addition to the change in potential over time at constant current input and output [56]. The response of various cell components to frequency and voltage is shown by spectroscopy of electrochemical impedance (Nyquist and Bode plots).

PRINCIPLES OF SUPERCAPACITOR

a. Atomistic and multiscale modeling permits the assessment, selection, and even design of novel materials, topologies, production processes, and energy storage strategies.

b. A general recognize of the functioning, breakdown, and deterioration of modern energy storage materials and systems, as well as charge transport and storage at nanoscale electrochemical surfaces.

c. Research on nanoscale processes, asymmetric supercapacitors, double-layer charge storage, pseudo-capacitance, and the creation and use of novel techniques, both in situ and in operation [57].

NANOMATERIALS FOR SUPERCAPACITORS

Porous carbon with a large surface area. Since they have a large surface area, are inexpensive, and are simple to produce, ACs have historically been the most popular material for EDLCs. To create the EDLCs that are currently on the market, numerous materials are pyrolyzed and activated, including wood, petroleum, plant matter, and phenolic resins (such are cryogels and aerogels). Due to the electrolyte's deterioration, the performance of EDLC can be compromised by the high impurity or ash content of naturally generated electrodes. Additionally, the pore size varies from source to source and is constrained by what nature offers.

Naturally generated electrodes contain high levels of contaminants or ash content, which might impair EDLC performance due to electrolyte deterioration. Pore size is also constrained by nature and may differ from source to source. Synthetic carbons, on the other hand, offer outstanding conductivity, a high surface area, regulated bulk and surface chemistry, and an adjustable porous structure, making them attractive candidates as electrodes for high-energy-density supercapacitors with long cyclic stability [53]. It is possible to improve capacitance by carefully performing thermal, chemical, and electrochemical processes to increase the accessible area of the surface and functional groups on the surface, or by extending the voltage that operates range past the limit of a water-based electrolyte solution.

An organic precursor impregnated in an inorganic template is carbonized, and the template is then taken out to generate the carbon template. Two typical template materials are mesoporous molecular sieves, which are often made of mesoporous silica and clay, where the carbon precursor is infiltrated between the lamellae. Another material that has long-range order, multidimensional channels, and an open microporous structure is zeolites. It is anticipated that all of the holes will be accessible to the electrolyte because the pores of the resulting carbons are homogenous and well-organized. Furthermore, carbons with microporosity and not much mesoporosity can have surface areas of over $4,000\,m^2/g$ when the size of the pores is perfectly controlled.

Fuertes et al. converted the carbon precursor furfuryl alcohol into porous carbon using SBA-16 silica as a template. The carbon that results from this process has a unimodal pore size distribution and the same form as the silica template. Due to the optimized pore size that facilitates simple transfer of the ions in the electrolyte, the template carbon performed exceptionally well in supercapacitors at high charge–discharge rates [58]. Using a zeolite template, Xu et al. created hierarchical porous carbon. Pore structures were regulated by altering the period of the carbon precursor's chemical vapor deposition. The pore structure of the final template carbon was also altered by pretreatment of zeolite with NaOH prior to precursor deposition. Samples with the highest surface area accessibility, which reached 215 F/g in a 6 M KOH aqueous electrolyte, demonstrated the best electrochemical performance. Carbide-derived carbon (CDC), a new class of synthetic mesoporous carbon, was recently created by carefully eliminating non-carbon atoms from carbides during high-temperature processing (i.e., chlorination) [1,59]. By selecting the precursor (i.e., the composition and structure of the carbide) and the chlorination conditions, it is possible to tailor the porous structure of CDC, including the average size of pores, distribution of pore size, volume of pores, and specific area of the surface. A material for supercapacitor electrodes has been constructed by CDC with a specific surface area of $43,100\,m^2/g$ with holes that are B0.3–10 nm in size [60].

According to Presser et al., electrospun titanium carbide (TiC) nanofelts were used as a precursor to create nano-fibrous felts (nanofelts) of CDC. The main characteristics of the precursor, such as high interconnectivity and structural integrity, are preserved during the conformal transformation of TiC into CDC. The created TiC-CDC nanofelts are mechanically flexible and resilient, and can be used as an electrode material for supercapacitor applications without the use of a binder [61]. After being created by chlorinating the precursor at 600°C, the TiC-CDC nanofelts have an average pore size of B1 nm, a high specific surface area of 1,390 m^2/g, and graphitic carbon ribbons contained in a highly disordered carbon matrix.

The aqueous electrolyte has a high gravimetric capacitance of 110 F/g, while the organic electrolyte has a gravimetric capacitance of 65 F/g, according to electrochemical measurements. Korenblit et al. were able to successfully prepare CDC from SiC. Polycarbosilane was infiltrated into an ordered mesoporous SiO$_2$ template, which was subsequently etched away to reveal mesoporous SiC. The technique produced porous CDC nanorods with aligned mesopores between the particles [62]. Pore widths were less than 4 nm, and the BET-specific surface area was larger than in previous SiC-CDC tests, at 2,250–2,430 m^2/g. The specific capacitance of TEATFB/AN (tetraethylammonium tetrafluoroborate in acetonitrile) was 170 F/g. The structured mesopores enable quick ion conductivity into the bulk CDC particles, resulting in a fast frequency response and capacitance retention at high current densities [63].

Graphene and Carbon Nanotubes (CNTs)

Carbon nanotubes and graphene were additionally employed in the manufacturing of supercapacitors due to their high conductivity. They are highly conductive and have a mesopore network, which allows for rapid ion transport in EDLCs. The lack of micropores, on the other hand, results in a comparatively small surface area of 100–400 m^2/g. Higher capacitance can, however, be obtained with faultier outer CNT walls, and the inclusion of a tiny covering of amorphous carbon on the outer walls results in enhanced charge accumulation. Furthermore, increasing surface area through activation mechanisms may increase capacitance. An et al. achieved a maximum specific capacitance of 180 F/g with a high power density of 20 kW/kg by heating CNTs to enhance their specific surface area and pore dispersion [64]. However, as compared to other AC electrodes (up to 3,000 m^2/g) or mesoporous carbons (up to 1,730 m^2/g), CNTs have a low specific surface area.

Researchers have also used ammonia, an aqueous NaOH-KOH solution, or nitric acid on the surfaces of CNTs to introduce oxygen functional groups (carboxyl, lactone, phenolic, ether groups, aldehyde, and so on) and therefore improve the electrochemical properties of the electrodes. For example, Niu et al. treated MWNTs with HNO$_3$ and measured a specific capacitance of 113 F/g with an electrolyte solution of 38 wt% H$_2$SO$_4$ [65].

By surface treating the CNT electrode with ammonia plasma, Yoon et al. increased its capacitance from 38.7 to 207.3 F/g. The oxygen groups, on the other hand, might promote capacitor instability, resulting in increased resistance and capacitance deterioration. Adding surface oxygen groups to the CNT electrode would also be detrimental in an organic electrolyte [66].

The formula for supercapacitor specific power is $P_{max} = V^2/4R$, where V is the working voltage and R is the equivalent series resistance. The resistance is made up of the electronic resistance of the electrode material, the resistance at the interface between the electrode and the current collector, the resistance generated by ions diffusing through the microscopic pores of the electrode material, and other resistances [67]. That is, for high performance, such as the density of CNT arrays, the characteristics of the electrode materials and the type of electrode generated become critical. CNTs can be directly deposited as electrodes on the current collector (Ni, Al, alloy, or other metal) to improve energy storage while decreasing contact resistance. Dorfler et al., for example, designed a scalable process for directly manufacturing vertically aligned carbon nanotubes with thicknesses ranging from 5 to 20 nm on metal foil substrates [68]. The obtained capacitance was close to the theoretical value expected for single-walled carbon nanotubes, but the resistance was exceptionally low (53.6–232 m/cm^2). Because the polymer binders produce less impurities, the electrochemical performance of this CNT film electrode would improve as a result of its production. Some research groups even developed tiny, lightweight electrodes that employed dense CNT networks as both the active electrode material and the current collector. Graphene has a theoretical surface area of 2,675 m^2/g, which might result in a specific capacitance of 550 F/g. However, aggregation of the graphene sheets, which reduces the available surface area and results in capacitances of just over 100 F/g, prevents this from being accomplished. To avoid aggregation, Wang et al. used a gas-solid reduction approach to achieve a maximum specific capacitance of 205 F/g at 1.0 V in an aqueous electrolyte with an energy density of 28.5 Wh/kg. Liu et al. developed curved graphene supercapacitors to prevent the graphene sheets from stacking again. They achieved extraordinarily high energy densities by using an ionic liquid electrolyte (85.6 Wh/kg at 1 A/g at room temperature or 136 Whk/g at 80°C). Several organizations have reported using CNTs as a spacer between graphene sheets to prevent agglomeration and act as a conductive additive and binder. The specific capacitance of a capacitor can reach 326 F/g [69] with a weight ratio of 9 to 1 between the graphene layer and the CNT.

Porous Conducting Polymers

Conducting polymers are a desirable pseudocapacitive material because they are simple to make, have a high charge density, are relatively inexpensive, and work well as thin films. These substances can undergo reversible reduction and oxidation. The majority of the material's redox processes provide high energy storage and low self-discharge, while power density typically decreases due to slow diffusion [70]. Polyaniline is likely the most investigated conducting polymer due to its various redox states. However, other prevalent polymers include polypyrrole and its derivatives. The preparation characteristics, especially the synthetic process and monomer structure, frequently have a considerable impact on capacitance [71].

Sharma et al. created highly porous poly-pyrrole films by employing a distinctive pulsed polymerization method. 10 mm flakes with short polymer chain lengths and high levels of conjugation were produced using a shorter pulse period. Due to fewer chain defects and a high level of doping, the resultant capacitance was 400 F/g with an energy density of 250 Wh/kg and strong cycle stability. A high power density was discovered in poly(3,4-ethylenedioxythiophene) nanotubes grown on an alumina template because of their rapid charge–discharge capabilities [72]. While the thin nanotube walls provide a short diffusion distance, the hollow structure makes it easy for counter-ions to be carried into the polymer. Recently, Wang et al. developed aligned polyaniline nanowire arrays with high specific surfaces and organized nanostructures. In comparison to other results for polyaniline nanowire networks (742 F/g) and polyaniline nanowire arrays (700 F/g) [73–74], the capacitance in aqueous solution is significantly higher at 950 F/g. Because of how quickly electrolyte ions may penetrate the polymer due to the narrow nanowire width, a capacitance of 780 F/g was maintained at high current densities.

Nanostructured Metal Oxides

Hydrous RuO$_2$, which has a reported capacitance of 850 F/g, is the best example of redox pseudo-capacitance. Strong conductivity, consistent redox activity across a wide voltage range, and high reversibility are all characteristics of the material. However, usefulness of RuO$_2$ is constrained by its high price. The development of nanostructures from these materials, and substituting RuO$_2$ with less expensive oxides of transition metals and nitrides including Mn, Fe, Co, Ni, V, Mo, and In, have been the main topics of recent research. For instance, cobalt oxide, which has high redox activity and good reversibility at a low price, makes a good substitute for RuO$_2$. Co$_3$O$_4$ nanowire arrays were produced on nickel foam by Gao et al. [75] utilizing a template-free growth technique. The nanowires, which had a diameter of 250 nm and a length of 15 mm, were constructed from stacked nanoplatelets. Greater nanowire loading was made possible by a longer growth time. The binder less electrodes had a capacitance of 746 F/g in an aqueous KOH electrolyte and saw a capacitance decrease of less than 15% after 500 cycles.

Using a plasma spray technique, Tummala et al. produced Co_3O_4 electrodes that were porous, bendable, and nanostructured. Without the use of a binder or carbon addition, the material was simply deposited directly on the stainless-steel current collector. In contrast to bulk cobalt oxide, which has a capacitance of 209 F/g, Deng et al. discovered that nanostructured cobalt oxide can achieve capacitances as high as 2,200 F/g [76]. A current collector film constructed of nickel nanopetals is covered with a thin coating of cobalt oxide to create a large surface area nanostructure with more active sites for pseudocapacitive reactions and shorter ionic and electronic transport distances, which enhances electrode kinetics. Other metal oxides' weaker conductivity than RuO_2 is one drawback. To overcome this restriction, many groups have focused on nanocomposites. For instance, conductive polymers or composites of MnO_2 and CNT can have a temperature range of 210–415 F/g. Lang et al. improved the conductivity by plating nanocrystalline MnO_2 onto nano-porous gold substrates, producing capacitances as high as 601 F/g at slow scan rates and 170 F/g at greater current density [77]. Kim et al. created NiO-TiO_2 nanotube arrays as electrodes that had a significant surface area, high packing density, and well-organized pore networks. According to calculations, the NiO component of the nanotube arrays has a capacitance between 100 and 300 F/g. Due to the ordered film design, which is thought to enable quick electron and ion movement, these electrodes also displayed higher rate capability than nanoparticle film electrodes. V_2O_5 nanowire and CNT composites can attain capacitances of 300 F/g and energy densities of 40 Whkg/g at power densities of 210 W/kg in an aqueous electrolyte. Fe_3O_4 has a low electronic conductivity but is a promising electrode material due to its low cost and minimal environmental impact. On the other hand, composites of reduced graphene oxide sheets and Fe_3O_4 nanoparticles show a capacitance of 480 F/g, 67 Wh/kg of energy density and 5,506 W/kg of power density. Due to the combination of dual storing mechanisms from EDLC and pseudo-capacitance, the combination of metal oxides and carbon-based materials further raises the possibility of a large capacitance [78,79].

REMARKS

Supercapacitors perform better and store more energy due to the higher surface area, simpler electrolyte access, and shorter diffusion distances that are generally provided by nanostructures. In contrast to PC, which rely on fast redox reactions between the electrolyte and the active material, EDLCs physically segregate charge at the active material/electrolyte interface to form a double-layer structure. The most often used carbon materials for EDLCs include AC, carbon aerogels, carbon generated from carbides, carbon nanotubes (CNTs), and graphene. This is because they are cheap, simple to make, chemically inert, stable, and have a low electrical resistance. Increasing the surface area by activation processes, templating, or preventing agglomeration increases the capacitance in the case of graphene. The surface area can only be increased so much, though, because the pore structure needs to be precisely created to allow electrolyte ion diffusion. Conducting polymers like polypyrrole, polythiophene, and polyaniline as well as metal oxides like RuO_2, MnO_2, and Co_3O_4 are a few examples of common pseudocapacitive substances. Pseudocapacitive materials have a far higher capacitance than EDLCs, but their cycle stability is poorer. One emerging research strategy for utilizing the capacitance contributions from both double-layer growth and pseudo-capacitance is to combine pseudocapacitive materials with carbon-based materials.

BATTERIES

The demand for eco-friendly living and the dwindling availability of conventional fossil fuels have accelerated the search for clean, renewable energy, which has further fueled the development of secondary ion batteries—such as lithium and sodium ion batteries—as energy storage technologies [80]. Since M. Stanley Whittingham initially developed the concept of rechargeable LIBs in 1973, liquid-state electrolyte (LSE) has dominated the commercial business for more than 30 years [81,82]. Michel Armand developed the SSB concept in 1978 as a result of the LSE-based LIBs' development challenges (such as low safety and low energy density) [83,84]. SSBs confront more challenging and complicated interface issues than liquid-state batteries (LSBs) [85]. In comparison to LSBs, SSBs face more difficult and complex interface challenges [85]. Reduced interfacial Li+ transport resistance but increased interfacial side reactions are the results of the flowable LSE in LSBs being able to fully penetrate the electrode and make full contact with the surface of the active materials [84]. The interfacial transport of Li+ is significantly hampered in SSBs by the rigid SSE's insufficient contact with the electrode surface, which results in extraordinarily high interfacial impedance. The SEI of SSBs, on the other hand, is extremely strong and a mixed conductor of ions and electrons [86]. The electrical conductivity of SEI in SSBs typically accounts for the steadily increasing interfacial resistance because of the prolonged electron transfer-induced consumption of SSE and the thickening of SEI. SSBs also include contact failures between electrode particles, failures connected with SSE particles, and failures associated with electrode particles and ion- or electron-conductive particles in addition to these interface issues between the electrode and the electrolyte [84,85]. The aforementioned interface failure of SSBs severely limits the electrochemical activity of electrode materials, negating the advantage of high energy density even if they are more secure than LSBs. It is critical to address their interface failure in order to promote the widespread development of SSBs in the impending global energy Internet and electric-driven future.

Nanotechnology is frequently employed in the design and control of various functional materials due to its vast range of physicochemical properties when compared to conventional scale. Without a doubt, nanotechnology is essential to the development of battery materials and the solution of significant scientific issues in them. Khalil Amine and his coworkers conducted a thorough assessment of the critical function of nanotechnology in the representative cathodes (such as graphite, $Li_4Ti_5O_{12}$, and Si) and anodes (such as $LiNixCoyMn1-x-yO_2$) for liquid electrolyte–based LIBs in 2016. The impact of nanotechnology on electrode materials for liquid LIBs in 2020 was then discussed by Kisuk Kang's group based on aspects of thermodynamics and kinetics. It was noted that the nanoscale strategy is an important way to overcome the physical–chemical limitations of the existing conventional-scale electrodes and address their development roadblocks. It has been demonstrated that nanotechnology enhances the mechanical characteristics, increases the surface activity of electrode materials, and facilitates the flow of lithium ions and electrons in liquid LIBs. Examples of this often include nanoscale active materials and surface modification layers. The switch from LSBs to SSBs has resulted in more severe and complex failure difficulties represented by interfacial problems, which presents new hurdles for the application of the nano-sizing technique for LSBs to SSBs. As a result, it is critical to reevaluate how nanotechnology functions in significant SSB materials, with a focus on the issue of interface failure. This will help to better support the role that nanotechnology plays in furthering the industrialization of SSBs [86–90].

Materials science is crucial to battery research and technology because electrode materials and their interfaces need to be precisely designed and built for ideal electrochemical transformations and long-term stability. From mobile devices to the electrical grid, the need for high-energy-density or high-power-density energy storage materials is always rising. Improved energy storage is possible with materials with at least one nanometer-scale dimension. However, there are also challenges with manufacturing and stability. Over the past 20 years, control of materials at the nanoscale has been a crucial enabler in the advancement of battery technology. Engineered materials with nanoscale dimensions have showed promise and have been carefully studied for batteries. Two benefits of nanoscale materials designed specifically for batteries include shorter ion diffusion lengths and higher resistance to mechanical deterioration [91–94].

As a result of charge and discharge, the materials inside a battery cell go through a complicated evolution that involves structural changes to the active materials, ion diffusion, modifications to local strain and stress, and the appearance of new phases at interfaces. Characterizing dynamic processes in batteries is necessary to comprehend how materials change and deteriorate over time, which can result in a reduction in charge storage capacity, as well as how their structure, chemistry, and morphology affect electrochemical behavior. When paired with synthetic techniques that allow for nanoscale control over material structure and chemistry, knowledge of how materials form in batteries can be used to create new materials that provide greater energy, power, and long-term stability.

Over the past ten years, there has been a surge in the number of specialized experimental methods for studying the dynamic evolution of battery materials at the nanoscale. These procedures were created thanks to advancements in instrumentation, the adoption of methodologies from other fields, and pure scientific inventiveness. Batteries are closed systems and frequently contain components that are air-sensitive, specialized methodologies must be created for their evaluation [95–98]. Therefore, numerous X-ray imaging methods, including X-ray tomography for observing particle transformations, Bragg coherent diffraction imaging for tracking the movement of particular dislocations in active particles, scanning X-ray spectroscopy, and imaging and scattering techniques for observing strain in particular particles during battery reactions, have been used to study nanoscale dynamics in battery materials [99–101].

Transmission electron microscopy (TEM) is a vital tool for analyzing battery materials. In situ TEM has been used to reveal nanoscale reaction mechanisms in solid and liquid environments for a wide variety of technologically significant active materials, and the more recent development of cryogenic TEM techniques has been useful for understanding the atomic-scale structure of delicate battery materials like lithium metal and its interfaces [102–106]. Other important techniques for understanding at the nanoscale include nuclear magnetic resonance and X-ray spectroscopy [107–110]. The major needs for batteries in electrified transportation, along with better energy content and cheaper cost, are improved safety, longevity, and durability. Current efforts to replace the electrode materials used in Li-ion batteries with ones that can store more Li in a given volume or weight will likely continue in the near future. Among these new materials are alloy anodes, lithium metal anodes, conversion cathodes, and high-Ni or Li-rich oxide cathodes. Recent years have seen a substantial focus on nanoscale characterization as it relates to understanding the transformation mechanisms and degrading behavior associated to these novel materials. This considerable research and development led to the effective application of several of these materials in current commercial Li-ion batteries. Additionally, sodium ion-based methods and materials are being developed, and these may be more cost-effective. Beyond these activities, there is a considerable possibility to progress battery technology in the development of new battery topologies for energy storage [111,112]. An example of such an architecture that could offer higher energy density while boosting dependability and endurance is the solid-state battery. These batteries lack the flammable liquid electrolyte necessary for Li-ion batteries because they are entirely constructed of solid materials. The development of the materials and interfaces required to produce cells that can match or even outperform Li-ion batteries in terms of their energy and power characteristics has received more attention recently, despite the fact that this technology has been around for a while. The all-solid nature of these systems presents a fundamental issue, however, as the required material changes during charge and discharge may worsen capacity loss. Many of the same active materials that are used in Li-ion batteries are also used in solid-state batteries. However, due to the radically different environment that the solid-state design creates, certain material and interface transformations occur. Fundamental understanding of transformation and degradation mechanisms in materials and at interfaces will become increasingly important in the future years, and nanoscale characterization will be a key component of this effort. In reality, characterization of modern materials has lately revealed different interfacial alterations in comparison to conventional liquid-based batteries, and more fascinating breakthroughs are imminent [113–116].

One of the main goals of battery material characterization is to be able to directly link the measured output of a battery cell (such as voltage or current) to the underlying material changes and degradation mechanisms. This connection is necessary because the materials inside a working battery may be easily assessed by measuring the output of the battery. Understanding how minute fluctuations in output voltage, temperature, cell volume/pressure, or other externally detectable characteristics are related to the beginning of material degradation throughout a range of length scales in a battery is the next step after understanding this fundamental association. Thanks to this insight, a pack comprising thousands of individual cells could be more precisely monitored during usage, and the health of a cell based on these measurables could be connected to the health of the materials inside. This is a challenging notion given that the disintegration of a single cell frequently involves extraordinary occurrences, such as the deposition of Li metal at the anode or excessive local interphase development. To understand the circumstances that result in these uncommon events at the nanoscale level, a combination of experimentation, data analytics, and modeling will be necessary. Early studies in this area have already shown how important it is to recognize early deterioration mechanisms by performing a systematic examination of the voltage outputs from a number of cells [117].

These efforts have been tremendously supported by the nanoscale engineering of materials, which has enhanced our ability to characterize and understand materials at the nanoscale. The foundations of battery functioning, which form the cornerstone for the creation and dissemination of this technology, will continue to be studied by the communities of materials and nanoscience. In our increasingly electrified society, electrochemical energy storage is ready to play a significant role. Nanomaterials offer much superior ionic transport and electrical conductivity as compared to conventional battery and supercapacitor materials. Additionally, they enable full intercalation of the particle volume's intercalation sites, resulting in high specific capacities and rapid ion diffusion. Due to these properties, nanomaterial electrodes are a promising way to store high-energy and high-power energy since they can withstand high currents. There are still several challenges in employing nanomaterials in energy storage technologies, with the exception of multi-wall carbon nanotube additives and carbon coatings on silicon particles used in lithium-ion battery electrodes. After decades of study and development, a variety of nanomaterials, including oxides, chalcogenides, carbides, carbon, and elements that interact with lithium to form alloys, are now available. One-dimensional (1D) nanowires, tubes, and belts, two-dimensional (2D) nanoflakes and sheets, and three-dimensional (3D) porous nanonetworks are just a few of the many particle morphologies that are covered in this library. These chemically varied nanoscale building blocks can be used to create energy storage technologies that are not feasible with traditional materials, such as wearable and structural energy storage technology. In addition to lithium ions, they can be mixed.

Nanomaterials have a limited application in energy storage devices due to their high surface area, which causes parasitic interactions with the electrolyte, particularly during the first cycle, known as the first cycle irreversibility, and their propensity to aggregate. The development of intelligent nanomaterial assembly into controlled-geometry architectures is therefore a goal of future strategies. Nanomaterial-based devices that can't be built with traditional slurry-based procedures should be made with the aid of advanced manufacturing processes including 3D printing, roll-to-roll manufacturing, self-assembly from solutions, atomic-layer deposition, and others. The long-desired flexible, stretchable, wearable, and structural energy storage and harvesting solutions for the Internet of Things and other disruptive technologies can also be made possible by such manufacturing approaches.

Remarks

Sectors of the global economy that have a strong demand for dependable and reasonably priced electrical energy storage technologies include transportation, portable gadgets, and sustainable intermittent electricity generation. Among the different electrochemical energy storage technologies under investigation, rechargeable lithium-sulfur (Li-S) batteries remain to be the most promising platform for reversibly storing significant amounts of electrical energy at a reasonable cost confined by the underlying cell chemistry. In order to fulfill its potential, Li-S storage technology must discover solutions to a number of fundamental problems, including the complicated solution chemistry of lithiated sulfur compounds in frequently used electrolytes and the naturally low electrical conductivity of sulfur and sulfides. These problems appear to be well suited for both basic solutions assisted by the tools of nanotechnology and innovative nanomaterial-based solutions. Due to the Li-S battery's potential to store significant amounts of electrochemical energy on a low-cost platform, researchers from all over the world are working very hard to develop answers. A solid set of design "rules" based on the knowledge gathered from both mechanistic and applications-oriented studies may be used to create the sulfur cathode and electrolyte for increased capacity and cycling performance.

REFERENCES

[1] Li Li, Changchang Xu, Chengcheng Chen, Yijing Wang, Lifang Jiao, Huatang Yuan, Sodium alanate system for efficient hydrogen storage, *International Journal of Hydrogen and Energy*, 38(2013): 8798–8812. https://doi.org/10.1016/j.ijhydene.2013.04.109

[2] L. Zaluski, A. Zaluska and J. O. Strom-Olsen, Nanocrystalline metal hydrides, *Journal of Alloys and Compounds*, 253(1997): 70–79. https://doi.org/10.1016/S0925-8388(96)02985-4

[3] Craig M. Jensen, Ragaiy Zidan, Nathan Mariels, Allan Hee, Chrystel Hagen, Advanced titanium doping of sodium aluminum hydride: Segue to a practical hydrogen storage material. *International Journal of Hydrogen and Energy*, 24(1999): 461–465. https://doi.org/10.1016/S0360-3199(98)00092-5

[4] G. P. Bud Peterson, Chen Li, Moran Wang, and Gang Chen, Micro/nano transport phenomena in renewable energy and energy efficiency. *Advanced Mechanical Engineering*, (2010): 1–2. https://doi.org/10.1155/2010/170590

[5] Chang-jun Liu, Uwe Burghaus, Flemming Besenbacher and Zhong Lin Wang, Preparation and characterization of nanomaterials for sustainable energy production, *Nano Focus*, 4(2010): 5517–5526. https://doi.org/10.1021/nn102420c

[6] A. Nikitin, X.L. Li, Z.Y. Zhang, H. Ogasawara, H.J. Dai, and A. Nilsson, A Hydrogen storage in carbon nanotubes through the formation of stable C-H bonds, *Nano Letters*, 8(2008): 162–167. https://doi.org/10.1021/nl072325k

[7] G. Girishkumar, Matthew Rettker, Robert Underhile, David Binz, K. Vinodgopal, Paul McGinn, Prashant Kamat, Single-wall carbon nanotube-based proton exchange membrane assembly for hydrogen fuel cells. *Langmuir*, 21(18); (2005): 8487–8494. https://doi.org/10.1021/la051499j

[8] Renata Orinakova, Andrej Orinak, Recent applications of carbon nanotubes in hydrogen production and storage. *Fuel*, 90(2011): 3123–3140. https://doi.org/10.1016/j.fuel.2011.06.051

[9] R.S. Amin, R.M. Abdel Hameed, K.M. El-Khatib, M. Elsayed Youssef, Electrocatalytic activity of nanostructured Ni and PdeNi on Vulcan XC-72Rcarbon black for methanol oxidation in alkaline medium. *International Journal of Hydrogen and Energy*, 39(2014): 2026–2041. https://doi.org/10.1016/j.ijhydene.2013.11.033

[10] Dena Z. Khater, R. S. Amin, Amani E. Fetohi, K. M. El-Khatib, Mohamed Mahmoud, Bifunctional manganese oxide-silver nanocomposites anchored on graphitic mesoporous carbon to promote oxygen reduction and inhibit cathodicbiofilm growth for long-term operation of microbial fuel cells fed with sewage. *Sustain Energy Fuels*, 6(2022): 430–439. https://doi.org/10.1039/D1SE01479J.

[11] Yaoyi Xie, Hanzhi Yu, Liming Deng, R. S. Amin, Deshuang Yu, Amani E. Fetohi, Maxim Yu. Maximov, Linlin Li, K. M. El-Khatib, Shengjie Peng, Anchoring stable FeS2 nanoparticles on MXene nanosheets via interface engineering for efficient water splitting, *Inorganic Chemistry Frontiers*, 9(2022): 662. https://doi.org/10.1039/D1QI01465J

[12] Nor Fatina Raduwan, Norazuwana Shaari, Siti Kartom Kamarudin, Mohd Shabudin Masdar, Rozan Mohamad Yunus, An overview of nanomaterials in fuel cells: Synthesis method and application, *International Journal of Hydrogen Energy*, 47(2022)42: 18468–18495. https://doi.org/10.1016/j.ijhydene.2022.03.035

[13] Schmidt, et al. Electro-catalytic activity of Pt-Ru alloy colloids for CO and CO/H_2 electro-oxidation: Stripping voltammetry and rotating-disk measurements, *Langmuir*, 14 (1997): 2591–2595.

[14] W. Chrzanowski, H. Kim, A. Wieckowski, Enhancement in methanol oxidation by spontaneously deposited ruthenium on low-index platinum-electrodes, *Catalysis Letters*, 50(1998): 69–75. https://doi.org/10.1023/A:1019042329841

[15] S. R. Brankovic, J. X. Wand, R. R. Adzic, Pt submonolayers on Ru nanoparticles-a novel low Pt loading, high CO tolerance fuel-cell electrocatalyst, *Electrochemical and Solid-State Letters*, 4(2001): A217–220. https://doi.org/10.1149/1.1414943

[16] Hockaday et al. In Proc. Fuel Cell Seminar, Portland, Oregon, USA (2000) 791–794 (Courtesy Associates, Washington DC).

[17] G. Q. Sun, J. T. Wang, R. F. Savinell, Iron(III) tetramethoxyphenylporphyrin (Fetmpp) as methanol tolerant electrocatalyst for oxygen reduction in direct methanol fuel-cells, *Journal of Applied Electrochemistry*, 28(1998): 1087–1093.

[18] R. W. Reeve, P. A. Christensen, A. Hamnett, S. A. Haydock, S. C. Roy, Methanol tolerant oxygen reduction catalysts based on transition metal sulphides, *Journal of the Electrochemical Society*, 145(1998): 3463–3471. https://doi.org/10.1149/1.1838828

[19] Supramaniam Srinivasan, Renaut Mosdale, Philippe Stevens, Christopher Yang, Fuel cells: Reaching the era of clean and efficient power generation in the twenty-first century, *Annual Review of Environment and Resources*, 24(1999): 281–238.

[20] N. Giordano, E. Passalacqua, L. Pino, A.S. Arico, V. Antonucci, M. Vivaldi, K. Kinoshita, Analysis of platinum particle size and oxygen reduction in phosphoric acid, *Electrochimica Acta*, 36(1991): 1979–1984. https://doi.org/10.1016/0013-4686(91)85082-I

[21] A.S. Arico, S. Srinivasan, V. Antonucci, DMFCs: From fundamental aspects to technology development, *Fuel Cells*, 1(2001)133–161. https://doi.org/10.1002/16156854(200107)1:2<133::AID-FUCE133>3.0.CO;2-5

[22] R. Nolte, K. Ledjeff, M. Bauer, R. Mülhaupt, Partially sulfonated poly(arylene ether sulfone)-a versatile proton conducting membrane material for modern energy conversion technologies, *Journal of Membrane Science*, 83(1993): 211–220. https://doi.org/10.1016/0376-7388(93)85268-2

[23] Jochen Kerres, Andreas Ullrich, Frank Meier, Thomas Haring, Synthesis and characterization of novel acid-base polymer blends for application in membrane fuel cells. *Solid State Ionics*, 125(1999): 243–249. https://doi.org/10.1016/S0167-2738(99)00181-2

[24] G. Alberti, M. Casciola, Composite membranes for medium-temperature PEM fuel cells. *Annual Review of Materials Research*, 33(2003): 129–154. https://doi.org/10.1146/annurev.matsci.33.022702.154702

[25] Oumarou Savadogo, Emerging membranes for electrochemical systems: (I) solid polymer electrolyte membranes for fuel cell systems, *Journal of New Materials for Electrochemical Systems*, 1(1998): 47–66.

[26] Qingfeng Li, Ronghuan He, Jens Oluf Jensen and Niels J. Bjerrum, Approaches and recent development of polymer electrolyte membranes for fuel cells operating above 100°C, *Chemistry of Materials*, 15, (2003): 4896–4815. https://doi.org/10.1021/cm0310519

[27] K.D. Kreuer, On the development of proton conducting polymer membranes for hydrogen and methanol fuel cells, *Journal of Membrane Science*, 185(2001): 29–39. https://doi.org/10.1016/S0376-7388(00)00632-3

[28] Masahiro Watanabe, Hiroyuki Uchida, Yasuhiro Seki, Masaomi Emori, Paul Stonehart, Self-humidifying polymer electrolyte membranes for fuel cells, *Journal of the Electrochemical Society*, 143(1996): 3847–3852. https://doi.org/10.1149/1.1837307

[29] A. S. Arico, P. Creti, P. L. Antonucci, V. Antonucci, Comparison of ethanol and methanol oxidation in a liquid feed solid polymer electrolyte fuel cells at high temperature, *Electrochemical and Solid-State Letters*, 1(1998): 66–68. https://doi.org/10.1149/1.1390638

[30] A.S. Arico, V. Baglio, A. Di Blasi, V. Antonucci, FTIR spectroscopic investigation of inorganic fillers for composite DMFC membranes, *Electrochemistry Communication*, 5(2003): 862–866. https://doi.org/10.1016/j.elecom.2003.08.007

[31] Klaus-Dieter Kreuer, Stephen J. Paddison, Eckhard Spohr, Michael Schuster, Transport in proton conductors for fuel cell applications: Simulation, elementary reactions and phenomenology, *Chemical Reviews*, 104(2004): 4637–4678. https://doi.org/10.1021/cr020715f

[32] K.D. Kreuer, On the development of proton conducting materials for technological applications, *Solid State Ionics*, 97(1997): 1–15. https://doi.org/10.1016/S0167-2738(97)00082-9

[33] M.K. Debe, *Handbook of Fuel Cells-Fundamentals, Technology and Applications* Vol. 3 (eds. Vielstich, W., Gasteiger, H. A. & Lamm, A.) Ch. 45(2003): 576-589 (Wiley, Chichester, UK, 2003).

[34] R. Atanasoski, In 4th Int. Conf. Applications of Conducting Polymers, ICCP4, Como, Italy Abstract 22(2004).

[35] J. Schoonman, Nanoionics, *Solid State Ionics*, 157(2003): 319–326. https://doi.org/10.1016/S0167-2738(02)00228-X

[36] B. Steele, A. Heinzel, Materials for fuel-cell technologies. *Nature,* 414 (2001): 345–352. https://doi.org/10.1038/35104620

[37] Joachim Maier, Defect chemistry and ion transport in nanostructured materials, Part II Aspects of nanoionics, *Solid State Ionics*, 157(2003): 327–334. https://doi.org/10.1016/S0167-2738(02)00229-1

[38] Philippe Knauth, Harry L. Tuller, Solid-state ionics: roots, status, and futureprospects. *Journal of the American Ceramic Society*, 85(2002): 1654–1680. https://doi.org/10.1111/j.1151-2916.2002.tb00334.x

[39] E. Perry Murray, T. Tsai, S.A. Barnett, A direct-methane fuel cell with a ceria-based anode. *Nature*, 400(1999): 649–651. https://doi.org/10.1038/23220

[40] Seungdoo Park, John M. Vohs, Raymond J. Gorte, Direct oxidation of hydrocarbons in a solid-oxide fuel cell, *Nature*, 404(2000): 265–267. https://doi.org/10.1038/35005040

[41] Y.P. Zhou, K. Feng, Y. Sun, L. Zhou, A brief review on the study of hydrogen storage in terms of carbon nanotubes, *Progress in Chemistry*, 15(2003): 345–350.

[42] Louis Schlappach, Andreas Zuttel, Hydrogen-storage materials for mobile applications, *Nature*, 414(2001): 353–358. https://doi.org/10.1038/35104634

[43] A. C. Dillon, K. M. Jones, T. A. Bekkedahl, C. H. Kiang, D. S. Bethune, M. J. Heben, Storage of hydrogen in single-walled carbon nanotubes. *Nature*, 386(1997): 377–379. https://doi.org/10.1038/386377a0

[44] A. Chambers, C. Park, R.T.K. Baker, N.M. Rodriguez, Hydrogen storage in graphite nanofibers, *The Journal of Physical Chemistry*, 102(1998): 4253–4256. https://doi.org/10.1021/jp980114l

[45] R. Yang, Hydrogen storage by alkali-doped carbon nanotubes-revisited, *Carbon*, 38(2000): 623–626. https://doi.org/10.1016/S0008-6223(99)00273-0

[46] A.C. Dillon, M.J. Heben, Hydrogen Storage using carbon adsorbents- past, present and future, *Applied Physics A*, 72(2001): 133–142.

[47] M. Hirscher, M. Becher, Hydrogen storage in carbon nanotubes, *Journal of Nanoscience and Nanotechnology*, 3(2003): 3–17. https://doi.org/10.1166/jnn.2003.172

[48] C.X. Shang, M. Bououdina, Y. Song, Z.X. Guo, Mechanical alloying and electronic simulation of MgH_2-M systems (M-Al, Ti, Fe, Ni, Cu, and Nb) for hydrogen storage, *International Journal of Hydrogen Energy*, 29(2004): 73–80. https://doi.org/10.1016/S0360-3199(03)00045-4

[49] J.L. Bobet, E. Grigorova, M. Khrussanova, M. Khristov, P., Peshev, Hydrogen sorption properties of the nanocomposite 90wt.% Mg2Ni-10 wt.% V, *Journal of Alloys and Compounds,* 356(2003): 593–597. https://doi.org/10.1016/S0925-8388(03)00684-4

[50] Renzhi Ma, Yoshio Bando, Tadao Sato, Dmitri Golberg, Hongwei Zhu, Cailu Xu, Dehai Wu, Synthesis of boron-nitride nanofibers and measurement of their hydrogen uptake capacity. *Applied Physics Letters,* 81(2002): 5225–5227. https://doi.org/10.1063/1.1534415

[51] Jae-Hun Kim, Kai Zhu, Yanfa Yan, Craig L. Perkins, Arthur J. Frank, Microstructure and pseudocapacitive properties of electrodes constructed of oriented NiO-TiO_2 nanotube arrays, *Journal of Nano Letters*, 10(2010): 4099–4104. https://doi.org/10.1021/nl102203s

[52] Marcus Rose, Yair Korenblit, Emanuel Kockrick, Lars Borchardt, Martin Oschatz, Stefan Kaskel, Gleb Yushin, Hierarchical micro- and mesoporous carbide-derived carbon as a high-performance electrode material in supercapacitors, *ACS Nano*, 4(2010): 1337–1344. https://doi.org/10.1002/smll.201001898

[53] Chang Liu, Feng Li, Lai-Peng Ma, Hui-Ming Cheng, Advanced Materials for Energy Storage, *Advanced Materials*, 22(2010): E28. https://doi.org/10.1002/adma.200903328

[54] John Miller, Introduction to electrochemical capacitor technology, *IEEE Electrical Insulation Magazine*, 26(2010): 40–47. https://doi.org/10.1109/MEI.2010.5511188

[55] Patrice Simon, Yury Gogotsi, Materials for electrochemical capacitors, *Nature Materials*, 7(2008): 845–854. https://doi.org/10.1038/nmat2297

[56] Elzbieta Frackowiak, Carbon materials for supercapacitor application, *Physical Chemistry Chemical Physics*, 9(2007): 1774–1785. https://doi.org/10.1039/B618139M

[57] Peter Hall, Mojtaba Mirzaeian, Isobel Fletcher, Fiona Sillars, Anthony Rennie, Gbolahan Shitta-Bey, Grant Wilson, Andrew Cruden, Rebecca Carter, Energy storage in electrochemical capacitors: designing functional materials to improve performance, *Energy & Environmental Science*, 3(2010): 1238–1251. https://doi.org/10.1039/C0EE00004C

[58] Chuang Peng, Shengwen Zhang, Daniel Jewell, George Z. Chen, Carbon nanotube and conducting polymer composites for supercapacitors, *Progress in Natural Science*, 18(2008): 777–788. https://doi.org/10.1016/j.pnsc.2008.03.002

[59] Reudiger Kotz, Martin Carlen, Principles and applications of electrochemical capacitors, *Electrochimica Acta*, 45(2000): 2483–2498. https://doi.org/10.1016/S0013-4686(00)00354-6

[60] Yonghee Kim, O Young Kweon, Yousang Won, Joon Hak Oh, Deformable and stretchable electrodes for soft electronic devices, *Macromolecular Research*, 27(2019): 625–639. https://doi.org/10.1007/s13233-019-7175-4

[61] Kai Wang, Jiyong Huang, Zhixiang Wei, Conducting polyaniline nanowire arrays for high performance supercapacitors, *The Journal of Physical Chemistry C*, 114(2010): 8062–8067. https://doi.org/10.1021/jp9113255

[62] Yinyi Gao, Shuli Chen, Dianxue Cao, Guiling Wang, Jinling Yin, Electrochemical Capacitance of Co_3O_4 nanowire arrays supported on nickel foam, *Journal of Power Sources*, 195(2010): 1757–1760. https://doi.org/10.1016/j.jpowsour.2009.09.048

[63] Qifeng Zhang, Evan Uchaker, Stephanie L. Candelaria, Guozhong Cao, Nanomaterials for energy conversion and storage, *Nanotechnology*, 20(2009): 175602–175606. https://doi.org/10.1039/C3CS00009E

[64] Raghavender Tummala, Ramesh Guduru, Pravansu Mohanty, Nanostructured Co_3O_4 electrodes for supercapacitor applications from plasma spray technique, *Journal of Power Sources,* 209(2012): 44–51. https://doi.org/10.1016/j.jpowsour.2012.02.071.

[65] John Chmiola, Gleb Yushin, Yury Gogotsi, Cristelle Portet, Simon, Pierre-Louis Taberna, Anomalous increase in carbon capacitance at pore sizes less than 1 nanometer, *Science*, 313(2006): 1760–1763. https://doi.org/10.1126/science.113219

[66] Barbieri, Hahn Matthias, Herzog Andrea, Reudiger Kotz, Capacitance limits of high surface area activated carbons for double layer capacitors, *Carbon*, 43(2005): 1303–1310. https://doi.org/10.1016/j.carbon.2005.01.001

[67] Juergen Biener, Michael Stadermann, Matthew Suss, Marcus Worsley, Monika Biener, Klint Rose, Theodore Baumann, Advanced carbon aerogels for energy applications, *Energy & Environmental Science*, 4(2011): 656–667. https://doi.org/10.1039/C0EE00627K

[68] Jun Liu, Guozhong Cao, Zhenguo Yang, Donghai Wang, Dan Dubois, Xiaodong Zhou, Gordon L. Graff, Larry R. Pederson, Ji-Guang Zhang, Oriented nanostructures for energy conversion and storage, *Chem Sus Chem*, 1(2008): 676–697. https://doi.org/10.1002/cssc.200800087

[69] Li Li Zhang, Zhao, Carbon-based materials as supercapacitor electrodes, *Chemical Society Reviews*, 20(2010): 5983–5992. https://doi.org/10.1039/B813846J

[70] Andrew F. Burke, Batteries and ultracapacitors for electric, hybrid, and fuel Cell vehicles, Proceedings of the IEEE, 95(2007): 806-820. https://doi.org/10.1109/JPROC.2007.892490.

[71] Vasile V.N. Obreja, On the performance of supercapacitors with electrodes based on carbon nanotubes and carbon activated material-A review, *Physica E: Low-dimensional Systems and Nanostructures*, 40(2008): 2596–2605. https://doi.org/10.1016/j.physe.2007.09.044.

[72] Conway B E, *Electrochemical Super Capacitors: Scientific Fundamentals and Technological Applications*, Plenum Press, (1999). https://doi.org/10.1007/978-1-4757-3058-6.

[73] Garcia-Gomez Alejandra, Miles P, Centeno Teresa A, Rojo Jose M, Uniaxially oriented carbon monoliths as supercapacitor electrodes, *Electrochimica Acta*, 55(2010): 8539–8544. https://doi.org/10.1016/j.electacta.2010.07.072

[74] E.G Calvo, Conchi O Ania, L. Zubizarreta, J. Angel Menendez, Ana Arenillas, Exploring new routes in the synthesis of carbon xerogels for their application in electric double-layer capacitors, *Energy Fuels*, 24(2010) 3334–3339. https://doi.org/10.1021/ef901465j

[75] Szczurek Andrzej, Jurewicz Krzysztof, Amaral-Labat Gisèle Aparecida, Fierro Vanessa, Pizzi Antonio Li, Celzard Alain, Structure and electrochemical capacitance of carbon cryogels derived from phenol-formaldehyde resins, *Carbon*, 48(2010): 3874–3883. https://doi.org/10.1016/j.carbon.2010.06.053

[76] Michio Inagaki, Hidetaka Konno, Osamu Tanaike, carbon materials for electrochemical capacitors, *Journal Power Sources*, 195(2010): 7880–7903. https://doi.org/10.1016/j.jpowsour.2010.06.036

[77] J. Lee, J. Kim, T. Hyeon, Recent progress in the synthesis of porous carbon materials, *Advanced Materials*, 18(2006): 2073–2094. https://doi.org/10.1002/adma.200501576

[78] Hirotomo Nishihara, Takashi Kyotani, Zeolite-templated carbons - Three-dimensional microporous graphene frameworks, *Chemical Communications*, 54(45); (2018): 876–879. https://doi.org/10.1039/C8CC01932K

[79] Fuertes, Lota, Centeno, Frackowiak, Templated mesoporous carbons for supercapacitor application, *Electrochimica Acta*, 50(2005): 2799–2805. https://doi.org/10.1016/j.electacta.2004.11.027

[80] Jiang-Kui Hu, Hong Yuan, Shi-Jie Yang, Yang Lu, Shuo Sun, Jia Liu, Yu-Long Liao, Shuai Li, Chen-Zi Zhao, Jia-Qi Huang, Dry electrode technology for scalable and flexible high-energy sulfur cathodes in all-solid-state lithium-sulfur batteries, *Journal of Energy Chemistry*, 71(2022): 612–618. https://doi.org/10.1016/j.jechem.2022.04.048

[81] John Goodenough, How we made the Li-ion rechargeable battery, *Nature Electronics*, 1 (2018): 204–204. https://doi.org/10.1038/s41928-018-0048-6

[82] Jing Xie, Yi-Chun Lu, A retrospective on lithium-ion batteries, *Nature Communications*, 11 (2020): 2499. https://doi.org/https://doi.org/10.1038/s41467-020-16259-9.

[83] Hong Li, Solid state battery, what's next?, *Next Energy*, 1(2023): 100007. https://doi.org/10.1016/j.nxener.2023.100007

[84] Lin Xu, Shun Tang, Yu Cheng, Kangyan Wang, Jiyuan Liang, Cui Liu, Yuan-Cheng Cao, Feng Wei, Liqiang Mai, Interfaces in solid-state lithium batteries, *Joule*, 2(2018): 1991–2015. https://doi.org/10.1016/j.joule.2018.07.009

[85] Jürgen Janek, Wolfgang Zeier, Challenges in speeding up solid-state battery development, *Nature Energy*, 8(2023): 230–240. https://doi.org/10.1038/s41560-023-01208-9

[86] Shuixin Xia, Xinsheng Wu, Zhichu Zhang, Yi Cui, Wei Liu, Practical challenges and future perspectives of all-solid-state lithium-metal batteries, *Chem*, 5(2019): 753–785. https://doi.org/10.1016/j.chempr.2018.11.013

[87] Chae Oh, Brett Lucht, Interfacial issues and modification of solid electrolyte interphase for Li metal anode in liquid and solid electrolytes, *Advanced Energy Materials*, 13(2023): 2203791. https://doi.org/10.1002/aenm.202203791

[88] Jian-Feng Li, Zhangquan Peng, Xin Xu, Functional nanomaterials for energy and catalysis, what's next? *Next Nanotechnology*, (2023): 100001. https://doi.org/10.1016/j.nxnano.2023.100011

[89] Jun Lu, Zonghai Chen, Zifeng Ma, Feng Pan, Larry A. Curtiss, Khalil Amine, The role of nanotechnology in the development of battery materials for electric vehicles, *Nature Nanotechnology*, 11(2016): 1031–1038. https://doi.org/10.1038/nnano.2016.207

[90] Sung-Kyun Jung, Insang Hwang, Donghee Chang, Kyu-Young Park, Sung Joo Kim, Won Mo Seong, Donggun Eum et al, Nanoscale phenomena in lithium-ion batteries, *Chemical Reviews*, 120(14); (2019): 6684-6737. https://doi.org/10.1021/acs.chemrev.9b00405

[91] Candace K. Chan, Hailin Peng, Gao Liu, Kevin McIlwrath, Xiao Feng Zhang, Robert A. Huggins, Yi Cui, High-performance lithium battery anodes using silicon nanowires. *Nature Nanotechnology*, 3 (1); (2008): 31–35. https://doi.org/10.1038/nnano.2007.411

[92] Sung-Yoon Chung, Jason T Bloking, Yet-Ming Chiang, Electronically conductive phospho-olivines as lithium storage electrodes. *Nature. Materials*, 1(2); (2002): 123–128. https://doi.org/10.1038/nmat732

[93] M. Stanley Whittingham, Lithium batteries: 50 years of advances to address the next 20 years of climate issues. *Nano Letters*, 20(12); (2020): 8435–8437. https://doi.org/10.1021/acs.nanolett.0c04347

[94] Yongming Sun, Nian Liu, Yi Cui, Promises and challenges of nanomaterials for lithium-based rechargeable batteries. *Nature Energy*, 1(7); (2016): 16071. https://doi.org/10.1038/nenergy.2016.71

[95] Matthew G. Boebinger, John A. Lewis, Stephanie E. Sandoval, Matthew T. McDowell, Understanding transformations in battery materials using in situ and operando experiments: Progress and outlook. *ACS Energy Letters*, 5(1); (2020): 335–345. https://doi.org/10.1021/acsenergylett.9b02514

[96] Simon Müller, Manuel Lippuner, Mariana Verezhak, Vincent De Andrade, Francesco De Carlo, Vanessa Wood, Multimodal nanoscale tomographic imaging for battery electrodes. *Advanced Energy Materials*, 10(28); (2020): 1904119. https://doi.org/10.1002/aenm.201904119

[97] Martin Ebner, Federica Marone, Marco Stampanoni, Vanessa Wood, Visualization and quantification of electrochemical and mechanical degradation in Li ion batteries *Science*, 342(6159); (2013): 716. https://doi.org/10.1126/science.1241882

[98] Andrew Ulvestad, Andrej Singer, Jesse N Clark, Hyung-Man Cho, Jong Woo Kim, Ross Harder, Jorg Maser, Ying shirley Meng, Oleg G Shpyrko, Topological defect dynamics in operando battery nanoparticles. *Science*, 348(6241); (2015): 1344. https://doi.org/10.1126/science.aaa1313

[99] Jongwoo Lim, Yiyang Li, Daan Hein Alsem, Hongyun So, Sang Chul Lee, Peng Bai, Daniel A Cogswell, Xuzhao Liu, Norman Jin, Young-sang Yu, Norman J Salmon, David A Shapiro, Martin Z Bazant, Tolek Tyliszczak, William C Chueh, Origin and hysteresis of lithium compositional spatiodynamics within battery primary particles. *Science*, 353(6299); (2016): 566. https://doi.org/10.1126/science.aaf4914

[100] Andrew Ulvestad, Andrej Singer, Hyung-Man Cho, Jesse N. Clark, Ross Harder, Jorg Maser, Ying Shirley Meng, Oleg G. Shpyrko, Single particle nanomechanics in operando batteries via lensless strain mapping. *Nano Letters*, 14 (9); (2014): 5123–5127. https://doi.org/10.1021/nl501858u

[101] Francisco Javier Quintero Cortes, Matthew G. Boebinger, Michael Xu, Andrew Ulvestad, Matthew T. McDowell, Operando synchrotron measurement of strain evolution in individual alloying anode particles within lithium batteries. *ACS Energy Letters*, 3(2); (2018): 349–355. https://doi.org/10.1021/acsenergylett.7b01185

[102] Chong-Min Wang, Wu Xu, Jun Liu, Ji-Guang Zhang, Lax V. Saraf, Bruce W. Arey, Daiwon Choi, Zhen-Guo Yang, Jie Xiao, Suntharampillai Thevuthasan, and Donald R. Baer, In situ transmission electron microscopy observation of microstructure and phase evolution in a SnO_2 nanowire during lithium intercalation. *Nano Letters*, 11(5); (2011): 1874–1880. https://doi.org/10.1021/nl200272n

[103] Jian Yu Huang, Li Zhong, Chong Min Wang, John P Sullivan, Wu Xu, Li Qiang Zhang, Scott X Mao, Nicholas S Hudak, Xiao Hua Liu, Arunkumar Subramanian, Hongyou Fan, Liang Qi, Akihiro Kushima, Ju Li, In situ observation of the electrochemical lithiation of a single SnO_2 nanowire electrode. *Science*, 330(6010); (2010): 1515. https://doi.org/10.1126/science.1195628

[104] Robert L Sacci, Jennifer M. Black, Nina Balke, Nancy J. Dudney, Karren L. More and Raymond R. Unocic, Nanoscale imaging of fundamental Li battery chemistry: Solid-electrolyte interphase formation and preferential growth of lithium metal nanoclusters. *Nano Letters*, 15(3); (2015): 2011–2018. https://doi.org/10.1021/nl5048626

[105] Matthew T. McDowell, Seok Woo Lee, Justin T. Harris, Brian A. Korgel, Chongmin Wang, William D. Nix and Yi Cui, In situ TEM of two-phase lithiation of amorphous silicon nanospheres, *Nano Letters*, 13(2); (2013): 758–764. https://doi.org/10.1021/nl3044508

[106] Cheng Ma, Yongqiang Cheng, Kuibo Yin, Jian Luo, Asma Sharafi, Jeff Sakamoto, Juchuan Li, Karren L More, Nancy J Dudney, Miaofang Chi, Interfacial stability of Li metal-solid electrolyte elucidated via in situ electron microscopy. *Nano Letters*, 16(11); (2016): 7030–7036. https://doi.org/10.1021/acs.nanolett.6b03223

[107] Yuzhang Li, Yanbin Li, Allen Pei, Kai Yan, Yongming Sun, Chun-Lan Wu, Lydia-Marie Joubert, Richard Chin, Ai Leen Koh, Yi Yu, John Perrino, Benjamin Butz, Steven Chu, Yi Cui, Atomic structure of sensitive battery materials and interfaces revealed by cryo-electron microscopy. *Science*, 358(6362); (2017): 506. https://doi.org/10.1126/science.aam6014

[108] Xuefeng Wang, Minghao Zhang, Judith Alvarado, Shen Wang, Mahsa Sina, Bingyu Lu, James Bouwer, Wu Xu, Jie Xiao, Ji-Guang Zhang, Jun Liu, Ying Shirley Meng, New insights on the structure of electrochemically deposited lithium metal and its solid electrolyte interphases via cryogenic TEM. *Nano Letters*, 17(12); (2017): 7606–7612. https://doi.org/10.1021/acs.nanolett.7b03606

[109] Baris Key, Mathieu Morcrette, Jean-Marie Tarascon, Clare P Grey, Pair distribution function analysis and solid state NMR studies of silicon electrodes for lithium ion batteries: Understanding the (de) lithiation mechanisms. *Journal of American Chemical Society*, 133(3); (2011): 503–512. https://doi.org/10.1021/ja108085d.

[110] Jie Gao, Michael A. Lowe, Yasuyuki Kiya and Hector D. Abruna, Effects of liquid electrolytes on the charge-discharge performance of rechargeable lithium/sulfur batteries: Electrochemical and in-situ X-ray absorption spectroscopic studies. *Journal of Physical Chemistry C*, 115(50); (2011): 25132–25137. https://doi.org/10.1021/jp207714c.

[111] Xiaoqiao Zeng, Matthew Li, Deia Abd El-Hady, Wael Alshitari, Abdullah S. Al-Bogami, Jun Lu, Khalil Amine, Commercialization of lithium battery technologies for electric vehicles. *Advanced Energy Materials*, 27(2019): 1900161, https://doi.org/10.1002/aenm.201900161

[112] A cost and resource analysis of sodium-ion battChristoph Vaalma, Daniel Buchholz, Marcel Weil & Stefano Passerinieries. *Nature Reviews Materials*, 3(4); (2018): 18013, https://doi.org/10.1038/natrevmats.2018.13.

[113] Jurgen Janek & Wolfgang G. Zeier, A solid future for battery development. *Nature Energy*, 1(9); (2016): 16141, https://doi.org/10.1038/nenergy.2016.141

[114] Ziying Wang, Dhamodaran Santhanagopalan, Wei Zhang, Feng Wang, Huolin L. Xin, Kai He, Juchuan Li, Nancy Dudney and Ying Shirley Meng, In situ STEM-EELS observation of nanoscale interfacial phenomena in all-solid-state batteries. *Nano Letters*, 16(6); (2016): 3760–3767. https://doi.org/10.1021/acs.nanolett.6b01119

[115] John A Lewis, Francisco Javier Quintero Cortes, Yuhgene Liu, John C Miers, Ankit Verma, Bairav S Vishnugopi, Jared Tippens, Dhruv Prakash, Thomas S Marchese, Sang Yun Han, Chanhee Lee, Pralav P Shetty, Hyun-Wook Lee, Pavel Shevchenko, Francesco De Carlo, Christopher Saldana, Partha P Mukherjee, Matthew T McDowell, Linking void and interphase evolution to electrochemistry in solid-state batteries using operando X-ray tomography. *Nature Materials*, 20(4); (2021): 503–510. https://doi.org/10.1038/s41563-020-00903-2

[116] Shuai Hao, Sohrab R. Daemi, Thomas M. M. Heenan, Wenjia Du, Chun Tan, Malte Storm, Christoph Rau, Dan J.L. Brett, Paul R. Shearing, Tracking lithium penetration in solid electrolytes in 3D by in-situ synchrotron X-ray computed tomography. *Nano Energy*, 82(2021): 105744. https://doi.org/10.1016/j.nanoen.2021.105744

[117] Kristen A. Severson, Peter M. Attia, Norman Jin, Nicholas Perkins, Benben Jiang, Zi Yang, Michael H. Chen, Muratahan Aykol, Patrick K. Herring, Dimitrios Fraggedakis, Martin Z. Bazant, Stephen J. Harris, William C. Chueh and Richard D. Braatz, Data-driven prediction of battery cycle life before capacity degradation. *Nature Energy*, 4(5); (2019): 383–391. https://doi.org/10.1038/s41560-019-0356-8

10 Overview of Metal Oxide Nanocomposites for Supercapacitor Applications
Perspectives and Progress

*Deepak R. Kasai, H. Shanavaz, M.K. Prashanth,
K. Yogesh Kumar, and M.S. Raghu*

INTRODUCTION

Rapid economic growth has increased the demand for fossil fuels such as coal, petroleum, and natural gas, as well as contribute to worsening climate change and pollution around the world. Clean energy sources like solar, wind, and tidal power can help reduce the effects of today's energy use and pollution. With the introduction of cutting-edge energy storage technology, the problem with green energy's (solar, wind, hydroelectric, and geothermal) intermittent output could be resolved. Keeping clean energy on hand is essential for meeting human requirements, and conventional energy storage systems like batteries, supercapacitors, and fuel cells are necessary for this. All of these alternative energy sources rely on the electrochemical energy conversion principle, including the next-generation batteries, supercapacitors (SCs), and fuel cells [1]. Supercapacitors (SCs) have become increasingly popular due to their superior specific capacitance (Cs), high power density (Pd), and long cycle life. These characteristics have helped to establish SCs as a promising new class of storage device [2–3]. Solar cells (SCs) and batteries can distribute electricity generated from renewable sources without connecting to the system [4]. Electrode substances need to be able to endure strong electric fields and keep a high dielectric constant in order for capacitive energy storage to be as effective as rechargeable battery packs at storing large amounts of power and energy. To improve their dielectric permittivity, researchers have experimented with incorporating ferroelectric metal oxides (MOs) into dielectric composites. These oxides include $Pb(Zr, Ti)O_3$ (PZT), $Pb(Mg_{0.33}Nb_{0.77})O_3$-$PbTiO_3$ (PMNT), and $BaTiO_3$ (BT) [5]. As electrodes for supercapacitors (also known as pseudocapacitors), electroactive MOs like ruthenium oxide (RuO_2), manganese oxide (MnOx), nickel oxide (NiO), etc. offer fast and reversed redox reactions, leading to a higher energy density capacity at the expense of power density and life cycle. Research into developing next-generation supercapacitors, which incorporate a range of MOs in carbon nanomaterials, has made strides toward overcoming this limitation and improving performance in recent years.

To better facilitate ion transport at the electrode–electrolyte interface, electrochemical electrodes can benefit from the incorporation of nanomaterials due to their increased porosity and surface area compared to those of bulk phase structures. MO nanocomposites are able to store a lot of charge since they are thermally stable and have many electroactive areas. Supercapacitors can make great use of the distinctive size and surface reliant (e.g., morphological) characteristics of carbon nanomaterials (CNM) like carbon nanotubes (CNTs), graphene and reduced graphene oxide (rGO), graphene nanofoam (GF), and carbon nanofibers (CNFs) [6–7]. Assembling an electrode for an EDLC capacitor was made possible by CNM's extraordinary high surface areas and moderate electronic conductivity in the first generation of CNM-based supercapacitors.

DOI: 10.1201/9781003479239-10

This chapter summarizes the significant developments in the field of supercapacitors during the past few years, with an emphasis on MO-based nanocomposites, by analyzing a selection of typical research publications from the preceding years. The various MO-based nanocomposites used for supercapacitors are categorized, and a summary table of recent outcomes reported on their electrochemical energy storage performances is provided. Finally, some potential perspectives and obstacles for MO nanocomposites are highlighted. With the goal of improving supercapacitor applications, several novel MO nanocomposites have been introduced. This chapter examines the similarities and differences between MO nanocomposites and the fundamental concepts of supercapacitors, and gives a succinct comparison and analysis of their possible applications. This chapter concludes with a brief review of recent developments in supercapacitor energy storage and the numerous MO nanocomposites used as electrode materials for supercapacitors.

ORIGIN OF SUPERCAPACITORS

Beginning with fuel cells and rechargeable batteries, general electric engineers experimented with porous carbon electrodes in the early 1950s with respect to design of capacitors. In the realm of electrochemical energy storage devices, supercapacitors—also known as "electrochemical capacitors" or "ultra-capacitors"—are a key player due to their capacity to store electrical energy at a much higher quantity and rate than traditional capacitors and batteries. Supercapacitors are superior to other energy storage devices in terms of both the amount of energy they can store and the speed with which they can release that energy. There is a lot of room for the growth of supercapacitors in mobile electronics, renewable energy systems, and hybrid electric vehicles [8]. Research and development in the field has led to the discovery that there are three broad categories of supercapacitors, namely, electrochemical double-layer capacitors, pseudocapacitors, and hybrid capacitors shown in Figure 10.1 [9]. The ability to store electric charge is what distinguishes one class from another. Generally, energy is stored in EDLCs through the buildup of charge at the interface between the electrode and the electrolyte. Pseudocapacitors store energy via a quick and reversible faradaic redox process. As it stands, supercapacitors' limited energy storage capacity is their biggest drawback. It is essential to remember that the design and preparation of the electrode materials are the primary factors in determining the electrochemical performance of supercapacitors. To clarify, the materials used for the electrodes have a significant impact on the performance of supercapacitors [10–11].

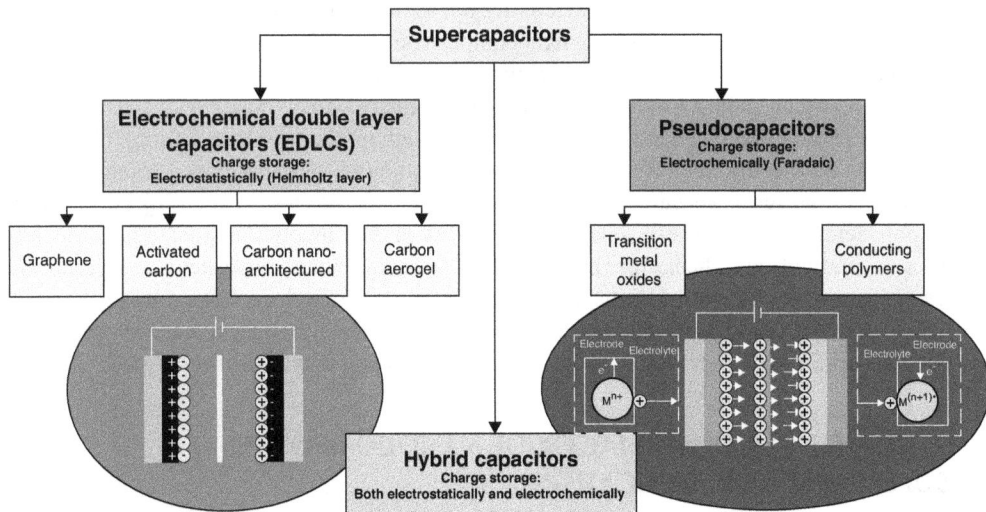

FIGURE 10.1 Different types of supercapacitors [9].

OVERVIEW OF METAL OXIDE NANOCOMPOSITES

MOs play a crucial role in the creation of high-tech, high-performance energy storage systems. There has been a lot of interest in using MO in supercapacitors because of their high theoretical specific capacitance due to the faradaic charge transfer process, but their low energy density is a big limitation. Electrode materials such as ruthenium oxide (RuO_2), manganese oxide (MnO_2), nickel oxide (NiO), cobalt oxide (Co_3O_4), and vanadium pentoxide (V_2O_5) have been extensively investigated as potential solutions to the problem of low energy density. Due to their large surface area, the flexibility of synthesis in different component morphologies, and their many oxidation states (+1 to +7), MO materials have been recognized as interesting possibilities for electrodes in energy storage devices [12]. In addition, they play a crucial role in the electrodes of electrochemical supercapacitors, and it may be possible to greatly increase capacitance by changing and regulating their defects and surface/interfaces on a nanoscale. They have enhanced energy density, but their low electrical conductivity, unpredictable volume expansion, and delayed ions movement in the bulk phase have severely restricted their usefulness. MOs, such as RuO_2, MnO_2, NiO, Co_3O_4, and V_2O_5, are the most popular choices for usage in supercapacitors [13].

Conducting polymers (CPs) and MOs are two materials that, like any others, have their advantages and disadvantages. In spite of their high conductivity, simple synthesis procedures, and inexpensive cost, the CP-like PANI have limited utility in supercapacitors due to their poor mechanical stability and propensity for degradation over repeated charge–discharge cycles. This is because CP expands and contracts at the CP chain/electrolyte interface as a result of related ion and solvent transport. By contrast, MOs have a large specific capacitance, are inexpensive, and can be found in large quantities despite their small footprint. Incorporating the benefits of both CP and MO into the design and optimization of a nanoscale material often results in novel nanocomposites. Nanocomposites are materials that enhance a product's macroscopic qualities through their nanoscale structure. Special mechanical, barrier, and weight-saving capabilities, as well as improved long-term heat, wear, and scratch resistance, are all features of nanocomposite materials. Nanocomposites are composites in which at least one of the components' size is 100 nm or smaller. Changes in electrical breakdown strength, melting temperature, color, magnetism, charge capacity, and the size of the interaction zone are only some of the ways in which CP and MO nanocomposites can stand apart. Characteristics of nanocomposite electrodes are affected not only by the constituent materials but also by the shape and interfacial qualities of combination. The benefits of CP-MO nanocomposites include malleability, toughness, coatability of polymers, and the strength and longevity of MO. They also possess unique synergetic characteristics not found in the individual ingredients.

It has been shown that by incorporating nanocomposite electroactive materials into supercapacitors, a number of attributes, including specific surface area, electrical and ionic conductivities, specific capacitance, cyclic stability, and energy and power density, can be greatly enhanced. To create nanocomposites with novel characteristics, scientists throughout the world are refining novel methods of synthesis. As a result, MO nanocomposites have become a prominent choice among the materials employed in the supercapacitor industry [14–15]. Nanocomposites are the subject of research aimed at enhancing the efficacy, efficiency, and durability of their practical applications through the integration of optimal ingredients for the synthesis of specific nanocomposites. Nanocomposites were created by doping CP with MO, as demonstrated by the research conducted by Shahabuddin et al. [16]. Supercapacitor and photocatalytic performances of the synthesized material were evaluated. The data tended to support the idea that the nanocomposite outperformed pure CP.

EVOLUTION OF VARIOUS METAL OXIDE NANOCOMPOSITES FOR SUPERCAPACITOR APPLICATION

Nanoparticles based on MOs can be modified to take on novel features. These can include, but are not limited to, increased surface area and surface activities, larger pores, lower electron-hole recombination, and higher conductivities [17]. Asymmetric supercapacitors play a crucial role in amplifying the properties of the included material when it comes to the use of MOs in a super-capacitor. In most cases, a bigger potential difference may be held between electrodes when one electrode is made of carbon-based materials and the other is made of an MO. Because of this, the supercapacitor is able to store more energy and produce more power. Because of this larger potential window, a variety of electrolytes can be used in asymmetric supercapacitors.

Recent research has shown that nanocomposites based on MOs are a very desirable material across a wide variety of scientific disciplines. Carbon nanotube–ZnO nanocomposite was described as electrode material for supercapacitor with the highest computed specific capacitance of 323.9 F/g, and many other nanocomposites based on transition MOs have been reported in the literature for their beneficial, crucial practical uses [18]. CNTs or CNFs and finely tailored microporous carbons, graphene, etc. have been proposed as new types of carbon materials having a large device surface area [19]. One-atom-thick carbon sheets are rolled at different angles to create the walls of CNTs. The final identity of each nanotube shell as a metal or a semiconductor is determined by these dis-tinct angles. Due to their unique one-dimensional (1D) structure, CNTs can be used to create macro-scopically useful one-dimensional (1D) fibers, two-dimensional (2D) films, and three-dimensional (3D) foams. CNTs have excellent conductivity, a huge surface area, and favorable mechanical quali-ties at the macroscopic level, allowing oxides to be employed as flexible SC electrodes or sustaining pseudocapacitive electrode components of varying proportions [20]. The performance of CNTs as semiconductors is predicated on the one-atom-thick sheets of carbon that are rolled at precise angles to form hollow cylinder walls. There has been little research on the electrochemical performances of SCs after incorporating new materials into CNTs, as opposed to the preparation methodology, structural design, flexibility, or application of nanocomposite materials, to name a few. Capacitance values of >300 F/g were consistently observed across all nanocomposites, which is typical for CNT nanocomposites and the focus of the present study. To try to provide a concise summary of a com-parative study involving 1D CNT fibers, 2D CNT film, and 3D CNT sponges and aerogels, Zhu et al. evaluated several CNT kinds [21]. They demonstrated various SC architectures and discussed how they affected the electrochemical make-up of the electrode. CNTs enable nanoparticles to easily contact their surface and create a matrix network for ion transfer. Various electrolytes and anode/cathode combinations can increase the achievable window, scan rate, and energy density of SCs. Managing pH with electrolytes can also help electrodes retain their retention cycle. Some nanomaterials show tremendous SC potential but struggle in application. CNT provides valuable substrate and backbone support for optimal performance. Ingredients with optimal electrolyte and methods for improved electrochemical properties are also included here. Electrodes made from carbon nanotubes (CNTs) and MOs have showed a lot of promise in previous studies. They have shown improved characteristics on a wide range of MOs and metals. This is because the higher spe-cific capacitance and longer cycle life of carbon-based nanoparticles, MO-based nanocomposites, transition MO–based particles, and nanocomposites (TMONPs and TMONCs) have widespread importance. Many scientists have tried TMONCs in ultra-capacitor and supercapacitor applications.

The performance metrics of several CNT-based SCs are summarized in Table 10.1.

Guangyu and colleagues prepared Co_3O_4/graphene nanocomposites for supercapacitor applica-tions. They found that Co_3O_4/graphene nanocomposites exhibited specific capacitance around 30 F/g at 1 A/g for three-electrode systems and 215 F/g at 0.4 A/g for two-electrode systems. When compared to the three-electrode system, the specific capacitance measured using only two elec-trodes is lower. They were able to achieve good cyclic stability of Co_3O_4/graphene in both the three-electrode and two-electrode systems [72]. Furthermore, specifically for use in superca-pacitors, Leontyeva et al. synthesized a carbon-supported NiO (NiO/C) nanocomposite and ana-lyzed its electrochemical properties. Based on the cyclic voltammetry scan rate increasing from 5 to 40 mV/s, the authors determined a specific capacitance of 1,100 to 777 F/g. Later, the gal-vanostatic charge–discharge curve was used to determine that the mass-normalized current is required to generate 970 F/g. After a thousand cycles, they observed that the cycle life was very

TABLE 10.1

Various MO Nanocomposites Used in Supercapacitor and Their Performance

	Electrodes	Specific Capacitance (Fg^{-1})	Electrolyte	Current Density (Ag^{-1})	References
Graphene-based	RuO_2/GO	512	$1 M H_2SO_4$	2.5	[22]
	Mn_3O_4/rGO	228	$1 M Na_2SO_4$	5	[23]
	MnO_2/Gr F	672	$0.5 M Na_2SO_4$	1	[24]
	Fe_3O_4/rGO	220	1 M KOH	0.5	[25]
	Fe_3O_4/GA	440	6 M KOH	1.5	[26]
	$NiCo_2O_4$/RGO	1388	6 M KOH	0.5	[27]
	$LaAlO_3$/RGO	721	1.0 M KOH	0.5	[28]
	Ta_2O_5/NRGO	808	$1 M H_2SO_4$	2	[29]
	RGO/V_2O_5	906	1.0 M KOH	2	[30]
	Ni_2MnO_4–x@rGO	1344.7	2 M KOH	1	[31]
	Bi_2O_3@rGO	560	1.0 M KOH	5	[32]
MXene-based	$MnO_2/CNT/Ti_3C_2Tx$	181	$1 M Na_2SO_4$	1	[33]
	MnO_2/d-Ti3C$_2$Tx	242	1 M KOH	1	[34]
	α-Fe_2O_3/Ti_3C_2Tx	405.4	5 M LiCl	2	[35]
	Fe_3O_4/rGO/Ti_3C_2Tx	45.8	$1 M Na_2SO_4$	0.2	[36]
	CGO/PDAAQ@MXene	346	$1 M H_2SO_4$	1	[37]
TMDs-based	RuO_2/MoS_2	719	1 M KOH	1	[38]
	Mn_3O_4/MoS_2	119.3	$1 M Na_2SO_4$	1	[39]
	Mn_3O_4/MoS_2/HGRs	608	$1 M Na_2SO_4$	1	[40]
	Fe_2O_3/Graphene/MoS_2	98.2	2 M KOH	1	[41]
	Fe_3O_4/PHG/MoS_2	830	$1 M H_2SO_4$	1	[42]
	$ZnWO_4/WS_2$	1280.7	3 M KOH	3	[43]
	$MoTe_2$/rGO	1196.4	4 M KOH	1	[44]
LDHs-based	MnO_2/CoAl-LDH	1088	6 M KOH	1	[45]
	CoNiFe/LDHs	3130	3 M KOH	1	[46]
	Fe_3O_4@NC/CoMn–LDH	313	1 M KOH	1	[47]
	$NiCo_2O_4$@NiCoAl-LDH	1814.2	2 M KOH	1	[48]
	$NiFe_2O_4$-NiCo-LDH@rGO	750	6 M KOH	1	[49]
	TMOs/rGO/NiCo LDH	2763	1 M KOH	1	[50]
PANI	Ag@MnO_2/PANI	1028.66	2 M KOH	1	[51]
	MnO_2/PANI	428.6	6 M KOH	1	[52]
	PAni/MnO_2/Porous Carbon Nanofiber	289	$1 M H_2SO_4$	5	[53]
	RuO_2/TiO_2/pani	67.4	$0.1 M H_2SO_4$	0.2	[54]
	Co_3O_4/PANI	1308	$1 M H_2SO_4$	1	[55]
	Co_3O_4/PANI	3105.46	6 M KOH	1	[56]
	Co_3O_4 (FNCO, Fe-Ni co-doped)	1171	$1 M H_2SO_4$	1	[57]
	$PANI/SnO_2$	636	$1 M H_2SO_4$	2	[58]
	Polyaniline/MoS_2	687	$1 M H_2SO_4$	5	[59]
	PANI/MWCNTs	1551	$1 M H_2SO_4$	2	[60]
	PANI/α-Fe_2O_3	974	$1 M H_2SO_4$	2	[61]

(Continued)

TABLE 10.1 (*Continued*)
Various MO Nanocomposites Used in Supercapacitor and Their Performance

	Electrodes	Specific Capacitance (Fg^{-1})	Electrolyte	Current Density (Ag^{-1})	References
PPy	Ni foam-MnO$_2$//Ni foam-PPy	59.29	LiClO$_4$	1	[62]
	PPy-MnO$_2$-CC	270	LiClO$_4$	1	[63]
	C$_3$N$_4$/PPy/MnO$_2$	509.4	1 M Na$_2$SO$_4$	1	[64]
	CC/N-CNWs/Ni@MnO$_2$	571.4	4 mM NaHCO$_3$	1	[65]
	CB/MnO$_2$/PPy	273.2	1 M H$_2$SO$_4$	1	[66]
	Ti$_3$C$_2$Tx and PPy/MnO$_2$	61.5	(PVA/H$_2$SO$_4$)	2	[67]
	MnO$_2$/CNT//PPy/CNT WASC	10.7	LiCl/PVA	1	[68]
	Ppy/NiO	679	0.1 M LiClO$_4$	1	[69]
	NiO-CoO-PPy	1123	2 M KOH	1	[70]
	NiO/Gr/PPy	970.85	2 M KOH	1	[71]

good (with 840 F/g) [73]. Nanocomposites of poly(3,4-ethylenedioxythiophene) (PEDOT) and poly(aniline) (PANI)-manganese dioxide (MnO$_2$) were created by Pintu et al. These authors looked at their electrochemical characteristics and compared them to those of pure MnO$_2$. Maximum calculated specific capacitances of PEDOT-MnO$_2$, PANI-MnO$_2$, and MnO$_2$ alone are 315, 221, and 158 F/g, respectively [74].

CeO$_2$ demonstrates potential redox characteristics when combined with an alkaline electrolyte. CeO$_2$ is a unique and exciting substance used in hybrid SCs due to its abundance, ease of manufacture, and fastest electrochemical redox chemistry of Ce^{3+} and Ce^{4+} ions [75]. However, CeO$_2$ electrode cannot be employed as a SC electrode due to its low capacitance, low electrical conductivity, and tiny surface area [76]. The hydrothermal approach was used by Luo et al. to integrate CeO$_2$ nanoparticles into CNTs, yielding CeO$_2$-4 wt.% CNTs nanocomposites, whose SPc was 818 F/g in 2 M KOH electrolyte and whose capacity was retained at a high level (95.3% after 2,000 cycles) [77]. The Schottky barrier that forms at the CeO$_2$/CNTs interface in these nanocomposites provides a convincing description of the energy storage theory and suggests that they may be a promising candidate for supercapacitors. Composites based on cerium oxide (CeO$_2$) and activated carbon (AC) were developed by Aravinda et al. [78]. They confirmed an SPc of 162 F/g in a two-electrode system, and the electrode has a power density of 3,500 W/kg even at a high current density of 18 mA/cm^2, with outstanding cycle stability. Deng and co-workers elaborated that a CeO$_2$/ZZ nanocomposite prepared via simple hydrothermal method (at a mole ratio of 1:1) exhibited an specific conductance (SPs) of 455.6 F/g at a current density of 1 A/g and an SPc retention of 81.1% after 2,000 continuous charge–discharge cycles in 6 ml KOH electrolyte [79]. The SPc of CeO$_2$/MWCNTs is about 246% higher than that of MWCNT electrodes that are also made using chemical processes.

Masaki Ujihara et al. have prepared ternary MO nanocomposites for high-efficiency supercapacitor applications. Binder-free electrodes made of complicated TMO on Ni foam were made at the same time from mixtures of Ni, Co, and Mn nitrates using a simple and inexpensive method called electrodeposition [80]. There were some distinct differences in the nanostructures of the deposited materials between the cathode (which needed a heat treatment to oxidize to TMO) and the anode blocks on the anode and nanosheets on the cathode shown in Figure 10.2. To evaluate the real-world performance of supercapacitors, a symmetric electrode system was built by joining two cathode-3 electrodes with a filter paper separator presented in Figure 10.3. In the potential range of 0.4–0.5 V, the cyclic voltammetry (CV) curves took on a square form (Figure 10.3a). Broad anodic and cathodic peaks, with centers at 0.12 and 0.056 V, were observed in the CV curves as the potential spectrum was widened from 0.6 to 0.8 V. These features were explained by reversible redox

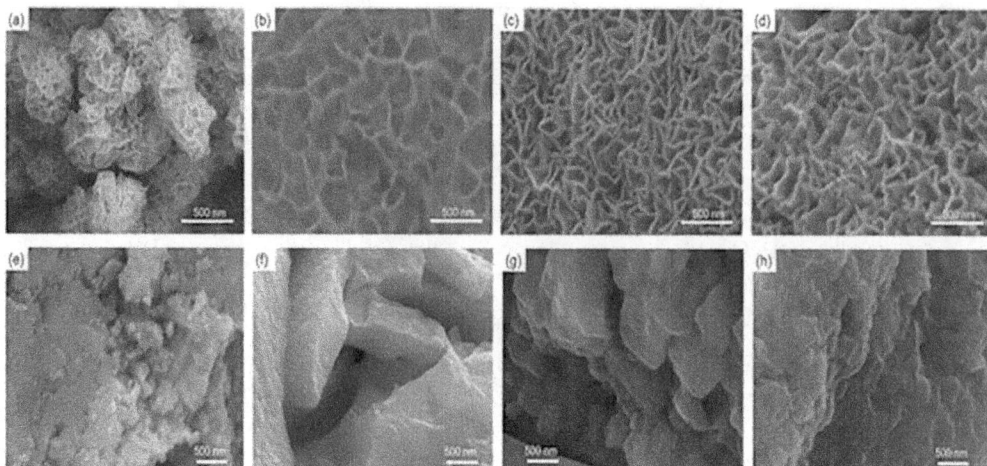

FIGURE 10.2 SEM image of cathode materials: (a) cathode-1, (b) cathode-2, (c) cathode-3, (d) cathode-4, (e) anode-1, (f) anode-2, (g) anode-3, and (h) anode-4 [80].

processes. Nearly identical CV curve morphologies were seen over a range of scan speeds, with an increase in current responsiveness accompanied by a shift in the peak locations toward greater positivity (oxidation peak) and greater negativity (Figure 10.3b). Current densities ranged from 1.67 to 5.56 A/g during GCD tests performed in the potential range of 0.2 to 0.8 V (Figure 10.3c). A capacitance retention of 75% was achieved with a specific capacitance of 440 F/g derived from the GCD curve at 1.67 A/g, and a reduction to 330 F/g at 5.56 A/g (Figure 10.3d). As mentioned previously,

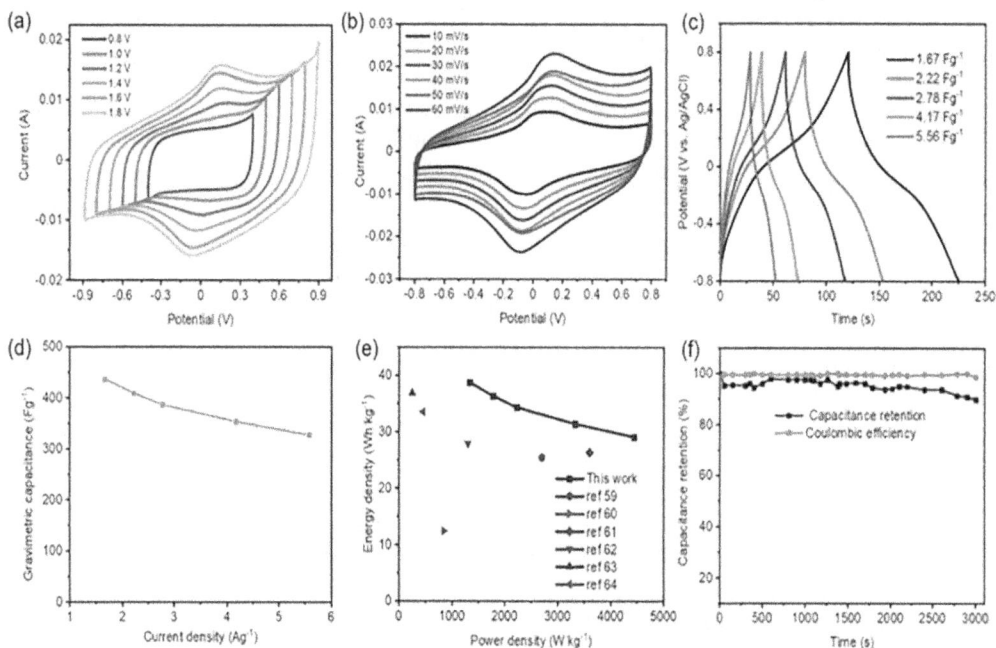

FIGURE 10.3 Electrochemical properties observed in symmetric electrode system using cathode-3: Curves (a) observed at various potential windows; (b) different scan rates; (c) GCD curves observed at various current densities; (d) gravimetric capacitance vs current density; and (e) Ragone plot, cyclic stability, and Coulombic efficiency tested at 11.11 Ag/f [80].

the specific capacitance of cathode-3 decreased with increasing scan rate due to the restriction of ion diffusion as per Ragone diagrams (Figure 10.3e). Over the course of 3,000 GCD cycles, the cycle stability was examined at a high current density of 11 A/g. As an electrode material for supercapacitor devices, manganese dioxide-graphene (MnO_2) has received a lot of attention. MnO_2 electrode stores electrical potential by allowing manganese to undergo an oxidation state transition from III to IV. Reversible addition and removal of electrolyte cations to achieve charge neutrality during reduction and oxidation of Mn^{+3}/Mn^{+4} gives MnO_2 thin films their pseudocapacitive behaviors of around 100 nm, and the latter was a nanocage with a diameter between 100 and 120 nm [81].

Both 110°C for 2 hours and 120°C for 3 hours were used in the hydrothermal synthesis of MnO_2. After 500 cycles, NS-MnO_2 exhibits its initial specific capacity of 121 F/g at 0.1 A/g. Nanocage-MnO_2 exhibits a specific starting power of 278 F/g, and even after 2,000 cycles at 0.1 A/g, the specific capacitance of the nanocage-MnO_2 remains to be 200 F/g. The different nanocage that was still there and undamaged contributed to the higher specific capacitance [82]. Using varying amounts of $KMnO_4$, hydrothermal synthesis was used to create morphologically distinct MnO_2 nanoparticles, including micro-sandwiches, micro-sticks, nanowires, and micro-flowers. Electrochemically, micro-flowers excel, with a specific capacitance of 185 F/g at 0.5 A/g and 85% retention after 10,000 cycles. The improved electrochemical characteristics were the result of the increased surface area, which explained the high results [83]. In order to generate higher capacitance, the polymer-assisted hydrothermal process with PEDOT/PSS was implemented. A capacitance of 365.5 F/g at 1 A/g was achieved in nanorods using this approach, with roughly 80% retention after 2,000 cycles. The polymer's enhanced capacity is due to the effective binding of its organic framework to its inorganic substrate when heated to high temperatures [84].

Adding an intercalating ingredient, such as reduced graphene oxide (rGO), can significantly increase a nanocomposite's capacity. Having desirable properties such as high conductivity, superior heat resistance, and little maintenance is icing on the cake. Using these items, one's abilities can be greatly enhanced. The nanorods with tunnel-like shape were synthesized with ease at low cost using hydrothermal synthesis from MnO_2 and reduced graphene oxide (rGO), yielding an extraordinary specific capacitance of 759 F/g at 2 A/g power density. It also showed excellent charge storage capability, with an impressive 88% capacity retention after 3,500 cycles and a maximum specific energy of 64.6 Wh/kg [85]. Similar experiments with varying concentrations of $KMnO_4$ yielded a specific capacity of 255 F/g at a current density of 0.5 A/g, with retention of roughly 85% after 10,000 cycles. This was the case for MnO_2/rGO-25. Large specific surface area, which aids in energy storage, was responsible for these remarkable qualities [86]. In order to create an asymmetric supercapacitor, manganese dioxide nano-flowers were electrochemically formed on graphene oxide sheets. The graphene oxide (GO) sheets measured to have a capacitance of 385.2 F/g at 0.5 A/g, with excellent retention up to 5,000 cycles. Their capacitance improved after we tweaked the MnO_2 deposition cycles [87].

It has been found that ruthenium oxides have a high redox state and high SPc throughout a broad potential window. Nano-crystalline hydrous ruthenium dioxide ($RuO_2 \cdot xH_2O$) is commonly utilized because of its ultra-high pseudocapacitance and exceptionally long cycle life, and it demonstrates good thermal stability, long life cycle, and reversible redox processes [88]. The easy hydrothermal process for the synthesis of GO is enhanced by nano-string clusters of ruthenium oxide (GO-RuO_2), as described by Raghu and colleagues [89]. The authors reported that specific capacitance using CV on a GO-RuO_2 nano-string composite has showed values as high as 859 F/g at a scan rate of 5 mV/s, while measurements using chronopotentiometry (CP) have revealed values as high as 512 F/g with a current density of 2.5 A/g. The maximal power density of this material is 1,028 W/kg, and its energy density is 246 Wh/kg. The authors confirmed that the GO-RuO_2 nano-string composite electrode had excellent cyclic stability, retaining approximately 99% of its initial capacitance after 500 cycles. The advantages of vanadium oxide over other MOs are its high capacity, numerous valence states, lower cost, and layered structure that allows for simple access to electrolytes. Among the electrode materials used in supercapacitor devices, it is among the most promising [90]. Hybrid aerogels comprising vanadium pentoxide nano-belts and rGO were designed in a 3D structure, and their electrochemical performance was 225.6 F/g at a current density of 0.5 A/g. They observed that their composite had a higher capacitance than either of the two individual performances (92.3 F/g for pure V_2O_5 and 102.3 F/g for rGO), with a strong cycling stability of 90.2% over 5,000 cycles [90].

With its exceptional qualities, tin oxide (SnO_2) is a ceramic material that finds extensive application in gas sensors, displays and solar cell anode material, catalysis, and secondary lithium battery usage. Tin oxides have been suggested as a promising material for SC applications because of their inexpensive cost, minimal environmental effect, and thermal stability in air. Nonetheless, there has been a dearth of research on this topic. The elastic quasi-solid-state SnO_2/CNT nanocomposite SCs were fabricated using a dc-pulsed nitrogen atmospheric-pressure plasma jet (APPJ) [91]. It is possible to run a highly reactive and energy-dense APPJ at a reasonable cost. APPJ processing can be completed in as little as 15 seconds. Based on capacitance-voltage (CV) measurements (potential scan rate=2 mV/s), the strong bending capability is indicated by the flat value of 3.86 mF/cm and the curved value of 4.42 mF/cm^2 (bending radius R=0.55 cm). A capacitance retention of 98.3% after 1,000 cycles of CV testing in bending at R=0.55 cm was within acceptable limits, suggesting reliable functioning. In the past, a dc-pulse nitrogen APPJ has been used to successfully build a gel-electrolyte SnO_2/CNT nanocomposite SCs [92]. Because furnace calcination has less of an effect on the reaction with CNTs, carbonaceous binders with nanocomposites (SnO_2 and CNTs) are used instead. This is done for a period of 10 minutes at 400°C [91]. After that, gel electrolytes are mixed together to create gel-electrolyte SCs, which are functional when bent despite their seemingly solid form.

Nickel oxide, one of the transition MOs, is widely utilized as an electrode material for SCs due to its low cost, high theoretical capacitance, abundant supply, and pseudocapacitive behavior [93]. There are typically two methods utilized to improve the nickel electrode materials' electrochemical efficiency. Nickel oxides have low electrical conductivity, but the CNT network helps transmit electrons during charge/discharge cycles, leading to a high SPc. It was stated by Amar Prasad Yadav et al. that a nickel oxide and polyaniline (PANI) nanocomposites–based electrode material could be easily, cheaply, and controllably synthesized for use in supercapacitors [94]. Supercapacitors were evaluated in 1 M H_2SO_4 using CV, galvanostatic charge–discharge (GCD), and electrochemical impedance spectroscopy (EIS). The numerous phases of the nanocomposite electrode material improved the overall performance of the energy storage behavior. When compared to other nanocomposites, PANI-NiO_3 stands out for having the highest specific capacitance, with 623 F/g at 1 A/g current density. Significant cyclic stability was also demonstrated by the PANI-NiO_3 electrode, as 89.4% of its initial capacitance was still present after 5,000 cycles of GCD at 20 A/g. They laid out PANI-NiO_3 nanocomposite as a potential electrode material for use in supercapacitors. Basheer and co-workers prepared a nickel-vanadium–based bimetallic precursor using the polymerization process by urea-formaldehyde as copolymers [95]. First, a $Ni_3V_2O_8$-NC magnetic nanocomposite was formed by calcining the precursor at 800°C in an argon atmosphere. The synthesized $Ni_3V_2O_8$-NC electrode was studied electrochemically (EC) in a three-electrode EC workstation. An increased capacitance (CS) of 915 F/g was observed at 50 mV/s when the cyclic voltmeter was immersed in a 5 M potassium hydroxide electrolyte. A capacitive enhancement of 1,045 F/g was also observed in a galvanic charge–discharge (GCD) investigation at a current density (It) of 10 A/g. Increased ion accessibility and large charge storage also characterize the synthesized $Ni_3V_2O_8$-NC nanocomposite, which has a good power density (Pt) of 356.67 W/kg. The maximum power transfer (Pt) of 285.17 W/kg was achieved at an energy density (Et) of 67.34 Wh/kg. $Ni_3V_2O_8$-NC is a magnetic nanocomposite that has improved GCD rate, cycle stability, and Et, making it a candidate for a high-quality supercapacitor electrode.

Since cobalt oxides are cheap, abundant, and safe for the environment, they have been proposed as a good electrode material for (SC) applications. Using a straightforward hydrothermal synthesis process, GO and cobalt oxide (Co_3O_4) nanocomposites were successfully created, with the resulting material showing great promise as a sustainable energy storage/conversion device component [96]. The efficiency of a supercapacitor was measured after a fabricated GO/Co_3O_4 nanocomposite electrode was used with a PVA/KOH gel electrolyte as a separator. This allows the GO/Co$_3$O$_4$-5 electrode to reach its maximum specific capacitance (Csp) of 1,012 F/g at a current density (CD) of 2 A/g. The constructed GO/Co$_3$O$_4$-5 electrode supercapacitor maintains an energy density of 18.9 Wh/kg and a high power density of 1364.8 W/kg. Capacitance was found to be preserved at 70.3% after 4,000 cycles, while Coulombic efficiency was measured at 96.9%. These findings prove that the GO/Co$_3$O$_4$ nanocomposite material is a promising candidate for use in supercapacitors.

SUPERCAPACITORS AND THEIR MECHANISM

Supercapacitors, owing to their exceptionally high capacitance, are also called ultra-capacitors. Farad (F) units of capacitance are high for SCs. The SC uses the method of physically accumulating the charge itself in order to control the charge and discharge timings, as well as to accomplish the long life and high energy density. Electric charge is physically adsorbed and desorbs from the surface of activated carbon, which is the basis of the SC principle [97]. Typically, the essential parts of SC are two electrodes: an electrolyte and a separator [98]. Typically, SCs have separators between the anode and cathode electrodes that allow ions and electrolytes (both aqueous and non-aqueous) to flow freely. Given that electrostatic adsorption at the electrode/electrolyte interface is the basis for EDLCs and faradic reaction at the electrolyte/electrode interface is the basis for pseudo capacitors, it is possible to classify SCs into these two categories [99]. The process of charge storage is facilitated by the electrostatic accumulation of charges at the interface between the electrode and the electrolyte [100]. The charge/discharge process in EDLCs is identical to the dielectric behavior of conventional capacitors because no faradaic reaction takes place during energy storage. Charge adsorption/desorption is typically accomplished in EDLCs with the help of materials such as graphene, activated carbon, and carbon aerogel.

One type of supercapacitor is pseudocapacitor, also called a redox capacitor, which stores energy via a fast and reversible faradaic reaction on the surface of the active components. The charge storage mechanism of pseudocapacitors relies on faradaic charge transfer, which occurs on the electrode's surface or a few nearby surface layers. Pseudocapacitors outperform EDLCs in both energy density and capacitance by a factor of 10–100 [101]. Pseudocapacitors often make use of MOs and conductive polymers. With high specific capacitance, conductivity, and surface area, these materials are able to store and transfer large amounts of energy and electricity. That's why scientists considered them as potentially useful materials for electrochemical energy storage [102]. Hybrid capacitors incorporate both EDLCs and pseudocapacitors, combining their best features. The primary drawback of TMOs is their poor conductivity, which makes it difficult for them to reach the high specific capacitance value predicted by theory. There have been numerous attempts to make HCs, but they have all ultimately failed due to the inherent limitations of the various materials involved.

It is possible to compute the theoretical pseudocapacitance of MO electrode materials by applying Equation (10.1):

$$C_i = \frac{n \times F}{M \times V} \tag{10.1}$$

where n refers to the number of electrons that were given to the oxidation reduction reaction, F stands for the value of the Faraday constant, M stands for the molar mass of the materials that were used, and V is the potential window. In addition, energy and power densities illustrate the two important frameworks of MO electrodes, which might be determined utilizing Equations (10.2) and (10.3):

$$E = \frac{CV^2}{2} \tag{10.2}$$

$$P = \frac{V^2}{4R} \tag{10.3}$$

where C is the capacitance measured in farads (F), V stands for the voltage, and R stands for the equivalent series resistance (ESR) measured in ohms.

The SPc is determined using the equation $C_s=2\times I\times \Delta t/(\Delta V\times m)$, and the necessary parameters are required to be collected from the discharge process in order to complete the calculation. In this case, I denotes the constant current, t stands for the reaction's infinite duration, and V appears to be the working potential range. The capacitance of EDLC is enhanced because of the increased specific surface area caused by the large pore volume of the electrode. However, when electrolyte ions are too tiny to allow unimpeded diffusion in and out, microspores can create pore resistance [103]. The significance of a well-defined pore size distribution becomes particularly obvious in the context of utilizing organic electrolytes, since it is vital for meeting the demanding criteria of high energy density. Therefore, the enhancement of power density necessitates the development of materials possessing high capacitance and low resistance [104].

FUTURE PERSPECTIVES AND CHALLENGES

Considerable research efforts have been dedicated to the modification of the production technique and the improvement of the electrochemical performances of MO nanocomposites. However, they continue to encounter significant challenges that must be resolved. Future prospects for enhancing the electrochemical performance of MO nanocomposites as a supercapacitor electrode will be defined by the extent to which these obstacles can be addressed. (i) Despite the many published studies on MO nanocomposites electrode material for supercapacitor applications, little is known about the effect their structure has on the electrochemical performances. Developing a deeper comprehension of the material's structure and chemical property relationships is essential to enhancing electrochemical performance. Consequently, knowledge of the intermolecular interactions between the components must be gained by experimentation and theory. (ii) Green synthesis approaches aid in the development of eco-friendly, high-conductivity electrodes and electrolytes, and a porous design enhances the performance of MO nanocomposites as electrodes. This has the potential to increase the energy density and voltage range of operation. (iii) The ability to mass-produce MO nanocomposites is crucial for their widespread use in industry. As a result, there is a dearth of MO nanocomposites for widespread use, and it is recommended that we develop more effective alternate approaches to increase their industrial production. Thus, a great deal of study is required to discover highly effective, sustainable, and low-cost ways for industrial-scale synthesis of exceptionally well MO nanocomposites. Based on the survey findings, there are several potential avenues for future research that might be explored.

  i. The prioritization of composites is crucial in the realm of design and production methodologies.
 ii. The preference for lightweight and flexible materials is anticipated among future generations, thereby necessitating their consideration.
iii. The essential necessity across various industries is the assurance that products do not pose any environmental hazards.

CONCLUSION

In the current context of cost-effective and practical devices, there is significant research being conducted on MO nanocomposites. This is due to their ability to easily relax strain during electrochemical cycling, their large specific surface areas, and their capacity to provide pathways for both ion and electron transport and de/intercalation. In summary, the utilization of MO nanocomposites presents opportunities for enhancing the energy density of supercapacitors (SCs). Furthermore, work presented theoretical insights derived from density functional theory on the electrochemical characteristics of the electrode materials as documented in the published literature. Several developing MO nanocomposites have been introduced with the aim of enhancing supercapacitor applications. This chapter provides a concise comparison and analysis of the potential applications of MO nanocomposites and the underlying principles of supercapacitors. It also provides a succinct overview of the progress made in the field of supercapacitor energy storage, as well as the various MO nanocomposites utilized as electrode materials for supercapacitors. Finally, it addresses the current challenges, the future prospects, and the areas that need to be the focus of future research.

REFERENCES

[1] A. Kanwade, S. Gupta, A. Kankane, A. Srivastava, S.C. Yadav, P.M. Shirage, Phosphate-based cathode materials to boost the electrochemical performance of sodium-ion batteries, *Sustain. Energy Fuels* 6 (2022) 3114–3147, https://doi.org/10.1039/D2SE00475E.

[2] K.S. Kumar, N. Choudhary, D. Pandey, L. Hurtado, H.-S. Chung, L. Tetard, Y.W. Jung, J. Tomas, High-performance flexible asymmetric supercapacitor based on rGO anode and WO3/WS2 core/shell nanowire cathode, *Nanotechnology* 31 (2020) 435405.

[3] S.Y. Shin, M.W. Shin, Nickel Metal-organic Framework (Ni-MOF) derived NiO/C@CNF composite for the application of high performance self-standing supercapacitor electrode, *Appl. Surf. Sci.* 540 (2021) 148295.

[4] Z. Gao, Y. Zhang, N. Song, X. Li, Biomass-derived renewable carbon materials for electrochemical energy storage, *Mater. Res. Lett.* 5 (2) (2017) 69–88.

[5] P. Barber, S. Balasubramanian, Y. Anguchamy, S. Gong, A. Wibowo, H. Gao, H.J. Ploehn, H.C. Zur Loye, Polymer composite and nanocomposite dielectric materials for pulse power energy storage, *Materials* 2 (2009) 1697–1733.

[6] T. Cottineau, M. Toupin, T. Delahaye, T. Brousse, D. Bélanger, Nanostructured transition metal oxides for aqueous hybrid electrochemical supercapacitors, *Appl. Phys. A* 82 (2006) 599–606.

[7] M. Pusty, P.M. Shirage, Gold nanoparticle-cellulose/PDMS nanocomposite: A flexible dielectric material for harvesting mechanical energy, *RSC Adv.* 10 (2020) 10097–10112.

[8] B. Chameh, M. Moradi, S. Hajati, F.A. Hessari, Design and construction of ZIF (8 and 67) supported Fe3O4 composite as advanced materials of high performance supercapacitor, *Phys. E Syst. Nanostruct.* 126 (2021) 114442.

[9] W.S. Niu, Z.Y. Xiao, S.F. Wang, S.R. Zhai, L.F. Qin, Z.Y. Zhao, Q.D. An, Synthesis of nickel sulfide-supported on porous carbon from a natural seaweed-derived polysaccharide for high-performance supercapacitors, *J. Alloys Compd.* 853 (2021) 157123.

[10] Y. Shi, L. Sun, Y.X. Zhang, H.C. Si, C. Sun, J.L. Gu, Y. Gong, X.W. Li, Y.H. Zhang, SnS2 nanodots decorated on RGO sheets with enhanced pseudocapacitive performance for asymmetric supercapacitors, *J. Alloys Compd.* 853 (2021) 156903.

[11] J.Y. Hao, X.F. Zou, L. Feng, W.P. Li, B. Xiang, Q. Hu, X.Y. Liang, Q.B. Wu, Facile fabrication of core-shell structured Ni(OH)2/Ni(PO3)2 composite via one-step electrodeposition for high performance asymmetric supercapacitor, *J. Colloid Interface Sci.* 583 (2021) 243–254.

[12] M.A.A. Mohd Abdah, N.H.N. Azman, S. Kulandaivalu, Y. Sulaiman, Review of the use of transition-metal-oxide and conducting polymer-based fibres for high-performance supercapacitors, *Mater. Des.* 186 (2020) 108199.

[13] C. An, Y. Zhang, H. Guo, Y. Wang, Metal oxide-based supercapacitors: Progress and prospectives, *Nanoscale Adv.* 1 (2019) 4644.

[14] H.S. Roy, M.M. Islam, M.Y.A. Mollah, M.A.B.H. Susan, Polyaniline-MnO2 composites prepared in-situ during oxidative polymerization of aniline for supercapacitor applications, *Mater. Today Proc.* 29 (2020) 1013.

[15] W.H. Low, P.S. Khiew, S.S. Lim, C.W. Siong, E.R. Ezeigwe, Recent development of mixed transition metal oxide and graphene/mixed transition metal oxide based hybrid nanostructures for advanced supercapacitors, *J. Alloys Compd.* 775 (2019) 1324.

[16] S. Shahabuddin, R. Gaur, N. Mukherjee, P. Chandra, R. Khanam, Conducting polymers-based nanocomposites: Innovative materials for wastewater treatment and energy storage, *Mater. Today Proc.* 62 (2022) 6950.

[17] S. Thambidurai, P. Gowthaman, M. Venkatachalam, S. Suresh, Enhanced Bactericidal performance of nickel oxide-zinc oxide nanocomposites synthesized by facile chemical co-precipitation method, *J. Alloys Compd.* 830 (2020) 154642.

[18] Y. Zhang, X. Sun, L. Pan, H. Li, Z. Sun, C. Sun, et al. Carbon nanotube-ZnO nanocomposite electrodes for supercapacitors, *Solid State Ion.* 180 (2009) 1525–1528.

[19] C.G. Cameron, Electrochemical capacitors 17, in: Breitkopf, S-Lyons (Ed.), *Springer Handbook of Electrochemical Energy*, Springer Handbooks, 2017, pp. 563–589, https://doi.org/10.1007/978-3-662-46 657-5_17.

[20] L. Liu, Z. Niu, J. Chen, Flexible supercapacitors based on carbon nanotubes, *Chin. Chem. Lett.* 29 (2018) 571–581, https://doi.org/10.1016/j.cclet.2018.01.013.

[21] S. Zhu, J. Ni, Y. Li, Carbon nanotube-based electrodes for flexible supercapacitors, *Nano Res.* 13 (2020) 1825–1841, https://doi.org/10.1007/s12274-020-2729-5.

[22] K.Y. Kumar, S. Archana, R. Namitha, B.P. Prasanna, S.C. Sharma, M.S. Raghu, Ruthenium oxide nanostring clusters anchored graphene oxide nanocomposites for high-performance supercapacitors application, *Mater. Res. Bull.* 107 (2018) 347–354.

[23] K.M. Anilkumar, M. Manoj, B. Jinisha, V.S. Pradeep, S. Jayalekshmi, Mn3O4/reduced graphene oxide nanocomposite electrodes with tailored morphology for high power supercapacitor applications, *Electrochim. Acta* 236 (2017) 424–433.

[24] V. Gupta, A.M. Kannan, S. Kumar, Graphene foam (GF)/manganese oxide (MnO2) nanocomposites for high performance supercapacitors, *J. Energy Storage* 30 (2020) 101575.

[25] Q. Wang, L. Jiao, H. Du, Y. Wang, H. Yuan, Fe3O4 nanoparticles grown on graphene as advanced electrode materials for supercapacitors, *J. Power Sources* 245 (2014) 101–106.

[26] A.M. Khattak, H. Yin, Z.A. Ghazi, B. Liang, A. Iqbal, N.A. Khan, Z. Tang, Three dimensional iron oxide/graphene aerogel hybrids as all-solid-state flexible supercapacitor electrodes, *RSC Adv.* 6 (64) (2016) 58994–59000.

[27] Q. Li, C. Lu, C. Chen, L. Xie, Y. Li, Q. Kong, H. Wang, Layered NiCo2O4/reduced graphene oxide composite as an advanced electrode for supercapacitor, *Energy Storage Mater.* 8 (2017) 59–67.

[28] T.N.V. Raj, P.A. Hoskeri, H.B. Muralidhara, C.R. Manjunatha, K.Y. Kumar, M.S. Raghu, Facile synthesis of perovskite lanthanum aluminate and its green reduced graphene oxide composite for high performance supercapacitors, *J. Electroanal. Chem.* 858 (2020) 113830.

[29] T.N.V. Raj, P.A. Hoskeri, H.B. Muralidhara, B.P. Prasanna, K.Y. Kumar, F.A. Alharthi, M.S. Raghu, Tantalum pentoxide functionalized nitrogen-doped reduced graphene oxide as a competent electrode material for enhanced specific capacitance in a hybrid supercapacitor device, *J. Alloys Compd.* 861 (2021) 158572.

[30] T.N.V. Raj, P.A. Hoskeri, S. Hamzad, M.S. Anantha, C.M. Joseph, H.B. Muralidhara, K.Y. Kumar, F.A. Alharti, B.H. Jeon, M.S. Raghu, Moringa Oleifera leaf extract mediated synthesis of reduced graphene oxide-vanadium pentoxide nanocomposite for enhanced specific capacitance in supercapacitors, *Inorg. Chem. Commun.* 142 (2022) 109648.

[31] X. Zhang, X. Gan, T. Wang, H. Li, W. Shi, X. Zhao, X. Yan, Y. Liu, B. Liu, P-induced oxygen-deficient P-Ni2MnO4–x@rGO with enhanced energy density for supercapacitor, *J. Alloys Compd.* 937 (2023) 168321.

[32] S.M. Mbam, R.M. Obodo, O.O. Apeh, A.C. Nwanya, A.B.C. Ekwealor, N. Nwulu, F.I. Ezema, Performance evaluation of Bi2O3@GO and Bi2O3@rGO composites electrode for supercapacitor application, *J. Mater. Sci. Mater. Electron.* 34 (2023) 1405.

[33] Q. Liu, J. Yang, X. Luo, Y. Miao, Y. Zhang, W. Xu, M. Zhu, Fabrication of a fibrous MnO2@ MXene/CNT electrode for high-performance flexible supercapacitor, *Ceram. Int.* 46 (8) (2020) 11874–11881.

[34] Q. Wang, Z. Zhang, Z. Zhang, X. Zhou, G. Ma, Facile synthesis of MXene/MnO2 composite with high specific capacitance, *J. Solid State Electrochem.* 23 (2) (2019) 361–365.

[35] R. Zou, H. Quan, M. Pan, S. Zhou, D. Chen, X. Luo, Self-assembled MXene (Ti3C2TX)/α-Fe2O3 nanocomposite as negative electrode material for supercapacitors, *Electrochim. Acta* 292 (2018) 31–38.

[36] T. Arun, A. Mohanty, A. Rosenkranz, B. Wang, J. Yu, M.J. Morel, A. Ramadoss, Role of electrolytes on the electrochemical characteristics of Fe3O4/MXene/RGO composites for supercapacitor applications, *Electrochim. Acta* 367 (2021), 137473.

[37] N. An, W. Li, Z. Shao, L. Zhou, Y. He, D. Sun, X. Dong, Z. Hu, Graphene oxide coated polyaminoanthraquinone@MXene based flexible film electrode for high-performance supercapacitor, *J. Energy Storage* 57, (2023) 106180.

[38] M.Y. Li, C.H. Chen, Y. Shi, L.J. Li, Heterostructures based on two-dimensional layered materials and their potential applications, *Mater. Today* 19 (6) (2016) 322–335.

[39] M. Wang, H. Fei, P. Zhang, L. Yin, Hierarchically layered MoS2/Mn3O4 hybrid architectures for electrochemical supercapacitors with enhanced performance, *Electrochim. Acta* 209 (2016) 389–398.

[40] Z. Chang, X. Zhu, X. Ju, X. Li, X. Zheng, W. Zhang, Z. Ren, Synthesis of hierarchical hollow urchin-like HGRs/MoS2/MnO2 composite and its excellent supercapacitor performance, *J. Alloys Compd.* 775 (2019) 241–247.

[41] R. Palanisamy, D. Karuppiah, S. Venkatesan, S. Mani, M. Kuppusamy, S. Marimuthu, R. Perumalsamy, High-performance asymmetric supercapacitor fabricated with a novel MoS2/Fe2O3/Graphene composite electrode, *Colloid Interface Sci. Commun.* 46 (2022), 100573.

[42] M. Sarno, A. Troisi, Supercapacitors based on high surface area MoS2 and MoS2-Fe3O4 nanostructures supported on physical exfoliated graphite, *J. Nanosci. Nanotechnol.* 17 (6) (2017) 3735–3743.

[43] T. Anitha, A.E. Reddy, I.K. Durga, S.S. Rao, H.W. Nam, H.J. Kim, Facile synthesis of ZnWO4@ WS2 cauliflower-like structures for supercapacitors with enhanced electrochemical performance, *J. Electroanal. Chem.* 841 (2019) 86–93.

[44] M. Abdullah, S. Khan, K. Jabbour, M. Imran, M.F. Ashiq, P. John, S. Manzoor, T. Munawar, M.N. Ashiq, Development of binder-free MoTe2/rGO electrode via hydrothermal route for supercapacitor application, *Electrochim. Acta* 466 (2023) 143020.

[45] Z.P. Diao, Y.X. Zhang, X.D. Hao, Z.Q. Wen, Facile synthesis of CoAl-LDH/MnO2 hierarchical nanocomposites for high-performance supercapacitors, *Ceram. Int.* 40 (1) (2014) 2115–2120.

[46] H. Pourfarzad, M. Shabani-Nooshabadi, M.R. Ganjali, H. Kashani, Synthesis of Ni-Co-fe layered double hydroXide and Fe2O3/Graphene nanocomposites as actively materials for high electrochemical performance supercapacitors, *Electrochim. Acta* 317 (2019) 83–92.

[47] J. Zhou, S. Xu, L. Ni, N. Chen, X. Li, C. Lu, W. Hou, Iron oxide encapsulated in nitrogen-doped carbon as high energy anode material for asymmetric supercapacitors, *J. Power Sources* 438 (2019) 227047.

[48] X. He, Q. Liu, J. Liu, R. Li, H. Zhang, R. Chen, J. Wang, Hierarchical NiCo2O4@ NiCoAl-layered double hydroXide core/shell nanoforest arrays as advanced electrodes for high-performance asymmetric supercapacitors, *J. Alloys Compd.* 724 (2017) 130–138.

[49] D. Chu, F. Li, X. Song, H. Ma, L. Tan, H. Pang, B. Xiao, A novel dual-tasking hollow cube NiFe2O4-NiCo-LDH@ rGO hierarchical material for high performance supercapacitor and glucose sensor, *J. Colloid Interface Sci.* 568 (2020) 130–138.

[50] A.M. Ghadimi, S. Ghasemi, A. Omrani, F. Mousavi, Nickel cobalt LDH/graphene film on nickel-foam-supported ternary transition metal oxides for supercapacitor applications, *Energy Fuels* 37 (4) (2023) 3121–3133.

[51] M.B. Poudel, M. Shin, H.J. Kim, Polyaniline-silver-manganese dioxide nanorod ternary composite for asymmetric supercapacitor with remarkable electrochemical performance, *Int. J. Hydrogen Energy* 46 (2021) 474.

[52] Y. Huang, S. Bao, J. Lu, Flower-like MnO2/polyaniline/hollow mesoporous silica as electrode for high-performance all-solidState supercapacitors, *J. Alloys Compd.* 845 (2020) 156192.

[53] M. Dirican, M. Yanilmaz, A.M. Asiri, X. Zhang, Polyaniline/MnO2/porous carbon nanofiber electrodes for supercapacitors, *J. Electroanal. Chem.* 861 (2020) 113995.

[54] M.A. Arvizu, F.J. Gonzalez, A. Romero-Galarza, F.J. Rodriguez-Varela, C.R. Garcia, M.A. Garcia-Lobato, Symmetric supercapacitors of PANI coated RuO2/TiO2 macroporous structures prepared by electrostatic spray deposition, *J. Electrochem. Soc.* 169 (2022) 020564.

[55] A.R. Athira, T.C. Bhagya, A.H. Riyas, T.S. Xavier, S.M.A. Shibli, Design and fabrication of Co3O4 anchored PANI binary composite supercapacitors with enhanced electrochemical performance and stability, *J. Mater. Sci. Mater. Electron.* 33 (2022) 2829.

[56] Y. Fan, H. Chen, Y. Li, D. Cui, Z. Fan, C. Xue, PANI-Co3O4 with excellent specific capacitance as an electrode for supercapacitors, *Ceram. Int.* 47 (2021) 8433.

[57] M. Usman, N. Adnan, M.T. Ahsan, S. Javed, M.S. Butt, M.A. Akran, In situ synthesis of a polyaniline/ Fe-Ni co-doped Fe3O4 composite for the electrode material of supercapacitors with improved cycling stability, *ACS Omega* 6 (2021) 1190.

[58] B.P. Prasanna, D.N. Avadhani, V. Raj, K.Y. Kumar, M.S. Raghu, Fabrication of PANI/SnO2 hybrid nanocomposites via interfacial polymerization for high performance supercapacitors applications, *Surf. Eng. Appl. Electrochem.* 55 (2019) 463–471.

[59] M.S. Raghu, K.Y. Kumar, S. Rao, T. Aravinda, B.P. Prasanna, M.K. Prashanth, Fabrication of polyaniline-few-layer MoS2 nanocomposite for high energy density supercapacitors, *Polym. Bull.* 75 (2018) 4359–4375.

[60] B.P. Prasanna, D.N. Avadhani, K. Chaitra, N. Nagaraju, N. Kathyayini, Synthesis of polyaniline/ MWCNTs by interfacial polymerization for superior hybrid supercapacitance performance, *J. Polym. Res.* 25 (5) (2018) 123.

[61] B.P. Prasanna, D.N. Avadhani, M.S. Raghu, K.Y. Kumar, Synthesis of polyaniline/α-Fe2O3 nanocomposite electrode material for supercapacitor applications, *Mater. Today Commun.* 12 (2017) 72–78.

[62] Y. Xie, J. Zhang, H. Xu, T. Zhou, Laser-assisted mask-free patterning strategy for high-performance hybrid micro-supercapacitors with 3D current collectors, *Chem. Eng. J.* 437 (2022) 135493.

[63] W.J. Zhuo, Y.H. Wang, C.T. Huang, M.J. Deng, Enhanced pseudocapacitive performance of symmetric polypyrrole-MnO2 electrode and polymer gel electrolyte, *Polymers* 13 (2021) 3577.

[64] H. Pourfarzad, R. Badrnezhad, M. Ghaemmaghami, M. Saremi, In situ synthesis of C3N4/PPy/MnO2 nanocomposite as high-performance active material for assymetric supercapacitor, *Ionics* 27 (2021) 4057.

[65] C. Kang, J. Fang, L. Fu, S. Li, Q. Liu, Hierarchical carbon nanowire/Ni@MnO2 nanocomposites for high-performance asymmetric supercapacitors, *Chem. Eur. J.* 26 (2020) 16392.

[66] M. Ates, O. Kuzgun, Modified carbon black, CB/MnO2 and CB/MnO2/PPy nanocomposites synthesised by microwave-assisted method for energy storage devices with high electrochemical performances, *Plast. Rubber Compos.* 49 (2020) 342.

[67] X. Li, Y. Ma, P. Shen, C. Zhang, M. Cao, S. Xiao, J. Yan, S. Luo, Y. Gao, An ultrahigh energy density flexible asymmetric microsupercapacitor based on Ti3C2Tx and PPy/MnO2 with wide voltage window, *Adv. Mater. Technol.* 5 (2020) 2000272.

[68] C. Ren, Y. Yan, B. Sun, B. Gu, T.W. Chou, Wet-spinning assembly and in situ electrodeposition of carbon nanotube-based composite fibers for high energy density wire-shaped asymmetric supercapacitor, *J. Colloid Interface Sci.* 569 (2020) 298.

[69] J.E. Nady, A. Shokry, M. Khalil, S. Ebrahim, A.M. Elshaer, M. Anas, One-step electrodeposition of a polypyrrole/NiO nanocomposite as a supercapacitor electrode, *Sci. Rep.* 12 (2022) 3611.

[70] Z.M. Shen, X.J. Luo, Y.Y. Zhu, Y.S. Liu, Facile co-deposition of NiO-CoO-PPy composite for asymmetric supercapacitors, *J. Energy Storage* 51 (2022) 104475.

[71] S.Z. Golkhatmi, A. Sedghi, H.N. Miankushki, M. Khalaj, Structural properties and supercapacitive performance evaluation of the nickel oxide/graphene/polypyrrole hybrid ternary nanocomposite in aqueous and organic electrolytes, *Energy* 214 (2021) 118950.

[72] G. He, J. Li, H. Chen, J. Shi, X. Sun, S. Chen, X. Wang, Hydrothermal preparation of Co3O4@ graphene nanocomposite for supercapacitor with enhanced capacitive performance, *Mater. Lett.* 82 (2012) 61–63.

[73] D.V. Leontyeva, I.N. Leontyev, M.V. Avramenko, Y.I. Yuzyuk, Y.A. Kukushkina, N.V. Smirnova, Electrochemical dispergation as a simple and effective technique toward preparation of NiO based nanocomposite for supercapacitor application, *Electrochim. Acta* 114 (2013) 356–362.

[74] P. Sen, A. De, A.D. Chowdhury, S.K. Bandyopadhyay, N. Agnihotri, M. Mukherjee, Conducting polymer based manganese dioxide nanocomposite as supercapacitor, *Electrochim. Acta* 108 (2013) 265–273.

[75] F. Xiong, D. Zhou, Z. Xie, Y. Chen, A study of the Ce3+/Ce4+ redox couple in sulfamic acid for redox battery application, *Appl. Energy* 99 (2012) 291–296, https://doi.org/10.1016/j.apenergy.2012.05.021.

[76] F. Paquin, J. Rivnay, A. Salleo, N. Stingelin, C. Silva, Multi-phase semicrystalline microstructures drive exciton dissociation in neat plastic semiconductors, *J. Mater. Chem. C* 3 (2015) 10715–10722, https://doi.org/10.1039/C5TC02043C.

[77] Y. Luo, T. Yang, Q. Zhao, M. Zhang, CeO2/CNTs hybrid with high performance as electrode materials for supercapacitor, *J. Alloys Compd.* 729 (2017) 64–70, https://doi.org/10.1016/j.jallcom.2017.09.165.

[78] L.S. Aravinda, K.U. Bhat, B.R. Bhat, Nano CeO2/activated carbon based composite electrodes for high performance supercapacitor, *Mater. Lett.* 112 (2013) 158–161, https://doi.org/10.1016/j.matlet.2013.09.009.

[79] D. Deng, N. Chen, Y. Li, X. Xing, X. Liu, X. Xiao, Y. Wang, Cerium oxide nanoparticles/multi-wall carbon nanotubes composites: Facile synthesis and electrochemical performances as supercapacitor electrode materials, *Phys. E: Low-Dimens. Syst. Nanostruct.*, 86 (2017) 284–291, 2017, https://doi.org/10.1016/j.physe.2016.10.031.

[80] E.M. Abebe, M. Ujihara, Simultaneous electrodeposition of ternary metal oxide nanocomposites for high-efficiency supercapacitor applications, *ACS Omega* 7 (2022) 17161–17174.

[81] S.-L. Kuo, N.-L. Wu, Investigation of pseudocapacitive charge-storage reaction of MnO2·nH2O supercapacitors in aqueous electrolytes, *J. Electrochem. Soc.* 153 (7) (2006) A1317, https://doi.org/10.1149/1.2197667.

[82] Y. Zheng, X. Zheng, Hydrothermal synthesis of MnO2 with different morphological characteristics as electrode material for high electrochemical performance supercapacitors, *Int. J. Electrochem. Sci.* 15 (2020) 1465–1473, https://doi.org/10.20964/2020.02.57.

[83] X. Wu, F. Yang, H. Dong, J. Sui, Q. Zhang, J. Yu, Q. Zhang, L. Dong, Controllable synthesis of MnO2 with different structures for supercapacitor electrodes, *J. Electroanal. Chem.* 848 (2019), 113332, https://doi.org/10.1016/j.jelechem.2019.113332.

[84] C. Yin, H. Zhou, J. Li, Facile one-step hydrothermal synthesis of PEDOT: PSS/ MnO2 nanorod hybrids for high-rate supercapacitor electrode materials, *Ionics* 25 (2019) 685–695, https://doi.org/10.1007/s11581-018-2680-6.

[85] S. Jadhav, R.S. Kalubarme, C. Terashima, B.B. Kale, V. Godbole, A. Fujishima, S.W. Gosavi, Manganese dioxide/reduced graphene oxide composite an electrode material for high-performance solid-state supercapacitor, *Electrochim. Acta* 299 (2019) 34–44, https://doi.org/10.1016/ j.electacta.2018.12.182.

[86] Q. Zhanga, X. Wua, Q. Zhanga, F. Yanga, H. Donga, J. Suia, L. Dong, One-step hydrothermal synthesis of MnO2/graphene composite for electrochemical energy storage, *J. Electroanal. Chem.* 837 (2019) 108–115, https://doi.org/10.1016/j.jelechem.2019.02.031.

[87] O. Sadak, W. Wang, J. Guan, A.K. Sundramoorthy, S. Gunasekaran, MnO2 nanoflowers deposited on graphene paper as electrode materials for supercapacitors, *ACS Appl. Nano Mater.* 2 (7) (2019) 4386–4394, https://doi.org/10.1021/acsanm.9b00797.

[88] L.Y. Chen, Y. Hou, J.L. Kang, A. Hirata, T. Fujita, M.W. Chen, Toward the theoretical capacitance of RuO2 reinforced by highly conductive nanoporous gold, *Adv. Energy Mater.* 3 (2013) 851–856, https:// doi.org/10.1002/aenm.201300024.

[89] K.Y. Kumara, S. Archanaa, R. Namithaa, B.P. Prasannaa, S.C. Sharmab, M.S. Raghu, Ruthenium oxide nanostring clusters anchored Graphene oxide nanocomposites for high-performance supercapacitors application, *Mater. Res. Bull.* 107 (2018) 347–354.

[90] L. Yao, C. Zhang, N. Hu, L. Zhang, Z. Zhou, Y. Zhang, Three-dimensional skeleton networks of reduced graphene oxide nanosheets/vanadium pentoxide nanobelts hybrid for high-performance supercapacitors, *Electrochim. Acta* 295 (2019) 14–21, https://doi.org/10.1016/j.electacta.2018.10.134.

[91] C.Y. Liao, F.H. Kuok, C.W. Chen, C.C. Hsu, J.Z. Chen, Flexible quasi-solid-state SnO2/CNT supercapacitor processed by a dc-pulse nitrogen atmospheric-pressure plasma jet, *J. Energy Storage* 11 (2017) 237–241, https://doi.org/10.1016/j. est.2017.03.007.

[92] H.W. Liu, S.P. Liang, T.J. Wu, H. Chang, P.K. Kao, C.C. Hsu, J.Z. Chen, P.T. Chou, I.C. Cheng, Rapid atmospheric pressure plasma jet processed reduced graphene oxide counter electrodes for dye-sensitized solar cells, *ACS Appl. Mater. Interfaces* 6 (2014) 15105–15112, https://doi.org/10.1021/am503217f.

[93] G. Zhang, L. Wang, Y. Liu, W. Li, F. Yu, W. Lu, H. Huang, Cracks bring robustness: A pre-cracked NiO nanosponge electrode with greatly enhanced cycle stability and rate performance, *J. Mater. Chem. A Mater.* 4 (2016) 8211–8218, https://doi.org/10.1039/c6ta02568d.

[94] K.P. Gautam, D. Acharya, I. Bhatta, V. Subedi, M. Das, S. Neupane, J. Kunwar, K. Chhetri, A.P. Yadav, Nickel oxide-incorporated polyaniline nanocomposites as an efficient electrode material for supercapacitor application, *Inorganics* 10 (2022) 86. https://doi.org/10.3390/inorganics10060086.

[95] A.R.Z. Almotairy, B.M. Al-Maswari, K. Alkanad, N.K. Lokanath, R.T. Radhika, B.M. Venkatesha, Nickel vanadate nitrogen-doped carbon nanocomposites for high-performance supercapacitor electrode, *Heliyon* 9 (2023) e18496.

[96] S. Veeresh, H. Ganesha, Y.S. Nagaraju, H. Vijeth, M. Vandana, M. Basappa, H. Devendrappa, Graphene oxide/cobalt oxide nanocomposite for high-performance electrode for supercapacitor application, *J. Energy Storage* 52 (2022) 104715.

[97] W. Wu, D. Niu, J. Zhu, Y. Gao, D. Wei, C. Zhao, C. Wang, F. Wang, L. Wang, L. Yang, Hierarchical architecture of Ti3C2@PDA/NiCo2S4 composite electrode as high-performance supercapacitors, *Ceram. Int.* 45 (2019) 16261–16269, https://doi.org/10.1016/j.ceramint.2019.05.149.

[98] D. Wu, H. Yu, C. Hou, H. Du, X. Song, T. Shi, X. Sun, B. Wang, NiS nanoparticles assembled on biological cell walls-derived porous hollow carbon spheres as a novel battery-type electrode for hybrid supercapacitor, *J. Mater. Sci.* 55 (2020) 14431–14446, https://doi.org/10.1007/s10853-020-05022-6.

[99] Y. Deng, Y. Xie, K. Zou, X. Ji, Review on recent advances in nitrogen-doped carbons: preparations and applications in supercapacitors, *J. Mater. Chem. A* 4 (2015) 1144–1173, https://doi.org/10.1039/ c5ta08620e.

[100] S.S. Siwal, Q. Zhang, N. Devi, V.K. Thakur, Carbon-based polymer nanocomposite for high-performance energy storage applications, *Polymer* 12 (2020), https://doi.org/10.3390/polym12030505.

[101] L. Feng, Y. Zhu, H. Ding, C. Ni, Recent progress in nickel based materials for high performance pseudocapacitor electrodes, *J. Power Sources* 267 (2014) 430–444, https://doi.org/10.1016/j.jpowsour.2014.05.092.

[102] D. Yang, M.I. Ionescu, 8- metal oxide-carbon hybrid materials for application in supercapacitors, in: D.P. Dubal, P. Gomez-Romero (Eds.), *Metal Oxides*, Elsevier, 2017, pp. 193–218, https://doi.org/10.1016/ B978-0-12-810464-4.00008-5.

[103] S. Joseph, D.M. Kempaiah, M.R. Benzigar, H. Ilbeygi, G. Singh, S.N. Talapaneni, D.H. Park, A. Vinu, Highly ordered mesoporous carbons with high specific surface area from carbonated soft drink for supercapacitor application, *Microporous Mesoporous Mater.* 280 (2019) 337–346, https://doi.org/10.1016/j. micromeso.2019.02.020.

[104] H. Pan, J. Li, Y.P. Feng, Carbon nanotubes for supercapacitor, *Nanoscale Res. Lett.* 5 (2010) 654–668, https://doi.org/10.1007/s11671-009-9508-2.

11 CeO$_2$:Tb^{3+} Nanophosphor for Display Applications

R. Vijay Kumar, H.J. Amith Yadav, V.S. Patil,
G.H. Nagaveni, H.P. Nagaswarupa, and
Ramachandra Naik

INTRODUCTION

In recent years, very much attention has been paid to the development of display applications in the industry. As we know, nanoparticles lie in the size range between 1 and 100 nm [1]. Due their size, they have unique properties compared to bulk materials. These particles are used increasingly in catalysis to boost chemical reactions. This reduces the quality of catalytic materials necessary to produce desired results, saving money and reducing pollutants in environment. All these different properties and applications were incorporated in the technical aspects which led to the development of a field called nanotechnology. Due to their high surface to volume ratio, nanoparticles have a diverse range of physical and chemical characteristics [2]. Applications of nanoscience and nanotechnology have quickly flourished over the last few decades, and still the trend continues.

Metal oxide nanoparticles play important roles in physical, chemical, and material sciences. Indeed, the metal elements are able to form a large diversity of oxides. These oxide nanoparticles can adopt a large number of physical and structural geometries with an electronic structure that can exhibit semiconductor or insulator, technological applications. Metal oxide nanoparticles are used in fabrication of microelectronic circuits, sensors, piezoelectric devices, fuel cells, coatings for the passivation of surfaces against corrosion, and as modification of different materials. In the emerging field of nanotechnology, the primary goal is to make nanostructures and nanofibers of various properties with the help bulk or single particle species. Cerium oxide nanoparticle, also known as ceric oxide or ceric dioxide, is a metal oxide nanoparticle of the rare earth metal cerium. It is a pale yellow white powder with the chemical formula CeO$_2$. Cerium oxide (CeO$_2$) nanoparticles have been confirmed to be capable of serving as photo catalysts for environmental remediation because of their strong redox ability, long-term stability, nontoxicity, and cost effectiveness [3]. These CeO$_2$ metal oxide nanoparticles are used as ultraviolet absorber, polishing agent, and gas sensor, as well as for commercial purposes in day-to-day life like cosmetics products and also fabrication and latent fingerprint application.

The chemical element terbium has the atomic number 65 and the symbol Tb. It is a brilliant white, rear earth metal that is malleable and flexible. Terbium, the ninth lanthanide, is an electropositive metal that reacts with water to produce hydrogen gas. Terbium is found in many minerals, including cerite, gadolinite, monazite, xenotime, and euxenite, but it never occurs naturally as a free element. The bright lemon-yellow color of the terbium (III) cation is the result of a strong green emission line combined with other lines in orange and red. The yttrofluorite assortment of the mineral fluorite owes its rich yellow fluorescence to some extent to terbium. Since terbium is susceptible to oxidation, it is primarily utilized in its elemental state for research. Single terbium atoms have been secluded by embedding them into fullerene molecules.

Gomutra or cow urine is a natural liquid byproduct of metabolism. It has no side effects or disadvantages and is used in Ayurveda from the ancient era to now because of its medicinal properties. It has many years of its own history of common usage in human life such as Poojas in Indian traditional

DOI: 10.1201/9781003479239-11

culture, with the belief that it manipulates human life. It contains metabolism products including 95% of water, 2.5% of urea, 2.5% enzymes, minerals, and 24 different types of salts. It is freely available, low cost, and ecofriendly. It also has its applications in fertilizers, medicine, etc. CeO$_2$-doped Tb nanoparticles can be prepared from cow urine by combustion method.

These CeO$_2$: Tb^{3+} nanoparticles extraction will possible to incorporate their properties and nature further applied to various applications in industries, engineering material science works, biomedicines, nanotechnology fields knowingly its very scalable purposes to make next generations explore their ideas to be identified and to be develop theirs essential needs which is low cost, high durability, environmental friendly most prominence materials [4]. By the help of production it also reduce abundance to everyone's need by taking with action of combustion method reduced time and cost of the process and highly pure outcome products easy to handle whole process of experimental studies.

SYNTHESIS

The starting materials for the fabrication of CeO$_2$:Tb^{3+} nanophoshor are cerium nitrate, terbium nitrate, and cow urine as a fuel. Cerium nitrate and cow urine were mixed and dissolved in distilled water. Terbium nitrate was used as the Tb source in the solution. The concentration of Tb was varied from 1 to 5 mol%. A homogeneous solution was obtained after stirring for 20 min. The solution was transferred to a pre-heated muffle furnace maintained at a temperature of 450 ± 10°C. The obtained product is calcinated at 700°C for 2h and applied to different techniques of characterization.

RESULT AND DISCUSSIONS

Figure 11.1 shows the PXRD pattern of CeO$_2$:Tb^{3+} (1–5 mol%) nanophosphor. Prominent PXRD peaks are observed at 2θ values 28.5, 33, 47.5, 56, 69, and 77 which can be assigned to (hkl) values

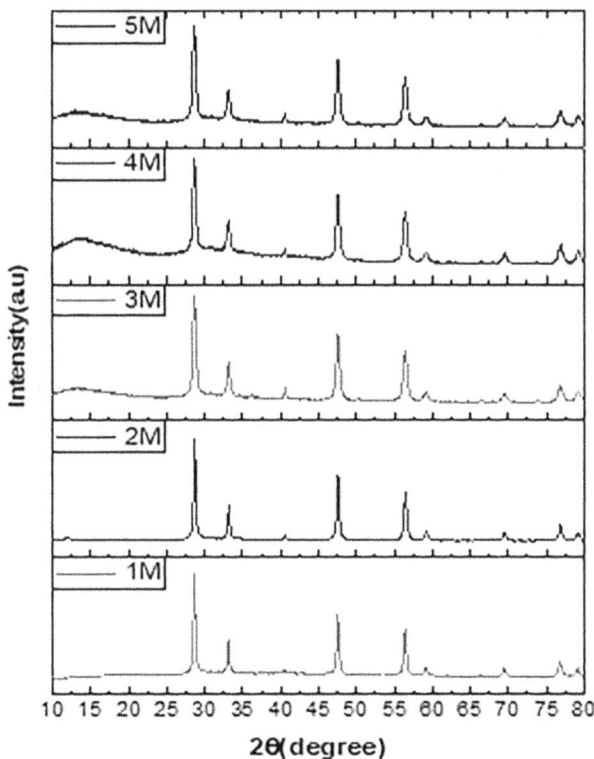

FIGURE 11.1　PXRD pattern of (1–5 mol%) Tb^{3+}-doped CeO$_2$ nanoparticles.

(1 1 1) (2 0 0) (2 1 1) (2 2 0) (3 1 1) (4 2 0), respectively as per the JCPDS card No. 81-0792 [5]. The obtained PXRD pattern confirms the cubic fluorite structure.

The optical properties and energy gap (E_g) of CeO_2:Tb^{3+} (1–5 mol%) were studied by UV-visible spectra as shown in Figure 11.2. UV–Vis spectroscopic measurements were carried out at room temperature to study the effect of Tb doping on CeO_2 in the range 200–800 nm. Figure 11.2a was observed that a broad absorption peak located at around ~300 nm for 2 mol% of Tb doping on CeO_2. Except 2 mol% of Tb doping on CeO_2 nanophosphor, other shows strong absorption from 300 to 800 nm in Figure 11.2a. The band gap of the CeO_2:Tb^{3+} (1–5 mol%) Nps was estimated from this absorption spectra using Wood–Tauc's relation $\left[\alpha h\gamma = c\left(h\gamma - E_g\right)^K \right]$ shown in Figure 11.2b, where h - Planks constant, α - absorption coefficient, γ - frequency, E_g - optical band gap [6].

$$\alpha\alpha \frac{\left(h\gamma - E_p - E_i\right)^2}{\left(e^{\frac{h\gamma}{kt}} - 1\right)} + \frac{\left(h\gamma - E_p - E_i\right)2e^{\frac{h\gamma}{kt}}}{\left(e^{\frac{h\gamma}{kt}} - 1\right)}$$

The E_g values were estimated by extrapolating the linear portion of the curve or tail in the UV–V is absorbance spectra and their values were found in the range 4.6–5.8 eV.

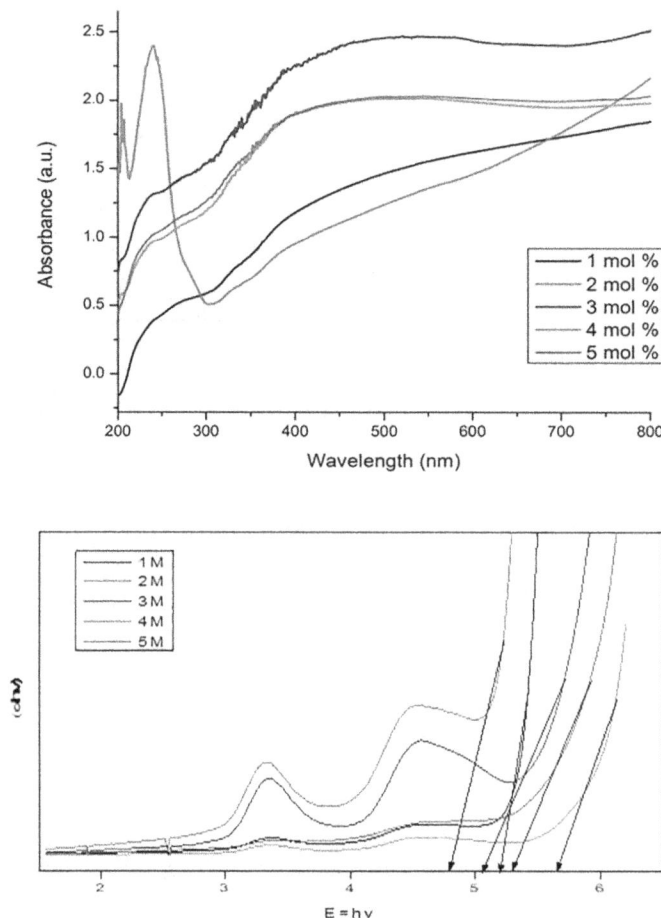

FIGURE 11.2 UV-visible spectrum of (1–5 mol%) Tb^{3+}-doped CeO_2 nanophosphor.

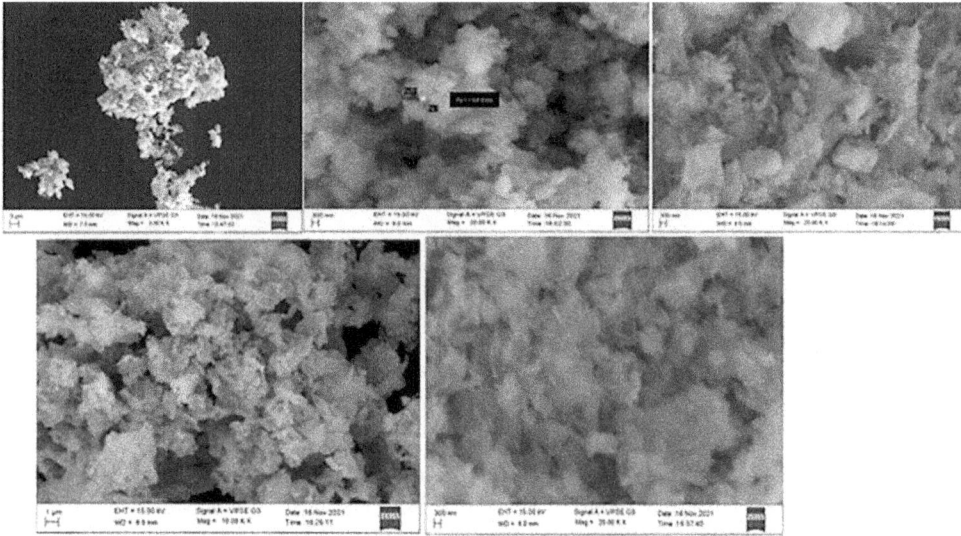

FIGURE 11.3   SEM micrographs of doped Tb^{3+} (1–5 mol%) CeO$_2$ nanophosphors.

Figure 11.3 shows SEM images of Tb^{3+} (1–5 mol%) doped CeO$_2$ nanophosphors. The morphology was strongly dependent on the doping concentration of Tb. SEM images show that in the initial concentration of Tb^{3+} are agglomerated, loosely packed, and void pores. When the concentration of Tb^{3+} increases, atoms are nearly packed with less pores. In the morphology studies shown in final concentration, atoms are packed with less pores [7].

Figure 11.4 shows the emission spectra of prepared CeO$_2$:Tb^{3+} (1–5 mol%) nanophosphor keeping excitation wavelength at 250 nm. In the figure, the peak is maximum at 320, 328, 340, 360, 500, and 750 nm [8].

The color coordinates of CIE chromaticity diagram are shown in the XY graph in Figure 11.5. The abbreviation CIE stands for "commission internationale de l'eclairage". The graph shows the distribution of different characteristic wavelengths in the visible spectrum. In the CIE diagram, the

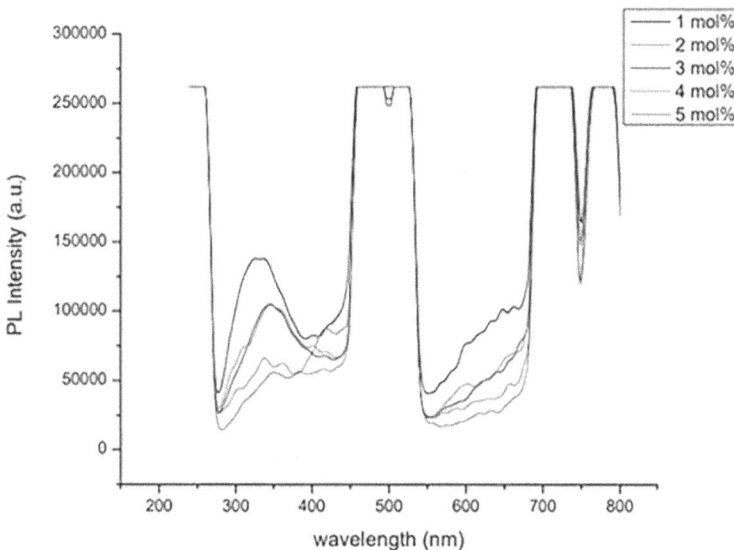

FIGURE 11.4   PL intensity shows emission spectra of CeO$_2$:Tb^{3+} (1–5 mol%) nanophosphor.

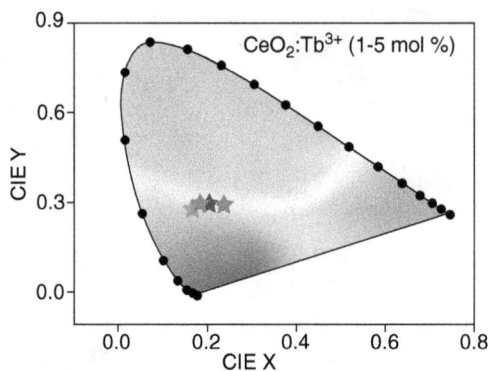

FIGURE 11.5   Color chromaticity diagram of $CeO_2:Tb^{3+}$ (1–5 mol%) nanophosphor.

three fundamental colors are green, blue, red with their particular standard wavelengths, and they produce different coordinates of different types of color shades. The CIE coordinates of $CeO_2:Tb^{3+}$ (1–5 mol%) nanophosphor are correlated with blue shades of colors [9].

CONCLUSION

In this chapter, we have prepared a $CeO_2:Tb^{3+}$ (1–5 mol%) nanophosphor by the combustion method using cow urine as fuel. The PXRD outcomes specify a cubic fluorite structure. The energy band gap values range from 4.6 to 5.8 eV. The product has an agglomerated morphology. The chromaticity results indicate that the CIE coordinates fall in the blue region.

ACKNOWLEDGMENT

The author Dr. Amith Yadav, H.J. thanks IQAC Davangere University, Davangere, India for the sanction of seed money.

REFERENCES

[1] N. Latha, D.R. Lavanya, G.P. Darshan, B.R. Radha Krushna, H.B. Premkumar, H.C. Prameela, H. Nagabhushana, Green emanating BiOCl: Tb^{3+} phosphors for strategic development of dermatoglyphics and anti-counterfeiting applications, *Inorganic Chemistry Communications*, 138(2022) 109266. https://doi.org/10.1016/j.inoche.2022.109266.

[2] V.N. Hegde, K.R. Jyothi, K.R. Bhagya, J. Lumbini, R. Somashekar, H. Nagabhushana, V.V. Manju, Structural, morphological and mechanical properties of Dy3+ doped $Sr_2MgSi_2O_7$ nanocomposites, *Journal of Solid State Chemistry*, 315(2022) 123501. https://doi.org/10.1016/j.jssc.2022.123501.

[3] N.B. Raj, N.T. Pavithra Gowda, O.S. Pooja, S.K. Sukrutha, B. Purushotham, H.P. Nagaswarupa, M.R. Anil Kumar, B.S. Surendra, S.S. TR, S.C. Prashantha, Eco-friendly synthesis of CeO_2 NPs using Aloe barbadensis Mill extract: Its biological and photocatalytic activities for industrial dye treatment applications, *Journal of Photochemistry and Photobiology*, 7(2021) 100038. https://doi.org/10.1016/j.jpap.2021.100038.

[4] B. Gopalakrishna Pillai Leela, D.N. Rajendran, Effect of lanthanide ion co-doping on the luminescence in the cerium-doped zinc oxide-phosphor system, *Spectroscopy Letters*, 52(2019) 431–440. https://doi.org/10.1080/00387010.2019.1659824.

[5] J. Malleshappa, H. Nagabhushana, S.C. Sharma, D.V. Sunitha, N. Dhananjaya, C. Shivakumara, B.M. Nagabhushana, Self propagating combustion synthesis and luminescent properties of nanocrystalline $CeO_2:Tb^{3+}$ (1-10 mol%) phosphors, *Journal of Alloys and Compounds*, 590(2014) 131–139. https://doi.org/10.1016/j.jallcom.2013.11.213.

[6] S. Balaraman, B. Iruson, S. Krishnmoorthy, M. Elayaperumal, S. Sangaraju, Synthesis of Er$_2$O$_3$ blended CeO$_2$ nanocomposites and investigation of their biomedical applications, *Chemical Physics Impact*, 6(2023) 100167. https://doi.org/10.1016/j.chphi.2023.100167.

[7] H.J.A. Yadav, B. Eraiah, H. Nagabhushana, G.P. Darshan, B.D. Prasad, S.C. Sharma, H.B. Premkumar, K.S. Anantharaju, G.R. Vijayakumar, Facile ultrasound route to prepare micro/nano superstructures for multifunctional applications, *ACS Sustainable Chemistry & Engineering*, 5(2017) 2061–2074. https://doi.org/10.1021/acssuschemeng.6b01693.

[8] H.J.A. Yadav, B. Eraiah, M.N. Kalasad, M. Thippeswamy, V. Rajasreelatha, Synthesis, characterization of ZrO$_2$: Tb^{3+}(1-9 mol%) nanophosphors for blue lighting applications and antibacterial property, *Biointerface Research in Applied Chemistry*, 12(2022) 7147–7158. https://doi.org/10.33263/BRIAC126.71477158.

[9] H.R. Girisha, B.R. Radha Krushna, D.R. Lavanya, P.B. Daruka, S.C. Sharma, H. Nagabhushana, Anti-counterfeiting, latent fingerprint detection and optical thermometry using a multi-stimulus down-converting La$_2$CaZnO$_5$:Er$_{3+}$ phosphor, *Optical Materials*, 134(2022) 113053. https://doi.org/10.1016/j.optmat.2022.113053.

12 Synthesis of Novel Nanocomposites for the Development of 3-G Dye-Sensitized Solar Cells

H.A. Deepa, M. Rudresh, G.M. Madhu, and R. Ravishankar

INTRODUCTION

Energy is the prime mover of economic growth and is vital to the sustenance of the modern economy. Future economic growth crucially depends on the long-term availability of energy from sources that are renewable, affordable, accessible, and environmental-friendly. Inexhaustible energies are sources of clean and increasingly competitive energy. They differ from fossil fuels principally in their diversity, abundance, and potential for use anywhere on the planet. Above all, they produce neither greenhouse gases which cause climate change nor polluting emissions [1–3]. Renewable energy sources and technologies have the potential to provide solutions to the longstanding energy crisis faced by developing countries. The most abundantly available renewable energy is solar energy, and sustainable energy production plays a vital role in the overall development of the country. Inexhaustible energy sources have accounted for over half of the total electricity in the last decade [4–6]. The most efficient method to obtain electricity is the conversion of quantum energy utilizing the solar cell. There are a set of three generations of solar cells, namely, first generation (1G), second generation (2G), and third generation (3G). Dye-sensitized solar cells (DSCs) belong to 3G solar cells which are an efficient substitute to conventional silicon-based solar cells due to their low manufacturing cost and ease of fabrication. A DSC is made up of a photoanode, a counterelectrode, and an electrolyte sandwiched between two electrodes. Usually, wide-bandgap nanocrystalline semiconductors of TiO_2 are incorporated as photoanodes on which light-sensitizing dyes are adsorbed. The nanocrystalline material used in developing the photoanode of DSC plays a vital role in the efficient functioning of DSC as it performs the function of adsorbing the dye which in turn absorbs energy from solar radiation to produce electricity [7–12].

The work achieved by several researchers in the field of DSC includes the development of DSCs using various photo-electrode semiconductor materials and finding the key factor influencing the performance of DSC. Though most of the investigations are done using a specific semiconductor metal oxide as the photoanode material, not many works have been accomplished by using the combination of nano metal oxides or the nanocomposites in the development of DSC. Few researchers have utilized nanocomposites in their research, but the efficiency achieved by them is not up to the mark. In the present chapter, fabrication of DSCs was achieved by incorporating novel nanocomposites as photo-electrodes, and their photovoltaic performance has been analyzed and elaborated.

DOI: 10.1201/9781003479239-12

SYNTHESIS OF NOVEL NANOCOMPOSITES

Synthesis of TiO₂-ZnO Composite by the Solution Combustion Method

The solution combustion method was employed to synthesize TiO_2-ZnO composite (molar ratio 1:1) as described by Chung et al. [13]. The stepwise synthesis procedure is demonstrated in Figure 12.1. The precursors used were $Ti(C_4H_9O)_4$ and $Zn (NO_3)_2.6H_2O$ with glycine as a fuel. The reactants were subjected to hydrolysis followed by nitration reaction. A blend of 6.8 ml of titanium (IV) butoxide and 2.35 g of $ZnNO_3.6H_2O$ was placed in a petri plate with the addition of 3.60 ml of distilled water. 1:1 HNO_3 was added to this blend under sonication. A stoichiometric amount of glycine was added, and the contents were placed in a muffle furnace at 400°C and ignited. The completion of the reaction was indicated by the formation of a froth-type precipitate. The obtained product was calcined at 450°C for 2 hours. A pinkish-white TiO_2-ZnO composite was formed, which was ground well in a mortar. Similarly, the TiO_2-ZnO composites were formed in different molar ratios of 2:1 and 3:1, respectively.

Synthesis of ZnO-CeO₂ Composite by the Solution Combustion Method

ZnO-CeO_2 composite (molar ratio 0.8:0.2) was prepared using the combustion technique as reported by Venkatesham et al. [14]. The stepwise synthesis procedure is represented in Figure 12.2. Cerium nitrate hexahydrate and zinc nitrate hexahydrate were the precursors, and sugar was utilized as a

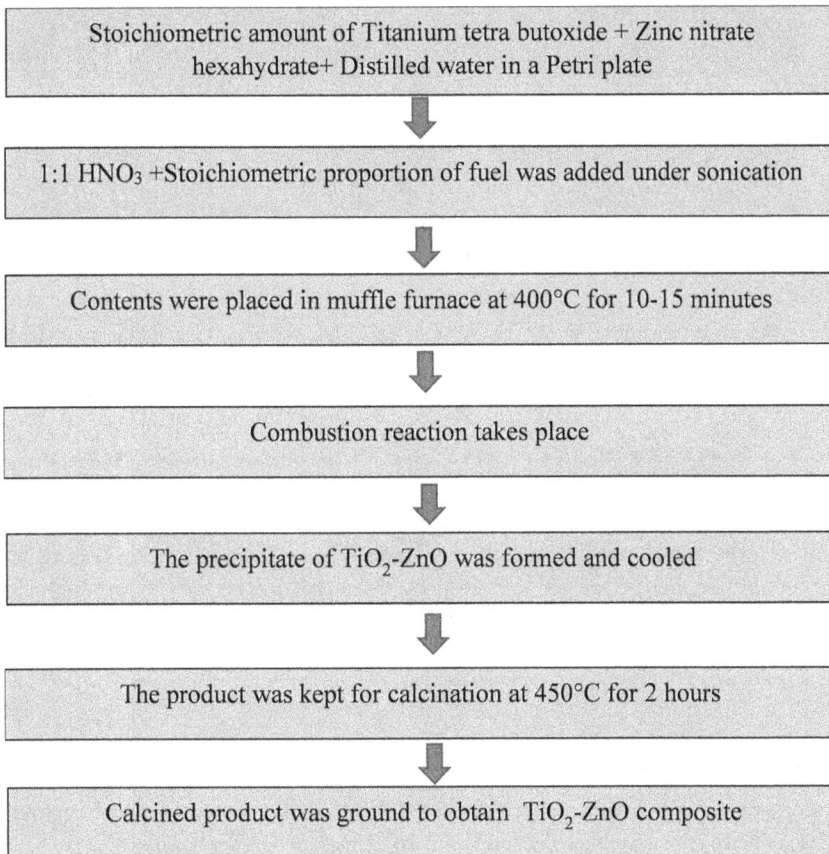

FIGURE 12.1 Stepwise procedure for the synthesis of TiO_2-ZnO composite by the solution combustion method.

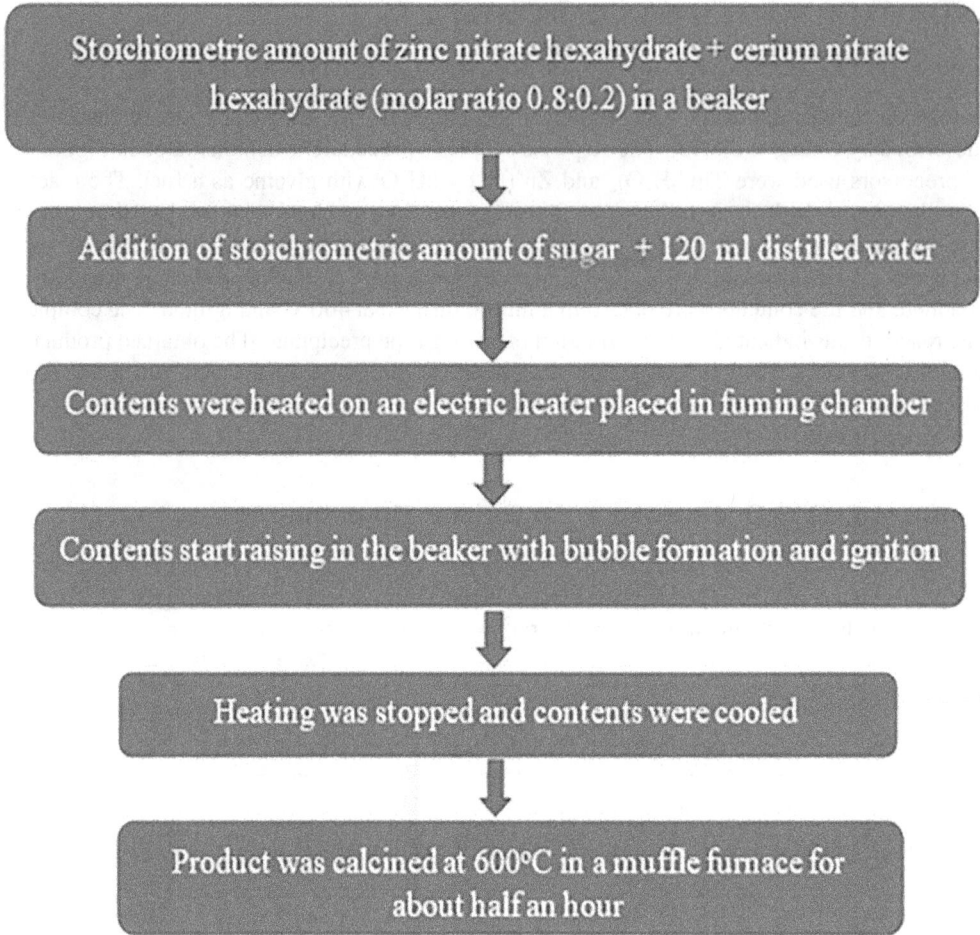

FIGURE 12.2 Stepwise procedure for the synthesis of $ZnO\text{-}CeO_2$ composite using the solution combustion method.

fuel. 16.0 g of Zn $(NO_3)_2$, 5.839 g of Ce $(NO_3)_3$, and 12.459 g of sugar were taken in a glass container. The precursors and the sugar were dissolved in 120 ml distilled water, placed on a hot plate, and heated in a fuming chamber. As the heating continued, the contents in the glass vessel raised with the bubble formation and the fumes were emanated. The fuming chamber was switched on to exhaust the fumes liberated. This was followed by the formation of a fluffy product with sudden ignition. The product formed was calcined at 600°C in a muffle furnace for half an hour to obtain a pale-yellow–colored composite of $ZnO\text{-}CeO_2$. Similarly, the $ZnO\text{-}CeO_2$ composites were formed in different molar ratios of 0.7:0.3 and 0.9:0.1, respectively.

CHARACTERIZATION OF NANOMATERIALS

In this section, the characterization results of the synthesized nanomaterials employing XRD, EDAX, SEM, UV-vis spectroscopic analysis, and BET analysis are explicated.

CHARACTERIZATION OF TiO₂-ZnO COMPOSITES FOR DSC APPLICATIONS

XRD Analysis

The crystallinity of the TiO_2-ZnO composite was analyzed employing the XRD. Figure 12.3 represents the XRD patterns of TiO_2-ZnO composites synthesized using the solution combustion method with different molar ratios of 1:1 (TZ-C1), 2:1 (TZ-C2), and 3:1 (TZ-C3), respectively.

The XRD pattern represented in Figure 12.3a emphasizes the formation of TZ-C1 with anatase and zinc phases. The formation of diffraction peaks at $2\theta = 25.32°$, 37.91°, 47.95°, 54.31°, 56.76°, and 62.27° θ corresponding to the standard planes (101), (112), (200), (105), (211), and (204) of ICDD #84-1286 for anatase TiO_2 was observed. The diffraction peaks at $2\theta = 32.75°$, 35.30°, 37.91°, 47.95°, 56.76°, and 62.27° corresponding to the standard planes (100), (002), (101), (102), (110), and (103) of ICDD #36-1451 for wurtzite ZnO was observed.

The XRD graph represented in Figure 12.3b confirmed the development of TZ-C2 in anatase and zinc phases. The formation of characteristic diffraction peaks at $2\theta = 25.360°$, 47.970°, 53.610°, and 62.23° corresponding to the standard planes (101), (200), (105), and (118) of ICDD #84-1286 for

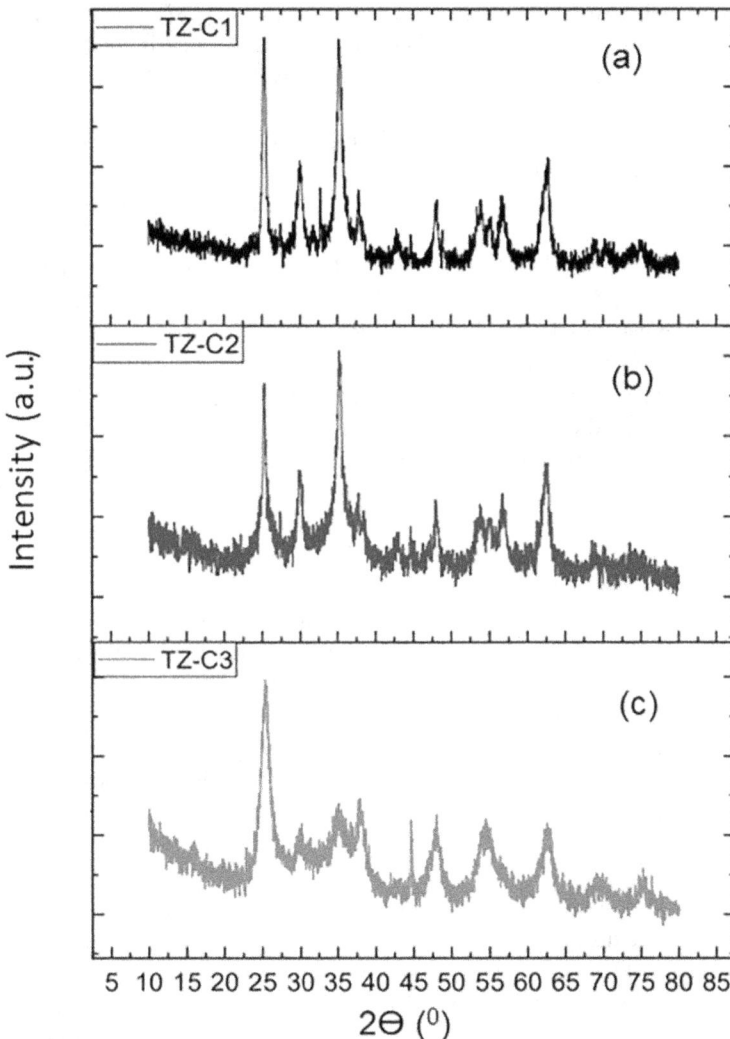

FIGURE 12.3 XRD patterns of combustion-derived composites: (a) TZ-C1, (b) TZ-C2, and (c) TZ-C3.

anatase TiO_2 was noticed. For wurtzite ZnO, the creation of diffraction peaks at $2\theta = 35.24°$, $38.03°$, $47.97°$, and $56.73°$ correlating to the standard planes (002), (101), (102), and (110) of ICDD #36-1451 was noted [15,16].

For the composite TZ-C3 (Figure 12.3c), a slight deviation in crystallinity was observed from the XRD analysis with the formation of unwanted peaks at $2\theta = 30.45°$, $35.52°$, and $44.64°$. The change in molar concentration resulted in deviation from crystallinity of the composite TZ-C3. The average crystallite sizes of TZ-C1, TZ-C2, and TZ-C3 were determined to be 24.57, 30.35, and 28.04 nm, respectively. The change in molar concentration resulted in varied crystallite sizes of the composites.

EDAX Analysis

EDAX was carried out to determine the elemental composition of TiO_2-ZnO composites. Figure 12.4 represents the EDAX analysis of TiO_2-ZnO composites produced by the solution combustion method with molar ratios of 1:1 (TZ-C1), 2:1 (TZ-C2), and 3:1 (TZ-C3).

It was observed that the composite TZ-C1 was formed in the stoichiometric ratio (by weight) as Ti = 19.660%, O = 26.560%, and Zn = 53.780%, respectively. The elemental composition of the TZ-C2 with stoichiometric weight % was determined to be 45.30, 37.52, and 17.180 for O, Ti, and Zn, respectively, using the EDAX analysis. TZ-C3 composite was formed in the stoichiometric weight ratio as Ti = 39.68%, O = 41.49%, and Zn = 18.83%, respectively. The EDAX analysis performed for the combustion-derived TiO_2-ZnO composites (TZ-C1, TZ-C2, and TZ-C3) demonstrated that no other impurities were formed except Ti, O, and Zn elements. The EDAX measurements of TZ-C1, TZ-C2 and TZ-C3 are briefed in Table 12.1.

SEM Analysis

The SEM images of the combustion-derived TiO_2-ZnO composites with magnifications 20, 66, and 20 kx for the composites TZ-C1, TZ-C2, and TZ-C3, are represented in Figure 12.5, respectively. The TZ-C1 composite was nearly spherical and aggregated. The particles were uniformly distributed. The TZ-C2 composite was noticed to be slightly bigger and almost spherical with evenly distributed particles. They were seen as more agglomerated compared to TZ-C1. The composite TZ-C3 was more aggregated compared to TZ-C1 and TZ-C2. The composite particles of TZ-C3 were not having a definite shape and were unevenly distributed, unlike the other two composites.

UV Analysis

TiO_2-ZnO composites were analyzed employing a UV-visible spectrophotometer which is presented in Figure 12.6. The bandgap energy of TZ-C1, TZ-C2, and TZ-C3 composites were computed to be 3.04, 3.06, and 3.07 eV, respectively. The bandgap energy values obtained in the present work are lower than that stated by Manikandan et al. (3.18 eV) [17]. The decrease in bandgap energy may be attributed to the aggregation and higher crystallize size of the composite particles which may have resulted in red-shift of the cut-off wavelength [17]. The variation in bandgap energy may also be attributed to the change in the molar concentrations of TiO_2 and ZnO nanoparticles. Hence, the composite with the molar ratio of 3:1 (TZ-C3) has exhibited the highest bandgap energy compared to the other two composites.

BET Analysis

The BET analysis of the synthesized TiO_2-ZnO composites was conducted to ascertain the specific surface area. The surface areas of TZ-C1, TZ-C2, and TZ-C3 were found to be 18.25, 18.54, and 22.45 m²/g, respectively. The surface area of the TiO_2-ZnO composites in the current work is lower than the results testified by Vlazan et al. (44.48 m²/g) [18] and Song et al. (66.3 m²/g) [19], but found to be higher than the results reported by Ahmad et al. (9.41 m²/g) [20]. Figure 12.7 exhibits the N_2 adsorption–desorption isotherms of TZ-C1, TZ-C2, and TZ-C3, respectively. The surface area, pore diameter, and average pore volume of the TiO_2-ZnO composites are summarized in Table 12.2.

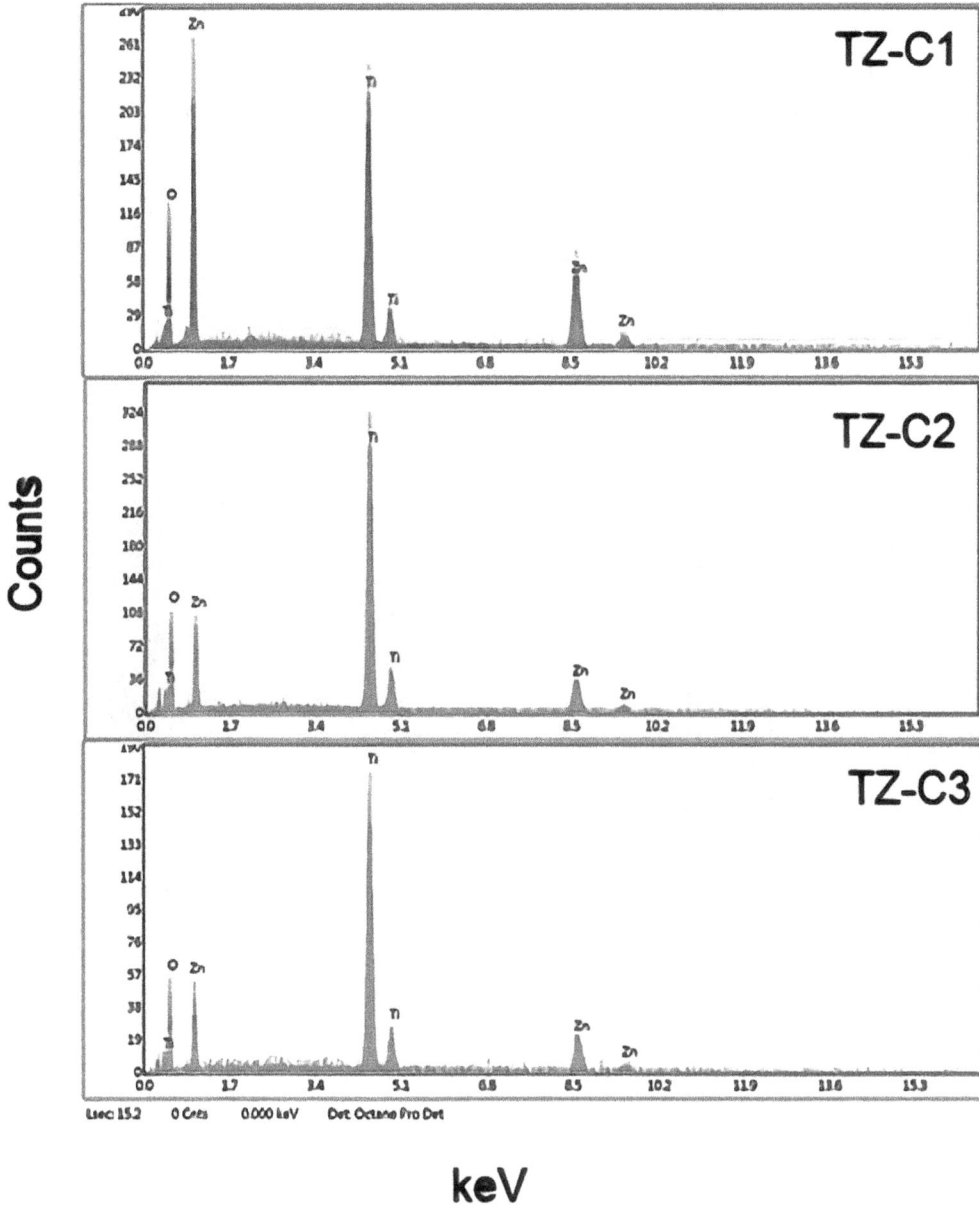

FIGURE 12.4 EDAX analysis of combustion-derived composites: TZ-C1, TZ-C2, and TZ-C3.

CHARACTERIZATION OF ZNO-CEO$_2$ COMPOSITES FOR DSC APPLICATIONS

XRD ANALYSIS

The crystallinity of the ZnO-CeO$_2$ composites was analyzed employing the XRD analysis. Figure 12.8 represents the XRD patterns of ZnO-CeO$_2$ composites synthesized using the solution combustion method with different molar ratios of 0.7:0.3, 0.8:0.2, and 0.9:0.1 designated as ZC-1, ZC-2, and ZC-3, respectively.

Figure 12.8a represents the XRD pattern of ZC-1 with the creation diffraction peaks at $2\theta = 28.52°$, $32.79°$, $47.56°$, and $56.45°$ corresponding to the standard cerium oxide planes of (111), (200), (220),

TABLE 12.1
EDAX Analysis of TZ-C1, TZ-C2, and TZ-C3

Nanomaterial	Element	Weight %	Atomic %
TZ-C1	O	26.56	57.38
	Zn	53.78	28.44
	Ti	19.66	14.18
TZ-C2	O	45.30	73.02
	Ti	37.52	20.20
	Zn	17.18	6.78
TZ-C3	O	41.49	69.91
	Ti	39.68	22.33
	Zn	18.83	7.76

FIGURE 12.5   SEM images of combustion-derived composites: (a) TZ-C1, (b) TZ-C2, and (c) TZ-C3.

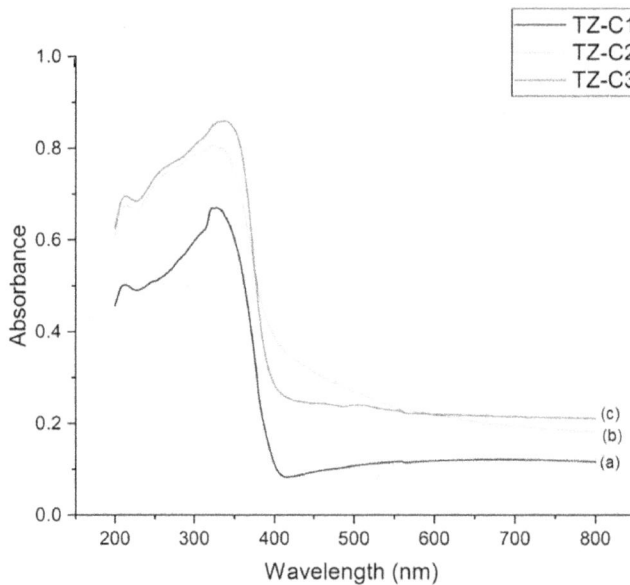

FIGURE 12.6   UV analysis of combustion-derived TZ-C1, TZ-C2, and TZ-C3 composites.

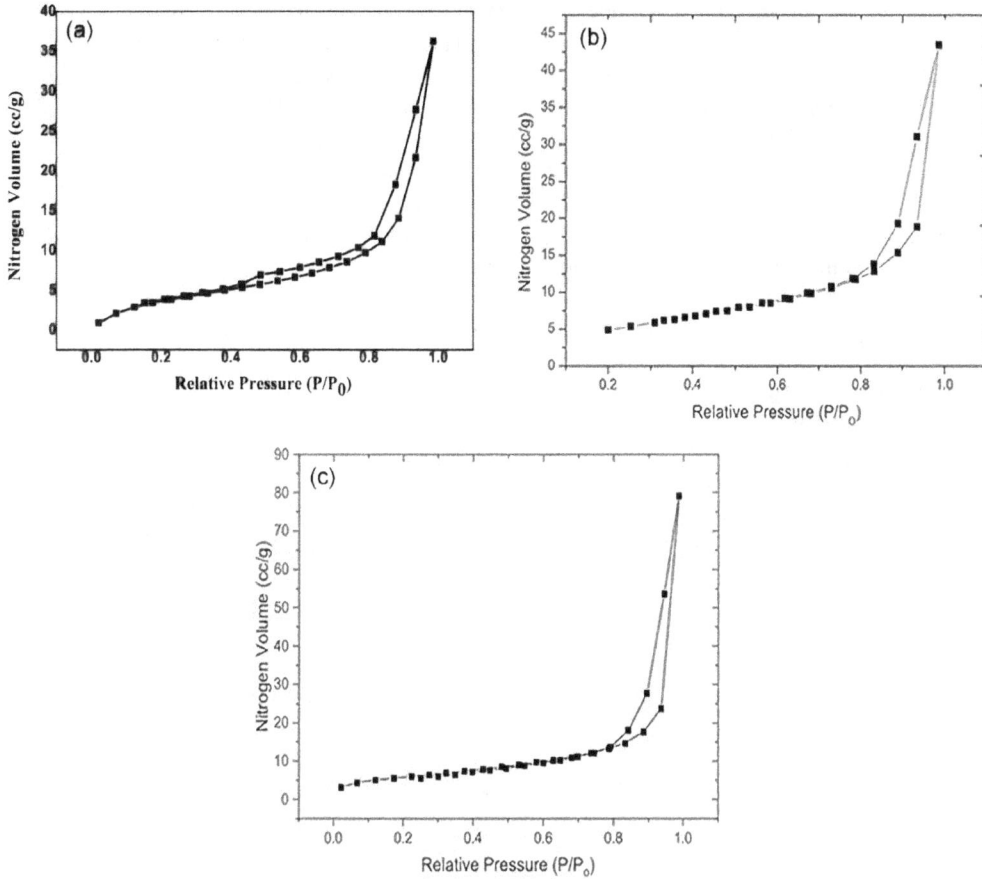

FIGURE 12.7 Nitrogen adsorption–desorption plots of (a) TZ-C1, (b) TZ-C2, and (c) TZ-C3.

TABLE 12.2

Surface Area, Pore Diameter, and Pore Volume of TZ-C1, TZ-C2, and TZ-C3

Nanomaterial	Surface Area (m^2/g)	Pore Diameter (10^{-9} m)	Pore Volume (cm^3/g)
TZ-C1	18.25	2.87	0.058
TZ-C2	18.54	3.46	0.067
TZ-C3	22.45	3.360	0.095

and (311), respectively. The diffraction peaks formed at angle $2\theta=34.44°$, $36.25°$, $47.56°$, $56.45°$, $67.99°$, and $69.06°$ corresponded to the standard zinc oxide planes of (002), (101), (102), (110), (112), and (201), respectively, thus confirming the development of ZnO-CeO_2 composite.

Figure 12.8b demonstrates the XRD analysis of ZC-2 composite with the development of characteristic diffraction peaks at $2\theta=28.61°$, $32.23°$, $47.48°$, and $56.48°$ corresponding to the standard cerium oxide planes (111), (200), (220), and (311), respectively. The diffraction peaks formed at angle $2\theta=34.50°$, $36.31°$, $47.48°$, $56.48°$, $67.90°$, and $69.20°$ correlated to the standard zinc oxide planes of (002), (101), (102), (110), (112), and (201), respectively, thus confirming the formation of ZnO-CeO_2 composite. The diffraction peaks of ZC-1 and ZC-2 matched with the ICDD #81-0792 [21].

Figure 12.8c represents the XRD analysis of ZC-3 with the development of diffraction peaks at $2\theta = 47.44°$ and $56.53°$ correlating to the standard cerium oxide planes of (220) and (311), respectively. The diffraction peaks were formed at angle $2\theta = 34.45°$, $36.27°$, $47.44°$, $56.53°$, and $68.31°$ correlating to the standard zinc oxide planes of (002), (101), (102), (110), and (112), respectively, thus confirming the formation of $ZnO\text{-}CeO_2$ composite. From Figure 12.8c, we could observe that there was a slight deviation in crystallinity for the formation of composite ZC-3. The deviation in crystallinity may be attributed to the aggregation of composite particles and variation in the molar concentration of ZnO and CeO_2. The average crystallite sizes of the composites ZC-1, ZC-2, and ZC-3 were computed to be 15.78, 17.49, and 20.52 nm, respectively.

EDAX ANALYSIS

EDAX was performed to determine the elemental composition of $ZnO\text{-}CeO_2$ composites. Figure 12.9 represents the EDAX measurements of $ZnO\text{-}CeO_2$ composites with molar ratios of 0.7:0.3, 0.8:0.2, and 0.9:0.1 designated as ZC-1, ZC-2, and ZC-3, respectively.

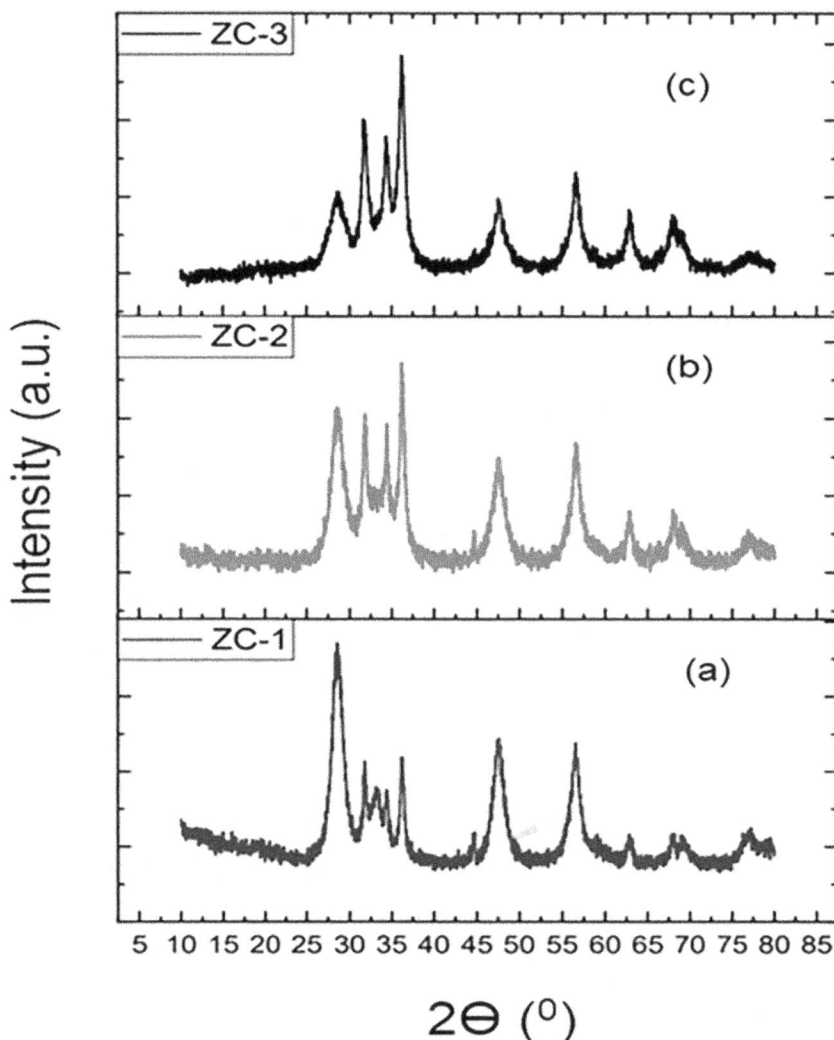

FIGURE 12.8   XRD patterns of $ZnO\text{-}CeO_2$ composites: (a) ZC-1, (b) ZC-2, and (c) ZC-3.

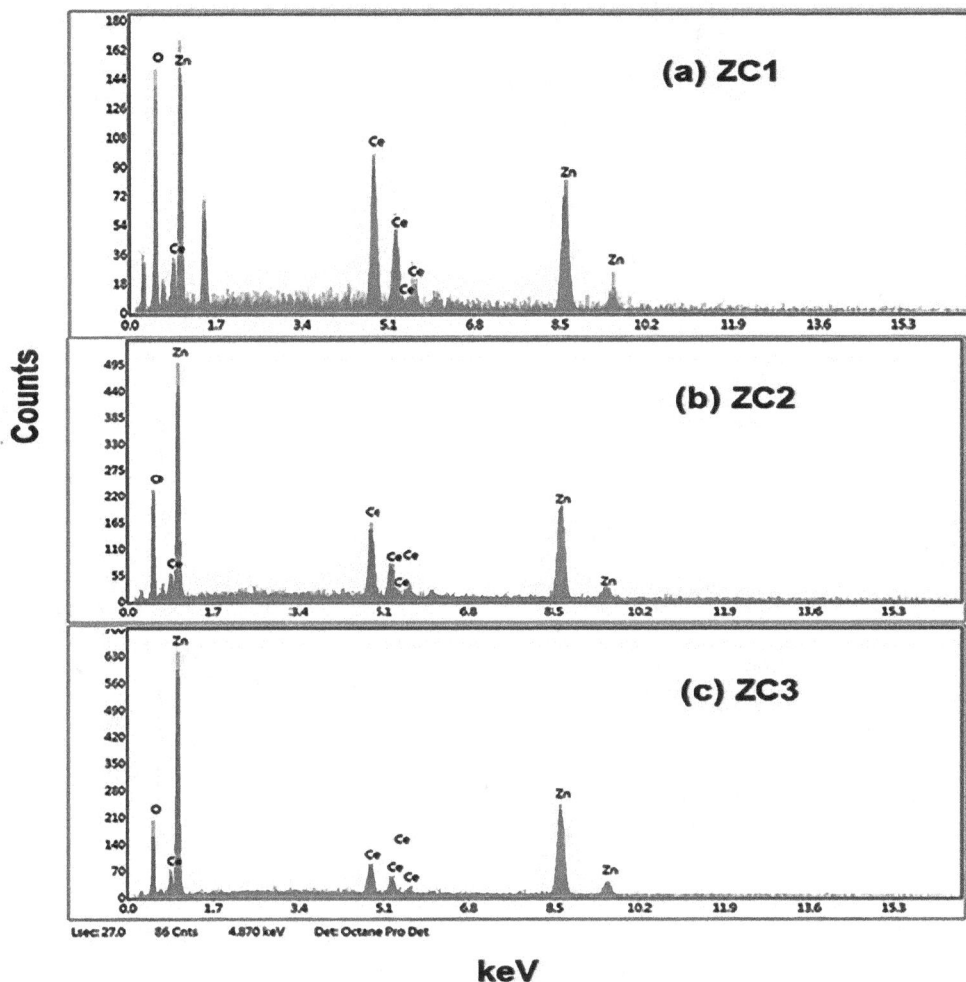

FIGURE 12.9 EDAX measurement of ZnO-CeO$_2$ composites: (a) ZC-1, (b) ZC-2, and (c) ZC-3.

It was noted that the composite ZC-1 was formed in the stoichiometric ratio (by weight) of Zn = 39.42%, Ce = 36.29%, and O = 24.29%. The elemental composition of the ZC-2 composite was determined (by weight %) as 50.42, 22.36, and 27.22 for Zn, O, and Ce, respectively. The composite ZC-3 was formed in the stoichiometric weight ratio of Zn = 63.15%, Ce = 17.03%, and O = 19.82%. From the EDAX measurements of composites, it was confirmed that impurities (like other elements) were not produced while synthesizing the ZnO-CeO$_2$ composites. The EDAX measurements of ZC-1, ZC-2, and ZC-3 are presented in Table 12.3.

SEM ANALYSIS

The SEM analysis was performed to ascertain the morphology of the ZnO-CeO$_2$ composites, which are represented in Figure 12.10. We could observe that the ZC-1 composites were roughly spherical, aggregated, and uniformly distributed as shown in Figure 12.10a with a magnification of 40 kx. From Figure 12.10b, we could infer that the ZC-2 composites (with magnification 67.5 kx) were nearly spherical, bigger, and more agglomerated compared to ZC-1 composites. The morphology of ZC-3 composites (with magnification 66.8 kx) is represented in Figure 12.10c. From the SEM image, we could observe that the ZC-3 composites were not having a definite shape and they were

TABLE 12.3
EDAX Analysis of ZC-1, ZC-2, and ZC-3

Nanomaterial	Element	Weight %	Atomic %
ZC-1	O	24.29	63.78
	Ce	36.29	10.88
	Zn	39.42	25.34
ZC-2	O	22.36	59.14
	Ce	27.22	8.22
	Zn	50.42	32.63
ZC-3	O	19.82	53.25
	Ce	17.03	5.22
	Zn	63.15	41.53

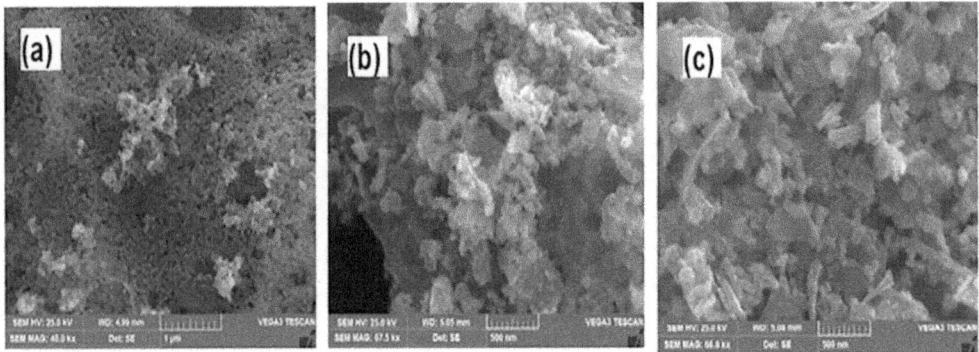

FIGURE 12.10   SEM images of ZnO-CeO$_2$ composites: (a) ZC-1, (b) ZC-2, and (c) ZC-3.

more agglomerated compared to the other two composites, namely, ZC-1, and ZC-2. The aggregation of composite particles might be due to the higher processing temperatures utilized in the solution combustion method.

UV Analysis

ZnO-CeO$_2$ composites were analyzed employing a UV-vis spectrophotometer to determine the bandgap energy (Eg). Figure 12.11 represents the UV analysis of ZC-1, ZC-2, and ZC-3, respectively. The bandgap energy for ZC-1, ZC-2, and ZC-3 composites were determined to be 3.02, 2.96, and 2.94 eV, respectively. The Eg results obtained in the present work were lower than the figures reported by He et al. (3.25 eV) [22] and Liu et al. (3.72 eV) [23]. He [22] and his team had produced ZnO-CeO$_2$ composites with varied molar ratios using the one-pot hydrothermal method. Liu et al. [23] had synthesized ZnO-CeO$_2$ composites with varrying molar ratios using the precipitation technique. It was noticed that the bandgap energy varied with the increase in the molar concentration of cerium oxide nanoparticles. This could be attributed to the increase in the number of oxygen vacancies created owing to the reduction of cerium particles [22].

BET Analysis

The BET analysis of the synthesized ZnO-CeO$_2$ composites was performed to ascertain the specific surface area. The surface area of ZC-1, ZC-2, and ZC-3 composites was determined to be

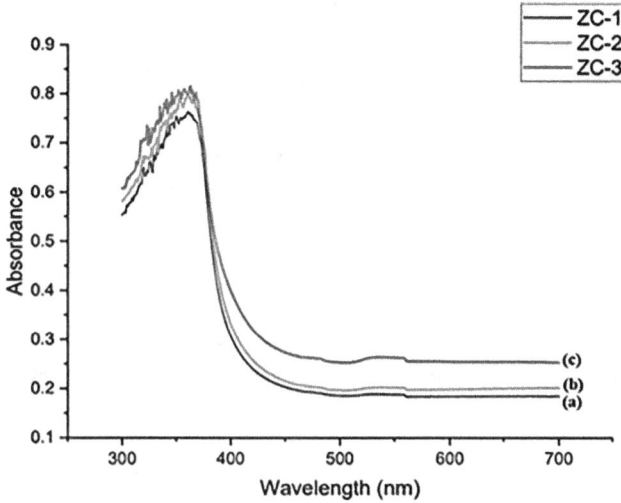

FIGURE 12.11   UV-vis analysis of ZnO-CeO$_2$ composites: (a) ZC-1, (b) ZC-2, and (c) ZC-3.

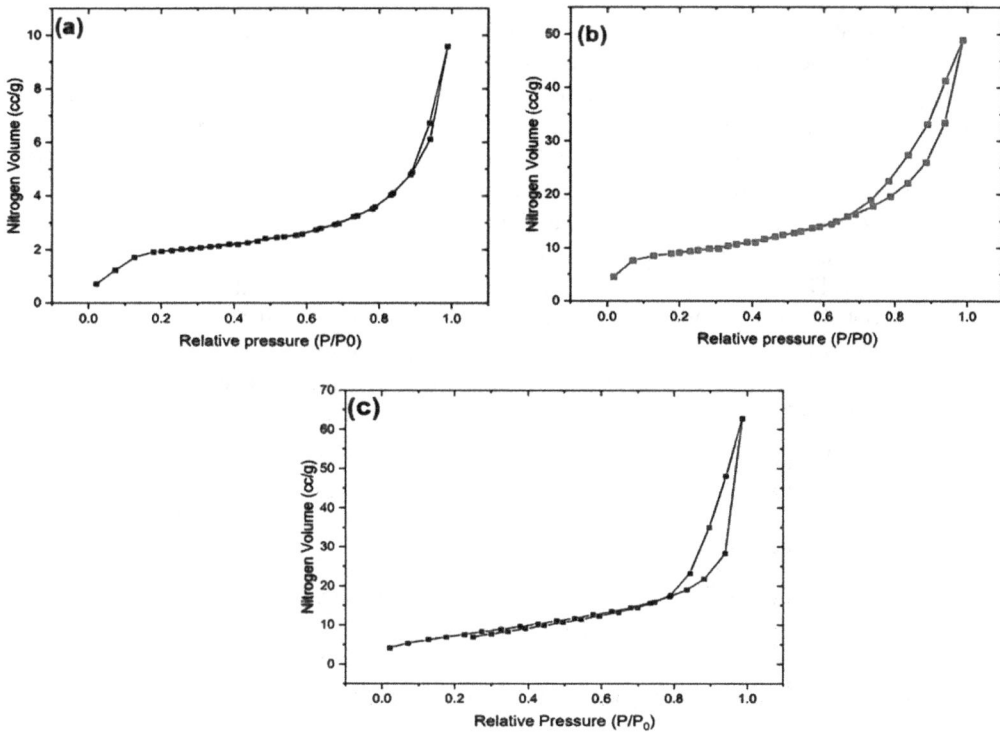

FIGURE 12.12   Nitrogen adsorption–desorption plots of (a) ZC-1, (b) ZC-2, and (c) ZC-3.

7.167, 32.874, and 19.337 m^2/g, respectively. Figure 12.12 exhibits the N$_2$ adsorption–desorption isotherms of ZnO-CeO$_2$ composites. The surface area, pore diameter, and average pore volume of the ZnO-CeO$_2$ composites are summarized in Table 12.4.

The surface area of the ZnO-CeO$_2$ composites in the present work is lesser than the results stated by Zhang et al. (37.056 m^2/g) [24]. Sol-gel route was employed to synthesize the ZnO-CeO$_2$

TABLE 12.4

Specific Surface Area, Pore Diameter, and Pore Volume of ZC-1, ZC-2, and ZC-3 Composites

Nanomaterial	Specific Surface Area (m²/g)	Pore Diameter (nm)	Pore Volume (cm³/g)
ZC-1	7.167	3.117	0.017
ZC-2	32.874	3.971	0.080
ZC-3	19.337	2.814	0.122

composites by Zhang [24] and his team. The lower surface area of the ZC composites in the present work could be due to the higher processing temperature used during the solution combustion method [13].

APPLICATION STUDIES

FABRICATION OF DYE-SENSITIZED SOLAR CELLS (DSCs)

DSCs were formulated employing a similar method as described by Govindaraj et al. [25]. Fluorine-doped tin oxide (FTO)–coated glass slides, with 2.2 mm thickness, resistivity 7 ohms/cm², and transmittance greater than 85%, were used as glass substrates for developing photoanode and counterelectrode. The FTO glass slides were rinsed with deionized water followed by ultrasonicating in ethanol for 10 minutes and dried. The synthesized novel nanocomposites (in the form of a paste) were deposited on the conductive side of the FTO glass slides employing doctor blade technique. These photoanodes were dried at room temperature and sintered at 450°C for half an hour. The glass substrates thus prepared were soaked in 5 mmol/l solution of N719 dye for 24 hours for dye sensitization. The nanomaterial-coated films were then rinsed with ethanol and dried in air after dye sensitization. Platisol T/SP was coated on another FTO glass slide which was used as a counterelectrode. The FTO glass substrate coated with Platisol T/SP was sintered at 400°C for half an hour and cooled to room temperature. The counterelectrode was placed on top of the photoanode, and the two electrodes are assembled using a binder clip. The electrolyte (Iodolyte HI-30) was introduced amid the photoanode and the counterelectrode by capillary movement. The active area of DSC was nearly 0.25 cm². The fabrication process of DSC is demonstrated in Figure 12.13. The fabricated DSCs were subjected to a solar simulator analysis under an irradiation of 100 mW/cm² to determine the photovoltaic performance.

PHOTOVOLTAIC PERFORMANCE OF DSC

Determining the I-V (current-voltage) characteristic is the key technique utilized in the estimation of photovoltaic performance of DSC. Under the standard cell temperature of 25°C and a standard solar irradiation of 100 mW/cm² (AM 1.5), the photovoltaic parameters such as open-circuit voltage (V_{oc}), short-circuit current (I_{sc}), short-circuit current density (j_{sc}), efficiency (η), and fill factor (FF) are evaluated. Figure 12.14 demonstrates the I-V characteristics of a typical DSC [26,27].

The FF is a vital factor as it determines the quality of a cell. The fill factor can be determined by computing the ratio of maximum power (P_{max}) output and the product of V_{oc} and I_{sc}. Therefore,

$$FF = \frac{Power_{(max)}}{V_{oc} \cdot I_{sc}} = \frac{I_{max} \cdot V_{max}}{V_{oc} \cdot I_{sc}} \tag{12.1}$$

The power conversion efficiency (η) of a DSC can be evaluated by computing the ratio of maximum power output (P_{max}) from a cell to the solar power input (P_{in}) to the cell. It can be computed using Equation (12.2):

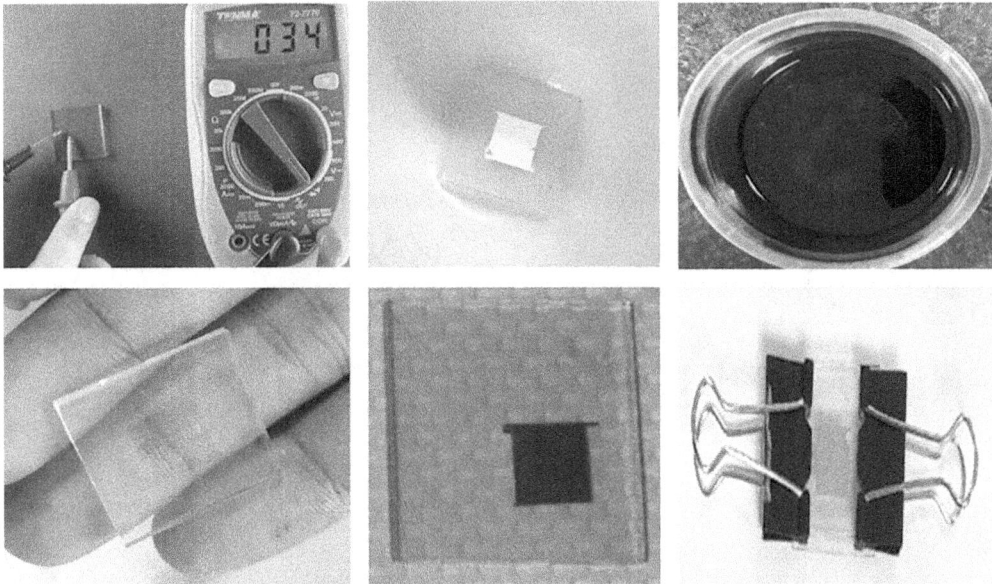

FIGURE 12.13 Fabrication process of dye-sensitized solar cell (DSC).

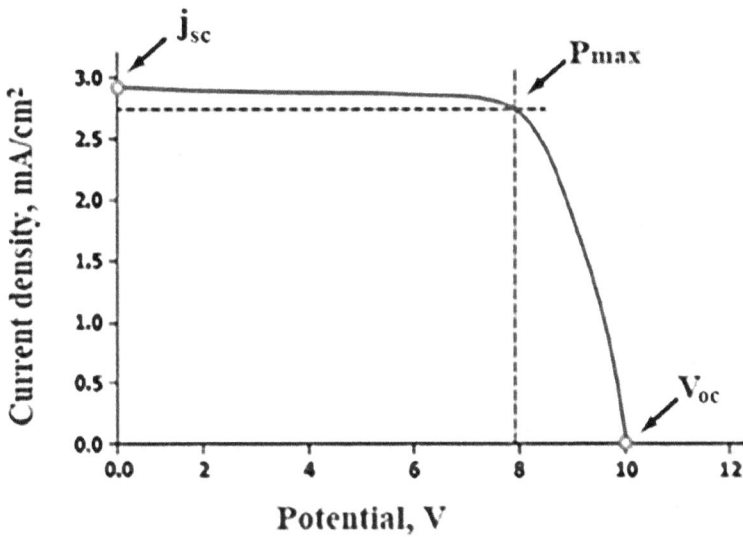

FIGURE 12.14 I-V characteristics of a typical DSC [26].

$$\eta = \frac{P_{max}}{P_{in}} = \frac{V_{oc} \cdot j_{sc} \cdot FF}{P_{in}} \tag{12.2}$$

where j_{sc} represents the short-circuit current density (mA/cm²) [27,28].

I-V CHARACTERIZATION OF DSCs FABRICATED WITH TiO₂-ZnO COMPOSITES

The DSCs fabricated with TiO₂-ZnO composites were subjected to a solar simulator analysis under an illumination of 100 mW/cm². Figure 12.15 represents the I-V characteristics of DSCs formed using combustion-derived TiO₂-ZnO composites as the semiconductor photoanode materials.

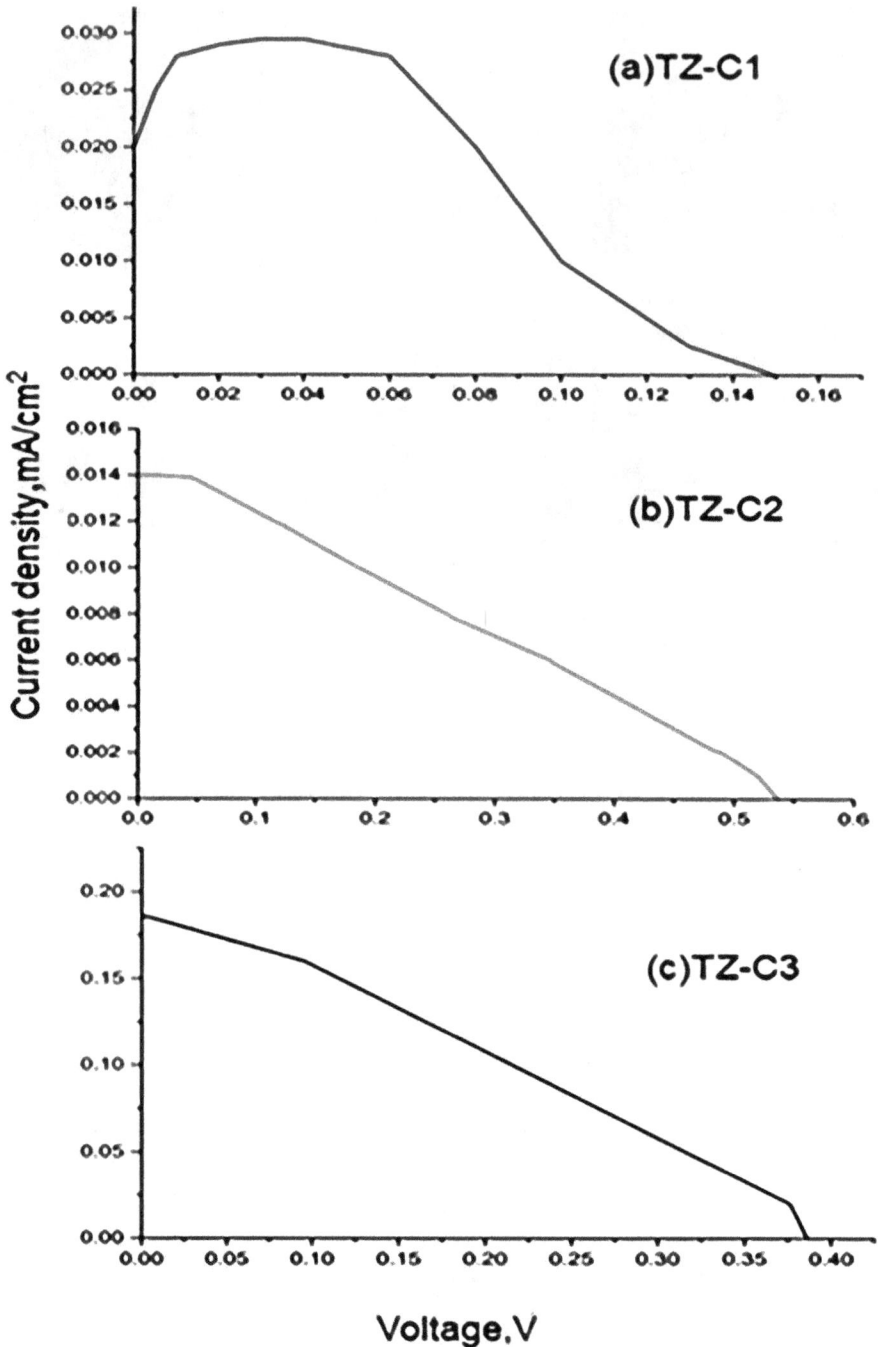

FIGURE 12.15   I-V characteristics of the DSCs fabricated with (a) TZ-C1, (b) TZ-C2, and (c) TZ-C3.

The I-V parameters of the DSCs formulated with the composites TZ-C1, TZ-C2, and TZ-C3 are represented in Table 12.5.

The open-circuit voltages of the devices formed with the composites TZ-C1, TZ-C2, and TZ-C3 were determined to be 0.151, 0.534, and 0.384 V, respectively. The V_{oc} values were lower than the results described by Giannouli et al. (0.43 V) [29] for the composites TZ-C1 and TZ-C3 but higher

TABLE 12.5
I-V Parameters of the DSCs Formulated with the Composites TZ-C1,
TZ-C2, and TZ-C3 Photoanodes

Photoanode	j_{sc} (mA/cm²)	V_{oc} (V)	FF (%)	Efficiency (%)
TZ-C1	0.0221	0.151	75.44	0.00175
TZ-C2	0.014	0.534	26.71	0.0022
TZ-C3	0.185	0.384	27.81	0.0198

for the composite TZ-C2. Giannouli et al. [29] had utilized Coumarine 343 and Rose Bengal as the sensitizers in their work.

The j_{sc} values for the DSCs fabricated with the composites TZ-C1, TZ-C2, and TZ-C3 were determined to be 0.0221, 0.014, and 0.185 mA/cm², respectively. The current density results obtained in the current study are higher than the results presented by Wahyuningsih et al. (0.204 μA/cm²) [30], where mechanochemical methods were utilized for producing TiO_2-ZnO nanorods, and N3 dye was employed as the sensitizer in the formation of the cell. The difference in current densities (j_{sc}) and voltages (V_{oc}) recorded might have resulted due to the diverse size and morphology of the prepared nanostructures, several synthesis techniques utilized, and the nature of dye employed for sensitization.

The efficiency of the DSCs developed using the composites TZ-C1, TZ-C2, and TZ-C3 was evaluated as 0.00175%, 0.0022%, and 0.0198%, respectively. The drastic decrease in efficacy might be due to the lower surface area and aggregation of particles of TiO_2-ZnO, which could have led to poor adsorption of the sensitizer on the composite semiconducting layer and higher resistance for the electron transfer across the photo-electrode resulting in decreased efficacy [13].

I-V Characterization of DSCs Fabricated with ZnO-CeO$_2$ Composites

The DSCs fabricated with ZnO-CeO$_2$ composites were subjected to a solar simulator analysis under an illumination of 100 mW/cm². Figure 12.16 represents the I-V characteristics of DSCs formed using ZC-1, ZC-2, and ZC-3 as the semiconductor photoanode materials, respectively. The I-V parameters of the DSCs formed with the ZnO-CeO$_2$ composites are presented in Table 12.6.

The DSCs fabricated using the ZnO-CeO$_2$ composites as the photoanode did not yield competent power conversion efficiencies (PCE). Values of PCE for the DSCs were found to be 0.101, 0.792, and 0.525 for ZC-1, ZC-2, and ZC-3, respectively, with ZC-2 exhibiting the highest efficiency of 0.792% and ZC-1 the least. The j_{sc} values were recorded as 0.38, 1.39, and 1.07 mA/cm² for ZC-1, ZC-2, and ZC-3, respectively. The overall decrease in the PCE and current density values may be due to the following factors. Firstly, a restriction in the mobility of the electrons may be caused by the aggregation in the ZnO-CeO$_2$ composites [31,32]. Integration of cerium oxide nanoparticles with ZnO has lowered the efficiency and current density values. This may be attributed to the rise in the number of oxygen vacancies created owing to the reduction of cerium particles which may perhaps have resulted in the obstruction of electron transfer across the photoanode [22]. In the current study, although a substantial reduction in the overall photovoltaic performance was observed in the DSCs synthesized with the ZnO-CeO$_2$ composites as anode, ZC-2 has shown a satisfactory performance due to the integration of CeO$_2$ with ZnO in an optimum ratio of 0.8:0.2 [33].

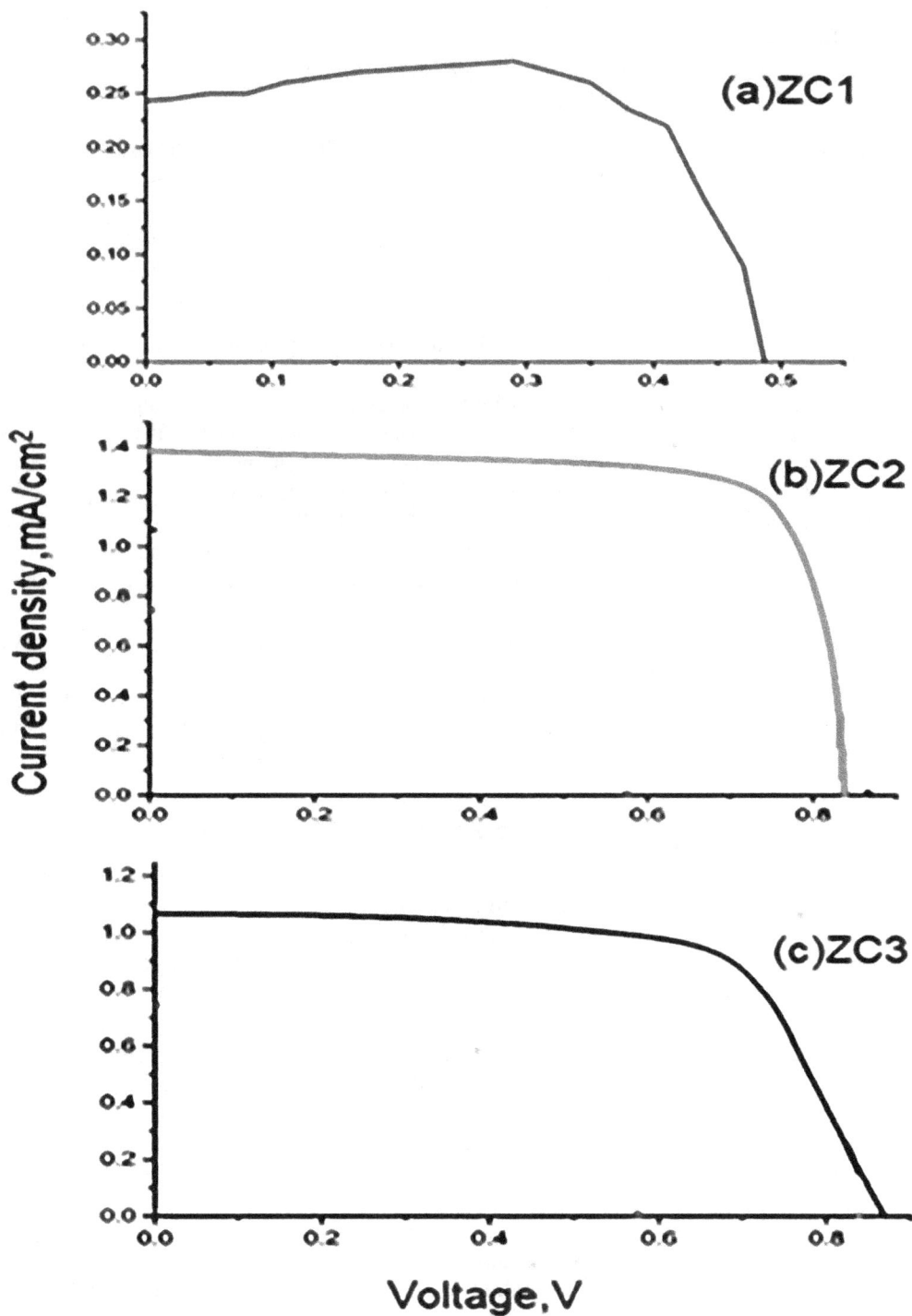

FIGURE 12.16   I-V characteristics of the DSCs formed with (a) ZC-1, (b) ZC-2, and (c) ZC-3 composites.

TABLE 12.6
I-V Parameters of the DSCs Formulated with ZC-1, ZC-2, and ZC-3
Composites

Photoanode	j_{sc} (mA/cm²)	V_{oc} (V)	FF (%)	Efficiency (%)
ZC-1	0.38	0.519	51.77	0.101
ZC-2	1.39	0.772	73.87	0.792
ZC-3	1.07	0.797	61.50	0.525

CONCLUSION

The outcome of the work presented in this chapter can be summarized as follows. Novel nano-composites of TiO_2-ZnO and ZnO-CeO_2 with different molar ratios were synthesized using the solution combustion technique. The synthesized nanomaterials were characterized by XRD, SEM, EDAX, UV, and BET analysis. DSCs were fabricated incorporating the synthesized nanomaterials as the photoanode semiconducting materials of DSC. The photovoltaic performance of the DSCs fabricated was assessed under a solar simulator of radiation 100 mW/cm². The change in the molar concentration of the composites resulted in varied morphology of the nanomaterials. The higher processing and calcination temperatures led to the slight deviation in crystallinity and aggregation of nanostructures. The integration of ZnO nanoparticles with TiO_2 nanoparticles resulted in the decline of photovoltaic performance of DSCs due to the formation of Zn^{2+} and dye complex, which causes hindrance to the electron mobility across the photoanode. The incorporation of CeO_2 nanoparticles with nano ZnO resulted in the reduced photovoltaic performance of DSCs owing to the formation of oxygen vacancies which is due to the reduction of cerium particles. The bandgap energy of the composites was observed to be lower than that of the pure nanoparticles. The incorporation of ZnO nanoparticles with TiO_2 nanoparticles and CeO_2 nanoparticles with nano ZnO resulted in a decline of bandgap energy. It can be observed that the composites of ZnO-CeO_2 which were utilized as photo-electrode material of DSC have displayed better photovoltaic performance compared to that of TiO_2-ZnO composites. This may be attributed to the higher surface area of nanomaterials which have facilitated better dye loading, higher electron mobility, and lower recombination reactions across the interfaces. It could be inferred that electron transfer via composites to the transparent conducting layer would be effectual and the mobility of electrons may be augmented by optimizing the characteristics of semiconducting photoanode materials, that is, utilizing optimized molar concentrations for the synthesis of composites, increasing the surface area of the composites, and decreasing the processing temperature, which may lead to improved efficacy.

REFERENCES

[1] Ahmad, Tanveer, and Dongdong Zhang. "A critical review of comparative global historical energy consumption and future demand: The story told so far." *Energy Reports* 6 (2020): 1973–1991. https://doi.org/10.1016/j.egyr.2020.07.020

[2] Hagfeldt, Anders, Gerrit Boschloo, Licheng Sun, Lars Kloo, and Henrik Pettersson. "Dye-sensitized solar cells." *Chemical Reviews* 110, no. 11 (2010): 6595–6663. https://doi.org/10.1021/cr900356p

[3] Reddy, K. Govardhan, T. G. Deepak, G. S. Anjusree, Sara Thomas, Sajini Vadukumpully, K. R. V. Subramanian, Shantikumar V. Nair, and A. Sreekumaran Nair. "On global energy scenario, dye-sensitized solar cells and the promise of nanotechnology." *Physical Chemistry Chemical Physics* 16, no. 15 (2014): 6838–6858. https://doi.org/10.1039/c3cp55448a www.rsc.org/pccp

[4] Gielen, Dolf, Francisco Boshell, Deger Saygin, Morgan D. Bazilian, Nicholas Wagner, and Ricardo Gorini. "The role of renewable energy in the global energy transformation." *Energy Strategy Reviews* 24 (2019): 38–50. https://doi.org/10.1016/j.esr.2019.01.006

[5] Moriarty, Patrick, and Damon Honnery. "What is the global potential for renewable energy?" *Renewable and Sustainable Energy Reviews* 16, no. 1 (2012): 244–252. https://doi.org/10.1016/j.rser.2011.07.151

[6] Kannan, Nadarajah, and Divagar Vakeesan. "Solar energy for future world: A review." *Renewable and Sustainable Energy Reviews* 62 (2016): 1092–1105. https://doi.org/10.1016/j.rser.2016.05.022

[7] Goetzberger, Adolf, Joachim Luther, and Gerhard Willeke. "Solar cells: Past, present, future." *Solar Energy Materials and Solar Cells* 74, no. 1–4 (2002): 1–11. https://doi.org/10.1016/S0927-0248(02)00042-9

[8] Ahmad, Khuram Shahzad, Syeda Naima Naqvi, and Shaan Bibi Jaffri. "Systematic review elucidating the generations and classifications of solar cells contributing towards environmental sustainability integration." *Reviews in Inorganic Chemistry* 41, no. 1 (2021): 21–39. https://doi.org/10.1515/revic-2020-0009

[9] Bagher, Askari Mohammad, Mirzaei Mahmoud Abadi Vahid, and Mirhabibi Mohsen. "Types of solar cells and application." *American Journal of Optics and Photonics* 3, no. 5 (2015): 94–113. https://doi.org/10.11648/j.ajop.20150305.17

[10] Gonçalves, Luís Moreira, Verónica de Zea Bermudez, Helena Aguilar Ribeiro, and Adélio Magalhães Mendes. "Dye-sensitized solar cells: A safe bet for the future." *Energy & Environmental Science* 1, no. 6 (2008): 655–667. https://doi.org/10.1039/B807236A

[11] Sharma, Khushboo, Vinay Sharma, and S. S. Sharma. "Dye-sensitized solar cells: Fundamentals and current status." *Nanoscale Research Letters* 13 (2018): 1–46. https://doi.org/10.1186/s11671-018-2760-6

[12] Gong, Jiawei, K. Sumathy, Qiquan Qiao, and Zhengping Zhou. "Review on dye-sensitized solar cells (DSSCs): Advanced techniques and research trends." *Renewable and Sustainable Energy Reviews* 68 (2017): 234–246. https://doi.org/10.1016/j.rser.2016.09.097

[13] Chung, Shyan-Lung, and Ching-Mei Wang. "Solution combustion synthesis of TiO_2 and its use for fabrication of photoelectrode for dye-sensitized solar cell." *Journal of Materials Science & Technology* 28, no. 8 (2012): 713–722. https://doi.org/10.1016/S1005-0302(12)60120-0

[14] Venkatesham, Vuppala, G. M. Madhu, S. V. Satyanarayana, and H. S. Preetham. "Adsorption of lead on gel combustion derived nano ZnO." *Procedia Engineering* 51 (2013): 308–313. https://doi.org/10.1016/j.proeng.2013.01.041

[15] Chen, Xiaobo, and Samuel S. Mao. "Titanium dioxide nanomaterials: Synthesis, properties, modifications, and applications." *Chemical Reviews* 107, no. 7 (2007): 2891–2959. https://doi.org/10.1021/cr0500535

[16] Pugazhendhi, K., Steven D'Almeida, P. Naveen Kumar, J. Sahaya Selva Mary, Tenzin Tenkyong, D. J. Sharmila, J. Madhavan, and J. Merline Shyla. "Hybrid TiO_2/ZnO and TiO_2/Al plasmon impregnated ZnO nanocomposite photoanodes for DSSCs: Synthesis and characterisation." *Materials Research Express* 5, no. 4 (2018): 045053. https://doi.org/10.1088/2053-1591/aab7af

[17] Manikandan, V. S., Akshaya K. Palai, Smita Mohanty, and Sanjay K. Nayak. "Eosin-Y sensitized core-shell TiO_2-ZnO nano-structured photoanodes for dye-sensitized solar cell applications." *Journal of Photochemistry and Photobiology B: Biology* 183 (2018): 397–404. https://doi.org/10.1016/j.jphotobiol.2018.05.001

[18] Vlazan, P., D. H. Ursu, C. Irina-Moisescu, I. Miron, P. Sfirloaga, and E. Rusu. "Structural and electrical properties of TiO_2/ZnO core-shell nanoparticles synthesized by hydrothermal method." *Materials Characterization* 101 (2015): 153–158. https://doi.org/10.1016/j.matchar.2015.01.017

[19] Song, Lixin, Qingxu Jiang, Pingfan Du, Yefeng Yang, Jie Xiong, and Can Cui. "Novel structure of TiO2-ZnO core shell rice grain for photoanode of dye-sensitized solar cells." *Journal of Power Sources* 261 (2014): 1–6. https://doi.org/10.1016/j.jpowsour.2014.03.030

[20] Ahmad, Waqar, Umer Mehmood, Amir Al-Ahmed, Fahad A. Al-Sulaiman, M. Zaheer Aslam, Muhammad Shahzad Kamal, and R. A. Shawabkeh. "Synthesis of zinc oxide/titanium dioxide (ZnO/TiO_2) nanocomposites by wet incipient wetness impregnation method and preparation of ZnO/TiO_2 paste using poly (vinylpyrrolidone) for efficient dye-sensitized solar cells." *Electrochimica Acta* 222 (2016): 473–480. https://doi.org/10.1016/j.electacta.2016.10.200

[21] Sani, Z. Khosousi, F. E. Ghodsi, and J. Mazloom. "Photoluminescence and electrochemical properties of transparent CeO 2-ZnO nanocomposite thin films prepared by Pechini method." *Applied Physics A* 123 (2017): 1–10. https://doi.org/10.1007/s00339-016-0729-9

[22] He, Geping, Huiqing Fan, and Zhiwei Wang. "Enhanced optical properties of heterostructured ZnO/CeO_2 nanocomposite fabricated by one-pot hydrothermal method: Fluorescence and ultraviolet absorption and visible light transparency." *Optical Materials* 38 (2014): 145–153. https://doi.org/10.1016/j.optmat.2014.09.037

[23] Liu, I. Tsan, Min-Hsiung Hon, and Lay Gaik Teoh. "The preparation, characterization and photocatalytic activity of radical-shaped CeO_2/ZnO microstructures." *Ceramics International* 40, no. 3 (2014): 4019–4024. https://doi.org/10.1016/j.ceramint.2013.08.053

[24] Zhang, Qian, Xiaoru Zhao, Libing Duan, Hao Shen, and Ruidi Liu. "Controlling oxygen vacancies and enhanced visible light photocatalysis of CeO_2/ZnO nanocomposites." *Journal of Photochemistry and Photobiology A: Chemistry* 392 (2020): 112156. https://doi.org/10.1016/j.jphotochem.2019.112156

[25] Govindaraj, Rajamanickam, M. Senthil Pandian, G. Senthil Murugan, P. Ramasamy, and Sumita Mukhopadhyay. "Synthesis of porous titanium dioxide nanorods/nanoparticles and their properties for dye sensitized solar cells." *Journal of Materials Science: Materials in Electronics* 26 (2015): 2609–2613. https://doi.org/10.1007/s10854-015-2731-y

[26] Lee, J. Joon, Md Mahbubur Rahman, Subrata Sarker, NC Deb Nath, AJ Saleh Ahammad, and Jae Kwan Lee. "Metal oxides and their composites for the photoelectrode of dye sensitized solar cells." *Advances in Composite Materials for Medicine and Nanotechnology* 1 (2011): 181–210. https://www.researchgate.net/publication/283405123_Metal_oxides_and_their_composites_for_dye-sensitized_solar_cells

[27] Singh, Pooja, T. R. Vinay, Archana Balyan, and Sandeep Prabhu M. Gangadhara. "PV and IV characteristics of solar cell." *Design Engineering* (2021): 520–528. https://www.researchgate.net/publication/352381037_PV_and_IV_Characteristics_of_Solar_Cell

[28] Gong, Jiawei, K. Sumathy, Qiquan Qiao, and Zhengping Zhou. "Review on dye-sensitized solar cells (DSSCs): Advanced techniques and research trends." *Renewable and Sustainable Energy Reviews* 68 (2017): 234–246. https://doi.org/10.1016/j.rser.2016.09.097

[29] Giannouli, Myrsini. "Nanostructured ZnO, TiO_2, and composite ZnO/TiO_2 films for application in dye-sensitized solar cells." *International Journal of Photoenergy* 2013 (2013). https://doi.org/10.1155/2013/612095

[30] Wahyuningsih, S., A. H. Ramelan, R. Hidayat, G. Fadillah, H. Munawaroh, and L. N. M. Z. Saputri. "Synthesis of TiO_2 NRs-ZnO Composite for Dye Sensitized Solar Cell Photoanodes." In *IOP Conference Series: Earth and Environmental Science*, vol. 75, no. 1, p. 012006. IOP Publishing, 2017. https://doi.org/10.1088/1755-1315/75/1/012006

[31] Lai, Fang I., Jui-Fu Yang, and Shou-Yi Kuo. "Efficiency enhancement of dye-sensitized solar cells' performance with ZnO nanorods grown by low-temperature hydrothermal reaction." *Materials* 8, no. 12 (2015): 8860–8867. https://doi.org/10.3390/ma8125499

[32] Suliman, Ali Elkhidir, Yiwen Tang, and Liang Xu. "Preparation of ZnO nanoparticles and nanosheets and their application to dye-sensitized solar cells." *Solar Energy Materials and Solar Cells* 91, no. 18 (2007): 1658–1662. https://doi.org/10.1016/j.solmat.2007.05.014

[33] Hossain, Md Ashraf, Changjin Son, and Sangwoo Lim. "Improvement in the photovoltaic performance of a dye-sensitized solar cell by the addition of CeO_2: Gd nanoparticles in the photoanode." *Journal of Industrial and Engineering Chemistry* 65 (2018): 418–422. https://doi.org/10.1016/j.jiec.2018.05.015

13 Perovskite-Spinel Oxide Nanocomposites for Optoelectronic Applications

Anand S. Kakde, Vishwajit M. Gaikwad,
Ajay B. Lad, and Kishor G. Rewatkar

INTRODUCTION

BACKGROUND OF COMPLEX OXIDE NANOCOMPOSITES

Nowadays, nanomaterials are widely used in numerous applications only because of their considerable dominance over the bulk materials [1]. Nanomaterials and its nanocomposites are of great interest only due to unique structural, optical, electrical, magnetic, etc. properties. The metal oxide–based nanocomposites are much popular among the researchers as they plays a crucial role in the enhancement of functional properties [2]. These oxide nanocomposites are of special class. Here, the simultaneous spin, charge, and orbital interactions are existing that lead to their multiple applications [3]. The complex oxide nanocomposites have exhibited exceptional behaviours, viz. ferromagnetic, anti-ferromagnetic, ferrimagnetic, multiferroic, superconducting, etc. [4]. The magneto-electric phenomena in nanocomposites have also drawn the attention. However, this effect is yet to be used for technical applications at large scale [5]. At present, different nanocomposites are explored to study the magneto-electric nature. The existence of magnetic and optical/electrical ordering in oxide causes the magneto-optic, optoelectronic, and magneto-electric effect. The nanocomposites made of particular crystal structure (perovskite and spinel) are in extensive demand because of their exceptional optoelectronic properties. Generally, the magneto-electric effect is the coupling between the magnetic and dielectric dipoles, which weakly exists and hence limits the applications in data storage, multifunctional devices, etc. The oxide composites with magnetic and electric ordering are rarely available. Thus, most of the researchers are centred their research work on the proper selection of dopant so as to synthesise oxide composites that must have strong ordering between different order parameters. The selection of materials from perovskite and spinel class involves consideration of many properties such as piezo-electric, dielectric, permeability, and magnetostriction. Unlike nanomaterials, nanocomposites are a combination of two host materials as a result it is bi-phase in nature at nanoscale. The composites are made up of two or more material phases where one material acts as the host matrix and the other acts as the particle phase. Here the particle phase is embedded on the host phase or vice versa. As depicted in Figure 13.1, nanocomposites are quite well recognised among the researchers and are broadly studied over the decades [6]. The oxide nanocomposites possess distinct lattice structures with an interface separating them, which made them potential materials for the advanced optical, electrical, optoelectronic, and catalyst applications [7].

Nanocomposites with the spinel structure along with perovskite structure are the most promising for the optoelectronic/magneto-electric applications [8]. The structural representation of perovskite component (LFO) and spinel component (NFO) along with the structure of PS composite is shown in Figure 13.2 In these oxide nanocomposites, only by changing the concentration of particle phase of composites, the required combination for a specific application can be synthesised.

DOI: 10.1201/9781003479239-13

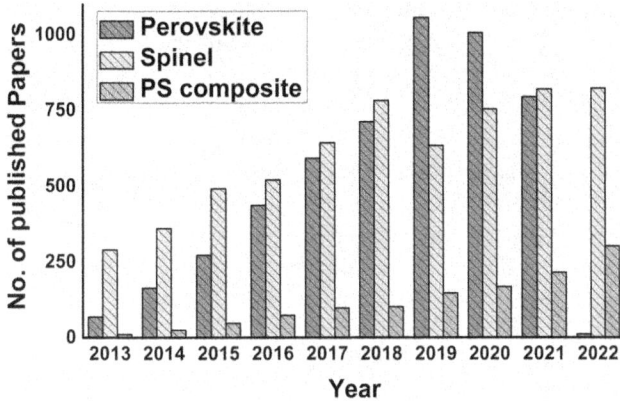

FIGURE 13.1   Number of wise-year publications for perovskite (P), spinel (S), and PS composites.

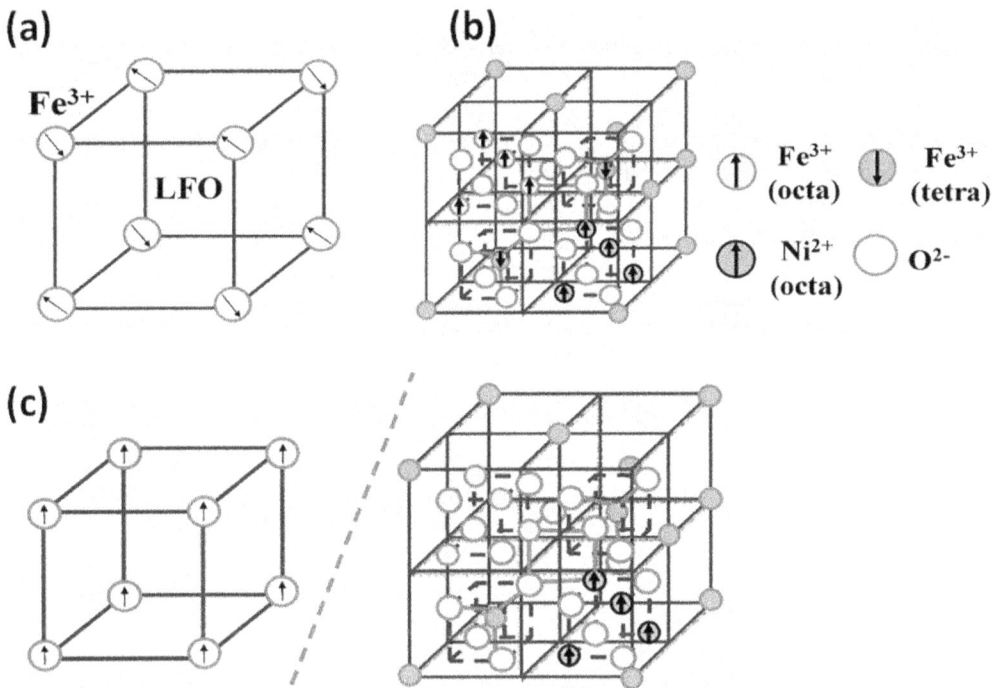

FIGURE 13.2   Unit cell representation of (a) perovskite material, (b) spinel material, and (c) PS composite [9].

PEROVSKITE-SPINEL (PS) NANOCOMPOSITES IN DIFFERENT FORMS

PEROVSKITE-SPINEL (PS) IN BULK FORM

The bulk composites of $LaFeO_3$ (perovskite)-$NiFe_2O_4$ (spinel), $BiFeO_3$-$CoFe_2O_4$, $ZnTiO_3$-$ZnTi_2O_4$, and $BiFeO_3$ (bulk form)-$ZnFe_2O_4$ (nano form) were synthesised using physical and chemical methods. At room temperature, these bulk composites are in a stable magnetic and dielectric ordering. The observed improvement in magnetic behaviour is credited to spin re-orientation of magnetic moments of perovskites by inducing strain at the interface of perovskite and spinel composites. The synthesised composite possesses a high dielectric constant with negligible loss, and hence in the room temperature, perovskite-spinel bulk composite is favourable for various technical applications [10].

To figure out the effect of interfacial strain on electric and magnetic properties, a different PS composite was studied. The examination of the composite reveals that due reorganisation of spins at the interface of composite alters the strain energy and magnetocrystalline anisotropy, which in turn improves magnetisation. As compared to pure $BiFeO_3$ perovskite, in composite, the leakage current drastically drops due to decrease in dielectric loss. Hence, at room temperature, this composite can be used for multifunctional applications [11]. In general, $ZnTi_2O_4$ and $ZnTiO_3$ have the cubic crystal symmetry. $ZnTi_2O_4$ and $ZnTiO_3$, respectively, possess direct and indirect wide bandgaps. Spin-dependent electronic structure can be used to tune the energy bandgap of PS composites. For an optoelectronic application, the materials with direct bandgap play a vital role as that of indirect bandgap. The structure–property relationship of this bulk composite confirmed its applications as optoelectronic device in the UV region [12].

Perovskite-Spinel (PS) in Nano Form

Many perovskite-spinel composites, viz. $La_{0.8}Sr_{0.2}CoO_3$-$CoFe_2O_4$, $ZnSnO_3$-$ZnFe_2O_4$, $BaTiO_3$-$CrFe_2O_4$, $BiFeO_3$-$CoFe_2O_4$, and $BiFeO_3$-$ZnFe_2O_4$, were synthesised in nano form using different methods [13,14]. These multicomponent nanocomposites hold low resistivity, great thermal conductivity, and very low toxicity, and they are chemically firm and excellently homogenous. The spinel (magnetic) and perovskite (ferroelectric) phases contribute equally to the formation of magneto-electric nanocomposites. At room temperature, these nanocomposites own the magneto-electric effect which is much stronger as that of bulk composites. The structural, composition, lattice strain, and plane orientation properties of PS nanocomposites can be modified by proper substitution of dopants at 'A' and 'B' sites. Nowadays, the coating over the nanocomposite materials forms a core-shell structure that has undue advantages because of their applications in wastewater management, biomedicine, and energy storage devices [15]. The PS nanocomposites in core-shell form confirms that with the increase in spinel content, (i) the decrease in crystallite size and (ii) enhancement of ferroelectric and magnetic properties were observed, while (iii) the dielectric behaviour remains constant. This indicates that core-shell PS nanocomposites are very handy for the fabrication of multipurpose devices. The optical properties of glass-coated PS nanocomposites confirm the optical bandgap in the range of 1.063–3.977 eV, which was good with earlier reported values [16]. Subsequently, PS nanocomposites can be recycled with a very low loss of photocatalytic activities, which is associated to low e^-–h^+ recombination. Thus, the unique structural and optoelectronic properties of glass-coated PS nanocomposites suggest that these materials are of n-type semiconductor. The dielectric properties also play a decisive role in improving the magneto-electric effect in PS nanocomposites. At higher frequency, dielectric constant shows a dispersion which is the typical behaviour of PS composites. The deficiency of oxygen atoms is always a drawback in perovskite materials, and as a result, there is the existence of space-charge polarisation [17]. Thus, at high temperatures, there is a significant rise in these charges, and hence, PS nanocomposites show a good conductivity. The variation in dielectric constant with change in applied magnetic field occurs not only because of magneto-electric phenomena but also because of magnetostriction effect. In the presence of magnetic field, this magnetostriction effect leads to change in the lattice constant, and hence, magnetocapacitance was induced in PS nanocomposites. At room temperature, the presence of magnetocapacitance confirms the strong magneto-electric effect.

Perovskite-Spinel (PS) Composites in Thin-Film Form

The various thin-film nanocomposites, such as $BiFeO_3$-$NiFe_2O_4$, $BiFeO_3$-$CoFe_2O_4$, $SrMnO_3$-$Mn_{1.56}Co_{0.96}Ni_{0.48}O_4$, and $LaMnO_3$-$Mn_{1.56}Co_{0.96}Ni_{0.48}O_4$, were synthesised using pulsed laser deposition and solution method. The PS bi-phase epitaxial thin film is of great interest. Composition and morphology control the bi-phase of these nanocomposites, which will be useful to design PS thin films at nanoscale [18,19]. The morphology of composite thin film has its influence on crystal orientation (miller plane) of the substrate and magnetic parameters, viz. magnetisation saturation, magnetocrystalline anisotropy, etc. The substrate is also a vital factor as it controls the strain component in the growth of PS nanocomposite thin films. The PS thin films are magnetically very hard due to magnetic anisotropy of spinel material, but it can be changed with the strain. Hence, by altering the orientation of growth, it is possible to modify morphology and limit the strain in PS nanocomposite thin films. For any epitaxial axis $(00l)$, the surface energy of spinel phase is much more than that of perovskite, and hence, perovskite makes the substrate moist. Generally, low resistance and high thermal constant (B) value play a vital role for negative temperature coefficient (NTC) films. The research scholars successfully improved the value of B much further by synthesising the PS nanocomposite thin films. Moreover, PS thin film that is bi-phase in nature exhibits much better firmness of surface, morphology, and texture as that of single-phase spinel thin film. If the temperature increases, then the resistivity of PS thin film reduces drastically, which indicates the NTC behaviour. Notably, if the content of perovskite is increased in PS nanocomposite thin films, then the value of B enhanced. The temperature coefficient resistance (TCR) clearly has its dependence on morphology, and the morphology has its influence on optical properties [20]. At room temperature, with the increase in fraction of perovskite, the grain boundaries are also increased, which leads to enhancement in TCR and expected shift in absorption peak to higher frequencies. This urges to increase the optical bandgap in PS nanocomposite thin films.

PROGRESS OF PEROVSKITE-SPINEL (PS) NANOCOMPOSITES IN THE CURRENT PERSPECTIVE

The bi-phase PS nanocomposites exist with exceptional properties that were never shown by any individual pure phase material. As stated earlier, the PS composites are in different forms, namely, bulk, nano, and thin-film forms, and have potential applications in and as transducers, memory devices, switches, sensors, microwave absorption devices, medical instruments, data storage devices, and homemade appliances [11,20,21]. Currently, many materials are engineered to have novel PS nanocomposites with multifunctional advanced properties that can be controlled with precision in terms of shape and size at nanometric scale. Recently, the carbon incorporated in PS nanocomposites, which are the best alternatives, are considered for energy-conversion devices for oxygen evolution (ORE), hydrogen evolution reaction (HER) and oxygen reduction (ORR). The carbon materials (carbon nanotubes or graphene) with PS nanocomposites effectively control the catalytic properties. The cations at octahedral sites of perovskite and spinel materials, where high-energy orbitals can easily get overlapped with oxygen molecules, have better catalytic activities specifically for OER and ORR evolution than those at tetrahedral sites. The carbon-combined PS nanocomposites are considered to be the best environmental and sustainable substitute for electro-chemical energy devices [22]. The spinel oxide derived from perovskite hydroxides is the most prominent substance for gas-sensing application. These days, the need of gas sensors is more appealing, and hence, the research scholars are continuously searching for a suitable gas-sensing material that possesses quick response to gas detection and is highly sensitive to gas. In PS nanocomposites, electron mobility is very high, and they have an extremely narrow bandgap that forms oxygen ions on the surface, which reveals the gas sensing interest. The divalent cations in spinel and perovskite materials are placed at octahedral sites, which improves catalytic behaviour, thereby improving the gas-sensing ability in PS nanocomposites. Further, the nanosized particles, high surface area, chemical composition, charge state, etc. of PS nanocomposites make them a more suitable and prominent material for gas sensor applications [23].

MATERIALS FABRICATION AND CONNECTIVITY SCHEMES

PREPARATION METHODS AND THEIR ROLE

The PS nanocomposites can be synthesised in different routes including both physical and chemical methods. Most often, the perovskite and spinel phases are mixed physically to the PS composites. But due to the physical mixing of two different phases, the particles are aggregated to form a cluster that damages the overall properties of PS nanocomposite. To overcome this problem, many research scholars approach the chemical-solution method. Initially, the individual phases are prepared using a chemical technique to have excellent chemical stability and homogeneity, and then the physical mixing takes place for better coupling between the perovskite and spinel phases. The different synthesis methods generally used to synthesise the PS nanocomposites are discussed in the following.

SOL-GEL COMBUSTION METHOD

The sol-gel combustion method is a very renowned chemical technique to synthesise nanoparticles. The precursors are dissolved in distilled water to form a transparent solution referred to as 'SOL'. Then the SOL is placed on a hot plate to convert it into 'GEL', and then the GEL is fired in hot furnace or in microwave oven for self-combustion to get desired nanoparticles [24]. Initially, perovskite $BiFeO_3$ was formed independently, where bismuth and ferric ions in the nitrate form are dissolved in 30 ml of distilled water and 2 ml of nitric acid to form SOL. Similar process was repeated to form spinel ferrite $NiFe_2O_4$ with nickel and ferric ions. For the synthesis of a PS composite, the SOL was mixed with citric acid in a ratio of 1:1 to be used as fuel. The combined SOL was now put on a hot plate at 70°C to form GEL. The obtained GEL was then placed in a microwave oven for self-combustion, and the resultant ash was obtained. In a mortar and pestle, the ash was grinded about 30 minutes and then placed in muffle furnace at 600°C to form the PS composite. Basically, the sol-gel method is based on the hydrolysis or condensation phenomena. The exact role of this method is obtaining a high specific surface area and a steady surface composition so as to have a better control on the overall properties of synthesised PS nanocomposites. The sol-gel method has many advantages such as low cost, better distribution of particles, flexible control on composite phase, and simple setup [25]. Hence, the sol-gel combustion method is one of the most economical and preferred techniques to synthesise PS nanocomposites, where one phase can easily be fixed on to another phase.

PULSED LASER DEPOSITION METHOD

The pulsed laser deposition method is one of the most emerging physical techniques to synthesise PS nanocomposites. Usually, this method is used to prepare thin films of composites materials. The method involves three basic steps: (i) to remove plume for solid target, (ii) to generate high-energy laser beam, and (iii) to deposit the vapour on the substrate to form thin film [26]. The whole setup of this method is placed inside the vacuum. In the pulsed laser deposition technique, the high energetic laser beam strikes the solid target and vaporizes it by strong heating. Now by condensation of the vapour, it forms a thin film on the substrate. The PS nanocomposites of $LaSrCoO_3$ and $CoFe_2O_4$ were synthesised by combinatorial pulsed laser deposition (cPLD) on $SrTiO_3$ substrate. In this method, separate perovskite ($LaSrCoO_3$) and spinel ($CoFe_2O_4$) targets are required. Both the targets were alternately ablated to generate the single layer which was deposited on the substrate placed at 600°C [27]. As shown in Figure 13.3, if the substrate is not rotated, then uniform formation of PS nanocomposite thin film is probably not possible (Positions 1–3), but if the substrate is rotated with the help of a holder, then a uniform thin film can be formed. The expected role of the method is to synthesise the PS nanocomposite with a high degree of crystallinity, excellent morphology, better chemical composition, etc. Additionally, if the deposition conditions of this method are better controlled, then the morphology, thickness, and chemical composition of the film can be changed. The advantages of the method are that many composites can be synthesised within short time in controlled conditions, and the transfer from the target to the substrate is easy.

SOLID-STATE REACTION

In spite of huge development in the chemical-based synthesis methods, still a wide range of composites are prepared via solid-state reaction. In this method, different raw materials that are already available in the solid form are mixed together vigorously to form a solid final product which will be

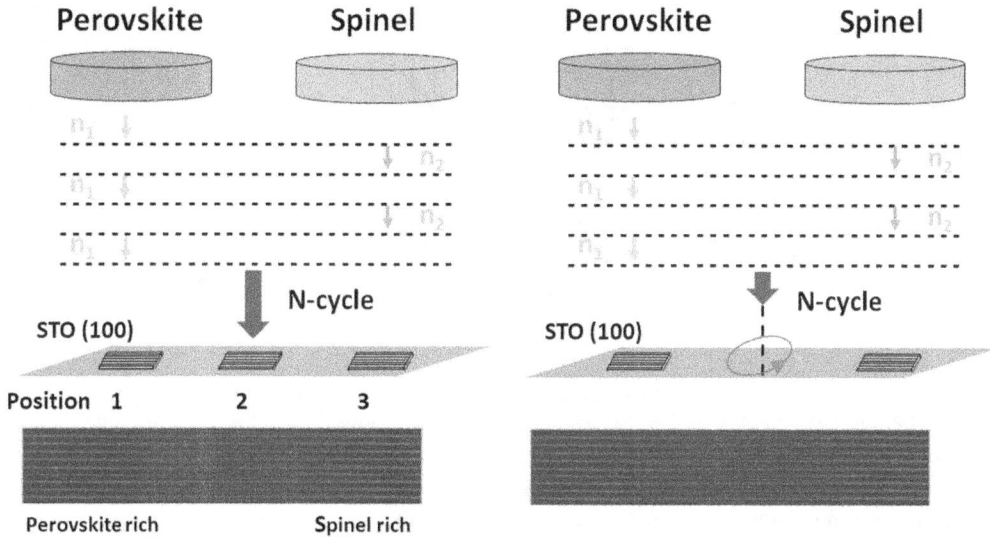

FIGURE 13.3 Growth process of PS nanocomposites: (a) fixed substrate and (b) rotating substrate [27].

treated in high temperature to remove all the residues from the final product. The PS composite of $BaTiO_3$-$Mn_{0.5}Zn_{0.5}Fe_2O_4$ was synthesised using barium acetate, titanium dioxide, manganese oxide, zinc oxide, iron oxide, sodium chloride, and nonyphenyl ether (NP-30). These materials were mixed together in a ratio of 4:4:1:1:2:120:20 and ground for half an hour to make the mixture homogenous. Here, NP-30 is used as non-ionic surfactant for better mixing and NaCl for reaction purpose. The mixture was sonicated and then put in furnace at 850°C for 5 hours. The mixture was then washed rigorously with distilled water and placed on a hot plate at 80°C for over-night drying to obtain the PS composite [28]. The solid-state reaction method is completely solvent free, which makes it more convenient and easily controllable method to obtain the nanoscale properties of synthesised materials [29]. The method is very cost effective and simple, and may be used for large production, but it requires high-temperature treatment.

3-0, 2-2, AND 3-1 CONNECTIVITY SCHEMES IN PS NANOCOMPOSITES

The connectivity schemes are important to fabricate composites in various forms (bulk, layered, film, and 2D). Any physiochemical properties can be altered only through the interconnection made between the various phases of composite materials. Connectivity defines the dimension of components connected to one another. The stability and efficiency of component phases are strictly dependent on the connectivity. Here, we restrict our focus on 3-0, 2-2, and 3-1 connectivity schemes in PS nanocomposites and interpret the stated connectivity from the literature [30,31]. Figure 13.4 represents the connectivity patterns in the two-phase system where the blue-green–coloured box indicates three-dimensional connection, while the white box indicates one-dimensional connection.

The 3-0 connectivity is often referred to as particulate composites where particles of one phase dispersed in the three-dimensional matrix of other phases. The 3-0-type PS nanocomposites have a high dispersion of magnetostrictive (spinel phase) dispersed in piezoelectric (perovskite phase), and hence have a piezoelectric–magnetostrictive interface which is required for magneto-electric effect. This interface results in space-charge polarisation, penetration of charges from piezoelectric medium due to magnetostrictive effect, and leakage current that weakens the magneto-electric effect.

The 2-2 connectivity is referred to as laminate composites where two-dimensional layers are arranged horizontally. As compared to 3-0 type, in 2-2 connectivity, the piezoelectric–magnetostrictive interface is certainly minimised. However, the reduced interface is unable to

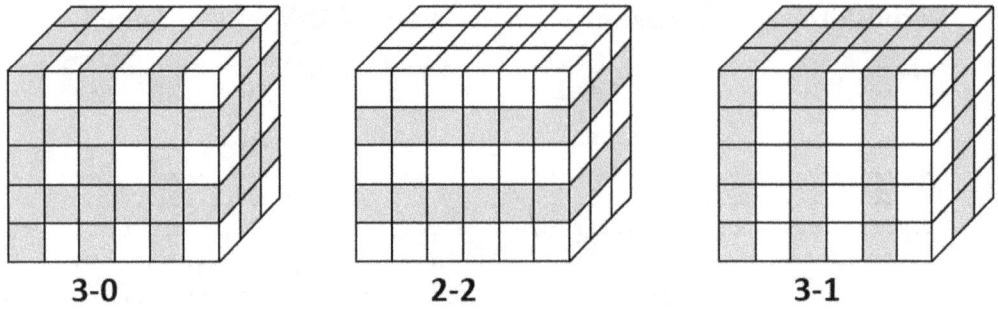

FIGURE 13.4   The connectivity patterns in two-phase system [26].

enhance the magneto-electric effect due to the substrate clamping effect. The only advantage of 2-2 type over the 3-0-type connectivity is that it can be easily handled and controlled.

The 3-1 connectivity is referred to as fibre composite where one-dimensional rods or nano-pillars embedded in a three-dimensional matrix structure. As compared to 3-0 and 2-2 types, 3-1 type have a high interface–volume ratio, and hence, it gains the recognition of researchers straightway. In 3-1 type, the nano-pillars of spinel phase are inserted in the perovskite phase. As a result, the penetration of charges and the agglomeration of small clusters are reduced to a large scale that increases the interface–volume ratio.

ROLE OF INTERFACE IN THE MODIFICATION OF STRUCTURE AND PROPERTIES

The interface is the area, where composites are in contact with each other. The improvement in any property of nanocomposites basically depends on the interface between the component phases. At interface, many interactions occur such as realignment of chemical bonds, spin-orbital rearrangement, and changes in electronic structure. The control of interaction at interface among the different phases is found to be difficult. The structure and chemical composition of interface are necessary to understand the coupling effect between ferroelectric and magnetostriction. The manner different component phase bonds interact at the interface is the most serious step, and it also helps us understand the coupling phenomena among magnetism, elasticity, and ferroelectricity. Any complicated surface interface can be easily resolved by considering 1 to 1 atomic correlation. Figure 13.5 reveals the bonding between $BiFeO_3$ layer of perovskite phase and $(Ni, Fe)O_2$ layer of spinel phase.

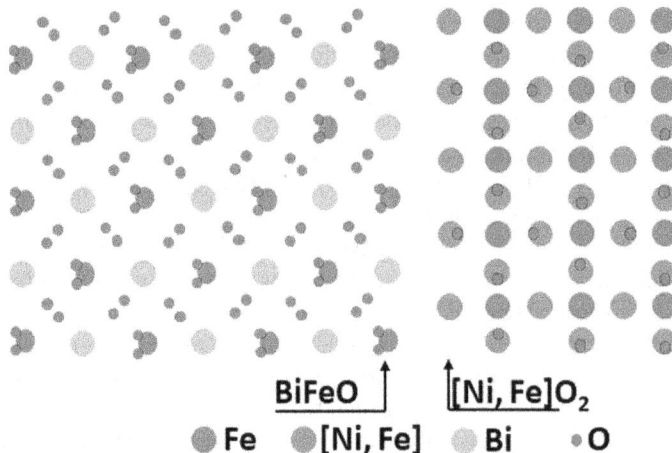

FIGURE 13.5   Structure model of the perovskite-spinel interface [32].

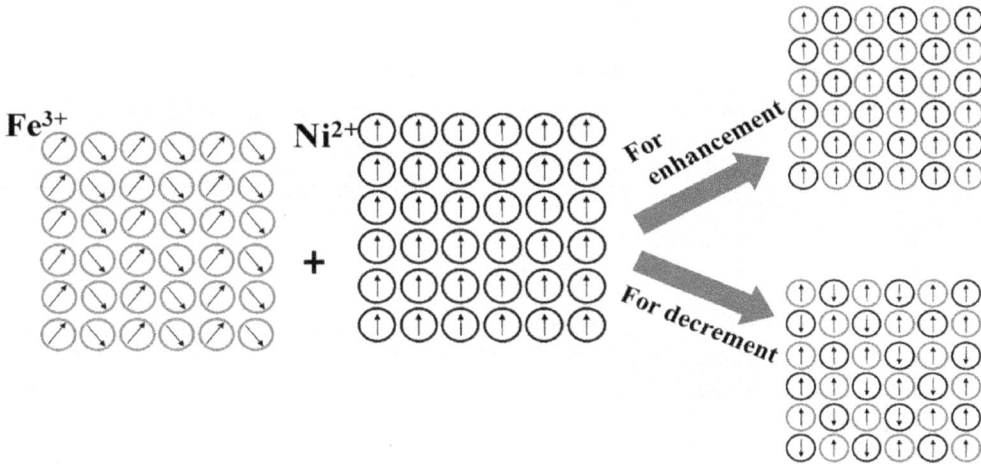

FIGURE 13.6 Spin alignment of perovskite (LFO) and spinel (NFO) materials [9].

As mentioned layers have octahedral sites at interface, the structure continuity is too good which needs less interface energy and hence requires small rearrangement of chemical bonds. At the interface, maximum structure continuity results in steady and firm elastic-coupling [32].

Let's discuss the spin alignment mechanism at the interface of LFO ($LaFeO_3$)–NFO ($NiFe_2O_4$) nanocomposites. The spinel phase has a ferrimagnetic order where the magnetic spins of Fe^{+3} at octahedral and tetrahedral sites are antiparallel and hence cancel out each other. In spinel material, only magnetic spins of Ni^{+2} can contribute to the magnetisation while the perovskite materials possess minimum magnetisation, due to spin canted effect. Now the magnetic spins of perovskite (Fe^{+3}) and spinel (Ni^{+2}) phase interact at the interface, which leads to change the overall magnetisation of PS composites. Figure 13.6 explains how the magnetic spins of LFO-NFO phase reconstruct themselves to increase or decrease magnetisation of composite.

PROPERTIES OF PS NANOCOMPOSITES

MAGNETIC AND ELECTRIC ORDERS

Varying the electric field by the external magnetic field or vice versa is called the magneto-electric effect. This effect exists between polarisation of ferroelectric and magnetisation of ferromagnetic phase coupled with each other through strain. At room temperature, the magneto-electric phenomena in PS nanocomposites have gained a valuable consideration due to their wide range of applications. Comparatively, magneto-electric phenomena are very much effective in PS nanocomposites as that exist in any individual single phase. At first, the magneto-electric effect was observed in $BaTiO_3$-$CoFe_2O_4$ (BTO-CFO 2-2-type) nanocomposite thin film. In this PS nanocomposite thin films, the magnetic study reveals the structural changes at ferroelectric Curie temperature. The perovskite phase changes from cubic to tetragonal which leads to distort the spinel phase. As a result, saturation magnetisation of spinel phase at this Curie temperature was found to be decreased by the magnetostriction effect. Alternately, in the absence of magnetic field, the voltage of about −16 V is applied and nearly half of magnetisation of spinel nano-pillars are rotated with a magnetic field of about 800 G. Almost complete magnetisation is rotated only because of the magneto-electric effect in PS composites [33]. Generally, due to low clamping effect of substrate, 3-1-type PS composites are more promising to exhibit magneto-electric effect than that of 2-2-type PS nanocomposites. The same PS nanocomposite BTO-CFO in (3-1 type) was studied to explore more about the magneto-electric effect. This nanocomposite is represented by a vertical aligned nanostructure

TABLE 13.1

The Connectivity Scheme with Its Magneto-Electric Coefficient [26]

Connectivity Type	ME Coefficient (mV/cmOe)
3-0	155
2-2	10,300
3-1	500

(VAN), where ferromagnetic nano-pillars have an epitaxial growth in ferromagnetic medium. The (3-1)-type BFO-CFO nanocomposite has shown both magnetisation along easy axis and polarisation. Similarly, like 2-2-type BFO-CFO nanocomposite, the change in magnetisation occurs at the ferroelectric transition temperature of perovskite phase, which shows a significantly improved magneto-electric effect. The 3-1-type PS nanocomposites with low clamping effect and VAN possess better magneto-electric coupling than 2-2-type PS nanocomposites [34]. Because of large leakage current, it is difficult to polarise the ferroelectric phase of (3-0)-type PS composites. In 3-0 type, the clamping effect as well as leakage current which restricts the strain are concerned issues that hammer the magneto-electric effect in PS nanocomposites. For $(PbFe_{0.5}Nb_{0.5}O_3)–(Co_{0.6}Zn_{0.4}Fe_{1.7}Mn_{0.3}O_4)$, 3-0-type PS composites are prominent materials for low-power memory devices and magnetic sensors [35] (Table 13.1).

Optical Characteristics

For optoelectronic applications, the optical properties of the materials provide the basic and necessary information. Dielectric function, absorption coefficient, conductivity, etc. are the other optical variables that can be used to explain how PS composites behave with electromagnetic radiation. The electronic structure of both perovskite and spinel phases can be modified at the interface of their composite. Modification of electronic structure affects their band structures and ultimately the distance between upper valence and lower conduction band. Therefore, the energy bandgap is greatly affected by the interface phenomenon that exists in PS composites.

Optoelectronic Functionalities

Recently, PS nanocomposites have gained a lot of attention only due to its optical and electronic properties. On the basis of structural, optical, and electronic properties, various PS composites have been studied in different circumstances to find their most suitable applications in the optoelectronic viewpoint. The electronic energy band and density of states are the most important parameters to explain optical properties. In PS nanocomposites, perovskite phase deals with low energy of s-orbital, big size of cations, and anti-bonding state in conduction band minimum. But the complex structural environment of PS nanocomposites reduces the stability of optoelectronic devices. Now, researchers try to find how to improve the stability of PS composite optoelectronic devices. PS composites are important for high-speed optical communications. Their thin-film forms can be integrated into the VLC technology, which has advantages like high data transmission rate, good security, and no electromagnetic interference. PS composites can be exploited for information encryption and decryption. They increase the security of information and perform repeated encryption and decryption. PS composites are suggested for making potential strategy for information encoding and decoding. Metal halide perovskite with spinel produces outstanding novel properties which can be promising in the field of artificial intelligence. Perovskite-based composites can be very flexible to combine light and electricity, which can be fruitful for overcoming the limitations of traditional electrical devices. PS composites can simulate the signal learning, processing, and memory of the nervous system. Overall, PS composite materials have multiple modern applications. In future, devices based on PS materials will become potential candidates for especially integrated chip fusing sensing, storage, and computing [36].

CONCLUSION

PS composites are the most prominent nanomaterials for numerous applications in the field of science, technology, biomedical, industry, etc. Various chemical and physical methods are available to synthesise PS composites, and hence, they are existing in different forms (bulk, nano, and thin film) that significantly improve the structural, physical, and other properties. Nowadays, carbon materials are effectively incorporated in PS nanocomposites for their catalytic properties and energy devices. The connectivity schemes 3-0, 2-2, and 3-1 are important as they interconnect the different phases in PS nanocomposites. The interface reveals the different coupling mechanism that exists between ferroelectric and magnetic phenomena. The different properties stated that the PS nanocomposites are suitable materials for optoelectronic devices. These nanocomposites are useful in optical communication, VLC technology, and high-speed data transmission. This chapter basically summarises the brief idea about the synthesis, properties, and utility of PS nanocomposites as optoelectronic devices.

REFERENCES

[1] Tetiana A. Dontsova, Svitlana V. Nahirniak, and Ihor M. Astrelin, Metaloxide nanomaterials and nanocomposites of ecological purpose, *Journal of Nanomaterials*, (2019), 1–31, https://doi.org/10.1155/2019/5942194.

[2] Elizabeth Gager, William Halbert, and Juan C. Nino, Complex oxide nanoparticle synthesis: Where to begin to do it right?, *Ceramics*, 5(4), (2022), 1019–1034, https://doi.org/10.3390/ceramics5040073.

[3] Fatima Sayed, Ganesh Kotnana, Giuseppe Muscas, Federico Locardi, Antonio Comite, Gaspare Varvaro, Davide Peddis, Roland Mathieu, and Tapati Sarkar, Symbiotic, low-temperature, and scalable synthesis of bi-mag- netic complex oxide nanocomposites, Nanoscale Advances, 2, (2020), 851, https://doi.org/10.1039/ c9na00619b.

[4] Aiping Chen, Qing Su, Hyungkyu Han, Erik Enriquez, and Quanxi Jia, Metal oxide nanocomposites: A perspective from strain, defect, and interface, *Advanced Materials*, 31, (2019), 1803241, https://doi.org/10.1002/adma.201803241.

[5] S. M. Yusuf, P. K. Manna, M. M. Shirolkar, S. K. Kulkarni, R. Tewari, and G. K. Dey, A study of exchange bias in $BiFeO_3$ core/$NiFe_2O_4$ shell nanoparticles, *Journal of Applied Physics*, 113(17), (2013), 1–6, https://doi.org/10.1063/1.4803549.

[6] P. Pramanik, D. C. Joshi, N. Tiwari, T. Sarkar, S. Pittala, O. O. Salman, M.-M. Manga, and S. Thota, Cationic distribution, exchange interactions, and relaxation dynamics in Zn-diluted $MnCo_2O_4$ nanostructures, *Journal of Applied Physics*, 125, (2019), 124302, https://doi.org/10.1063/1.5079800.

[7] Mohammad Hossein Habibi and Maryam Mardani, Synthesis and characterization of bi-component $ZnSnO_3$/Zn_2SnO_4 (perovskite/spinel) nano-composites for photocatalytic degradation of Intracron Blue: Structural, opto-electronic and morphology study, *Journal of Molecular Liquids*, 238, (2017), 397–340, https://doi.org/10.1016/j.molliq.2017.05.011.

[8] Xiang-chun Liu, Junhui Li., F. Gao, H. Du, Molten salt synthesis of Zn2SnO4 micro crystals with regularoctahedral morphology, Journal of Materials Science: Materials in Electronics, 28, (2017), 3860–3864, https://doi. org/10.1007/s10854-016-5998-8.

[9] Vishwajit M. Gaikwad and Smita A. Acharya, Novel perovskite-spinel composite approach toenhance the magnetization of $LaFeO_3$, *RSC Advances*, 5, (2015), 14366, https://doi.org/10.1039/c4ra11619d.

[10] Vishwajit M. Gaikwad and Smita A. Acharya, Exploration of magnetically stable $BiFeO_3$-$CoFe_2O_4$ composites with significant dielectric ordering at room temperature, *Journal of Alloys and Compounds*, 755, (2018), 168–176, https://doi.org/10.1016/j.jallcom.2018.04.273.

[11] Vishwajit M. Gaikwad and Smita A. Acharya, Perovskite-spinel composite approach to modify room temperature structural, magnetic and dielectric behavior of $BiFeO_3$, *Journal of Alloys and Compounds*, 695, (2017), 3689–3703, https://doi.org/10.1016/j.jallcom.2016.11.367.

[12] Zahid Ali, Sajad Ali, Iftikhar Ahmad, Imad Khan, and H. A. Rahnamaye Aliabad, Structural and opto-electronic properties of the zinc titanate perovskite and spinel by modified Becke-Johnson potential, *Physica B*, 420, (2013), 54–57, https://doi.org/10.1016/j.physb.2013.03.042.

[13] Preethy Augustine, Y. Narayana, and Nandakumar Kalarickal, Enhancement of room temperature magneto-electric coupling effect in perovskite-spinel $(1-x)BiFeO_{3-x}ZnFe_2O_4$ nanocomposites, *Materials Today: Proceedings*, 35, (2021), 436–439, https://doi.org/10.1016/j.matpr.2020.02.948.

[14] H. Zheng, J. Wang, L. Mohaddes-Ardabili, M. Wuttig, L. Salamanca-Riba, D. G. Schlom, and R. Ramesh, Three-dimensional heteroepitaxy in self-assembled $BaTiO_3$ -$CoFe_2O_4$ nanostructures, *Applied Physics Letters*, 85, (2004), 2035, https://doi.org/10.1063/1.1786653.

[15] Ayesha Khalid, Murtaza Saleem, Shahzad Naseem, Shahid M. Ramay, Hamid M. Shaikh, and Shahid Atiq, Magneto-electric coupling and multifunctionality in $BiFeO_3$-$CoFe_2O_4$ coreshell nano-composites, *Ceramics International*, 46, (2020), 12828–12836, https://doi.org/10.1016/j.ceramint.2020.02.053.

[16] Mohammad Hossein Habibi and Maryam Mardani, Synthesis and characterization of bi-component $ZnSnO_3/Zn_2SnO_4$ (perovskite/spinel) nano-composites for photocatalytic degradation of Intracron Blue: Structural, opto-electronic and morphology study, *Journal of Molecular Liquids*, 238, (2017), 397–401, https://doi.org/10.1016/j.molliq.2017.05.011.

[17] Poonam Uniyal and K. L. Yadav, Synthesis and study of multiferroic properties of $ZnFe_2O_4$-$BiFeO_3$ nanocomposites, *Journal of Alloys and Compounds*, 492, (2010), 406–410, https://doi.org/10.1016/j.jallcom.2009.10.275.

[18] Sung Park, Min Seok Kim, Junho Yang, Tae Cheol Kim, Seung Ho Han, and Dong Hun Kim, Spinel-perovskite nanocomposite thin films on various substrates, *Journal of Nanoscience and Nanotechnology*, 17, (2017), 3523–3527, https://doi.org/10.1166/jnn.2017.14085.

[19] Li Yan, Feiming Bai, Jiefan Li, and D. Viehland, Nanobelt structure in perovskite-spinel composite thin films, *Journal of American Ceramic Society*, 92(1), (2009), 17–20, https://doi.org/10.1111/j.1551-2916.2008.02825.x.

[20] Jun Zhang, Wenwen Kong, and Aimin Chang, Fabrication and properties of high B value $[Mn_{1.56}Co_{0.96}Ni_{0.48}O_4]_{1-x}[SrMnO_3]_x$ ($0 \leq x \leq 0.5$) spinel-perovskite composite NTC films, *Journal of Materials Science: Materials in Electronics*, 29, (2018), 9613–9620, https://doi.org/10.1007/s10854-018-8997-0.

[21] Amit Kumar, K. L. Yadav, Hemant Singh, Ratnakar Pandu, and P. Ravinder Reddy, Structural, magnetic and dielectric properties of $xCrFe_2O_4$-$(1-x)$ $BiFeO_3$ multiferroic nanocomposites, *Physica B*, 405, (2010), 2362–2366, https://dio.org/10.1016/j.physb.2010.02.038.

[22] Jhony Xavier Flores Lasluisa, Francisco J. Huerta, D. Cazorla-Amorós, Emilia Morallon, Transition metal oxides with perovskite and spinel structures forelectrochemical energy production applications, Environmental Research, 214, (2022), 113731, https://doi.org/10.1016/j.envres.2022.113731.

[23] Kuan Tian, Wenhui Zhang, Su-Ning Sun, Lu Xing, Zi-Yuan Li, Tong-Tong Zhang, Niu-Niu Yang, Bei-Bei Kuang, and Hua-Yao Li, Design and fabrication of spinel nanocomposites derived from perovskite-hydroxides as gas sensing layer for volatile organic compounds detection, *Sensors and Actuators B: Chemical*, 329, (2021), 129076, https://doi.org/10.1016/j.snb.2020.129076.

[24] Anand. S. Kakde, Raju. M. Belekar, Gautam. C. Wakde, Mohan. A. Borikar, Kishor. G. Rewatkar, and Buddhaghosh. A. Shingade, Evidence of magnetic dilution due to unusual occupancy of zinc on B-site inNiFe2O4 spinel nano-ferrite, Journal of Solid State Chemistry, 300, (2021), 122279, https://doi.org/10.1016/j.jssc.2021.122279.

[25] Anand. S. Kakde, Gautam. C. Wakde, Mashuq. A. Wani, Vishwajit. M. Gaikwad, Nandkishor. S. Meshram, Ajay. B. Lad, Kishor. G. Rewatkar, and Raju. M. Belekar, Exploration of Ce+3 substitution on electron density distribution, optical, and magnetic properties of Ni-Co-Zn spinel nano-ferrites, Journal of Sol-Gel Science and Technology, 107, (2023), 401–416, https://doi.org/10.1007/s10971-023-06121-x

[26] Mohsin Rafique, Study of the Magnetoelectric Properties of Multiferroic Thin Films and Composites for Device Applications, Ph.D Thesis, CIIT/FA09-PPH-010/ISB, (2014).

[27] Yan Chen, Shuchi Ojha, Nikolai Tsvetkov, Dong Hun Kim, Bilge Yildiz, and C. A. Ross, Spinel/perovskite cobaltite nanocomposites synthesized by combinatorial pulsed laser deposition, *CrystEngComm*, 18(40), (2016), 7745–7752, https://doi.org/10.1039/c6ce01445c.

[28] Yaodong Yang, Shashank Priya, Jie-Fang Li, and Dwight Viehland, Two-phase coexistence in single-grain $BaTiO_3$-$(Mn_{0.5}Zn_{0.5})Fe_2O_4$ composites, via solid-state reaction, *Journal of American Chemical Society*, 92(7), (2009), 1552–1555, https://doi.org/10.1111/j.1551-2916.2009.03068.x.

[29] Amit Kumar, Soumen Dutt, Seonock Kim, Taewan Kwon, Santosh S. Patil, Nitee Kumari, Sampathkumar Jeevanandham, and In Su Lee, Solid-state reaction synthesis of nanoscale materials: Strategies and applications, *Chemical Reviews*, 122, (2022), 12748–12863, https://doi.org/10.1021/acs.chemrev.1c00637.

[30] R. Newnham, D. Skinner, and L. Cross, Connectivity and piezoelectric-pyroelectric composites, *Materials Research Bulletin*, 13, (1978), 525–536, https://doi.org/10.1016/0025-5408(78)90161-7.

[31] Heng-Jui Liu, Wen-I Liang, Ying-Hao Chu, Haimei Zheng, and Ramamoorthy Ramesh, Self-assembled vertical heteroepitaxial nanostructures: From growth to functionalities, *MRS Communications*, 4, (2014), 31–44, https://doi.org/10.1557/mrc.2014.13.

[32] Q. Zhan, R. Yu, S. P. Crane, H. Zheng, C. Kisielowski, and R. Ramesh, Structure and interface chemistry of perovskite-spinel nanocomposite thin films, *Applied Physics Letters*, 89, (2006), 172902, https://doi.org/10.1063/1.2364692.

[33] Li Yan, Yaodong Yang, Zhiguang Wang, Zengping Xing, Jiefang Li, and D. Viehland, Review of magnetoelectric perovskite-spinel self-assembled nano-composite thin films, *Journal of Material Science*, 44, (2009), 5080–5094, https://doi.org/10.1007/s10853-009-3679-1.

[34] Dong Hun Kim, Shuai Ning, and Caroline A. Ross, Self-assembled multiferroic perovskite-spinel nanocomposite thin films: Epitaxial growth, templating and integration on Silicon, *Journal of Materials Chemistry C*, 7, (2019), 9128–9148, https://doi.org/10.1039/C9TC02033K.

[35] Krishnamayee Bhoi, H. S. Mohanty, Ravikant, Md. F. Abdullah, Dhiren K. Pradhan, S. Narendra Babu, A. K. Singh, P. N. Vishwakarma, A. Kumar, R. Thomas, and Dillip K. Pradhan, Unravelling the nature of magneto-electric coupling in room temperature multiferroic particulate $(PbFe_{0.5}Nb_{0.5}O_3)$-$(Co_{0.6}Zn_{0.4}Fe_{1.7}Mn_{0.3}O_4)$ composites, *Scientific Reports*, 11, (2021), 3149, https://doi.org/10.1038/s41598-021-82399-7.

[36] X. Liu, Y. Wang, Y. Wang, Y. Zhao, J. Yu, X. Shan, Y. Tong, X. Lian, X. Wan, L. Wang, P. Tian, and H. Kuo, Recent advances in perovskites-based optoelectronics, *Nanotechnology Reviews*, 11(1), (2022), 3063–3094, https://doi.org/10.1515/ntrev-2022-0494.

14 Synthesis and Characterization of Pure and Doped Titanium Oxide Nanoparticles for Catalysis

Bullapura Matt Santhosh, S. Manjunatha, and Shivarudrappa Honnali Pattanashetty

INTRODUCTION

Nanoscale materials are in the size between 1 and 100 nm. The properties exhibited by nanoparticles (NPs) lie in between its molecular and bulk phases. Several studies have proved that the properties of nanocrystalline semiconductor particles are different from those of bulk. In recent days, nanoscience plays an important role in many applications due to its tunable electrical, optical, and magnetic properties. Pure and doped metal oxide NPs such as ZnO, TiO_2, Fe_2O_3, CeO_2, Mo_2O_3, V_2O_5, Mn_2O_3, Co_2O_3, NiO, CuO, ZrO_2, and other transition and inner transition metal oxides are well studied for various applications. Particularly, TiO_2 NPs are the most studied materials for various applications including catalysis and photocatalysis. TiO_2 is similar to an ideal semiconductor for photocatalysis, but it has poor absorption in the visible region and shows quick recombination of photogenerated electron/hole pairs [1]. TiO_2 was discovered in 1791 from ilmenite ore, and the photocatalytic activity was first observed in 1929. Further studies evidence that TiO_2 exists in three different polymorphs, namely, anatase, rutile, and brookite. Investigations show that TiO_2 is the most efficient photocatalyst.

TiO_2 is a low cost, highly stable, and nontoxic material. The photocatalytic property of TiO_2 has been explored enormously for various applications such as purification of waste water, catalysis and photocatalysis for water purification, degradation of organic molecules, photochemical water splitting reaction for hydrogen generation, and as a catalyst for transformations of organic molecules [2–6]. Nano TiO_2 can be obtained in different sizes and morphology and various shapes in the form of nanowires, nanoribbons, nanotubes, nanosheets, and arrays. These particles can be synthesized through different methods such as sol-gel, hydrothermal, and co-precipitation.

SYNTHETIC METHODS

SYNTHESIS OF TiO_2 NANOPARTICLES

Two important methodologies are followed for the synthesis of metal oxide NPs:

a. Top-down approach
b. Bottom-up approach

DOI: 10.1201/9781003479239-14

Top-Down Approach

A top-down approach is used to transform a bulk material into a nanoproduct. Both physical and chemical methods were used to reduce size. Some of the techniques used in the top-down approach include sputtering, laser pulse ablation, milling/ball milling, etching, and lithography evaporation–condensation reaction. The top-down approach does have several drawbacks, and the most prominent of them is that surface imperfections are imposed on the product. This might have an impact on the product's physical characteristics and surface defects.

Bottom-Up Approach

The bottom-up approach is employed by joining atom by atom, molecule by molecule, and cluster by cluster. This method is used to create the majority of nanostructures that have the capacity to vary in uniformity, size, and morphology. A wide variety of techniques are available for chemical synthesis, including solvothermal/hydrothermal synthesis, polyol condensation, chemical vapor deposition (CVD), sol-gel method, aerosol approach, pyrolysis, electrochemical method, thermal decomposition, and use of metal oxide frameworks. Particularly in green synthesis, the process in bottom-up synthesis is controlled to small particle formation. Therefore, researchers might claim that the bottom-up approach is essential for preparation of nanostructures and nanomaterials.

Hydrothermal/Solvothermal Synthesis

With the reaction taking place in aqueous solutions, the hydrothermal synthesis is often carried out in steel pressure containers called autoclaves with or without Teflon liners under controlled temperature or pressure. When the temperature rises above the water's boiling point, the pressure reaches saturation with vapor. The internal pressure created in the autoclave is mostly influenced by the temperature, nature of the solvent, and volume of solution in the vessel.

The precursors of titanium are treated hydrothermally to obtain precipitates of TiO_2 NPs. The precipitates were obtained by mixing a stoichiometric amount of titanium butoxide solution in isopropanol with deionized water ($[H_2O]/[Ti] = 150$), then heated at 70°C–80°C for 1 hour, and then subjected to a heat treatment at 200°C–300°C for 24–48 hours [7]. TiO_2 nanorods and NPs have both been produced via the hydrothermal technique using $TiCl_4$ solution at 350–450 K for 12 hours.

A $100 \, cm^3$ Teflon-lined autoclave was filled with 2–3 g of P_{25}-TiO_2 white power as part of a standard preparation procedure. The autoclave was then filled with 80 ml of 10 M NaOH aqueous solution, placed inside a stainless-steel vessel, and kept at 130°C for 24 hours. The acquired sample was filtered and repeatedly rinsed with distilled water once the autoclave had naturally cooled to room temperature. The finished products were then collected, washed with dil. HCl solution (pH 1–2) for 24 hours, and repeatedly rinsed with distilled water until the pH value reached 7. The items were finally annealed at 400°C in air for 2 hours [8].

Sol-Gel Method

TiO_2 NPs have been created through the hydrolysis of a titanium precursor using the sol-gel technique. Typically, this procedure uses titanium (IV) hydrolysis process. The low-temperature growth of Ti-O-Ti chains is advantageous. 3D, closely packed polymeric skeletons are formed due to Ti-O-Ti chain formation. The $Ti(OH)_4$ formation is promoted by high rates of hydrolysis for a required volume of water. The existence of a significant Ti-OH concentration and inadequate three-dimensional growth of particles are loosely packed as a result of polymeric skeletons formed in the presence of a large excess of water [9].

Chemical Vapor Deposition

Any process in which vaporized components are condensed to form solid-phase materials is referred to as vapor deposition. These procedures are typically employed to produce coatings that change the mechanical, electrical, thermal, optical, corrosion-resistance, and wear-resistance characteristics of various substrates. Thermal energy is used in CVD techniques to heat the gases in the coating chamber and drive the deposition reaction. The CVD process factors that can be manipulated to produce the desired size of the materials include gas composition, flow rate, pressure, deposition temperature, deposition chamber geometry. Nanocrystalline TiO_2 films by CVD on various substrates using $TiCl_4$ as a precursor at a relatively low temperature and found that, the type of substrate influences on the size and distribution of nanograins in the films [10,11].

Sonochemical Method

A variety of nanostructured materials, including high surface area of transition metals oxides, alloys, and colloids, are produced enormously from the use of ultrasound. Ultrasound does not directly interact with molecular entities to produce its chemical effects. Instead, acoustic cavitation, the growth, expansion, and impulsive collapse of bubbles in a liquid, is the source of sonochemistry [12].

TiO_2 NPs are typically prepared by dissolving TiO_2 pellets in NaOH solution and stirring vigorously for 2–3 hours at room temperature. The yellowish solution was then exposed to ultrasonic radiation for 2 hours at room temperature. The resulting precipitates were centrifuged, several times washed, decanted, and dried with deionized water at 60°C for 24 hours [12].

Microwave Method

High-frequency electromagnetic waves have the ability to process dielectric materials. The main microwave heating frequencies range from 900 to 2,450 MHz. At lower microwave frequencies, energy can be transferred to the material by conductive currents that are flowing because of the movement of ionic components. The energy absorption at higher frequencies is mostly caused by molecules with a permanent dipole that tends to reorient when exposed to a microwave electric field.

Different TiO_2 nanostructures are generated using microwave radiation. For instance, using a solution of $TiOCl_2$ and both a conventional and a microwave thermal treatment, it was possible to synthesize TiO_2 by forced hydrolysis under hydrothermal conditions. The microwave digesting system was used to carry out the microwave-assisted syntheses. Temperature and pressure ($P_{max} = 14.1$ atm) are used to regulate the system, which employs microwaves operating at 2.45 GHz. Microwave hydrothermal treatments are carried out at 195°C for varying lengths of time, ranging from 5 minutes to 1 hour [13]. TiO_2 nanotubes were produced using microwave radiation by reacting anatase, rutile, or mixed-phase TiO_2 crystals with aqueous NaOH solutions at a certain microwave power output.

CHARACTERIZATION OF TiO_2 NANOPARTICLES

Nanomaterials after synthesis through chemical, or physical techniques, high-tech equipments to be used for their physico-chemical characterization. The characterization techniques are greatly support for their further technological applications.

SURFACE MORPHOLOGY AND SHAPE OF TiO_2 NANOPARTICLES

The SEM, TEM, FE-SEM, and EDAX methods for structure and surface analysis can be used to characterize the TiO_2 NPs synthesized using different methods. Various results confirmed the spherical shape and agglomerate form of the TiO_2 NPs. TiO_2 NPs have been observed to typically be formed by clusters that form the bunch-type surface and have a spherical shape with small pores, according to SEM examination. Anatase, brookite, and rutile are smaller, tetragonal crystals than their bulk materials like anatase and brookite, respectively. TiO_2 NPs have relatively large crystals with porous structures that exhibit unique and distinctive surface morphology. TiO_2 nanotubes have significant uses in the environmental sector because of their porous characteristics.

A number of investigations have shown that TiO_2 NPs may or may not contain companion atoms in the final product. For instance, in the green synthesis of TiO_2 NPs, Ti and O atom percentages were 40% and 60%, respectively, while Fe atom percentage was <1%, and Ca and K peaks were only faintly visible, which showed that TiO_2 NPs included trace amounts of various contaminants.

XRD ANALYSIS TiO_2 NANOPARTICLES

XRD patterns were frequently used to characterize the crystallinity and crystallite size of the produced material.

For the purpose of comparing the X-ray diffraction pattern with the pattern of the substance-specified plane angles, the structural data were gathered. The collected data are compared to the Joint Committee on Powder Diffraction Standards to confirm crystallinity. The feature facets of the tetragonal anatase phase of the crystallite with space group I41/amd and lattice parameters of a=b=3.792, c=9.554 are the standards for TiO_2 NPs. TiO_2 NPs display an intense peak at the 2 value, especially at 25.6, which corresponds to the anatase (101) plane, between 24° and 28°. Planes at (101), (004), (200), and (211) correspond to anatase. Crystalline planes of (110), (101), (111), (211), and (220) correspond to the rutile phase of TiO_2 [14].

PHOTOCATALYTIC ACTIVITY OF TiO_2 NANOPARTICLES

TiO_2 is considered a semiconductor particle in which the conduction band is separated from the valence band and has an empty energy level. The valence band has a filled energy level and is filled with electrons. An electron from the valence band enters a vacant hole in the conduction band. TiO_2 NPs participate in photocatalysis reactions due to the movement of electrons from the valence band to the conduction band. TiO_2 is a semiconductor, and therefore, photons with enough energy will cause electron–hole pairs to form. When ultra-violet (UV) light was absorbed by the TiO_2 NPs, the electrons in the valence band transitioned to the conduction band and filled the holes. Reactive oxygen species (ROS), hydroxyl radicals, and holes are produced when oxygen and water in the environment interact with each other. Hydrogen peroxide, singlet oxygen, and photocatalytic oxidation of NPs are produced in addition to hydroxyl and superoxide radicals. All of these radicals are known to be incredibly reactive and, when in touch with organic substances, can rapidly degrade them. Many hazardous pollutants, such as fatal dyes and nitroarene compounds, are now present in domestic and industrial wastes, which affect the environment and contaminate water sources. Due to their limited solubility and high stability, hazardous dyes and other noxious compounds are pervasive and dangerous to aquatic life. Metallic NPs with a unique structure and recently synthesized properties were developed. Additionally, the large surface area of these metallic NPs makes them effective heterogeneous catalysts.

Photocatalytic mechanism of TiO_2 can be described as an electron will move from the valence band to the conduction band when the TiO_2 semiconductor is irradiated with more energy than the band gap, resulting in holes in the valence band and electrons in the conduction band. While the holes interact with the nearby water to form hydroxyl radicals, the electrons interact with the surrounding oxygen to produce superoxide radicals. Both of these free radicals will react with nearby substances present in water, particularly organic substances or polymers. The organic substance or polymer will be broken down by this hydroxyl radical into water and carbon dioxide. The superoxide radicals will react with the water to form hydrogen peroxide, which will next react with electrons to produce hydroxyl radicals. Finally, the hydrogen peroxide will interact with light irradiated to produce another hydroxyl radical. The organic polymer (R) will be broken down by this hydroxyl radical into water and carbon dioxide [15].

Metallic NPs with a unique structure and properties were developed. Additionally, the large surface area of these metallic NPs makes them effective heterogeneous catalysts. The ability to quickly recover and recycle the nanostructured catalysts together with the reaction mixture is another benefit. Important factors include the toxicity and aggregation of the NPs. TiO_2 NPs have mostly been used in catalysis due to their excellent stability, low toxicity, optical characteristics, and photocatalytic potential. According to a number of studies, green-mediated TiO_2 NPs can be used to photocatalytically degrade a variety of dyes and chemicals. The composites showed increased photocatalytic activity, maximum photocatalytic activity, and degradation of the rhodamine B dye. Green-mediated NPs outperformed chemically prepared TiO_2 NPs when their photocatalytic potential was compared to that of chemically produced TiO_2 NPs. The ability to decrease is influenced by the temperature, the type of dye employed, and the phytochemicals found in various plant species. Other metallic NPs were added to TiO_2 to increase their catalytic capability.

PHOTOCATALYTIC ACTIVITY OF DOPED TiO_2 NANOPARTICLES

Metal ion or non-metal dopants can be deposited into TiO_2 to enhance the photocatalyst's performance. In order to get over the limitations of nano TiO_2 such as wideband gap, poor photocatalysis in direct sunlight, and thermal instability, doping approaches have been used in photocatalysis. Most dopants have the ability to boost nano-doped TiO_2 photocatalytic effectiveness. Dopants can change the electronic structure of nano-TiO_2 to increase the effective range of photon sensitivity of photocatalyst from the UV region to the visible light region. Physicochemical characteristics such as high crystallinity (high proportion of anatase phase), high specific surface area, and small crystallite size are some of the good physicochemical qualities that dopants can impart. Dopants can generate a charge space carrier area on the surface of TiO_2 that prevents the recombination of photogenerated electron–hole pairs, accelerating the production of hydroxyl radicals and so increasing the rate of photocatalytic reaction. Additionally, dopants can speed up photodegradation and serve as an active site for the adsorption of contaminants or pollutants.

Various parameters that doping requires to accomplish excellent photocatalytic activity of NPs are doping concentration and dosage of doped TiO_2 NPs. Apart from these, the activity of catalyst depends on the calcination temperature and initial reactant concentration. Different metal- and non-metal-doped TiO_2 NPs give a good photocatalytic activity.

PHOTOCATALYTIC ACTIVITY OF METAL-DOPED TiO_2 NPS

Transition metal–doped TiO_2 NPs are the promising photocatalysts, especially metals having variable valency. Metal atoms such as Fe, Zn, Cu, Pd, Sn, Ni, Mn, Co, Ag, and Au are the most studied dopants for TiO_2.

For instance, if Fe is a dopant, the electron–hole pairs are trapped by iron, which prevents their recombination. On the other hand, Ti^{4+} ionic radius (0.75 A°) and Fe^{3+} is (0.79 A°) are comparable. This property makes it easier for Fe^{3+} ions to bind to the TiO_2 crystal lattice. Fe^{3+}-doped TiO_2, which forms nanocrystalline particles with high surface areas and ensures good photocatalytic efficiency with a band gap of 2.6 eV and Fe, was found to prevent the agglomeration of the particles [16]. The performance of the synthesized TiO_2 films for photocatalysis and their wetting characteristics were examined as a result of the co-addition of Zn^{2+} and sodium dodecyl benzene sulfonate. It was noted that the addition of Zn^{2+} enhanced both the photocatalytic activity and the hydrophilicity, which can be attributed to surface oxygen defects.

According to a paper, under UV and visible light irradiation, the sulfated Mo-doped TiO_2@ fumed SiO_2 composite exhibits good photocatalytic activity for the degradation of methyl orange (MO).

Malathion was employed as a model compound to reveal the breakdown of an organophosphorus insecticide by the solar photocatalytic activity of WO_3/TiO_2 photocatalysts. The improved photocatalytic activity and long-term stability of TiO_2 NPs were seen with WO_3 doping. For the photocatalytic degradation of formic acid, La-doped nanorods and nanotubes that were calcined at 900 K have been employed. La was found to impede the development of TiO_2 crystallites and increases the lifetime of photogenerated electron–hole pairs.

PHOTOCATALYTIC ACTIVITY OF NON-METAL-DOPED TiO_2 NANOPARTICLES

Degradation of MB aqueous solution under the illumination of visible light, the photocatalytic activity of nitrogen-doped TiO_2 NPs was examined. It was found that the nitrogen-doped TiO_2 wavelength range was moved toward the visible light.

Ionic radius of nitrogen (1.71) is much larger than that of oxygen (1.4), and it is challenging to replace O with N. Therefore, two nitrogen atoms should be added in place of three oxygen atoms in order to maintain electroneutrality and create an oxygen vacancy. Nitrogen appears to facilitate the production of oxygen vacancies as evidenced by the energy of oxygen vacancies in titania decreasing from 4.2 to 0.6 eV in the presence of nitrogen. The oxygen vacancies ensure that N-doped TiO_2 is activated and promote absorption in the visible range (400 and 600 nm). The hydrothermal and post-impregnation techniques were used to produce Cr- and N-co-doped TiO_2 nanotubes in two steps. It has been observed that there has been an increase in photocatalytic activity for the breakdown of MO up to 97.16%.

To ensure the photocatalytic activity of nanocrystalline S, N-co-doped TiO_2 thin films, and powders under visible and solar light irradiation, degradation of MO was employed. It was discovered that MO solution totally changed color in 75 minutes when modified TiO_2 powder was present. According to certain studies, when coupled with graphene oxide (GO), GO-TiO_2 exhibited significantly increased photocatalytic activity for the decomposition of Acid Orange 7 when exposed to UV radiation.

CONCLUSION

In recent years, water pollution has received significant attention. For the treatment of wastewater, a variety of approaches and techniques have been used, but advanced oxidation processes (AOPs) have emerged as the most successful and promising option. Compared to other approaches, they offer a number of benefits. This series examines the photocatalytic activity of nanomaterials. They have a greater amount of available reactive surface area, are less expensive, and are resistant to corrosion in the presence of water and other chemicals. These NPs can be used as efficient photocatalysts for the degradation of a range of organic pollutants that pose a serious threat to the environment. By adding certain metal and non-metal ions as dopants, their photocatalytic activity can be enhanced significantly.

REFERENCES

[1] Jenny Schneider, Masaya Matsuoka, Masato Takeuchi, Jinlong Zhang, Yu Horiuchi, Masakazu Anpo, Detlef W. Bahnemann, Understanding TiO_2 Photocatalysis: Mechanisms and Materials, *Chemical Reviews*, 2014, 114(19), 9919–9986.

[2] Akira Fujishima, Kenichi Honda, Electrochemical Photolysis of Water at a Semiconductor Electrode, *Nature*, 1972, 238, 37–38.

[3] Takahiro Kaida, Kota Kobayashi, Maoya Adachi, Fukuji Suzuki, Optical Characteristics of Titanium Oxide Interference Film and the Film Laminated with Oxides and Their Applications for Cosmetics, *Journal of Cosmetic Science*, 2004, 55(2), 219–220.

[4] Sinisa Ivankovic, Marijan Gotic, Mislav Jurin, Svetozar Music, Photokilling Squamous Carcinoma Cells SCCVII with Ultrafine Particles of Selected Metal Oxides, *Journal of Sol-Gel Science and Technology*, 2003, 27, 225–233.

[5] Michael Gratzel, Sol-Gel Processed TiO_2 Films for Photovoltaic Applications, *Journal of Sol-Gel Science and Technology*, 2001, 22, 7–13.

[6] Michael Gratzel, Dye-Sensitized Solar Cells, *Journal of Photochemistry and Photobiology C: Photochemistry Reviews*, 2003, 4(2), 145–153.

[7] Howraa Ghassan Hameed, Nadia A. Abdulrahman, Synthesis of TiO_2 Nanoparticles by Hydrothermal Method and Characterization of their Antibacterial Activity: Investigation of the Impact of Magnetism on the Photocatalytic Properties of the Nanoparticles, *Physical Chemistry Research*, 2023, 11(4), 771–782.

[8] P. Maheswari, S. Ponnusamy, S. Harish, M. R. Ganesh, Y. Hayakawa, Hydrothermal Synthesis of Pure and Bio Modified TiO_2: Characterization, Evaluation of Antibacterial Activity against Gram Positive and Gram-negative Bacteria and Anticancer Activity against KB Oral Cancer Cell Line, *Arabian Journal of Chemistry*, 2020, 13(1), 3484–3497.

[9] Mohamad M. Ahmad, Shehla Mushtaq, Hassan S. Al Qahtani, A. Sedky, Mir Waqas Alam, Investigation of TiO_2 Nanoparticles Synthesized by Sol-Gel Method for Effectual Photodegradation, *Oxidation and Reduction Reaction, Crystals*, 2021, 11(12), 1456.

[10] Jie Wu, Guo-Ren Bai, Jeffrey A. Eastman, Guangwen Zhou, Vijay K. Vasudevan, Synthesis of TiO_2 Nanoparticles Using Chemical Vapor Condensation, *MRS Online Proceedings Library*, 2005, 879, Article number: 712.

[11] Abdullah M. Alotaibi, Sanjayan Sathasivam, Benjamin A. D. Williamson, Andreas Kafizas, Carlos Sotelo-Vazquez, Alaric Taylor, David O. Scanlon, Ivan P. Parkin, Chemical Vapor Deposition of Photocatalytically Active Pure Brookite TiO_2 Thin Films, *Chemistry of Materials*, 2018, 30(4), 1353–1361.

[12] Jingjing Guo, Shenmin Zhu, Zhixin Chen, Yao Li, Ziyong Yu, Qinglei Liu, Jingbo Li, Chuanliang Feng, Di Zhang, Sonochemical Synthesis of TiO_2 Nanoparticles on Graphene for Use as Photocatalyst, *Ultrasonics Sonochemistry*, 2011, 18(5), 1082–1090.

[13] A. K. Singh, Umesh T. Nakate, Photocatalytic Properties of Microwave-Synthesized TiO_2 and ZnO Nanoparticles Using Malachite Green Dye, *Journal of Nanoparticles*, 2013, Article number: 310809.

[14] P. K. Singh, Soumya Mukherjee, C. K. Ghosh, S. Maitra, Influence of Precursor Type on Structural, Morphological, Dielectric and Magnetic Properties of TiO_2 Nanoparticles, *Cerâmica*, 2017, 63, 549–556.

[15] Nasikhudin, M. Diantoro, A. Kusumaatmaja, K. Triyana, Study on Photocatalytic Properties of TiO_2 Nanoparticle in Various pH Condition, *Journal of Physics: Conference Series*, 2018, vol. 1011, 012069, IOP Publishing.

[16] Md. Kamrul Hossain, Md. Mufazzal Hossain, Shamim Akhtar, Studies on Synthesis, Characterization, and Photo-catalytic Activity of TiO_2 and Cr-Doped TiO_2 for the Degradation of *p*-Chlorophenol, *Omega*, 2023, 8, 1979–1988.

15 Lanthanum-Based Compounds Promoted by Carbon Substrates for Catalytic Applications

I. Prabha, J. Hemalatha, C. Senthamil, J.J. Umashankar, K. Preethi, and S. Nandhabala

INTRODUCTION

In 2015, the United Nations put forward 17 Sustainable Development Goals (SDGs) to protect the planet by contributing to green energy and practicing environmental remediation. Research is ongoing all over the world in the search of efficient materials that can perform the energy generation reaction as well as in retreating the environment. Carbon is one such material with a lot of potential and can be used in a wide range of applications. The advantage of carbon materials is that they are very flexible and can be modified into wearable electronics with improved comfort for encouraging human activity by maintaining sustainability. With these advantages, carbon and its functionalized forms can even act as sensors to measure temperature, pressure, humidity, and health-related ailments. Carbon-based materials are well known for their electrochemical behaviors attributed to their accessible large surface areas [1]. Recent advances are made by functionalizing the carbon materials using other active materials, especially metals and metal-based advanced structures. Furthermore, increasing interest in using lanthanum (the second largest available rare earth metal) with carbon materials is particularly studied for its exceptional characteristics. In lanthanum, the electrons in the 4f orbital facilitate electron transition so that the conductivity of the metal is increased which in turn increases the catalytic performances [2]. For instance, theophylline is a drug used to treat chronic obstructive pulmonary disease which can be detected efficiently using lanthanum and carbon nanofiber (LV@CNF) nanocomposites [3]. It is also reported that lanthanum can also be used to eliminate inorganic heavy metal pollutants from water through adsorption. The attribution of its performance is highly related to the surface morphology and the active surface functional groups. Here, in this chapter, we discuss about different carbon-based materials, lanthanum composites supported by carbon substrates, and their role in photo-and electrocatalytic applications such as catalytic reduction and degradation.

TYPES OF CARBON MATERIALS

Due to its unique features such as high surface area, good electron mobility, and flexible surface chemistry, carbon materials are used in photocatalysis. These characteristics make carbon materials useful platforms for powering numerous photocatalytic reactions. Among the various types of carbon materials, carbon nitride facilitates water splitting and pollutant degradation. In photocatalytic devices, carbon nanotubes improved pollutant dissolution and electron transport. In the process of purifying water and air, activated carbon expands its surface area and helps remove pollutants through adsorption. Fullerene not only helps increase the charge separation, but it also aids in accelerating the light-driven reactions in photocatalysis. The various sources of this carbon

DOI: 10.1201/9781003479239-15

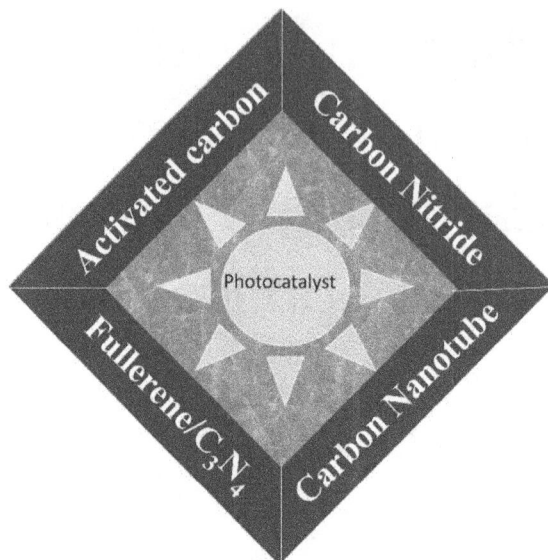

FIGURE 15.1   Functionalized carbon materials.

material and its characteristics are given in the following. Figure 15.1 shows different functionalized carbon materials.

ACTIVATED CARBON

Activated carbon (AC) is a well-known adsorbent used in various applications such as purification of water resources and removal of contaminants from the air. The main advantage of AC is its easy preparation from carbonaceous sources. There are notable shreds of evidence that the AC prepared by burning coconuts, peat, and wood are used for the removal of pollutants from water. One of the most crucial components for cleaning up organic and inorganic harmful chemicals from industrial wastewater is AC. This is made from agricultural waste that contains a lot of carbon. For instance, the *Manikara zapota* peel could be used effectively as a raw material for the manufacture of AC using the KOH chemical activation method. According to XRD analysis, MnO_2 and phosphorous-doped $MnO_2(P-MnO_2/AC)$ exhibited an excellent tetragonal structure and are crystalline in contrast to the *M. zapota* peel AC, which consists of an amorphous as well as graphitic carbon phase. The CN stretching vibrations mode produced the unique carbon peak in the FT-IR spectrum at $1,560 cm^{-1}$, the phosphorus peak at $1,040 cm^{-1}$ (which is produced by phosphorus-oxy compounds found in aliphatic phosphones), and the main amine CN stretching modes as well as vibrations with $P-MnO^2/AC$ structure. With P-ions doped into the MnO_2/AC structure, the peaks are traveling in the direction of lower wave number. UV-vis analysis demonstrated that MnO_2 and $P-MnO_2/AC$ nanorods display exceptional catalytic degradation capability for methyl orange and Rhodamine B dyes in the presence of H^+ion. According to the result of the characterization, building a MnO_2 and $P-MnO_2/AC$ heterojunction using AC may improve optical absorption and charge transfer properties of MnO_2, which would lead to improved photocatalytic activity for the photodegradation of methylene blue and Rhodamine B dyes [4].

BAMBOO CHARCOAL (BC)/C3N4

Bamboo charcoal was made by calcinating the mature bamboo plants refuse between 600°C and 1,200°C. Bamboo charcoals are known to have a very large surface area which is many times greater than normal charcoal, and in turn, its adsorptive power is four times greater than charcoal. When the BC is composed with other materials like BC/g-C_3N_4 composite, it can act as a photocatalyst. This is due to a Schottky barrier which typically forms at the interface between a photocatalyst and a co-catalyst, which is eliminated at the BC/g-C_3N_4 interface thanks to a greater Femi level of BC than g-C_3N_4. The addition of BC speeds electron transfers and increases charge carrier separation, which dramatically raises photocatalytic activity of g-C_3N_4. By utilizing natural resources and eco-friendly materials for photocatalytic H_2 production, the design of the composite opens up new possibilities. Other semiconductor photocatalysts can also be modified using the BC loading method [5]. In comparison to pristine g-C_3N_4 and other modified g-C_3N_4, the H_2 evolution in 0.2% BC/g-C_3N_4 is approximately 2.3 times higher. Compared to pure BC and g-C_3N_4, the BET surface area of 0.2% BC/g-C_3N_4 was significantly higher. According to the BJH pore size distribution, the average pore diameter for 0.2% BC/g-C_3N_4 was 12.1 nm and for g-C_3N_4 was 16.7 nm. In general, 0.2% BC/g-C_3N_4 can offer more active sites as well as reaction space for photocatalysis due to its higher surface area and greater pore volume. The addition of BC to g-C_3N_4 improves the conductivity of catalysts and preventing the recombination of charge carriers produced by photosynthesis.

Biopolymer/CNT

Because of their biodegradability, non-toxicity, and reusable nature, biopolymer-based catalysts are viable materials for environmental treatment. These catalysts are often made from organic functional groups, which include metal ions, enzymes, or chitosan ions added to natural polymers like cellulose, starch, and chitosan to create catalytic properties. Application areas for biopolymer-based catalysts in environmental treatment include soil remediation, wastewater treatment, and air pollution control [6]. For instance, the multi-walled carbon nanotubes (MWCNT) are made into a membrane with chitosan biopolymer. The engineered MWCNTs/chitosan nanofibrous membrane possessed the ability to adsorb 60.34 mg/g of Cu(II), 86.18% of methylene blue, and 83.60% of methyl orange. Chitosan/magnetic MWCNT can remove >322 mg/g of nitrate from the water resources. Chitosan/MWCNT/$CoFe_2$-NH_2 removed 42.48 mg/g of TBBPA (tetrabromobisphenol A). Chitosan/CNT removed 222.85 mg/g of diazinon. Chitosan/MCNT removed 36.10 mg/g of phosphate. Chitosan-coated MWCNT and magnetic Chitosan@MWCNT removed 205 mg/g and 256.4 mg/g of acetaminophen, respectively. Chitosan hydrogel scaffold modified with CNT removed 404.2 mg/g of phenol, and chitosan/carboxylated CNT composite removed 307.5% of U(VI). A thermally stable nanocomposite consisting of MWCNTs and chitosan (Cs) (25:75 wt.%) was described by Abdel Salam et al. and utilized to adsorb Cu(II), NI(II), Zn(II), and Cd(II) ions from aqueous solutions by packing within a glass column. The experimental results for the removal percentages of these ions were 100%, 100%, 99%, and 98%, respectively.

Encapsulating Fullerenes into C_3N_4

Several C_3N_4 and C60 composites that resemble graphite were created using an in situ generation technique. Simply changing the weight ratios between the C60 and the graphite-like C_3N_4 allowed the efficiency charge separation and photocatalysis for the resulting composites to be modified. It was discovered that when comparing graphite-like C_3N_4 and the C60, the C_3N_4/C60@6:1 composite exhibited the strongest built-in electric field, realizing an effective charge separation and providing improved photocatalytic efficiency for formaldehyde degradation. Overall, this study has provided insightful information about designing and creating unique synergistic systems for real-world uses, notably in the area of photocatalysis. While the two strong bands at 1,243 and 1,398 cm^{-1} were attributed to C-N stretching, the bands positioned at 1,610 cm^{-1} were assigned to C=N stretching vibration. It was a weak interaction among C60 and graphite-like C_3N_4, as evidenced by the shifting of the prominent peaks of graphite-like C_3N_4 within the C60/graphite-like C_3N_4 mixture, which ranged from 1,200 to 1,700 cm^{-1}. Compared to the graphite-like C_3N_4, this may have facilitated electron transport and enhanced the photocatalytic activity of the composites. The binding energies of N (N-C=N) in graphite-like carbon nitride and C_3N_4/C60 shifted from 400.5 to 399.8 eV according to a comparison of the N 1s XPS spectrum of the graphite-like C_3N_4 and that of the graphite-like C_3N_4/C60. Given that graphite-like carbon nitride and C_3N_4/C60 had BET surface areas of 40.0 and 29.6 m^2/g, respectively, this change implies that the graphite-like C_3N_4 and the C60 interacted. Although significantly less than that of graphite-like C_3N_4(0.093 cm^3/g), the pore volume of the graphite-like C_3N_4/C60@6:1 compound was nevertheless quite high. The C60 nanoparticles may have been easily absorbed into the graphite-like C_3N_4 nanosheets due to small diameter, producing a well-dispersed composite. After the addition of C60, the graphite-like C_3N_4/C60@6:1 composite had clear red-shift edges indicating that it could widen the spectrum to the visible range in comparison to the bare graphite-like C_3N_4. The bandgap values of the graphite-like C_3N_4/C60@6:1 and the graphite-like C_3N_4 were calculated to be 2.62 and 2.85 eV, respectively, using the equation $Eg = 1,240/\lambda$. The graphite-like C_3N_4/C60@6:1 has a lower bandgap in which it can be excited more readily to produce free holes and electrons. The photocatalytic effectiveness increased from 40.2% to 96.4% with an increase in the loading of C60 on the graphite-like C_3N_4, showing that the integration of C60 can significantly improve graphite-like C_3N_4 during photocatalytic degradation of formaldehyde [7].

LANTHANUM COMPOUNDS SUPPORTED BY CARBON SUBSTRATE

One of the most notable research accomplishments is thought to be the development of engineered nanomaterials. A significant class of designed nanomaterials utilized in agriculture, industry, and medicine are metal oxide nanoparticles (NPs) [8]. Lanthanum nanoparticles (La$_2$ NPs), which are naturally occurring as a rare earth metal oxide and possess a spherical structure, are the basis for a wide variety of potential applications [9]. For instance, Weiwei Gu et al. synthesized $La_2O_2CO_3$-encapsulated La_2O_3 nanoparticles on carbon ($La_2O_2CO_3$@La_2O_3/C) via chemical precipitation in an aqueous solution containing various concentrations of CTAB, followed by calcination at 750°C. The obtained lanthanum compound nanoparticles on carbon strongly depend on the CTAB concentrations in terms of particle size, loading, and dispersion. The minimum particle size (7.1 nm) and maximum actual loading (44.5 wt.%) of the nanoparticles were attained at the CTAB concentration of 24.8 mmol/L. The synthesized hybrid catalyst displayed a noticeably increased oxygen reduction reaction (ORR) electrocatalytic activity that is comparable to commercial Pt/C (20 wt.%) when compared to pure La_2O_3, $La_2O_2CO_3$, and carbon support. Additionally, the hybrid catalyst outperforms Pt/C catalysts in terms of durability over the long run. The improved ORR electrocatalytic activity is considerably associated with $La_2O_2CO_3$ layers stuffed at the interface between carbon and La_2O_3, where the $La_2O_2CO_3$ chemically links the carbon support through –C–O–C(=O)–O–La–O– bonds. Four steps are followed in the process of oxygen reduction, which is catalyzed by the hybrid catalyst, and they are as follows: (i) surface-adsorbed hydroxide (–OH$^-_{ad}$) substitution, (ii) surface peroxide formation (–OOH$^-$) formation, (iii) surface oxide (–O$^-$) formation, and (iv) surface-adsorbed hydroxide generation. The presence of the specific covalent –C–O–C(=O)–O–La–O– bonds in hybrid catalysts would favorably influence the surface-adsorbed hydroxide (OH ad) desorption process and promote active oxygen adsorption, which significantly boosts ORR performance. The development of lanthanum compound–carbon hybrid as the next generation for ORR electrocatalysts benefited from a precise understanding of the ORR process catalyzed by hybrid catalysts [10].

Additionally, Saikat Kumar Kulia et al. worked on La^{3+} ions–decorated 2D-g-C_3N_4 using an easy and scalable chemisorption technique. To reach the maximal adsorption capacity (657.32 mg/g), process variables like temperature, pH, and duration are tuned. The interaction of La^{3+} ions with 2D-g-C_3N_4 has been confirmed by several structural, optical, and morphological observations. Defects brought on by the adsorption of La^{3+} are further supported by low bandgap energy like 2.21 eV and a greater I_D/I_G ratio (0.82). The HR-TEM images revealed the lateral size to be ~ 1 μm, and SAED pattern showed the obtained graphitic carbon nitride crystalline plane experimentally. AFM images revealed the average thickness of the exfoliated 2D sheet estimated to be ~ 6 nm before La^{3+} adsorption, but after La^{3+} adsorption, thickness increases up to ~ 8 nm. The fabricated material was tested for the photocatalytic degradation of ciprofloxacin, which shows 93% efficiency at 1g/L of catalyst dose. The photodegradation efficiency and reaction rate were higher for La^{3+}-2D-g-C_3N_4 than 2D-g-C_3N_4 (~11% and ~32.3%, respectively). Therefore, the scavenging investigation demonstrated that O_2 and OH play substantial roles in the photodegradation of ciprofloxacin. The stability of the recycled La^{3+}-2D-g-C_3N_4 has been confirmed by FT-IR and UV-Vin absorption spectra. Therefore, in the realm of pollutant removal, the now-produced nanocatalyst has exploited for environmental remediation of the developing antibiotics present in water [11].

Using a modified GCE electrode with GQDs@La^{3+}@ZrO_2, Thota Trinadh et al. have created a unique electrochemical sensor for the detection of flutamide. To create GQDs, a simple bottom-up method has been used. GQDs were created by pyrolyzing citric acid in a single step and then dispersing the carbonized byproducts in an alkaline solution. By combining precursors such as zirconium chloride and lanthanum nitrate, the compound La^{3+}@ZrO_2 was created. GQDs@La^{3+}@ZrO_2 nanocomposite was synthesized by dispersing GQDs and La^{3+}@ZrO_2in distilled water and sonicating for 30 minutes for better suspension and drop cast in the pre-casted surface of GCE. To increase electrode sensitivity and electrochemical signals, the aforementioned nanocomposite was organized, described, and used. The techniques such as XRD, FT-IR, UV, FE-SEM, and XPS supported the synthesis of GQDs@La^{3+}@ZrO_2. Moreover, the synergistic effects of GQDs@La^{3+}@ZrO_2 nanostructures led to the good electrochemical activities of sensors toward flutamide. The suggested sensor demonstrated an excellent trace-level LOD, wide linear range, repeatability, selectivity, and accuracy. The fabrication method is also very quick and affordable. The findings suggested that flutamide in biological samples may be detectable using the suggested sensor [12].

La_2O_3 nanoparticles were evenly dispersed across the surface of the chitosan matrix to form a hybrid nanocomposite film to perform admirably as a heterogeneous for the production of pyridines and pyrazoles using triethylamine as a traditional catalyst. The chitosan-La_2O_3 nanocomposite was found to be a more efficient catalyst in these processes than triethylamine, in addition to having a better influence on the environment. According to the catalytic experiment, the higher catalytic potency of chitosan-La_2O_3 nanocomposite when compared to individual components was mostly due to the synergistic impact caused by the combination of the basic character of both La_2O_3 nanoparticles and chitosan itself. The nanocatalyst coating had a positive environmental impact and could be easily removed, restored, and reused without losing its catalytic activity. Last but not least, more research into the chitosan-metal oxide hybrid nanocomposite is required in a number of organic processes [13].

SYNTHESIS METHODS OF CARBON-BASED NANOMATERIALS

Sirajudheen et al. followed the simple method for the synthesis of chitosan-lanthanum-graphite (CS-La-GR) composite. In a 2% acetic acid solution, 2 g of chitosan was completely dissolved with dropwise addition of 5% (W/V) $LaCl_3$, and it was agitated for 2.5 hours at 50°C. 0.5 g of graphite powder was slowly added to the above hot solution and stirred once again for 3 hours at 50°C. To achieve the complete formation of the product as CS-La-GR, 2% NaOH was added drop by drop to the reaction mixture. The prepared CS-La-GR composite was washed with deionized water and dried at 30°C. Hereafter, the material was cross-linked with glutaraldehyde and dried at 30°C where glutaraldehyde was used as a cross-linking agent [14]. Similarly, Bao Liu et al. described the non-cross-linked lanthanum-chitosan (La-CTS-0X) and cross-linked lanthanum-chitosan (La-CTS-1X/2X) nanocomposites via the simple co-precipitation method [15]. Bin Wang et al. synthesized the lanthanum cross-linked polyvinyl alcohol/alginate/palygorskite (LPAP) composite hydrogel [16]. Fei Yu et al. introduced lanthanum-modified κ-carrageenan/sodium alginate aerogels

(La/κ-car/SA) via the sol-gel method [17]. Table 4.1 describes the various synthesis methods of carbon-based substrates supported by lanthanum nanocomposites.

Gaurav Sharma et al. synthesized the oxidized graphite/lanthanum oxide/zirconium oxide nanocomposite (OG/La$_2$O$_3$/ZrO$_2$) using the simple sonication method. For obtaining OG/La$_2$O$_3$/ZrO$_2$, first La$_2$O$_3$ was synthesized by following the modified Hummers method. Subsequently, La$_2$O$_3$ was converted into OG/La$_2$O$_3$ using the simple precipitation method. Then, 30 mg of OG/La$_2$O$_3$ was added to 40 mL of 0.2 M zirconium oxychloride solution under constant stirring for 2 hours followed by ultrasonicating for 15 minutes. Afterward, 30 mL of NH$_4$OH was added drop by drop to the reaction mixture to maintain the pH of 9.0 and stirred for 50 minutes. Then it was filtered and washed with deionized water followed by drying at 50°C [18]. Vinuth Raj et al. prepared the RGO-supported lanthanum aluminate (RGO-LaAlO$_3$). RGO-LaAlO$_3$ nanocomposite was synthesized in three stages. In the first stage, LaAlO$_3$ was synthesized via a gel route and low-temperature pathway. By using modified Hummers process, graphene oxide (GO) was synthesized from natural graphite flakes. Then, it was converted to reduced GO with green tea extract (GTE) using the membrane method. In order to prepare RGO-LaAlO$_3$ nanocomposite, 20 mg of RGO was dissolved in 20 mL of ethanol followed by sonication for 15 minutes. In the above-mentioned solution, 0.1 g of LaAlO$_3$ powder was added with continuous stirring at room temperature for an additional 1 hour. This reaction mixture was transferred into the 100 mL hydrothermal autoclave for 5 hours at 120°C. The obtained product was washed with double-distilled water followed by ethanol using centrifugation and dried in a vacuum oven [19].

Ruochong Zhang et al. introduced the soft-nanocomposites (HSA-LaF$_3$/OA) in three stages via the gelation method. First, the LaF$_3$ nanoparticles were synthesized by the co-precipitation method. In the second stage, the oleic acid–supported LaF$_3$ (LaF$_3$-OA) was prepared by co-precipitation process like LaF$_3$ nanoparticles preparation. Finally, in the third stage, the soft-nanocomposites such as 12-hydroxystearic acid–linked LaF$_3$/OA (HSA-LaF$_3$/OA-U) were prepared by the simple sonication method. In a 10 mL sample bottle of liquid parafilm (LP) base oil, the gelator (12-HSA) and LaF$_3$ nanoparticles were introduced in the concentration of 1 wt.%. For the synthesis of LaF$_3$/12-HSA soft-nanocomposite, three stages were followed. Steps I and II involved ultrasonication for 5 minutes and followed by heating at 90°C for 3 minutes. Both processed under room temperature and ultrasonication were two options for Step III. The soft-nanocomposite was made by treating the LP dispersed on 12-HSA and LaF$_3$/OA with ultrasound for 5 minutes, and the mixture was treated on a hot plate at 90°C and maintained for 3 minutes and then cooled under room temperature sequentially. The "HSA-LaF$_3$/OA-U" indicated the ultrasonication method in the final cooling process [20].

Muthumariappan Akilarasan et al. introduced the lanthanum molybdates–covered reduced graphene oxide (La$_2$(MoO$_4$)$_3$@rGO) nanocomposites. The La$_2$(MoO$_4$)$_3$@rGO nanocomposites were successfully achieved by two stages, namely, hydrothermal and simple sonication methods. In the first step, La$_2$(MoO$_4$)$_3$ was prepared using 25 mL of 1.0×10^{-3}M La(NO$_3$)$_3$.6 H$_2$O and 5.0×10^{-5}M (NH$_4$)$_6$Mo$_7$O$_{24}$.4 H$_2$O solutions prepared using deionized water. The 25 mL of prepared lanthanum nitrate solution was dropwise added to the ammonium molybdate solution with constant stirring. The pH of the reaction mixture was maintained at pH = 9.0 using 1 M NaOH solution for the white-colored precipitate formation. This reaction mixture was transferred into the 100 mL autoclave and kept in the heating at 180°C for 4 hours. In order to remove the water-soluble and organic impurities, the synthesized La$_2$(MoO$_4$)$_3$ was centrifuged and then washed with DD water and ethanol. Then 2:5 ratio of rGO:La$_2$(MoO$_4$)$_3$ was dispersed in 20 mL of deionized water and continued stirring for 30 minutes. The resulting mixture was then subjected to ultrasonication for 1.5 hours. Finally, the resulting product was collected and dried for 18 hours in a hot-air oven at 70°C [21]. Selvarasu Maheshwaran et al. used the hydrothermal technique to fabricate the GO@LaVO$_4$ nanocomposites. Lanthanum nitrate (La(NO$_3$)$_3$) and sodium vanadate (Na$_3$VO$_4$) solutions with an equimolar ratio of 5 mL were prepared using deionized water and mixed followed by stirring for 10 minutes. 3 M nitric acid (HNO$_3$) was utilized for maintaining the pH of 2.0. Now the solution was

TABLE 15.1

Synthesis Methods of Lanthanum Nanocomposites Supported by Carbon-Based Substrates

S. No.	Type of Material	Synthesis Method	Reference
1	Layered double hydroxides (LDH) embedded polyvinyl alcohol/ lanthanum alginate hydrogels	Cross-linking method	[23]
2	Lanthanum-modified nanochitosan-hierarchical ZSM-5 zeolite nanocomposite	Gelation method	[24]
3	Lanthanum-encapsulated chitosan-kaolin clay (LCK) hybrid composite	Cross-linking method	[25]
4	Gelatin-polyvinyl alcohol/lanthanum oxide composite	Reversed micelles method	[26]
5	Lanthanum (III)-encapsulated chitosan-montmorillonite composite	Cross-linking method	[27]
6	Lanthanum hydroxide embedded interpenetrating network poly (vinyl alcohol)/sodium alginate hydrogel beads	Reversed micelles method	[28]
7	Lanthanum-modified chitosan-attapulgite (La-CTS-ATP) composite	Gelation method	[29]
8	3D porous lanthanum-modified attapulgite chitosan hydrogel bead	Cross-linking method	[30]
9	Zirconium-/lanthanum-modified chitosan/polyvinyl alcohol composite	Reversed micelles method	[31]
10	Lanthanum-doped aminated graphene oxide@aminated chitosan microspheres	Cross-linking method	[32]

turned to golden yellow color from white color. Hereafter, the 1 M sodium hydroxide (NaOH) solution was dropwise added to the reaction mixture until the solution gets 30 mL of volume. Then, the GO was prepared by the modified Hummersmethod. Now, the prepared 10 mg of GO was added to the above reaction mixture with constant stirring for 10 minutes. This resultant solution was shifted to a Teflon-lined stainless autoclave which is kept in the heating process at 180°C for 24 hours. After the reaction was finished, the reaction mixture was centrifuged and washed with water and ethanol before being allowed to cool at ambient temperature. Finally, the product was dried at 80°C [22]. Table 15.1 explains the synthesis methods of lanthanum nanocomposites supported by carbon-based substrates.

APPLICATIONS

PHOTOCATALYTIC REDUCTION

A photocatalyst has been used in the process of photocatalytic reaction by capturing the light energy. In this reaction, a target chemical received electrons from the photocatalyst, changing it from a higher-valence state to a lower one. Numerous industries, including environmental remediation, water purification, and the creation of renewable fuels, could benefit from the use of this technology. In photocatalytic reduction, light energy has been captured and converted into chemical energy using photocatalysts. The photocatalyst produced electron–hole pairs when exposed to light. The electrons become extremely reactive and might participate in reduction reactions, whereas the holes might take part in oxidation reactions. The photocatalyst aided in the transfer of electrons to the targeted molecules during reduction, transforming them into desired compounds [33]. The creation of renewable energy and the remediation of the environment are two uses for this technique. To improve the surface area in the case of solids, the solid photocatalyst was ground into a fine powder. In the case of suspension, the photocatalyst was dissolved in a solvent to create a colloidal solution. To obtain a uniform dispersion, the mixer was stirred. By combining the reactant with the proper solvent, the reactant solution was prepared. Depending on the requirements of the experiment, the concentration could be changed. The reaction vessel that contained the photocatalyst and reactant solution was exposed to light. To activate the photocatalyst, UV lamps were turned on or the reaction vessel was exposed to sunlight.

Reactive sites were produced on the photocatalyst's surface as a result of the light energy being absorbed and converted into electron–hole pairs. The reduction took place within minutes to hours, depending on the reactants and circumstances. The mixture was stirred to ensure that the photocatalyst was exposed to the reactants uniformly [34]. The light source was cut off after the specified reaction time. Using the proper methods such as filtration or centrifugation, the products from the reaction mixture were separated. The products were analyzed using appropriate techniques (e.g., spectroscopy and chromatography) to confirm the reduction process. The products can be purified later. The electrons (e^-) in the valence band were excited to an empty conduction band when a photocatalyst was exposed to light with energy equal to or greater than its bandgap energy, leaving a positive hole (h^+) in the valence band. Photocatalytic reduction generated positive holes and electrons. Due to their electronic structures, they can serve as intermediaries in chemical redox reactions where conduction band electrons are reductants [35]. Figure 15.2 shows the photocatalytic mechanism followed by the catalyst.

$$Photocatalyst + h\nu \rightarrow Photocatalyst\ (h^+ + e^-)$$

The blue fluorescent NP-carbon dots photocatalyst was used to reduce Cr (VI) to Cr (III) in the aqueous phase while under the influence of visible light. Natural sunlight demonstrates its large contribution to the photoreduction of Cr (VI) to Cr (III) in comparison to artificial light. Additionally, it was discovered that the metal-doped NP-carbon dots enhanced the photocatalyst's efficiency even more [36]. $LaMnO_3$ nanoparticles were found to significantly increase the ability of $g\text{-}C_3N_4$ nanosheets to absorb visible light, which in turn increased the number of photo-induced electron–hole pairs needed for photocatalytic reactions and increased photocatalytic activity. An increase in photocatalytic activity caused by the addition of $LaMnO_3$ resulted in the formation of a direct Z-scheme $LaMnO_3/g\text{-}C_3N_4$ framework that not only made it easier to separate the electrons from holes in both $LaMnO_3$ and $g\text{-}C_3N_4$ but also provided the photocatalyst with a strong redox ability and was found to reduce organic pollutants [37]. Figure 15.3 shows the mechanism of degradation of organic pollutants by photocatalyst.

The CO yield from the g-CNT was 150 μmole/g-cat/h, which was 1.1 and 1.9 times higher than the yield from the graphitic carbon nanosheets and graphitic carbon nitride, respectively. Similar to this, 15.60 μmole/g-cat/h of H_2 was generated over g-CNT, which was 1.10 and 1.64 times more than

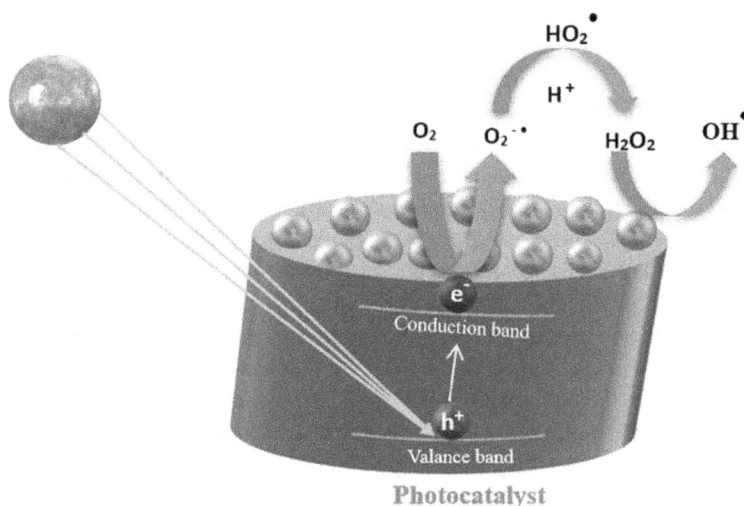

FIGURE 15.2 Photocatalytic mechanism.

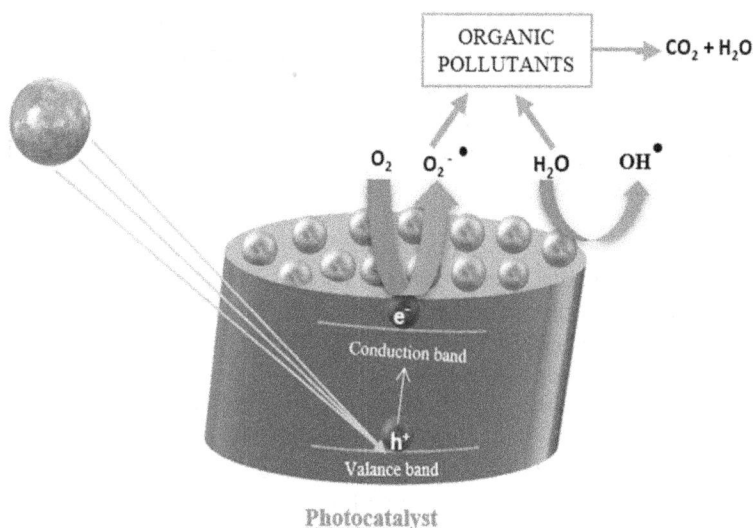

FIGURE 15.3 Degradation mechanism of organic pollutants by photocatalyst.

that achieved with g-CNT and g-CN, respectively. The maximum photoactivity of all the samples was found in 5 wt.% La-modified g-CNT, with a CO generation rate of 491 μmole/g-cat/h, which is 3.27 times greater than utilizing a g-CNT sample. Similar to this, 46 μmole/g-cat/h of H_2 production over 5 wt.% La/g-CNT was achieved, which was 2.95 times greater than utilizing graphitic carbon nanotubes. Due to their two-dimensional (2D) structure, g-CN nanosheets would allow for more charge production and separation. To reduce and oxidize CO_2 and CH_4, these charges must first travel a great distance due to surface interactions in photocatalysis. As a result, the catalyst surface would experience greater recombination rates, which would lower photoactivity. Comparatively, charges would go linearly to the catalytic surface in nanotubes because of their one-dimensional structure. As a result, employing g-CNT resulted in more CO and H_2 generation when compared with g-CN and g-CNS [38].

PHOTOCATALYTIC DEGRADATION OF PESTICIDES AND DYES

It is typically anticipated that persistent organic contaminants, such as different organic dyes and pesticides, will have a negative impact on the ecosystem. Modernization of an industry and its excessive use to meet human demands result in environmental contamination, particularly water pollution. In addition to being carcinogenic, dyes and pesticides have a number of negative impacts on the environment, numerous ecosystems, and human health. This makes it important to eliminate these dangerous contaminants from natural water resources as well as industrial effluents. Many conventional technologies, including electrochemical and ozone-based methods with ion exchange precipitation, and solvent extraction, are available to remove such organic contaminants from wastewater. The disadvantages of these traditional technologies include hazardous sludge, insufficient removal, other secondary waste that necessitates additional purification, raising the overall cost, and taking a lot of time. Using semiconductors in the nanodomain, photocatalytic degradation of various contaminants, such as dyes and pesticides, has evolved as a viable and efficient method for environmental cleanup over the last few decades. One of the most modern and popular methods of treating wastewater effectively involved the use of photocatalysts in the advanced oxidation process (AOP), which uses semiconductor metal oxides, chalcogenides, metal nanoparticles, and composites to degrade organic pollutants. Recent research has shown that many semiconductor metal oxides including TiO_2, SnO_2, ZnO, and Nb_2O_5 are investigated as potential heterogeneous photocatalysts.

In heterogeneous photocatalysis, reactive oxygen species (ROS) like hydroxyl (OH) and superoxide (O_2) radicals are produced after specific amount of UV or visible radiation is absorbed, resulting in the generation of photo-induced charge carriers, namely electrons and hole (e^-/h^+). By the process of mineralization, the ROS are completely capable of transforming large, complex organic pollutants into smaller, harmless, and less hazardous compounds like water, NH_3, and carbon dioxide [39].

PESTICIDE DEGRADATION

Due to the numerous diseases that it causes in both humans and animals, water pollution is a serious global issue. Water contamination occurred from a wide range of sources and in many different structures. Water pollution has largely caused by agriculture as well. Pesticides come in a variety of forms and used to protect crops from pests. Pests are proven to harm a sizable portion of the crop yield each year, costing farmers much. Pesticides are substances that are applied to crops to prevent pests or insects from destroying them. These pesticides can contaminate water in a number of ways, including direct runoff during spraying and during rainy seasons, when pesticides from crops mix in rainy water and stored in ponds and lakes, making the environment unfit for plants and living things.

By using the co-precipitation approach, ZnO and La-doped ZnO nanoparticles were successfully created. Under UV light irradiation, the photocatalytic activity was investigated on monocrotophos (insecticide that belongs to the organophosphates family). According to the test results, monocrotophos containing La-doped ZnO nanoparticles degrades most rapidly when exposed to UV radiation. The adsorption rate of monocrotophos on the surface of nanoparticles accelerates as lanthanum is uniformly distributed on the ZnO surface and greater degradation rate is visible. The highest levels of degradation were seen in lanthanum-doped ZnO as a result of an increase in surface area, which also increased the rate at which monocrotophos adsorbs to nanoparticle surfaces since degradation takes place there. When ZnO is doped with lanthanum, the material's surface shape changes, allowing monocrotophos to adsorb more quickly and in turn, speeding up the pace of deterioration to a greater extent. As a result, the goal of successfully removing organic contaminants from water was attained [40].

The simple chemical solution deposition approach might be able to deposit La into Bi_2O_3. Due to the meticulously controlled manufacturing of thin films, the integration of La^{3+} ions into the lattice structure of Bi_2O_3 exhibited the change in the XRD signature (201) and the creation of very lovely marigold flower-like morphology. La-doped Bi_2O_3's bandgap (2.8 eV) and increased visible light absorption both contributed to the breakdown of Carbol Fuchsin and Chlorpyrifos (organophosphate pesticide) organic pollutants. More Carbol Fuchsin than Chlorpyrifos degrading efficiency was discovered. By analyzing the metabolites during the degradation process, LC-MS was able to demonstrate the efficacy of the degradation. The primary ROS in the breakdown of both compounds was the hydroxide ion. La-doped Bi_2O_3 film was found to be a better catalyst than a pure Bi_2O_3 film by degradation study, minimization of electron–hole recombination by La doping induced the lattice contraction of Bi_2O_3. Secondly, La acted as an electron acceptor by being a p-type dopant, and thirdly, the lowered conduction band position, which has operated as a driving force for the electron transportation process during degradation under illumination, is suggested by the observed reduced bandgap [41].

Utilizing La-doped ZnO nanorods with PAN (poly acetonitrile) nanofiber, the degradation of methyl paraoxon (MP), an organophosphate pesticide, was investigated. One of the materials that needs more research for pesticide breakdown is ZnO. La-doped ZnO nanocrystals were produced on the surface of each PAN fiber to maintain their stability and enable their recovery after the degradation process. Shape of La-doped ZnO nanorods and crystallinity of Wurtzite were both shown by FE-SEM and XRD experiments, respectively. The defect-opulent material was identified by its photoluminescence property, which is crucial in the degradation. The degradation study identified the best condition for fully degrading MP in an aqueous solution under UV radiation. Under ideal circumstances, MP degradation followed second-order reaction kinetics with a high tenacity coefficient. Under ideal circumstance, MP completely broke down into the tertiary degradation product when it was subjected to degradation in a sample that had been spiked with MP. To determine if the degraded sample could be discharged into the aquatic ecosystem, an acute toxicity assessment was done. Up to three subsequent cycles, the photocatalytic material exhibited exceptional stability. It can be said that the catalytic compound La-doped ZnO/PAN nanofiber is very innovative in the process of environmental remediation [42].

A resilient, reusable dual photocatalyst made of g-C_3N_4/GO/La_2O_3 has been created and utilized to break down the carbamate insecticide carbofuran. The studies like XRD, FT-IR, and SEM were utilized to describe the produced nanomaterial. The developed nanocomposite exhibited significant photocatalytic activity for the oxidation of the insecticide carbofuran when exposed to light. Carbofuran photocatalytic degradation was accomplished after 120 minutes of light irradiation by 80% and employing H_2O_2 at 85.6%. When utilizing g-C_3N_4, g-C_3N_4/GO, and g-C_3N_4/La_2O_3, the photocatalytic degradation was only 53.75%, 62.5%, and 67.5%, respectively. This ternary nanocomposite can absorb more visible light while also efficiently reducing the recombination of photo-induced electron–hole pairs. Additionally, with 48% effectiveness in the photocatalytic degradation of carbofuran after five cycles, ternary photocatalysts demonstrated exceptional sustainability and recyclability. However, adding the dosage had no further effect on elimination efficiency. This result showed that a substantial quantity of photocatalysts decreased the light penetration due to the photocatalyst's greater suspension [43]. Figure 15.4 shows the graphical illustration of photodegradation of dyes by carbon substrate–supported lanthanum nanoparticles.

ELECTROCATALYTIC ACTIVITY OF LANTHANUM COMPOUNDS

Modern research focuses on the generation of intermittent energy from renewable energy sources such as wind, solar, geothermal, hydrodynamic, and others, with the conversion of energy into chemical energy. One of the most pure and renewable energy sources is hydrogen fuel, which is produced by electrocatalytic water-splitting devices by employing a chemical process as an energy storage system. The main difficulty is in designing and creating stable, highly effective

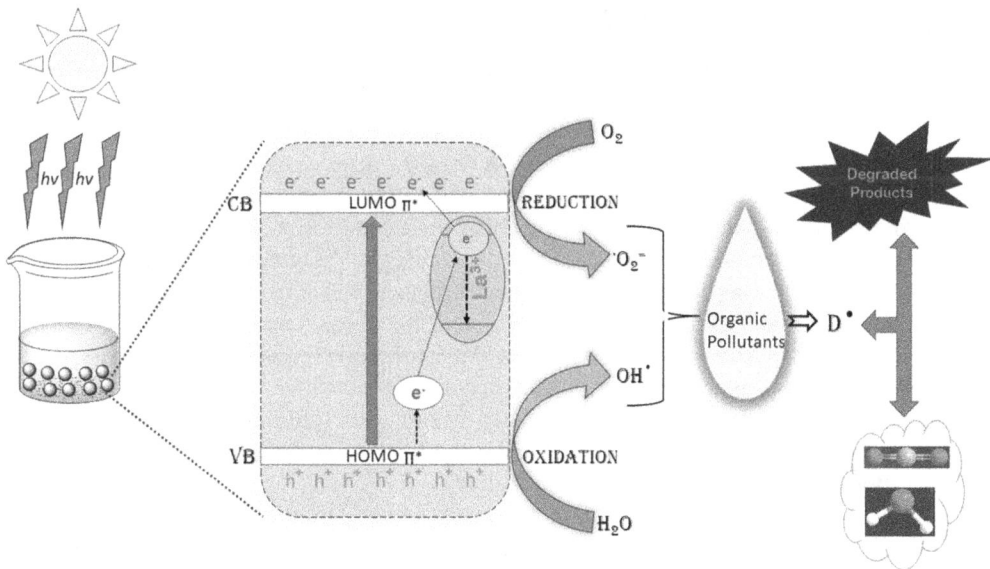

FIGURE 15.4 Graphical illustration of photodegradation of dyes by carbon substrate–supported lanthanum nanoparticles.

electrocatalysts that are also affordable for the oxygen evolution reaction (OER) and hydrogen evolution reaction (HER) processes. The noble metals are utilized to increase total water splitting and catalytic activity. Noble metals such as Ru/Pt and Ir have higher OER evolution catalytic activity, but their high price limits their use as catalysts in large-scale water-splitting reactions. It is required to investigate new, affordable, and highly effective materials as potential replacement catalysts for large-scale water splitting. The production of hydrogen and oxygen is crucial for the OER and HER of the water-splitting devices. Recent research on water splitting has focused on bifunctional OER and HER catalysts. Due to their excellent performance in redox reactions, many 3D-orbital transition metal oxide and sulfide electrocatalysts have drawn great deal of attention [44]. Alternative energy sources are required due to environmental concern as well as the quick depletion of fossil fuels. Therefore, it is vitally desirable to have modern, economical, and reliable energy technologies such as batteries, fuel cells (FCs), and water electrolyzers. The development of sustainable energy storage and conversion systems depends on three electrochemical processes: ORR, OER, and HER. A fuel cell produces the mixure of hydrogen and oxygen (renewable energy sources) by the electrolysis of water. Hydrogen has been used to produce electrochemical cells, especially fuel cells, by a process known as the water-splitting reaction ($2H_2O \rightarrow H_2+O_2$). Two half reactions like HER (water reduction) and OER (water oxidation) reactions are necessary for the electrochemical water-splitting process. This process is hindered by the large overpotentials of both half reactions and slow transport kinetics. To lower the overpotentials for OER and HER which improve the electrolytic efficiency, active and affordable catalysts are needed [45].

Recently, lanthanum compounds have drawn attention as a potential electrocatalyst for electrochemical reactions such as OER, HER, ORR, and methanol oxidation reactions (MORs) due to the properties of these lanthanum-based compounds as inexpensive and effective catalysts can be attributed to two factors based on their distinctive electronic structures. First of all, lanthanum ions are found in the f-block, and they have unoccupied 4f and 5d orbitals that have characteristic f-f and f-d electronic configurations to generate special bonding properties. As a result, the intra-atomic hybridization between the 4f orbitals and the 5d orbitals via the electron transfer inside lanthanum-based compounds determines the fundamental bandgap of lanthanum compounds, which is dependent on both the 4f and 5d orbitals of lanthanum ions. Secondly, the La 4f orbitals have energy levels that are below the fermi level by about 5 eV, which causes the 4f orbitals to partially delocalize and couple with the compounds' conduction band (CB) because of the strong interaction between the localized 4f orbitals and the delocalized conduction band, whereas certain electrons in 4f orbitals may be able to transfer into the conduction band, further changing the electrical structure of these lanthanum compounds. The incredible electrocatalytic activity of lanthanum-based compounds may be a result of their unique electronic structures [46].

Various concentrations of lanthanum-doped CuO nanoparticles were created by a synthetic process, and they served as an electrocatalyst that was highly effective for general water splitting. The particle size is shown to noticeably grow as the dopant concentration is changed above 1%, and the surface-to-volume ratio is seen to rapidly decline. According to the study, there was drop in series resistance as 1% dopant was added to the CuO nanostructure, which increased the active sites and improved the conductivity of the electrode. However, as the dopant concentration rises, resistance rises as well, with complicated distortions resulting in significantly fewer active sites. In an alkaline electrolyte, the 1% La-doped CuO performed better than expected in the OER and HER activity parameters. As a result, 10 mA/cm² was attained for overall water splitting at a cell voltage of 1.6 V, which is in line with benchmark standards. The rare earth-doped transition metal oxides can have a new application in the mass generation of hydrogen [47].

In this study, varied thicknesses of La_2S_3-MnS/GO (LMS/GO) thin films were deposited on a stainless steel (SS) substrate utilizing the binder-free SILAR (successive ionic layer adsorption reaction) process. It has been found that the LMS-90/GO thin film is extremely effective and has great potential for OER activities. Enhanced OER catalytic activity of La_2S_3-MnS-90/GO thin film is mostly due to its nanoflake-like shape, hydrophilic nature, high surface area, and direct contact with the SS substrate through GO film. With a micro-/mesoporous structure, the ultrathin nanosheets of the LMS-90/GO thin film displayed a high specific surface area of 170 m²/g. With an over potential of 263 mV and a current density of 10 mA/cm², the LMS-90/GO electrode displayed porous thin nanosheets and hydrophilic character. The LMS-90/GO thin film displayed a high electroactive surface area (1,725 F/cm²). In conclusion, the electrocatalytic performance of the LMS-90/GO composite thin film was good, and its 50-hour stability suggested that it might be used in water splitting as industrial applications [48]. Sundaresan et al. successfully showed the formation of lanthanum tungstate nanoparticles (NPs) using a sonochemical approach as a novel nanomaterial for the sensitive and selective electrochemical detection of anti-scald-inhibitor diphenylamine (DPA). It's interesting to note that La₂(WO₄)₃/SPCE showed a great sensitivity of 1.021 µA/µM/cm², a broad linear range of 0.01–58.06 µM, and a very low detection limit of 0.0024 µM. Additionally, La₂(WO₄)₃/SPCE was used to determine DPA in apple juice as the real sample analysis and showed remarkable selectivity even in the presence of interfering compounds. Therefore, the current work offered a novel method for synthesizing and fabricating rare earth metal tungstate in order to create unique nanostructures that serve as an effective electrocatalyst for electrochemical sensor applications [49].

Li Goa et al. have successfully shown the pyrolytic disintegration of the dual-metal La/ZIF-67 embedded crystalline cobalt/amorphous LaCoOx hybrid nanoparticles in porous nitrogen-doped carbon for hydrazine oxidation reaction (HzOR) due to high energy density, low onset potential, several uses in hydrazine-assisted hydrogen production and direct hydrazine fuel cells (DHFC), hydrazine electro-oxidation has drawn lot of attention. The highly conductive and porous N-C species could act as a robust skeleton for anchoring the catalytically active hybrid nanoparticles with a simple charge transport, resulting in synergistically realizing enhanced HzOR activity. The crystalline cobalt/amorphous LaCoOx hybrid nanoparticles could provide a wealth of active sites for HzOR. Promoted HzOR performance was attained by the synergistic effects of the enhanced active sites, wide surface area, and simple charge transport. This material is a promising electrocatalyst for effective hydrazine-assisted hydrogen production because the optimized catalyst typically displayed an ultralow onset potential of −0.17 V vs. RHE, a high HzOR current density of 69.2 mA/cm² at 0.3 V vs. RHE, and superior stability for 20-hour continuous catalysis [49]. Gao et al. synthesized a 2D Bi₂WO₆/lanthanum titanate (LTO) heterojunction, which served as a carrier to modify Pt NPS. Using the MOR, the electrochemical performance of the Pt-Bi₂WO₆/LTO sample was assessed. The Pt-Bi₂WO₆/LTO sample demonstrated outstanding electrocatalytic performance and stability with visible light exposure in comparison to conventional electrocatalytic procedure because of the synergistic effect of electrocatalysis and photocatalysis. Increased photo-electrocatalytic activity in the MOR is the result of the introduction of 2D LTO to efficiently hybridize with Bi₂WO₆ nanosheets and improve the interface charge separation of Bi₂WO₆ under visible light. The result suggested that a suitable photo-functional carrier for fuel cell processes is the 2D-Bi₂WO₆/LTO heterojunction. In order to improve the electrocatalytic oxidation of methanol with the aid of light irradiation, the current investigations offered a new framework for building a 2D semiconductor heterojunction as the conventional electrocatalysts' carrier [50]. Figure 15.5 depicts a graphical illustration of the mechanism of water splitting.

EFFECTIVE CATALYTIC ADSORPTION

Adsorption is the phenomenon of gases or solutes adhering to solid or liquid surfaces, and it is a mass transfer process. Some substances that impact solid surfaces are drawn to the surface by these imbalanced forces and remain there. The adsorption process can be separated into two groups: physical adsorption and chemical adsorption, depending on the various adsorption forces. Van der Waals' forces interact between molecules to cause physical adsorption, which can happen when AC is used to adsorb gas. Physical adsorption often takes place at low temperatures with a quick adsorption rate, little heat produced during the process, and without selection. Due to weak intermolecular interaction the structure of the molecules scarcely adsorb the substance. So the adsorption energy is

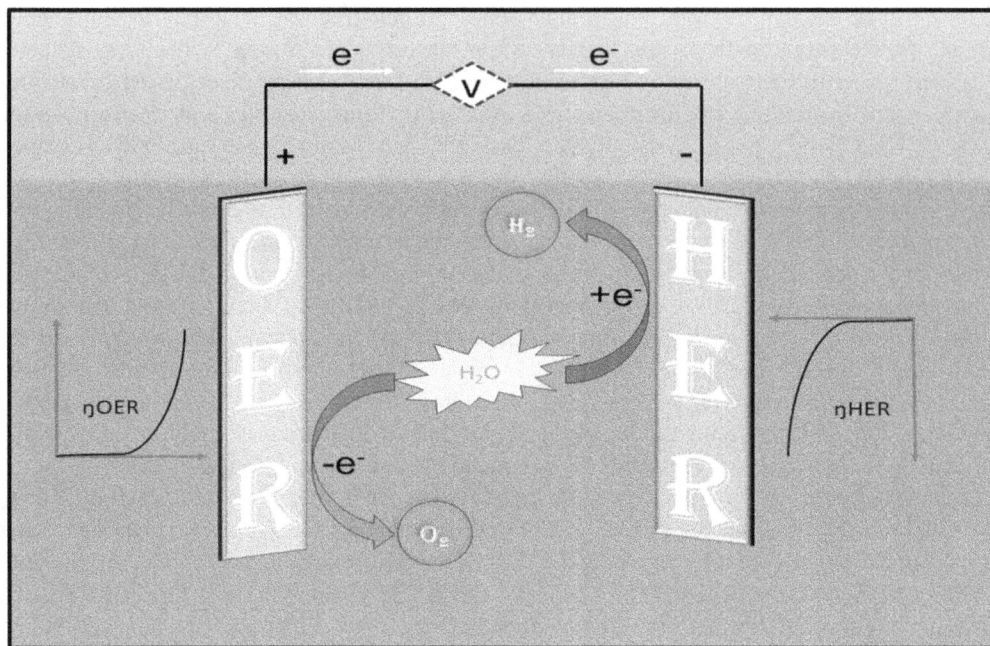

FIGURE 15.5   Graphical illustration of mechanism of water splitting.

low, and the adsorbed substance is easily detachable. Chemical adsorption is adsorption that occurs as a result of chemical bonding. Chemical bonds were created and broken during the process of chemical adsorption. The processes of physical and chemical adsorption frequently co-occur [51]. Figure 15.6 represents adsorption mechanism followed by photocatalysts.

AC is the most frequently employed adsorbent in wastewater treatment. Due to its large specific surface area, many microporous structures, and high hydrophobicity, AC has excellent adsorption capability for the majority of pollutants [52]. The efficiency of photocatalytic degradation can be increased by the adsorption process, which can also reduce the accumulation of pollutants from wastewater on the catalyst surface [53]. AC has become a popular adsorbent for air and water filtration among carbon compounds. High surface area, the existence of surface functional groups, and the advantageous porous structure are the benefits. For phosphate removal, different kinds of metal-loaded BC were synthesized. La-modified biochar among them showed the highest ability for phosphate adsorption (93.91 mg/g) [54]. Wu and associates (2017) fabricated $La(OH)_3/Fe_3O_4$ nanocomposites as an extremely efficient phosphate adsorbent made by adding chemicals to the sludge charcoal which was produced and tested. The result indicated that the La-600SS-OH has the highest phosphate adsorption capability (93.91 mg/g). The phosphate adsorption was unaffected by other common wastewater anions, and the adsorbent functioned best in the pH range from 3.0 to 6.0 [55]. The removal of trace metals present in the contaminated groundwater analyzed was shown to be best accomplished by the granular activated carbons like GAC383 and GACLa1073. A useful tuning procedure is to chemically activate GAC with nanoscale La_2O_3 metal oxide. Due to the inclusion of increased microporosity, treatment that generated porosity, more surface groups, and surface area upon activation and doping significantly improved the adsorption potential of the resultant carbons [56]. Compared to Granular Activated Carbons 383, GACLa1073 was found to be more effective at removing hazardous metal pollutants. GACLa1073 can contribute more to the removal of 99.3% of Fe, 96.6% of Co, 99.3% of Mn, 76.3% of As, 63.0% of Ba, 50.4% of Li, 1% of Mg, 71.3% of Al, 47.7% of Cr, 99.2% of Ni, 7.11% of Cu, 24.3% of Zn, 55.1% of Cd, 32.6% of Tl, and 94.9% of Pb, but GAC contributes comparatively less to the removal of all the other metals

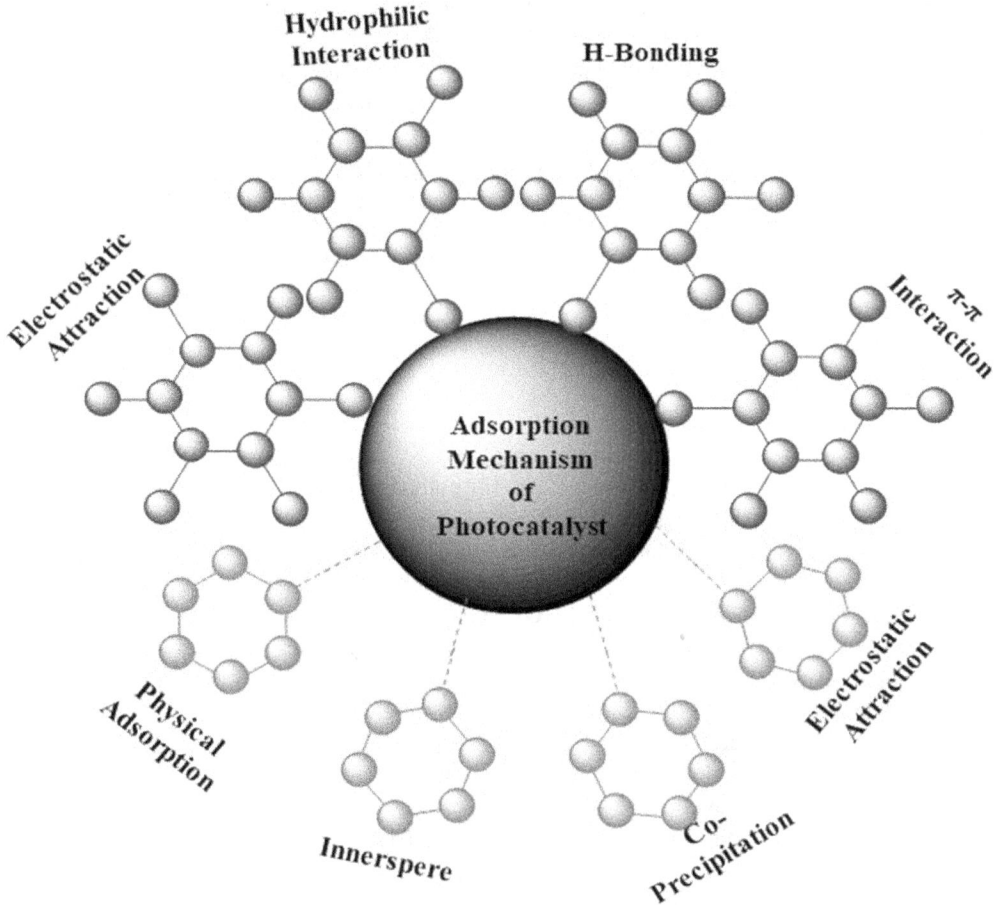

FIGURE 15.6 Adsorption mechanism enhanced by photocatalysts.

than Mg, Cu, Zn, Cd, Tl, Cr, Al, Ba, and As [57]. For the first time, lanthanum as well as cobalt species were added to porous carbon made of sugar. According to the result, La/Co-SC's surface morphology has undergone significant modification, and both its specific surface area and the size of its pores are larger than those of virgin SC, indicating that it is suitable for use as an effective adsorbent. According to the results, La/Co-SC performs well at adsorbing methyl orange, with an adsorption capacity of 285.71 mg/g, outperforming La-SC, Co-SC, and other types of same-type adsorbents. Additionally, the La/Co-SC demonstrated strong regeneration capabilities, pointing to its potential as an environmentally friendly adsorbent. La/Co-SC is suitable for treating wastewater contaminated with dyes, especially to get rid of methyl orange [58].

FUTURE PERSPECTIVE

La-MnS/GO thin film possesses enhanced charge separation which could boost efficiency for water purification, CO_2 reduction, and solar energy conversion. Further development can be carried out by doping it with reduced graphene oxide, and the efficiency of the material can be studied on the basis of energy storage. There aren't many studies done on lanthanum tungstate ($La_2(WO_4)_3$) for drug delivery. However, it has energy-absorbing characteristics that can be improved upon by doping it with gold nanoparticles. Additionally, the material's capabilities for photoacoustic imaging, medication delivery, heat creation, and targeted destruction of cancer cells can be evaluated in the future. It is anticipated that 2D Bi_2WO_6/LTO will improve ion diffusion in advanced batteries; nevertheless, the same material may potentially be investigated for energy storage in the future. It can be examined for gas-sensing applications due to its large surface area. The presence of g-C_3N_4 and La_2O_3 could improve the photothermal efficiency. Since GO can convert light energy into heat, which can be used to target drug delivery by altering the surface or attaching ligands to it, it can be used to selectively destroy cancer cells. As a result, g-C_3N_4/GO/La_2O_3 will be further studied for applications in drug administration, imaging, biosensing, and photothermal therapy for cancer cells. Additionally, improvements in fuel cell performance and efficiency will be investigated for La-doped Bi_2O_3, as well as heat to electricity transformation, degradation of pollutants, and sensors that can detect a number of gases including hydrogen, carbon monoxide, and volatile organic compounds.

CONCLUSION

This study discovered that adsorbents made from low-cost lanthanum-based carbon are extremely effective at removing direct organic and inorganic pollutants from organic contaminants in the environment. It also demonstrated that the efficacy of such adsorption and degradation approaches was highly reliant on a variety of parameters. Emerging pollutants in water will be a growing worry in the next years, demanding research into the development of new methods to preserve potable water supplies. To strengthen the industrial production processes of wastewater, we need a sophisticated water treatment technology to remove various types of impurities existing in water resources. Several research organizations are now working on the removal of organic and inorganic pollutants from water and soil. The stages of AOP development follow a cyclical pattern. Furthermore, laboratory investigations focus solely on the kinetics of target organic pollutants, ignoring the toxicity of degradation intermediates. Many researchers, according to several publications, developed flawed models for photocatalytic oxidation reactors. This review has shown that the lanthanum-based carbon materials are very efficient for photocatalytic degradation and reduction.

REFERENCES

[1] Chin Wei Tan, Kok Hong Tan, Yit Thai Ong, Abdul Rahman Mohamed, Sharif Hussein Sharif Zein, Soon Huat Tan, Energy and environmental applications of carbon nanotubes, *Environmental Chemistry Letters*, 10(2012): 265–273. https://doi.org/10.1007/s10311-012-0356-4.

[2] Koh Yuen Koh, Yi Yang, J. Paul Chen, Critical review on lanthanum-based materials used for water purification through adsorption of inorganic contaminants, *Critical Reviews in Environmental Science and Technology*, 52(2022): 1773–1823. https://doi.org/10.1080/10643389.2020.1864958.

[3] Subramaniyan Vinoth, Sea-Fue Wang, Lanthanum vanadate-based carbon nanocomposite as an electrochemical probe for amperometric detection of theophylline in real food samples, *Food Chemistry*, 427(2023): 136623. https://doi.org/10.1016/j.foodchem.2023.136623.

[4] S. Manikandan, D. Sasikumar, Improving sunlight-photocatalytic activity of undoped and phosphorous doped MnO_2 with activated carbon from bio-waste with nanorods morphology, *Inorganic Chemistry Communications*, 144(2022): 109942. https://doi.org/10.1016/j.inoche.2022.109942.

[5] Yongfeng Lu, Wensong Wang, Hongrui Cheng, Haijiang Qiu, Wenhao Sun, Xiao Fang, Jiefeng Zhu, Yuanhui Zheng, Bamboo-charcoal-loaded graphitic carbon nitride for photocatalytic hydrogen evolution, *International Journal of Hydrogen Energy*, 47(2022): 3733–3740. https://doi.org/10.1016/j.ijhydene.2021.10.267.

[6] Xuan Wang, Mehrasa Tarahomi, Reza Sheibani, Changlei Xia, Weidong Wang, Progresses in lignin, cellulose, starch, chitosan, chitin, alginate, and gun/carbon nanotube (nano)composite for environmental application: A review, *International Journal of Biological Macromolecules*, 241(2023): 124472. https://doi.org/10.1016/j.ijbiomac.2023.124472.

[7] Dongmei Peng, Zhongfeng Zhang, Jijuan Zhang, Yang Yang, Improving photocatalytic activity for formaldehyde degradation by encapsulating C60 fullerenes into graphite-like C3N4 through the enhancement of built-in electric fields, *Molecules*, 28(2023): 5815. https://doi.org/10.3390/molecules28155815.

[8] Anna Gnach, Tomasz Lipinski, Artur Bednarkiewicz, Jacek Rybka, John Capobianco, Upconverting nanoparticles: Assessing the toxicity, *Chemical Society Reviews*, 44(2015): 1561–1584. https://doi.org/10.1039/C4CS00177J.

[9] Brabu Balusamy, Burcu Ertit Tastan, Seyda Fikirdesici Ergen, Tamer Uyar, Turgay Tekinay, Toxicity of lanthanum oxide (La_2O_3) nanoparticles in aquatic environments, *Environmental Science: Processes & Impacts*, 17(2015): 1265–1270. https://doi.org/10.1039/C5EM00035A.

[10] Weiwei Gu, Ye Song, Jingjun Liu, Feng Wang, Lanthanum-based compounds: electronic band-gap-dependent electrocatalytic materials for oxygen reduction reaction, *A European Journal*, 23(2017): 10126–10132. https://doi.org/10.1002/chem.201701136.

[11] Saikat Kumar Kuila, Deepak Kumar Gorai, Bramha Gupta, Ashok Kumar Gupta, Chandra Sekhar Tiwary, Tarun Kumar Kundu, Lanthanum ions decorated 2-dimensional g-C_3N_4 for ciprofloxacin photodegradation, *Chemosphere*, 268(2021): 128780. https://doi.org/10.1016/j.chemosphere.2020.128780.

[12] Nivedita Shukla, Dinesh K. Verma, Alok K. Singh, Bharat Kumar, Kavita, Rashmi B. Rastogi, Ternary composite of methionine-functionalized graphene oxide, lanthanum-doped yttria nanoparticles, and molybdenum disulfide nanosheets for thin-film lubrication, *ACS Applied Nanomaterials*, 8(2020): 8012. https://doi.org/10.1021/acsanm.0c01468.

[13] Thota Trinadh, Harisankar Khuntia, Tummala Anusha, Kali Sai Bhavani, J. V. Shanmukha Kumar, Pradeep Kumar Brahman, Synthesis and characterization of nanocomposite material based on graphene quantum dots and lanthanum doped zirconia nanoparticles: An electrochemical sensing application towards flutamide in urine sample, *Dimond and Related Materials*, 110(2020): 108143. https://doi.org/10.1016/j.diamond.2020.108143.

[14] Khaled D. Khalil, Sayed M. Riyadh, Mariusz Jaremko, Thoraya A. Farghaly, Mohamed Hagar, Synthesis of chitosan-La_2O_3 nanocomposite and its utility as a powerful catalyst in the synthesis of pyridines and pyrazoles, *Molecules*, 26(2021): 3689. https://doi.org/10.3390/molecules26123689.

[15] P. Sirajudheen, Sankaran Meenakshi, Facile synthesis of chitosan-La^{3+}-graphite composite and its influence in photocatalytic degradation of methylene blue, *International Journal of Biological Macromolecules*, 133(2019): 253–261. https://doi.org/j.ijbiomac.2019.04.073.

[16] Bao Liu, Yueying Yu, Qifeng Han, Sichao Lou, Lingfan Zhang, Wenqing Zhang, Fast and efficient phosphate removal on lanthanum-chitosan composite synthesized by controlling the amount of cross-linking agent, *International Journal of Biological Macromolecules*, 157(2020): 247–258. https://doi.org/j.ijbiomac.2020.04.159.

[17] Bin Wang, Xiaoling Hu, Dao Zhou, Heng Zhang, Rongfan Chen, Wenbin Guo, Hongyu Wang, Wei Zhang, Zhenzhen Hong, Wanlin Lyu, Highly selective and sustainable clean-up of phosphate from aqueous phase by eco-friendly lanthanum cross-linked polyvinyl alcohol/alginate/palygorskite composite hydrogel beads, *Journal of Cleaner Production*, 298(2021): 126878. https://doi.org/10.1016/j.jclepro.2021.126878.

[18] Fei Yu, Zhengqu Yang, Xiachen Zhang, Peiyu Yang, Jie Ma, Lanthanum modification k-carrageenan/sodium alginate dual-network aerogels for efficient adsorption of ciprofloxacin hydrochloride, *Environmental Technology and Innovation*, 24(2021): 102052. https://doi.org/j.eti.2021.102052.

[19] Gaurav Sharma, Amit Kumar, Shweta Sharma, Sameerah I. Al-Saeedi, Ghadah M. Al-Senan, Ayman Nafady, Tansri Ahamad, Mu Naushad, Florian J. Stadler, Fabrication of oxidized graphite supported La_2O_3/ZrO_2 nanocomposite for the photoremediation of toxic fast green dye, *Journal of Molecular Liquids*, 277(2019): 738–748. https://doi.org/10.1016/j.molliq.2018.12.126.

[20] T. N. Vinuth Raj, Priya A. Hoskeri, H. B. Muralidhara, C. R. Manjunatha, K. Yogesh Kumar, M. S. Raghu, Facile synthesis of perovskite lanthanum aluminate and its green reduced graphene oxide composite for high performance supercapacitors, *Journal of Electroanalytical Chemistry*, 858(2020): 113830. https://doi.org/10.1016/j.jelechem.2020.113830.

[21] Ruochong Zhang, Qi Ding, Songwei Zhang, Yi Li, Qingho Niu, Lingling Yang, Jun Ye, Litian Hu, The effects of ultrasonication on the microstructure, gelling and tribological properties of 12-HAS soft-nanocomposite with LaF_3 nanoparticles, *Colloids and Surfaces A: Physicochemical and Engineering Aspects*, 586(2020): 124247. https://doi.org/10.1016/j.colsurfa.2019.124247.

[22] Muthumariappan Akilarasan, Elayappan Tamilalagan, Shen-Ming Chen, Selvarasu Maheshwaran, Chih-Hsuan Fan, Mohamed A. Habila, Mika Sillanpaa, Rational synthesis of rare-earth lanthanum molybdate covered reduced graphene oxide nanocomposites for the voltammetric detection of moxifloxacin hydrochloride, *Bioelctrochemistry*, 146(2022): 108145. https://doi.org/10.1016/j.bioelechem.2022.108145.

[23] Selvarasu Maheshwaran, Muthumariappan Akilarasan, Tse-Wei Chen, Shen-Ming Chen, Elayappan Tamilalagan, Ting-Yu Jiang, Eman A. Alabdullkarem, Mustafa Soylak, Electrocatalytic evaluation of graphene oxide warped tetragonal t-lanthanum vanadate (GO@LaVO$_4$) nanocomposites for the voltammetric detection of antifungal and anthiprotozoal drug (clioquinol), *Microchimica Acta*, 188(2021): 1–9. https://doi.org/10.1007/s00604-021-04758-5.

[24] Lihua Feng, Qian Zhang, Fangying Ji, Lei Jiang, Caocong Liu, Quishi Shen, Qian Liu, Phosphate removal performances of layered double hydroxides (LDH) embedded polyvinyl alcohol/lanthanum alginate hydrogels, *Chemical Engineering Journal*, 430(2022): 132754. https://doi.org/10.1016/j.cej.2021.132754.

[25] Samira Salehi, Mojtaba Hosseinifard, Highly efficient removal of phosphate by lanthanum modified nanochitosan-hierarchical ZSM-5 zeolite nanocomposite: Characteristics and mechanism, *Cellulose*, 27(2020): 4637–4664. https://doi.org/10.1007/s10570-020-03094-w.

[26] Hyder Ali Thagira Banu, Perumal Karthikeyan, Sivakumar Vigneshwaran, Sankaran Meenakshi, Adsorptive performance of lanthanum encapsulated biopolymer chitosan-kaolin clay hybrid composite for the recovery of nitrite and phosphate from water, *International Journal of Biological Macromolecules*, 154(2020): 188–197. https://doi.org/10.1016/j.ijbiomac.2020.03.074.

[27] Mohammad Fuzail Siddiqui, Tabrez Alam Khan, Gelatin-polyvinyl alcohol/lanthanum oxide composite: A novel adsorbent for sequestration of arsenic species from aqueous environment, *Journal of Water Process Engineering*, 34(2020): 101071. https://doi.org/10.1016/j.jwpe.2019.101071.

[28] H. Thangira Banu, P. Karthikeyan, Sankaran Meenakshi, Lanthanum (III) encapsulated chitosan-montmorillonite composite for the adsorptive removal of phosphate ions from aqueous solution, *International Journal of Biological Macromolecules*, 112(2018): 284–293. https://doi.org/10.1016/j.ijbiomac.2018.01.138.

[29] Aijiao Zhou, Chang Zhu, Wangwei Chen, Jun Wan, Tao Tao, Tian C. Zhang, Pengchao Xie, Phosphorus recovery from water by lanthanum hydroxide embedded interpenetrating network poly (vinyl alcohol)/sodium alginate hydrogel beads, *Colloids and Surfaces A: Physicochemical and Engineering Aspects*, 554(2018): 237–244. https://doi.org/10.1016/j.colsurfa.2018.05.086.

[30] Hao Kong, Qian Li, Xiangqun Zhang, Peizhen Chen, Gengtao Zhang, Zhiping Huang, Lanthanum modified chitosan-attapulgite composite for phosphate removal from water: Performance, mechanisms and applicability, *International Journal of Biological Macromolecules*, 224(2023): 984–997. https://doi.org/10.1016/j.ijbiomac.2022.10.183.

[31] Hao Kong, Jiarui Wang, Gentao Zhang, Feng Shen, Qian Li, Zhiping Huang, Synthesis of three-dimensional porous lanthanum modified attapulgite chitosan hydrogel bead for phosphate removal: performance, mechanism, cost-benefit analysis, *Separation and Purification Technology*, 320(2023): 124098. https://doi.org/10.1016/j.seppur.2023.124098.

[32] Liping Mei, Jiao Wei, Ruirui Yang, Fei Ke, Chuanyi Peng, Ruyan Hou, Junsheng Liu, Xiaochun Wan, Huimei Cai, Zirconium/lanthanum-modified chitosan/polyvinyl alcohol composite adsorbent for rapid removal of fluoride, *International Journal of Biological Macromolecules*, 243(2023): 125155. https://doi.org/10.1016/j.ijbiomac.2023.125155.

[33] Abdelazeem S. Eltaweli, Karim Ibrahim, Eman M. Abd El-Monaem, Gehan M. El-Subruiti, Ahmed M. Omer, Phosphate removal by lanthanum-doped aminated graphene oxide@aminated chitosan microspheres: insights into the adsorption mechanism, *Journal of Cleaner Production*, 385(2023): 135640. https://doi.org/10.1016/j.jclepro.2022.135650.

[34] Peng Zhang, Jijie Zhang, Jinlong Gong, Tantalum-based semiconductors for solar water splitting, *RSC Chemical Society Reviews*, 43(2014): 4395–4422. https://doi.org/10.1039/C3CS60438A.

[35] Thomas Simon, Michael T. Carlson, Jacek K. Stolarczyk, Jochen Feldmann, Electron transfer rate vs recombination losses in photocatalytic H$_2$ generation on Pt-decorated CdS nanorods, *ACS Energy Letters*, 6(2016): 1137–1142. https://doi.org/10.1021/acaenergylett.6b00468.

[36] Quanjun Xiang, Bei Cheng, Jiaguo Yu, Graphene-based photocatalysts for solar-fuel generation, *Angewandte Chemie International Edition*, 54(2015): 11350–11366. https://doi.org/10.1002/anie.201411096.

[37] Shasha Zhu, Dunwei Wang, Photocatalysis: basic principles, diverse forms of implementations and emerging scientific opportunities, *Advanced Energy Materials*, 7(2017): 1700841. https://doi.org/10.1002/anem.201700841.

[38] Rajendra C. Pawar, Da-Hyun Choi, Caroline S. Lee, Reduced graphene oxide composites with MWCNTs and single crystalline hematite nanorhombohedra for applicationsin water purification, *International Journal of Hydrogen Energy*, 40(2015): 767–778. https://doi.org/10.1016/j.ijhydene.2014.08.084.

[39] Md. Ashraful Islam Molla, Abrar Zadeed Ahmed, Satoshi Kaneco, Chapter 3 Reaction mechanism for photocatalytic degradation of organic pollutants, *Nanostructured Photocatalysts*, (2021): 63–84. https://doi.org/10.1016/B978-0-12-823007-7.00011-0.

[40] Shivendu Rajan, Nandita Dasgupta, Eric Lichtfouse, *Nanoscience in Food and Agriculture 5*, Springer, 26(2017): 366. https://doi.org/10.1007/978-3-319-58496-6.

[41] Danilo Spasiano, Raffaele Marotta, Sixto Malato, Pilar Fernandez-Ibanez, Ilaria Di Somma, Solar photocatalysis: Materials, reactors, some commercial, and pre-industrialized applications. A comprehensive approach, *Applied Catalysis B: Environmental*, 170–171(2015): 90–123. https://doi.org/10.1016/j.apcatb.2014.12.050.

[42] Anshu Bhati, Satyesh Raj Anand, Deepika Saini, Gunture, Sumit Kumar Sonkar, Sunlight-induced photoreduction of $Cr(VI)$ to $Cr(III)$ in wastewater by nitrogen-phosphorus-doped carbon dots, *npj Clean Water*, 2(2019): 12. https://doi.org/10.1038/s41545-019-0036-z.

[43] Jin Luo, Jinfen Chen, Rongting Guo, Yuelien Qiu, Wen Li, Xiaosong Zhou, Xiaomei Ning, Liang Zhan, Rational construction of direct Z-scheme $LaMnO_3/g\text{-}C_3N_4$ hybrid for improved visible-light photocatalytic tetracycline degradation, *Separation and Purification Technology*, 211(2019): 882–894. https://doi.org/10.1016/j.seppur.2018.10.062.

[44] Ayyaz Muhammad, Muhammad Tahir, Saad S. Al-Shanhrani, Arshid Mahmood Ali, Sami Ullah Rather, Template free synthesis of graphitic carbon nitride nanotubes mediated by lanthanum (La/g-CNT) for selective photocatalytic CO_2 reduction via dry reforming of methane (DRM) to fuels, *Applied Surface Science*, 504(2020): 144177. https://doi.org/10.1016/j.apsusc.2019.144177.

[45] Narges Omrani, Alireza Nezamzadeh-Ejhieh, Focus on scavengers' effects and GC-MASS analysis of photodegradation intermediates of sulfasalazine by Cu_2O/CdS nanocomposite, *Separation and Purification Technology*, 235(2020): 116228. https://doi.org/10.1016/j.seppur.2019.116228.

[46] Subhash D. Khairnar, Anil N. Kulkarni, Sachin G. Shinde, Sunil D. Marathe, Yogesh V. Marathe, Sanjay D. Dhole, Vinod S. Shrivastava, Synthesis and characterization of 2-D La-doped Bi_2O_3 for photocatalytic degradation of organic dye and pesticide, *Journal of Photochemistry and Photobiology*, 6(2021): 100030. https://doi.org/10.1016/j.jpap.2021.100030.

[47] Hadis Derikvandi, Alireza Nezamzadeh-Ejhieh, A comprehensive study on enhancement and optimization of photocatalytic activity of ZnS and SnS_2: Response Surface Methodology (RSM), n-n heterojunction, supporting and nanoparticles study, *Journal of Photochemistry and Photobiology A: Chemistry*, 348(2017): 68–78. https://doi.org/10.1016/j.jphotochem.2017.08.007.

[48] Hadis Derikvandi, Alireza Nezamzadeh-Ejhieh, Increased photocatalytic activity of NiO and ZnO in photodegradation of a model drug aqueous solution: Effect of coupling, supporting, particles size and calcination temperature, *Journal of Hazardous Materials*, 321(2017): 629–638. https://doi.org/10.1016/j.jhazmat.2016.09.056.

[49] Geetika Geetika, Vijay Luxmi, Ashavani Kumar, Effect of lanthanum doping on structural and optical properties of ZnO along with photocatalytic activity in degradating toxic pesticide (monocrotophos), *AIP Conference Proceedings*, 2142(2019): 040009. https://doi.org/10.1063/1.5122346.

[50] Krishnasamy Lakshmi, Krishna Kadirvelu, Palathurai Subramaniam Mohan, Reclaimable La: ZnO/PAN nanofiber catalyst for photodegradation of methyl paraoxon and its toxicological evaluation utilizing early life stages of zebra fish (Danio rerio), *Chemical Engineering Journal*, 357(2019): 724–736. https://doi.org/10.1016/j.cej.2018.09.201.

[51] S. Tabasum, A. Sharma, S. Rani, S. Chaudhary, A. Q. Malik, D. Kumar, T. Deshpande, Prolific fabrication of lanthanum oxide with graphitic carbon/graphene oxide for enhancing photocatalytic degradation of carbofuran from aqueous solution, *Rasayan Journal of Chemistry*, 16(2023): 740–745. https://doi.org/10.31788/RJC.2023.1628266.

[52] Vikas J. Mane, Shital B. Kale, Shivaji B. Ubale, Vaibhav C. Lokhande, Umakant M. Patil, Chandrkant D. Lokhande, Lanthanum sulphide-manganese sulphide/graphene oxide (La2S3-MnS/GO) composite thin film as an electrocatalyst for oxygen evolution reactions, *Journal of Solid-State Electrochemistry*, 25(2021): 1775–1788. https://doi.org/10.1007/s10008-021-04945-7.

[53] Tehmeena Maryum Butt, Naveed Kausar Janjua, Ayesha Mujtaba, Shabana Ali Zaman, Rotaba Ansir, Ammara Rafique, Poshmal Sumreen, Maria Mukhtar, Maria Pervaiz, Azra Yaqub, Zareen Akhter, Tariq Yasin, Ghazanfar Abbas, Rizwan Raza and Dmitry Medvedev, B-site doping in lanthanum cerate nanomaterials for water electrocatalysis, *Journal of the Electrochemical Society*, 167(2020): 026503. https://doi.org/10.1149/1945-7111/ab63c0.

[54] Weiwei Gu, Ye Song, Jingjun Liu, Feng Wang, Lanthanum-based compounds: Electronic bandgap-dependent electro-catalytic materials toward oxygen reduction reaction, *Chemistry - A European Journal*, 23(2017): 10126–10132. https://doi.org/10.1002/chem.201701136.

[55] John D. Rodney, S. Deepa Priya, M. Cyril Robinson, C. Justin Raj, Suresh Perumal, Byung Chul Kim, S. Jerome Das, Lanthanum doped copper oxide nanoparticle enabled proficient bi-functional electro-catalyst for overall water splitting, *International Journal of Hydrogen Energy*, 45(2020): 24684–24696. https://doi.org/10.1016/j.ijhydene.2020.06.240.

[56] Periyasamy Sundaresan, Periyasamy Gnanaprakasam, Shen-Ming Chen, Ramalinga Viswanathan Mangalaraja, Wu Lei, Qingli Hao, Simple sonochemical synthesis of lanthanum tungstate ($La_2(WO_4)_3$) nanoparticles as an enhanced electrocatalyst for the selective electrochemical determination of anti-scald-inhibitor diphenylamine, *Ultrasonics Sonochemistry*, 58(2019): 104647. https://doi.org/10.1016/j.ultsonch.2019.104647.

[57] Li Gao, Junfeng Xie, Shan-Shan Liu, Shanshan Lou, Zimeng Wei, Xiaojiao Zhu, and Bo Tang, Crystalline cobalt/amorphous $LaCoO_x$ hybrid nanoparticles embedded in porous nitrogen-doped carbon as efficient electrocatalysts for hydrazine-assisted hydrogen production, *ACS Applied Materials & Interfaces*, 12(2020): 24701–24709. https://doi.org/10.1021/acsami.0c02124.

[58] Haifeng Gao, Chunyang Zhai, Nianqing Fu, Yukou Du, Kevin Yu, and Mingshan Zhu, Synthesis of Pt nanoparticles supported on a novel 2D Bismuth tungstate/lanthanum titanate heterojunction for photo-electrocatalytic oxidation of methanol, *Journal of Colloid and Interface Science*, 561(2019): 338–347. https://doi.org/10.1016/j.jcis.2019.10.114.

16 Catalysis for Environmental and Energy Applications
The Potential of Multifunctional Nanomaterials

*Anjana Vinod, Jyothi Vaz, K. Divyarani, Praveen Martis,
S. Sreenivasa, Vinayak Adimule, and L. Parashuram*

INTRODUCTION

Due to the abundant energy from the sun, sunlight has become a sustainable energy for the menacing challenges in the environment [1,2]. The environmental benefits of photocatalysis can address the consequences of contemporary problems. The mechanics of natural photosynthesis serve as a model for photocatalysis. The Z-scheme is widely used to explain light-driven processes [3–5]. Photocatalysis has more advantages than thermocatalysis owing to their selectivity in both oxidation and reduction process. The photocatalytic process is extensively attractive due to greater efficiency and economic viability and hence used in pollutant degradation, hydrogen evolution, and carbon capture [6,7]. Also, nanostructured photocatalysts due to their higher active sites, surface defects, and higher surface area attributed towards boosting photocatalysis compared to their bulk counterparts. These materials due to quantum size effects enhance the feasibility of water splitting. When their configurations and dimensions are altered, the redox potentials and bandgaps also align and harness photocatalysis [8]. To address environmental issues and advance sustainable energy solutions, electrocatalysis is an essential tool [9,10]. The electrochemical processes that turn CO_2 into useful chemicals and fuels have attracted a lot of interest in the effort to reduce CO_2 emissions. The capacity of metal-based electrochemical systems, in particular Cd, Pb, Ag, In, and Cu, to enhance CO_2 reduction is well established. These systems depend on electrode materials with characteristics that promote charge transfer and have lower overpotentials. For the production of high-purity hydrogen [11,12], electrochemical water splitting is essential. Research has focused on non-noble-metal HER catalysts, with transition metal–based materials including Fe, Bi, Co, Ni, W, and Mo and their compounds appearing as attractive alternatives to Pt-based groups to overcome their cost and availability limitations [13]. The nitrogen reduction reaction (NRR) involves electrocatalysis to function. Noble metals with favourable NRR catalytic activity, such as Au, Pt, Pd, and Ru, are rare and expensive [14]. As they can exhibit catalytic activity in the NRR, transition-metal oxides provide affordable and environmentally favourable alternatives. Because of their large surface area, permeability, and internal free space, various nanohybrids, including different metal-organic frameworks (MOFs), are becoming more popular. These materials provide an abundance of electrochemical active sites and facilitate mass transfer during catalytic reactions. The combination of light and catalysis has become a ground-breaking strategy in the search for environmentally benign and sustainable technology [15]. Hydrogen evolution [16–18], CO_2 reduction, environmental degradation [19,20], and oxygen evolution are a few examples of the many uses for photo electrocatalysis, which combines photochemistry and electrocatalysis. This introduction gives us an overview of these revolutionary procedures and how they may revolutionise energy conversion, environmental cleanup, and the fight against climate change. Photo electrocatalysis represents a convergence of scientific

DOI: 10.1201/9781003479239-16

principles, from semiconductor physics to electrochemistry and photochemistry. It capitalises on the ability of semiconductors to absorb photons and generate charge carriers, which are then used to drive electrochemical reactions.

In this introduction, we will delve deeper into the fundamental mechanisms, challenges, and recent breakthroughs in the various applications of photocatalysis, electrocatalysis, and photo elec-trocatalysis, shedding light on their potential to reshape our energy landscape and address pressing environmental concerns in the years ahead.

NANOMATERIALS FOR CATALYSIS

PHOTOCATALYTIC APPLICATIONS OF NANOMATERIALS

Photocatalytic CO_2 Reduction

Jie-Yinn Tang et al. developed a new approach to engineer the heterostructure of BCN/CN_x by forming p-n junction and isotype simultaneously. The binary composite showed exceptional quali-ties due to strong interfacial contacts and engineered bandgap, and was reliable for photocatalytic CO_2 reduction. The charge dynamics was enhanced by built-in electric field along with p-n junction (Figure 16.1c). Also, the BCN/CN_x showed 11.70 mol/g CH_4 generation with high stability, which was higher than the individual components [21]. Young Ho Park et al. achieved reduction of CO_2 to CO by using 1T/2H-MoS_2@RT by using reaction system with gas phase. The CO_2 reduction was enhanced due to the formation of heterojunction (Figure 16.1d), presence of Ti^{3+} sites and also annealing MoS_2 has resulted in better light absorption. A greater performance and an increased stability were observed for 1T/2H-MoS_2@RT compared to pristine samples in the CO production rate [22]. Peng Wang et al. developed a hybrid of PCN-222/$CsPbBr_3$ for the transformation of CO_2 to HCOOH. The better charge separation and transfer abilities in the heterojunction and the forma-tion of Z-scheme resulted in higher rate of CO_2 reduction. From the DFT studies, a preference was foreseen for HCOOH than the carbonyl pathway, and these results resonated with the experimen-tal results. The optimised catalyst shows a higher HCOOH production of 189.9 μmol/g/h having a selectivity of 100% [23]. Jingxue Wang et al. developed a UiO-type MOF, incorporating Co(II) sites and precisely controlling the coordination environment of Co (II) by altering coordination count of nitrogen. It was found that compared to other UiO-Co-N_x samples, UiO-Co-N_3 exhibited higher CO_2 reduction. Hence, it was found that the number of coordinating atoms of nitrogen sur-rounding the Co site has a significant impact on the photocatalytic performance. In addition, the theoretical calculations revealed the lowest energy barrier in RDS and favourable CO* desorption energy for UiO-Co-N_3 in comparison to its counterparts [24]. Shaoqi Zhang et al. doped Fe to Bi_2O_3, represented it as $Fe_xBi_{2-x}O_3$, and employed them in CO_2 reduction. The optimal introduction of Fe to Bi_2O_3 increased the electron density at the surface and boosted the light absorption. Thus, the potential driving force towards enhancing CO_2 reduction was the high density of electrons at its sur-face due to surface occupation of double metals. $Fe_xBi_{2-x}O_3$ outperformed Bi_2O_3 in CO_2 reduction leading to 30.06 μmol/g/h CO evolution [25].

Photocatalytic Hydrogen Evolution

Dehang Ma et al. developed two catalysts, where one was binary WS_2-WO_3 and the other was ternary WO_3-WS_2-MoS_2 for using them in hydrogen evolution. Out of these, the performance of WO_3-WS_2 exhibited a hydrogen evolution of 1,637 μmol/h/g outperforming the ternary composite. This was primarily due to highly effective Z-scheme WS_2/WO_3 than the heterojunction formed in WO_3-WS_2-MoS_2. The results also showed that when the temperature of calcination was varied from 400°C to 500°C, the photocatalysis decreased abruptly [26]. S. Jayachitra et al. developed $NiSe_2$ nanoparticles by a supercritical fluid–assisted method and then prepared $NiSe_2$/TiO_2 composite. A better performance in hydrogen evolution was seen for the $NiSe_2$/TiO_2 composite after introducing $NiSe_2$. Further studies of them done by DFT as shown in Figure 16.1a and b confirmed that the

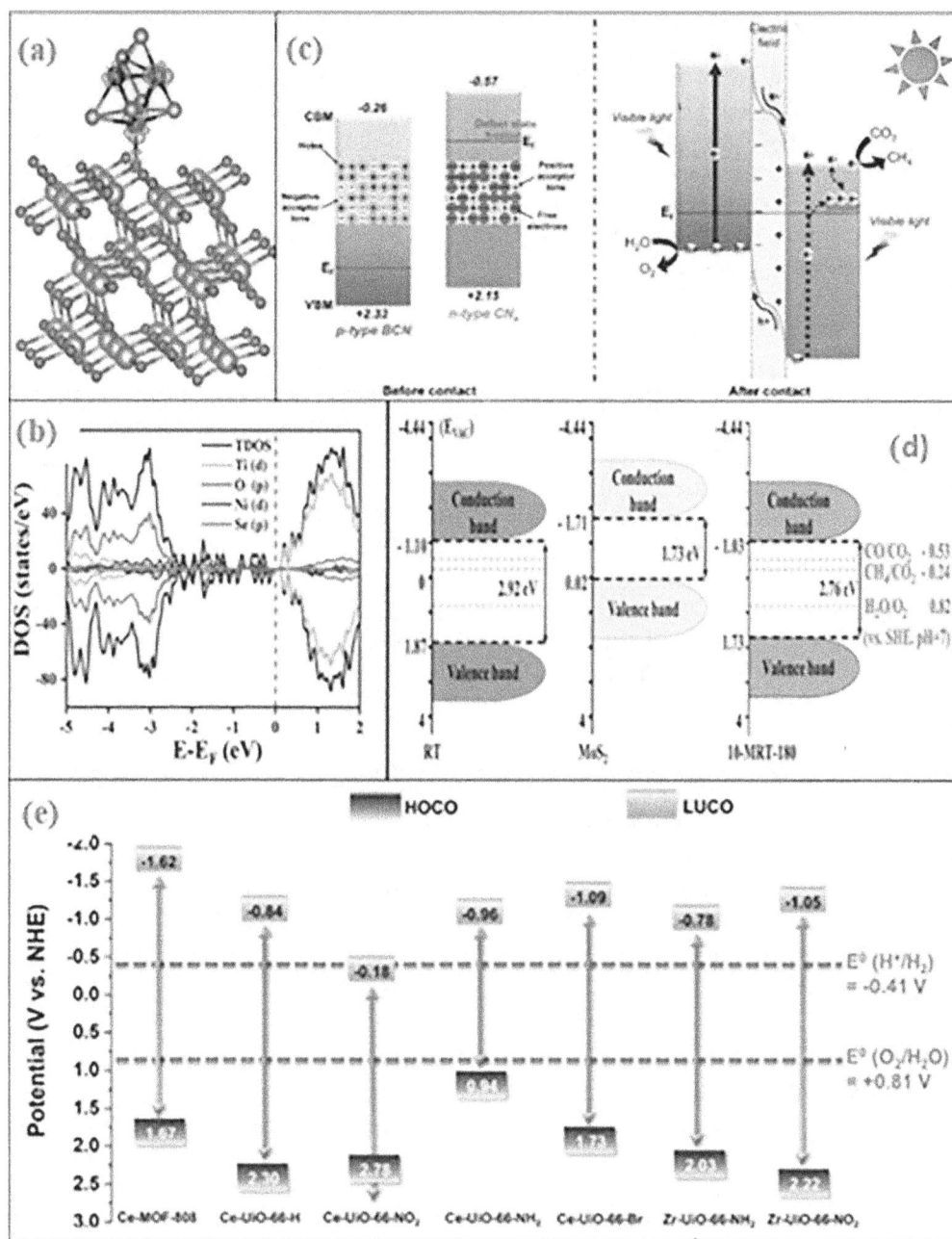

FIGURE 16.1 (a) The charge density differences of TNS nanocomposites are shown. Grey, red, blue, and yellow balls represent Ti, O, Ni, and Se atoms, respectively. The charge density is represented by excess charge light blue (depletion charge-pink colour) iso-surface (value is 0.01 e⁻/Å³). (b) DOS plot of TNS nanocomposite. (c) Schematic representation of the cascade charge transfer across the p-n junction in BCN/CNₓ isotype heterostructure for photocatalytic CO_2 reduction. (d) Band positions in RT, MoS_2, and 10-MRT-180 samples. (e) Energy-level diagram vs. NHE and vacuum of substituted Ce-UiO-66-X and MOF-808(Ce) materials as well as Zr-UiO-66-NH_2 and Zr-UiO-66-NO_2 taken as reference compounds and indication of the required potential for the HER or OER vs. NHE at pH 7.

metallic sites favoured the separation of e^-h^+ pair and electron transfer and elucidated the water splitting process [27]. In their study, Peyman Gholami et al. developed an NLDH-NOG catalyst for exploiting them in hydrogen evolution. Modifications done by oxygen and nitrogen plasma causes structural flaws, improves surface area and hydrophilicity. Due to plasma modifications and the combined effect of oxygen and nitrogen doping, the interfacial contact improved the efficiency in both hydrogen production and sulfanilamide degradation [28]. Fahad A. Alharthi et al. developed $Mg_3V_2O_8$-rGO and displayed a substantial hydrogen evolution of 97.45 µmol/g/h, which was higher than pristine $Mg_3V_2O_8$, and the synergistic effects of rGO together with $Mg_3V_2O_8$-rGO showed reusability up to four cycles [29].

Photocatalytic O_2 Reaction

Laxmi Prasad Rao Pala and Nageswara Rao Peela developed an oxygen evolution reaction (OER) catalyst from IrO_2/TiO_2 films. They observed a linear correlation of OER rate with its thickness. The results demonstrated that the films showed 1.84 µmol/h/cm². Due to adhesion of film to glass substrate, prolonged viability and sustainable functionality were seen for four cycles [30]. Yi Lu et al. developed an AMC heterojunction made of $Ag_3PO_4/MoS_2/g$-C_3N_4 by the electrostatic method combined with ion-exchange. The material was capable of showing 11 times higher OER rate than pristine Ag_3PO_4. The material followed a Z-scheme pathway. Due to the incorporation of MoS_2, it contributed to enhancement in OER by acting as a medium of electron transport and finally produced 232.1 µmol/L/g/h [31]. Shan Dai et al. developed an MOF catalyst based on cerium and showed greater activity in both OER and HER due to LMCT. Additionally, despite considerable cerium leaching, these Ce-based materials demonstrated recyclability, keeping their ability to catalyse and crystal structure for at least three successive applications. Ce-UiO-NH₂ was an outstanding choice because it had the best LUCO-HOCO band energy values for total water splitting (OWS) (see Figure 16.1e for an energy-level diagram). Under simulated solar irradiation, a highly stable, reusable, and active photocatalyst for OWS was produced when Pt NPs were used as a support [32]. Hugo A. Vignolo-González et al. used RONS, i.e., 2D Ru oxide nanosheets, by combining with WO_3 to design a semiconducting material for OER. They compared the activity of ruthenium nanoparticles impregnation on WO_3 and ruthenium nanosheet impregnation on WO_3, and found doubled OER activity in RONS/WO_3 composite due to regulated band edges in RONS and higher optical properties [33].

ELECTROCATALYTIC APPLICATIONS OF NANOMATERIALS

Electrocatalytic CO_2 Reduction

By densely coating sulphur vacancy-rich CdS nanoparticles on Ti_3C_2 using a basic solvothermal technique, Yuwei Wang and others created a variety of MXene composite catalysts for the use of CO_2 reduction in an aqueous electrolyte, which is illustrated in Figure 16.2d. Fast electron transfer was given by the high conductivity of two-dimensional MXene skeleton, which also improved electrolyte infiltration and increased the electrochemical surface area. On the surface of Ti_3C_2 MXene, CdS nanoparticles with many sulphur vacancies were attached, creating active sites for CO_2 reduction. Making use of these advantages, the ideal CdS/Ti_3C_2 exhibited quick CO_2 electroreduction reaction kinetics and a high CO Faraday efficiency of 94% in –1.0 V vs. RHE. According to SEM pictures, nanosheet structure of Ti_3C_2 MXene was preserved in VS-CdS/Ti_3C_2. This two-dimensional shape in electrocatalytic processes offered the advantages of reactant association to active sites and product diffusion [34]. A new g-C_3N_4/Cu_2O-FeO heterogeneous nanocomposite catalyst for electrochemical CO_2 reduction to CO was reported by Girma W. Woyessa and colleagues. This catalyst had a maximum faradaic efficiency of 84.4% at a low onset overpotential of –0.24 V vs. normal hydrogen electrode (NHE). With a high selectivity of 96.3% at –1.60 V vs. Ag/AgCl, the turnover frequency for the conversion of CO_2 to CO reached 10,300 h⁻¹, which translated to a thermodynamic

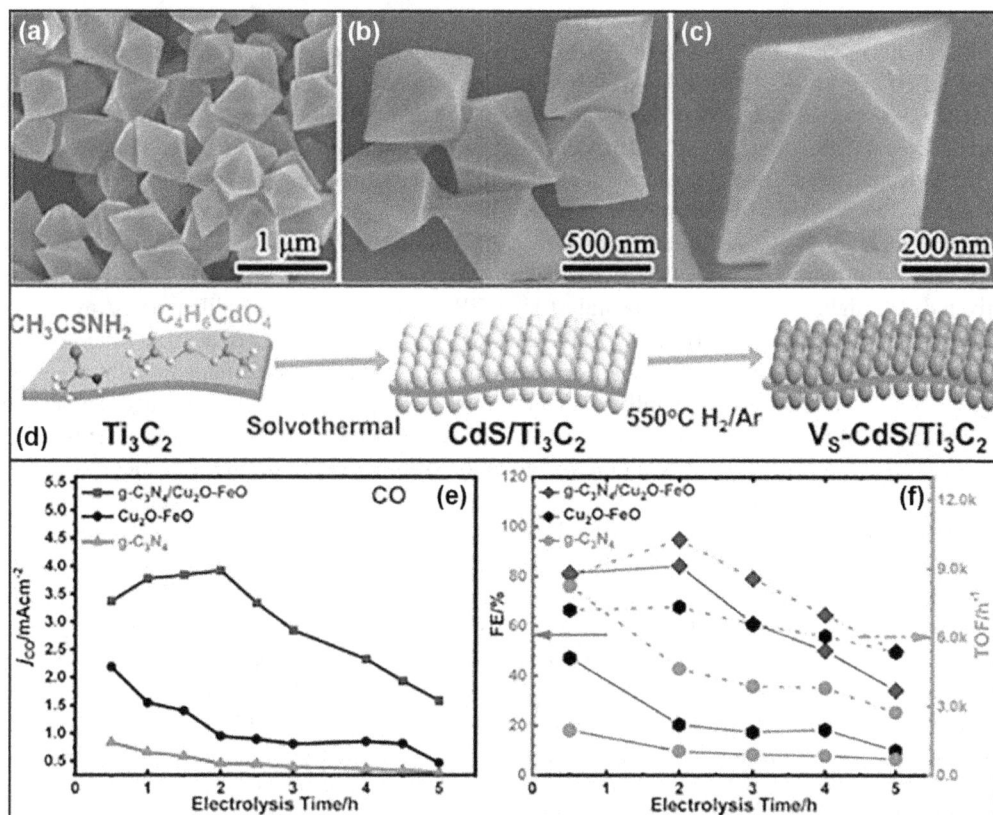

FIGURE 16.2 (a–c) MIL-53(Fe) template particles: SEM images. (d) Schematic illustration of the preparation of VS-CdS/Ti$_3$C$_2$. (e) Partial current density 'jCO'. (f) Plot of TOF and FE for 'CO' vs. electrolysis time using g-C$_3$N$_4$/Cu$_2$O-FeO, Cu$_2$O-FeO, and g-C$_3$N$_4$.

overpotential of −0.865 V vs. NHE. The superior CO$_2$ reduction to CO was due to both the increased surface region that is electrochemically active and the close interfacial interaction between g-C$_3$N$_4$ and the metal oxide (Cu$_2$O-FeO). Small Cu$_2$O and FeO composite particles (~10 nm) placed on the interfaces of g-C$_3$N$_4$ nanosheets, verifying the production of a heterostructure nanocomposite material, gave the pristine g-C$_3$N$_4$ matrix a transparent sheet-like appearance, showing the nanosheet's layered nature. Plot of TOF and FE for 'CO' vs. electrolysis time is given in Figure 16.2e and f [35]. Cu$_3$(BTC)$_2$ metal-organic framework (Cu-MOF) and graphene oxide were used in the research by Sun-Mi Hwang and others on the electroreduction of CO$_2$ to formic acid. The hydrothermal process was utilised to create the electrode with the synthesised material, which was then tested for CO$_2$ electroreduction at varied polarisation potentials. HCOOH was identified as the primary product of reduction by ion chromatography. The greatest HCOOH concentrations generated for various supporting electrolytes were 0.3050 mM (−0.45 V), 0.1404 mM (−0.1 V), 66.57 mM (−0.6 V), 0.2690 mM (−0.5 V), 0.2390 mM (−0.5 V), and 0.7784 mM (−0.4 V). The created catalyst successfully converted and reduced CO$_2$ with a notable degree of efficiency. With 0.1 M TBAB/DMF electrolyte, a high faradic efficacy of 58% was achieved. The MOF had a cubic crystal with a truncated octahedral shape, and their sizes ranged between 19 and 25 μm, according to SEM analysis [36].

Electrocatalytic H$_2$ Evolution

As improved multifunctional electrocatalysts, Dengfeng Wu and colleagues described a simple method for synthesising ternary PtNiCu nanostructures containing a low Pt content, which have accessible surfaces and hollow interiors (H-PtNiCu-AAT NPs). Due to a clever fusion of several structural advantages, H-PtNiCu-AAT NPs displayed remarkable activity and endurance towards HER, OER, and MOR. H-PtNiCu-AAT NPs had a mass activity and a specific activity towards ORR at 0.9 V (versus RHE) that were, respectively, 7.1 and 6.9 times greater than those of commercial Pt/C (0.138 A/mg$_{Pt}$ and 0.212 mA/cm^2). The production of hollow PtNiCu NPs was caused by the galvanic substitution reaction and atomic diffusion among in situ preformed CuNi nanocrystals and Pt species. H-PtNiCu NPs were shown to have hollow and open nanostructures using TEM examination. The face-centred-cubic (fcc) PtNiCu nanoalloys' (111) planes were indexed to the average interplanar spacing, which was determined to be 0.211 nm [37]. Titanium dioxide was activated as a highly effective electrocatalyst for HER by Bowen Ren and colleagues. Amorphous Cu-doped TiO$_2$ was successfully prepared. With a low overpotential of 92 mV at 10 mA/cm^2 in alkaline conditions, the activated TiO$_2$ demonstrated an exceptional HER performance that was significantly higher than that of the crystalline titanium dioxide (over 400 mV). The average diameter of Cu-A in the SEM image was 150–250 nm, and the wire-like shape was clearly visible after the TiO$_2$ deposit. The combined effects of Cu doping and amorphisation led to the catalytic activity, which not only improved electrical conductivity but also reinforced the orbital hybridisation of H1s and O2p, aiding in the stabilisation of the adsorbed water molecule [38]. Meso-Cu-BTC MOF was the focus of Ravi Nivetha and colleagues' investigation, and they used it to further understand its function in the electrocatalytic hydrogen evolution reaction, also called HER in 1 M NaOH solution. The solvothermal technique was not used to carry out the synthesis. The Meso-Cu-BTC electrocatalyst had an onset potential of 25 mV, an overpotential of 89.32 mV, a Tafel slope value of 33.41 mV/dec, an exchange current density of 6.0 mA/cm^2, and no discernible degradation even after 1,000 cycles. These outcomes were mostly attributed to a highly porous Meso-Cu-BTC octahedral MOF network for quick electronic and ionic transfer from the electrocatalyst particle surfaces. According to FESEM pictures, the MOF had cubic crystals with octahedral geometry, with each octahedral edge measuring between 10 and 20 μm in length [39].

Electrocatalytic N$_2$ Reduction

Fe-doped nanoparticles of TiO$_2$ were created by Tongwei Wu and others for electrochemical nitrogen (N$_2$) reduction. They concentrated on converting N$_2$ to NH$_3$. According to high-resolution TEM (HR-TEM) characterisation images of TiO$_2$ and Fe-TiO$_2$ nanoparticles, the lattice fringes of these particles had a distance of 0.349 and 0.352 nm, respectively, which is equivalent to the (101) plane of anatase TiO$_2$. The NRR activity of TiO$_2$ catalyst was increased by Fe, an efficient dopant. Fe-doped TiO$_2$ catalyst achieved a high FE of 25.6% in 0.5 M LiClO$_4$ and a significant NH$_3$ yield of 25.47 g/h/mgcat at −0.40 V, with excellent electrochemical and structural stabilities. According to the results of the DFT calculations, adding Fe to TiO$_2$ (101) caused an increase in oxygen vacancies that further encouraged N$_2$ activation. The strong catalytic performances were due to the synergistic interaction between bi-Ti^{3+} and oxygen vacancy [40]. Shijian Luo and co-workers developed MOF-derived Co$_3$O$_4$@NC with a core–shell structure for nitrogen (N$_2$) electrochemical reduction. A high NH$_3$ production of 42.58 μgh/mg$_{cat}$ and a faradaic efficiency of 8.5% at 0.2 V vs reversible hydrogen electrode in 0.05 M H$_2$SO$_4$ were displayed by the synthesised material. The results showed that the synergistic interactions between N-doped carbon and Co$_3$O$_4$ with considerable oxygen vacancy were the cause of the excellent N$_2$ reduction reaction performance of Co$_3$O$_4$@NCs. TEM pictures of Co$_3$O$_4$@NC samples revealed an entirely different internal structure. The interior structure of Co$_3$O$_4$@NC-5 was stable and resembled the rhombic dodecahedron of ZIF-67. Co$_3$O$_4$@NC-15 featured a porous and hollow inner structure, while Co$_3$O$_4$@NC-10 had a core-shell inner structure. It was hypothesised that the Kirkendall effect might have generated their internal structure [41]. Yuyao Ji and colleagues developed an effective electrocatalyst for ambient N$_2$ reduction using a nanoporous CeO$_2$ nanowire array on a Ti mesh (np-CeO$_2$/TM). This catalyst generated 38.6 μgh/mg$_{cat}$ of NH$_3$ at −0.3 V vs. reversible hydrogen electrode while achieving a high faradaic efficiency of 4.7% in 0.1 M HCl. The MnO$_2$-CeO$_2$ nanowire arrays were visible in the SEM pictures, showing that the np-CeO$_2$/TM structure preserved the nanowire array characteristic. HRTEM also supported an interplanar separation of 0.313 nm, which corresponded to the CeO$_2$ (111) plane [42].

Electrocatalytic O$_2$ Reduction

Using MIL-53(Fe) particles as sacrificial templates, Jingyun Wang and colleagues established an easy MOF-derived approach to create hollow FeOOH polyhedral, which was subsequently attached with dispersive Ni (OH)$_2$ nanosheets. Due to their synergistic effect and distinctive hollow structure, the developed hollow FeOOH@Ni(OH)$_2$ composite displayed improved OER performance compared to either Ni(OH)$_2$ or FeOOH. The composite had a honeycomb surface structure and an average length of about 900 nm and a diameter of around 600 nm as shown in Figure 16.2a–c. The composite displayed an effective OER activity, with a Tafel slope of 70 mV/dec and an overpotential of 310 mV@10 mAcm^{-2}. Due to their distinctive hollow architectures and the interplay between these two materials, the OER performance was improved [43]. In their ground-breaking research, Tae Yong Yoo and colleagues described a novel synthetic technique to directly generate distributed MPt alloy nanomaterials (M=Fe, Co, or Ni) on different carbon substrates with high catalytic loading. The intermetallic L1$_0$-FePt on rGO (37 wt.%-FePt/rGO) microscopic and magnetic experiments revealed homogenous bimetallic alloying with ordered atomic configuration. Following 20,000 potential cycles, the excellent ORR mass activity (1.96 A/mg$_{Pt}$) and specific activity (4.1 mA/cm$_{Pt}$^{2}) of 37 wt.%-FePt/rGO, which are 11.5 and 18.8 times greater than those of commercial Pt/C, respectively, were still present. It was also possible to create other 1:1 Pt-alloy nanoparticles on carbon substrates, such as NiPt/rGO and CoPt/rGO, which have ORR activities similar to those of FePt/rGO. Additionally, the STEM-EDS mapping revealed that an N-doped carbon layer covered the whole surface of FePt/rGO [44]. In order to create carbon fibre paper (CFP), Yun-Hyuk Choi worked on the production of single-phase VO$_2$ (M1 phase) nanoparticles that were uniformly covered on the surface of individual carbon fibres to employ as a highly effective electrocatalyst for the OER. The direct integration of VO$_2$ nanoparticles onto conductive carbon finer paper was accomplished using the vacuum annealing method. The created VO$_2$ nanoparticles, which had a diameter of about 300 nm and were uniform in size, were clearly visible in the FESEM pictures covering the surface of each carbon fibre. In a 1 M aqueous solution of KOH, the VO$_2$/CFP had the highest electrocatalytic OER activity and the lowest 10 (350 mV) and Tafel slope (46 mV/dec) values when compared to the vacuum-annealed V$_2$O$_5$ and the hydrothermally produced VO$_2$ (M1), α-V$_2$O$_5$, and γ'-V$_2$O$_5$. Considered V^{4+} components and V$^{4+/5+}$ redox pairs in VO$_2$ are the catalytically active site. Comparative investigations showed that the oxidation state of V^{4+} was more advantageous for the OER catalysis than that of V^{5+} in vanadium oxide [45].

PHOTO-ELECTROCATALYTIC APPLICATIONS OF NANOMATERIALS

Photo-Electrocatalytic CO$_2$ Reduction

Beatriz Costa e Silva et al. researched how film thickness influences CO$_2$ reduction reaction in the MOF Cu(BDC) that was electrochemically synthesised by the ligand 1,4-benzenedicarboxylate (1,4-BDC) on metallic copper, by anodic deposition for time periods of 30 and 6.5 minutes. Using ideal conditions, 234 mol/L of methanol was produced in 3 hours by photo-electrocatalytic reduction of CO$_2$ using Cu/Cu$_2$O-Cu(BDC) electrodes. The highest concentrations of methanol formed were obtained at applied potentials slightly lower than the flat band potential at +0.22 V, where a potential gradient over the electrode of Cu was modified by Cu (BDC) film. The proposed mechanism of the BDC electrode is shown in Figure 16.3a. This photo-electrocatalytic study shows that the prepared MOF increased the conversion of CO$_2$ to methanol 20 times [46]. Similarly, Alejandro Aranda-Aguirre et al. worked on photo-electrocatalytic conversion of CO$_2$ to methanol with alternative multilayer photoelectrode of fluoride-doped tin oxide FTO/Cu/Bi$_2$Se$_3$-Se/Cu$_2$O [47]. Thus, intralayer interface of n-type chalcogenide bismuth selenide (Bi$_2$Se$_3$) in contact with a thin film of

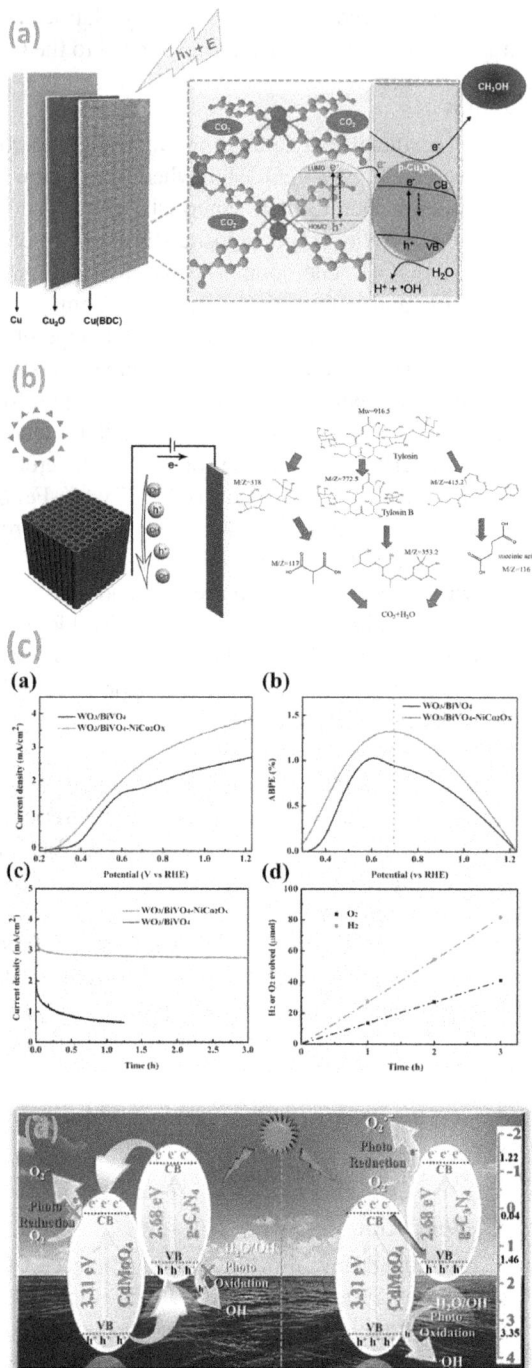

FIGURE 16.3 (a) Schematic representation of the proposed mechanism for CO_2 reduction on the Cu/Cu_2O-Cu (BDC) electrode. (b) Schematic representation of PEC degradation process of tylosin. (c) J-V of $WO_3/BiVO_4$ and $WO_3/BiVO_4þNiCo_2O_x$ photoanodes in 0.1 M potassium borate buffer solution. (a) The applied bias photon-to-current efficiency of $WO_3/BiVO_4$ and $WO_3/BiVO_4þNiCo_2O_x$ photoanode. (b) The J-t plot of $WO_3/BiVO_4þNiCo_2O_x$ photoanode for water oxidation at 0.7 V vs. RHE in 0.1 M potassium borate buffer solution under AM 1.5G illumination. (c) Faradaic efficiency of $WO_3/BiVO_4þNiCo_2O_x$ photoanode. (d) Effective H_2 and O_2 evolution over the time. (d) Photocatalytic dye degradation mechanism of CMO/CN-10 composite photocatalyst.

p-type photo electrocatalyst will create an n-p heterojunction. The characteristics of chalcogenide (Bi_2Se_3) was impacted by the displacement of intralayer deposition, which had an impact on the band edges' electrocatalytic performance. The Iph vs. E curves indicated that Bi_2Se_3 enriched with metallic Se (Bi_2Se_3 electrodeposition potential is −0.1 V) displayed a higher photo-electrocatalytic response with defined reduction peaks for CO_2 reduction with peak potential at −0.6 V. Under photo-electrocatalytic process of TO/Cu/Bi_2Se_3/Cu_2O electrode, there was an enhancement in methanol production of sixfold with an accumulation of 4.50 mM. These findings indicated that the combined effects of light exposure and applied potential enhance artificial photosynthesis processes for the reduction of CO_2 and valorise the photo-electrocatalytic system by coupling of topological insulator/p-type semiconductor. In another study, Wenchao Ma et al. worked on CO_2 reduction to formate by core-shell-structured non-noble-metal Ni@In cocatalyst loaded p-type silicon nanowire arrays (SiNWs) [48], where the Ni@In cocatalyst displayed a core-shell structure, with an approximate Ni@In nanoparticle size of 12.5 nm and an average in shell thickness of 1.8 nm, according to TEM images. The Ni@In core-shell structure was created for Ni@In/SiNWs with 1.0 hour of photo deposition time, and the FE of formate attained the highest values. The FE of formate and the current density on the Ni@In/SiNWs significantly increased to 87% and 3.6 mA/cm^2 at 1.2 V versus RHE, respectively, and were around 12 and 2.0 times higher than those on SiNWs after loading a core-shell-structured Ni@In cocatalyst on SiNWs.

Photo-Electrocatalytic Degradation

The performance of P25 nanoparticulate electrodes with WO_3 on FTO electrodes was compared for their ability for photoelectrochemical degradation of sulfamethoxazole (SMX), inactivation of the viral surrogate, and MS2 bacteriophage [49]. The photocatalytic degradation of sulfamethoxazole and inactivation of MS2 was studied with an electrochemical support. In comparison to P25, WO_3 photoanodes degraded SMX and eliminated MS2 more quickly—nearly 50% of SMX was converted after 2 hours and MS2 was removed three times more quickly. Several advantageous factors were credited for the WO_3 nanostructures' superior performance versus P25, including increased surface area due to the morphology and structure of WO_3. The uniformly developed layer of brick-like nanostructures was vertical, and the thickness of the nanoplates ranged from 40 to 300 nm. The thicknesses were uniform, but the P25 film was close to 50 μm thick, while the WO_3 coating was just around 0.5 μm thick, shown by SEM. Absorption of radiation found to be 480 nm (estimated through the bandgap values). It is also noted that the ratio of absorbed radiation to incident radiation for WO_3 is higher than that of P25, which was estimated to be 1.14 times higher at 365 nm due to the WO_3 nanostructures.

By growing CuO NPs in situ on the surface of BiOCl NSs, a 0D/2D CuO/BiOCl heterojunction was designed. For the photo-electrocatalytic degradation of AFB1 under light irradiation and 0.25 V of bias voltage, the CuO/BiOCl/ITO photoelectrode with an efficient degradation rate of 81.3% was identified. BiOCl NSs could absorb Cu^{2+} as they dispersed in the copper salt solution owing to the presence of many dangling bonds and the unsaturated coordination of surface oxygen. As the temperature of the DMF solution rose, the absorbed Cu^{2+} could further react with the oxygen atoms on the surface of the BiOCl to create CuO crystal nuclei. Finally, a stable and heterostructured CuO/BiOCl composite was produced by the slow development of these CuO crystal nuclei on the surface of BiOCl [50]. Gaolian Zhang et al. worked on phosphorus-doped TiO_2 nanotube arrays (TNTAs/P) by calcining TNTAs with amorphous red phosphorus. The P dopants increased the photo-induced carrier separation efficiency and optical light absorption, which improved the photo-electrocatalytic (PEC) degradation performance. The average tubular diameter and length of the vertically oriented TNTAs were around 50 nm and 10 m, respectively. The nanotubes were still aligned in order following phosphorous doping treatment by SEM images. Tylosin could be eliminated by 80% within the same reaction time, while PEC degradation could remove 63% of total organic carbon (TOC) after 4 hours (Figure 16.3b). Tylosin degradation was examined using the TNTAs/P (0.75) photoanode for the photolysis, photocatalysis (PC), electrocatalysis (EC), and photo electrocatalysis (PEC) processes. It was clear that out of these four methods, the PEC technique had the best removal efficiency. The process of degradation was a little bit slower at high tylosin concentrations (50 mg/L) than it was in tylosin dilutions. Within 250 minutes, the optimised TNTAs/P (0.75) could break down 79% of the tylosin [51].

Photo-Electrocatalytic Water Splitting

To study photo-electrochemical water splitting for the hydrogen evolution process (PEC-HER), a Ru-MoS$_2$ heterostructure catalyst was built and mounted upon a Si photoelectrode Ru-MoS$_2$ which enhanced charge separation and transport by enhancing the internal electric field and decreasing charge transfer resistance at the photoelectrode/electrolyte interface [52]. According to the subsequent HER tests and calculations analysis, the particular heterostructured Ru-MoS$_2$ offered an advantageous electronic structure for catalytic hydrogen evolution, notably for the active hetero-interface sites. Due to the heterostructured RuMoS$_2$ electrical characteristics, the catalyst on the photoelectrode promoted band bending and lowered the electron transport resistance at the junction of the photoelectrode and solution, increasing the efficiency of charge separation and transfer. Thus, considerably high half-cell solar-to-hydrogen conversion efficiency (HC-STH) of 7.28% was achieved at 0.2 V RHE. By electrodepositing 3D worm-like bismuth vanadate (BiVO$_4$) onto a 2D thin tungsten trioxide (WO$_3$) underlayer, a mixed-dimensional structured photoanode was created. This greatly enhanced the photocatalytic activity and the charge separation efficiency. BiVO$_4$ and WO$_3$ were in intimate contact in a 2D/3D heterojunction, which raised the photoanode conductivity and increased the lifespan of the photogenerated charge [53]. NiCo$_2$O$_x$ coated 2D/3D WO$_3$/BiVO$_4$ photoanode exhibited an enhanced photocurrent of 3.85 mA/cm^2 at 1.23 V vs. RHE. Also, it showed stability for over 3 hours, with a faraday efficiency of nearly 100%, and sulphite oxidation photocurrent of 3.85 m^{-2} at 1.23 V v/s. RHE. 92% of photo-induced hole reached the photoanode surface from the bulk in the 2D/3D WO$_3$/BiVO$_4$ heterojunction as shown in Figure 16.3c. S. Sadhasivam et al. employed hydrothermal and SILAR approaches for the synthesis of a 1D heterojunction comprising TiO$_2$ nanorod and metal chalcogenide nanoparticles, where a platform for bias-free solar water splitting with a top-open vertical photoanode structure was created. For fast carrier transport to 1D nanorods bonded to conducting substrates, hetero-epitaxial interface designs on arrays of 1D TiO$_2$ nanorods with CdS and CdSe nanoparticles were ideal. Since CdS/CdSe had a more positive conduction band edge in the bias potential between −600 and +200 mV vs. Ag/AgCl, 15 cycles of solar nanoarrays (TSE15) showed potentially higher water oxidation capability [54]. Throughout the experiment, the surface of the counterelectrode (Pt) evolved the constant H$_2$ bubbles (Figure 16.3c). In the chronoamperometric examination, CdS- and CdSe-deposited TiO$_2$ exhibited a significant photocurrent generation with a constant photocurrent density.

Photo-Electrocatalytic O$_2$ Evolution

On the g-C$_3$N$_4$ (CN) sheets, CdMoO$_4$ (CMO) microspheres were deposited and thus CMO microspheres on CN sheets were created using an easy one-pot in situ hydrothermal process. In a three-electrode system run by a potentiostat/galvanostat CHI, 6273D workstation, electrochemical and photo-electrochemical experiments were made. The CMO/CN composite photocatalyst that had been optimised exhibited remarkable efficiency in the PEC-OER [55]. After hybridising with CN, the absorption edge shifted towards a higher wavelength area in comparison to CMO. The absorption edge steadily changed towards the lower wavelength for the increase in the amount of CMO on CN will further support the establishment of a heterojunction between two semiconductors. The data clearly demonstrated that CMO/CN-10 had the lowest Tafel value, 283 mV/dec, followed by CMO/CN-15 (313 mV/dec), CMO/CN-5 (337 mV/dec), CMO (380 mV/dec), and CN (386 mV/dec). The prepared materials also showed excellent results for photocatalytic dye degradation of organic pollutants as given in Figure 16.3d. Ru$_4$(hmp)$_4$(CO)$_8$] is a cluster molecule synthesized using ruthenium and cubane. The phenomenal application of the same as an effective anodic material for solar water splitting in a photo-electrochemical cell (PEC) with four 2-pyCH$_2$O-bridged ligands is explained [56]. The first cubane Ru_4 ruthenium cluster for water oxidation exhibited low initial potential, quick kinetics, and exceptional photocurrent for the oxygen evolution (anodic) reaction in a photo-electrochemical cell without the need for external cocatalyst, which also had crystal unit cell voids large enough to accommodate a 1.2 Å radius spherical probe. The voids were visible in vertical form as such solvent-accessible gaps could have an impact on the oxidation or reduction reaction's kinetics. The hypothetical half-cell solar-to-hydrogen (HC-STH) efficiency was reported to be 91.92%, which was comparable to recent reports for the OER. Cobalt tungstate solid solutions with iron and manganese were hydrothermally synthesised as $Co_{1-(x+y)}$FexMnyWO$_4$ series. The OER activity also increased by iron and manganese doping. Through LSV measurements, the electrocatalytic activity towards the OER of each sample was assessed [57]. Bandgap reduction in $CoWO_4$ was initiated by co-doping iron and manganese in cobalt. The bandgap was dramatically reduced by cobalt doping of iron, reaching a level that was never before observed in tungstate. By contrast, sample C5 needed a much smaller overpotential of 410 mV to reach 5 mA/cm^2. On the surfaces of the working electrode and counterelectrode, respectively, there was also noticeable bubbling of both hydrogen and oxygen bubbles in large quantities.

CONCLUSION

Multifunctional nanohybrids have evolved as innovative frontiers in the field of nanotechnology, giving astounding potential for a wide range of applications through photocatalytic, electrocatalytic, and photo-electrocatalytic processes. These innovative materials have the potential to revolutionise a variety of industries, from environmental remediation to energy conversion and storage, owing to their customised characteristics and flexible capabilities. As we dive farther into the world of nanohybrids, we get access to an ever-expanding toolbox of answers to urgent global problems. Nanomaterials and catalytic processes can work together in a synergistic way to transform our planet, promoting sustainable growth and welcoming a new era of technological innovation. The revolutionary potential of multifunctional nanohybrids is poised to impact the future with sustained study and innovation.

REFERENCES

[1] W. Nabgan, A.A. Jalil, B. Nabgan, M. Ikram, M.W. Ali, A. Kumar, P. Lakshminarayana, A state of the art overview of carbon-based composites applications for detecting and eliminating pharmaceuticals containing wastewater, *Chemosphere* 288 (2022) 132535. https://doi.org/10.1016/J.CHEMOSPHERE. 2021.132535.

[2] M.S. Raghu, A.S. Alkorbi, K.Y. Kumar, M.K. Prashanth, L. Parashuram, A. Abate, F.A. Alharti, B.H. Jeon, Samarium vanadate affixed sulfur self doped g-C$_3$N$_4$ heterojunction; photocatalytic, photoelectro-catalytic hydrogen evolution and dye degradation, *Int. J. Hydrogen Energy* 47 (2022) 12988–13003.

[3] M.S. Raghu, L. Parashuram, M. Prashanth, K.Y. Kumar, C.P. Kumar, H. Alrobei, Simple in-situ functionalization of polyaniline with boroncarbonitride as potential multipurpose photocatalyst: Generation of hydrogen, organic and inorganic pollutant detoxification, *Nano-Struct. Nano-Objects* 25 (2022) 100667.

[4] M. Ubaidullah, A.M. Al-Enizi, A. Nafady, S.F. Shaikh, K.Y. Kumar, M.K. Prashanth, L. Parashuram, B.H. Jeon, M.S. Raghu, B. Pandit, Photocatalytic CO$_2$ reduction and pesticide degradation over g-C$_3$N$_4$/ Ce$_2$S$_3$ heterojunction, *J. Environ. Chem. Eng.* 11 (2023) 109675.

[5] A. Alsulami, Y.K. Kumarswamy, M.K. Prashanth, S. Hamzada, P. Lakshminarayana, C.B. Pradeep Kumar, B.H. Jeon, M.S. Raghu, Fabrication of FeVO4/RGO nanocomposite: An amperometric probe for sensitive detection of methyl parathion in green beans and solar light-induced degradation, *ACS Omega* 7 (2022) 45239–45252. https://doi.org/10.1021/acsomega.2c05729.

[6] K. Divyarani, S. Sreenivasa, T.M.C. Rao, W. Nabgan, F.A. Alharthi, B.H. Jeon, L. Parashuram, Boosting sulfate radical assisted photocatalytic advanced oxidative degradation of tetracycline via few-layered CoZn@MOF/GO nanosheets, *Colloids Surf. A Physicochem. Eng. Asp.* 671 (2023) 131606. https://doi. org/10.1016/j.colsurfa.2023.131606.

[7] S. Akshatha, V.S. Anusuya Devi, L. Parashuram, S. Sreenivasa, Synergistic effect of samarium doped magnesium zirconate photocatalyst for the degradation of methylene blue dye via efficient charge separation pathway and its photoluminescence studies, *Int. J. Sci. Res. Comput. Sci. Eng. Inf. Technol.* 4 (2019) 334–338.

[8] K.Y. Kumar, M.K. Prashanth, H. Shanavaz, L. Parashuram, F.A. Alharti, B.H. Jeon, M.S. Raghu, Green and facile synthesis of strontium doped Nb_2O_5/RGO photocatalyst: Efficacy towards H_2 evolution, benzophenone-3 degradation and Cr(VI) reduction, *Catal. Commun.* 173 (2023). https://doi.org/10.1016/j.catcom.2022.106560.

[9] L. Parashuram, S. Sreenivasa, S. Akshatha, V. Udayakumar, S.S. Kumar, A non-enzymatic electrochemical sensor based on ZrO_2: Cu(I) nanosphere modified carbon paste electrode for electro-catalytic oxidative detection of glucose in raw Citrus aurantium var. sinensis, *Food Chem.* 300 (2019) 125178. https://doi.org/10.1016/j.foodchem.2019.125178.

[10] C. Sun, X. Liao, P. Huang, G. Shan, X. Ma, L. Fu, L. Zhou, W. Kong, A self-assembled electrochemical immunosensor for ultra-sensitive detection of ochratoxin A in medicinal and edible malt, *Food Chem.* 315 (2020) 126289. https://doi.org/10.1016/j.foodchem.2020.126289.

[11] L. Parashuram, M.K. Prashanth, P. Krishnaiah, C.B.P. Kumar, F.A. Alharti, K.Y. Kumar, B.H. Jeon, M.S. Raghu, Nitrogen doped carbon spheres from Tamarindus indica shell decorated with vanadium pentoxide; photoelectrochemical water splitting, photochemical hydrogen evolution & degradation of Bisphenol A, *Chemosphere* 287 (2022) 132348. https://doi.org/10.1016/J.CHEMOSPHERE.2021.132348.

[12] S. Hamzad, K. Kumar, M.K. Prashanth, D. Radhika, L. Parashuram, F. Alharti, B. Jeon, M.S. Raghu, Boron doped RGO from discharged dry cells decorated Niobium pentoxide for enhanced visible light-induced hydrogen evolution and water decontamination, *Surf. Interfaces* 36 (2023) 102544. https://doi.org/10.1016/j.surfin.2022.102544.

[13] S. Akshatha, S. Sreenivasa, L. Parashuram, V.U. Kumar, S.C. Sharma, H. Nagabhushana, S. Kumar, T. Maiyalagan, Synergistic effect of hybrid Ce^{3+}/Ce^{4+} doped Bi_2O_3 nano-sphere photocatalyst for enhanced photocatalytic degradation of alizarin red S dye and its NUV excited photoluminescence studies, *J. Environ. Chem. Eng.* 7 (2019) 103053. https://doi.org/10.1016/j.jece.2019.103053.

[14] K.Y. Kumar, L. Parashuram, M.K. Prashanth, C.B. Pradeep Kumar, F.A. Alharti, P. Krishnaiah, B.H. Jeon, M. Govindasamy, M.S. Raghu, N-doped reduced graphene oxide anchored with δTa_2O_5 for energy and environmental remediation: Efficient light-driven hydrogen evolution and simultaneous degradation of textile dyes, *Adv. Powder Technol.* 32 (2021) 2202–2212. https://doi.org/10.1016/j.apt.2021.04.031.

[15] M.S. Raghu, L. Parashuram, K.Y. Kumar, B.P. Prasanna, S. Rao, P. Krishnaiah, K.N. Prashanth, C.B.P. Kumar, H. Alrobei, Facile green synthesis of boroncarbonitride using orange peel; Its application in high-performance supercapacitors and detection of levodopa in real samples, *Mater. Today Commun.* 24 (2020) 101033. https://doi.org/10.1016/j.mtcomm.2020.101033.

[16] S. Akshatha, S. Sreenivasa, L. Parashuram, V.U. Kumar, F.A. Alharthi, T.M. Chakrapani Rao, S. Kumar, Microwave assisted green synthesis of p-type Co_3O_4@Mesoporous carbon spheres for simultaneous degradation of dyes and photocatalytic hydrogen evolution reaction, *Mater. Sci. Semicond. Process.* 121 (2021) 105432. https://doi.org/10.1016/j.mssp.2020.105432.

[17] S. Rao Akshatha, S. Sreenivasa, L. Parashuram, M.S. Raghu, K. Yogesh Kumar, T. Madhu Chakrapani Rao, Visible-light-induced photochemical hydrogen evolution and degradation of crystal violet dye by interwoven layered MoS_2/wurtzite ZnS heterostructure photocatalyst, *ChemistrySelect* 5 (2020) 6918–6926. https://doi.org/10.1002/slct.202001914.

[18] W. Nabgan, B. Nabgan, A.A. Jalil, M. Ikram, I. Hussain, M.B. Bahari, T.V. Tran, M. Alhassan, A.H.K. Owgi, L. Parashuram, A.H. Nordin, F. Medina, A bibliometric examination and state-of-the-art overview of hydrogen generation from photoelectrochemical water splitting, *Int. J. Hydrogen Energy* (2023). https://doi.org/10.1016/j.ijhydene.2023.05.162.

[19] B.C. Yallur, V. Adimule, M.S. Raghu, F.A. Alharthi, B. Jeon, L. Parashuram, Solar-light-sensitive Zr/Cu-(H2BDC-BPD) metal organic framework for photocatalytic dye degradation and hydrogen evolution, *Surf. Interfaces* 36 (2023) 102587.

[20] A.G. Alhamzani, T.A. Yousef, M.M. Abou-krisha, K.Y. Kumar, M.K. Prashanth, L. Parashuram, B. Hun, M.S. Raghu, Fabrication of layered In 2 S 3/WS 2 heterostructure for enhanced and efficient photocatalytic CO_2 reduction and various paraben degradation in water, *Chemosphere* 322 (2023) 138235. https://doi.org/10.1016/j.chemosphere.2023.138235.

[21] J.Y. Tang, C.C. Er, X.Y. Kong, B.J. Ng, Y.H. Chew, L.L. Tan, A.R. Mohamed, S.P. Chai, Two-dimensional interface engineering of g-C_3N_4/g-C_3N_4 nanohybrid: Synergy between isotype and p-n heterojunctions for highly efficient photocatalytic CO_2 reduction, *Chem. Eng. J.* 466 (2023) 143287. https://doi.org/10.1016/j.cej.2023.143287.

[22] Y.H. Park, D. Kim, C.B. Hiragond, J. Lee, J.W. Jung, C.H. Cho, I. In, S.I. In, Phase-controlled 1T/2H-MoS$_2$ interaction with reduced TiO$_2$ for highly stable photocatalytic CO$_2$ reduction into CO, *J. CO$_2$ Util.* 67 (2023). https://doi.org/10.1016/j.jcou.2022.102324.

[23] P. Wang, X. Ba, X. Zhang, H. Gao, M. Han, Z. Zhao, X. Chen, L. Wang, X. Diao, G. Wang, Direct Z-scheme heterojunction of PCN-222/CsPbBr$_3$ for boosting photocatalytic CO$_2$ reduction to HCOOH, *Chem. Eng. J.* 457 (2023) 141248. https://doi.org/10.1016/j.cej.2022.141248.

[24] J. Wang, K. Sun, D. Wang, X. Niu, Z. Lin, S. Wang, W. Yang, J. Huang, H.-L. Jiang, Precise regulation of the coordination environment of single Co(II) sites in a metal-organic framework for boosting CO$_2$ photoreduction, *ACS Catal.* 13 (2023) 8760–8769. https://doi.org/10.1021/acscatal.3c01003.

[25] S. Zhang, H. Yu, Y. Wang, Y. Yan, J. Dai, D. Shu, X. Wu, Surface dual metal occupations in Fe-doped FexBi$_2$-xO$_3$ induce highly efficient photocatalytic CO$_2$ reduction, *ACS Appl. Mater. Interfaces* 15 (2023) 25049–25057. https://doi.org/10.1021/acsami.3c02784.

[26] D. Ma, M. Yin, K. Liang, M. Xue, Y. Fan, Z. Li, Simple synthesis and efficient photocatalytic hydrogen production of WO$_3$-WS$_2$ and WO$_3$-WS$_2$-MoS$_2$, *Mater. Sci. Semicond. Process.* 167 (2023) 107788. https://doi.org/10.1016/j.mssp.2023.107788.

[27] S. Jayachitra, D. Mahendiran, P. Ravi, P. Murugan, M. Sathish, Highly conductive NiSe$_2$ nanoparticle as a co-catalyst over TiO$_2$ for enhanced photocatalytic hydrogen production, *Appl. Catal. B Environ.* 307 (2022) 121159. https://doi.org/10.1016/j.apcatb.2022.121159.

[28] P. Gholami, A. Heidari, A. Khataee, M. Ritala, Oxygen and nitrogen plasma modifications of ZnCuCo LDH-graphene nanocomposites for photocatalytic hydrogen production and antibiotic degradation, *Sep. Purif. Technol.* 325 (2023) 124706. https://doi.org/10.1016/j.seppur.2023.124706.

[29] F.A. Alharthi, A. El Marghany, N.A.Y. Abduh, I. Hasan, Hydrothermal synthesis of a magnesium vanadate-functionalized reduced graphene oxide nanocomposite for an efficient photocatalytic hydrogen production, *ACS Omega* 8 (2023) 31493–31499. https://doi.org/10.1021/acsomega.3c04476.

[30] L.P.R. Pala, N.R. Peela, Visible light active IrO$_2$/TiO$_2$ films for oxygen evolution from photocatalytic water splitting in an optofluidic planar microreactor, *Renew. Energy* 197 (2022) 902–910. https://doi.org/10.1016/j.renene.2022.08.017.

[31] Y. Lu, X.K. Cui, C.X. Zhao, X.F. Yang, Highly efficient tandem Z-scheme heterojunctions for visible light-based photocatalytic oxygen evolution reaction, *Water Sci. Eng.* 13 (2020) 299–306. https://doi.org/10.1016/j.wse.2020.12.005.

[32] S. Dai, E. Montero-Lanzuela, A. Tissot, H.G. Baldoví, H. García, S. Navalón, C. Serre, Room temperature design of Ce(iv)-MOFs: From photocatalytic HER and OER to overall water splitting under simulated sunlight irradiation, *Chem. Sci.* 14 (2023) 3451–3461. https://doi.org/10.1039/d2sc05161c.

[33] H.A. Vignolo-González, A. Gouder, S. Laha, V. Duppel, S. Carretero-Palacios, A. Jiménez-Solano, T. Oshima, P. Schützendübe, B.V. Lotsch, Morphology matters: 0D/2D WO$_3$ nanoparticle-ruthenium oxide nanosheet composites for enhanced photocatalytic oxygen evolution reaction rates, *Adv. Energy Mater.* 13 (2023). https://doi.org/10.1002/aenm.202203315.

[34] Y. Wang, R. Du, Z. Li, H. Song, Z. Chao, D. Zu, D. Chong, N. Gao, C. Li, Rationally designed CdS/Ti$_3$C$_2$ MXene electrocatalysts for efficient CO$_2$ reduction in aqueous electrolyte, *Ceram. Int.* 47 (2021) 28321–28327. https://doi.org/10.1016/j.ceramint.2021.06.249.

[35] G.W. Woyessa, J.B. dela Cruz, M. Rameez, C.H. Hung, Nanocomposite catalyst of graphitic carbon nitride and Cu/Fe mixed metal oxide for electrochemical CO$_2$ reduction to CO, *Appl. Catal. B Environ.* 291 (2021). https://doi.org/10.1016/j.apcatb.2021.120052.

[36] S.M. Hwang, S.Y. Choi, M.H. Youn, W. Lee, K.T. Park, K. Gothandapani, A.N. Grace, S.K. Jeong, Investigation on electroreduction of CO$_2$ to formic acid using Cu$_3$(BTC)$_2$ metal-organic framework (Cu-MOF) and graphene oxide, *ACS Omega* 5 (2020) 23919–23930. https://doi.org/10.1021/acsomega.0c03170.

[37] D. Wu, W. Zhang, A. Lin, D. Cheng, Low Pt-content ternary PtNiCu nanoparticles with hollow interiors and accessible surfaces as enhanced multifunctional electrocatalysts, *ACS Appl. Mater. Interfaces* 12 (2020) 9600–9608. https://doi.org/10.1021/acsami.9b20076.

[38] B. Ren, Q. Jin, Y. Li, Y. Li, H. Cui, C. Wang, Activating titanium dioxide as a new efficient electrocatalyst: From theory to experiment, *ACS Appl. Mater. Interfaces* 12 (2020) 11607–11615. https://doi.org/10.1021/acsami.9b21575.

[39] R. Nivetha, A. Sajeev, A.M. Paul, K. Gothandapani, S. Gnanasekar, G. Jacob, R. Sellappan, V. Raghavan, N. Krishna Chandar, S. Pitchaimuthu, S.K. Jeong, A.N. Grace, Cu based Metal Organic Framework (Cu-MOF) for electrocatalytic hydrogen evolution reaction, *Mater. Res. Express* 7 (2020) 114001. https://doi.org/10.1088/2053-1591/abb056.

[40] T. Wu, X. Zhu, Z. Xing, S. Mou, C. Li, Y. Qiao, Q. Liu, Y. Luo, X. Shi, Y. Zhang, X. Sun, Greatly improving electrochemical N_2 reduction over TiO_2 nanoparticles by iron doping, *Angew. Chem. Int. Ed.* 58 (2019) 18449–18453. https://doi.org/10.1002/anie.201911153.

[41] S. Luo, X. Li, B. Zhang, Z. Luo, M. Luo, MOF-derived Co_3O_4@NC with core-shell structures for N_2 electrochemical reduction under ambient conditions, *ACS Appl. Mater. Interfaces* 11 (2019) 26891–26897. https://doi.org/10.1021/acsami.9b07100.

[42] Y. Ji, X. Liu, A nanoporous CeO_2 nanowire array by acid etching preparation: An efficient electrocatalyst for ambient N_2 reduction, *Mater. Adv.* 2 (2021) 3552–3555. https://doi.org/10.1039/d1ma00243k.

[43] J. Wang, S. Li, R. Lin, G. Tu, J. Wang, Z. Li, MOF-derived hollow β-FeOOH polyhedra anchored with α-Ni(OH) 2 nanosheets as efficient electrocatalysts for oxygen evolution, *Electrochim. Acta* 301 (2019) 258–266. https://doi.org/10.1016/j.electacta.2019.01.157.

[44] T.Y. Yoo, J.M. Yoo, A.K. Sinha, M.S. Bootharaju, E. Jung, H.S. Lee, B.H. Lee, J. Kim, W.H. Antink, Y.M. Kim, J. Lee, E. Lee, D.W. Lee, S.P. Cho, S.J. Yoo, Y.E. Sung, T. Hyeon, Direct synthesis of intermetallic platinum-alloy nanoparticles highly loaded on carbon supports for efficient electrocatalysis, *J. Am. Chem. Soc.* 142 (2020) 14190–14200. https://doi.org/10.1021/jacs.0c05140.

[45] Y.H. Choi, VO2 as a highly efficient electrocatalyst for the oxygen evolution reaction, *Nanomaterials* 12 (2022) 939. https://doi.org/10.3390/nano12060939.

[46] B.C.E. Silva, K. Irikura, J.B.S. Flor, R.M.M. Dos Santos, A. Lachgar, R.C.G. Frem, M.V.B. Zanoni, Electrochemical preparation of Cu/Cu_2O-Cu(BDC) metal-organic framework electrodes for photoelectrocatalytic reduction of CO_2, *J. CO_2 Util.* 42 (2020) 101299. https://doi.org/10.1016/j.jcou.2020.101299.

[47] A. Aranda-Aguirre, J. Ojeda, J.F. Brito, S. Garcia-Segura, M.V.B. Zanoni, H. Alarcon, Photoelectrodes of Cu_2O with interfacial structure of topological insulator Bi2Se3 contributes to selective photoelectrocatalytic reduction of CO_2 towards methanol, *J. CO_2 Util.* 39 (2020) 101154. https://doi.org/10.1016/j.jcou.2020.101154.

[48] W. Ma, M. Xie, S. Xie, L. Wei, Y. Cai, Q. Zhang, Y. Wang, Nickel and indium core-shell co-catalysts loaded silicon nanowire arrays for efficient photoelectrocatalytic reduction of CO_2 to formate, *J. Energy Chem.* 54 (2021) 422–428. https://doi.org/10.1016/j.jechem.2020.06.023.

[49] A. Tolosana-Moranchel, N. Pichel, H. Lubarsky, J.A. Byrne, P. Fernández-Ibañez, Photoelectrocatalytic degradation of pharmaceuticals and inactivation of viruses in water with tungsten oxide electrodes, *J. Environ. Chem. Eng.* 10 (2022) 107955. https://doi.org/10.1016/j.jece.2022.107955.

[50] L. Mao, H. Liu, L. Yao, W. Wen, M.M. Chen, X. Zhang, S. Wang, Construction of a dual-functional CuO/BiOCl heterojunction for high-efficiently photoelectrochemical biosensing and photoelectrocatalytic degradation of aflatoxin B1, *Chem. Eng. J.* 429 (2022) 132297. https://doi.org/10.1016/j.cej.2021.132297.

[51] G. Zhang, G. Huang, C. Yang, S. Chen, Y. Xu, S. Zhang, P. Lu, J. Sun, Y. Zhu, D. Yang, Efficient photoelectrocatalytic degradation of tylosin on TiO_2 nanotube arrays with tunable phosphorus dopants, *J. Environ. Chem. Eng.* 9 (2021) 104742. https://doi.org/10.1016/j.jece.2020.104742.

[52] F. Zhang, X. Yu, J. Hu, L. Lei, Y. He, X. Zhang, Coupling Ru-MoS_2 heterostructure with silicon for efficient photoelectrocatalytic water splitting, *Chem. Eng. J.* 423 (2021). https://doi.org/10.1016/j.cej.2021.130231.

[53] P. Wei, Y. Wen, K. Lin, X. Li, 2D/3D WO_3/$BiVO_4$ heterostructures for efficient photoelectrocatalytic water splitting, *Int. J. Hydrogen Energy* 46 (2021) 27506–27515. https://doi.org/10.1016/j.ijhydene.2021.06.007.

[54] S. Sadhasivam, A. Gunasekaran, N. Anbarasan, N. Mukilan, K. Jeganathan, CdS and CdSe nanoparticles activated 1D TiO_2 heterostructure nanoarray photoelectrodes for enhanced photoelectrocatalytic water splitting, *Int. J. Hydrogen Energy* 46 (2021) 26381–26390. https://doi.org/10.1016/j.ijhydene.2021.05.144.

[55] A. Gandamalla, S. Manchala, P. Anand, Y.P. Fu, V. Shanker, Development of versatile $CdMoO_4$/g-C_3N_4 nanocomposite for enhanced photoelectrochemical oxygen evolution reaction and photocatalytic dye degradation applications, *Mater. Today Chem.* 19 (2021) 100392. https://doi.org/10.1016/j.mtchem.2020.100392.

[56] A. Singh, N. Choudhary, S.M. Mobin, P. Mathur, Cubane $Ru_4(CO)_8$ cluster containing 4 pyridine-methanol ligands as a highly efficient photoelectrocatalyst for oxygen evolution reaction from water, *J. Organomet. Chem.* 940 (2021) 121769. https://doi.org/10.1016/j.jorganchem.2021.121769.

[57] M. Athar, M. Fiaz, M.A. Farid, M. Tahir, M.A. Asghar, S. Ul Hassan, M. Hasan, Iron and manganese codoped cobalt tungstates $Co1_{-(x+y)}Fe_xMn_y WO_4$ as efficient photoelectrocatalysts for oxygen evolution reaction, *ACS Omega* 6 (2021) 7334–7341. https://doi.org/10.1021/acsomega.0c05412.

17 Review of Ferrite Nanocomposites as Adsorbents of Heavy Metal Ions from Aqueous Solutions

G.D. Prasanna, Ashwini Rayar, and C.S. Naveen

INTRODUCTION

Unwanted substances alter the quality of water, which is harmful to the environment and human health. Water is a significant source of infection since it is a universal solvent. The WHO estimates that 80% of diseases are water-borne. Numerous nations' drinking water does not adhere to WHO criteria. According to research, 3.1% of deaths are attributable to unclean, low-quality water. Commercial and residential effluent radioactive waste, waste discharge, marine dumping, water tank leaks, and air deposition are all factors that contribute to water pollution. Heavy metals and industrial trash that have been improperly disposed of can build up in lakes and rivers, harming both wildlife and people. The leading cause of immunological suppression, acute poisoning, and impaired reproduction is toxins in industrial waste. Infectious diseases, including typhoid, cholera, and other illnesses that cause vomiting, skin diseases, renal difficulties, and gastroenteritis, spread through contaminated water. Direct nutrient damage to animals has an impact on human health. Sea weed, marine birds, molluscs, marine birds, fish, crustaceans, and other sea animals that are used as food for humans are being killed by water contaminants. Along the food chain, the concentration of insecticides like DDT is rising. Insecticides can pose a danger to human health [1].

Water pollution can be caused by heavy metals, which are highly toxic, is persistent in the environment, and able to accumulate in living organisms. The harmful effects of heavy metal ions such as Pb(II), Cr(II), Mn(II), Ni(II), As(V), Cd(II), and Hg(II) on living organisms are well documented [2]. Heavy metals can enter wastewater from both natural and human-made sources. Natural sources include soil erosion, volcanic activities, and weathering of rocks and minerals. Anthropogenic sources include activities such as mineral processing, fuel combustion, street runoffs, landfills, agricultural activities, and industrial activities such as mining, printed board manufacturing, metal finishing and plating, semiconductor manufacturing, and textile dyes [3]. The global concern over water pollution has increased due to the introduction of pollutants such as dyes, heavy metals, and pharmaceutical drugs. Wastewater streams often contain a harmful combination of heavy metals and dyes, which can have toxic effects on both humans and living organisms. This can cause poisoning symptoms and accumulate within the food chain [4].

Several methods are employed to eliminate heavy metals from substances, such as ion exchange, chemical precipitation, chemical oxidation and reduction, reverse osmosis, ultrafiltration, electrodialysis, and adsorption [5]. Out of all the methods available, the adsorption method is widely acknowledged for effectively eliminating heavy metals from water [6]. The secondary waste generated by the adsorption method is negligible compared to the other techniques.

Effective adsorbents are crucial for efficient adsorption, and as such, researchers have become progressively attentive in developing adsorbents with superior properties [7]. Heavy metal ion elimination requires an adsorbent with sufficient binding sites. Activated carbon, clay, and metal

DOI: 10.1201/9781003479239-17

oxides are commonly employed, but they have limitations such as low adsorption capacities, poor reusability, and lack of functional tunability. To overcome these nano-dimensioned adsorbents, constraints are synthesised and utilised for decontamination. Nanomaterials are efficient due to their high surface area, enhanced active sites, and functional groups on their surfaces. Treating wastewater using the photocatalytic degradation of organic pollutants is a highly efficient method compared to traditional techniques such as desalination, adsorption, and reverse osmosis [8]. In addition, at the nanoscale level, the surface energy is high, and the surface structure is size-dependent, which can lead to the formation of highly active adsorption sites. This results in an increased capacity for adsorption when normalised to the surface area. Nanotechnology allows for the engineering and fabricating of materials with desired structures and functions using nano-sized building blocks, making it highly versatile. Additionally, most atoms on the nanoparticle surface are unsaturated, making it easy to bind with other particles. Therefore, many materials can be modified by changing the surface [9].

Recently, adsorbents with magnetic materials have attracted much consideration from researchers owing to their significant effect in accelerating separation and improving the efficiency of water treatment [10]. Ferrite magnetic nanoparticles are one common magnetic nanoparticle used as adsorbent in adsorption [11]. Magnetic nanoparticles are known for their unique features, which include magnetic separation, biocompatibility, ease of recycling, chemical stability, superparamagnetic, low toxicity and cost, high saturation magnetisation, uniform particle size distribution, and large surface area [12].

Adsorption is a prevalent technique recently owing to its effectiveness (also at low contaminant concentrations), selectivity, renewability of used adsorbents, and cost-efficiency. Due to these properties, the adsorption process is employed in treating industrial wastewater contaminated with heavy metal ions [13].

FERRITE NANOPARTICLES

The history of ferrites and their uses dates back several centuries. As early as 800 B.C., Greek writings described the loadstone (magnetite, Fe_3O_4) as the first natural non-metallic solid that could attract iron. Navigators in those times used magnetite to locate the magnetic North, considered a significant discovery [13]. William Gilbert published the first scientific study of magnetism called "De Magnete". In 1819, Hans Christian Oersted observed that an electric current in a wire could impact a magnetic compass needle. Magnetite, a naturally occurring mineral, is a weak hard ferrite. Hard ferrites have a permanent magnetism and are produced in a limited range of shapes and sizes. On the other hand, soft ferrites are available in countless sizes and shapes, making them suitable for various applications. In electronics, ferrites find their primary usage in three areas: power applications, sensing applications, and electromagnetic interference (EMI) suppression. Due to their possible geometries, material characteristics, and cost-effectiveness, ferrites are widely preferred for both innovative and conventional applications [13].

Ferrite materials have a general molecular formula of MFe_2O_4, where M can be Fe, Co, Ni, Mn, Zn, or Cu. These materials exhibit supermagnetic properties in the nanoscale of 20 nm. In a single unit Cell of ferrite, 64 tetrahedral and 32 octahedral positions are available for cations. However, only eight tetrahedral and 24 octahedral positions are filled by cations. The specific cation occupying each position depends on its affinity towards both sites, and its position affects the chemical and physical properties of the ferrite nanoparticles. Each cation occupies specific places according to its affinity for both sites, and these positions have an impact on the chemical and physical characteristics of nanoparticles. Their placements are additionally influenced by conditions, synthesis procedures, size of interstices, ionic radii, and stabilisation energy [14]. In the MFe_2O_4, the tetrahedral sites are occupied by the divalent ions (M^{2+}) and the trivalent sites are occupied by the majority of trivalent cations (Fe^{3+}).

CLASSIFICATIONS OF FERRITES

Ferrites are categorised based on crystal structure and response to the magnetic field as shown in the flow chart in Figure 17.1.

SPINEL FERRITE

Spinel ferrite has a chemical formula MFe_2O_4, where M denotes divalent metal ions. It encompasses two interstitial sites called tetrahedral (A) and octahedral (B). With a range of cations that can fit into the tetrahedral A site and the octahedral B site, a difference in the characteristics of ferrites can be achieved. Divalent metal ions can replace M to obtain a variety of spinel ferrites. Furthermore, trivalent ions, such as Cr^{3+} and Al^{3+}, can replace Fe^{3+} ions. Fe^{3+}, a combination of divalent and tetravalent ions, can also replace ions (Figure 17.2).

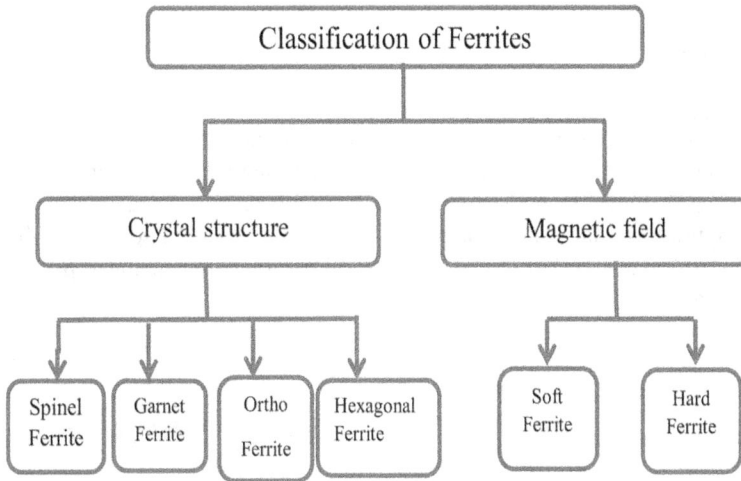

FIGURE 17.1   Classification of ferrites.

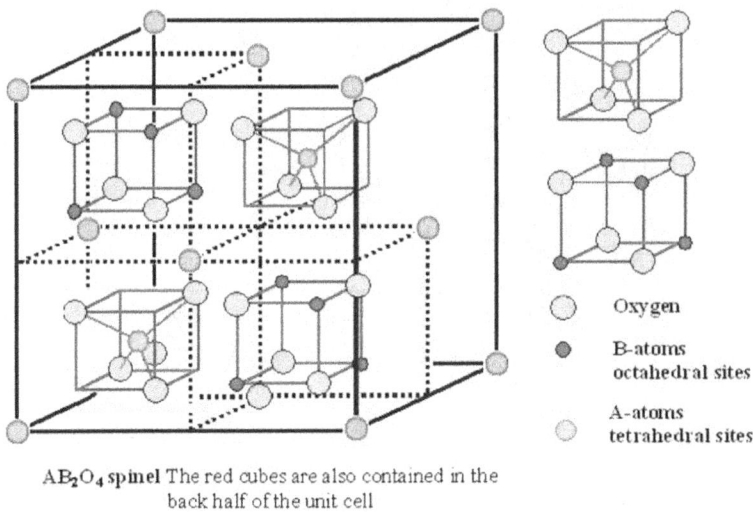

AB_2O_4 spinel The red cubes are also contained in the back half of the unit cell

FIGURE 17.2   Schematic representation of the spinel ferrite [15].

GARNET FERRITES

Ferrimagnetic garnet is represented by $Me_3Fe_5O_{12}$, with Me being a trivalent ion like yttrium or rare earth metals. A cubic unit cell contains eight molecules of $Me_3Fe_5O_{12}$ or 160 atoms. Me ions occupy C sites and are enclosed by eight oxygen ions, while Fe^{3+} ions are distributed over tetrahedral and octahedral sites. In spinels, magnetic alignment is achieved through superexchange interaction via oxygen ions (Figure 17.3).

ORTHO-FERRITES

$MeFeO_3$, Me a massive trivalent metal ion like a rare earth ion, is the general formula for ortho-ferrites. They form crystals with an orthorhombic unit cell and a deformed perovskite structure. Due to the slight canting that occurs when two anti-ferromagnetically connected lattices are aligned, these ortho-ferrites exhibit weak ferromagnetism (Figure 17.4).

HEXAGONAL FERRITES

$MeFe_{12}O_{19}$ is the typical formula for hexagonal ferrites. Me is a divalent ion with a large ionic radius, such as Ba^{2+}, Sr^{+2}, or Pb^{2+}. These ferrites are further divided into the compounds M, W, Y, Z, and U. These have unique crystal structures yet connected. The structure of the M compounds is the least complex. This class includes the barium ferrite. Barium ferrite has a hexagonal structure, and each unit cell has two unit formulae. The structure is comparable to that of spinel, in which the oxygen lattice, FCC, comprises several hexagonal layers of oxygen that are orientated perpendicular to the (111) direction (Figure 17.5 and Table 17.1).

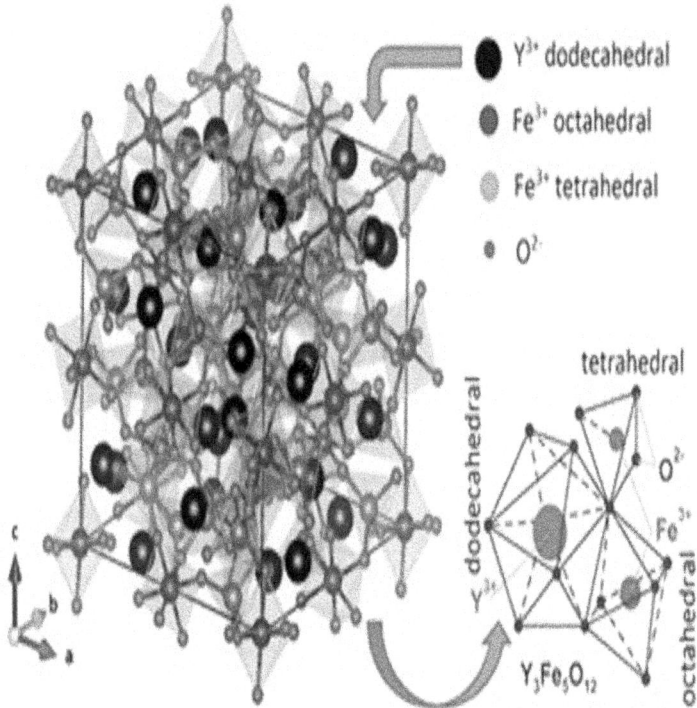

FIGURE 17.3 Unit cell of yttrium iron garnet ($Y_3Fe_5O_{12}$) with three distinct cation crystallographic sub-lattices [16].

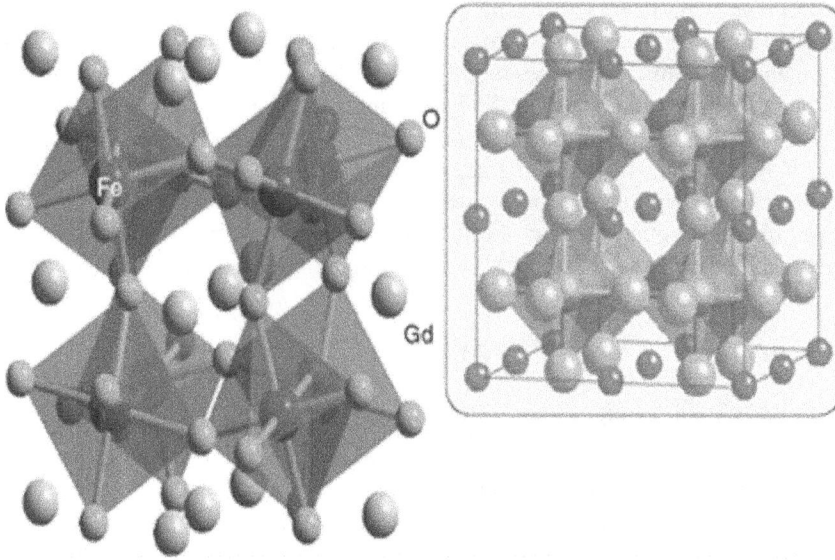

FIGURE 17.4 Structure of $GdFeO_3$ as an example of ortho-ferrite [17].

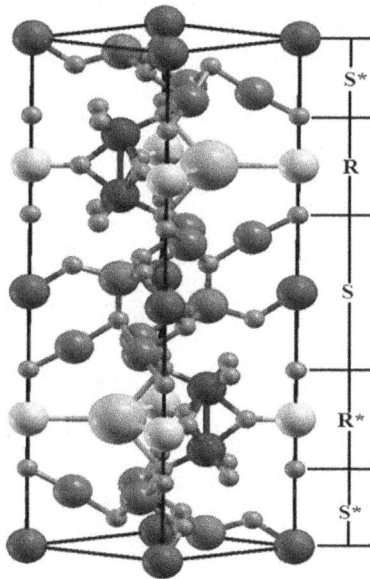

FIGURE 17.5 Crystal structure of M-type barium hexaferrite $BaFe_{12}O_{19}$ [18].

TABLE 17.1
Structure and General Formula of Ferrites

Structure of Ferrites	General Formula	Examples
Spinel ferrite	MFe_2O_4	$ZnFe_2O_4$, $MnFe_2O_4$
Garnet ferrite	$Me_3Fe_5O_{12}$	$Y_3Fe_5O_{12}$, $Gd_3Ga_5O_{12}$, $Tb_3Ga_5O_{12}$
Ortho-ferrite	$MeFeO_3$	$GdFeO_3$, $LaFe_2O_3$, $DyFeO_3$
Hexagonal ferrite	$MeFe_{12}O_{19}$	$BaFe_{12}O_{19}$, $Ba_2Me_2Fe_{12}O_2$, $BaCo_2Fe_{16}O_2$

Ferrites are known as superior magnetic materials compared to pure metals as they have high resistivity, easier manufacture, lower cost, and more excellent magnetisation characteristics. Ferrites are broadly employed in microwave devices, digital recording and audio–video, memory cores of computers, bubble devices, radar, and satellite communication.

Ferrites also have many applications in wastewater treatment, catalysts, electronic devices, and biomedical. The requirements of the characteristics of the ferrites differ in different fields. For effective removal of heavy metals from wastewater, ferrites should exhibit exceptional chemical reactivity and high adsorption capacity. For effective removal of heavy metals from wastewater, ferrites should exhibit exceptional chemical reactivity and high adsorption capacity [13].

SYNTHESIS OF FERRITE NANOPARTICLES

CO-PRECIPITATION METHOD

The co-precipitation technique is extensively used in the preparation of ferrite nanoparticles. This chemical co-precipitation process contains the stoichiometric mixture of ferric and ferrous salts in the ratio of 2:1 (Fe^{3+}/Fe^{2+}) without oxygen and the presence of a base [19].

$$Fe^{2+} + 2Fe^{3+} + 8OH^- \rightarrow Fe_3O_4$$

The pH value should be 8–14 for the formation of Fe_3O_4. The co-precipitation method makes producing a large amount of nanomaterial and governing the particle size possible [19].

The identical-size molecularly imprinted ferrite nanoparticles are synthesised by the co-precipitation method. The Fe_3O_4 MNPs were dispersed ultrasonically in distilled toluene. In molecularly imprinted ferrite, the spinel structure of iron oxide nanoparticles was synthesised by taking the aqueous ferrous and ferric salts in a 1:2 molar ratio and stirring at a fixed temperature. After stirring, it is centrifuged, and the brown residue was obtained and is dried at a specific temperature [20].

The Fe_3O_4/reduced graphene oxide (rGO) nanocomposite was synthesised by dispersing rGO in distilled water under ultrasonication. Then, $FeCl_3.6H_2O$ and $FeSO_4.7H_2O$ were mixed with the help of ultrasonication. Then, the ammonium hydroxide solution was added in drops with vigorous stirring. A black-coloured solid was obtained. It was then filtered and washed several times using ethanol and water [21].

The studies confirm that the 3-aminopropyltriethoxysilane (APS)-modified Fe_3O_4 (Fe_3O_4@APS) was synthesised via a co-precipitation technique with some modifications. The prepared ferrite nanoparticles were disseminated in distilled toluene followed by heating with mechanical agitation under nitrogen while APS was added. The obtained product is washed and dried. Crotonic acid and acrylic acid copolymer-modified ferrite nanoparticles (Fe_3O_4@APS@AA-co-CA) were also synthesised in a similar way.

Nickel ferrite nanoparticles (NFNs) are the finest nanosorbents because of their ability to eliminate heavy metals in terms of their simplicity of operation, low cost, and high efficiency for removing trace levels of heavy metal ions. The co-precipitation technique contains the nickel nitrate and ferric chloride salts mixed in a 1:2 molar ratio. The nickel nitrate hexahydrate was dissolved in distilled water to make the solution of ferric chloride ($FeCl_3$). The sodium hydroxide solution was added dropwise as a precipitating agent to the solution to gain the required pH of 8–9. After constant stirring and keeping at a temperature of 70°C for 2 hours, brown precipitates were formed. Then the obtained precipitate is cooled to room temperature and is washed several times with ethanol (C_2H_5OH) and distilled water to eliminate the unwanted impurities. The nanopower was ground and calcinated at 350°C–550°C for 3 hours in a muffle furnace, and then crushed to get the fine powder of NFNs [22].

Another novel ternary nanocomposite was recorded by using NFNs to increase the adsorption of lead from an aqueous solution, i.e., bentonite/chitosan/$NiFe_2O_4$ ternary nanocomposite. The binary composite of chitosan and nickel ferrite was made. Then bentonite is added to prepare the ternary nanocomposite. The finely powdered nickel ferrite was added to the chitosan and acetic acid solution. To obtain uniform dispersion, the solution is sonicated for 30 minutes. To obtain binary solution, the standard solution of bentonite is added and stirred. Lastly, using the centrifugation method, the ternary composites are collected [23].

The adsorption of Cr(III) and Pb(II) is improved using reduced graphene oxide nickel ferrite nanocomposite (rGONF). The rGONF was synthesised via the co-precipitation method. Graphene oxide (GO) is prepared by Hummer's method. The hydrazine hydrate was gradually added to the solution containing GO and a 2:1 molar ratio of $Fe^{3+}:Ni^{2+}$. Then the temperature of the solution is raised to 120°C with vigorous stirring. The precipitate was washed with distilled water and ethanol. The obtained black rGONF was collected by magnetic separation and dried at certain temperature [24].

The manganese ferrite nanoparticles were also prepared by the co-precipitation technique. The hexahydrate ferric chloride and tetrahydrate manganese chloride were mixed in water, and after 30 minutes, the prepared solution was introduced into a sodium hydroxide solution at 100°C. Then the obtained nanoparticles were separated and washed many times with distilled water. Lastly, the product of manganese ferrite oxide nanoparticles were filtered and dried at higher temperature [25].

To rise the adsorption efficiency, the manganese ferrites are prepared by modified co-precipitation Massart's method. In this process, the manganese sulphate was dissolved in water and hydrochloric acid and stirred. Similarly ferric sulphate is dissolved in deionised water and HCl. Then iron nitrate solution $(Fe(NO_3)_3.9H_2O)$ was mixed with vigorous stirring. Then NaOH solution was added dropwise and stirred until brown precipitate was observed. The mixture was thermally heated under vigorous agitation to obtain black precipitate. The precipitate is then rinsed with deionised water and acetone, and is dried at 45°C [26,27].

The modification of $MnFe_2O_4$ with GO was reported [28]. The $MnFe_2O_4$/GO nanocomposites enhance the adsorption of heavy metals. Here the GO was synthesised via the tour method. The prepared GO suspension was mixed with manganese chloride tetrahydrate and ferric chloride hexahydrate. Then the sonication process is carried out followed by adjustment of the pH to 10. The mixture is stirred vigorously, and the obtained product was washed and dried [29].

The studies conducted by Bharti Verma et al. [30] demonstrate that the cobalt ferrite nanoparticles can be prepared via co-precipitation method. The aqueous solution of citric acid is adjusted to the pH value of 5.7. This solution is added to ammonia, and it is centrifuged. It removes any inefficient stabilised particle and finally obtains the stable magnetic aqueous colloidal solution (Figure 17.6).

SOL-GEL METHOD

In this process, "sol" is the solution containing hydroxylation and condensation of molecular precursors, while "gel" is the three-dimensional network of nanoparticle from the evaporation of solvents. The porous structure of the NFNs was done by the sol-gel method. In the first step of the sol-gel method, the chitosan sol-gel solution was prepared by dissolving the chitosan in 2% (v/v) acetic acid solution. Then the solutions of $Ni(NO_3)_2.6H_2O$ and $Fe(NO_3)_3.9H_2O$ are mixed with 5 ml of water, and the mixture is added slowly into the chitosan solution with constant stirring. The chitosan gel template voids are occupied by the metal solution. The solution is then heated to 80°C. To form the complex precursors, the active organic groups of chitosan could react with the metal ions (Ni^{2+} and Fe^{3+}). The product is calcinated at 400°C for 5 hours. The organic part is then removed and is quenched using ice and water, which results in the 3D porous spinel ferrite. The obtained product was washed with water and is named as PNA (porous $NiFe_2O_4$ adsorbent) [31].

According to the studies conducted by Bharti Verma et al. [32], the NFNs modified by multi-walled carbon nanotubes (MWCNTs) will enhance the adsorption capacity of Cr(VI). The MWCNTs were synthesised by chemical vapour deposition technique. To grow the carbon nanotubes, the iron

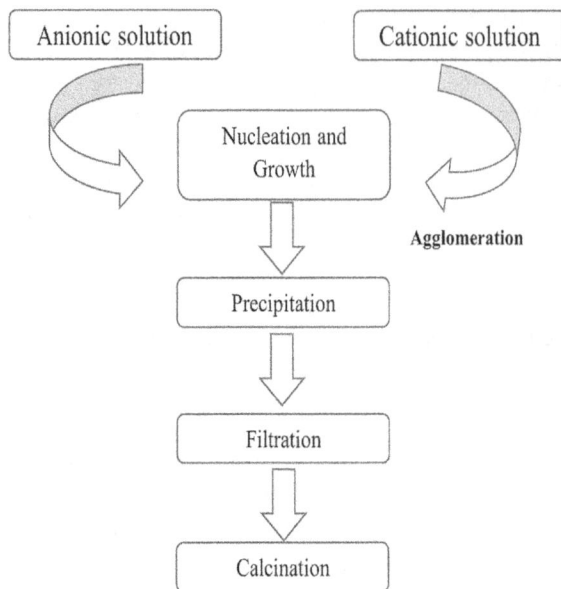

FIGURE 17.6   Schematic representation of co-precipitation method.

catalyst with 5% Fe was loaded. On the surface of the MWCNTs, a carboxyl group was embedded. Acid digestion of MWCNTs with HNO_3 was performed for functionalisation. Then to make the solution pH neutral, it was undergone vacuum filtration. The prepared NFNs are ultrasonically impregnated onto the SW-MWCNTs to synthesise the nickel ferrite nanocomposites (NFNCs).

The studies show that the sol-gel method can prepare the cobalt ferrite nanoparticles. In the preparation, the nitrate salts of cobalt (II) and iron (III) were dissolved in deionised water in the ratio of 1:2. D-Glucose is added to the solution with constant stirring at 60°C. By using a mortar and pestle, the residue can be obtained. Then the residue is calcinated in air. This calcinated product was crushed to make fine power and is called the cobalt ferrite [33]. The sol-gel auto-combustion is also used to make the cobalt ferrite nanoparticles. The specific molar ratio of cobalt nitrate and iron nitrate and citric acid were dissolved in water which forms the "sol". On heating the product to 80°C, a gel-type material was obtained. The $CoFe_2O_4$ nanoparticle is obtained as colour ashes after exposure to auto-ignition (Figure 17.7).

SOLVOTHERMAL METHOD

Another method to synthesise the magnetic nanomaterials is the solvothermal method. It is reported that the cobalt ferrite nanoparticles were synthesised by a solvothermal process in which the ethylene glycol, the combination of cobalt chloride and ferric chloride, is kept in ultrasonication. Polyethylene glycol and sodium acetate were added to the solution. The obtained solution is transferred to the Teflon-scaled autoclave at 200°C. The mixture is washed with ethanol and deionised water [34]. The synthesised cobalt ferrite is modified with GO and SiO_2 to enhance the adsorption capacity (Figure 17.8).

HYDROTHERMAL TECHNIQUE

Hydrothermal technique is another solvothermal method for generating mono-dispersed catalyst particles with controlled size and shape. It is prepared using high-temperature aqueous solution at a high vapour pressure level; hence, it is termed hydrothermal. This process needs a hydrothermal

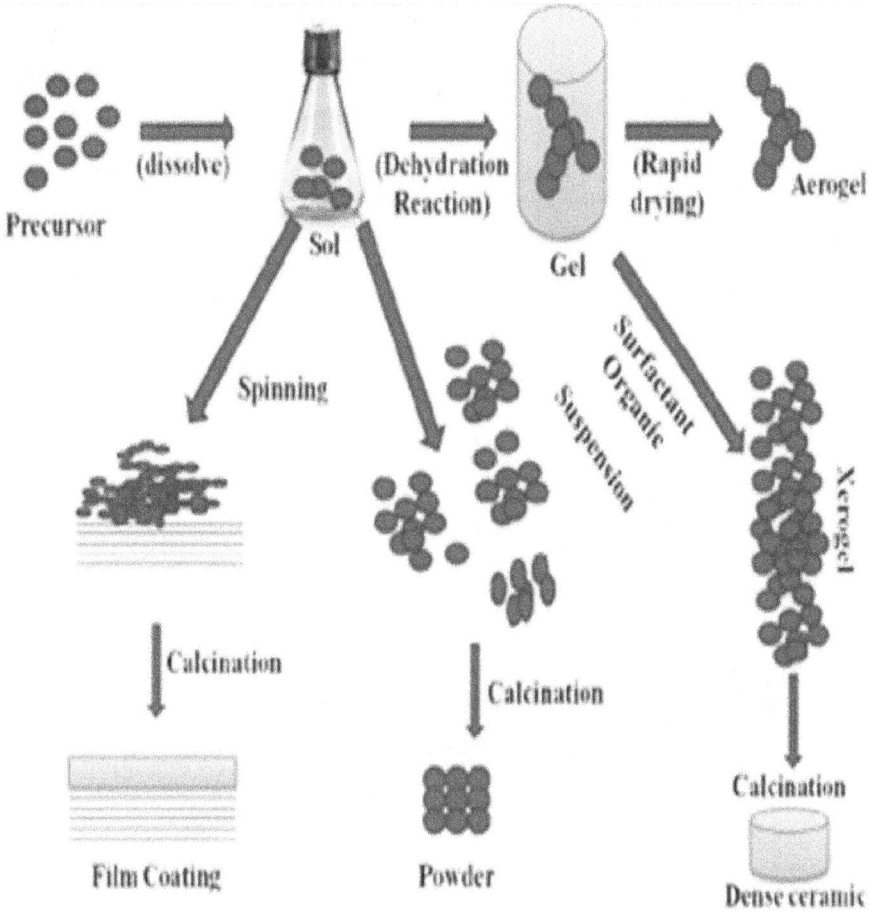

FIGURE 17.7 Schematic representation of sol-gel process.

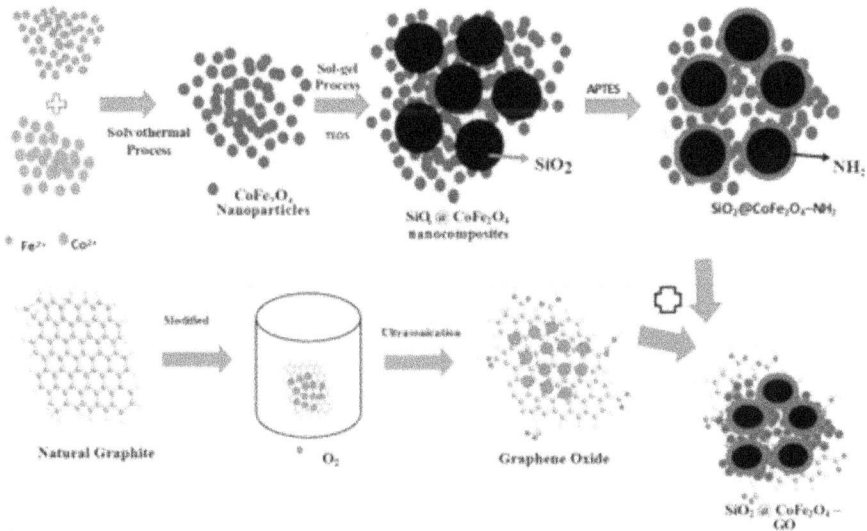

FIGURE 17.8 Schematic representation of solvothermal process.

autoclave reactor. Depending on temperature levels and the solubility of aqueous solution under hot water, the hydrothermal process is defined as an artificial mode to prepare single crystals (nanoparticles). The temperature difference between opposing ends of the autoclave's crystallising compartment requires maintaining a constant temperature. At one end, with a higher temperature where the solvent is dissolved, while the other is moderately cooler, nanoparticle growth occurs.

The magnetic nanoparticles modified with ammonia were synthesised via the hydrothermal reduction technique with ferric chloride solution. The solution is mixed with ethylene glycol to form a clear solution. Then, 1,6-hexanediamine is added to the solution with vigorous stirring. The mixture was poured into Teflon-linked stainless steel autoclave. The final product was exposed to the magnetic field and washed severally with water and ethanol [35] (Figure 17.9 and Table 17.2).

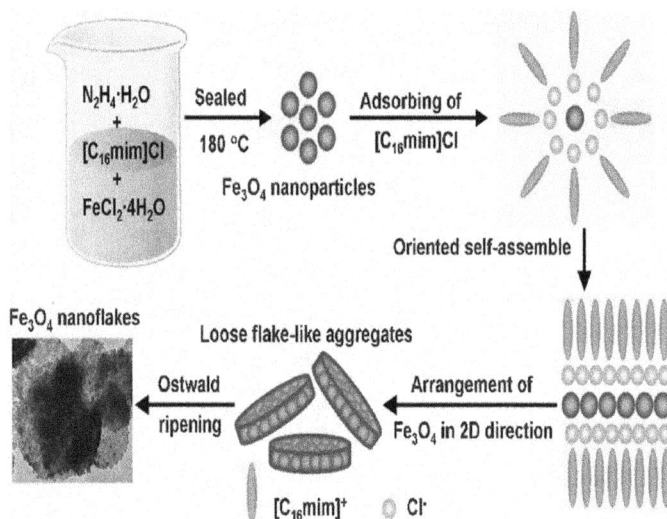

FIGURE 17.9   Representation of hydrothermal process.

TABLE 17.2
Advantages and Disadvantages of Various Synthesis Methods of MNPs

Preparation Techniques	Advantages	Disadvantages
Co-precipitation	• The synthesis method is simple, cheap, and eco-friendly. • High-yield and high-purity product. • Surface contains a large number of hydrophilic ligands.	• The response time is lengthy. • Poor crystalline nanoparticles. • The iron oxide formed under these circumstances displays high polydispersity.
Sol-gel method	• Desired shape and length can be obtained. • Homogeneous and high-adhesion products. • Can be operated at low temperatures. • Can create very fine powers.	• It is expensive. • It is important to take safety precautions during the calcination process because a significant amount of alcohol is released.
Solvothermal method	• Aqueous media is used. • High yield and reasonable control of size.	• High temperature and pressure are used. • The time of reaction is long.
Hydrothermal method	• Eco-friendly. • Reactants have raised reactivity. • Morphology can be altered easily.	• A costly autoclave is needed. • Safety concerns. • Samples of varying particle sizes are collected.

ADSORPTION CAPACITY

The adsorption capacity of the prepared NFNs depends on several factors. Once the adsorption process was complete, a magnet separated the adsorbent from the solution [36].

An equation can be used to calculate the amount of metal ions adsorbed on an adsorbent by finding the difference between the initial concentration of the metal solution and the equilibrium concentration of the metal ions [39,40]:

$$q_e = \frac{V(C_o - C_e)}{m}$$

where q_e is the equilibrium uptake; C_o (mg/L) and C_e (mg/L) are the metal ions' initial concentration and equilibrium concentration, respectively; V (L) is the solution volume; and m (g) is the adsorbent dosage [41,42].

The removal efficiency of the metal ions is calculated by using the following equation:

$$\text{Removal efficiency } (\%) = \frac{(C_o - C_e)}{C_o} \times 100$$

where C_o is the initial concentration and C_e is the equilibrium metal concentration.

ADSORPTION ISOTHERM MODELS

Adsorption isotherm is considered a critical tool to elucidate the interaction between the solutes and adsorbent, and is helpful for optimising adsorbent usage. By studying adsorption kinetics, the rate of adsorption is evaluated. The dependency of the adsorption mechanism on the physical and chemical properties of the adsorbent is determined by the adsorption mechanism [41].

Pseudo-first-order, pseudo-second-order, and intra-particle diffusion models are three non-linear kinetic models that can be applied to calculate the adsorption mechanism [42]. An adsorption isotherm determines the amount of adsorbed ions per unit mass of the adsorbent. The three isotherms are Freundlich, Langmuir, and Dubinin-Radushkevich (D-R). Adsorption isotherms are applied to the equilibrium data.

The Langmuir and Freundlich adsorption isotherm models were selected to investigate the chromium ions adsorption isotherms at a specific optimum pH. The pseudo-second-order model exhibited a good fit in the adsorption of Cr(VI). It is also revealed that the Langmuir isotherm model is best suited for the adsorption of Cr(VI) by cobalt ferrite nanoparticles, and the maximum adsorption capacity evaluated from the Langamuir isotherm model is found to be 10.35 mgg⁻¹ [33,45].

Various nanocomposites synthesised by distinct methods show different adsorption capacities for discrete target ions. Table 17.3 shows the adsorption capacities of other nanocomposites.

FACTORS DEPENDING ON ADSORPTION

There are several factors that depend on adsorption. The factors include the effect of adsorbent dosage, pH, and contact time, which are discussed in detail.

EFFECT OF pH ON THE REMOVAL OF HEAVY METALS

The removal of heavy metals from aqueous solution through adsorption critically depends on the effect of the pH value of the solution, as it controls the surface charge of the adsorbent. To maintain pH for maximum removal efficiency, the experiments are conducted from 2 to 14 under precise conditions [31]. It is observed that the removal efficiency of Pb(II) and Cd(II) ions increased with the increase in pH value [22]. This is due to the higher concentration of H^+ ions in the lower pH value.

TABLE 17.3

Adsorption Capacity of Various Ferrite Nanocomposites

Nanoparticles	Synthesis Method	Target Ions	Adsorption Capacity	References
Fe_3O_4	Co-precipitation	Pb(II), Cu(II), Zn(II), and Mn(II)	Pb^{2+} – 0.180 mmol/g Cu^{2+} – 0.170 mmol/g Zn^{2+} – 0.160 mmol/g Mn^{2+} – 0.140 mmol/g	[43]
$SiO_2@Fe_2O_3$	Co-precipitation	Ni^{2+}	2.64 mol/g	[20]
$Fe_3O_4@ZrP$	Co-precipitation	Hg^{2+}	181.8 mg/g	[44]
Fe_3O_4/rGO	Facile one-step process; modified Hammer's method	As(V), Pb(II), Ni(II)	As(V) – 58.48 mg/g Pb(II) – 65.79 mg/g Ni(II) – 76.34 mg/g	[21]
Fe_3O_4/MnO_2	Hydrothermal process	Cd(II)	53.2 mg/g	[46]
Fe_3O_4/HA	Co-precipitation	Hg(II), Pb(II), Cd(II), and Cu(II)	Hg(II) – 97.7 mg/g Pb(II) – 92.4 mg/g Cd(II) – 50.4 mg/g Cu(II) – 46.3 mg/g	[47]
$Fe_3O_4@APS@AA-CO-CA$	Co-precipitation	Cd(II), Zn(II), Pb(II), and Cu(II)	Cd^{2+} – 29.6 mg/g Zn^{2+} – 43.4 mg/g Pb^{2+} – 166.1 mg/g Cu^{2+} – 126.9 mg/g	[40]
Amino-functionalised Fe_3O_4	Simple one-pot synthesis	Pb(II)	40.10 mg/g	[36]
$NiFe_2O_4$	Co-precipitation	Cr(VI), Pb(II), and Cd(II)	Cr(VI) – 21.5437 mg/g Pb(II) – 19.8773 mg/g Cd(II) – 21.5343 mg/g	[22]
3-D porous $NiFe_2O_4$	Sol-gel method	Pb(II)	48.98 mg/g	[31]
$NiFe_2O_4$ @ SM-MWCNTs	Sol-gel auto-combustion	Cr(VI)	121.9 mg/g	[32]
$NiFe_2O_4$/chitosan/bentonite	Co-precipitation followed by calcination	Pb(II)	50 mg/100 ml	[23]
rGONF	Co-precipitation	Pb(II) and Cr(III)	Pb(II) – 121.95 mg/g Cr(III) – 126.58 mg/g	[24]
$CoFe_2O_4$	Co-precipitation	As(III), As(V)	As(III) – 100 mg/g As(V) – 74 mg/g	[48]
Zn^{2+}-Si-$CoFe_2O_4$	Wet chemical method	Pb(II)	19.8 mg/g	[49]
$CoFe_2O_4/g-C_3N_4$	Hydrothermal process	Pb(II)	32.49 mg/g	[50]
$MgFe_2O_4@CNT$	Sol-gel process	Cr(VI)	175.43 mg/g	[51]
$MnFe_2O_4$	Co-precipitation	As(III) and As(V)	As(III) – 94 mg/g As(V) – 90 mg/g	[48]
$MnFe_2O_4/GO$	"In situ" method	Ni^{2+}	152.67 mg/g	[29]
$HAP/ZnFe_2O_4$	Co-precipitation	Cd(II)	120.33 mg/g	[52]

As the pH of the solution increases, the concentration of H^+ ions decreases, increasing the adsorption of metal ions on the NFNs [22]. Cr(VI) adsorption falls at the rate as pH increases. The maximum removal efficiency is at a pH of 3. As the pH value increases, the removal efficiency of chromium ions gradually decreases.

In the case of NFNs modified by SM-MWCNTs, it is noted that the removal efficiency of Cr(VI) started declining after a pH of 6. There is a complex formation between chromium ions and ferric ions of iron oxide [32]. Surprisingly, iron oxide has more affinity for chromate ions than hydroxyl ions. Therefore, the governing mechanism of Cr(VI) adsorption is electrostatic attraction for acidic conditions and ion exchange for alkaline conditions. The point of zero charge (pH_{pzc}) is the point of pH value at which the electrical charge on the adsorbent surface becomes zero. This pH_{pzc} is less than 7.22 for NFNCs; hence, they become negatively charged and attract negatively charged dichromate easily [32]. Pb(II) adsorption in the case of reduced GO nickel ferrite nanocomposites rapidly increases from pH 2 to pH 3, and then slowly increases at pH 5. However, the adsorption of Cr(III) did not rapidly increase, then slowly rose, and reached a maximum at pH 4 [24].

In the manganese ferrite nanoparticles, the increase in the pH from 2 to 5leads to increase in trivalent chromium (Cr(III)) adsorption. However, Cr(III) adsorption decreases when pH increases from 5 to 11 [25]—in Cr(VI), increasing the pH from 2 to 11 leads to drastically reducing Cr(VI) adsorption. For Cr(III), in acidic pH up to 4, Cr^{3+} was the major ion species in the solution. Some chromium ions were hydrolysed to $Cr_3(OH)_4^{5+}$, $CrOH^{2+}$, and $Cr(OH)_2^+$ by increasing pH up to 6. In alkaline condition, $Cr(OH)_3$, $HCrO_4^-$, and CrO_4^{2-} were the major ion species in the solution [25].

In manganese ferrite modified with graphene oxide (FGO_2) nanocomposites, the adsorption capacity of Ni^{2+} is enhanced with increased pH values. When the pH level is below 5.5, the surface of FGO2 becomes positive. This leads to the formation of nickel ions in the form of Ni^{2+}. As a result, it becomes challenging for the nanocomposite material to bind with these ions [28]. The amount of H^+ ions increases in the solution at low pH, and –OH groups become positively charged to form –OH_2^+, which decreases the adsorption capacity of Ni^{2+} ions [28]. As the pH increases above the pH_{pzc}, the FGO_2 surface is highly negatively charged.

As(III) was better attracted to manganese ferrite NPs at pH 2, depicting improved adsorption capacity. Arsenic adsorption on $MnFe_2O_4$ nanoadsorbent was also almost independent of pH between 2 and 6 [27].

Cobalt ferrite shows heightened Cr(VI) adsorption tendency at pH 3 (76%), and adsorption performance eventually decreases with a rise in the solution basicity pH 11 (26%). Cr(VI) ions occur in the aqueous media as oxy anionic species such as $HCrO_4^-$, CrO_4^{2-}, and $Cr_2O_7^{-2}$ [33]. $HCrO_4^-$ and CrO_4^{2-} are the predominant species up to pH 5, whereas CrO_4^{-2} is dominant at pH > 6.5 [34].

$CoFe_2O_4$/SM-MWCNTs nanocomposites show the maximal removal of Cr(VI) at low pH. The presence of hydronium ions at low pH makes the surface of the adsorbent positive [30]. The adsorption of heavy metals and the interaction between Pb^{2+} species and thiol groups of adsorbent also influence functionalised cobalt ferrite nanoparticles. The adsorption capacity gradually increased with increasing initial pH value of the solution up to 7, while further increase in pH causes hydrolysis of Pb^{2+} [37].

In copper ferrite nanoparticles, the increase in the pH will reach a maximum at pH 6 and then decreases when the removal efficiency is increased. The solution pH is augmented to reduce competition between metal ions and protons, and enhance metal ion adsorption. The removal efficiency decreases at a pH higher than 6, which is owing to the precipitation of Pb(II) in the form of $Pb(OH)_2$ [38] (Figures 17.10 and 17.11).

EFFECT OF CONTACT TIME FOR THE REMOVAL OF HEAVY METALS

The adsorption of heavy metal ions from the aqueous solution increases with the increase in the contact time. It attains an equilibrium level, after which further increase in time does not improve the removal efficiency. NFNs show the maximum removal efficiency of Cr(VI) and Cd(II) ions as 85.21% and 84.45%, respectively, at an equilibrium level of 90 minutes. Pb(II) removal efficiency is 77.41% at 120 minutes equilibrium level [22]. However, the removal efficiency of Pb(II) and Cr(III) reaches 99% at the contact time of 90 minutes in the case of rGONF [24]. In another study, the NFNCs modified by SM-MWCNTs, for the adsorption of Cr(VI) ions reaches 91.6% in the first 60 minutes. This is because, as the contact time increases, the available number of sites reduces [32]. Hence, it is concluded that the different metal ions attain equilibrium at other times.

In manganese ferrite/GO nanocomposites, the Ni^{2+} adsorption capacity increased along with adsorption time. The Ni^{2+} uptake surged for the first 50 minutes and reached the equilibrium

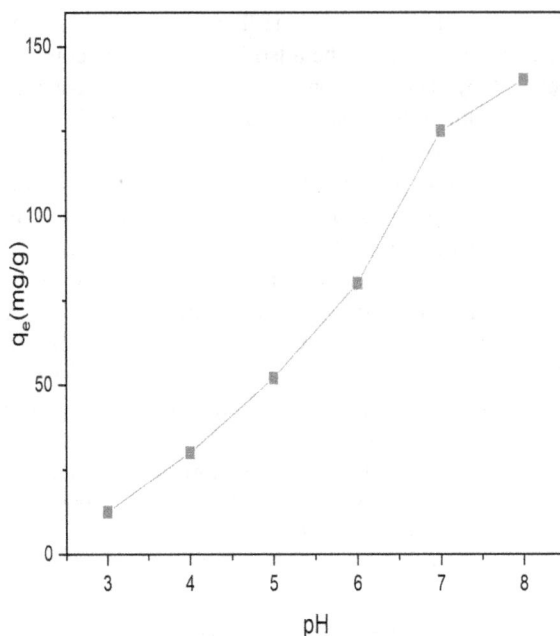

FIGURE 17.10 Effect of pH on the adsorption capacities of FGO$_2$ for Ni^{2+}.

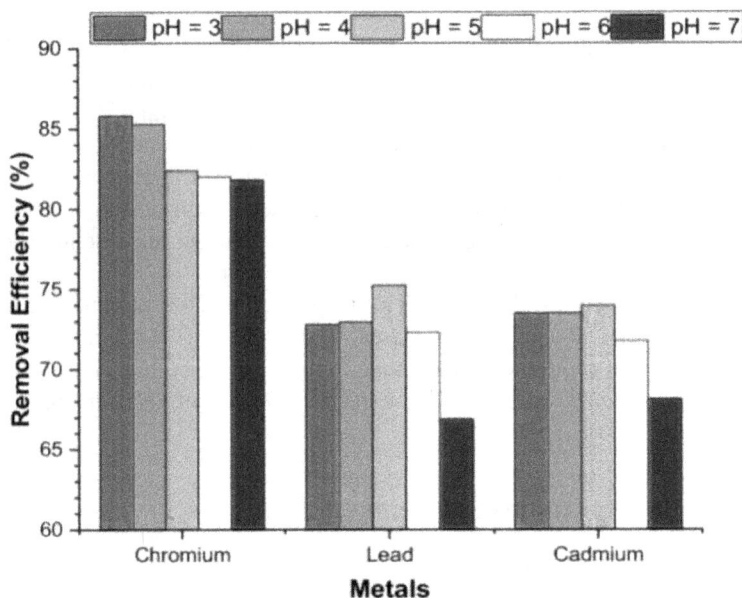

FIGURE 17.11 Effect of pH for removal of Cr(VI), Pb(II), and Cd(II) by adsorption in nickel ferrite nanoparticles.

after 270 minutes. The Ni^{2+} ions gradually filled the nanostructure active sites, leading to a decrease in adsorption capacity when the time increases. Thus, the adequate contact time of manganese ferrite/graphene oxide (FGO$_2$) was 270 minutes [28].

The adsorption of magnetic SiO$_2$@CoFe$_2$O$_4$ nanoparticles decorated on GO also depends on the contact time. The adsorption of Cr(VI) ions increased with time, and after 2 hours, equilibrium was achieved. One hundred and forty minutes was selected as the equilibrium time. This might be owing to the immediate utilisation of most available binding sites on the adsorbent surface [34] (Figures 17.12 and 17.13).

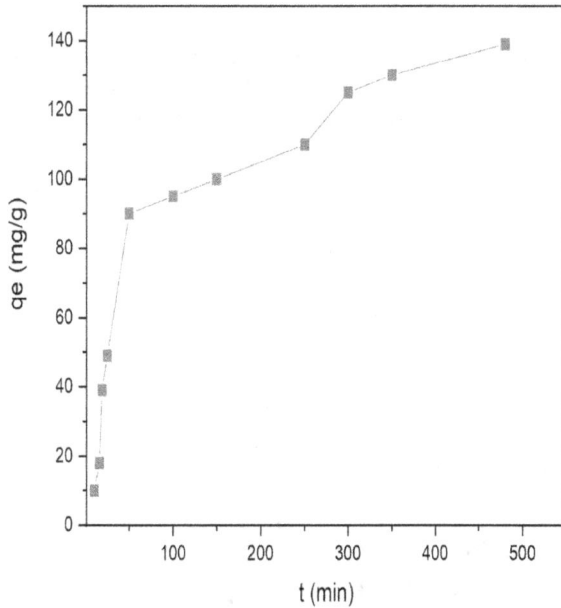

FIGURE 17.12 Effect of contact time on the adsorption capacities of FGO_2 for Ni^{2+}.

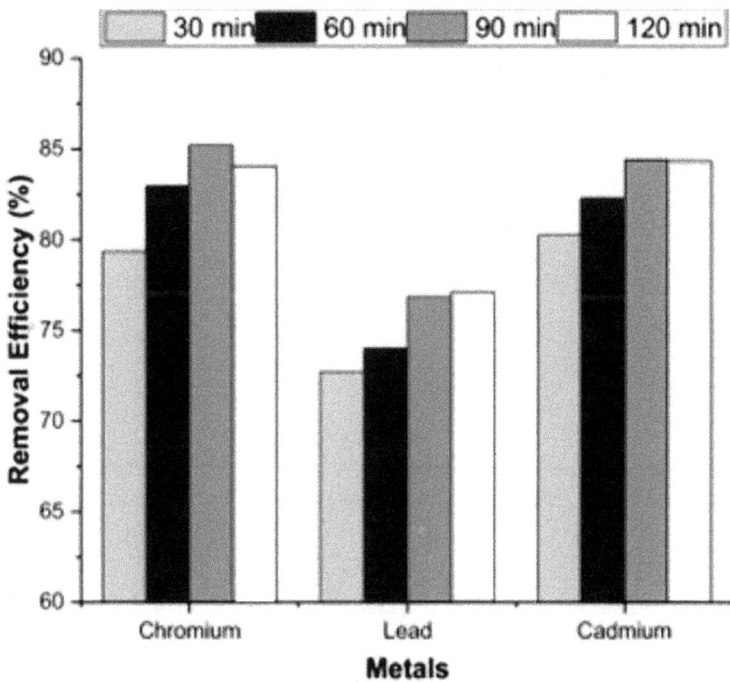

FIGURE 17.13 Effect of contact time for removal of Cr(VI), Pb(II), and Cd(II) by adsorption from nickel ferrite nanoparticles.

EFFECT OF ADSORBENT DOSAGE FOR THE REMOVAL OF HEAVY METALS

The increase in the adsorbent dosage increases the removal efficiency of the metal ions. The NFNs show a maximum removal efficiency of 89.8% for Cr(VI) ions on 30 mg of adsorbent dosage; in comparison, it is 79.47% and 87.24% for Pb(II) and Cd(II) ions on 40 mg of adsorbent dosage, respectively. As the adsorbent dosage increases, the percentage of adsorption also increases, which is due to the increase in adsorbent surface area and availability of more adsorption sites. However, after the optimal value of adsorbent dosage, the adsorption of metal ions decreases due to the overlapping of adsorption sites because of the overcrowding of the adsorbent particles [22]. The reduced GO nickel ferrite adsorbent dosage study revealed an inverse relationship between removal efficiency and adsorption capacity. It has been concluded that the maximum adsorption efficiency is recorded as 99.7% for Pb(II) and Cr(III) with 0.4 g/L rGONF dosage [24]. The removal of As(III) and As(V) form manganese ferrite nanoparticles shows both the active and passive adsorptions and significant differences at each concentration of nanoparticles, although the dynamic adsorption of arsenic was higher than passive adsorption. The increasing adsorbent concentration leads to a decreased arsenic ion adsorption [26].

The adsorption capacity of cobalt ferrite decreases gradually with rising adsorbent dosage. This is due to the availability of enough active surface sites at lower adsorbents [33]. On increasing the adsorbent dose, dynamic sites are available for the same concentration of Cr(VI). As a result, adsorption capacity decreases at a higher adsorbent dosage (Figure 17.14).

ADSORPTION MECHANISM FOR METAL AND THEIR OXIDE-DOPED NANOFERRITES

The ion exchange mechanism provides a clear explanation for the adsorption behaviour of oxide-doped and metal nanoferrites. The concentration of dopants in the nanoferrite sample directly impacts the size and crystalline structure of the nanoparticles. This is due to the variation in atomic

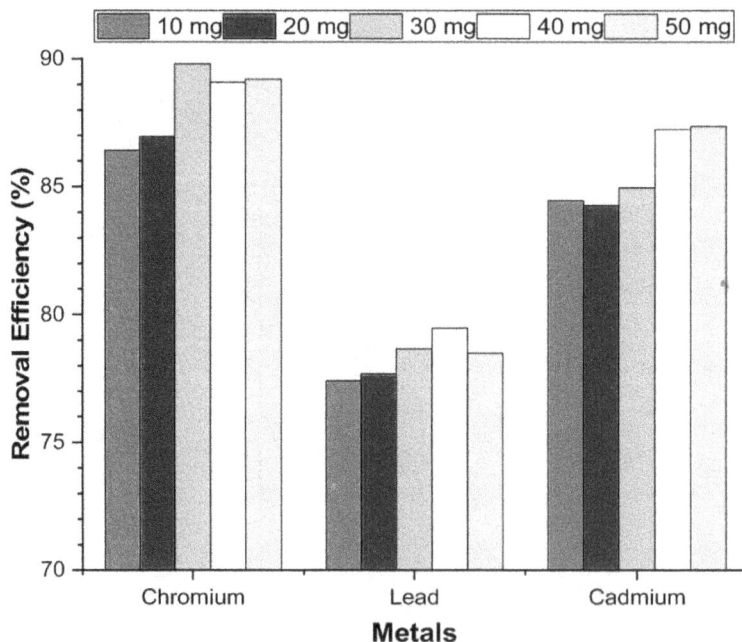

FIGURE 17.14 Effect of adsorbent dosage for removal of Cr(VI), Pb(II), and Cd(II) on nickel ferrite nanoparticles.

size between the central metal ions and the substituted dopant ions on the crystal lattice surface, directly influencing the lattice constant, interplanar spacing, and concentration of free electrons in the crystal lattice. These variations in the crystal lattice structure increase the active sites on the doped surface, which alters the adsorption behaviour of doped nanoferrites. Free electrons on the crystal lattice surface aids in reducing the pollutant's high oxidation state to a lower one, making it less stable and more easily absorbed on the surface of doped surface nanoferrites.

An experiment conducted by Ahalya et al. [53] involved synthesising $MnFe_2O_4$ and $Co_{1-x}Mn_xFe_2O_4$ using a wet chemical technique at a temperature of 900°C. The results of the adsorption behaviour studies showed that $Co_{1-x}Mn_xFe_2O_4$ had higher adsorption efficiency towards Cr(VI) compared to $MnFe_2O_4$. The adsorption of Cr(VI) increased as the cobalt content in $MnFe_2O_4$ risen from 0.2 to 0.6. The maximum adsorption efficiency was observed at a cobalt concentration of 0.6, indicating saturation in the crystal lattice. Additionally, researchers Kuai et al. [54] found that the size of nanoparticles in $Ce_{1-x}Zn_xFe_2O_4$ decreased as the content of Ce(III) ions increased. They prepared the material using the solvothermal technique. $Ni_{1-x}Zn_xFe_2O_4$ was synthesised by Liu et al. [55] via a sol-gel process. They observed that the magnetic composite had a higher adsorption rate for Congo Red (CR) compared to pure $ZnFe_2O_4$, indicating the potential of metal-doped ferrite to enhance adsorption behaviour [56,57–61]. The process and benefits of metal and oxide doping can be seen in Figure 17.15, which highlights the general mechanism of this technique.

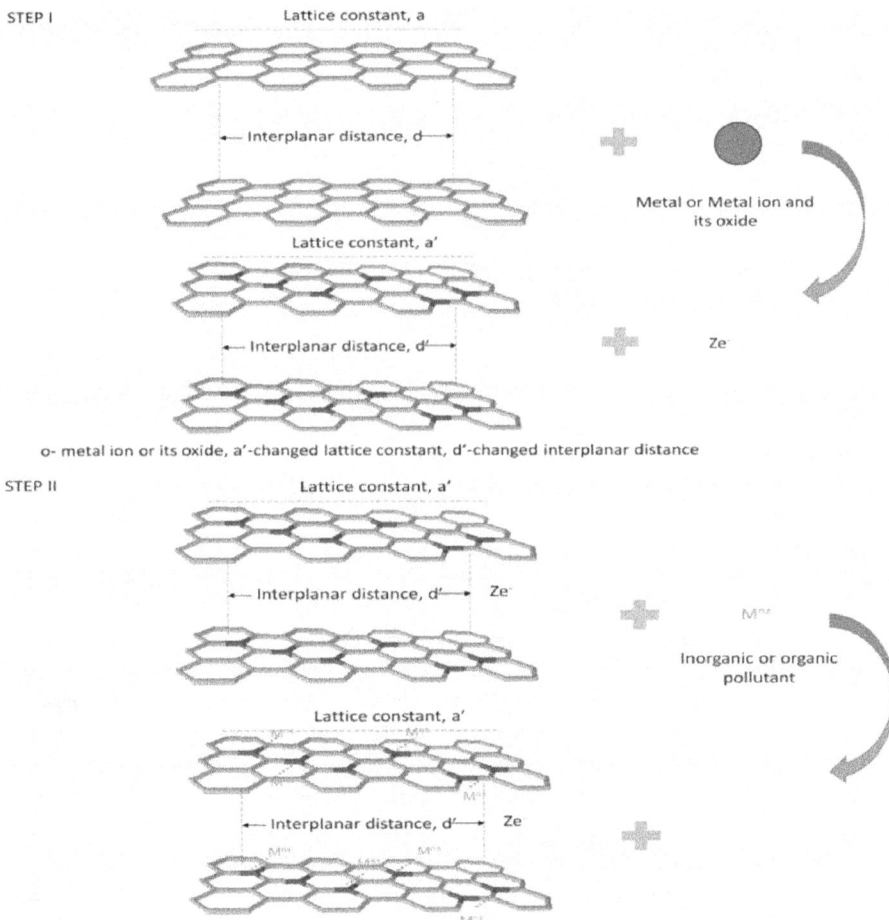

FIGURE 17.15   General adsorption mechanism of metal- and their oxide-doped nanoferrite.

TABLE 17.4

Adsorption Behaviour of Nanoferrites and Their Composites for Removal of Different Inorganic and Organic Pollutants

Adsorbent	Pollutant Adsorbed	% Removal	Adsorption Capacity (Q_{max} in mg/g)	References
$MnFe_2O_4$	Nitrobenzene	-	64.37	[67]
$CuFe_2O_4$	Cd(II)	-	13.9	[63]
$MnFe_2O_4$	Co(II)	99.2	67.4	[68]
$CaFe_2O_4$	Crystal Violet	72.0	10.7	[69]
	Congo Red	95.0	18.2	
$ZnFe_2O_4$	Acid Red 88	-	111.1	[70]
$\gamma - Fe_2O_3$	Cu(II)	-	44.6	[71]
	Cd(II)		42.0	
	Cr(VI)		40.5	
$BiFeO_3$	Rhodamine B	-	64.2	[72]
$CoFe_2O_4$	Congo Red	-	244.5	[73]
$Fe/BaFe_{12}O_{19}$	Methylene Blue	-	189.3	[66]
$ZnFe_2O_4/SDS$	Basic Blue 41	90	42	[74]
	Basic Red 18	80	61	
	Basic Violet 16	61	16	
$Ni_{0.5}Zn_{0.5}Fe_2O_4$	Methyl Blue	-	54.7	[75]
$NiFe_2O_4$/zeolite/sodium alginate	Methylene Blue	92	54	[76]
$NiFe_2O_4$/activated carbon	Direct Red 31	74	229.6	[77]
$ZnFe_2O_4$/chitosan	Brilliant Green	-	20	[78]
	Crystal Violet		14.3	

APPLICATIONS OF NANOFERRITES AND THEIR COMPOSITES IN THE ADSORPTION FIELD

Nanoferrites and their composites are widely used as adsorbents to remove different inorganic and organic pollutants from wastewater [62–65]. Results of literature studies reveal that nanoferrites and their composites show a high tendency to remove different types of organic and inorganic pollutants from wastewater [66]. The detailed review of metal ferrites and their composites as adsorbents for removing other inorganic and organic contaminants is mentioned in Table 17.4.

CONCLUSION

The ferrite nanocomposites exhibit an outstanding high capturing capacity, fast adsorption kinetics, extraordinary affinity, and excellent selectivity that can preferentially adsorb softer heavy metals in various ions in the water. The present study highlights the importance of ferrite nanocomposites as adsorbents for removing heavy metal ions from aqueous solutions. The different types of ferrite nanoparticles and their structures have been studied. The adsorption capacity of varying ferrite nanocomposites and the modifications needed to enhance the adsorption capacity are checked. It has also been investigated that other factors affect the adsorption efficiency, including solution pH, contact time, and adsorption dosage. Other ferrite nanocomposites show additional adsorption capacity for various heavy metal ions. As adsorption capacity increases, the removal of heavy metal ions from aqueous solution also increases. Among the other synthesis methods, the co-precipitation method of synthesis is the simplest as it is easy to control the particle size and composition. It gives high purity and high yield. The ferrites prepared using the co-precipitation method readily absorb the heavy metals in the solution.

The ferrite nanoparticles, such as $MnFe_2O_4$, $CoFe_2O_4$, Fe_3O_4@APS@AA-CO-CA, $MnFe_2O_4$/GO, HAP/$ZnFe_2O_4$, and rGONF, show adsorption of As(V) (90 mgg^{-1}), As(III) (100 mgg^{-1}), Pb(II) (166.1 mgg^{-1}), Ni^{2+} (152.67 mgg^{-1}), Cd(II) (120.33 mgg^{-1}), and Cr(III) (126.58 mgg^{-1}), respectively. Among these, Fe_3O_4@APS@AA-CO-CA, with an adsorption capacity of 166.1 mgg^{-1} for lead ions, exhibits a significant removal capacity. This implies that the modified ferrites adsorb the heavy metal ions to a greater extent. This suggests that the ferrite nanocomposites are one of the best adsorbents of heavy metal ions from an aqueous solution.

REFERENCES

[1] Shweta Wadhawan, Ayushi Jain, Jasamrit Nayyar, Surinder Mehta, Role of nanomaterials as adsorbents in heavy metal ion removal from waste water: A review, *Journal of Water Process Engineering*, 33(2020) 101038. https://doi.org/10.1016/j.jwpe.2019.101038.

[2] Mehtab Haseena, Muhammad Faheem Malik, Asma Javed, Sidra Arshad, Nayab Asif, Sharon Zulfiqar, Jaweria Hanif, Water pollution and human health, *Environmental Risk Assessment and Remediation*, 1(2017) 3. https://doi.org/10.4066/2529-8046.100020.

[3] Alexander E. Burakov, Evgeny V. Galunin, Irina V. Burakova, Anastassia E. Kucherova, Shilpi Agarwal, Alexey G. Tkachev, Vinod K. Gupta, Adsorption of heavy metals on conventional and nanostructured materials for wastewater treatment purposes: A review, *Ecotoxicology and Environmental Safety*, 148(2018) 702–712. https://doi.org/10.1016/j.ecoenv.2017.11.034.

[4] Sonal Agrawal, Nakshatra Bahadur Singh, Removal of arsenic from aqueous solution by an adsorbent nickel ferrite-polyaniline nanocomposite, *Indian Journal of Chemical Technology*, 23(2016) 374–383.

[5] Fenglian Fu, Qi Wang, Removal of heavy metal ions from wastewaters: A review, *Journal of Environmental Management*, 92(2011) 407–418. https://doi.org/10.1016/j.jenvman.2010.11.011.

[6] Gautam Kumar Sarma, Susmita Sen Guptha, Krishna G. Bhattacharyya, Nanomaterials as versatile adsorbents for heavy metal ions in water: A review, *Environmental Science and Pollution Research*, 26(2019) 6245–6278. https://doi.org/10.1007/s11356-018-04093-y.

[7] Sajad Tamjidi, Hossein Esmaeili, Bahareh Kamyab Moghadas, Application of magnetic adsorbents for removal of heavy metals from wastewater: A review study, *Material Research Express*, 6(2019) 102004. https://doi.org/10.1088/2053-1591/ab3ffb.

[8] Gurpinder Singh, Manpreet Kaur Ubhi, Kiran Jeet, Chetan Singla, Manpreet Kaur, A review on impacting parameters for photocatalytic degradation of organic effluents by ferrites and their nanocomposites, *Processes*, 11(2023) 1727. https://doi.org/10.3390/pr11061727.

[9] Renu, Madhu Agarwal, K. Singh, Heavy metal removal from wastewater using various adsorbents: A review, *Journal of Water Reuse and Desalination*, 7(2017) 387–419. https://doi.org/10.2166/wrd.2016.104.

[10] Simona Liliana Iconaru, Régis Guégan, Cristina Liana Popa, Mikael Motelica-Heino, Carmen Steluta Ciobanu, Daniela Predoi, Magnetite (Fe3O4) nanoparticles as adsorbents for As and Cu removal, *Applied Clay Science*, 134(2016) 128–135. https://doi.org/10.1016/j.clay.2016.08.019.

[11] Wahyu Waskito Aji, Edi Suharyadi, Study of heavy metal ions Mn(II), Zn(II), Fe(II), Ni(II), Cu(II), and Co(II) adsorption using MFe_2O_4 (M= Co^{2+}, Mg^{2+}, Zn^{2+}, Fe^{2+}, Mn^{2+}, and Ni^{2+}) magnetic nanoparticles as adsorbent, *Material Science Forum*, 901(2017) 142–148. https://doi.org/4028/www.scientific.net/MSF.901.142.

[12] Dun Chen, Tunsagnl Awut, Bin Liu, Yali Ma, Tao Wang, Ismayil Nurulla, Functionalized magnetic Fe_3O_4 nanoparticles for removal of heavy metal ions from aqueous solutions, *e-Polymers*, 16(2016) 0043. https://doi.org/10.1515/epoly-2016-004.

[13] Zafar Iqbal, Mohd Saquib Tanweer, Masood Alam, Reduced graphene oxide-modified spinel cobalt ferrite nanocomposite: Synthesis, characterization, and its superior adsorption performance for dyes and heavy metals, *ACS Omega*, 8(2023) 6376–6390. https://doi.org/10.1021/acsomega.2c06636.

[14] Kebede K. Kefeni, Titus A.M. Msagati, Bhekie B. Mamba, Ferrite nanoparticles: Synthesis, characterization and applications in electronic device, *Material Science and Engineering*, 215(2017) 37–55. https://doi.org/10.1016/j.mseb.2016.11.002.

[15] Ngonidzashe Masunga, Olga Kelebogile Mmelesi, Kebede K. Kefeni, Bhekie B. Mamba, Recent advances in copper ferrite nanoparticles and nanocomposites synthesis, magnetic properties and application in water treatment: Review, *Journal of Environmental Chemical Engineering*, 7(2019) 103179. https://doi.org/10.1016/j.jece.2019.103179.

[16] Lakshmikanta Aditya, R. Meena, et al., Development of indium doped calcium vanadium garnets for high power circulators at 505.8 MHz for Indus-2, *Indian Particle Accelerator Conference*, 9–12, 2018, RRCAT, Indore.

[17] A. Birajdar, Sagar E. Shirsath, Ram. H. Kadam, Sunil. M. Patange, Lohar Kishan Shankarrao, Dhanraj. R. Mane, A. R. Shitre, Role of Cr^{3+} ions on the microstructure development, and magnetic phase evolution of $Ni0.7Zn0.3Fe_2O_4$ ferrite nanoparticles, *Journal of Alloys and Compounds*, 512(2012) 316–322. https://doi.org/10.1016/j.jallcom.2011.09.087.

[18] Amithava Moitra, Sungho Kim, Seong-Gon Kim, Steven. C. Erwin, Yang-Ki Hong, Jihoon Park, Defect formation energy and magnetic properties of aluminum-substituted M-type barium hexaferrite, *Computational Condensed Matter*, 1(2014) 45–50. https://doi.org/10.1016/j.cocom.2014.11.001.

[19] Pragnesh N. Dave, Lakhan V. Chopda, Application of iron oxide nanomaterials for the removal of heavy metals, *Journal of Nanotechnology*, 14(2014). https://doi.org/10.1155/2014/398569.

[20] Irshad Ahmad, Weqar Ahmad Siddiqui, Tokeer Ahmad, Vasi Uddin Siddiqui, Synthesis and characterization of molecularly imprinted ferrite (SiO_2 @ Fe_3O_4) nanoparticles for the removal of nickel (Ni2+ ions) from aqueous solution, *Journal of Materials Research and Technology*, 8(2019) 1400–1411. https://doi.org/10.1016/j.jmrt.2018.09.011.

[21] Nguyen Thi Vuong Hoan, Nguyen Thi Anh Thu, Hoang Van Duc, Nguyen Duc Cuong, Dinh Quang Khieu, Vien Vo, Fe_3O_4/reduced graphene oxide nanocomposite: Synthesis and its application for toxic metal ion removal, *Journal of Chemistry, Hindawi*, 2016. https://doi.org/10.1155/2016/2418172.

[22] Waheed Ali Khoso, Noor Haleem, Muhammad Anwar Baig, Yousuf Jamal, Synthesis, characterization and heavy metal removal efficiency of nickel ferrite nanoparticles (NFNs), *Scientific Report*, 11(2021) 3790. https://doi.org/10.1038/s41598-021-83363-1.

[23] Debasish Guha Thakurata, Arijita Paul, Krishna Chandra Das, Siddhartha Sankar Dhar. Fabrication of hierarchical Bentonite/Chitosan/$NiFe_2O_4$ ternary nanocomposite and its efficiency in the removal of Pb(II) from aqueous medium, *Environmental Engineering Research*, 26(2021) 200216. https://doi.org/10.4491/eer.2020.216.

[24] Lakshmi Prasanna Lingamdinne, Im-Soon Kim, Jeong-Hyub Ha, Yoon-Young Chang, Janardhan Reddy Koduru, Jae-Kyu Yang, Enhanced adsorption removal of Pb(II) and Cr(III) by using nickel ferrite-reduced graphene oxide nanocomposites, *Journal Metals*, 7(2017) 6. https://doi.org/10.3390/met7060225.

[25] Behzad Eyvazi, Ahmad Jamshidi-Zanjani, Ahmad Khodadadi Darban, Synthesis of nano-magnetic $MnFe_2O_4$ to remove Cr(III) and Cr(VI) from aqueous solution: A comprehensive study, *Environmental Pollution*, 265(2020) 113685. https://doi.org/10.1016/j.envpol.2019.113685.

[26] Sergio Martinez-Vargas, Arturo I. Martínez, Elias E. Hernández-Beteta, Oscar F. Mijangos-Ricardez, Virgilio Vázquez-Hipólito, Cristóbal Patiño-Carachure, Jaime López-Luna, As(III) and As(V) adsorption on manganese ferrite nanoparticles, *Journal of Molecular Structure*, 1154(2018) 524–534. https://doi.org/10.1016/j.molstruc.2017.10.076.

[27] Jaime López-Luna, Loida E. Ramírez-Montes, Sergio Martinez-Vargas, et al. Linear and nonlinear kinetic and isotherm adsorption models for arsenic removal by manganese ferrite nanoparticles, *SN Applied Sciences*, 1(2019) 950. https://doi.org/10.1007/s42452-019-0977-3.

[28] Lu Thi Mong Thy, Nguyen Hoan Kiem, Tran Hoang Tu, Lu Minh Phu, Doan Thi Yen Oanh, Hoang Minh Nam, Mai Thanh Phong, Nguyen Huu Hieu, Fabrication of manganese ferrite/graphene oxide nanocomposites for removal of nickel ions, methylene blue from water, *Chemical Physics*, 533(2020) 110700. https://doi.org/10.1016/j.chemphys.2020.110700.

[29] C. N. Chinnasamy, Aria Yang, S. D. Yoon, Kailin Hsu, Size-dependent magnetic properties and cation inversion in chemically synthesized $MnFe_2O_4$ nanoparticles, *Journal of Applied Physics*, 1010(2007) 509. https://doi.org/10.1063/1.2710218.

[30] Bharti Verma, Chandrajit Balomajumder, Fabrication of magnetic cobalt ferrite nanocomposites: An advanced method of removal of toxic dichromate ions from electroplating wastewater, *Korean Journal of Chemical Engineering*, 37(2020) 1157–1165. https://doi.org/10.1007/s11814-020-0516-3.

[31] D. Harikishore Kumar Reddy, Seung-Mok Lee, Three-dimensional porous spinel ferrite as an adsorbent for Pb(II) removal from aqueous solutions, *Industrial & Engineering Chemistry Research*, 52(2013) 15789–15800. https://doi.org/10.1021/ie303359e.

[32] Bharti Verma, Chandrajit Balomajumder, Synthesis of magnetic nickel ferrites nanocomposites: An advanced remediation of electroplating wastewater, *Journal of the Taiwan Institute of Chemical Engineers*, 112(2020) 106–115. https://doi.org/10.1016/j.jtice.2020.07.006.

[33] Suraj Prakash Tripathy, Raghunath Acharya, Mira Das, Rashmi Acharya, Kulamani Parida, Adsorptive remediation of Cr (VI) from aqueous solution using cobalt ferrite: Kinetics and isotherm studies, *Materials Today: Proceedings*, 30(2020) 289–293. https://doi.org/10.1016/j.matpr.2020.01.534.

[34] Chella Santhosh, Ehsan Daneshvar, Pratap Kollu, Sirpa Peräniemi, Andrews Nirmala Grace, Amit Bhatnagar, Magnetic SiO_2@$CoFe_2O_4$ nanoparticles decorated on graphene oxide as efficient adsorbents for the removal of anionic pollutants from water, *Chemical Engineering Journal*, 322(2017) 472–487. https://doi.org/10.1016/j.cej.2017.03.144.

[35] Yanqing Tan, Man Chen, Yongmei Hao, High efficient removal of Pb (II) by amino-functionalized Fe_3O_4 magnetic nano-particles, *Chemical Engineering Journal*, 191(2012) 104–111. https://doi.org/10.1016/j.cej.2012.02.075.

[36] Fei Ge, Meng-Meng Li, Hui Ye, Bao-Xiang Zhao, Effective removal of heavy metal ions Cd^{2+}, Zn^{2+}, Pb^{2+}, Cu^{2+} from aqueous solution by polymer-modified magnetic nanoparticles, *Journal of Hazardous Materials*, 211(2012) 366–372. https://doi.org/10.1016/j.jhazmat.2011.12.013.

[37] Branka Viltuznik, Aljosa Kosak, Yuriy Zub, Aleksandra Lobnik, Removal of Pb(II) ions from aqueous systems using thiol-functionalized cobalt-ferrite magnetic nanoparticles, *Journal of Sol-Gel Science and Technology*, 68(2013) 3. https://doi.org/10.1007/s10971-013-3072-z.

[38] G. Sreekala, A. Fathima Beevi, R. Resmi, B. Beena, Removal of lead (II) ions from water using copper ferrite nanoparticles synthesized by green method, *Materials Today: Proceedings*, 45(2020) 3986–3990. https://doi.org/10.1016/j.matpr.2020.09.087.

[39] Naim Sezgin, Musa Şahin, Arzu Yalcin, Yuksel Koseoglu, Synthesis, characterization and, the heavy metal removal efficiency of MFe_2O_4 (M=Ni, Cu), *Ekoloji*, 89(2013) 89–96. https://doi.org/10.5053/ekoloji.2013.8911.

[40] K.H. Ranaweera, M.N.C. Grainger, A.D. French, M. R. Mucalo, Construction and demolition waste repurposed for heavy metal ion removal from wastewater: A review of current approaches, *International Journal of Environmental Science and Technology*, 20(2023) 9393–9422. https://doi.org/10.1007/s13762-023-05029-x.

[41] Xiang Cai, Ying Gao, Qian Sun, Zuliang Chen, Removal of co-contaminants Cu (II) and nitrate from aqueous solution using kaolin-Fe/Ni nanoparticles, *Chemical Engineering Journal*, 244(2014) 19–26. https://doi.org/10.1016/j.cej.2014.01.040.

[42] Sippy Kalra, Ranu Gadi, Vijay Bahadur Yadav, Clay based nanocomposites for the removal of heavy metals from water: A review, *Journal of Environmental Management*, 232(2018) 803–817. https://doi.org/10.1016/j.jenvman.2018.11.120.

[43] Liliana Giraldo, Alessandro Erto, Juan Carlos Moreno-Piraján, Magnetite nanoparticles for removal of heavy metals from aqueous solutions: Synthesis and characterization, *Adsorption, Environmental Technologies*, 19(2013) 465–474. https://doi.org/10.1007/s10450-012-9468-1.

[44] Tansir Ahamad, Mu Naushad, Basheer M. Al-Maswari, Jahangeer Ahmed, Zeid A. ALOthman, Saad M. Alshehri, Ayoub Abdullah Alqadami, Synthesis of a recyclable mesoporous nanocomposite for efficient removal of toxic Hg^{2+} from aqueous medium, *Journal of Industrial and Engineering Chemistry*, 53(2017) 268–275. https://doi.org/10.1016/j.jiec.2017.04.035.

[45] Kaichuang Zhang, Xinbao Gao, Qian Zhang, Hao Chen, Xuefang Chen, Fe_3O_4 nanoparticles decorated MWCNTs @ C ferrite nanocomposites and their enhanced microwave absorption properties, *Journal of Magnetism and Magnetic Materials*, 452(2018) 55–63. https://doi.org/10.1016/j.jmmm.2017.12.039.

[46] Eun-Ju Kim, Chung-Seop Lee, Yoon-Young Chang, Yoon-Seok Chang, Hierarchically structured manganese oxide-coated magnetic nanocomposites for the efficient removal of heavy metal ions from aqueous systems, *ACS Applied Matter Interfaces*, 5(2013) 9628–9634. https://doi.org/10.1021/am402615m.

[47] Jing-fu Liu, Zong-shan Zhao, Gui-bin Jiang, Coating Fe_3O_4 magnetic nanoparticles with humic acid for high efficient removal of heavy metals in water, *Environmental Science Technology*, 18(2008) 49–54. https://doi.org/10.1021/es800924c.

[48] Shengxiao Zhang, Hongyun Niu, Yaqi Cai, Xiaoli Zhao, Yali Shi, Arsenite and arsenate adsorption on coprecipitated bimetal oxide magnetic nanomaterials: $MnFe_2O_4$ and $CoFe_2O_4$, *Chemical Engineering Journal*, 158(2010) 599–607. https://doi.org/10.1016/j.cej.2010.02.013.

[49] Arundhati Sengupta, Rohan Rao, D. Bahadur, Zn^{2+}-silica modified cobalt ferrite magnetic nanostructured composite for efficient adsorption of cationic pollutants from water, *ACS Sustainable Chemistry & Engineering*, 5(2017) 1280–1286. https://doi.org/10.1021/acssuschemeng.6b01186.

[50] Davis Jacob Inbaraj, Bagavath Chandran and Chitra Mangalaraj, Synthesis of $CoFe_2O_4$ and $CoFe_2$/g-C_3N_4 nanocomposite via honey mediated sol-gel auto combustion method and hydrothermal method with enhanced photocatalytic and efficient Pb^{2+} adsorption property, *Materials Science and Engineering: B*, 177(2012) 269–27. https://doi.org/10.1088/2053-1591/aafd5d.

[51] Bharti Verma, Chandrajit Balomajumder, Magnetic magnesium ferrite-doped multi-walled carbon nanotubes: An advanced treatment of chromium-containing wastewater, *Environmental Science and Pollution Research International*, 12(2020) 13844–13854. https://doi.org/10.1007/s11356-020-07988-x.

[52] Krishna Chandra Das, Siddhartha S. Dhar, Removal of cadmium(II) from aqueous solution by hydroxyapatite-encapsulated zinc ferrite (HAP/ZnFe$_2$O$_4$) nanocomposite: Kinetics and isotherm study, *Environmental Science and Pollution Research*, 27(2020) 37977–37988. https://doi.org/10.1007/s11356-020-09832-8.

[53] K. Ahalya, N. Suriyanarayanan, V. Ranjithkumar, Effect of cobalt substitution on structural and magnetic properties and chromium adsorption of manganese ferrite nano Particles, *Journal of Magnetism and Magnetic Materials*, 372(2014) 208–213. https://doi.org/10.1016/j.jmmm.2014.07.030.

[54] Sanke Kuai, Zhibin Zhang, Zhaodong Nan, Synthesis of Ce^{3+} doped ZnFe$_2$O$_4$ self-assembled clusters and adsorption of chromium(VI), *Journal of Hazardous Materials*, 250(2013) 229–237. https://doi.org/10.1016/j.jhazmat.2013.01.074.

[55] Ruijiang Liu, Hongxia Fu, Hengbo Yin, Peng Wang, Lu Lu, Yuting Tao, A facile sol combustion and calcinations process for the preparation of magnetic Ni$_{0.5}$Zn$_{0.5}$Fe$_2$O$_4$ nanopowders and their adsorption behaviors of Congo red, *Powder Technologies*, 274(2015) 418–425. https://doi.org/10.1016/j.powtec.2015.01.045.

[56] Mamata. Mohapatra, S. Anand, Synthesis and applications of nano-structured iron oxides/hydroxides-a review, *International Journal of Engineering Science and Technologies*, 2(2010) 127–146. https://doi.org/10.4314/ijest.v2i8.63846.

[57] Grégorio Crini, Recent developments in polysaccharide-based materials used as adsorbents in wastewater treatment, *Progress in Polymers Science*, 30(2005) 38–70. https://doi.org/10.1016/j.progpolymsci.2004.11.002.

[58] Muhammad Rahim and Mas Rosemal Hakim Mas Haris, Application of biopolymer composites in arsenic removal from aqueous medium: A review, *Journal of Radiation Research and Applied Sciences*, 8(2015) 255–263. https://doi.org/10.1016/j.jrras.2015.03.001.

[59] Yao-Jen Tu, Chen-Feng You and Chien-Kuei Chang, Kinetics and thermodynamics of adsorption for Cd on green manufactured nano-particles, *Journal of Hazard Materials*, 235(2012) 116–122. https://doi.org/10.1016/j.jhazmat.2012.07.030.

[60] Naef A. A. Qasem, Ramy H. Mohammed, Dahiru U. Lawal, Removal of heavy metal ions from wastewater: A comprehensive and critical review, *Npj Clean Water*, 4(2021) 36. https://doi.org/10.1038/s41545-021-00127-0T.

[61] Ruhma Rashid, Iqrash Shafiq, Parveen Akhter, Muhammad Javid Iqbal, Murid Hussain A state-of-the-art review on wastewater treatment techniques: The effectiveness of adsorption method, *Environmental Science and Pollution Research*, 28(2021) 9050–9066. https://doi.org/10.1007/s11356-021-12395-x.

[62] Xinchun Yang, Zhou Wang, Maoxiang Jing, Ruijiang Liu, Fuzhan Song and Xiangqian Shen, Magnetic nanocomposite Ba-ferrite/α-iron hollow microfiber: A multifunctional 1D space platform for dyes removal and microwave absorption, *Ceramic International*, 40 (2014) 15585–15594. https://doi.org/10.1016/j.ceramint.2014.07.035.

[63] Muhammad Aslam Malana, Raheela Beenish Qureshi and Muhammad Naeem Ashiq, Adsorption studies of arsenic on nano aluminium doped manganese copper ferrite polymer (MA, VA, AA) composite: Kinetics and mechanism, *Chemical Engineering Journal*, 172(2011) 721–727. https://doi.org/10.1016/j.cej.2011.06.041.

[64] Neelam Kumar, Naveen Chandra Joshi, Potential of PTH-Fe$_3$O$_4$ based nanomaterial for the removal of Pb (II), Cd (II), and Cr (VI) ions, *Journal of Inorganic Organometallic Polymers*, 32(2022) 1234–1245. https://doi.org/10.1007/s10904-021-02173-0.

[65] Abbas Afkhami, Samira Aghajani, Milad Mohseni and Tayyebeh Madrakian, Effectiveness of Ni$_{0.5}$Zn$_{0.5}$Fe$_2$O$_4$ for the removal and preconcentration of Cr(VI), Mo(VI), V(V) and W(VI) oxyanions from water and wastewater samples, *Journal of the Iranian Chemical Society*, 12(2015) 2007–2013. https://doi.org/10.1007/s13738-015-0675-z.

[66] Mukesh Kumar, Harmanjit Singh Dosanjh, Sonika, Jandeep Singh, Kamarul Monir, Harminder Singh, Review on magnetic nanoferrites and their composites as alternatives in waste water treatment: Synthesis, modifications and applications, *Environmental Science: Water Research & Technology*, 6(2020) 491–514. https://doi.org/10.1039/C9EW00858F.

[67] Roshni Rathore, Aakash Waghmare, Sarita Rai, and Vimlesh Chandra, Removal of nitrobenzene from aqueous solution using manganese ferrite nanoparticles, *Inorganic Chemistry Communications*, 153(2023) 110848. https://doi.org/10.1016/j.inoche.2023.110848.

[68] Varsha Srivastava, Sharma Yogesh. Chandra and Milka Sillanpää, Application of nano-magnesso ferrite (n-MgFe$_2$O$_4$) for the removal of Co^{2+} ions from synthetic wastewater: Kinetic, equilibrium and thermodynamic studies, *Applications of Surface Science*, 338(2015) 42–54. https://doi.org/10.1016/j.apsusc.2015.02.072.

[69] Shuai An, Xueyan Liu, Lijun Yang and Lei Zhang, Enhancement removal of crystal violet dye using magnetic calcium ferrite nanoparticle: Study in single- and binary-solute systems, *Chemical Engineering Research and Design*, 94(2015) 726–735. https://doi.org/10.1016/j.cherd.2014.10.013.

[70] Wojciech Konicki, Daniel Sibera, Ewa Mijowska, Zofia Lendzion-Bieluń and Urszula Narkiewicz, Equilibrium and kinetic studies on acid dye Acid Red 88 adsorption by magnetic $ZnFe_2O_4$ spinel ferrite nanoparticles, *Journal of Colloid and Interface Science*, 398(2013) 152–160. https://doi.org/10.1016/j.jcis.2013.02.021.

[71] Andra Predescu and Avram Nicolae, Adsorption of Zn, Cu and Cd from waste waters by means of maghemite nanoparticles, *Scientific Bulletin, Series B: Chemistry and Materials Science*, 74(2012) 255–264.

[72] Tayyebe Soltani and Mohammad H. Entezari, Sono-synthesis of bismuth ferrite nanoparticles with high photocatalytic activity in degradation of Rhodamine B under solar light irradiation, *Chemical Engineering Journal*, 223(2013) 145–154. https://doi.org/10.1016/j.cej.2013.02.124.

[73] Lixia Wang, Jianchen Li, Yingqi Wang, Lijun Zhao and Qing Jiang, Adsorption capability for Congo red on nanocrystalline MFe_2O_4 (M= Mn, Fe, Co, Ni) spinel ferrites, *Chemical Engineering Journal*, 181(2012) 72–79. https://doi.org/10.1016/j.cej.2011.10.088.

[74] Niyaz Mohammad Mahmoodi, Surface modification of magnetic nanoparticle and dye removal from ternary systems, *Journal of Industrial and Engineering Chemistry*, 27(2015) 251–259. https://doi.org/10.1016/j.jiec.2014.12.042.

[75] Ruijiang Liu, Xiangqian Shen, Xinchun Yang, Qiuji Wang and Fang Yang, Adsorption characteristics of methyl blue onto magnetic $Ni_{0.5}Zn_{0.5}Fe_2O_4$ nanoparticles prepared by the rapid combustion process, *Journal of Nanoparticle Research*, 15(2013) 1679. https://doi.org/10.1007/s11051-013-1679-1.

[76] Mahsa Bayat, Vahid Javanbakht and Javad Esmaili, Synthesis of zeolite/nickel ferrite/sodium alginate bionanocomposite via a co-precipitation technique for efficient removal of water-soluble methylene blue dye, *International Journal of Biological Macromolecules*, 116(2018) 607–619. https://doi.org/10.1016/j.ijbiomac.2018.05.012.

[77] Milad Jamal Livani and Mohsen Ghorbani, Fabrication of $NiFe_2O_4$ magnetic nanoparticles loaded on activated carbon as novel nanoadsorbent for Direct Red 31 and Direct Blue 78 adsorption, *Environmental Technologies*, 39(2018) 2977–2993. https://doi.org/10.1080/09593330.2017.1370024.

[78] Mukesh Kumar, Harmanjit Singh Dosanjh and Harminder Singh, Magnetic zinc ferrite-chitosan bio-composite: synthesis, characterization and adsorption behavior studies for cationic dyes in single and binary systems, *Journal of Inorganic Organometallic Polymer Materials*, 28(2018) 880–898. https://doi.org/10.1007/s10904-017-0752-0.

18 Iron-Based Catalysts for Fischer–Tropsch Synthesis for Light Olefins Production from Syngas

Rajender Boddula, Ramyakrishna Pothu,
Ramachandra Naik, Ahmed Bahgat Radwan,
and Noora Al-Qahtani

INTRODUCTION

Despite the rapid and considerable advancement of new energy sources, such as solar energy and wind energy, traditional fossil fuels remain the dominant force in global energy consumption, accounting for a substantial 86% of the total energy consumption in 2019 as reported by BP [1]. One of the critical challenges associated with fossil energy, particularly crude oil, lies in the instability of its prices and scarcity. The erratic nature of crude oil prices exerts a direct and significant impact on various sectors, encompassing energy, fuels, and chemical products. This volatility in prices creates a ripple effect throughout economies and industries, influencing everything from consumer costs to investment decisions. Moreover, the persistent high cost of petroleum-derived fuels and chemical products is a concern that looms large on the horizon. This cost burden is poised to intensify due to the escalating demand for these products, even as the industry grapples with the limitations presented by dwindling oil reserves and the challenges tied to crude oil production infrastructure [2]. This enduring prevalence of fossil fuels persists despite the notable progress in harnessing alternative and cleaner energy options. In the midst of these global dynamics, an alternative to oil has emerged in the form of coal, biomass, and natural gas. This resource has garnered recognition for its relatively cleaner combustion characteristics compared to conventional fossil fuels. Moreover, these possess the dual advantage of being capable of responding to climate change concerns by serving as a lower-carbon energy source [3]. In-addition, the extraction of natural gas from shale formations has not only expanded supply but has also contributed to making this resource more economically viable. The implications of this extend beyond immediate energy consumption to align with broader environmental goals. Additionally, it has proven to be price-competitive, thanks in part to the transformative impact of the shale revolution.

FISCHER–TROPSCH SYNTHESIS (FTS)

A noteworthy avenue that has garnered attention within the realm of using natural gas as an alternative to oil is the realm of gas-to-liquids (GTL) technologies (Figure 18.1). The GTL process, rooted in Fischer–Tropsch chemistry, offers a transformative approach. These technologies have emerged as a promising solution for converting natural gas into a range of valuable products, including fuel and petrochemical derivatives [5]. The process involves the transformation of methane, the primary constituent of natural gas, into products such as diesel, naphtha, and paraffin, etc. [4]. This presents a sustainable and versatile alternative to conventional crude-based products, aligning

DOI: 10.1201/9781003479239-18

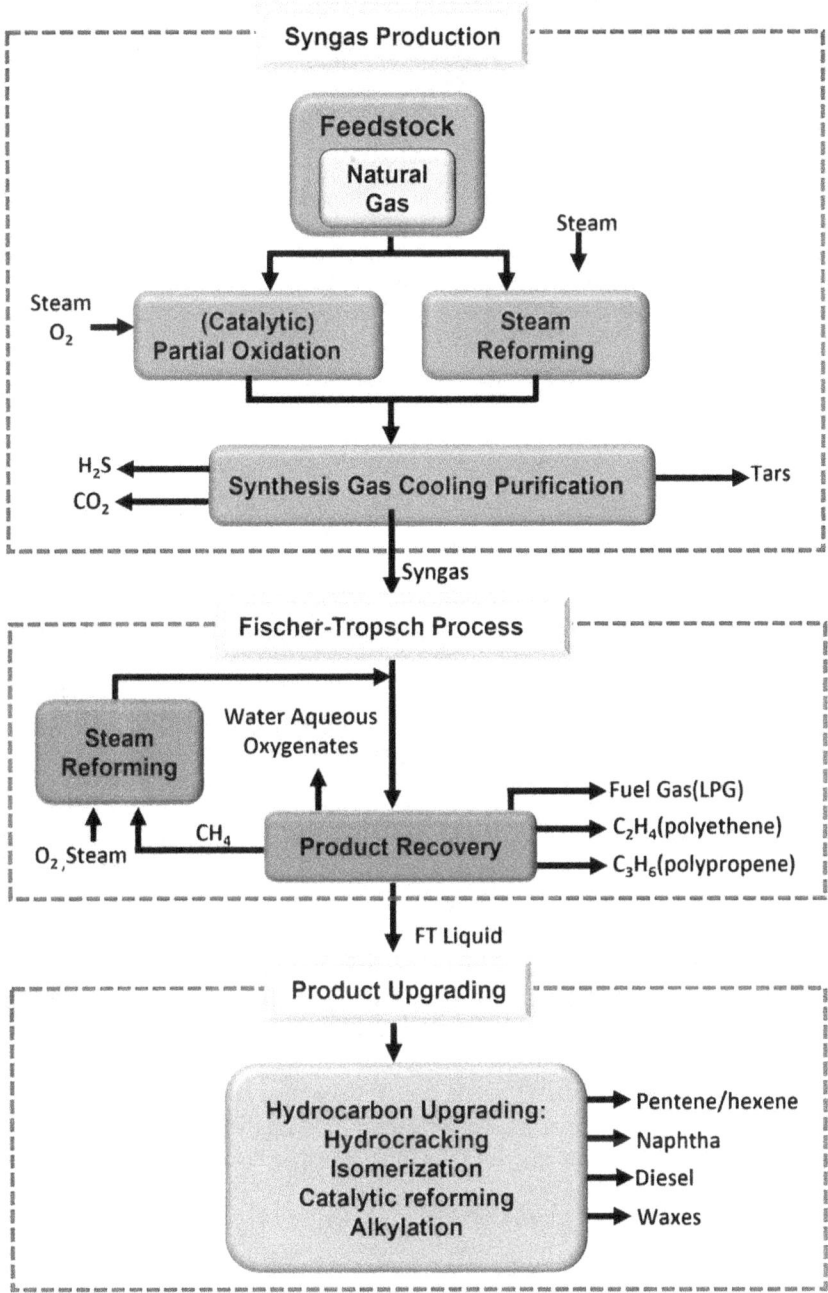

FIGURE 18.1 Whole process of the GTL technology [4].

with the evolving landscape of energy and resource utilization. The significance of this process lies in its potential to create high-quality liquid products that exhibit fewer impurities compared to those derived from crude oil. The GTL process follows a sequence of three essential steps. It commences with methane reforming, a process that generates syngas, which subsequently serves as the precursor for the FTS stage. This latter phase is responsible for the conversion of syngas into a spectrum of valuable hydrocarbon products. The final stage of the GTL process pertains to product upgrading, a crucial step that elevates the quality and utility of the end products. The inception of

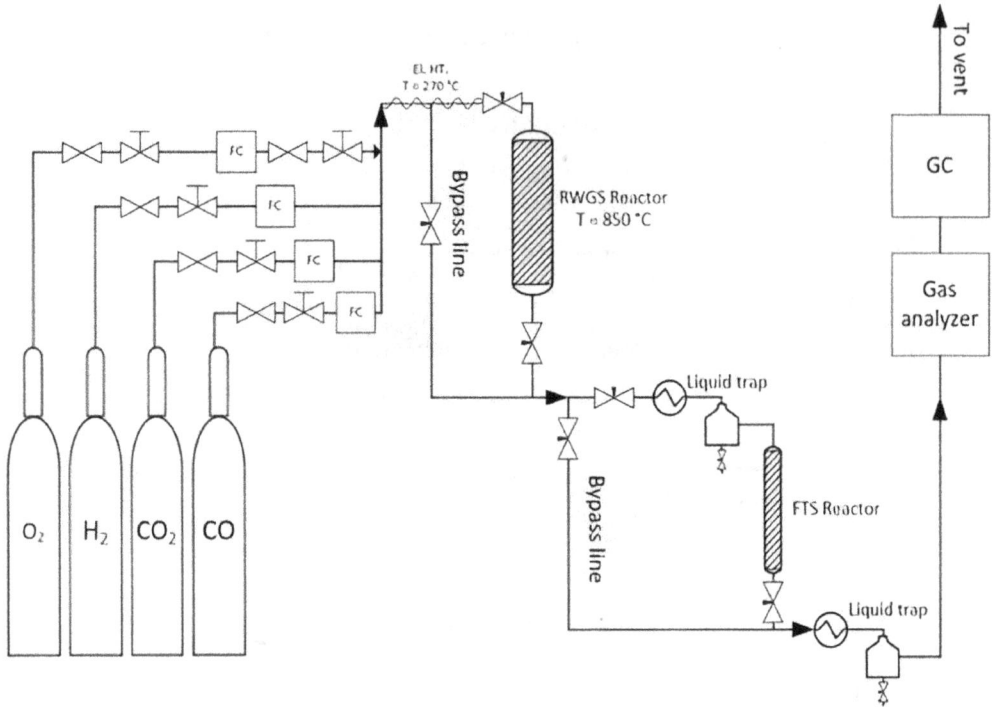

FIGURE 18.2   Schematic of the experimental setup for FTS process [7].

Fischer–Tropsch synthesis (FTS) dates to 1926 when it was first reported by the Kaiser Wilhelm Institut für Kohlenforschung [6]. Since then, this process has garnered remarkable attention, particularly from regions characterized by an abundance of coal or natural gas but lacking in substantial petroleum reserves.

FTS is a highly exothermic reaction commonly employed to catalytically convert syngas into higher hydrocarbons and oxygenates, which can be subsequently upgraded to produce clean chemicals and transportation fuels. The reactor model is presented in Figure 18.2. This reaction pathway involves catalytic polymerization reactions that transform syngas into an array of products encompassing paraffins, olefins, alcohols, and aldehydes with varying carbon numbers, spanning from gasoline to diesel, contingent on the utilization of appropriate catalysts. Among these, paraffins and olefins stand out as the primary outcomes of FTS (mentioned in below equations) [2].

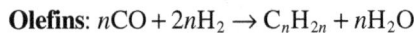

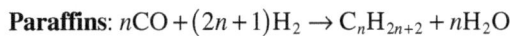

$$\textbf{Olefins}: nCO + 2nH_2 \rightarrow C_nH_{2n} + nH_2O$$

$$\textbf{Paraffins}: nCO + (2n+1)H_2 \rightarrow C_nH_{2n+2} + nH_2O$$

LIGHT OLEFINS

The demand for light olefins ($C_2^=$–$C_4^=$), including ethylene (C_2H_4), propylene (C_3H_6), and butylene (C_4H_8), is on the rise, driven by their significance as essential raw materials for a wide range of products such as plastics, synthetic fibers, dyes, and synthetic rubber. The global olefin market is projected to reach a substantial value of approximately USD 329.30 billion by 2028, reflecting a growth rate of 4.5% starting from 2020. In 2019, the market witnessed the production of 208 million tons of ethylene, 110 million tons of propylene, and 4.6 million tons of light olefins [8].

The expansion of economies and the wave of industrialization in numerous developing nations have contributed to the escalating demand for light olefins. Consequently, there has been a surge of interest in advancing olefin production technologies. Traditionally, light olefins are generated through processes such as steam-cracking naphtha or liquefied petroleum gas and, more recently, via direct propane dehydrogenation [9]. However, the reliance on petrochemical-based approaches is waning due to the imperative for energy conservation and emission reduction. In response to this paradigm shift, the utilization of syngas (a combination of hydrogen and carbon monoxide) has emerged as a prominent non-petroleum route for light olefin production. This avenue has garnered significant attention and is considered a focal point of research. Nonetheless, this production method presents challenges characterized by a heavy reliance on specific raw materials, high energy consumption during production, and modest product selectivity. Given the limitations inherent in conventional production methods, there is an urgent call to pioneer clean energy production routes that capitalize on non-petroleum resources as feedstocks. The use of syngas as a foundational platform for synthesizing compounds and liquid fuels is a pivotal pathway for harnessing non-petroleum carbon sources [2]. The evolution of this approach, aimed at supplanting the traditional petroleum-based production route, holds paramount importance in the context of realizing a sustainable global development strategy and securing national energy resilience. The successful execution of this transition represents a critical agenda for driving sustainable industrial progress and safeguarding energy security on a global scale [3].

CATALYSTS

In the realm of FTS, transition metals capable of adsorbing syngas (a combination of hydrogen (H_2) and carbon monoxide (CO)) and reducing metal oxides play a pivotal role as catalysts. These transition metals, found within groups III and VI of the periodic table, possess distinct characteristics that influence their efficacy in FTS processes. Notably, these metals tend to form oxides that exhibit remarkable stability, often proving challenging to reduce during FTS. However, they find utility in facilitating the dissociative adsorption of carbon monoxide (CO), a crucial aspect of the overall process. Certain transition metals, including copper (Cu), palladium (Pd), platinum (Pt), and iridium (Ir), exhibit a notable difficulty in achieving CO dissociation. This inherent limitation often results in elevated methanol selectivity during FTS, underscoring the intricate nature of the catalytic reactions involved. In contrast, transition metals such as iron (Fe), cobalt (Co), nickel (Ni), and ruthenium (Ru) stand out as promising candidates for FTS due to their adeptness in CO dissociation. This capability subsequently translates to higher rates of chain growth, a pivotal step in the synthesis process. Research has also explored the potential of rhodium (Rh) in this context, identifying it as a feasible option in select cases. However, the economic viability of utilizing certain metals demands consideration. For instance, ruthenium (Ru) is limited by its scarcity and elevated cost, making it less practical as a catalyst candidate. Additionally, nickel (Ni) showcases a propensity for methane selectivity over the desired products, thus presenting a limitation. As such, Fe and Co have emerged as the most prevalent and pragmatic choices for FTS catalysts. Interestingly, Fe-based catalysts exhibit a greater capacity to tolerate sulfur impurities when compared to their cobalt counterparts. This attribute suggests their suitability for processes involving biomass- and coal-derived syngas. Furthermore, Fe demonstrates a lower hydrogenation activity relative to Co, contributing to heightened selectivity toward olefins.

IRON (Fe)-BASED CATALYSTS

Various synthesis methods have been employed to produce catalysts with diverse properties. Methods such as wet-chemical, hydrothermal synthesis followed by thermal treatment method have shown high potential. Iron catalysts exhibit distinct selectivity patterns at varying temperatures. At lower temperatures, they tend to favor the formation of paraffins, while raising the temperature shifts their selectivity toward olefins. This shift is attributed to the temperature-dependent alteration of reaction pathways. The activation energy required for CO hydrogenation is notably similar to that of iron carbide formation, i.e., active sites within Fe-based FTS catalysts are often attributed to distinct ferric carbides, including ε-Fe$_2$C, θ-Fe$_3$C, Fe$_7$C$_3$, and χ-Fe$_5$C$_2$ [10].

Briefly, the primary compounds found in Fe-based catalysts include iron carbide, metallic iron, and magnetite (Fe$_3$O$_4$). Understanding the phase transformation of iron species within both bulk and surface regions under H$_2$ atmosphere reveals a sequence of transitions involving a-Fe$_2$O$_3$, Fe$_3$O$_4$, FeO, and a-Fe. Interactions between metal and support result in the creation of metastable FeO due to this strong interaction. Reduced iron phases, such as Fe$_3$O$_4$, FeO, and a-Fe, can transform into iron carbides in the presence of CO or syngas. Carburization ability follows the order a-Fe > FeO > Fe$_3$O$_4$ [10]. During FTS, a balance between iron carbides and Fe(II) oxide species is reached, and hydrocarbon species form primarily on the catalyst surface layers as reduced iron phases undergo complete carburization to iron carbides. This conversion of reduced iron phases to iron carbides on surface layers positively impacts hydrocarbon species formation. In CO or syngas atmospheres, diverse iron carbide types such as α-Fe$_2$C, ε-Fe$_{2.2}$C, Hägg χ-Fe$_5$C$_2$, Fe$_7$C$_3$, or θ-Fe$_3$C can arise from these reduced iron species. To determine the active phase in Fe catalysts, researchers often employ in situ magnetization measurements to ascertain the Curie temperature, which is the temperature where the first derivative of magnetization changes. This allows for the identification of the active phase based on the observed Curie temperature. In FTS processes targeting light olefins, the active phases of Fe-based catalysts are frequently reported to be iron carbides such as Hägg χ-Fe$_5$C$_2$, α-Fe$_2$C, ε-Fe$_{2.2}$C, and Fe$_7$C$_3$ [11]. The stability of α-phase carbides lies below 250°C, while Hägg carbide remains stable within the temperature range of 250°C–350°C. Consequently, among the iron carbides, Hägg χ-Fe$_5$C$_2$ is widely recognized as the active phase within the typical FTS temperature range (240°C–360°C), featuring a Curie temperature between 205°C and 238°C [12]. Fe7C3 emerges as a carbide phase during CO treatment of the catalyst at moderate temperatures, and θ-Fe3C remains stable above 350°C [13]. These different carbides exhibit nuanced catalytic properties that influence the outcomes of the FTS process. For instance, Liu et al.'s research demonstrated that controlling the silicon shell thickness allowed for the stabilization of Fe$_7$C$_3$ phases, thus offering insights into the iron phase selection during iron-based FTS catalysis [14]. Consequently, during the FTS process, iron carbides are formed as a consequential aspect of the reaction mechanism. The process gives rise to diverse forms of iron carbides, each impacting the reaction outcomes. Taking a microscopic perspective, the catalyst's reduction, carbonation, and the interaction between iron and its support play pivotal roles in the performance of Fe-based FTS catalysts. Furthermore, aspects like the dissociation and adsorption of syngas, the immediate desorption of light olefins, the control of CO$_2$, and the extent of olefin hydrogenation all profoundly impact the catalyst's efficacy in FTS. Scaling up, attributes such as pore characteristics, stability, catalyst reaction kinetics, transformation behaviors, and product component analysis serve as crucial benchmarks for catalyst evaluation. Subsequently, this discussion will delve into how these key perspectives of catalyst performance are influenced by promoters and support materials. The nuanced effects of metals, promoters, and supports, which are among the most influential parameters, are dissected in detail concerning their impact on catalyst performance. Therefore, in the realm of FTS catalysts, the incorporation of promoters holds considerable significance, enhancing the catalyst's reactivity. Simultaneously, the choice of support materials exerts a vital influence on the catalyst's overall performance. A comprehensive summary of iron-based catalysts with incorporating promoters and supports pertinent to the production of light olefins via FTS is provided in Table 18.1.

SUPPORTS

Support materials, utilized to load active components, often possess large specific surface areas and unique pore structures. Analogous to promoters, the influence of support materials varies across different scales. On the micro level, support materials can interact with catalyst metals, shaping heat

TABLE 18.1
Fe-Based Catalysts for Syngas Conversion to Light Olefins

Catalyst	Promoter	Support	Reaction Conditions			Conversion of CO (%)	Selectivity of Light Olefins (%)	References
			Temp (°C)	Pressure (MPa)	H_2/CO Ratio			
Fe	–	$mSiO_2$	300	2	2.1	15.4	12.8	[12]
Fe	–	SiO_2	300	2	2.1	28.5	15.2	[12]
Fe_2N	–	Al_2O_3	280	2	1	33	22.3	[35]
FeMn	–	HZSM-5	280	1	2	78.5	41.4	[36]
α-Fe_2O_3	–	MnO_2	280	1	1	48.2	51.1	[37]
Fe	–	MgO nanosheets	300	1	1	35.5	29.6	[18]
Fe	–	MgO nanocubes	300	1	1	35.7	21.5	[18]
Fe	–	AC	300	0.1	1	4.8	30.6	[19]
Fe	–	CNT	350	1	1	14.4	32.4	[38]
Fe	–	t-CNT	300	0.1	1	9.1	36.4	[19]
Fe	–	NCNTs	300	0.1	1	14.4	46.7	[19]
ε-Fe_2C	–	graphene	300	1	1	95	18	[22]
FeCu	–	graphite	260	2	1	44.9	37.8	[39]
Fe	–	N-doped graphene	350	2	2	97.2	41.2	[40]
35Fe	–	hNCNC-3	350	0.1	1	3.5	54.1	[41]
Fe_5C_2	–	NC600	250	1.8	2	22.3	55.6	[42]
Fe	–	PANI	350	2	1	79	47	[43]
FeMnK	–	H-S-1 zeolite	280	0.5	1.9	13.9	50.6	[44]
FeMnK	–	ZSM-5	280	0.5	1.9	11.7	28.7	[45]
Fe	Cu	–	300	2	1	76	29	[32]
γ-Fe_2O_3	0.5Mn	–	320	1	1	55.1	61.2	[46]
Fe	100Zr	–	280	1	1	40.6	57	[26]
Fe	Zr/K	–	270	1	2	82.1	34.6	[47]
Fe	Zr/Na	–	320	1.5	2	95.2	35.8	[48]
FeMn	Cu3.0	–	300	2	2	96.9	40.1	[31]
Fe	K/B	–	340	0.5	1	18.4	66.4	[49]
FeMn	1.5Li	–	320	1.5	2	85.6	36.7	[50]
$FeMnO_x$	Na/S	–	280	0.1	2	19	18	[51]
Fe	Sn	SiO_2	350	1	2	53	17	[52]
Fe	Sb	SiO_2	350	1	2	47	17	[52]
Fe	Mn	SiO_2	300	1	1	50.5	54.6	[17]
FeMn	Cu	rGO	300	2	1	87	30	[32]
MgFe	K	HSG	340	2	1	65.5	57.8	[53]
Fe	Pb	CNT	350	0.1	1	18.6	57.7	[33]
Fe	Bi	CNT	350	0.1	1	10	60.9	[33]
Fe	Bi	CNT	350	1	1	28.9	37.5	[38]
Fe	Pb	CNT	350	1	1	34.3	35.3	[38]
Fe	Pb/K	CNT	350	1	1	76.2	52.6	[38]
Fe	Mn/K	CNT	300	0.1	1	22.7	50.3	[20]
Fe	K	NCNTs	300	0.1	1	16.5	54.6	[19]
Fe	K	NGNFs	300	2	1	16.1	11.7	[34]
Fe	K	NCNTS	300	2	1	28.1	10.8	[34]

hCNC=hierarchical carbon nanocage; HSG=honeycomb-like structured graphene; PANI=polyaniline; NC=N-doped carbon; H-S-1=silicalite-1-zeolite; AC=activated carbon; t-CNTs=pretreated carbons; $mSiO_2$=mesoporous silica; NCNTs=nitrogen-doped CNTs.

and mass transfer dynamics during catalysis. On a macro level, support materials can enhance the dispersion of active phases, maximizing catalyst surface area and even exerting control over active particle size to a certain extent. Moreover, the mechanical stability and strength of catalysts can be affected by support materials. Notably, c-Fe_5C_2 catalysts exhibited a preference for synthesizing longer-chain hydrocarbons and alkanes. Effective promotion strategies further improved selectivity toward C_2–C_4 olefins. Feyzi and colleagues employed a Fe-Ni/Al_2O_3 catalyst for light olefin production from synthesis gas. The investigation revealed that the catalyst demonstrated the highest selectivity toward C_2–C_4 olefins (77.8%), alongside low selectivity toward methane (9.1%) and CO_2 (0.3%) under specific conditions. The introduction of K_2S as a catalyst modifier amplified the selectivity toward C_2–C_4 olefins [15]. The particle size of iron catalysts has intriguing ramifications on their selectivity behavior. Smaller Fe nanoparticles, typically less than 7–9 nm, exhibited heightened CH_4 selectivity compared to their larger counterparts. This particle size dependency is linked to the specific sites on the catalyst crystals, with CH_4 formation enhanced at corners and edges, which are more prevalent in smaller particles [16]. In contrast, larger Fe nanoparticles favor the production of olefins due to the prominence of terrace sites that foster olefin formation. The support material also holds sway in the catalytic process. Larger pore sizes in the support structure promote the generation of heavy hydrocarbons and elevate the selectivity toward light olefins. Research indicated that Fe-based catalysts (specifically FeMn) characterized by pore sizes within the range of 50–80 nm facilitated the efficient diffusion of reactants and products [17]. This dynamic also curtailed secondary reactions of light olefins, ultimately leading to heightened selectivity toward light olefins. Moving to catalyst support, the incorporation of magnesium oxide (MgO) as a basic support and structural promoter within Fe-based catalysts exhibited the potential to elevate the olefin-to-paraffin ratio. Researchers synthesized Fe/MgO nanosheet and MgO cube catalysts using diverse methods, such as incipient wetness impregnation and deposition–precipitation. Of note, ultrasonic impregnation led to the development of Fe/MgO nanosheet catalysts with robust basicity sites of MgO [18]. This distinct characteristic correlated with improved dissociative CO adsorption and heightened olefin selectivity.

Traditionally, oxide materials with high temperature resistance such as alumina and silicon oxide serve as common support materials. However, interactions between oxide supports and iron particles can influence iron oxide reduction and iron carbide formation. This has driven recent progress in using inert carbon materials like graphene and carbon nanotubes (CNTs) as catalyst supports. The role of support materials was further underscored by the advantageous impact of carbon nanotubes (CNTs) and carbon nanofibers (CNFs) as catalyst supports. Lu and collaborators [19] established iron catalysts immobilized onto N-doped carbon nanotubes, revealing remarkable catalytic selectivity, activity, and stability for producing lower olefins. This performance was attributed to factors including high dissociative CO adsorption, inhibition of secondary hydrogenation of lower olefins, and the presence of the active-phase χ-Fe_5C_2. Nitrogen within N-doped CNTs facilitated an anchoring effect and intrinsic basicity, ensuring the catalyst's robustness during the FTS process. Wang and associates [20] extended this concept by employing CNTs-supported Fe catalysts with manganese and potassium as promoters. Two synthesis methods were utilized, showcasing the superiority of the Fe/MnK-CNTs catalyst over FeMnK/CNTs in terms of activity and stability for light olefins production. The advantageous attributes of small-sized and uniform nanoparticles, weak metal–support interaction, uniform promoter distribution, and enhanced support defects contributed to the heightened performance of the former catalyst. Additionally, Roe and team [21] explored FTS using CNTs-supported catalysts based on iron in both gas phase and supercritical hexane operating conditions. Carbon-supported catalysts demonstrated elevated activity, reduced CH_4 formation, and enhanced selectivity toward olefins and oxygenates. The adoption of supercritical conditions facilitated the extraction of olefins through improved heat management, limiting methanation and allowing intermediates to readhere and continue propagation. Supercritical operation also highlighted the potential for enhancing factors such as chain growth, CO conversion, and selectivity toward unhydrogenated products.

Fe-based catalysts in FT reactions have encountered a formidable challenge – the formation of an amorphous carbon or carbide layer that hampers catalytic activity. Encapsulation of pure-phase ε-Fe$_2$C nanocrystals within graphene layers synthesized by pyrolysis method has yielded remarkable results [22]. Breaking free from this limitation was no small task, but the introduction of a few graphene layers emerged as a transformative solution, leading to the development of ε-Fe$_2$C@graphene catalysts (Figure 18.3a–d). These graphene layers effectively confine the Fe-based catalyst, preventing the formation of the troublesome amorphous carbon layer, and bolstering the stability of the highly active ε-Fe$_2$C phase (Figure 18.3b). Crucially, the ε-Fe$_2$C@graphene catalysts exhibited consistently high activities across a range of Fe loadings, from 10 to 50 wt.% (Figure 18.3c). In stark contrast, reference Fe/C catalysts suffered a sharp decrease in performance with increasing Fe loading, a phenomenon attributed to Fe species aggregation and active carbide phase oxidation under reaction conditions, as depicted in Figure 18.3a. Under realistic FT synthesis conditions, this innovative ε-Fe$_2$C@graphene catalyst achieves an iron-time yield that astounds at 1,258 $\mu mol_{CO}g_{Fe}$/s.

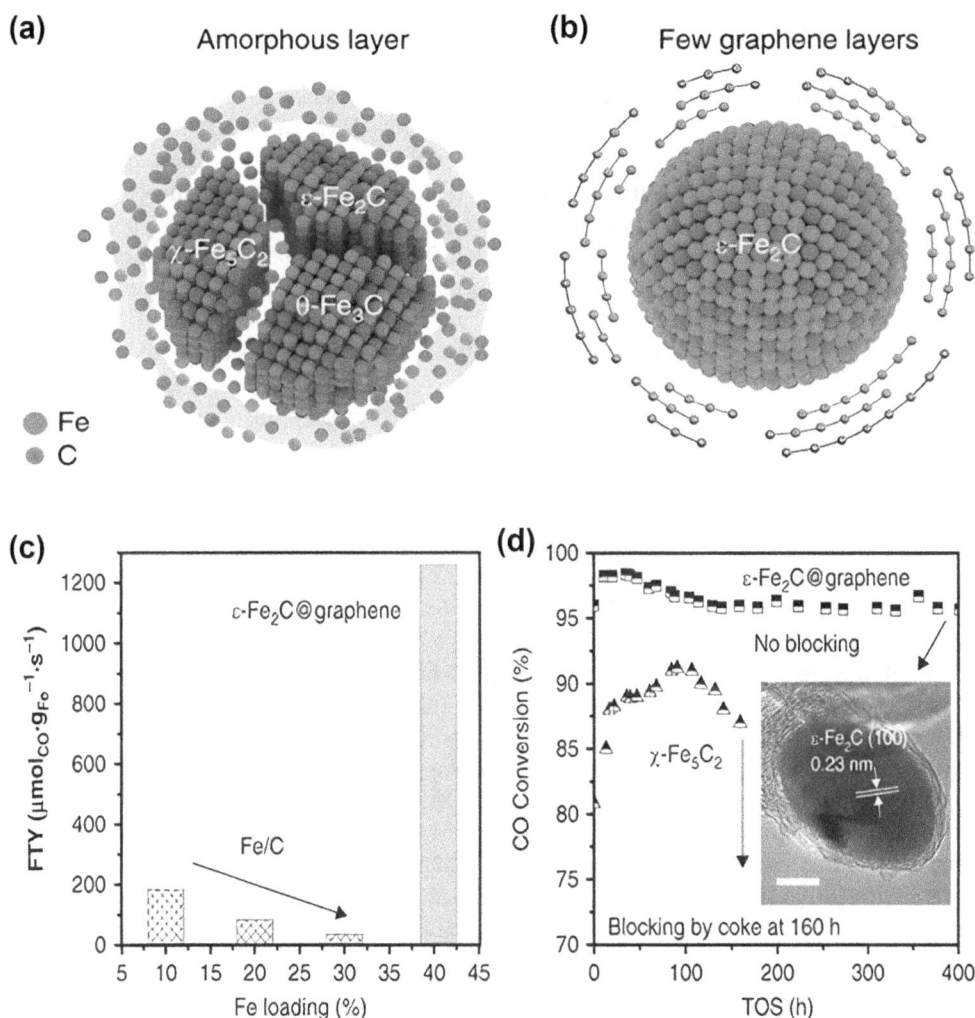

FIGURE 18.3 Schematic models of (a) unconfined iron carbide (ε-Fe$_2$C), (b) graphene layer-confined ε-Fe$_2$C, (c) comparison of the Fe-time yield (FTY) values between the iron catalyst loaded on active carbon with various iron loadings and the ε-Fe$_2$C@graphene catalyst, and (d) stability long-run (inset: TEM image of spent catalyst) [22].

This remarkable performance stands as a true game-changer, boasting an order of magnitude improvement over conventional carbon-supported Fe catalysts. It maintained its efficiency for over 400 hours (Figure 18.3d), even at higher CO conversion rates (~95%) indicating the prevention of coke deposition on the catalyst surface by graphene layers. Conversely, the unencapsulated χ-Fe$_5$C$_2$ catalyst faced significant deactivation, with carbon deposition ultimately blocking gas flow after 160 hours of operation (Figure 18.3d). Delving into the science behind this achievement, density functional theory (DFT) calculations unveil the feasibility of ε-Fe$_2$C formation through the carburization of α-Fe precursor, facilitated by the interfacial interactions within the ε-Fe$_2$C@graphene structure [22].

PROMOTERS

The introduction of promoters and meticulous support optimization is pivotal for achieving various objectives, including effective CO adsorption and dissociation, moderation of water-gas shift reactions (WGSR), controlled hydrogenation, carbon–carbon coupling, and ensuring reaction stability and durability. Consequently, recent studies have been increasingly dedicated to the addition of promoters and the alteration of support materials. The efficacy and selectivity of these catalysts are significantly influenced by the nature and concentration of promoters. Promoters play a significant role in catalyst performance, and commonly employed ones include alkali metals (Na and K), alkaline earth metals (Mg), transition metals (Mn, Cu, Zn), and non-metallic elements (S and N). These promoters can be categorized into electronic and structural promoters. Electronic promoters enhance catalyst element electronic states to bolster catalytic performance, whereas structural promoters optimize catalyst structure for uniform active ingredient distribution. The type, quantity, and addition method of promoters profoundly affect the distribution of active phases, dissociative adsorption efficiency, and reductive carbonization in modified FTO catalysts. These factors cascade into varying CO conversion efficiency, hydrocarbon selectivity, and catalyst stability. Galvis and colleagues explored the introduction of sulfur and sodium in low concentrations to an Fe/α-Al$_2$O$_3$ catalyst. This strategic modification aimed to enhance catalytic activity and reduce methane production while elevating C$_2$–C$_4$ olefins selectivity [23]. The researchers observed a decline in methane selectivity due to chain growth probability following sodium addition. In contrast, sulfur addition stimulated higher olefin production.

Among the commonly employed catalyst promoters, alkali metals (such as Na and K), alkaline earth metals (like Mg), transition metals (including Mn, Cu, and Zn), and non-metallic elements (such as S and N) have gained prominence. These promoters play distinct roles, classifiable as electronic and structural promoters. Electronic promoters augment the electronic state of catalyst elements, consequently enhancing catalytic performance. In contrast, structural promoters optimize the catalyst structure and facilitate the uniform dispersion of active components. The type, quantity, and manner of adding promoters critically influence the distribution of active phases, CO adsorption capabilities, and reductive carbonization of modified FTS catalysts. These, in turn, translate into varying CO conversion performances, hydrocarbon selectivity, and catalyst stability. Interestingly, additional transition metals, such as Zn, Zr, and Cu, have been investigated as potential promoters in FTO processes. Zn, in comparison to Mn, has demonstrated the ability to facilitate the reduction of iron oxide, leading to significantly enhanced catalytic activity. The formation of ZnFe$_2$O$_4$ indicated the strong interaction between Zn and Fe, promoting smaller crystals of Fe$_5$C$_2$ through a spinel structure [24,25]. Zirconia, when highly dispersed on the iron surface, counteracted the formation of Fe–Zr mixed oxide. This effect led to increased C$_2$–C$_4$ olefin selectivity [26]. In the exploration of catalytic performances in FTS, Xiong and colleagues delved into the effects of alkali metal promoters, including Li, Na, and K, on iron catalysts supported on carbon nanotubes. Their investigation illuminated how the introduction of alkali metals influenced both the size of the catalyst crystallites and the corresponding reduction in surface area. Alkali metals displayed varying degrees of basicity, with an ascending order of Li < Na < K [27]. Notably, increased loading of Na and K correlated with a heightened olefin/paraffin ratio and a greater formation of long-chain hydrocarbons. Another study [28] explored the impact of alkali addition to Fe/SiO_2 catalysts. This alteration brought about changes in catalyst reduction due to robust interactions between alkali and iron metal. Potassium hindered the initial reduction of iron oxide while simultaneously enhancing the formation of metallic iron from FeO, thereby influencing the FTS activity of iron catalysts. Additionally, an augmentation in alkali atomic number correlated with an intensified carbonization of catalysts. Li and colleagues focused on alkali metals (Li, Na, K, Rb, and Cs) as promoters for iron-based FTS catalysts. This study revealed that while Li and Na could effectively penetrate the catalyst surface, K, Rb, and Cs faced challenges in diffusion. The inclusion of alkali metals led to a significant improvement in the selectivity toward olefins and heavier hydrocarbons, concomitantly reducing the selectivity toward methane and alkane. A noteworthy finding was the diffusion of Li out of the catalyst, with K exhibiting limited mobility in the iron catalyst post-FTS reactions. Research has shown that potassium introduced into Fe/C catalysts increases the surface area and average pore diameter. This facilitates the formation of iron carbides, amplifying the adsorption and dissociation of CO. Furthermore, computational studies demonstrated that potassium reduces the dissociation barrier of CO on the metal surface in K/Rh catalysts. The increased potassium concentration on Fe_5C_2 surfaces correlates with enhanced CO adsorption energy, leading to C–O bond elongation. Potassium's promoter effect was evidenced by transient response curves, showcasing altered CH_4 desorption kinetics and suggesting the establishment of a new, slower pathway for CH_4 formation. These alterations were attributed to the influence of alkalis on surface adsorption, particularly in enhancing the adsorption and dissociation of CO [29]. Sodium carbonate is a commonly employed precipitator during catalyst preparation due to its cost-effectiveness. Activation monitoring of Na-promoted Fe–Zn catalysts revealed electron-rich Fe_5C_2 surfaces, which inhibited C_3H_6 hydrogenation and was confirmed through pulse experiments [25]. Na also influences the FeNaMg catalyst's surface electronic environment, suppressing methane formation while enhancing CO adsorption and ultimately increasing selectivity toward aromatics [30]. Copper (Cu) can reduce catalysts, enhance CO dissociative adsorption with H_2 assistance, and alter the interaction between metal and support, leading to improved activity. Lower surface alkalinity due to Cu prevents carbon chain elongation, enhancing light olefin selectivity by inhibiting long-chain hydrocarbon formation. For example, the Cu promoter enhances CO dissociative adsorption through alterations in interactions between Fe, Mn, and the SiO_2 binder. This adjustment led to the Cu-3.0 catalyst displaying lower olefins selectivity [31]. Copper, by acting as an active site for H_2 dissociation, lowers the reduction temperature of iron oxide. This facilitates the removal of oxygen atoms from the surface and alters the catalyst's surface state, thereby promoting overall catalytic activity [32]. Remarkably, even main group metals like bismuth exhibit promoter effects on iron catalysts, resulting in a substantial rate increase and heightened light olefin selectivity due to continuous oxidation and reduction cycles experienced by metallic bismuth [33]. Spark-plasma sintering technique allowed for the production of potassium-promoted iron catalysts supported on both N-doped graphene nanoflakes (FeKNGNF700) and N-doped carbon nanotubes (FeKNCNT800) [34]. Notably, the catalyst developed using N-doped graphene nanoflakes exhibited a significant enhancement in FTS activity, achieving an impressive rate of 196 $\mu mol_{CO}/g_{Fe}/s$. What's particularly noteworthy is that this enhanced activity was achieved while maintaining a high C_{5+} selectivity of approximately 65%–70%, signifying the production of valuable higher C_2–C_4 olefins (30%–40%). In the realm of iron-containing FTS catalysts, it is well established that iron carbide plays a pivotal role in the active phase. In the X-ray diffraction profile of FeKNGNF700, as illustrated in Figure 18.4, only the reflections corresponding to the cementite (Fe_3C) phase were observed. This is indicative of the absence of iron oxide phases within this sample, likely attributed to the small size of oxide particles. On the other hand, the XRD analysis of FeKNCNT800 revealed a more complex composition. It included a combination of the body-centered cubic (bcc-Fe) and austenite (fcc-Fe, C) phases, alongside residual traces of the Mo_2C phase inherited from the growth catalyst. This diverse phase composition highlights the nuanced structures that can be achieved and tailored in the pursuit of optimizing FTS catalysts.

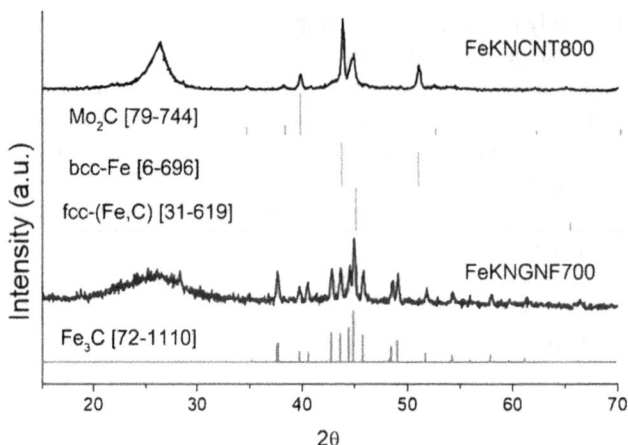

FIGURE 18.4   XRD profiles of fresh FeKNGNF700 and FeKNCNT800 FTS catalysts [34].

Overall, the intricate interplay of catalyst composition, promoter incorporation, and support materials orchestrates the complex realm of FTS for light olefins. The manipulation of these factors continues to be a focal point in the quest for optimized catalytic processes that yield valuable olefinic products.

CHALLENGES AND PERSPECTIVES

Despite the significant progress made, challenges persist in the development and application of iron-based catalysts for FTS of light olefins. The intricate interplay among catalyst composition, phase transitions, and reaction conditions requires further exploration to optimize catalytic outcomes. Achieving high selectivity for desired products, such as light olefins, while minimizing undesirable byproducts remains a challenge, necessitating the design of catalysts that can tailor product distributions. The integration of iron-based catalysts with scalable and cost-effective processes poses a challenge. Addressing the potential for catalyst deactivation due to factors such as carbon deposition, sintering, and impurities in syngas is vital for achieving prolonged catalyst lifetimes. Moreover, bridging the gap between laboratory-scale success and real-world industrial applications demands robust catalyst engineering, scale-up, and economic viability.

The field of iron-based catalysts for FTS of light olefins holds exciting perspectives and avenues for further exploration. Continued advancements in catalyst design, supported by computational modeling and experimental insights, will provide deeper insights into catalyst behavior and guide the development of tailored catalysts for specific applications. Exploring novel catalyst supports, promoters, and synthetic methods will contribute to fine-tuning catalytic performance. The integration of catalytic processes with renewable and low-carbon feedstocks is an emerging avenue. Harnessing renewable energy sources for syngas production and coupling FTS with carbon capture and utilization could lead to more sustainable and environmentally friendly hydrocarbon production. The pursuit of multifunctional catalysts that combine FTS with other catalytic reactions, such as upgrading of biomass-derived feedstocks, holds potential for process intensification and diversification.

CONCLUSIONS

In summary, the utilization of iron-based catalysts for Fischer–Tropsch synthesis (FTS) to produce light olefins from syngas presents a promising pathway for sustainable hydrocarbon production. The research landscape explored in this chapter underscores the significance of this catalytic process in addressing the increasing demand for light olefins, vital components in a range of industrial applications. The comprehensive understanding of catalyst composition, phase transitions, and active sites provides a foundation for optimizing catalytic performance. The identification of key active phases, such as iron carbides, and their transformation under different reaction conditions deepen our insights into the mechanisms governing hydrocarbon production. Catalyst promoters and supports contribute to fine-tuning catalyst behavior, enabling enhanced selectivity, conversion rates, and stability. Alkali metals, alkaline earth metals, transition metals, and non-metallic elements are commonly employed as promoters, impacting the electronic and structural properties of catalysts. Supports, such as graphene and carbon nanotubes, are investigated for their potential to enhance dispersion, stability, and catalytic activity of the iron-based catalysts. The active phases of Fe-based catalysts in FTS, namely iron carbides and metallic iron, are scrutinized for their role in hydrocarbon production. The transformation of these phases under different reaction conditions significantly influences the selectivity and performance of the catalysts. A thorough understanding of phase transitions and interactions between catalyst components is crucial for optimizing catalyst design and performance. The utilization of iron-based catalysts for FTS to produce light olefins from syngas holds promise for addressing the growing demand for hydrocarbon products while aligning with sustainability goals. By addressing challenges and leveraging emerging perspectives, researchers and industries can drive innovation in catalytic processes and contribute to a cleaner and more efficient energy landscape.

ACKNOWLEDGMENTS

This work was supported by Qatar University through a National Capacity Building Program Grant (NCBP), [QUCP-CAM-20/23-463]. Statements made herein are solely the responsibility of the authors.

REFERENCES

[1] BP Energy Outlook 2019 edition, (n.d.). https://www.bp.com/content/dam/bp/business-sites/en/global/corporate/pdfs/energy-economics/energy-outlook/bp-energy-outlook-2019.pdf.

[2] D. Wang, Y. Gu, Q. Chen, Z. Tang, Direct conversion of syngas to alpha olefins via Fischer-Tropsch synthesis: Process development and comparative techno-economic-environmental analysis, *Energy*. 263 (2023). https://doi.org/10.1016/j.energy.2022.125991.

[3] H. Yu, C. Wang, T. Lin, Y. An, Y. Wang, Q. Chang, F. Yu, Y. Wei, F. Sun, Z. Jiang, S. Li, Y. Sun, L. Zhong, Direct production of olefins from syngas with ultrahigh carbon efficiency, *Nat Commun*. 13 (2022) 5987. https://doi.org/10.1038/s41467-022-33715-w.

[4] F.T. Alsudani, A.N. Saeed, N.S. Ali, H.S. Majdi, H.G. Salih, T.M. Albayati, N.M.C. Saady, Z.M. Shakor, Fisher-Tropsch synthesis for conversion of methane into liquid hydrocarbons through Gas-to-Liquids (GTL) process: A review, *Methane*. 2 (2023) 24–43. https://doi.org/10.3390/methane2010002.

[5] K.J. Kim, K.Y. Kim, G.B. Rhim, M.H. Youn, Y.L. Lee, D.H. Chun, H.S. Roh, Nano-catalysts for gas to liquids: A concise review, *Chem Eng J*. 468 (2023). https://doi.org/10.1016/j.cej.2023.143632.

[6] P. Zhai, Y. Li, M. Wang, J. Liu, Z. Cao, J. Zhang, Y. Xu, X. Liu, Y.W. Li, Q. Zhu, D. Xiao, X.D. Wen, D. Ma, Development of direct conversion of syngas to unsaturated hydrocarbons based on Fischer-Tropsch route, *Chem*. 7 (2021) 3027–3051. https://doi.org/10.1016/j.chempr.2021.08.019.

[7] M. Reinikainen, A. Braunschweiler, S. Korpilo, P. Simell, V. Alopaeus, Two-step conversion of CO_2 to light olefins: Laboratory-scale demonstration and scale-up considerations, *ChemEngineering*. 6 (2022) 96. https://doi.org/10.3390/chemengineering6060096.

[8] S.A. Chernyak, M. Corda, J.P. Dath, V.V. Ordomsky, A.Y. Khodakov, Light olefin synthesis from a diversity of renewable and fossil feedstocks: State-of the-art and outlook, *Chem Soc Rev*. 51 (2022) 7994–8044. https://doi.org/10.1039/d1cs01036k.

[9] B.V. Vora, Development of catalytic processes for the production of olefins, *Trans Indian Natl Acad Eng*. 8 (2023) 201–219. https://doi.org/10.1007/s41403-023-00401-2.

[10] M. Ding, Y. Yang, B. Wu, Y. Li, T. Wang, L. Ma, Study on reduction and carburization behaviors of iron phases for iron-based Fischer-Tropsch synthesis catalyst, *Appl Energy*. 160 (2015) 982–989. https://doi.org/10.1016/j.apenergy.2014.12.042.

[11] A. Yahyazadeh, A.K. Dalai, W. Ma, L. Zhang, Fischer-Tropsch synthesis for light olefins from syngas: A review of catalyst development, *Reactions*. 2 (2021) 227–257. https://doi.org/10.3390/reactions2030015.

[12] K. Cheng, M. Virginie, V.V. Ordomsky, C. Cordier, P.A. Chernavskii, M.I. Ivantsov, S. Paul, Y. Wang, A.Y. Khodakov, Pore size effects in higherature Fischer-Tropsch synthesis over supported iron catalysts, *J Catal*. 328 (2015) 139–150. https://doi.org/10.1016/j.jcat.2014.12.007.

[13] Q. Chang, C. Zhang, C. Liu, Y. Wei, A.V. Cheruvathur, A.I. Dugulan, J.W. Niemantsverdriet, X. Liu, Y. He, M. Qing, L. Zheng, Y. Yun, Y. Yang, Y. Li, Relationship between iron carbide phases (ε-Fe_2C, Fe_7C_3, and χ-Fe_5C_2) and catalytic performances of Fe/SiO2 Fischer-Tropsch catalysts, *ACS Catal*. 8 (2018) 3304–3316. https://doi.org/10.1021/acscatal.7b04085.

[14] X. Liu, T. Lin, P. Liu, L. Zhong, Hydrophobic interfaces regulate iron carbide phases and catalytic performance of $FeZnO_x$ nanoparticles for Fischer-Tropsch to olefins, *Appl Catal B*. 331 (2023). https://doi.org/10.1016/j.apcatb.2023.122697.

[15] M. Feyzi, M.M. Khodaei, J. Shahmoradi, Effect of sulfur on the catalytic performance of Fe-Ni/Al_2O_3 catalysts for light olefins production, *J Taiwan Inst Chem Eng*. 45 (2014) 452–460. https://doi.org/10.1016/j.jtice.2013.05.017.

[16] J.X. Liu, P. Wang, W. Xu, E.J.M. Hensen, Particle size and crystal phase effects in Fischer-Tropsch catalysts, *Engineering*. 3 (2017) 467–476. https://doi.org/10.1016/J.ENG.2017.04.012.

[17] Y. Liu, J.-F. Chen, Y. Zhang, The effect of pore size or iron particle size on the formation of light olefins in Fischer-Tropsch synthesis, *RSC Adv*. 5 (2015) 29002–29007. https://doi.org/10.1039/C5RA02319J.

[18] S.-Y. Li, L.Ü. Shuai, Y.-H. Zhang, J.-L. Li, L. Zhong-Neng, W. Li, Syngas-derived olefins over iron-based catalysts: Effects of basic properties of MgO nanocrystals, *J. Fuel Chem. Technol*. 46 (2018) 1342–1351. https://doi.org/10.1016/S1872-5813(18)30054-9.

[19] J. Lu, L. Yang, B. Xu, Q. Wu, D. Zhang, S. Yuan, Y. Zhai, X. Wang, Y. Fan, Z. Hu, Promotion effects of nitrogen doping into carbon nanotubes on supported iron fischer-tropsch catalysts for lower olefins, *ACS Catal*. 4 (2014) 613–621. https://doi.org/10.1021/cs400931z.

[20] D. Wang, X. Zhou, J. Ji, X. Duan, G. Qian, X. Zhou, D. Chen, W. Yuan, Modified carbon nanotubes by $KMnO_4$ supported iron Fischer-Tropsch catalyst for the direct conversion of syngas to lower olefins, *J Mater Chem A*. 3 (2015) 4560–4567. https://doi.org/10.1039/c4ta05202a.

[21] D.P. Roe, R. Xu, C.B. Roberts, Influence of a carbon nanotube support and supercritical fluid reaction medium on Fe-catalyzed Fischer-Tropsch synthesis, *Appl Catal A Gen*. 543 (2017) 141–149. https://doi.org/10.1016/j.apcata.2017.06.020.

[22] S. Lyu, L. Wang, Z. Li, S. Yin, J. Chen, Y. Zhang, J. Li, Y. Wang, Stabilization of ε-iron carbide as high-temperature catalyst under realistic Fischer-Tropsch synthesis conditions, *Nat Commun*. 11 (2020). https://doi.org/10.1038/s41467-020-20068-5.

[23] H.M. Torres Galvis, A.C.J. Koeken, J.H. Bitter, T. Davidian, M. Ruitenbeek, A.I. Dugulan, K.P. De Jong, Effects of sodium and sulfur on catalytic performance of supported iron catalysts for the Fischer-Tropsch synthesis of lower olefins, *J Catal*. 303 (2013) 22–30. https://doi.org/10.1016/j.jcat.2013.03.010.

[24] M. Zhao, C. Yan, S. Jinchang, Z. Qianwen, Modified iron catalyst for direct synthesis of light olefin from syngas, *Catal Today*. 316 (2018) 142–148. https://doi.org/10.1016/j.cattod.2018.05.018.

[25] P. Zhai, C. Xu, R. Gao, X. Liu, M. Li, W. Li, X. Fu, C. Jia, J. Xie, M. Zhao, X. Wang, Y.-W. Li, Q. Zhang, X.-D. Wen, D. Ma, Highly tunable selectivity for syngas-derived alkenes over zinc and sodium-modulated Fe_5C_2 catalyst, *Angew Chem*. 128 (2016) 10056–10061. https://doi.org/10.1002/ange.201603556.

[26] S. Zhang, D. Li, Y. Liu, Y. Zhang, Q. Wu, Zirconium doped precipitated Fe-based catalyst for Fischer-Tropsch synthesis to light olefins at industrially relevant conditions, *Catal Letters*. 149 (2019) 1486–1495. https://doi.org/10.1007/s10562-019-02775-x.

[27] H. Xiong, M.A. Motchelaho, M. Moyo, L.L. Jewell, N.J. Coville, Effect of Group I alkali metal promoters on Fe/CNT catalysts in Fischer-Tropsch synthesis, *Fuel*. 150 (2015) 687–696. https://doi.org/10.1016/j.fuel.2015.02.099.

[28] J. Li, X. Cheng, C. Zhang, Q. Chang, J. Wang, X. Wang, Z. Lv, W. Dong, Y. Yang, Y. Li, Effect of alkalis on iron-based Fischer-Tropsch synthesis catalysts: Alkali-FeO_x interaction, reduction, and catalytic performance, *Appl Catal A Gen*. 528 (2016) 131–141. https://doi.org/10.1016/j.apcata.2016.10.006.

[29] J. Li, X. Cheng, C. Zhang, J. Wang, W. Dong, Y. Yang, Y. Li, Alkalis in iron-based Fischer-Tropsch synthesis catalysts: Distribution, migration and promotion, *J Chem Technol Biotechnol*. 92 (2017) 1472–1480. https://doi.org/10.1002/jctb.5152.

[30] S. Yang, M. Li, M.A. Nawaz, G. Song, W. Xiao, Z. Wang, D. Liu, High selectivity to aromatics by a Mg and Na Co-modified catalyst in direct conversion of syngas, *ACS Omega*. 5 (2020) 11701–11709. https://doi.org/10.1021/acsomega.0c01007.

[31] W. Gong, R.-P. Ye, J. Ding, T. Wang, X. Shi, C.K. Russell, J. Tang, E.G. Eddings, Y. Zhang, M. Fan, Effect of copper on highly effective Fe-Mn based catalysts during production of light olefins via Fischer-Tropsch process with low CO_2 emission, *Appl Catal B.* 278 (2020) 119302. https://doi.org/10.1016/j.apcatb.2020.119302.

[32] A.H. Nasser, H.M. El-Bery, H. ELnaggar, I.K. Basha, A.A. El-Moneim, Selective conversion of syngas to olefins via novel Cu-promoted Fe/RGO and Fe-Mn/RGO Fischer-Tropsch catalysts: Fixed-bed reactor vs Slurry-bed reactor, *ACS Omega.* 6 (2021) 31099–31111. https://doi.org/10.1021/acsomega.1c04476.

[33] B. Gu, V.V. Ordomsky, M. Bahri, O. Ersen, P.A. Chernavskii, D. Filimonov, A.Y. Khodakov, Effects of the promotion with bismuth and lead on direct synthesis of light olefins from syngas over carbon nanotube supported iron catalysts, *Appl Catal B.* 234 (2018) 153–166. https://doi.org/10.1016/j.apcatb.2018.04.025.

[34] S.A. Chernyak, D.N. Stolbov, K.I. Maslakov, R.V. Kazantsev, O.L. Eliseev, D.O. Moskovskikh, S.V. Savilov, Graphene nanoflake- and carbon nanotube-supported iron-potassium 3D-catalysts for hydrocarbon synthesis from syngas, *Nanomaterials.* 12 (2022). https://doi.org/10.3390/nano12244491.

[35] X.P. Fu, W.Z. Yu, C. Ma, J. Lin, S.Q. Sun, S.Q. Li, P.N. Ren, F.Y. Jia, M.Y. Li, W.W. Wang, X. Wang, C.J. Jia, K. Wu, R. Si, C.H. Yan, Supported Fe2C catalysts originated from Fe_2N phase and active for Fischer-Tropsch synthesis, *Appl Catal B.* 284 (2021). https://doi.org/10.1016/j.apcatb.2020.119702.

[36] F. Song, X. Yong, X. Wu, W. Zhang, Q. Ma, T. Zhao, M. Tan, Z. Guo, H. Zhao, G. Yang, N. Tsubaki, Y. Tan, FeMn@HZSM-5 capsule catalyst for light olefins direct synthesis via Fischer-Tropsch synthesis: Studies on depressing the CO_2 formation, *Appl Catal B.* 300 (2022). https://doi.org/10.1016/j.apcatb.2021.120713.

[37] F. Lu, J. Huang, Q. Wu, Y. Zhang, Mixture of α-Fe_2O_3 and MnO_2 powders for direct conversion of syngas to light olefins, *Appl Catal A Gen.* 621 (2021). https://doi.org/10.1016/j.apcata.2021.118213.

[38] B. Gu, S. He, D.V. Peron, D.R. Strossi Pedrolo, S. Moldovan, M.C. Ribeiro, B. Lobato, P.A. Chernavskii, V.V. Ordomsky, A.Y. Khodakov, Synergy of nanoconfinement and promotion in the design of efficient supported iron catalysts for direct olefin synthesis from syngas, *J Catal.* 376 (2019) 1–16. https://doi.org/10.1016/j.jcat.2019.06.035.

[39] C. Li, I. Sayaka, F. Chisato, K. Fujimoto, Development of high performance graphite-supported iron catalyst for Fischer-Tropsch synthesis, *Appl Catal A Gen.* 509 (2016) 123–129. https://doi.org/10.1016/j.apcata.2015.10.028.

[40] L. Guo, Z. Guo, J. Liang, X. Yong, S. Sun, W. Zhang, J. Sun, T. Zhao, J. Li, Y. Cui, B. Zhang, G. Yang, N. Tsubaki, Quick microwave assembling nitrogen-regulated graphene supported iron nanoparticles for Fischer-Tropsch synthesis, *Chem Eng J.* 429 (2022). https://doi.org/10.1016/j.cej.2021.132063.

[41] O. Zhuo, L. Yang, F. Gao, B. Xu, Q. Wu, Y. Fan, Y. Zhang, Y. Jiang, R. Huang, X. Wang, Z. Hu, Stabilizing the active phase of iron-based Fischer-Tropsch catalysts for lower olefins: Mechanism and strategy, *Chem Sci.* 10 (2019) 6083–6090. https://doi.org/10.1039/c9sc01210a.

[42] R. Li, Y. Li, Z. Li, Y. Wei, Q. Hao, Y. Shi, S. Ouyang, H. Yuan, T. Zhang, Electronically activated Fe_5C_2 via N-doped carbon to enhance photothermal syngas conversion to light olefins, *ACS Catal.* 12 (2022) 5316–5326. https://doi.org/10.1021/acscatal.2c00926.

[43] B. Gu, S. He, W. Zhou, J. Kang, K. Cheng, Q. Zhang, Y. Wang, Polyaniline-supported iron catalyst for selective synthesis of lower olefins from syngas, *J Energy Chem.* 26 (2017) 608–615. https://doi.org/10.1016/j.jechem.2017.04.009.

[44] C. Zhu, C. Huang, M. Zhang, Y. Han, K. Fang, Rational design of hierarchical zeolite encapsulating FeMnK architecture to enhance light olefins selectivity in Fischer-Tropsch synthesis, *Fuel.* 309 (2022). https://doi.org/10.1016/j.fuel.2021.122075.

[45] C. Zhu, C. Huang, M. Zhang, K. Fang, Design of ZSM-5 encapsulating FeMnK nanocatalysts for light olefins synthesis with enhanced carbon utilization efficiency, *Fuel.* 335 (2023). https://doi.org/10.1016/j.fuel.2022.126745.

[46] Y. Liu, F. Lu, Y. Tang, M. Liu, F.F. Tao, Y. Zhang, Effects of initial crystal structure of Fe_2O_3 and Mn promoter on effective active phase for syngas to light olefins, *Appl Catal B.* 261 (2020). https://doi.org/10.1016/j.apcatb.2019.118219.

[47] Z. Ma, C. Zhou, D. Wang, Y. Wang, W. He, Y. Tan, Q. Liu, Co-precipitated Fe-Zr catalysts for the Fischer-Tropsch synthesis of lower olefins ($C_2^O \sim C_4^O$): Synergistic effects of Fe and Zr, *J Catal.* 378 (2019) 209–219. https://doi.org/10.1016/j.jcat.2019.08.037.

[48] Z. Ma, H. Ma, H. Zhang, X. Wu, Q. Qian, Q. Sun, W. Ying, Direct conversion of syngas to light olefins through Fischer-Tropsch synthesis over Fe-Zr catalysts modified with sodium, *ACS Omega.* 6 (2021) 4968–4976. https://doi.org/10.1021/acsomega.0c06008.

[49] I.K. Ghosh, Z. Iqbal, S. Bhattacharya, A. Bordoloi, Insight of boron induced single-step synthesis of short-chain olefins from bio-derived syngas, *Fuel.* 263 (2020) 116663. https://doi.org/10.1016/j.fuel.2019.116663.

[50] X. Wu, W. Qian, H. Ma, H. Zhang, D. Liu, Q. Sun, W. Ying, Li-decorated Fe-Mn nanocatalyst for high-temperature Fischer-Tropsch synthesis of light olefins, *Fuel.* 257 (2019). https://doi.org/10.1016/j.fuel.2019.116101.

[51] X. Yang, J. Yang, Y. Wang, T. Zhao, H. Ben, X. Li, A. Holmen, Y. Huang, D. Chen, Promotional effects of sodium and sulfur on light olefins synthesis from syngas over iron-manganese catalyst, *Appl Catal B.* 300 (2022) 120716. https://doi.org/10.1016/j.apcatb.2021.120716.

[52] D.V. Peron, A.J. Barrios, A. Taschin, I. Dugulan, C. Marini, G. Gorni, S. Moldovan, S. Koneti, R. Wojcieszak, J.W. Thybaut, M. Virginie, A.Y. Khodakov, Active phases for high temperature Fischer-Tropsch synthesis in the silica supported iron catalysts promoted with antimony and tin, *Appl Catal B.* 292 (2021). https://doi.org/10.1016/j.apcatb.2021.120141.

[53] Y. Cheng, J. Tian, J. Lin, S. Wang, S. Xie, Y. Pei, S. Yan, M. Qiao, H. Xu, B. Zong, Potassium-promoted magnesium ferrite on 3D porous graphene as highly efficient catalyst for CO hydrogenation to lower olefins, *J Catal.* 374 (2019) 24–35. https://doi.org/10.1016/j.jcat.2019.04.024.

19 Wastewater Treatment Using Synthesised Metal Oxide– Based Nanocomposites

H.K. Jahnavi, S.V. Dhanyashree, S. Rajendra Prasad, H.V. Harini, and H.P. Nagaswarupa

INTRODUCTION

NANOCOMPOSITES

Composites are naturally occurring solid materials. When two or more constituent materials that have distinct physical or chemical properties are combined together, a new substance can be formed in a particular finished structure which has better properties than the original materials. The term "nanocomposite material" has greatly broadened over the years to include a wide range of systems, including dimensions 1, 2, and 3, and amorphous materials are made up of clearly different components and combined at the nanometer scale [1].

Nowadays, nanocomposites gained a lot of interest. Properties of nanocomposite materials are influenced by both their morphology and interfacial characteristics as well as their individual parents [2]. Numerous nanocomposites made up of oxides of metal have been described for their essential uses, such as the carbon nanotube–ZnO nanocomposite used for supercapacitors. CuO–ZnO nanocomposite was reported to be an effective antibacterial agent [3]. It was discovered that ZnO/ CuO nanocomposites had humidity-sensing capabilities [4]. GO/ZnO nanocomposite for photocatalytic degradation of basic fuchsin dye is only one example among many more [5]. The polyaniline/ ZnO nanocomposite has been utilized as a sensor for various applications [6]. Nanocomposite materials can be categorised into three distinct categories, like that of microcomposites based on their matrix materials:

1. Metal Matrix Nanocomposite (MMNC)
2. Polymer Matrix Nanocomposite (PMNC)
3. Ceramic Matrix Nanocomposite (CMNC)

Metal Matrix Nanocomposite (MMNC)

Like other composites, metal matrix nanocomposites (MMCs) constitute a combination of at least two physically and chemically distinct phases that are dispersed to give them properties. A metallic matrix usually consists of two phases: a fibrous or particulate phase [7]. Metal acts as the matrix in metal matrix nanocomposites, and ceramic acts as the reinforcement. Nowadays, the majority of metals and alloys can be used as matrix materials, which need for dispersant materials that are stable throughout a broad range of temperatures [8]. Metal matrix is distributed with various metal, ceramic, or organic materials in metal matrix composites. The purpose of reinforcements is typically to enhance various qualities of the base metal. The characteristics of MMCs are significantly influenced by the particle distribution. As basic metals in metal matrix composites, attention has been focused especially on copper, magnesium, and aluminium [9]. The production of materials

DOI: 10.1201/9781003479239-19

with high service temperatures and compression strengths is well suited for these nanocomposites. They have plenty of potential applications in a variety of fields, such as the manufacturing of structural materials and the automotive and aerospace sectors [7].

Polymer Matrix Nanocomposite (PMNC)

PMNC is the most common type of nanocomposites. It would consist of small particles evenly distributed all over a polymer. In the actual world, nanoparticles are dispersed using a polymer matrix [10]. In general, nanocomponents in PMNC are fillers known as nanofillers and are categorised as 1D linear, 2D layer, and 3D powder [11]. High-performance composite is a different term used to describe PMNC. When filler homogeneity is attained in PMNCs, the associated properties of the nanoscale filler are significantly distinct from the initial matrix. Nowadays, polymer matrix nanocomposites are created by utilising stiffer ceramic nanoparticles, including carbon nanotubes and clay [12]. The utilisation of polymer composites in various industrial applications has increased [13].

Ceramic Matrix Nanocomposite (CMNC)

The term "ceramic matrix composites" describes composites in which the ceramic matrix is reinforced. The ceramic matrix is in the continuous phase, while the reinforced elements are in the dispersed phase. Continuous fibres and discrete particles can make up the dispersed phase. The type of fibres employed, the matrix, the interface phase, and their interactions all affect the mechanical properties of fibre-reinforced CMNCs [14]. Ceramic matrix composites are yielding a substantial promise for enabling advanced technologies such as hypersonic engines and aircraft, lightweight and high lifetime prosthetics, and high-temperature electronic components [15]. Ceramic matrix composites have the benefits of high temperature resistance, low density, and low thermal expansion coefficient, which can greatly reduce engine construction quality and increase the temperature-bearing capacity for components utilised in high temperatures [14].

METAL OXIDE NANOCOMPOSITE FOR PHOTOCATALYSIS AND WASTEWATER TREATMENT

METAL OXIDE NANOCOMPOSITE FOR PHOTOCATALYSIS

Generally, photocatalysts are used to speed up chemical reactions when ultraviolet and visible light are present. The photocatalytic system has a sufficient bandgap, remains in a stable, adequate morphology, has a large surface area, and is reusable [16].

Based on the physical appearance of the reactants, two types of photocatalytic processes can be differentiated, i.e., homogeneous catalysis and heterogeneous catalysis [17]. By definition, homogeneous catalysis corresponds to a catalytic system in which the catalyst components and the reaction substrates are combined in one phase, most frequently the liquid phase. There are several known catalytic processes that occur when the catalyst and reacting substances do not exist in the same phase, and the term they are known as is heterogeneous catalyst. Systems using heterogeneous catalysts are known to be effective in treating wastewater. Recent studies are conducted on the usage of metal oxide nanocomposites as photocatalysts for their exceptional capabilities in water purification [18]. Different catalytic processes have made use of transition metals and their oxides. They exhibit numerous unique qualities and industrial uses.

In simple terms, photocatalysis is a process that utilises the absorption of light by a catalyst to start and speed up chemical reactions that lead to the mineralisation of the pollutants. As implied by the name, it involves photons and a catalyst. It includes the general equation [19].

$$\text{Organic pollutant} + O_2 \xrightarrow[\text{UV/ Solar light}]{\text{semiconductor}} CO_2 + H_2 + \text{mineral acids}$$

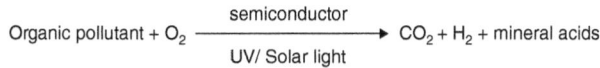

Photodegradation process mechanism is shown below,

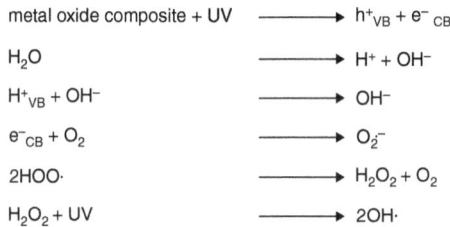

$$\text{metal oxide composite} + UV \longrightarrow h^+{}_{VB} + e^-{}_{CB}$$

$$H_2O \longrightarrow H^+ + OH^-$$

$$H^+{}_{VB} + OH^- \longrightarrow OH^-$$

$$e^-{}_{CB} + O_2 \longrightarrow O_2^-$$

$$2HOO\cdot \longrightarrow H_2O_2 + O_2$$

$$H_2O_2 + UV \longrightarrow 2OH\cdot$$

We will analyse various metal oxide–based nanocomposite materials (TiO_2, ZnO, and WOx) as improved photocatalysts in this review [20].

METAL OXIDE NANOCOMPOSITE FOR WASTEWATER TREATMENT

Wastewater is a liquid form of urban trash that is contaminated with harmful heavy metals, organic materials, bacteria, and inorganic soluble substances. These contaminants alter the chemical, biological, and physical properties of pure water [21]. Human health and development were seriously endangered by the harmful and biodegradable organic and inorganic pollutants that produce water pollution. Pollution of water has been a major issue recently, especially in locations where people utilise ground water and surface water for drinking and other household applications of water [22].

Zinc, chromium, nickel, lead, cadmium, and arsenic are examples of common heavy metal pollutants which cause possible dangers to the environment when present over the concentration limits set by the WHO. In severe circumstances, these can result in organ failures, body system dysfunction, and death [23]. Industrial wastewater is treated using a variety of approaches, including chemical, physical, and biological. Advanced methods of oxidation are attractive possibilities for the nonselective destruction of pollutants among chemical treatment techniques [24].

DYES AND THEIR STRUCTURE

The main factor influencing the colour of our outfits and fabrics is dye. Dyes are the substances with the capacity of binding to fabric. The longevity of the dye is greatly influenced by temperature and time. The majority of organic substances and dyes have colour because in the visible spectrum, (i) they absorb light, (ii) it contains a minimum of one chromophore, (iii) it contains conjugated system, and (iv) it shows resonance.

Dye is a substance made up of auxochrome and chromophore groups. Chromophore group's saturation determines dye shade due to its saturation. The pigment fibre reaction is caused by the auxochrome group [25].

Classification of dyes is as follows:

1. Natural Dye
2. Synthetic Dye

NATURAL DYE

The term "natural dye" refers to all dyes made from organic materials like plants, animals, and minerals. Natural dyes are typically not substantial and must be applied to fabrics with the aid of

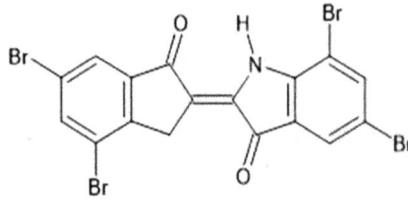

FIGURE 19.1 Structure of natural dye.

modifiers, typically a metallic salt that has an intense <u>attraction</u> for both the dyeing agent and the fibre itself [26] (Figure 19.1).

Synthetic Dye

Synthetic dyes are those made from either organic or inorganic compounds. Sir William Henry Perkin unintentionally discovered the first synthetic dye, mauveine or mauve, in 1856 [27]. In comparison to natural dyes, synthetic dyes are more resistant and more difficult to totally degrade with photolysis, the process involving biological and chemical decomposition, along with other conventional methods [28]. Synthetic dyes have high impacts on the aquatic ecology since they are non-bioabsorbable and decomposable [27] (Figure 19.2).

Further dyes can be classified based on their application.

Reactive Dyes

In reactive dyes, covalent bonds are formed when dyes react with fibres generally like cotton [30]. Reactive dyes are one of the widely used dyes because they are simple to use, have vibrant colours, and have high colour resistance [31].

Some examples of reactive dyes are reactive red 120 (bis-monochlorotriazine dye), reactive yellow 4 (dichlorotriazine dye), vat blue (indigoid vat dye), and sulphur black (Figure 19.3).

Direct Dyes

A dye is referred to as a direct dye when it dyes a fabric instantly without the aid of any binding agent. Congo red was the first direct dye, found in 1884 [32]. The ease of usage and low price of these dyes are two of their numerous advantages, and their low strength is their main flaw [33]. These dyes are more affordable. They come in a wide range of colours; however, their colour brightness is not very impressive [34].

Some examples for the direct dye are direct orange 40, direct orange 26, direct blue 6, direct blue 71, direct black 38, and direct red 2 (Figure 19.4).

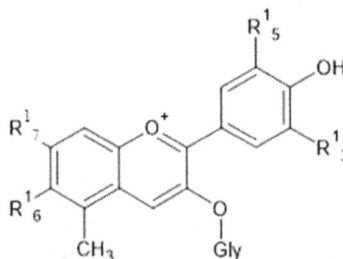

FIGURE 19.2 Structure of synthetic dye [29].

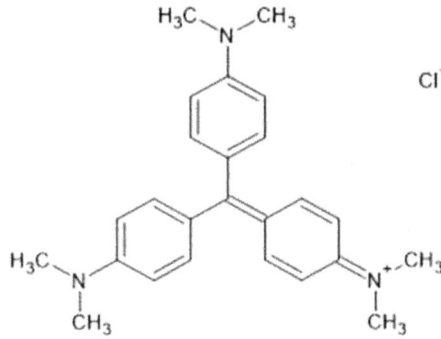

FIGURE 19.3   Structure of indigoid vat dye.

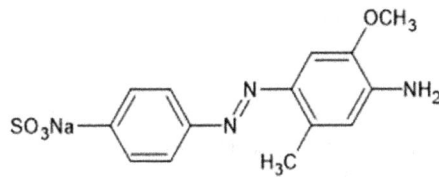

FIGURE 19.4   Structure of direct orange 40.

Vat Dyes

The term "vat dyes" refers to colour since they were once fermented in enormous wooden vats. Because of vat dye is insoluble in water, a non-renewable reducing substance is needed to convert the dye into the colourless form [35]. Due to its planar structure and superior durability properties, vat dyes have an intense attraction for cellulose fibres [36].

Some examples for vat dyes are vat blue 1 (synthetic indigo), vat black 25, indigo carmine, vat yellow 1, vat acid blue 74, and vat blue 4 (indanthrene) (Figure 19.5).

Sulphur Dyes

Sulphur dyes can be identified by the existence of sulphur. Each and every sulphur dye is reactive [37]. Organic substances called sulphur dyes can be produced by heating amines or phenolic compounds in any amount of sulphur. Because of their inexpensive cost, outstanding colour, and fastness capabilities, sulphur dyes are also employed [38].

Some examples for sulphur dyes are sulphur red 7, sulphur brilliant green, sulphur black, and sulphur blue 15 (Figure 19.6).

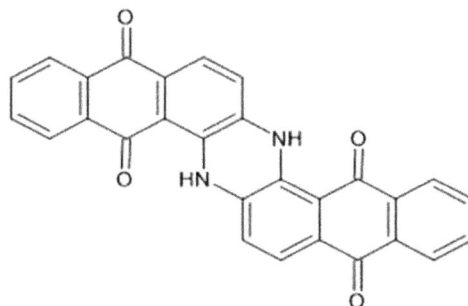

FIGURE 19.5   Structure of vat blue 4.

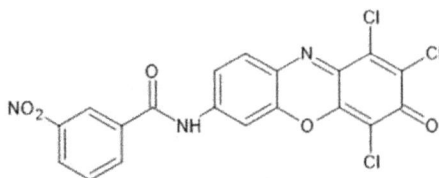

FIGURE 19.6 Structure of sulphur red 7.

Disperse Dye

Disperse dyes are largely water insoluble. In a wide range of sectors, including textiles, paper, and leather, disperse dyes are frequently utilised [39]. Dyeing process of polyester materials requires the addition of dispersants [40].

Some examples for disperse dyes are red disperse 60, blue disperse 7, disperse yellow 3, disperse red 9, and disperse violet 26 (Figure 19.7).

Acid Dyes

Acid dyes are anionic and dissolved in water that contains formic acid, acetic acid, and one or more sulphuric acid substituents. Affinities for the basic features of fibres like polyamides can be attributed to their acid nature. Due to their striking colours and excellent solubility, acid dyes, particularly sulfonic acid dyes, are frequently used in the textile, drugs, printing, leather goods, dye, paper, and other industry sectors [30].

Some examples for acid dyes are acid blue 74, acid blue 90, red acid 27, acid blue 349, acid yellow 36, and acid red 14 (Figure 19.8).

Basic Dye

Cationic dyes are another name for basic dyes. The basic dyes are less expensive, soluble in alcohol, but not readily soluble in water, and only a small number are very slightly soluble [30].

Some examples for basic dyes are basic blue 26, basic red 1, basic green 1, basic yellow 2, and basic blue 24 (Figure 19.9).

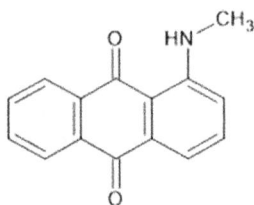

FIGURE 19.7 Structure of disperse red 9.

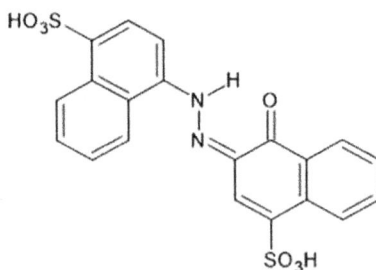

FIGURE 19.8 Structure of acid red 14.

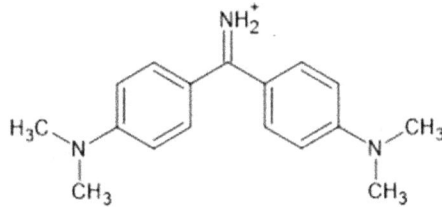

FIGURE 19.9   Structure of basic yellow 2.

SYNTHESIS OF METAL OXIDE–BASED COMPOSITES

A variety of methods, including chemical, physical, and biological methods, can be used to generate nanoparticles. In general, physical and chemical approaches work best to produce stable nanoparticles of similar size (Figure 19.10).

CHEMICAL METHOD

Due to the presence of some harmful chemicals absorbed on the surface, chemical synthesis techniques have been associated with a number of adverse side effects [41]. There are numerous methodologies of generating nanoparticles. Therefore, these topics are addressed in detail.

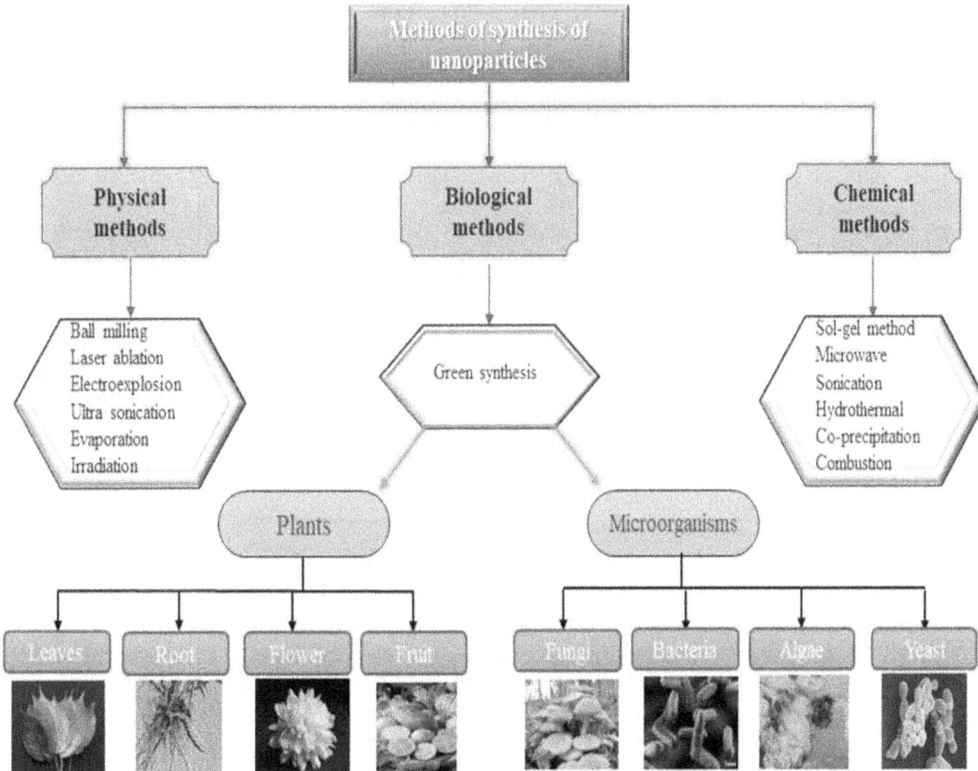

FIGURE 19.10   Schematic diagram of synthesis methods.

COMBUSTION METHOD

Solution combustion synthesis is a cost-effective method for producing simple and complex oxides, particularly in ceramics, due to its simplicity and efficiency [42,43]. The same method is used to create ZnO–TiO_2, which has applications as a gas sensor, where ZnO nanoplatelets with dispersed TiO_2 nanoparticles have been identified. Fe–Ni-doped TiO_2 was used to synthesise CNTs/Fe–Ni/TiO_2 by deposition of chemical vapours. MWCNT/ZnO were also synthesised [44].

SOL-GEL METHOD

Sol-gel is a versatile method for producing a wide variety of materials. It provides the capacity to create a wide variety of nano- and microstructures, and allows for the successful application of various processes [45]. A variety of transition metal oxide–based nanocomposites, including CNTs/TiO_2, WO_3/TiO_2, ZnO/reduced grapheme oxide, ZnO–SiO_2, CdO–ZnO, $CoTiO_3$/$CoFe_2O_4$, and ZnO–SnO_2, produced using the simple sol-gel synthesis method [46].

SONICATION METHOD

One of the most efficient ways to eliminate aggregation and enhance the distribution of nanostructures in the cement matrix is sonication. The best way for enhancing the dispersion of clay nanoparticles was discovered to be indirect sonication or a bath sonicator by means of the cement matrix [47]. Both a bath with ultrasonic waves and a probe with ultrasonic frequencies (sonicator) can be utilised for sonication [48].

CO-PRECIPITATION METHOD

It is a moist chemical procedure also known as a solvent displacement technique [49]. This process is a crucial tool to generate nanomaterials. When a substance's concentration reaches oversupply, a precipitate in solution occurs. Diffusion occurring on the surface of the precipitation will cause it to expand, resulting in the formation of nanoparticles [50]. Over the years, various combinations of transition metal oxides, including $CuWO_4$/ NiO, NiO/ZnO, CeO/CuO/ ZnO, CeO_2/ZnO/$ZnAl_2O_4$, ZnO/Ag and more have been observed [51].

MICROWAVE METHOD

For the production of nanomaterials and nanocomposites, microwave-assisted technologies are an exciting environmentally friendly method [52]. When fabricating Ag/ZnO/graphene for photocatalytic activity, Ag/ZnO/TiO_2 [53] and Mn_3O_4/Fe_2O_3/Fe_3O_4/rGO nanocomposites are used for supercapacitors and many more synthesised by using this method [54].

HYDROTHERMAL METHOD

Hydrothermal method was extensively utilised in the fabrication of transition metal oxide–based nanocomposites. The latest approach for separating materials from hot-water solutions at high vapour pressures is hydrothermal synthesis [55]. A variety of TMONCs via simple hydrothermal methods, including carbon nanotube/cubic ferrite, Fe_2O_3–ZnO, MnO_2–CNT, TiO_2/reduced ferrite, and CuO–ZnO, were synthesised [56].

GREEN METHOD

In the field of materials science, "green" synthesising has drawn a lot of attention as a trustworthy, durable, and environmentally friendly method to synthesise a variety of substances and nanomaterials [57]. Most microorganisms used in green synthesis are bacteria, algae, and fungi. Alternatively, extracts from leaves, plants, seeds, fruits, and peelings of various plants have also been used for the production of various nanoparticles due to the phytochemicals that are present in the extracts functioning as preserving and reducing agents [58].

Using Plant Extracts

Plants are an excellent option for nanoparticle production because the procedures utilising plant materials do not involve any chemicals that are dangerous. Plants are referred to as the inexpensive, low-maintenance manufacturing facilities of environment [59]. Plant extracts are simply combined with a solution of metallic salts at room temperature in order to generate nanoparticles. The reaction is finished in just a couple of minutes. Nanoparticles are synthesised by using a variety of plants like neem, *Aloe vera*, tea, *Cinnamomum camphora*, geranium leaf, and lemon grass [60].

Numerous plant extract nanocomposites have been synthesised including Ag–ZnO nanocomposite made from *Trigonella foenum-graecum* leaf extract, Fe_2O_3–Ag TMONCs made from *Psidium guajava* leaf extract, ZnO–NiO nanocomposite made from *Azadirachta indica* (neem) leaf extract, and ZnO/CuO nanocomposite made from cacao seed bark extract [61].

USING MICROORGANISMS

Microorganisms are significant nanofactories with enormous qualities such as environmentally friendliness, economically advantageousness, avoiding damaging toxic substances, and high energy demand required for physiochemical synthesis [44]. Microorganisms are being investigated for the production of metal nanoparticles. These organisms include bacteria, yeast, and fungi. Because of the pathogenic problem, utilising microorganisms is dangerous. Consequently, it is important to synthesise nanoparticles using greener techniques [62].

STATE OF THE ART IN METAL OXIDE COMPOSITES

1. Aluminium nitrate and zirconyl were used as oxidisers, while urea was selected as the fuel. The powders were heated for 2 hours at 1,200°C, 1,300°C, and 1,400°C. These powders were examined by X-ray diffraction, scanning electron microscopy, and transmission electron microscopy (Table 19.1).
2. Aluminium nitrate and magnesium nitrate were used as oxidisers. Urea was chosen as the fuel for our experiment. The furnace was set at a temperature between 450°C and 500°C. These powders were examined by XRD and TEM.
3. ZnO, CdO, and CdO–ZnO nanoparticles were synthesised using the sol-gel method. The sol-gels of ZnO and CdO–ZnO were calcined for 8 hours in air at 600°C and 400°C, respectively. These powders were characterised by X-ray diffraction (XRD), transmission electron microscopy (TEM), and diffuse reflectance spectroscopy (DRS).
4. Zn^{2+} and Cu^{2+} precursor salts were made using zinc chloride ($ZnCl_2$) and copper chloride ($CuCl_2$). Dry samples were ground and annealed for 2 hours at 300°C. These powders were examined by X-ray diffraction, and microscopic and spectroscopic techniques.
5. $FeCl_3.H_2O$ and $C_{12}H_{28}O_4Ti$ were the precursor substances utilised to prepare the samples. The gel was then gradually heated to 100°C for 7 days in a controlled oven. These powders were examined by XRD, TEM, DTA, and Mossbauer analysis.
6. The production of MnO_2–CNT nanocomposites was accomplished hydrothermally. The recovered product was then dried for 24 hours at 80°C in a vacuum oven. These powders were examined by X-ray diffraction, electron microscopy, and electrochemical methods.
7. The precursors employed in the microwave-assisted approach to create the CdO–ZnO nanocomposite were zinc chloride ($ZnCl_2$) and cadmium chloride ($CdCl_2.2H_2O$).

TABLE 19.1

Synthesis of Metal Oxide Composites by Chemical Methods

Metal Oxide Composite	Synthesis Methods	Application	References
Pd/Al_2O_3	Sol-combustion	Catalyst support	[63]
$Al_2O_3–ZrO_2$	Sol-combustion	Cutting tool	[64]
MAl_2O_4 (M=Mn and Zn)	Sol-combustion	Catalytic support	[65]
$MgAl_2O_4$	Sol-combustion	Structural material	[65]
$WO_3–ZrO_2$	Sol-combustion	Solvent-free synthesis of coumarins	[66]
$CeO_2–ZrO_2$	Sol-combustion	Oxygen storage capacitor	[67]
$LiMn_2O_4$	Sol-combustion	Lithium battery	[68]
$Y_2O_3–ZrO_2/YSZ$	Sol-combustion	SOFC electrolyte	[69]
ZrW_2O_8	Sol-combustion	Negative thermal expansion	[70]
$ZnO-SnO_2$	Co-precipitation	Promising materials	[71]
$ZnO-Fe_3O_4$	Co-precipitation	Novel catalyst	[72]
$NiO.CeO_2.ZnO$	Co-precipitation	Photolytic and antibacterial activities	[73]
$ZnO-CuO$	Co-precipitation	Antibacterial agent	[74]
$MgO-Al_2O_3$	Co-precipitation	Post-combustion capture	[75]
$MgO-CuO$	Co-precipitation	Glass ceramic for dental	[76]
$Zeolite-Y-Fe_3O_4$	Co-precipitation	Industry sector	[77]
ZnO-Activated carbon	Co-precipitation	Supercapacitor	[78]
Mullite-SiC	Sol-Gel	Coating for carbon materials	[79]
CNT's-TiO_2	Sol-Gel	Photo-assisted water electrolysis	[80]
TiO_2–MMT	Sol-Gel	Food packaging	[81]
TiO_2–Clay	Sol-Gel	Photocatalysis	[82]
ZnO–CdO	Sol-Gel	Gas sensing	[83]
ZnO/GO	Co-precipitation	Antibacterial properties	[84]
$TiO_2–Al_2O_3$	Sol-Gel	Improved sonocatalyst	[85]
$TiO_2–Fe_2O_3$	Sol-Gel	Dye decolorisation and supercapacitors	[86]
Ag–TiO_2	Sol-Gel	Photocatalytic and bactericidal behaviours	[87]
Al_2O_3–SiC	Sol-Gel	Slip casting and injection moulding	[88]
$WO_3–TiO_2$	Sol-Gel	PFAS removal under UVA/visible light	[89]
$Al_2O_3–TiO_2$	Hydrothermal	Dye-sensitised solar cells	[90]
ZnO-Kaolinite	Hydrothermal	Adsorption studies for tannery wastewater treatment	[91]
$CoFe_2O_4$/Graphite	Hydrothermal	Electrochemical and supercapacitor	[92]
MnO_2–CNT	Hydrothermal	Supercapacitor	[93]
ZnS–MMT	Hydrothermal	Degrading eosin B	[94]
CdS–TiO_2–Montmorillonite	Hydrothermal	Photocatalytic activity	[95]
Co–MgO	Hydrothermal	Oxidation of carbon monoxide	[96]
Mixed-phase TiO_2	Hydrothermal	Photo oxidation and photo reduction	[97]
Ag/ZnO–TiO_2	Microwave	Water purification	[53]
CdO–ZnO	Microwave	Gas sensing	[98]
CdO–NiO–ZnO	Microwave	Antibacterial activity	[99]
ZrO_2/ZnO	Sol-Gel	Optical uses	[99]
Al_2O_3/ZrO_2	Sol-Gel	Catalytic uses	[100]
ZrO_2/CeO_2	Sol-Gel	Photocatalytic uses	[101]
CeO_2/ZrO_2	Sol-Gel	Catalytic uses	[102]
ZrO_2/CeO_2	Co-precipitation	Thermal application	[103]
MgO/ZrO_2	Co-precipitation	Antibacterial uses	[104]

(Continued)

TABLE 19.1 (*Continued*)
Synthesis of Metal Oxide Composites by Chemical Methods

Metal Oxide Composite	Synthesis Methods	Application	References
$CeO_2/SeO_2/ZrO_2$	Co-precipitation	Fuel cells	[105]
Fe-Mn bimetal oxide	Co-precipitation	Contaminants' removal from the aquatic solutions	[106]
$NiFe_2O_4$	Co-precipitation	Targeted drug delivery	[107]
Graphene oxide/Fe_3O_4	Co-precipitation	Wastewater purification	[108]
ZnO-CuO	Co-precipitation	Antibacterial activity	[74]
$CuWO_4$/NiO	Co-precipitation	Photodegradation of organic dye in water	[51]
nickel oxide-zinc oxide	Co-precipitation	Enhanced bactericidal performance	[109]
CeO_2/CuO/ZnO	Co-precipitation	Photoluminescence, photocatalytic, and antibacterial activities	[110]
CeO_2/ZnO/$ZnAl_2O_4$	Co-precipitation	Photocatalytic activity	[111]
ZnO/Ag	Co-precipitation	Contact materials	[112]
ZnO/GO	Co-precipitation	Antibacterial properties	[113]
rGO/NiO	Microwave-assisted	Glucose detection	[114]
CuO/TiO_2/ZnO	Sol-Gel	Photocatalytic performance	[115]
Mn–Co–Fe oxide	Co-precipitation	Catalytic ozonation	[116]
NiO/TiO_2	Sol-Gel	Photodegradation of methylene blue dye	[117]
ZnO/CuO	Sol-Gel	Photocatalytic behaviour	[118]
CNTs/TiO_2	Sol-Gel	Photocatalytic activity	[80]
WO_3/TiO_2	Sol-Gel	Environmental purification	[89]
ZnO/Reduced graphene oxide	Sol-Gel	Photoelectrochemical water splitting	[119]
ZnO-SiO_2	Sol-Gel	Optoelectronic devices and as sensors	[120]
$CoTiO_3$/$CoFe_2O_4$	Sol-Gel	Photocatalytic properties	[121]
TiO_2/Ti_3C_2	Hydrothermal	Photocatalytic activity	[122]
$ZnFe_2O_4$	Hydrothermal	Photocatalytic activity	[56]
TiO_2-reduced graphene oxide	Hydrothermal	Optical limiting properties	[123]
Ag/ZnO/graphene	Microwave	Photocatalytic activity	[124]
Mn_3O_4–Fe_2O_3/Fe_3O_4@rGO	Microwave	Electrochemical performance	[54]
CuO–Fe_2O_3–MgO	Sol-gel method	Antibacterial properties	[125]
CuO/Fe_2O_3/ZnO	Sol-gel auto-combustion	Degradation pathways	[126]
NiO-CdO-ZnO n	Co-precipitation	Photocatalytic activity	[127]
CdO–NiO–ZnO	Microwave-assisted	Antibacterial activities	[99]
CdO–ZnO–MgO	Microwave-assisted	Antibacterial studies	[128]
CuO–Fe_2O_3–MgO	Sol-gel	Optical properties	[129]
CuO–Fe_2O_3–MgO–$CuFe_2O_4$	Auto-combustion	Antibacterial properties	[130]
MgO–$Bi_{2-x}Cr_xO_3$	Solvent-deficient	Antibacterial properties	[131]

The obtained solid powder was crushed into small pieces after being sintered at 773 K for 4 hours in air to create CdO–ZnO nanocomposite. These powders were examined by X-ray diffraction, scanning electron microscopy, and Fourier transform infrared spectroscopy [108].

8. 1 M of $Cd(CH_3COO)$, $Ni(CH_3COO)$, $H_2O_2.4H_2O$, $Zn(CH_3COO)_2.2H_2O$, and 2 M NaOH were used as starting materials. For 2 hours, it was calcined at 673 K to create a nanocomposite. These powders were examined by XRD, SEM, TEM, FTIR, EDS, and UV–Vis spectroscopy [74] (Table 19.2).

TABLE 19.2

Synthesis of Metal Oxide Composite by Green Methods

Metal Oxide Composite	Green Fuel	Synthesis Method	Application	References
Iron oxide–gold	*Pimenta dioica*	Green method	Photothermal therapeutic	[132]
Silver/iron oxide	L-Arginine	Green method	4-Nitrophenol reduction	[132]
Polystyrene/iron oxide	Green tea leaves (*Camellia sinensis*)	Green method	Wastewater treatment	[133]
Biochar/iron oxide	Banana peel extract	Green method	Methylene blue removal	[134]
$Fe_3N@C$	Fish skin	Green method	Fabricate carbon materials	[135]
ZnO-reduced graphene oxide	Sodium citrate	Green method	Supercapacitor and photocatalysis	[136]
Fe_2O_3–Ag	*Psidium guajava* leaf extract	Co-precipitation	Recyclable adsorbent for remediation of Cr(VI) from aqueous media	[137]
Zinc iron oxide ($ZnFe_2O_4$)	*Moringa oleifera* natural extract	Biological method	Electrochemical applications	[138]
ZnO/CuO	Seed bark extract of Theobroma cacao	Green method		[139]
ZnO/CuO	*Mentha longifolia* leaf extract	Microwave oven	Antibacterial agents	[140]
Fe, Cu oxide	Loquat leaf extracts	Green method	Removal of norfloxacin and ciprofloxacin	[141]
ZnO and ZnO/CuO	*Mentha longifolia* leaf extract	Green method	Antibacterial agents	[139]
Cu_2O/CuO–ZnO	*Alchornea cordifolia* leaf extract	Hydrothermal method	Anticancer	[61]
Au/TiO_2	*Averrhoa bilimbi* fruit and *Pandanus amaryllifolius* leaf extracts	Green method	Photocatalytic activity	[142]
Ag–ZnO	*Trigonella foenum-graecum* leaf extract	Green method	Antibacterial, antifungal, antioxidant, and photocatalytic properties	[143]
ZnO–NiO	Neem leaf extract	Hydrothermal method	Antifungal and antimicrobial	[144]
Chitosan-coated zinc oxide	Bioflavonoid *rutin*	Green chemistry	Antibacterial and photocatalytic activities	[145]
$MgFe_2O_4$	Lemon juice	Combustion	Biomedical application	[145]
$MgFe_2O_4$	Albumin	Auto-combustion	Physicochemical properties	[146]

1. Ferrous sulphate was used as the precursor for the synthesis of iron oxide nanoparticles. The mixture was agitated at 70°C for an additional 8 hours to produce silver/iron oxide composite nanoparticles. These powders were examined by TEM, XRD, and UV–vis spectroscopy.
2. The banana peel biochar was altered by combining 1.2 g to 200 ml of banana peel extract. The item was dried in an oven for 12 hours at 333 K. These powders were examined by XPS, SEM, XRD, and FTIR.
3. Fish was dried at 80°C after being repeatedly rinsed with deionised water. The dark powder was then placed in a porcelain boat and heated to 600°C for 4 hours. These powders were examined by XRD, TEM, TGA, and Raman shift.
4. $Zn(NO_3)_2.6H_2O$ aqueous solution (0.005 M) was mixed into the aqueous solution of GO (10 mg), and the mixture was subjected to sonication. The collected samples were dried at 60°C in a vacuum oven. Similar techniques were used to acquire graphene sheets. These powders were examined by SEM and XRD (Table 19.3).

TABLE 19.3

Photocatalytic Degradation of Metal Oxide Composites

Metal Oxide Composite	Synthesis Method	Dyes	Time	% of Degradation	Source of Light	References
Fe_2O_3/RGO	Green hydrothermal	4-Nitrophenol	50 minutes	98%	Visible light	[147]
Graphene–zinc oxide	Chemical precipitation	Rhodamine B	90 minutes	100%	Visible light	[148]
$LaFeO_3$/Ag_2CO_3	Co-precipitation	Rhodamine B and p-chlorophenol	45 minutes	99.5%	Sunlight	[149]
Ag–ZnO	Biogenic synthesis	Methyl Orange	5 hours	~100%	Visible light	[150]
ZnO/CuO	Microwave-assisted method	Acid Orange 7	120 minutes	80.5%	Sunlight	[151]
ZnO/Fe_2O_3	Solvothermal method	GRL dye	200 minutes	91.1%	Ultraviolet	[152]
Graphene–V_2O_5	Hydrothermal method	Methylene Blue	90 minutes	100%	Sunlight	[153]
Fe/Al/Ti oxide	Chemical route	Methylene blue	10 minutes	98.4%	Visible light	[154]
Zinc oxide–activated charcoal polyaniline	Precipitation method	Rhodamine B	120 minutes	95%	Visible light	[155]
Polyaniline/CdO	Chemical oxidation polymerisation	Malachite green	4 hours	99%	Sunlight	[156]
ZnO–RGO	Microwave-assisted	Methylene blue	260 minutes	88%		[157]
Graphene–ZnO	Chemical precipitation	Methylene blue	100 minutes	100%		[158]
ZnO NP–RGO	Sol-gel	Methylene blue	180 minutes	99.5%		[159]
ZnO NR–GO	Precipitation	Methylene blue	90 minutes	94%		[160]
ZnO–RGO	Hydrothermal	Methylene blue	130 minutes	99.5%		[161]
ZnO–RGO	Hydrothermal	Rhodamine B	150 minutes	92.9%		[162]
GD–ZnO	Hydrothermal	Methylene blue	120 minutes	68%		[163]
ZG	Solvothermal technique	Methyl orange	200 minutes	100%		[164]
G–ZnO	Chemical precipitation	Rhodamine B	90 minutes	100%		[148]
ZnO/Gr	Electrochemical exfoliation	Methyl orange	420 minutes	96%		[165]
ZnO–RGO	Electrostatic self-assembly	Rhodamine B	120 minutes	100%		[166]
rGO–ZnO	Hydrolysis-calcination	Methyl orange	150 minutes	100%		[167]
ZnO/l_2O_3		Methyl orange	60 minutes	100%	Sunlight	[168]
Graphene/SiO_2		Methyl orange	100 minutes	99%	UV irradiation	[169]
CeO_2–TiO_2		Crystal violet	60 minutes	98%	Visible light	[169]
AgO/Chitosan		Methylene blue	180 minutes	72%	Ultraviolet	[170]
Chitosan/ZnO		Methylene blue	180 minutes	64%	Ultraviolet	[171]

BIOSYNTHESIS OF METAL OXIDE COMPOSITES USING BACTERIA, ALGAE, AND FUNGI

BACTERIA

Due to their rapid development and relatively simple manipulation of genes, bacteria have received the most interest in the studies on the development of nanoparticles [172]. Bacteria have been extensively employed in biotechnological processes like biological remediation, genetic technology, and biological leaching. The disadvantages associated with antibiotics, such as difficulty in penetration and systemic excretion following treatment, have been solved through the use of metal oxide nanoparticles and green nanoparticles [173].

Antibacterial properties have been investigated by zinc oxide, iron oxide, titanium dioxide, copper oxide, nickel oxide, magnesium oxide, and calcium oxide, and many more.

ALGAE

Algae have been used in numerous experiments to produce nanoparticles. *Spirulina platensis*, a blue green alga, has purportedly been utilised in the protein-mediated production of gold nanoparticles [174]. Brown algae (*Sargassum muticum*) extract was effectively used to generate the nanocomposite, which was subsequently spread in polyethylene oxide polymer, and it contains several phenolic chemicals, fucoidan, steroids, terpenoids, and flavonoids, all of which serve as reducing and stabilising agents [175].

Algae of *Aegagropila linnaei* was selected as low-cost raw material to synthesise iron oxide composites [176]. Algae have been successfully used in nanotechnology to biosynthesise metal, metal oxide, and composite nanostructures, including Cu, Au, CuO, SiO_2, Ag, and ZnO [177].

FUNGI

The microbes known as fungi use energy by breaking down organic materials without using photosynthetic processes [178]. Due to their resistance and flexibility, fungi are dominating investigations on the biological synthesis of nanoparticles. Fungi are more useful for producing nanoparticles than other microorganisms because they grow faster and are easier to manage and generate in a laboratory environment than bacteria [179].

For the production of ZrO_2 nanoparticles, fungus can serve as excellent biotemplates. Fundamentally, the green production of ZrO_2 nanoparticles utilising fungi and bacteria has a very similar mechanism [180]. In comparison to other biological synthesis processes like those used by bacteria and plants, fungal species are better suitable for the production of ZrO_2 nanocomposites [181].

ANALYSIS OF METAL OXIDE NANOCOMPOSITES USING DIFFERENT TECHNIQUES

The many characteristics of metal oxide-based nanocomposites are examined through various kinds of characterization approaches. Some of the important and essential techniques are X-Ray diffraction (XRD), UV-Visible spectroscopy (UV), Scanning Electron Microscopy (SEM), Transmission Electron Microscopy (TEM), Fourier transform infrared spectroscopy (FTIR), Energy-Dispersive X-Ray Analysis (EDAX), Brunauer–Emmett–Teller (BET) Surface Area. (Figure 19.11).

X-RAY DIFFRACTION (XRD)

XRD studies of their synthesised transition metal oxide–based nanocomposites were conducted for their crystalline quality and phase study. An essential and popular method for material

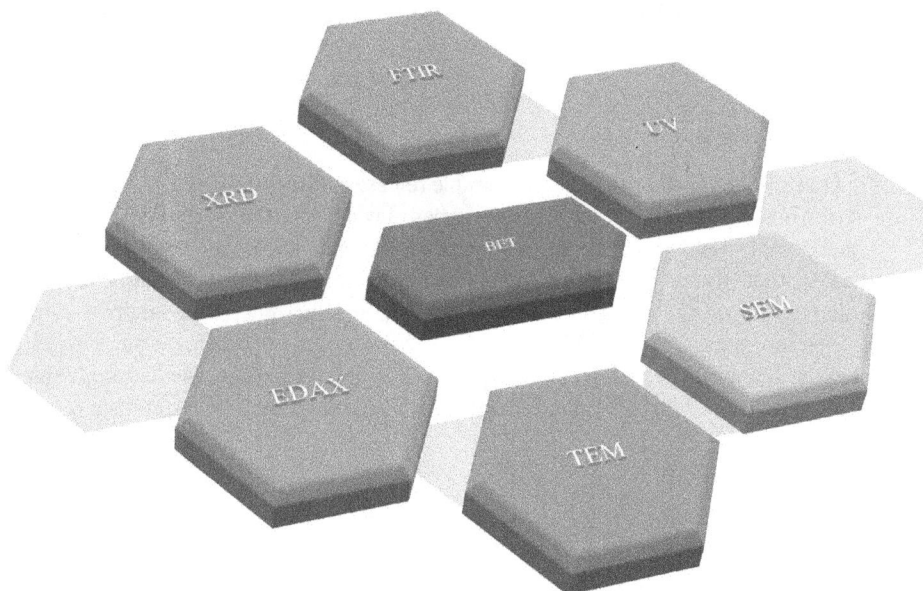

FIGURE 19.11   Schematic diagram of different techniques.

characterisation is X-ray diffraction [108]. We synthesised a graphene oxide/Fe_3O_4 nanocomposite, which was shown by the XRD data of their sample [106]. Also, the success of iron–manganese metal oxide nanocomposite production was confirmed by an XRD data analysis [107]. Similarly, other researchers synthesised nanocomposites based on transition metal oxides and studied the XRD results such as $NiFe_2O_4$ nanocomposite, ZnO nanocomposite, and ZnO–CuO nanocomposite [182]. Transmission electron microscopy and X-ray diffraction analyses are used to determine the sizes and structures of the nanoparticles in the nanocomposites. The fundamental and essential characterisation method for nanomaterials is XRD [183].

UV–Visible Spectroscopy

The most significant and frequently used spectrophotometric method for the examination of a wide range of chemicals is regarded to be the UV–Vis spectroscopy. This method is based on the measurement of the electromagnetic radiation's (EMR) interaction with a certain wavelength of matter [184]. Fabrication of CdO–MgO–Fe_2O_3 also reported UV–visible spectra for the same. We observed a wide absorption band in the UV–visible range of 340–500 nm. That nanocomposite showed red shift as compared to individual CdO, MgO, and Fe_2O_3 nanoparticles. He had also calculated the band gap of 1.76 eV for CdO–MgO–Fe_2O_3 nanocomposite [185].

Fourier Transform Infrared Spectroscopy (FTIR)

FTIR is a simple, quick, acceptable, harmless, and inexpensive technique. A kind of vibrational spectroscopy called Fourier transform infrared (FTIR) spectroscopy is helpful in the investigation of several soil chemical processes.

Researchers have studied functional groups in their synthesised nanomaterials by utilising FTIR spectroscopy in a number of published research publications for the synthesis and characterisation of transition metal oxide–based nanocomposites [98]. FTIR spectra for hexagonal ZnO is 505 cm^{-1} peak, and the 861 cm^{-1} peak is for cubic CdO [186]. When we studied CuO–ZnO composites and examined at their FTIR spectra, we saw that all of the 0.4% CuO–ZnO, 2% CuO–ZnO,

10% CuO–ZnO, and 50% CuO–ZnO displayed a band for metal–oxygen (M–O) stretching vibrations in the 400–600 cm^{-1} range [187].

Scanning Electron Microscopy (SEM)

By identifying the additional electrons that the sample releases after coming into interaction with the affecting electron beam, scanning electron microscopes may observe the sample surface [115]. CuO/TiO2/ZnO nanocomposites, which are semi-spheroidal nanoparticles with irregular shapes, have been synthesised and analysed [118]. ZnO/CuO nanocomposite was synthesised and analysed under SEM for finding the SEM image that shows irregular-shaped semi-spheroidal nanoparticles [188].

Scanning electron microscopy (SEM) is an important tool for characterisation of materials at the nanoscale. The scanning electron microscope (SEM) is based on the simple principle that when energetic electrons interact with a condensed substance, mainly solids, they produce a variety of measurable signals [189].

Transmission Electron Microscopy (TEM)

A high-resolution determining technique called transmission electron microscopy (TEM) records the movement of an electron beam through a material [190].

We found that all of the CuO–ZnO composites we tested and examined in their FTIR spectra had a band for metal–oxygen (M–O) stretching vibrations in the 400–600 cm^{-1} region [191]. NiO/TiO$_2$ with 1%, 2%, 5%, and 10% of NiO were synthesised. We examined TEM pictures and discovered that the amount of nickel oxide decreased as it increased. Additionally, the nanocomposite's size range was 4–32 nm. ZnO/CuO nanocomposite was synthesised and analysed under TEM for morphology and size study of this synthesised TiO$_2$–Ag nanocomposite. TEM confirmed agglomeration of spheroidal particles (10 nm) connected with the rod-shaped particles (50 nm).

Energy-Dispersive X-Ray Analysis (EDAX)

EDAX is useful to identifying elemental content present in synthesised transition metal oxide–based nanocomposites and also other nanomaterials. The method of energy-dispersive X-ray analysis (EDAX) is used to analyse nanoparticles using SEM [137].

Fe$_2$O$_3$–Ag nanocomposite had been synthesised, and its EDX evaluation confirmed that oxygen, silver, and iron were all present (weight percentages of 51.12, 23.25, and 25.63, respectively). EDX showed that the weight percentage of Ag was 5.60 and the weight of ZnO was 86.02 in the synthesised Ag–ZnO nanocomposite [143]. Elemental composition of synthesised Au/TiO$_2$ nanocomposite was identified by using EDX. EDX showed the presence of Au (wt.% of 9.53), Ti (wt.% of 32.66), and O (wt.% of 57.81) elements in Au/TiO$_2$ nanocomposite [142]. The basic concept behind EDAX is the generation of X-rays from a sample using an electron beam. The properties and type of the components contained in the sample are taken into consideration while producing the X-rays [192].

Raman Spectroscopy

Raman spectroscopy is a very flexible method that is frequently used to describe electrode materials [193]. Raman spectroscopy depends on an inelastic scattered light phenomenon for recognising analytical substances through vibrations in bonds between molecules. A tiny proportion of photons disperse when a sample is subjected to laser light. Elastically dispersed light with exactly the same frequency as the incoming light causes most of the scattering [194].

Brunauer–Emmett–Teller (BET) Surface Area

The BET is useful for surface area observations and pore properties analysis. In the BET studies of ZnO/CuO nanocomposite, we observed images with mesopores, and calculated specific surface area for ZnO/CuO (5%) and ZnO/CuO (10%) were 36.0 and $26.0\,m^2/g$, respectively. Also the calculated pore volumes for ZnO/CuO (5%) and ZnO/CuO (10%) were 0.0939 and $0.0984\,cm^3/g$, respectively [195].

Thermogravimetric Analysis (TGA)

Thermogravimetric analysis (TGA) is a form of quantitative analysis that measures the mass of a sample while it is heated in a furnace to a maximum of $1,600°C$ when the gas flow is either constant or fluctuating [139]. To determine the thermodynamic characteristics of each sample, a technique known as thermogravimetric analysis was used. TGA data have been used to compute kinetic variables such as thermal activation energy entropy, Gibbs energy, and enthalpy [196]. Synthesised $Fe_3O_4/CNC/Cu$ nanocomposite was studied, and its TGA/DTA results showed its thermal stability [197].

Differential Thermal Analysis (DTA)

The word "differential" denotes that the behavioural difference between the material for research and a reference material is studied in the thermal analysis known as differential thermal analysis [198].

When compared to other techniques, differential thermal analysis can be more advantageous because it allows for in situ testing of freezing acceptance, increases the number of tested species, tissue types, and sampling dates, and more accurately simulates the effects of freezing rate and duration [199].

Photocatalysis of Rhodamine B

Some dyes, notably organic dyes like rhodamine B, are toxic and represent a major danger, when they are released into the environment [200]. Rhodamine B is a dye that is commonly used to colour biological products as well as paints, acrylic, and other materials. When released into water directly, it is highly dangerous to organisms [201]. Photocatalytic activity of graphene–zinc oxide was evaluated by photocatalytic degradation of rhodamine B dye, and the solution was magnetically stirred in dark condition for about 90 minutes under visible light source. It shows 100% degradency.

Photocatalysis of Acid Orange 7

Deionised water was used to produce an acid orange 7 aqueous solution for the actual sample and textile wastewater for the control samples [202]. Photocatalytic activity of ZnO/CuO was evaluated by photocatalytic degradation of acid orange dye, and the solution was magnetically stirred for about 120 minutes under sunlight source. It shows 80.5% degradency.

Photocatalysis of Methyl Orange

One of the most widely used colours in the textile industry is methyl orange. Additionally, it serves as a titration's pH indicator [97]. An important dye that is frequently used as a colouring agent in the textile and leather industries is methyl orange, a synthetic azo dye [203]. Photocatalytic activity of Ag–ZnO was evaluated by photocatalytic degradation of methyl orange dye, and the solution was magnetically stirred in dark condition for about 5 hours under visible light source. It shows 100% degradency.

PHOTOCATALYSIS OF 4-NITROPHENOL

One such compound that is used to produce pharmaceuticals, pesticides, colours, and herbicide and insecticide is 4-nitrophenol. It has been identified as one among the organic contaminants, is soluble in water, and is harmful over 20 ppb [204]. Photocatalytic activity of Fe_2O_3 was evaluated by photocatalytic degradation of 4-nitrophenol dye, and the solution was magnetically stirred in dark condition for about 50 minutes under visible light source. It shows 98% degradency.

POSSIBLE OUTLOOK

Many practical approaches, including the sol-gel method, hydrothermal techniques, combustion method, microwave-aided method, sonication method, co-precipitation method, and green methods, have been studied for the synthesis of various transition metal oxide–based composites. Here, several synthetic methods have been used to synthesise and modify transition metal oxide–based nanocomposites. We already know that using a modified synthesis method, we can produce materials with special properties such as a large surface area, increased surface activity, a greater number of active sites on the surface, the introduction of mesoporosity, a reduction in electron hole recombination through doping, and an increase in the charge transfer process.

For structural and morphological research, we have also evaluated several issued characterisation methods of synthesised metal oxide–based nanocomposites, such as XRD, EDX, SEM, TEM, and FTIR. We have examined a special use for transition metal oxide nanocomposites based on their various topologies. For these transition metal oxide–based composites to be used in practical applications, morphologies and surface characteristics are crucial.

Transition metal oxide–based composites offer a wide range of reported possible real-world uses, including sensors, photocatalytic degradation of harmful dyes, medicinal applications including drug transport, antimicrobials, biosensors, and anticancer, energy generation through solar cells and H_2 synthesis, and energy storage from supercapacitors.

REFERENCES

[1] Sen, Mousumi. "Nanocomposite materials." *Nanotechnology and the Environment*. Technology and Engineering. IntechOpen (2020): 1–12.

[2] Okpala, Charles Chikwendu. "Nanocomposites-an overview." *International Journal of Engineering Research and Development* 8, no. 11 (2013): 17–23.

[3] Widiarti, N., J. K. Sae, and S. Wahyuni. "Synthesis CuO-ZnO nanocomposite and its application as an antibacterial agent." In *IOP Conference Series: Materials Science and Engineering*, vol. 172, no. 1, p. 012036. IOP Publishing, 2017.

[4] Ashok, C., K. V. Rao, and C. S. Chakra. "Synthesis and characterization of ZnO/CuO nanocomposite for humidity sensor application." *Journal of Advanced Chemical Sciences* 2 (2016): 223–226.

[5] Durmus, Zehra, Belma Zengin Kurt, and Ali Durmus. "Synthesis and characterization of graphene oxide/zinc oxide (GO/ZnO) nanocomposite and its utilization for photocatalytic degradation of basic fuchsin dye." *ChemistrySelect* 4, no. 1 (2019): 271–278. https://doi.org/10.1002/slct.201803635.

[6] Mehto, A., V. R. Mehto, J. Chauhan, I. Singh, and R. Pandey. "Preparation and characterization of polyaniline/ZnO composite sensor." *Journal of Nanomedicine Research* 5 (2017): 00104. https://doi.org/10.15406/jnmr.2017.05.00104.

[7] Al-Mutairi, N. H., A. H. Mehdi, and B. J. Kadhim. "Nanocomposites materials definitions, types and some of their applications: A review." *European Journal of Research Development and Sustainability* 3, no. 2 (2022): 102–108.

[8] Gupta, Pallav, Devendra Kumar, M. A. Quraishi, and Om Parkash. "Metal matrix nanocomposites and their application in corrosion control." *Advances in Nanomaterials*. Springer (2016): 231–246. https://doi.org/10.1007/978-81-322-2668-0_6.

[9] Srivastava, A. "Recent advances in metal matrix composites (MMCs): A review." *Biomedical Journal of Scientific & Technical Research* 1, no. 2 (2017): 520–522.

[10] Lossada, F., D. Hoenders, J. Guo, D. Jiao, and A. Walther. "Self-assembled bioinspired nanocomposites." *Accounts of Chemical Research* 53, no. 11 (2020): 2622–2635. https://doi.org/10.1021/acs.accounts.0c00448.

[11] Omanović-Mikličanin, E., A. Badnjević, A. Kazlagić, and M. Hajlovac. "Nanocomposites: A brief review." *Health and Technology* 10 (2020): 51–59. https://doi.org/10.1007/s12553-019-00380-x.

[12] Liu, Qianwen, Amin Zhang, Ruhao Wang, Qian Zhang, and Daxiang Cui. "A review on metal-and metal oxide-based nanozymes: Properties, mechanisms, and applications." *Nano-Micro Letters* 13 (2021): 1–53. https://doi.org/10.1007/s40820-021-00674-8.

[13] Naik Tejas Pramod,, Inderdeep Singh, and Apurbba Kumar Sharma. "Processing of polymer matrix composites using microwave energy: A review." *Composites Part A: Applied Science and Manufacturing* 156 (2022): 106870. https://doi.org/10.1016/j.compositesa.2022.106870.

[14] Li, Longbiao. "A micromechanical tension-tension fatigue hysteresis loops model of fiber-reinforced ceramic-matrix composites considering stochastic matrix fragmentation." *International Journal of Fatigue* 143 (2021): 106001. https://doi.org/10.1016/j.ijfatigue.2020.106001.

[15] Nieto, Andy, Ankita Bisht, Debrupa Lahiri, Cheng Zhang, and Arvind Agarwal. "Graphene reinforced metal and ceramic matrix composites: A review." *International Materials Reviews* 62, no. 5 (2017): 241–302. https://doi.org/10.1080/09506608.2016.1219481.

[16] Joshi, Naveen Chandra, Prateek Gururani, and Shiv Prasad Gairola. "Metal oxide nanoparticles and their nanocomposite-based materials as photocatalysts in the degradation of dyes." *Biointerface Research in Applied Chemistry* 12 (2022): 6557–6579. https://doi.org/10.33263/BRIAC125.65576579.

[17] Ameta, Rakshit, Meenakshi S. Solanki, Surbhi Benjamin, and Suresh C. Ameta. "Photocatalysis." In *Advanced Oxidation Processes for Waste Water Treatment*, pp. 135–175. Academic Press, 2018. https://doi.org/10.1016/B978-0-12-810499-6.00006-1.

[18] Jamjoum, Hayfa Alajilani Abraheem, Khalid Umar, Rohana Adnan, Mohd R. Razali, and Mohamad Nasir Mohamad Ibrahim. "Synthesis, characterization, and photocatalytic activities of graphene oxide/metal oxides nanocomposites: A review." *Frontiers in Chemistry* 9 (2021): 752276. https://doi.org/10.3389/fchem.2021.752276.

[19] Danish, Mir Sayed Shah, Liezel L. Estrella, Ivy Michelle A. Alemaida, Anton Lisin, Nikita Moiseev, Mikaeel Ahmadi, Massoma Nazari, Mohebullah Wali, Hameedullah Zaheb, and Tomonobu Senjyu. "Photocatalytic applications of metal oxides for sustainable environmental remediation." *Metals* 11, no. 1 (2021): 80. https://doi.org/10.3390/met11010080.

[20] Visakh, P. M., and B. Raneesh. "Metal oxide nanocomposites: State-of-the-art and new challenges." *Metal Oxide Nanocomposites: Synthesis and Applications*. Wiley (2020): 1–26. https://doi.org/10.1002/9781119364726.ch1.

[21] Naseem, Taiba, and Tayyiba Durrani. "The role of some important metal oxide nanoparticles for wastewater and antibacterial applications: A review." *Environmental Chemistry and Ecotoxicology* 3 (2021): 59–75. https://doi.org/10.1016/j.enceco.2020.12.001.

[22] Ramakoti, Ivaturi Siva, Achyut Kumar Panda, and Narayan Gouda. "A brief review on polymer nanocomposites: Current trends and prospects." *Journal of Polymer Engineering* 43, no. 8 (2023): 651–679. https://doi.org/10.1515/polyeng-2023-0103.

[23] Bashambu, Lavisha, Rasmeet Singh, and Jonita Verma. "Metal/metal oxide nanocomposite membranes for water purification." *Materials Today: Proceedings* 44 (2021): 538–545. https://doi.org/10.1016/j.matpr.2020.10.213.

[24] Tahir, Noor, Muhammad Zahid, Ijaz Ahmad Bhatti, Asim Mansha, Syed Ali Raza Naqvi, and Tajamal Hussain. "Metal oxide-based ternary nanocomposites for wastewater treatment." In *Aquananotechnology*, pp. 135–158. Elsevier, 2021. https://doi.org/10.1016/B978-0-12-821141-0.00022-7.

[25] Mujtahid, Fitriah, Paulus Lobo Gareso, Bidayatul Armynah, and Dahlang Tahir. "Review effect of various types of dyes and structures in supporting performance of dye-sensitized solar cell TiO_2-based nanocomposites." *International Journal of Energy Research* 46, no. 2 (2022): 726–742. https://doi.org/10.1002/er.7310.

[26] Samanta, Ashis Kumar, and Adwaita Konar. "Dyeing of textiles with natural dyes." *Natural Dyes* 3, no. 30–56 (2011).

[27] Shabbir, Mohd, and Masoom Naim. "Introduction to textiles and the environment." *Textiles and Clothing* (2019): 1–9. https://doi.org/10.1002/9781119526599.

[28] Javaid, Rahat, and Umair Yaqub Qazi. "Catalytic oxidation process for the degradation of synthetic dyes: An overview." *International Journal of Environmental Research and Public Health* 16, no. 11 (2019): 2066. https://doi.org/10.3390/ijerph16112066.

[29] Picos, Alain, and Juan M. Peralta-Hernández. "Genetic algorithm and artificial neural network model for prediction of discoloration dye from an electro-oxidation process in a press-type reactor." *Water Science and Technology* 78, no. 4 (2018): 925–935. https://doi.org/10.2166/wst.2018.370.

[30] Benkhaya, Said, Souad M'rabet, and Ahmed El Harfi. "A review on classifications, recent synthesis and applications of textile dyes." *Inorganic Chemistry Communications* 115 (2020): 107891. https://doi.org/10.1016/j.inoche.2020.107891.

[31] Değermenci, Gökçe Didar, Nejdet Değermenci, Vefa Ayvaoğlu, Ekrem Durmaz, Doğan Çakır, and Emre Akan. "Adsorption of reactive dyes on lignocellulosic waste; characterization, equilibrium, kinetic and thermodynamic studies." *Journal of Cleaner Production* 225 (2019): 1220–1229. https://doi.org/10.1016/j.jclepro.2019.03.260.

[32] Chen, Victor J., Robert E. Minto, Nicholas Manicke, and Gregory D. Smith. "Structural elucidation of two Congo red derivatives on dyed historical objects indicative of formaldehyde exposure and the potential for chemical fading." *Dyes and Pigments* 201 (2022): 110173. https://doi.org/10.1016/j.dyepig.2022.110173.

[33] Berradi, Mohamed, Rachid Hsissou, Mohammed Khudhair, Mohammed Assouag, Omar Cherkaoui, Abderrahim El Bachiri, and Ahmed El Harfi. "Textile finishing dyes and their impact on aquatic environs." *Heliyon* 5, no. 11 (2019):1–11.

[34] Zinatloo-Ajabshir, Sahar, Masoud Salavati-Niasari, and Zahra Zinatloo-Ajabshir. "Facile size-controlled preparation of highly photocatalytically active praseodymium zirconate nanostructures for degradation and removal of organic pollutants." *Separation and Purification Technology* 177 (2017): 110–120. https://doi.org/10.1016/j.seppur.2016.12.043.

[35] Yang, Zhuo, Wei Shen, Qianjin Chen, and Wei Wang. "Direct electrochemical reduction and dyeing properties of CI Vat Yellow 1 using carbon felt electrode." *Dyes and Pigments* 184 (2021): 108835. https://doi.org/10.1016/j.dyepig.2020.108835.

[36] Agarwal, Jyoti. "Dyes and dyeing processes for natural textiles and their key sustainability issues." In *Fundamentals of Natural Fibres and Textiles*, pp. 439–472. Woodhead Publishing, 2021. https://doi.org/10.1016/j.dyepig.2020.108835.

[37] Horobin, Richard W., Juan C. Stockert, and Hua Zhang. "Reactive dyes for living cells: Applications, artefacts, and some comparisons with textile dyeing." *Coloration Technology* 138, no. 1 (2022): 3–15. https://doi.org/10.1111/cote.12577.

[38] Atiq, Muhammad Sohaib, A. B. D. U. R. Rehman, Kashif Iqbal, Faiza Safdar, Abdul Basit, Munir Ashraf, Hafiz Shahzad Maqsood, and A. S. F. A. N. D. Y. A. R. Khan. "Salt free sulphur black dyeing of cotton fabric after cationization." *Cellulose Chemistry and Technology* 53, no. 1–2 (2019): 155–161.

[39] Rajabi, Ali Asghar, Yadollah Yamini, Mohammad Faraji, and Farahnaz Nourmohammadian. "Modified magnetite nanoparticles with cetyltrimethylammonium bromide as superior adsorbent for rapid removal of the disperse dyes from wastewater of textile companies." *Nanochemistry Research* 1, no. 1 (2016): 49–56. https://doi.org/10.7508/ncr.2016.01.006

[40] He, Jingjing, and Yan Luo. "Novel carboxylate comb-like dispersant used in disperse dyes." *Journal of Applied Polymer Science* 139, no. 20 (2022): 52147. https://doi.org/10.1002/app.52147.

[41] Hasan, Saba. "A review on nanoparticles: Their synthesis and types." *Research Journal of Recent Sciences* 2277 (2015): 2502.

[42] Reyes, Victoria Isabel. "Fabrication and Characterization of Iron-Based Catalysts for the Dehydrogenation of Fossil Fuels." PhD diss., The University of Texas at El Paso, 2022. https://doi.org/10.1002/chem.202000678.

[43] Parauha, Yatish R., Vaibhavi Sahu, and Sanjay J. Dhoble. "Prospective of combustion method for preparation of nanomaterials: A challenge." *Materials Science and Engineering: B* 267 (2021): 115054. https://doi.org/10.1016/j.mseb.2021.115054.

[44] Yadav, Satyapal, Neha Rani, and Kamal Saini. "A review on transition metal oxides based nanocomposites, their synthesis techniques, different morphologies and potential applications." In *IOP Conference Series: Materials Science and Engineering*, vol. 1225, no. 1, p. 012004. IOP Publishing, 2022. https://doi.org/10.1088/1757-899X/1225/1/012004.

[45] Owens, Gareth J., Rajendra K. Singh, Farzad Foroutan, Mustafa Alqaysi, Cheol-Min Han, Chinmaya Mahapatra, Hae-Won Kim, and Jonathan C. Knowles. "Sol-gel based materials for biomedical applications." *Progress in Materials Science* 77 (2016): 1–79. https://doi.org/10.1016/j.pmatsci.2015.12.001.

[46] Hamrouni, Abdessalem, Noomen Moussa, Agatino Di Paola, Leonardo Palmisano, Ammar Houas, and Francesco Parrino. "Photocatalytic activity of binary and ternary SnO_2-ZnO-$ZnWO_4$ nanocomposites." *Journal of Photochemistry and Photobiology A: Chemistry* 309 (2015): 47–54. https://doi.org/10.1016/j.molcata.2014.03.018.

[47] Chung, Deborah. D. L. "Carbon fibers, nanofibers, and nanotubes." *Carbon Composites* 2 (2017): 12–47.

[48] Altammar, Khadijah A. "A review on nanoparticles: Characteristics, synthesis, applications, and challenges." *Frontiers in Microbiology* 14 (2023): 1155622. https://doi.org/10.3389/fmicb.2023.1155622.

[49] Nam, Nguyen Hoang, and Nguyen Hoang Luong. "Nanoparticles: Synthesis and applications." In *Materials for Biomedical Engineering*, pp. 211–240. Elsevier, 2019. https://doi.org/10.1016/B978-0-08-102814-8.00008-1.

[50] Kumar, Anuj, Yun Kuang, Zheng Liang, and Xiaoming Sun. "Microwave chemistry, recent advancements, and eco-friendly microwave-assisted synthesis of nanoarchitectures and their applications: A review." *Materials Today Nano* 11 (2020): 100076. https://doi.org/10.1016/j.mtnano.2020.100076.

[51] Sedighi, Farideh, Mahdiyeh Esmaeili-Zare, Ali Sobhani-Nasab, and Mohsen Behpour. "Synthesis and characterization of $CuWO_4$ nanoparticle and $CuWO_4$/NiO nanocomposite using co-precipitation method; application in photodegradation of organic dye in water." *Journal of Materials Science: Materials in Electronics* 29 (2018): 13737–13745. https://doi.org/10.1007/s10854-018-9504-3.

[52] Dahiya, Manjeet S., Vijay K. Tomer, and Surender Duhan. "Metal-ferrite nanocomposites for targeted drug delivery." In *Applications of Nanocomposite Materials in Drug Delivery*, pp. 737–760. Woodhead Publishing, 2018. https://doi.org/10.1016/B978-0-12-813741-3.00032-7.

[53] Li, Li, Xiuli Zhang, Wenzhi Zhang, Lili Wang, Xi Chen, and Yu Gao. "Microwave-assisted synthesis of nanocomposite Ag/ZnO-TiO_2 and photocatalytic degradation Rhodamine B with different modes." *Colloids and Surfaces A: Physicochemical and Engineering Aspects* 457 (2014): 134–141. https://doi.org/10.1016/j.colsurfa.2014.05.060.

[54] Kumar, Rajesh, Sally M. Youssry, Kyaw Zay Ya, Wai Kian Tan, Go Kawamura, and Atsunori Matsuda. "Microwave-assisted synthesis of Mn_3O_4-Fe_2O_3/Fe_3O_4@ rGO ternary hybrids and electrochemical performance for supercapacitor electrode." *Diamond and Related Materials* 101 (2020): 107622. https://doi.org/10.1016/j.diamond.2019.107622.

[55] da Silva, Amandha Kaiser, Thiago Gomes Ricci, Ana Lúcia de Toffoli, Edvaldo Vasconcelos Soares Maciel, Carlos Eduardo Domingues Nazario, and Fernando Mauro Lanças. "The role of magnetic nanomaterials in miniaturized sample preparation techniques." In *Handbook on Miniaturization in Analytical Chemistry*, pp. 77–98. Elsevier, 2020. https://doi.org/10.1016/B978-0-12-819763-9.00004-0.

[56] Zhang, Guo-Ying, Ya-Qiu Sun, Dong-Zhao Gao, and Yan-Yan Xu. "Quasi-cube ZnFe2O4 nanocrystals: Hydrothermal synthesis and photocatalytic activity with TiO_2 (Degussa P25) as nanocomposite." *Materials Research Bulletin* 45, no. 7 (2010): 755–760. https://doi.org/10.1016/j.materresbull.2010.03.025.

[57] Arsiya, Farzaneh, Mohammad Hossein Sayadi, and Sara Sobhani. "Green synthesis of palladium nanoparticles using Chlorella vulgaris." *Materials Letters* 186 (2017): 113–115. https://doi.org/10.1016/j.matlet.2016.09.101.

[58] Jadoun, Sapana, Rizwan Arif, Nirmala Kumari Jangid, and Rajesh Kumar Meena. "Green synthesis of nanoparticles using plant extracts: A review." *Environmental Chemistry Letters* 19 (2021): 355–374. https://doi.org/10.1007/s10311-020-01074-x.

[59] Devi, Henam Sylvia, Muzaffar Ahmad Boda, Mohammad Ashraf Shah, Shazia Parveen, and Abdul Hamid Wani. "Green synthesis of iron oxide nanoparticles using Platanus orientalis leaf extract for antifungal activity." *Green Processing and Synthesis* 8, no. 1 (2019): 38–45. https://doi.org/10.1515/gps-2017-0145.

[60] Singh, Priyanka, Yu-Jin Kim, Dabing Zhang, and Deok-Chun Yang. "Biological synthesis of nanoparticles from plants and microorganisms." *Trends in Biotechnology* 34, no. 7 (2016): 588–599. https://doi.org/10.1016/j.tibtech.2016.02.006.

[61] Elemike, Elias E., Damian C. Onwudiwe, and Moganavelli Singh. "Eco-friendly synthesis of copper oxide, zinc oxide and copper oxide-zinc oxide nanocomposites, and their anticancer applications." *Journal of Inorganic and Organometallic Polymers and Materials* 30 (2020): 400–409. https://doi.org/10.1007/s10904-019-01198-w.

[62] Vithiya, K., and Sankar Sen. "Biosynthesis of nanoparticles." *International Journal of Pharmaceutical Sciences and Research* 2, no. 11 (2011): 2781.

[63] Greca, Maria Conceição, Caetano Moraes, and Ana Maria Segadães. "Palladium/alumina catalysts: Effect of the processing route on catalytic performance." *Applied Catalysis A: General* 216, no. 1–2 (2001): 267–276. https://doi.org/10.1016/S0926-860X(01)00571-3.

[64] Bhaduri, Sutapa B, and Sarita B Bhaduri. "Enhanced low temperature toughness of Al_2O_3-ZrO_2 nano/nano composites." *Nanostructured Materials* 8, no. 6 (1997): 755–763. https://doi.org/10.1016/S0965-9773(97)00215-8.

[65] Chen, Po Chou, Anindita Ganguly, Tata Sanjay Kanna Sharma, Kuan-Yu Chou, Shu-Mei Chang, and Kuo-Yuan Hwa. "Investigation of T site variation in spinel aluminates TAl_2O_4 (T= Mg, Zn & Cu), and formation of electrocatalyst $CuAl_2O_4$/carbon for efficient sensing application." *Chemosphere* 301 (2022): 134458. https://doi.org/10.1016/j.chemosphere.2022.134458.

[66] Mukasyan, Alexander S., Alexander S. Rogachev, and Singanahally Thippa Reddy Aruna. "Combustion synthesis in nanostructured reactive systems." *Advanced Powder Technology* 26, no. 3 (2015): 954–976. https://doi.org/10.1016/j.apt.2015.03.013.

[67] Devaiah, Damma, Lankela H. Reddy, Sang-Eon Park, and Benjaram M. Reddy. "Ceria-zirconia mixed oxides: Synthetic methods and applications." *Catalysis Reviews* 60, no. 2 (2018): 177–277. https://doi.org/10.1080/01614940.2017.1415058.

[68] Park, Hyu-Bum, Jeongsoo Kim, and Chi-Woo Lee. "Synthesis of $LiMn_2O_4$ powder by auto-ignited combustion of poly (acrylic acid)-metal nitrate precursor." *Journal of Power Sources* 92, no. 1–2 (2001): 124–130. https://doi.org/10.1016/S0378-7753(00)00512-7.

[69] Patil, Kashinath C., Singanahally Thippa Reddy Aruna, and Tanu Mimani. "Combustion synthesis: An update." *Current Opinion in Solid State and Materials Science* 6, no. 6 (2002): 507–512. https://doi.org/10.1016/S1359-0286(02)00123-7.

[70] KameswariU., Arthur W. Sleight, and Jahn S. O. Evans. "Rapid synthesis of ZrW_2O_8 and related phases, and structure refinement of $ZrWMoO_8$." *International Journal of Inorganic Materials* 2, no. 4 (2000): 333–337. https://doi.org/10.1016/S1466-6049(00)00029-5.

[71] Hamrouni, Abdessalem, Hinda Lachheb, and Ammar Houas. "Synthesis, characterization and photocatalytic activity of $ZnO-SnO_2$ nanocomposites." *Materials Science and Engineering: B* 178, no. 20 (2013): 1371–1379. https://doi.org/10.1016/j.mseb.2013.08.008.

[72] Farrokhi, Mehrdad, Seyydeh-Cobra Hosseini, Jae-Kyu Yang, and Mehdi Shirzad-Siboni. "Application of $ZnO-Fe_3O_4$ nanocomposite on the removal of azo dye from aqueous solutions: Kinetics and equilibrium studies." *Water, Air, & Soil Pollution* 225 (2014): 1–12. https://doi.org/10.1007/s11270-014-2113-8.

[73] Subhan, Md Abdus, Tanzir Ahmed, Nizam Uddin, Abdul Kalam Azad, and Kulsuma Begum. "Synthesis, characterization, PL properties, photocatalytic and antibacterial activities of nano multi-metal oxide $NiO·CeO_2·ZnO$." *Spectrochimica Acta Part A: Molecular and Biomolecular Spectroscopy* 136 (2015): 824–831. https://doi.org/10.1016/j.saa.2014.09.100.

[74] Jan, Tariq, Sohail Azmat, Qaisar Mansoor, H. M. Waqas, Muhammad Adil, Syed Zafar Ilyas, Ishaq Ahmad, and Muhammad Ismail. "Superior antibacterial activity of ZnO-CuO nanocomposite synthesized by a chemical co-precipitation approach." *Microbial Pathogenesis* 134 (2019): 103579. https://doi.org/10.1016/j.micpath.2019.103579.

[75] Nazari, Mahdi, and Rouein Halladj. "Adsorptive removal of fluoride ions from aqueous solution by using sonochemically synthesized nanomagnesia/alumina adsorbents: An experimental and modeling study." *Journal of the Taiwan Institute of Chemical Engineers* 45, no. 5 (2014): 2518–2525. https://doi.org/10.1016/j.jtice.2014.05.020.

[76] Kaviyarasu, K., C. Maria Magdalane, Krishnan Anand, E. Manikandan, and Maalik Maaza. "Synthesis and characterization studies of MgO: CuO nanocrystals by wet-chemical method." *Spectrochimica Acta Part A: Molecular and Biomolecular Spectroscopy* 142 (2015): 405–409. https://doi.org/10.1016/j.saa.2015.01.111.

[77] Nabiyouni, Gholamreza, Ali Shabani, Sajad Karimzadeh, Javad Ghasemi, and Hamid Ramazani. "Synthesis, characterization and magnetic investigations of Fe_3O_4 nanoparticles and zeolite-Y nanocomposites prepared by precipitation method." *Journal of Materials Science: Materials in Electronics* 26 (2015): 5677–5685. https://doi.org/10.1007/s10854-015-3118-9.

[78] Ghaedi, Mehrorang, Maryam Ghayedi, Syamak Nasiri Kokhdan, Reza Sahraei, and Ali Daneshfar. "Palladium, silver, and zinc oxide nanoparticles loaded on activated carbon as adsorbent for removal of bromophenol red from aqueous solution." *Journal of Industrial and Engineering Chemistry* 19, no. 4 (2013): 1209–1217. https://doi.org/10.1016/j.jiec.2012.12.020.

[79] Warrier, Krishna Gopakumar K, and G. M. Anilkumar. "Densification of mullite-SiC nanocomposite sol-gel precursors by pressureless sintering." *Materials Chemistry and Physics* 67, no. 1–3 (2001): 263–266. https://doi.org/10.1016/S0254-0584(00)00447-8.

[80] Gao, Bin, George Z. Chen, and Gianluca Li Puma. "Carbon nanotubes/titanium dioxide ($CNTs/TiO_2$) nanocomposites prepared by conventional and novel surfactant wrapping sol-gel methods exhibiting enhanced photocatalytic activity." *Applied Catalysis B: Environmental* 89, no. 3–4 (2009): 503–509. https://doi.org/10.1016/j.apcatb.2009.01.009.

[81] Dou, Binlin, Valerie Dupont, Weiguo Pan, and Bingbing Chen. "Removal of aqueous toxic Hg (II) by synthesized TiO_2 nanoparticles and TiO_2/montmorillonite." *Chemical Engineering Journal* 166, no. 2 (2011): 631–638. https://doi.org/10.1016/j.cej.2010.11.035.

[82] Belessi, Vassiliki C., Dimitra Lambropoulou, Ioannis Konstantinou, Alexandros Katsoulidis, Pomonis Pomonis, Dimitrios Petridis, and Triantafillos Albanis. "Structure and photocatalytic performance of TiO_2/clay nanocomposites for the degradation of dimethachlor." *Applied Catalysis B: Environmental* 73, no. 3–4 (2007): 292–299. https://doi.org/10.1016/j.apcatb.2006.12.011.

[83] Mosquera, Edgar, Ignacio del Pozo, and Mauricio Morel. "Structure and red shift of optical band gap in CdO-ZnO nanocomposite synthesized by the sol gel method." *Journal of Solid State Chemistry* 206 (2013): 265–271. https://doi.org/10.1016/j.jssc.2013.08.025.

[84] Ma, Wenxia, Meiling Lv, Feiping Cao, Zheng Fang, Yuqin Feng, Gang Zhang, Yongsheng Yang, and Hongjun Liu. "Synthesis and characterization of ZnO-GO composites with their piezoelectric catalytic and antibacterial properties." *Journal of Environmental Chemical Engineering* 10, no. 3 (2022): 107840. https://doi.org/10.1016/j.jece.2022.107840.

[85] Sivakumar, Sri V, C. P. Sibu, Poothayil Mukundan, P. Krishna Pillai, and Krishna Gopalkumar K. Warrier. "Nanoporous titania-alumina mixed oxides-an alkoxide free sol-gel synthesis." *Materials Letters* 58, no. 21 (2004): 2664–2669. https://doi.org/10.1016/j.matlet.2004.03.050.

[86] Bulin, Chaoke, Bo Li, Yanghuan Zhang, and Bangwen Zhang. "Removal performance and mechanism of nano α-Fe_2O_3/graphene oxide on aqueous Cr (VI)." *Journal of Physics and Chemistry of Solids* 147 (2020): 109659. https://doi.org/10.1016/j.jpcs.2020.109659.

[87] Zhang, Huanjun, and Guohua Chen. "Potent antibacterial activities of Ag/TiO_2 nanocomposite powders synthesized by a one-pot sol– gel method." *Environmental Science & Technology* 43, no. 8 (2009): 2905–2910. https://doi.org/10.1021/es803450f.

[88] Ananthakumar, Solaiappan, Kuttan Prabhakaran, U. S. Hareesh, P. Manohar, and Krishna Gopalkumar K. Warrier. "Gel casting process for Al_2O_3-SiC nanocomposites and its creep characteristics." *Materials Chemistry and Physics* 85, no. 1 (2004): 151–157. https://doi.org/10.1016/j.matchemphys.2003.12.022.

[89] Yang, Huaming, Rongrong Shi, Ke Zhang, Yuehua Hu, Aidong Tang, and Xianwei Li. "Synthesis of WO_3/TiO_2 nanocomposites via sol-gel method." *Journal of Alloys and Compounds* 398, no. 1–2 (2005): 200–202. https://doi.org/10.1016/j.jallcom.2005.02.002.

[90] Zou, Jin Feng, Hai Feng Chen, and Ting Yan. "Synthesis and adsorption properties of nano Al_2O_3/TiO_2 by hydrothermal method." *Advanced Materials Research* 580 (2012): 509–512. https://doi.org/10.4028/www.scientific.net/AMR.580.509.

[91] Kutláková, Kateřina Mamulová, Jonáš Tokarský, and Pavlína Peikertová. "Functional and eco-friendly nanocomposite kaolinite/ZnO with high photocatalytic activity." *Applied Catalysis B: Environmental* 162 (2015): 392–400. https://doi.org/10.1016/j.apcatb.2014.07.018.

[92] Gan, Lu, Songmin Shang, Chun Wah Marcus Yuen, Shou-xiang Jiang, and Enling Hu. "Hydrothermal synthesis of magnetic $CoFe_2O_4$/graphene nanocomposites with improved photocatalytic activity." *Applied Surface Science* 351 (2015): 140–147. https://doi.org/10.1016/j.apsusc.2015.05.130.

[93] Teng, Fei, Sunand Santhanagopalan, Ying Wang, and Dennis Desheng Meng. "In-situ hydrothermal synthesis of three-dimensional MnO_2-CNT nanocomposites and their electrochemical properties." *Journal of Alloys and Compounds* 499, no. 2 (2010): 259–264. https://doi.org/10.1016/j.jallcom.2010.03.181.

[94] Miao, Shiding, Zhimin Liu, Buxing Han, Haowen Yang, Zhenjiang Miao, and Zhenyu Sun. "Synthesis and characterization of ZnS-montmorillonite nanocomposites and their application for degrading eosin B." *Journal of Colloid and Interface Science* 301, no. 1 (2006): 116–122. https://doi.org/10.1016/j.jcis.2006.04.080.

[95] Zhou, Feng-shan, Dai-mei Chen, Bao-lin Cui, and Wei-heng Wang. "Synthesis and characterization of CdS/TiO_2-montmorillonite nanocomposite with enhanced visible-light absorption." *Journal of Spectroscopy* 2014 (2014). https://doi.org/10.1155/2014/961230.

[96] King Brink M., and J. G. Sebranek. "Combustion method for determination of crude protein in meat and meat products: collaborative study." *Journal of AOAC International* 76, no. 4 (1993): 787–793.

[97] Li, Gonghu, Shannon Ciston, Zoran V. Saponjic, Le Chen, Nada M. Dimitrijevic, Tijana Rajh, and Kimberly A. Gray. "Synthesizing mixed-phase TiO_2 nanocomposites using a hydrothermal method for photo-oxidation and photoreduction applications." *Journal of Catalysis* 253, no. 1 (2008): 105–110. https://doi.org/10.1016/j.jcat.2007.10.014.

[98] Karthik, Kannan, Sivasubramanian Dhanuskodi, Chandrakasan Gobinath, and Sivaperumal Sivaramakrishnan. "Microwave-assisted synthesis of CdO-ZnO nanocomposite and its antibacterial activity against human pathogens." *Spectrochimica Acta Part A: Molecular and Biomolecular Spectroscopy* 139 (2015): 7–12. https://doi.org/10.1016/j.saa.2014.11.079.

[99] Karthik, Kannan, Sivasubramanian Dhanuskodi, Chandrakasan Gobinath, S. Prabukumar, and Sivaperumal Sivaramakrishnan. "Multifunctional properties of microwave assisted CdO-NiO-ZnO mixed metal oxide nanocomposite: Enhanced photocatalytic and antibacterial activities." *Journal of Materials Science: Materials in Electronics* 29 (2018): 5459–5471. https://doi.org/10.1007/s10854-017-8513-y.

[100] Del Angel, Julio, Alberto F. Aguilera, Ignacio R. Galindo, Merced Martínez, and Tomas Viveros. "Synthesis and characterization of alumina-zirconia powders obtained by sol-gel method: Effect of solvent and water addition rate." *Scientific Research* 3 (2012): https://www.scirp.org/journal/PaperInformation.aspx?PaperID=22924.

[101] Wang, Xinyuan, Heriberto Pfeiffer, Jiangjiang Wei, Jianming Dan, Jinyu Wang, and Jinli Zhang. "3D porous Ca-modified Mg-Zr mixed metal oxide for fluoride adsorption." *Chemical Engineering Journal* 428 (2022): 131371. https://doi.org/10.1016/j.cej.2021.131371.

[102] Thammachart, Matina, Vissanu Meeyoo, Thirasak Risksomboon, and Somchai Osuwan. "Catalytic activity of CeO_2-ZrO_2 mixed oxide catalysts prepared via sol-gel technique: CO oxidation." *Catalysis Today* 68, no. 1–3 (2001): 53–61. https://doi.org/10.1016/S0920-5861(01)00322-4.

[103] Арсентьев, М. Ю., М. В. Калинина, П. А. Тихонов, Л. В. Морозова, А. С. Коваленко, Н. Ю. Ковалько, И. И. Хламов, and О. А. Шилова. "Синтез и свойства сенсорных оксидных наноразмерных пленок в системе ZrO_2-CeO_2." *Физика и химия стекла* 40, no. 3 (2014): 478–484.

[104] Precious Ayanwale, A. "Review of the synthesis, characterization and application of zirconia mixed metal oxide nanoparticles." *Instituto de Ciencias Biomédicas* 6 (2018): 1–10.

[105] Liu, Man, Changrong He, Jianxin Wang, Wei Guo Wang, and Zhenwei Wang. "Investigation of (CeO2) x $(Sc_2O_3)(0.1_{1-x})(ZrO_2)$ 0.89 (x = 0.01-0.10) electrolyte materials for intermediate-temperature solid oxide fuel cell." *Journal of Alloys and Compounds* 502, no. 2 (2010): 319–323. https://doi.org/10.1016/j.jallcom.2009.12.134.

[106] Eslami, Hadi, Mohammad Hassan Ehrampoush, Abbas Esmaeili, Ali Asghar Ebrahimi, Mohammad Taghi Ghaneian, Hossein Falahzadeh, and Mohammad Hossein Salmani. "Synthesis of mesoporous Fe-Mn bimetal oxide nanocomposite by aeration co-precipitation method: Physicochemical, structural, and optical properties." *Materials Chemistry and Physics* 224 (2019): 65–72. https://doi.org/10.1016/j.matchemphys.2018.11.067.

[107] Egizbek, Kamila B, A. L. Kozlovskiy, Katarzyna Ludzik, Maxim V Zdorovets, I. V. Korolkov, Beata Marciniak, Monika Jazdzewska, D. Chudoba, Assel Nazarova, and Renata Kontek. "Stability and cytotoxicity study of $NiFe_2O_4$ nanocomposites synthesized by co-precipitation and subsequent thermal annealing." *Ceramics International* 46, no. 10 (2020): 16548–16555. https://doi.org/10.1016/j.ceramint.2020.03.222.

[108] Pu, Shengyan, Shengyang Xue, Zeng Yang, Yaqi Hou, Rongxin Zhu, and Wei Chu. "In situ co-precipitation preparation of a superparamagnetic graphene oxide/Fe_3O_4 nanocomposite as an adsorbent for wastewater purification: Synthesis, characterization, kinetics, and isotherm studies." *Environmental Science and Pollution Research* 25 (2018): 17310–17320. https://doi.org/10.1007/s11356-018-1872-y.

[109] Thambidurai S., P. Gowthaman, M. Venkatachalam, and Sagadevan Suresh. "Enhanced bactericidal performance of nickel oxide-zinc oxide nanocomposites synthesized by facile chemical co-precipitation method." *Journal of Alloys and Compounds* 830 (2020): 154642. https://doi.org/10.1016/j.jallcom.2020.154642.

[110] Subhan, Md Abdus, Nizam Uddin, Prosenjit Sarker, Abul Kalam Azad, and Kulsuma Begum. "Photoluminescence, photocatalytic and antibacterial activities of $CeO_2 \cdot CuO \cdot ZnO$ nanocomposite fabricated by co-precipitation method." *Spectrochimica Acta Part A: Molecular and Biomolecular Spectroscopy* 149 (2015): 839–850. https://doi.org/10.1016/j.saa.2015.05.024.

[111] Somraksa, Wararat, Sumetha Suwanboon, Pongsaton Amornpitoksuk, and Chamnan Randorn. "Physical and photocatalytic properties of CeO 2/ZnO/ZnAl 2 O 4 ternary nanocomposite prepared by co-precipitation method." *Materials Research* 23 (2020). https://doi.org/10.1590/1980-5373-MR-2019-0627.

[112] Jazi, Fariborz Sharifian, Nader Parvin, Mohammadreza Rabiei, M. Tahriri, Z. Moshefi Shabestari, and Amir Reza Azadmehr. "Effect of the synthesis route on the grain size and morphology of ZnO/Ag nanocomposite." *Journal of Ceramic Processing Research* 13, no. 5 (2012): 523–526.

[113] Trinh, Le Thanh, Le Anh Bao Quynh, and Nguyen Huu Hieu. "Synthesis of zinc oxide/graphene oxide nanocomposite material for antibacterial application." *International Journal of Nanotechnology* 15, no. 1–3 (2018): 108–117. https://doi.org/10.1504/IJNT.2018.089542.

[114] Alghazzawi, Wedam, Ekram Danish, Hanan Alnahdi, and Mohamed Abdel Salam. "Rapid microwave-assisted hydrothermal green synthesis of rGO/NiO nanocomposite for glucose detection in diabetes." *Synthetic Metals* 267 (2020): 116401. https://doi.org/10.1016/j.synthmet.2020.116401.

[115] Taufik, Ardiansyah, Alfred Albert, and Rosari Saleh. "Sol-gel synthesis of ternary CuO/TiO$_2$/ZnO nano-composites for enhanced photocatalytic performance under UV and visible light irradiation." *Journal of Photochemistry and Photobiology A: Chemistry* 344 (2017): 149–162. https://doi.org/10.1016/j.jphotochem.2017.05.012.

[116] Ma, Zichuan, Lin Zhu, Xiaoyang Lu, Shengtao Xing, Yinsu Wu, and Yuanzhe Gao. "Catalytic ozonation of p-nitrophenol over mesoporous Mn-Co-Fe oxide." *Separation and Purification Technology* 133 (2014): 357–364. https://doi.org/10.1016/j.seppur.2014.07.011.

[117] Mohamed Aktham Ahmed. "Synthesis and structural features of mesoporous NiO/TiO$_2$ nanocomposites prepared by sol-gel method for photodegradation of methylene blue dye." *Journal of Photochemistry and Photobiology A: Chemistry* 238 (2012): 63–70. https://doi.org/10.1016/j.jphotochem.2012.04.010.

[118] Lavin, Alexis, Ramesh Sivasamy, Edgar Mosquera, and Mauricio J. Morel. "High proportion ZnO/CuO nanocomposites: Synthesis, structural and optical properties, and their photocatalytic behavior." *Surfaces and Interfaces* 17 (2019): 100367. https://doi.org/10.1016/j.surfin.2019.100367.

[119] Ghorbani, Mina, Hossein Abdizadeh, Mahtab Taheri, and Mohammad Reza Golobostanfard. "Enhanced photoelectrochemical water splitting in hierarchical porous ZnO/Reduced graphene oxide nanocomposite synthesized by sol-gel method." *International Journal of Hydrogen Energy* 43, no. 16 (2018): 7754–7763. https://doi.org/10.1016/j.ijhydene.2018.03.052.

[120] Grigorie, Alexandra Carmen, Cornelia Muntean, Titus Vlase, Cosmin Locovei, and Mircea Stefanescu. "ZnO-SiO$_2$ based nanocomposites prepared by a modified sol-gel method." *Materials Chemistry and Physics* 186 (2017): 399–406. https://doi.org/10.1016/j.matchemphys.2016.11.011.

[121] Ansari, Fatemeh, Azam Sobhani, and Masoud Salavati-Niasari. "Simple sol-gel synthesis and characterization of new CoTiO$_3$/CoFe$_2$O$_4$ nanocomposite by using liquid glucose, maltose and starch as fuel, capping and reducing agents." *Journal of Colloid and Interface Science* 514 (2018): 723–732. https://doi.org/10.1016/j.jcis.2017.12.083.

[122] Gao, Yupeng, Libo Wang, Aiguo Zhou, Zhengyang Li, Jingkuo Chen, Hari Bala, Qianku Hu, and Xinxin Cao. "Hydrothermal synthesis of TiO$_2$/Ti$_3$C$_2$ nanocomposites with enhanced photocatalytic activity." *Materials Letters* 150 (2015): 62–64. https://doi.org/10.1016/j.matlet.2015.02.135.

[123] Wang, Aijian, Wang Yu, Yu Fang, Yinglin Song, Ding Jia, Lingliang Long, Marie P. Cifuentes, Mark G. Humphrey, and Chi Zhang. "Facile hydrothermal synthesis and optical limiting properties of TiO$_2$-reduced graphene oxide nanocomposites." *Carbon* 89 (2015): 130–141. https://doi.org/10.1016/j.carbon.2015.03.037.

[124] Dou, Pingtao, Fatang Tan, Wei Wang, Ali Sarreshteh, Xueliang Qiao, Xiaolin Qiu, and Jianguo Chen. "One-step microwave-assisted synthesis of Ag/ZnO/graphene nanocomposites with enhanced photocatalytic activity." *Journal of Photochemistry and Photobiology A: Chemistry* 302 (2015): 17–22. https://doi.org/10.1016/j.jphotochem.2014.12.012.

[125] Alnahari, Hisham, Annas Al-Sharabi, A. H. Al-Hammadi, Abdel-Basit Al-Odayni, and Adnan Alnehia. "Synthesis of glycine-mediated CuO-Fe$_2$O$_3$-MgO nanocomposites: Structural, optical, and antibacterial properties." *Composites and Advanced Materials* 32 (2023): 26349833231176838. https://doi.org/10.1177/26349833231176838.

[126] Shekoohiyan, Sakine, Asieh Rahmania, Masoumeh Chamack, Gholamreza Moussavi, Omid Rahmanian, Vali Alipour, and Stefanos Giannakis. "A novel CuO/Fe$_2$O$_3$/ZnO composite for visible-light assisted photocatalytic oxidation of Bisphenol A: Kinetics, degradation pathways, and toxicity elimination." *Separation and Purification Technology* 242 (2020): 116821. https://doi.org/10.1016/j.seppur.2020.116821.

[127] Munawar, Tauseef, Faisal Iqbal, Sadaf Yasmeen, Khalid Mahmood, and Altaf Hussain. "Multi metal oxide NiO-CdO-ZnO nanocomposite-synthesis, structural, optical, electrical properties and enhanced sunlight driven photocatalytic activity." *Ceramics International* 46, no. 2 (2020): 2421–2437. https://doi.org/10.1016/j.ceramint.2019.09.236.

[128] Revathi, V., and K. Karthik. "Microwave assisted CdO-ZnO-MgO nanocomposite and its photocatalytic and antibacterial studies." *Journal of Materials Science: Materials in Electronics* 29 (2018): 18519–18530. https://doi.org/10.1007/s10854-018-9968-1.

[129] Alnahari, H., A. H. Al-Hammadi, A. Al-Sharabi, and A. Alnehia. "The effect of tartaric acid concentration on the structural, morphological and optical properties of CuO-Fe$_2$O$_3$-MgO nanocomposite." *JAST* 1, no. 2 (2023): 208–216.

[130] Alnahari, Hisham, A. H. Al-Hammadi, Annas Al-Sharabi, Adnan Alnehia, and Abdel-Basit Al-Odayni. "Structural, morphological, optical, and antibacterial properties of CuO-Fe$_2$O$_3$-MgO-CuFe$_2$O$_4$ nanocomposite synthesized via auto-combustion route." *Journal of Materials Science: Materials in Electronics* 34, no. 7 (2023): 682. https://doi.org/10.1007/s10854-023-10120-7.

[131] Al-Sharabi, Annas, Kholod S. S. Sada'a, Ahmed Al-Osta, and R. Abd-Shukor. "Structure, optical properties and antimicrobial activities of MgO-Bi$_{2-x}$Cr$_x$O$_3$ nanocomposites prepared via solvent-deficient method." *Scientific Reports* 12, no. 1 (2022): 10647. https://doi.org/10.1038/s41598-022-14811-9.

[132] Chiou, Jau-Rung, Bo-Hung Lai, Kai-Chih Hsu, and Dong-Hwang Chen. "One-pot green synthesis of silver/iron oxide composite nanoparticles for 4-nitrophenol reduction." *Journal of Hazardous Materials* 248 (2013): 394–400. https://doi.org/10.1016/j.jhazmat.2013.01.030.

[133] De Leon-Condes, Cristina A., Gabriela Roa-Morales, Gonzalo Martinez-Barrera, Patricia Balderas-Hernandez, Carmina Menchaca-Campos, and Fernando Urena-Nunez. "A novel sulfonated waste polystyrene/iron oxide nanoparticles composite: Green synthesis, characterization and applications." *Journal of Environmental Chemical Engineering* 7, no. 1 (2019): 102841. https://doi.org/10.1016/j.jece.2018.102841.

[134] Zhang, Ping, David O'Connor, Yinan Wang, Lin Jiang, Tianxiang Xia, Liuwei Wang, Daniel C. W. Tsang, Yong Sik Ok, and Deyi Hou. "A green biochar/iron oxide composite for methylene blue removal." *Journal of Hazardous Materials* 384 (2020): 121286. https://doi.org/10.1016/j.jhazmat.2019.121286.

[135] Zhou, Xinfeng, Bingbing Wang, Zirui Jia, Xinda Zhang, Xuehua Liu, Kuikui Wang, Binghui Xu, and Guanglei Wu. "Dielectric behavior of Fe$_3$N@ C composites with green synthesis and their remarkable electromagnetic wave absorption performance." *Journal of Colloid and Interface Science* 582 (2021): 515–525. https://doi.org/10.1016/j.jcis.2020.08.087.

[136] Zhang, Lidong, Guixiang Du, Bo Zhou, and Lei Wang. "Green synthesis of flower-like ZnO decorated reduced graphene oxide composites." *Ceramics International* 40, no. 1 (2014): 1241–1244. https://doi.org/10.1016/j.ceramint.2013.06.023.

[137] Biswal, Susanta Kumar, Gagan Kumar Panigrahi, and Shraban Kumar Sahoo. "Green synthesis of Fe$_2$O$_3$-Ag nanocomposite using Psidium guajava leaf extract: An eco-friendly and recyclable adsorbent for remediation of Cr (VI) from aqueous media." *Biophysical Chemistry* 263 (2020): 106392. https://doi.org/10.1016/j.bpc.2020.106392.

[138] Matinise, Nolubabalo, K. Kaviyarasu, Nametso Mongwaketsi, Saleh Khamlich, Lebogang Kotsedi, Noluthando Mayedwa, and Maalik Maaza. "Green synthesis of novel zinc iron oxide (ZnFe$_2$O$_4$) nanocomposite via Moringa Oleifera natural extract for electrochemical applications." *Applied Surface Science* 446 (2018): 66–73. https://doi.org/10.1016/j.apsusc.2018.02.187.

[139] Mohammadi-Aloucheh, Ramin, Aziz Habibi-Yangjeh, Abolfazl Bayrami, Saeid Latifi-Navid, and Asadollah Asadi. "Green synthesis of ZnO and ZnO/CuO nanocomposites in Mentha longifolia leaf extract: Characterization and their application as anti-bacterial agents." *Journal of Materials Science: Materials in Electronics* 29 (2018): 13596–13605. https://doi.org/10.1007/s10854-018-9487-0.

[140] Bordbar, Maryam, Neda Negahdar, and Mahmoud Nasrollahzadeh. "Melissa Officinalis L. leaf extract assisted green synthesis of CuO/ZnO nanocomposite for the reduction of 4-nitrophenol and Rhodamine B." *Separation and Purification Technology* 191 (2018): 295–300. https://doi.org/10.1016/j.seppur.2017.09.044.

[141] Liu, Qiujie, Peili Ma, Penglei Liu, Hongping Li, Xiuli Han, Lie Liu, and Weihua Zou. "Green synthesis of stable Fe, Cu oxide nanocomposites from loquat leaf extracts for removal of Norfloxacin and Ciprofloxacin." *Water Science and Technology* 81, no. 4 (2020): 694–708. https://doi.org/10.2166/wst.2020.152.

[142] Yulizar, Yoki, Dewangga Oky Bagus Apriandanu, and Aji Prasetiyo Wibowo. "Plant extract mediated synthesis of Au/TiO$_2$ nanocomposite and its photocatalytic activity under sodium light irradiation." *Composites Communications* 16 (2019): 50–56. https://doi.org/10.1016/j.coco.2019.08.006.

[143] Noohpisheh, Zahra, Hamzeh Amiri, Saeed Farhadi, and Abdolnaser Mohammadi-Gholami. "Green synthesis of Ag-ZnO nanocomposites using Trigonella foenum-graecum leaf extract and their antibacterial, antifungal, antioxidant and photocatalytic properties." *Spectrochimica Acta Part A: Molecular and Biomolecular Spectroscopy* 240 (2020): 118595. https://doi.org/10.1016/j.saa.2020.118595.

[144] Hessien, Manal, Enshirah Da'na, and Amel Taha. "Phytoextract assisted hydrothermal synthesis of ZnO-NiO nanocomposites using neem leaves extract." *Ceramics International* 47, no. 1 (2021): 811–816. https://doi.org/10.1016/j.ceramint.2020.08.192.

[145] Shunmuga Priya, R., E. Ranjith Kumar, A. Balamurugan, and C. Srinivas. "Green synthesized MgFe$_2$O$_4$ ferrites nanoparticles for biomedical applications." *Applied Physics A* 127, no. 7 (2021): 538. https://doi.org/10.1007/s00339-021-04699-z.

[146] Udhaya, P. Aji, M. Meena, and M. Abila Jeba Queen. "Green synthesis of $MgFe_2O_4$ nanoparticles using albumen as fuel and their physicochemical properties." *International Journal of Scientific Research in Physics and Applied Sciences* 7 (2019): 71–74.

[147] Mohan, B. Sathish, K. Ravi, R. Balaji Anjaneyulu, G. Satya Sree, and K. Basavaiah. "Fe_2O_3/RGO nanocomposite photocatalyst: Effective degradation of 4-Nitrophenol." *Physica B: Condensed Matter* 553 (2019): 190–194. https://doi.org/10.1016/j.physb.2018.10.033.

[148] Maruthupandy, Muthuchamy, Pan Qin, Thillaichidambaram Muneeswaran, Govindan Rajivgandhi, Franck Quero, and Ji-Ming Song. "Graphene-zinc oxide nanocomposites (G-ZnO NCs): Synthesis, characterization and their photocatalytic degradation of dye molecules." *Materials Science and Engineering: B* 254 (2020): 114516. https://doi.org/10.1016/j.mseb.2020.114516.

[149] Pirzada, Bilal M., Pushpendra, Ravi K. Kunchala, and Boddu S. Naidu. "Synthesis of $LaFeO_3$/Ag_2CO_3 nanocomposites for photocatalytic degradation of rhodamine B and p-chlorophenol under natural sunlight." *ACS Omega* 4, no. 2 (2019): 2618–2629. https://doi.org/10.1021/acsomega.8b02829.

[150] Ansari, Sajid Ali, Mohammad Mansoob Khan, Mohd Omaish Ansari, Jintae Lee, and Moo Hwan Cho. "Biogenic synthesis, photocatalytic, and photoelectrochemical performance of Ag-ZnO nanocomposite." *The Journal of Physical Chemistry C* 117, no. 51 (2013): 27023–27030. https://doi.org/10.1021/jp410063p.

[151] Ruan, Shuhong, Wenqian Huang, Mengjiu Zhao, Haiyan Song, and Zhihong Gao. "A Z-scheme mechanism of the novel ZnO/CuO nn heterojunction for photocatalytic degradation of Acid Orange 7." *Materials Science in Semiconductor Processing* 107 (2020): 104835. https://doi.org/10.1016/j.mssp.2019.104835.

[152] Hashim, Fouad Sh, Ayad F. Alkaim, Shymaa M. Mahdi, and Adel H. Omran Alkhayatt. "Photocatalytic degradation of GRL dye from aqueous solutions in the presence of ZnO/Fe_2O_3 nanocomposites." *Composites Communications* 16 (2019): 111–116. https://doi.org/10.1016/j.coco.2019.09.008.

[153] Shanmugam, Mahalingam, Ali Alsalme, Abdulaziz Alghamdi, and Ramasamy Jayavel. "Enhanced photocatalytic performance of the graphene-V_2O_5 nanocomposite in the degradation of methylene blue dye under direct sunlight." *ACS Applied Materials & Interfaces* 7, no. 27 (2015): 14905–14911. https://doi.org/10.1021/acsami.5b02715.

[154] Mukherjee, Arnab, Mrinal K. Adak, Sudipta Upadhyay, Julekha Khatun, Prasanta Dhak, Sadhana Khawas, Uttam Kumar Ghorai, and Debasis Dhak. "Efficient fluoride removal and dye degradation of contaminated water using Fe/Al/Ti oxide nanocomposite." *ACS Omega* 4, no. 6 (2019): 9686–9696. https://doi.org/10.1021/acsomega.9b00252.

[155] Steplin Paul Selvin, S., Agrawal Ganesh Kumar, L. Sarala, R. Rajaram, A. Sathiyan, Princy Merlin johnson, and I. Sharmila Lydia. "Photocatalytic degradation of rhodamine B using zinc oxide activated charcoal polyaniline nanocomposite and its survival assessment using aquatic animal model." *ACS Sustainable Chemistry & Engineering* 6, no. 1 (2018): 258–267. https://doi.org/10.1021/acssuschemeng.7b02335.

[156] Gulce, Handan, Volkan Eskizeybek, Bircan Haspulat, Fahriye Sarı, Ahmet Gülce, and Ahmet Avcı. "Preparation of a new polyaniline/CdO nanocomposite and investigation of its photocatalytic activity: Comparative study under UV light and natural sunlight irradiation." *Industrial & Engineering Chemistry Research* 52, no. 32 (2013): 10924–10934. https://doi.org/10.1021/ie401389e.

[157] Lv, Tian, Likun Pan, Xinjuan Liu, Ting Lu, Guang Zhu, and Zhuo Sun. "Enhanced photocatalytic degradation of methylene blue by ZnO-reduced graphene oxide composite synthesized via microwave-assisted reaction." *Journal of Alloys and Compounds* 509, no. 41 (2011): 10086–10091. https://doi.org/10.1016/j.jallcom.2011.08.045.

[158] Chen, Yan-Li, Zhong-Ai Hu, Yan-Qin Chang, Huan-Wen Wang, Zi-Yu Zhang, Yu-Ying Yang, and Hong-Ying Wu. "Zinc oxide/reduced graphene oxide composites and electrochemical capacitance enhanced by homogeneous incorporation of reduced graphene oxide sheets in zinc oxide matrix." *The Journal of Physical Chemistry C* 115, no. 5 (2011): 2563–2571. https://doi.org/10.1021/jp109597n.

[159] Yuan, Hong, Jianglin Ye, Chuanren Ye, Songsen Yin, Jieyun Li, Kai Su, Gang Fang et al. "Highly efficient preparation of graphite oxide without water enhanced oxidation." *Chemistry of Materials* 33, no. 5 (2021): 1731–1739. https://doi.org/10.1021/acs.chemmater.0c04505.

[160] Nipane, Sandip. V., Korake, Praksh. V., & Gokavi G. S. "Graphene-zinc oxide nanorod nanocomposite as photocatalyst for enhanced degradation of dyes under UV light irradiation." *Ceramics International* 41, no. 3, (2015): 4549–4557. https://doi.org/10.1016/j.ceramint.2014.11.151.

[161] Zhou, Xun, Tiejun Shi, and Haiou Zhou. "Hydrothermal preparation of ZnO-reduced graphene oxide hybrid with high performance in photocatalytic degradation." *Applied Surface Science* 258, no. 17 (2012): 6204–6211. https://doi.org/10.1016/j.apsusc.2012.02.131.

[162] Li, Xueshan, Qian Wang, Yibo Zhao, Wei Wu, Jianfeng Chen, and Hong Meng. "Green synthesis and photo-catalytic performances for ZnO-reduced graphene oxide nanocomposites." *Journal of Colloid and Interface Science* 411 (2013): 69–75. https://doi.org/10.1016/j.jcis.2013.08.050.

[163] Thangavel, Sakthivel, Karthikeyan Krishnamoorthy, Velmurugan Krishnaswamy, Nandhakumar Raju, Sang Jae Kim, and Gunasekaran Venugopal. "Graphdiyne-ZnO nanohybrids as an advanced photo-catalytic material." *The Journal of Physical Chemistry C* 119, no. 38 (2015): 22057–22065. https://doi.org/10.1021/acs.jpcc.5b06138.

[164] Leng Yuanpeng, Wang Wucong, Zhang Lin, Zabihi Fatemeh, & Zhao Yapin. "Fabrication and photocatalytical enhancement of ZnO-graphene hybrid using a continuous solvothermal technique." *The Journal of Supercritical Fluids* 91, (2014): 61–67. https://doi.org/10.1016/j.supflu.2014.04.012.

[165] Sawant, Sandesh Y., and Moo Hwan Cho. "Facile electrochemical assisted synthesis of ZnO/graphene nanosheets with enhanced photocatalytic activity." *RSC Advances* 5, no. 118 (2015): 97788–97797.

[166] Weng, Bo, Min-Quan Yang, Nan Zhang, and Yi-Jun Xu. "Toward the enhanced photoactivity and photostability of ZnO nanospheres via intimate surface coating with reduced graphene oxide." *Journal of Materials Chemistry A* 2, no. 24 (2014): 9380–9389.

[167] He, Jianjiang, Chunge Niu, Chao Yang, Jide Wang, and Xintai Su. "Reduced graphene oxide anchored with zinc oxide nanoparticles with enhanced photocatalytic activity and gas sensing properties." *RSC Advances* 4, no. 104 (2014): 60253–60259.

[168] Jabeen, Sobia, Javed Iqbal, Aqsa Arshad, Muhammad Saifullah Awan, and Muhammad Farooq Warsi. "$(In_{1-x}Fe_x)_2O_3$ nanostructures for photocatalytic degradation of various dyes." *Materials Chemistry and Physics* 243 (2020): 122516. https://doi.org/10.1016/j.matchemphys.2019.122516.

[169] Zahoor, Mehvish, Amara Arshad, Yaqoob Khan, Mazhar Iqbal, Sadia Zafar Bajwa, Razium Ali Soomro, Ishaq Ahmad et al. "Enhanced photocatalytic performance of CeO_2-TiO_2 nanocomposite for degradation of crystal violet dye and industrial waste effluent." *Applied Nanoscience* 8 (2018): 1091–1099. https://doi.org/10.1007/s13204-018-0730-z.

[170] Jusoh, R., H. D. Setiabudi, N. S. Kamarudin, and N. F. Sukor, "Chitosan-functionalized Ag nanoparticles for degradation of methylene blue: Effect of chitosan pretreatment." *The International Journal of Science & Technoledge* 5, no. 12 (2017): 103–109.

[171] Haldorai, Yuvaraj, and Jae-Jin Shim. "Chitosan-zinc oxide hybrid composite for enhanced dye degradation and antibacterial activity." *Composite Interfaces* 20, no. 5 (2013): 365–377. https://doi.org/10.1080/15685543.2013.806124.

[172] Singh, Jagpreet, Tanushree Dutta, Ki-Hyun Kim, Mohit Rawat, Pallabi Samddar, and Pawan Kumar. "'Green' synthesis of metals and their oxide nanoparticles: Applications for environmental remediation." *Journal of Nanobiotechnology* 16, no. 1 (2018): 1–24. https://doi.org/10.1186/s12951-018-0408-4.

[173] Pal, Gaurav, Priya Rai, and Anjana Pandey. "Green synthesis of nanoparticles: A greener approach for a cleaner future." In *Green Synthesis, Characterization and Applications of Nanoparticles*, pp. 1–26. Elsevier, 2019. https://doi.org/10.1016/B978-0-08-102579-6.00001-0.

[174] Sadek, Rawia Farag Hala A. Farrag, Shimaa Mohammad Abdelsalam, Z. M. H. Keiralla, Amany I. Raafat, and Eman Araby. "A powerful nanocomposite polymer prepared from metal oxide nanoparticles synthesized via brown algae as anti-corrosion and anti-biofilm." *Frontiers in Materials* 6 (2019): 140. https://doi.org/10.3389/fmats.2019.00140.

[175] Yu, Chen, Jingchun Tang, Hongji Su, Jingci Huang, Fangheng Liu, Lan Wang, and Hongwen Sun. "Development of a novel biochar/iron oxide composite from green algae for bisphenol-A removal: Adsorption and Fenton-like reaction." *Environmental Technology & Innovation* 28 (2022): 102647. https://doi.org/10.1016/j.eti.2022.102647.

[176] El-Refaey, Ahmed A., and Salem S. Salem. "Algae materials for bionanopesticides: nanoparticles and composites." In *Algae Materials*, pp. 219–230. Academic Press, 2023. https://doi.org/10.1016/B978-0-443-18816-9.00004-6.

[177] Chu, Ruoyu, Shuangxi Li, Liandong Zhu, Zhihong Yin, Dan Hu, Chenchen Liu, and Fan Mo. "A review on co-cultivation of microalgae with filamentous fungi: Efficient harvesting, wastewater treatment and biofuel production." *Renewable and Sustainable Energy Reviews* 139 (2021): 110689. https://doi.org/10.1016/j.rser.2020.110689.

[178] Tran, Thuan Van, Duyen Thi Cam Nguyen, Ponnusamy Senthil Kumar, Azam Taufik Mohd Din, Aishah Abdul Jalil, and Dai-Viet N. Vo. "Green synthesis of ZrO_2 nanoparticles and nanocomposites for biomedical and environmental applications: A review." *Environmental Chemistry Letters* (2022): 1–23. https://doi.org/10.1007/s10311-021-01367-9.

[179] Ghomi, Ahmad Reza Golnaraghi, Mohammad Mohammadi-Khanaposhti , Hossein Vahidi , Farzad Kobarfard, Mahdieh Ameri Shah Reza, and Hamed Barabadib. "Fungus-mediated extracellular biosynthesis and characterization of zirconium nanoparticles using standard penicillium species and their preliminary bactericidal potential: A novel biological approach to nanoparticle synthesis." *Iranian Journal of Pharmaceutical Research: IJPR* 18, no. 4 (2019): 2101. https://doi.org/10.22037%2Fijpr.2019.112382.13722.

[180] Ali, Asif, Yi Wai Chiang, and Rafael M. Santos. "X-ray diffraction techniques for mineral characterization: A review for engineers of the fundamentals, applications, and research directions." *Minerals* 12, no. 2 (2022): 205. https://doi.org/10.3390/min12020205.

[181] Ikim M. I., E. Yu Spiridonova, T. V. Belysheva, Vladimir F. Gromov, Genrik N. Gerasimov, and Leonid Lzrailevich Trakhtenberg. "Structural properties of metal oxide nanocomposites: Effect of preparation method." *Russian Journal of Physical Chemistry B* 10 (2016): 543–546. https://doi.org/10.1134/S1990793116030210.

[182] Akash, Muhammad Sajid Hamid, and Kanwal Rehman. *Essentials of Pharmaceutical Analysis.* Springer, 2020. https://doi.org/10.1007/978-981-15-1547-7.

[183] Fadlelmoula, Ahmed, Diana Pinho, Vitor Hugo Carvalho, Susana O. Catarino, and Graça Minas. "Fourier transform infrared (FTIR) spectroscopy to analyse human blood over the last 20 years: A review towards lab-on-a-chip devices." *Micromachines* 13, no. 2 (2022): 187. https://doi.org/10.3390/mi13020187.

[184] Rahman, Abdur, Humera Sabeeh, Sonia Zulfiqar, Philips Olaleye Agboola, Imran Shakir, and Muhammad Farooq Warsi. "Structural, optical and photocatalytic studies of trimetallic oxides nanostructures prepared via wet chemical approach." *Synthetic Metals* 259 (2020): 116228. https://doi.org/10.1016/j.synthmet.2019.116228.

[185] Vladár, András E., and Vasile-Dan Hodoroaba. "Characterization of nanoparticles by scanning electron microscopy." In *Characterization of Nanoparticles*, pp. 7–27. Elsevier, 2020. https://doi.org/10.1016/B978-0-12-814182-3.00002-X.

[186] Mansournia, Mohammadreza, and Leila Ghaderi. "CuO@ ZnO core-shell nanocomposites: Novel hydrothermal synthesis and enhancement in photocatalytic property." *Journal of Alloys and Compounds* 691 (2017): 171–177. https://doi.org/10.1016/j.jallcom.2016.08.267.

[187] Liang, Jing, Xixi Xiao, Tseng-Ming Chou, and Matthew Libera. "Analytical cryo-scanning electron microscopy of hydrated polymers and microgels." *Accounts of Chemical Research* 54, no. 10 (2021): 2386–2396. https://doi.org/10.1021/acs.accounts.1c00109.

[188] de Haan, Kevin, Zachary S. Ballard, Yair Rivenson, Yichen Wu, and Aydogan Ozcan. "Resolution enhancement in scanning electron microscopy using deep learning." *Scientific Reports* 9, no. 1 (2019): 12050. https://doi.org/10.1038/s41598-019-48444-2.

[189] Lin, Yue, Min Zhou, Xiaolin Tai, Hangfei Li, Xiao Han, and Jiaguo Yu. "Analytical transmission electron microscopy for emerging advanced materials." *Matter* 4, no. 7 (2021): 2309–2339.

[190] Hamrouni, Abdessalem, Noomen Moussa, Francesco Parrino, Agatino Di Paola, Ammar Houas, and Leonardo Palmisano. "Sol-gel synthesis and photocatalytic activity of ZnO-SnO$_2$ nanocomposites." *Journal of Molecular Catalysis A: Chemical* 390 (2014): 133–141. https://doi.org/10.1016/j.molcata.2014.03.018.

[191] Khan, Muhammad Saiful Islam, Se-Wook Oh, and Yun-Ji Kim. "Power of scanning electron microscopy and energy dispersive X-ray analysis in rapid microbial detection and identification at the single cell level." *Scientific Reports* 10, no. 1 (2020): 2368. https://doi.org/10.1038/s41598-020-59448-8.

[192] Kharat, Reval Singh, M. Saleem, Netram Kaurav, and H. S. Dagar. "Composites of ZnFe$_2$O$_4$ and GaFeO$_3$: Structural, optical bandgap and morphology studies." In *AIP Conference Proceedings*, vol. 2369, no. 1. AIP Publishing, 2021. https://doi.org/10.1063/5.0061848.

[193] Flores, Eibar, Petr Novák, and Erik J. Berg. "In situ and operando Raman spectroscopy of layered transition metal oxides for Li-ion battery cathodes." *Frontiers in Energy Research* 6 (2018): 82. https://doi.org/10.3389/fenrg.2018.00082.

[194] Petersen, Marlen, Zhilong Yu, and Xiaonan Lu. "Application of Raman spectroscopic methods in food safety: A review." *Biosensors* 11, no. 6 (2021): 187. https://doi.org/10.3390/bios11060187.

[195] Saadatkhah, Nooshin, Adrián Carillo Garcia, Sarah Ackermann, Philippe Leclerc, Mohammad Latifi, Said Samih, Gregory S. Patience, and Jamal Chaouki. "Experimental methods in chemical engineering: Thermogravimetric analysis-TGA." *The Canadian Journal of Chemical Engineering* 98, no. 1 (2020): 34–43. https://doi.org/10.1002/cjce.23673.

[196] Rami, J. M., C. D. Patel, C. M. Patel, and M. V. Patel. "Thermogravimetric analysis (TGA) of some synthesized metal oxide nanoparticles." *Materials Today: Proceedings* 43 (2021): 655–659. https://doi.org/10.1016/j.matpr.2020.12.554.

[197] Devi, Neeraj, Shobhit Srivastava, Bhumika Yogi, and Sujeet Kumar Gupta. "A review on differential thermal analysis." *Chemistry Research Journal* 6 (2021): 71–80.

[198] Malyshev, Andrey V., Ilka Beil, and Juergen Kreyling. "Differential thermal analysis: A fast alternative to frost tolerance measurements." *Plant Cold Acclimation: Methods and Protocols* (2020): 23–31. https://doi.org/10.1007/978-1-0716-0660-5_3.

[199] Abderrahmane Boughelout, Roberto Macaluso, Mohammad Kechouane, and Mohammad Trari. "Photocatalysis of rhodamine B and methyl orange degradation under solar light on ZnO and Cu$_2$O thin films." *Reaction Kinetics, Mechanisms and Catalysis* 129 (2020): 1115–1130. https://doi.org/10.1007/s11144-020-01741-8.

[200] Xu, Dong, and Hailing Ma. "Degradation of rhodamine B in water by ultrasound-assisted TiO$_2$ photocatalysis." *Journal of Cleaner Production* 313 (2021): 127758. https://doi.org/10.1016/j.jclepro.2021.127758.

[201] Liu, Junxia, Zhiwei Tang, Jialuo Zeng, Yiyun Zhong, Zhihong Long, Zhihong Wang, and Li Feng. "Degradation of acid orange 8 through photocatalysis in the presence of ZnO/polyaniline nanocomposite." *International Journal of Electrochemical Science* 17, no. 8 (2022): 220857. https://doi.org/10.20964/2022.08.57.

[202] Iwuozor, Kingsley O., Joshua O. Ighalo, Ebuka Chizitere Emenike, Lawal Adewale Ogunfowora, and Chinenye Adaobi Igwegbe. "Adsorption of methyl orange: A review on adsorbent performance." *Current Research in Green and Sustainable Chemistry* 4 (2021): 100179. https://doi.org/10.1016/j.crgsc.2021.100179.

[203] Hanafi, Muhammad Farhan, and Norzahir Sapawe. "A review on the water problem associate with organic pollutants derived from phenol, methyl orange, and remazol brilliant blue dyes." *Materials Today: Proceedings* 31 (2020): A141–A150. https://doi.org/10.1016/j.matpr.2021.01.258.

[204] Boruah, Bhanupriya, Rimzhim Gupta, Jayant M. Modak, and Giridhar Madras. "Novel insights into the properties of AgBiO$_3$ photocatalyst and its application in immobilized state for 4-nitrophenol degradation and bacteria inactivation." *Journal of Photochemistry and Photobiology A: Chemistry* 373 (2019): 105–115. https://doi.org/10.1016/j.jphotochem.2018.11.001.

20 ZnO-Based Nanostructured Materials for Photocatalytic Applications

Jayanti Mishra and Arabinda Baruah

NANOMATERIALS

Materials with at least one dimension falling within 1–100 nm are referred to as nanomaterials, e.g., zero-dimensional nanoparticles (quantum dots), one-dimensional nanorods and nanowires and two-dimensional thin films. Common approaches for the synthesis of nanoparticles are sol-gel, hydrothermal, colloidal and microemulsion methods. Bottom-up approach is preferred over top-down approach because of more homogeneous compositions, less defects, better short- and long-range ordering, etc. It also ends up reducing Gibbs free energy, so the nanomaterials produced stay close to thermodynamic equilibrium state. Nanomaterials are very popular in several areas [1–3] such as electronics, optics, therapeutics, drug delivery, dentistry, laser instruments, piezoelectric, catalytic and sensing due to their unique properties such as smaller size, higher surface area and larger band gap.

PHOTOCATALYSIS

Photocatalysis of hazardous substances by metal oxide nanomaterials is utilised in various ways, e.g., decomposition of harmful gases, H_2 gas production by photo splitting of water, cleaning of water and air, self-cleaning of huge buildings, airborne dust removal, killing of pathogens, anti-fogging action, odour control, cancer cell inactivation, oil spill cleaning and hazardous waste remediation [4–10] (Figure 20.1). Major dyes as pollutants from industries are Rhodamine B, Metanil Yellow, Amaranth, Quinoline Yellow, Carmine, Methyl Orange, Malachite Green, etc. [11,12].

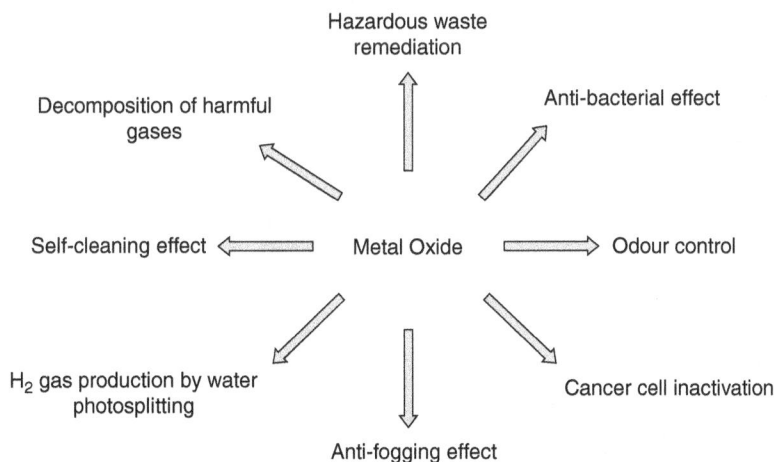

FIGURE 20.1 Application of metal oxides in various fields.

DOI: 10.1201/9781003479239-20

The dyes have chromophores which absorb sunlight and thus disturb the photosynthesis of aquatic flora. In addition, they increase the biological oxygen demand (BOD) and chemical oxygen demand (COD) rapidly [13]. The anaerobic degradation of suspended dyes produces carcinogenic amines which disturb ecological balance in environment [14,15]. A variety of industries including textile, paint, paper, cosmetic and food industries, e.g., fruit jam, sweets, candies and soft drinks [16], are using dyes, several of which are harmful. The conventional methods such as ozonisation, chlorination, activated charcoal treatment and enzyme treatment [17–23] are not sufficient for the removal of many chemicals such as drugs, dyes, cosmetic compounds, metabolites and disinfectants [24]. To overcome this problem, an efficient and environmentally friendly mechanisms for degradation of these harmful pollutants are needed.

A typical photocatalyst should be equipped with poor recombination rate of electrons and holes, because their fast recombination gives low quantum efficiency. In order to fulfil the favourable thermodynamic criteria, a photocatalytic material must possess the following qualities: a) to generate superoxide radical anions, the redox potential of conduction band electron should be negative, and b) to generate hydroxyl radicals, the redox potential of valence band hole should be positive. If photogenerated electrons and holes are located on different locations on the catalysts, the photocatalytic reaction rate may increase. It can be done by tuning band gap, morphology, particle size and surface area of the material [5,25,26]. Because of fulfilling the above criteria, several metal oxides, perovskite halides and metal sulphides [4,27,28] have been reported so far for their dye degradation ability [2,29,30].

MECHANISM OF PHOTOCATALYSIS

Chemical species such as hydroxyl radical (OH), superoxide radical anion (O_2^-) and HO_2 play vital roles in photocatalysis. Illumination of a photocatalyst by suitable wavelength of light along with aqueous medium from surrounding helps to generate hydroxyl radicals [5,7,26] (Figure 20.2).

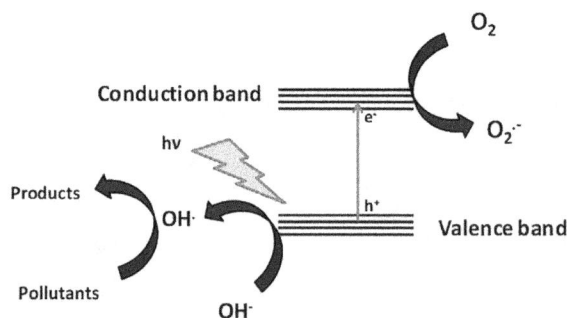

$$\text{Photocatalyst} + h\nu \rightarrow \text{Photocatalyst} \ (e^-_{CB} + h^+_{VB})$$

$$\text{Photocatalyst} \ (h^+_{VB}) + H_2O \rightarrow OH^\bullet$$

$$e^- + O_2 \rightarrow O_2^{\bullet-}$$

$$O_2^{\bullet-} + H_2O \rightarrow HO_2^{\bullet-} + OH$$

$$HO_2^\bullet + H_2O \rightarrow OH^\bullet + H_2O_2$$

$$H_2O_2 \rightarrow 2OH^\bullet$$

$$OH^\bullet + RhB \rightarrow CO_2 + H_2O$$

FIGURE 20.2 Schematic diagram of photocatalysis.

e^-_{CB}: Electron in the conduction band;
h^+_{VB}: Hole in the valence band

Upon illumination of the photocatalyst by suitable light source, holes in valence band and electrons in conduction band are separated. Eventually, water molecules are converted into hydroxyl radicals in valence band after reaction of hydroxide ions with holes, while oxygen molecules are converted into superoxide radical anions which are also converted into hydroxyl radicals finally. The hydroxyl radical degrades the dye to get carbon dioxide and water molecules.

ZnO NANOPARTICLES AS PHOTOCATALYST

ZnO is highly popular metal oxide known for its multifunctional applications in optics, electronics, sensing, photocatalysis [31–33], etc. It is a biocompatible and non-toxic semiconductor having a large band gap of 3.0–3.7 eV and a large free-exciton binding energy of 60 meV, which makes it suitable for room temperature exciton recombination [31,34–36]. ZnO is a type II–VI semiconductor which is borderline between ionic and covalent semiconductor with three forms, namely, zinc blende, wurtzite, and rock salt–type crystal structures with biocompatibility. Thermodynamically, the most favourable crystal structure of ZnO is wurtzite structure whose all octahedral sites are empty and possess as hexagonal closed packed lattice in which Zn atoms occupy half of the tetrahedral sites [36]. So, there is enormous possibility for ZnO to accommodate extrinsic dopants and intrinsic defects. Various synthetic routes for ZnO synthesis include the template-based synthesis by solution-phase synthesis, epitaxial growth, colloidal routes, electrochemical deposition and vapour deposition methods [37–44]. Solution-phase synthesis is very popular due to its mild reaction conditions and simple instrumentation requirements, in which hydrolysis of zinc precursors is done in the presence of any capping agent/surfactant/structure directing agents under alkaline conditions [45]. By tuning the size and morphology, the catalytic behaviour can be tuned as per the requirements.

VARIOUS CATEGORIES OF ZnO NANOPARTICLES FOR APPLICATION IN CATALYSIS

Based on their physical structure and chemical composition, ZnO-based materials can be classified into different categories as discussed in the following sections.

SURFACE-MODIFIED ZnO NANOPARTICLES

Functional groups regulate the surface defects by interacting with the electron deficiencies of the surface [46–48]. They help prevent agglomeration as well as restrict the size and increase the surface area [42,44–46]. Properties of capping agents such as the number and position of binding sites, chain length and steric positions govern the surface energy and the diffusion of ions of a catalyst. The formation of anisotropic ZnO can be achieved by a capping agent which regulates the growth of a crystal in different crystal planes. Various compounds such as polyvinyl pyrrolidone, citrate, thiols, oleylamine, dopamine, phosphonic acid, trioctylphosphine oxide, hydrazine, urea and thiourea derivatives [32,45,47,50,51–56] are reported as potential capping agents/surface directing agents to tune the size and morphology and hence properties of metal oxides. The presence of functional groups such as amines, carboxylic acids, alcohols, urea and thiourea helps the substrate achieve a range of binding efficiencies. They modify the surface properties by restricting or favouring a particular plane of the crystal, regulating the growth, reducing the surface defects and hence helpful in degradation of the desired compounds [49,57]. Mishra *et al.* reported surface-modified ZnO nanoparticles by urea- and thiourea-based tripodal ligands and explored their performance towards degradation of RhB dye [58–61]. Rao *et al.* observed that the growth of one dimension is faster and another dimension is slower in ZnO nanorods in the presence of PVP [62]. Synthesis of ZnO nanoparticles at room temperature using organic ligands [1,5-diphenyl-1,3,5-pentanetrione (pent), dimethyl-L-tartrate (dmlt) and citric acid] via sol-gel route was reported by Zobel and Neder *et al.* [63]. Vertically aligned 1D nanorods are obtained by using L-cysteine in basic medium by Liu, Gao and Tang *et al.* [64]. Jung *et al.* investigated the interaction of triethyl citrate with (0001) surface of ZnO over morphology change of ZnO nanostructures in order to achieve nanospheres from nano-flowers. Capping agents regulate the growth of structures in a particular direction by their different binding strengths towards different crystal planes to get the growth into favoured directions [65].

ZnO-Based Core-Shell Structures

In the past couple of decades, a tremendous amount of research work has been published on the synthesis of various core-shell nanostructures for photocatalytic and electrocatalytic applications [66]. In general, core-shell nanostructures exhibit remarkable traits of creating more efficient inter-facial combination and effective charge carriage possibility by dropping the rate of recombination at the grain boundary. Furthermore, the contacts between dissimilar semiconductor nanomaterials of the core and the shell can substantially enhance the catalytic performance owing to their favourable synergistic effects and greater photo stability [67].

Among such core-shell heterostructures, zinc oxide–based materials have been widely investigated owing to their advantageous features such as ease of synthesis, biocompatibility, enhanced catalytic efficiency and greater stability. Here, a few such fascinating reports have been briefly discussed highlighting their characteristic features and applications. Zeng *et al.* reported the synthesis of Zn/ZnO core/shell nanoparticles by laser ablation in a liquid medium for temperature-dependent blue photoluminescence properties having potential applications in optical and electronic devices [68]. For similar applications, Guidelli *et al.* made ZnO/Ag and ZnO/Au core-shell nanoparticles exhibiting strongly enhanced near-band-edge UV emission from the ZnO when excited at 325 nm [69]. They have observed that the enhancement is linearly proportional to the laser power under photoluminescence conditions and also increases with volume of nanoparticles before approaching an apparent saturation value. In another interesting report by Haldar *et al.*, gold-ZnO core-shell heterostructures have been prepared and used for the investigation of highly efficient fluorescence resonance energy transfer from Rhodamine 6G dye to Au@ZnO nanoparticle by steady-state and time-resolved spectroscopy [70]. A visible light–responsive, ZnO-based core-shell photocatalyst was reported by Kumar *et al.*, and they have employed graphitic carbon nitride with nitrogen-doped ZnO to design cost-effective, stable and novel material via an eco-friendly route. These organic–inorganic hybrid nanoplates have been used to degrade an organic dye under visible light with a degradation efficiency of more than 90% [71]. Zhai *et al.* have reported the synthesis of a hetero-structure having zinc oxide core with silica shell for photocatalytic degradation of Rhodamine B dye in water under acidic and alkaline conditions with very high efficiency [72]. Photocatalysis by ZnO@ZnS core-shell structures has been reported by Sadollahkhani *et al.* In their study, zinc oxide nanoparticles were prepared via the co-precipitation method, and using a simple chemical route, zinc sulphide shell was developed over ZnO nanoparticles [73]. They have studied photocatalytic decomposition of Rose Bengal dye at different pH values. There are plenty of reports in the litera-ture where core-shell heterostructures of ZnO with various other oxides, metals and sulphides have been successfully used as an efficient photocatalyst material [68–73].

Composites of ZnO with Other Metal Nanoparticles

ZnO rapidly forms composites with a wide variety of materials. Among them, composites of zinc oxide with different types of metals have been widely investigated in the literature for photocata-lytic applications. In order to enhance the photocatalytic performance of zinc oxide nanoparticles, one of the most promising strategies is to dope with noble metals. In such a study by Guy *et al.* [74], Au, Ag and Pd noble metals were separately doped on ZnO nanophotocatalysts by the borohydride reduction method, and their photocatalytic performance was investigated by the degradation of Congo red dye under ultraviolet radiation. The results have shown that among these three metals, Pd doping exhibited the best photocatalytic efficiency.

In another report by Liu and his co-workers [75], a two-step chemical method was employed for photocatalytic decontamination of water by synthesising worm-like morphology of silver/ZnO heterostructure. They have found that these novel composites are better photocatalysts than bare zinc oxide because of their greater light absorption. Zhang *et al.* also reported a similar silver/ZnO composite for photocatalytic applications [76]. Silver-decorated zinc oxide nanostructures were obtained by hydrothermal route, and these heterostructures were found to be very effective in photocatalytic degradation of Rhodamine B dye.

Apart from silver, gold and palladium, platinum has also been used to design new catalysts with zinc oxide. In a very interesting report by Yu *et al.* [77], hollow-structured Pt-ZnO microspheres have been engineered using mild solvothermal conditions using zinc acetate and $HPtCl_4$ as the precursors, and polyethylene glycol and ethylene glycol as the reducing agent and solvent, respectively. Photocatalytic inspection of the heterostructured microspheres for the breakdown of the dye acid orange II unveiled enormously high catalytic performance and solidity than that of pure zinc oxide and corresponding metal-doped ZnO. The extraordinary photocatalytic activity of hollow microspheres is attributed to their exceptional nanostructure and the reduced recombination rate exciton pairs by the Pt implanted on ZnO nanoparticles. Nanocrystalline gold-decorated zinc oxide nanocomposites have also been studied in the literature for visible light–assisted decomposition of methyl orange dye in water by Wu and his co-workers [78]. Here, initially zinc oxide nanorods and films were prepared using a chemical vapour deposition (CVD) technique. Later, nanocrystalline gold was deposited photo-catalytically on them using $HAuCl_4$/ethanol solutions under 365 nm irradiation. They have studied the effect of shell thickness on catalytic efficiency and found that at smaller shell thickness, the best results were obtained. Kanwal *et al.* have explored the potential of Ag-ZnO composites for total mineralisation of imidacloprid (a pesticide) with the aim to sustain the pollutant-free safe water supply [79]. The composites were prepared by the hydrothermal method and used for the degradation of imidacloprid under UV light while optimising the process parameters such as time of irradiation, pH of medium, pesticide concentration and composite loading. The results of the study revealed an increase in photodegradation of imidacloprid by Ag-ZnO composites than pure ZnO.

COMPOSITES WITH OTHER METAL OXIDE NANOPARTICLES

Liu *et al.*'s study shows that Bi_2WO_6 crystal seeds change the ratio of ZnO polar facets. Bi_2WO_6 controls the nucleation and growth rate of ZnO by changing its shape from nanorods to nanosheets. Higher proportion of Zn- terminated (0001) and O terminated (000$\bar{1}$) polar facets enhances the photocatalytic activity of ZnO [80]. Vo *et al.* established a heterojunction of p-type CuO semiconductor and n-type ZnO semiconductor by a one-step green chemical approach using rosin as a primary agent to fabricate a photocatalyst for removal of organic pollutants, i.e., phenol and methyl orange and nitrogen oxide (NO) gas under a visible light excitation [81].

Dhanalakshmi *et al.* synthesised $PrFeO_3$ (PFO), ZnO, and $PrFeO_3$/ZnO (PFO/ZnO) p-n heterostructure nanoparticles in situ by the hydrothermal method and demonstrated the effect of magnetic field on the simultaneous removal of phenol red and inorganic Cr (VI). They observed 99.6% and 96.4% removal of phenol red and inorganic Cr (VI) by PFO/ZnO nanocomposites by imposing 0.5T of external magnetic field [82]. Shen *et al.* developed magnetically separable Cu/ZnO/CoFe–CLDH composite by the Taguchi approach by hydrothermal coupling of Cu-doped ZnO and calcined CoFe–LDH. They integrated the visible light photocatalysis with persulfate activation to degrade bisphenol A (BPA). It is showing great optoelectronic, physicochemical and magnetic properties and good catalytic performance [83]. Lu *et al.* developed a series of ZnO-CoO$_x$-CeO$_2$ nanocomposites by the impregnation method which showed good adsorption activity and photocatalytic oxidation of Congo red dye [84]. Singh *et al.* synthesised undoped and co-doped ZnO nanoparticles by the co-precipitation method and investigated their photocatalytic properties towards reactive brown dye (RB-1) and antimicrobial potency towards two bacterial strains, respectively [85]. Mylarappa *et al.* prepared clay-doped ZnO composites by the solution combustion technique using citric acid as fuel and complexing agent, and investigated its ability for photocatalysis of malachite green dye in UV-induced environment. The synthesised compound has shown 90% degradation rate. Apart from that, clay/ZnO nanocomposite has shown electrochemical sensing of ascorbic acid, resorcinol and uric acid in 0.1 M KOH electrolyte [86].

Graphene-Based ZnO Nanoparticles

Busarello *et al.* have prepared nanocomposites between ZnO quantum dots (QDs) and graphene oxide (GO) by the sol-gel method in isopropanol with investigation of the effect of water addition over the size of QDs and photoluminescence. They evaluated the photocatalytic activity of these ZnO-based materials towards degradation of methylene blue (MB) and astrazon blue (AB) in isopropanol and water using fluorescence and Hg vapour lamps [87]. In situ generation of ZnO nanoparticles on the graphene surface using zinc benzoate dihydrazinate complex at 200°C was reported by Thangavelu *et al.*, which showed photodegradation of methylene blue in ethanol and electrochemical sensing of glucose [88]. Malekshoar *et al.* synthesised graphene-based titanium dioxide and zinc oxide composites by the hydrothermal process. The presence of graphene enhanced the photocatalytic degradation of phenol up to 30% [89].

ZnO Nanoparticles Synthesised via Green Route

Among the various methods to prepare nanoparticles, such as hydrothermal, microemulsion, sol-gel, co-precipitation, microwave, pulsed LASER, ultrasonication and CVD, the most lucrative one in the current scenario is the biosynthesis or green synthesis, which utilises plant resources, natural products, algae, fungi, bacteria or other biological precursors to obtain nanostructured materials without employing expensive and toxic chemicals, thereby making this route highly promising and sustainable considering the unprecedented rise in the global pollution level [90]. In this section, we have briefly summarised the recent progress in the domain of ZnO synthesis using green routes for photocatalytic applications.

Owing to the presence of different types of phytochemicals in plant extracts which can act as the reducing as well as capping agent in the formation of nanoparticles, there has been a tremendous growth in the number of articles reporting the use of plant resources for nanosynthesis. Osuntokun *et al.* have reported the production of ZnO nanoparticles having hexagonal phase and size in the range of 15–20 nm employing the aqueous broccoli extract [91]. These nanoparticles demonstrated greater than 70% degradation of methylene blue and phenol red dye under UV radiation. In another similar report by Aldeen *et al.* [92], photocatalyst ZnO nanoparticles with wurtzite structure and capable of degrading methylene blue dye were fruitfully produced utilising the extract of *Phoenix roebelenii* leaves. The extract of *Syzygium cumini* plant leaves was used for the green synthesis of zinc oxide nanoparticles having size around 12 nm by Sadiq *et al.* [93]. Synthesised photocatalysts exhibited 94% degradation of methylene blue dye in water.

The usage of microbes to make zinc oxide is by all accounts environmentally benign since no toxic or perilous compounds are employed in the synthesis, which makes this procedure more helpful in comparison to conventional techniques. Furthermore, fungi-mediated methods are mainly fascinating due to greater content of bioactive complexes in fungal species than other microorganisms. Ameen *et al.* [94] have reported the use of *Acremonium potronii*, a new fungal species found in fruits, soil, and marine environments, for the eco-friendly preparation of ZnO nanoparticles with sizes ranging between 13 and 15 nm. They have used these nanoparticles for photocatalytic decontamination of wastewater with a degradation efficiency of 93%.

MISCELLANEOUS

Akbari *et al.* synthesised ZnO@vanadium carbide for photocatalytic degradation of an antibiotic, i.e., sulfamethoxazole (SMX) [95]. Ranjith *et al.* have reported crosslinked electrospun polycyclodextrin (Poly-CD) nanofiber (NF) membrane coated with ZnO nanograins by atomic layer deposition (ALD) for the removal of Rhodamine B (RhB), methylene blue (MB), methyl orange (MO) and tetracycline (TC) from wastewater [96].

CONCLUSION

This chapter discussed about the various types of ZnO nanostructured materials for photocatalytic applications. Synthesis of ZnO nanostructured material through various ways using capping agents, surface directing agents, core-shell nanostructures and natural products for green synthesis is discussed here because of their importance in determining the photocatalytic properties of ZnO nanostructures. Apart from that, the role of graphene in the morphology and photocatalytic properties is also discussed. It was found that composites with metals and metal oxides such as silver, platinum, gold, palladium, Bi_2WO_6, CeO_2, clay and layered double hydroxides have remarkable impact over the catalytic properties of ZnO nanomaterials. This chapter covered the major aspects of the synthesis of ZnO nanomaterials for photocatalysis.

REFERENCES

[1] Logothetidis, S. "Nanotechnology: Principles and Applications." In *Nanostructured Materials and Their Applications*, pp. 1–22. Berlin, Heidelberg: Springer, 2011. https://doi.org/10.1007/978-3-642-222 27-6_1.

[2] De Jong, Wim H., and Paul JA Borm. "Drug Delivery and Nanoparticles: Applications and Hazards." *International Journal of Nanomedicine* 3, no. 2 (2008): 133–49. https://doi.org/10.2147/ijn.s596.

[3] Wang, Zhong Lin ZL. "Dr. Zhong Lin Wang gave an Invited Talk at Fall MRS Conference (Nov. 29, 2006). This is the First Introduction of Nano-PiezotronicsTM as a Field of Research in Academics. Here is the Abstract." *Science* 312 (2006): 242–46. https://doi.org/10.1126/science.1124005.

[4] Chen, Xiaobo, Shaohua Shen, Liejin Guo, and Samuel S. Mao. "Semiconductor-based Photocatalytic Hydrogen Generation." *Chemical Reviews* 110, no. 11 (2010): 6503–570. https://doi.org/10.1021/cr1001645.

[5] Hoffmann, Michael R., Scot T. Martin, Wonyong Choi, and Detlef W. Bahnemann. "Environmental Applications of Semiconductor Photocatalysis." *Chemical Reviews* 95, no. 1 (1995): 69–96. https://doi.org/10.1021/cr00033a004.

[6] Vinu, R., and Giridhar Madras. "Environmental Remediation by Photocatalysis." *Journal of the Indian Institute of Science* 90, no. 2 (2010): 189–230.

[7] Kabra, Kavita, Rubina Chaudhary, and Rameshwar L. Sawhney. "Treatment of Hazardous Organic and Inorganic Compounds through Aqueous-Phase Photocatalysis: A Review." *Industrial & Engineering Chemistry Research* 43, no. 24 (2004): 7683–96. https://doi.org/10.1021/ie0498551.

[8] Jiang, Lijuan, Yajun Wang, and Changgen Feng. "Application of Photocatalytic Technology in Environmental Safety." *Procedia Engineering* 45 (2012): 993–97. https://doi.org/10.1016/j.proeng.2012.08.271.

[9] Gamage, Joanne, and Zisheng Zhang. "Applications of Photocatalytic Disinfection." *International Journal of Photoenergy* 2010 (2010).https://doi.org/10.1155/2010/764870.

[10] Chaturvedi, Shalini, and Pragnesh N. Dave. "Environmental Application of Photocatalysis." In *Materials Science Forum*, vol. 734, pp. 273–94. Trans Tech Publications Ltd, 2013. https://doi.org/10.4028/www.scientific.net/msf.734.273.

[11] Włodarczyk, Elżbieta, and Paweł K. Zarzycki. "Chromatographic Behavior of Selected Dyes on Silica and Cellulose Micro-TLC Plates: Potential Application as Target Substances for Extraction, Chromatographic, and/or Microfluidic Systems." *Journal of Liquid Chromatography & Related Technologies* 40, no. 5–6 (2017): 259–81. https://doi.org/10.1080/10826076.2017.1298028.

[12] Dixit, Sumita, Subhash K. Khanna, and Mukul Das. "All India Survey for Analyses of Colors in Sweets and Savories: Exposure Risk in Indian Population." *Journal of Food Science* 78, no. 4 (2013): T642–47. https://doi.org/10.1111/1750-3841.12068.

[13] Masten, Susan J., and Simon HR Davies. "The Use of Ozonation to Degrade Organic Contaminants in Wastewaters." *Environmental Science & Technology* 28, no. 4 (1994): 180A–185A. https://doi.org/10.1021/es00053a718.

[14] Fu, Hongbo, Chengshi Pan, Wenqing Yao, and Yongfa Zhu. "Visible-Light-Induced Degradation of Rhodamine B by Nanosized Bi_2WO_6." *The Journal of Physical Chemistry B* 109, no. 47 (2005): 22432–9. https://doi.org/10.1021/jp052995j

[15] Horikoshi, Satoshi, Fukuyo Hojo, Hisao Hidaka, and Nick Serpone. "Environmental Remediation by an Integrated Microwave/UV Illumination Technique. 8. Fate of Carboxylic Acids, Aldehydes, Alkoxycarbonyl and Phenolic Substrates in a Microwave Radiation Field in the Presence of TiO_2 Particles under UV Irradiation." *Environmental Science & Technology* 38, no. 7 (2004): 2198–208. https://doi.org/10.1021/es034823a.

[16] Natarajan, Subramanian, Hari C. Bajaj, and Rajesh J. Tayade. "Recent Advances based on the Synergetic Effect of Adsorption for Removal of Dyes from Waste Water Using Photocatalytic Process." *Journal of Environmental Sciences* 65 (2018): 201–22. https://doi.org/10.1016/j.jes.2017.03.011.

[17] Satyawali, Y., and M. Balakrishnan. "Wastewater Treatment in Molasses-Based Alcohol Distilleries for COD and Color Removal: A Review." *Journal of Environmental Management* 86, no. 3 (2008): 481–97. https://doi.org/10.1016/j.jenvman.2006.12.024.

[18] Ambatkar Mugdha, and Mukundan Usha, Enzymatic Treatment of Wastewater Containing Dyestuffs Using Different Delivery Systems, *Scientific Reviews and Chemical Communications* 2 (2012): 31–40.

[19] Geethakarthi, A., and B. R. Phanikumar. "Adsorption of Reactive Dyes from Aqueous Solutions by Tannery Sludge Developed Activated Carbon: Kinetic and Equilibrium Studies." *International Journal of Environmental Science & Technology* 8, no. 3 (2011): 561–70. https://doi.org/10.1007/bf03326242.

[20] S. A. Abo-Farha. "Comparative Study of Oxidation of Some Azo Dyes by Different Advanced Oxidation Processes: Fenton, Fenton-Like, Photo-Fenton and Photo-Fenton-Like." *Journal of American Science* 6 (2010): 128–42.

[21] Tchobanoglous, George, H. David Stensel, Franklin Louis Burton. *Wastewater Engineering: Treatment and Resource Recovery*. McGraw-Hill, New York, 2014. ISBN: 1259010791.

[22] Sano, Larissa Lubomudrov. *Aquatic Toxicity, Ecological Risks, and Risk Tradeoff Analysis of Biocide Treatment for Unballasted Vessels*. University of Michigan, Michigan 2005.

[23] Karia, G. L., and R.A. Christian. *Wastewater Treatment*. PHI Learning Pvt. Ltd., Delhi, 2013. ISBN: 8120347358.

[24] Stackelberg, Paul E., Edward T. Furlong, Michael T. Meyer, Steven D. Zaugg, Alden K. Henderson, and Dori B. Reissman. "Persistence of Pharmaceutical Compounds and Other Organic Wastewater Contaminants in a Conventional Drinking-Water-Treatment Plant." *Science of the Total Environment* 329, no. 1–3 (2004): 99–113. https://doi.org/10.1016/j.scitotenv.2004.03.015.

[25] Turchi, C. "Photocatalytic Degradation of Organic Water Contaminants: Mechanisms Involving Hydroxyl Radical Attack." *Journal of Catalysis* 122, no. 1 (1990): 178–92. https://doi.org/10.1016/0021-9517(90)90269-p.

[26] (A) Fox, Marye Anne, and Maria T. Dulay. "Heterogeneous Photocatalysis." *Chemical Reviews* 93, no. 1 (1993): 341–57. https://doi.org/10.1021/cr00017a016; (B) Nosaka, Yoshio, and Atsuko Y. Nosaka. "Generation and Detection of Reactive Oxygen Species in Photocatalysis." *Chemical Reviews* 117, no. 17 (2017): 11302–36. https://doi.org/10.1021/acs.chemrev.7b00161.

[27] Khanchandani, Sunita, Simanta Kundu, Amitava Patra, and Ashok K. Ganguli. "Band Gap Tuning of ZnO/In_2S_3 Core/Shell Nanorod Arrays for Enhanced Visible-Light-Driven Photocatalysis." *The Journal of Physical Chemistry C* 117, no. 11 (2013): 5558–67. https://doi.org/10.1021/jp310495j.

[28] Kumar, Sandeep, Sunita Khanchandani, Meganathan Thirumal, and Ashok K. Ganguli. "Achieving Enhanced Visible-Light-Driven Photocatalysis Using Type-II $NaNbO_3/CdS$ Core/Shell Heterostructures." *ACS Applied Materials & Interfaces* 6, no. 15 (2014): 13221–33. https://doi.org/10.1021/am503055n.

[29] Geys, Jorina, Abderrahim Nemmar, Erik Verbeken, Erik Smolders, Monica Ratoi, Marc F. Hoylaerts, Benoit Nemery, and Peter H.M. Hoet. "Acute Toxicity and Prothrombotic Effects of Quantum Dots: Impact of Surface Charge." *Environmental Health Perspectives* 116, no. 12 (2008): 1607–13. https://doi.org/10.1289/ehp.11566.

[30] Hardman, Ron. "A Toxicologic Review of Quantum Dots: Toxicity Depends on Physicochemical and Environmental factors." *Environmental Health Perspectives* 114, no. 2 (2006): 165–72. https://doi.org/10.1289/ehp.8284.

[31] Janotti, Anderson, and Chris G Van de Walle. "Fundamentals of Zinc Oxide as a Semiconductor." *Reports on Progress in Physics* 72, no. 12 (2009): 126501. https://doi.org/10.1088/0034-4885/72/12/126501.

[32] Shim, Moonsub, and Philippe Guyot-Sionnest. "Organic-Capped ZnO Nanocrystals: Synthesis and n-Type Character." *Journal of the American Chemical Society* 123, no. 47 (2001): 11651–54. https://doi.org/10.1021/ja0163321.

[33] Hoffmann, Michael R., Scot T. Martin, Wonyong Choi, and Detlef W. Bahnemann. "Environmental Applications of Semiconductor Photocatalysis." *Chemical Reviews* 95, no. 1 (1995): 69–96. https://doi.org/10.1021/cr00033a004.

[34] Zhang, Yin, T. R Nayak, Hao Hong, and Weibo Cai. "Biomedical Applications of Zinc Oxide Nanomaterials." *Current Molecular Medicine* 13, no. 10 (2013): 1633–45. https://doi.org/10.2174/1566524013666131111130058.

[35] Lee, Kian Mun, Chin Wei Lai, Koh Sing Ngai, and Joon Ching Juan. "Recent Developments of Zinc Oxide Based Photocatalyst in Water Treatment Technology: A Review." *Water Research* 88 (2016): 428–48. https://doi.org/10.1016/j.watres.2015.09.045.

[36] Özgür, Ümit, Ya. I. Alivov, Chunli Liu, Ali Teke, Michael A. Reshchikov, S. Doğan, V. Avrutin, S.-J. Cho, and H. Morkoç. "A Comprehensive Review of ZnO Materials and Devices." *Journal of Applied Physics* 98, no. 4 (2005). https://doi.org/10.1063/1.1992666.

[37] Sounart, Thomas L., Jun Liu, James A. Voigt, Mae Huo, Erik D. Spoerke, and Bonnie McKenzie. "Secondary Nucleation and Growth of ZnO." *Journal of the American Chemical Society* 129, no. 51 (2007): 15786–93. https://doi.org/10.1021/ja071209g.

[38] Spanhel, Lubomir, and Marc A. Anderson. "Semiconductor Clusters in the Sol-Gel Process: Quantized Aggregation, Gelation, and Crystal Growth in Concentrated Zinc Oxide Colloids." *Journal of the American Chemical Society* 113, no. 8 (1991): 2826–33. https://doi.org/10.1021/ja00008a004.

[39] Zhang, Luyuan, Longwei Yin, Chengxiang Wang, Ning Lun, and Yongxin Qi. "Sol–Gel Growth of Hexagonal Faceted ZnO Prism Quantum Dots with Polar Surfaces for Enhanced Photocatalytic Activity." *ACS Applied Materials & Interfaces* 2, no. 6 (2010): 1769–73. https://doi.org/10.1021/am100274d.

[40] Zhang, Tierui, Wenjun Dong, Mary Keeter-Brewer, Sanjit Konar, Roland N. Njabon, and Z. Ryan Tian. "Site-Specific Nucleation and Growth Kinetics in Hierarchical Nanosyntheses of Branched ZnO Crystallites." *Journal of the American Chemical Society* 128, no. 33 (2006): 10960–68. https://doi.org/10.1021/ja0631596.

[41] Liu, Bin, and Hua Chun Zeng. "Hydrothermal Synthesis of ZnO Nanorods in the Diameter Regime of 50 Nm." *Journal of the American Chemical Society* 125, no. 15 (2003): 4430–31. https://doi.org/10.1021/ja0299452.

[42] Zeng, Haibo, Weiping Cai, Peisheng Liu, Xiaoxia Xu, Huijuan Zhou, Claus Klingshirn, and Heinz Kalt. "ZnO-Based Hollow Nanoparticles by Selective Etching: Elimination and Reconstruction of Metal–Semiconductor Interface, Improvement of Blue Emission and Photocatalysis." *ACS Nano* 2, no. 8 (2008): 1661–70. https://doi.org/10.1021/nn800353q.

[43] Gao, X. P., Z. F. Zheng, H. Y. Zhu, G. L. Pan, J. L. Bao, F. Wu, and D. Y. Song. "Rotor-like ZnO by Epitaxial Growth under Hydrothermal ConditionsElectronic Supplementary Information (ESI) Available: TEM Images of the Rod-like and Rotor-like ZnO in the Bright Field and Dark Field. See https://www.rsc.org/suppdata/cc/b4/b403252g/." *Chemical Communications* 12 (2004): 1428. https://doi.org/10.1039/b403252g.

[44] Bilgili, Ecevit, Rhye Hamey, and Brian Scarlett. "Nano-Milling of Pigment Agglomerates Using a Wet Stirred Media Mill: Elucidation of the Kinetics and Breakage Mechanisms." *Chemical Engineering Science* 61, no. 1 (2006): 149–57. https://doi.org/10.1016/j.ces.2004.11.063.

[45] Wong, Eva M., Paul G. Hoertz, Cindy J. Liang, Bai-Ming Shi, Gerald J. Meyer, and Peter C. Searson. "Influence of Organic Capping Ligands on the Growth Kinetics of ZnO Nanoparticles." *Langmuir* 17, no. 26 (2001): 8362–67. https://doi.org/10.1021/la010944h.

[46] Schneider, Jenny, Masaya Matsuoka, Masato Takeuchi, Jinlong Zhang, Yu Horiuchi, Masakazu Anpo, and Detlef W. Bahnemann. "Understanding TiO$_2$ Photocatalysis: Mechanisms and Materials." *Chemical Reviews* 114, 19 (2014): 9919–86. https://doi.org/10.1021/cr5001892.

[47] Mourdikoudis, Stefanos, and Luis M. Liz-Marzán. "Oleylamine in Nanoparticle Synthesis." *Chemistry of Materials* 25, no. 9 (2013): 1465–76. https://doi.org/10.1021/cm4000476.

[48] Comparelli, Roberto, Elisabetta Fanizza, M. L. Curri, Pantaleo Davide Cozzoli, G. Mascolo, and A. Agostiano. "UV-Induced Photocatalytic Degradation Of Azo Dyes By Organic-Capped ZnO Nanocrystals Immobilized Onto Substrates." *Applied Catalysis B: Environmental* 60, no. 1–2 (2005): 1–11. https://doi.org/10.1016/j.apcatb.2005.02.013.

[49] Sharma, Hemant, Navneet Kaur, Thangarasu Pandiyan, and Narinder Singh. "Surface Decoration of ZnO Nanoparticles: A New Strategy to Fine Tune the Recognition Properties of Imine Linked Receptor." *Sensors and Actuators B: Chemical* 166–167 (2012): 467–72. https://doi.org/10.1016/j.snb.2012.01.076.

[50] Medeiros Borsagli, Fernanda G.L., and Aislan E. Paiva. "Eco-Friendly Luminescent ZnO Nanoconjugates with Thiol Group for Potential Environmental Photocatalytic Activity." *Journal of Environmental Chemical Engineering* 9, no. 4 (August 2021): 105491. https://doi.org/10.1016/j.jece.2021.105491.

[51] Füldner, Stefan, Tatiana Mitkina, Tobias Trottmann, Alexandra Frimberger, Michael Gruber, and Burkhard König. "Urea Derivatives Enhance the Photocatalytic Activity of Dye-Modified Titanium Dioxide." *Photochemical & Photobiological Sciences* 10, no. 4 (2011): 623–25. https://doi.org/10.1039/c0pp00374c.

[52] Gui, Zhou, Jian Liu, Zhengzhou Wang, Lei Song, Yuan Hu, Weicheng Fan, and Daoyong Chen. "From Muticomponent Precursor to Nanoparticle Nanoribbons of ZnO." *The Journal of Physical Chemistry B* 109, no. 3 (2005): 1113–17. https://doi.org/10.1021/jp047088d.

[53] Jing, Liqiang, Wei Zhou, Guohui Tian, and Honggang Fu. "Surface Tuning for Oxide-Based Nanomaterials as Efficient Photocatalysts." *Chemical Society Reviews* 42, no. 24 (2013): 9509. https://doi.org/10.1039/c3cs60176e.

[54] Tian, Zhengrong R., James A. Voigt, Jun Liu, Bonnie Mckenzie, and Matthew J. Mcdermott. "Biomimetic Arrays of Oriented Helical ZnO Nanorods and Columns." *Journal of the American Chemical Society* 124, no. 44 (2002): 12954–55. https://doi.org/10.1021/ja0279545.

[55] Yin, Yadong, and A. Paul Alivisatos. "Colloidal Nanocrystal Synthesis and the Organic-Inorganic Interface." *Nature* 437, no. 7059 (2004): 664–70. https://doi.org/10.1038/nature04165.

[56] Sun, Hongqi, Yuan Bai, Youping Cheng, Wanqin Jin, and Nanping Xu. "Preparation and Characterization of Visible-Light-Driven Carbon–Sulfur-Codoped TiO$_2$ Photocatalysts." *Industrial & Engineering Chemistry Research* 45, no. 14 (2006): 4971–76. https://doi.org/10.1021/ie060350f.

[57] Sharma, Hemant, Ajnesh Singh, Navneet Kaur, and Narinder Singh. "ZnO-Based Imine-Linked Coupled Biocompatible Chemosensor for Nanomolar Detection of Co^{2+}." *ACS Sustainable Chemistry & Engineering* 1, no. 12 (2013): 1600–1608. https://doi.org/10.1021/sc400250s.

[58] Mishra, Jayanti, Navneet Kaur, and Ashok K Ganguli. "Morphological Changes of ZnO Nanoparticles, Directed by Urea/Thiourea-Based Tripodal Organic Ligands and Their Photocatalytic Properties." *Bulletin of Materials Science* 46, no. 2 (2023). https://doi.org/10.1007/s12034-023-02914-6.

[59] Mishra, Jayanti, Gagandeep Singh, Navneet Kaur, and Ashok K. Ganguli. "Role of Linker Molecules on Morphology of Tripodal Ligands Based Functionalized ZnO Nanoparticles and Its Effect on Photocatalysis." *Inorganic Chemistry Communications* 148 (2023): 110333. https://doi.org/10.1016/j.inoche.2022.110333.

[60] Mishra, Jayanti, Randeep Kaur, Navneet Kaur, and Ashok K. Ganguli. "Surface Functionalization of ZnO Nanoparticles by Schiff Base Oriented Tetrapodal Ligands at Room Temperature and Their Photocatalytic Applications: A Comparative Account on Effect of Structure Directing Agent on Photocatalysis." *Inorganic Chemistry Communications* 146 (2022): 110130. https://doi.org/10.1016/j.inoche.2022.110130.

[61] Mishra, Jayanti, Menaka Jha, Navneet Kaur, and Ashok K. Ganguli. "Room Temperature Synthesis of Urea Based Imidazole Functionalised ZnO Nanorods and Their Photocatalytic Application." *Materials Research Bulletin* 102 (2018): 311–18. https://doi.org/10.1016/j.materresbull.2018.02.045.

[62] Biswas, Kanishka, Barun Das, and C. N. R. Rao. "Growth Kinetics of ZnO Nanorods: Capping-Dependent Mechanism and Other Interesting Features." *The Journal of Physical Chemistry C* 112, no. 7 (2008): 2404–11. https://doi.org/10.1021/jp077506p.

[63] Zobel, Mirijam, Haimantee Chatterjee, Galina Matveeva, Ute Kolb, and Reinhard B. Neder. "Room-Temperature Sol-Gel Synthesis of Organic Ligand-Capped ZnO Nanoparticles." *Journal of Nanoparticle Research* 17, no. 5 (2015). https://doi.org/10.1007/s11051-015-3006-5.

[64] Liu, Lan, Lei Fu, Yong Liu, Yaling Liu, Peng Jiang, Shaoqin Liu, Mingyuan Gao, and Zhiyong Tang. "Bioinspired Synthesis of Vertically Aligned ZnO Nanorod Arrays: Toward Greener Chemistry." *Crystal Growth & Design* 9, no. 11 (2009): 4793–96. https://doi.org/10.1021/cg900634t.

[65] Jung, Seung-Ho, Eugene Oh, Kun-Hong Lee, Yosep Yang, Chan Gyung Park, Wanjun Park, and Soo-Hwan Jeong. "Sonochemical Preparation of Shape-Selective ZnO Nanostructures." *Crystal Growth & Design* 8, no. 1 (2007): 265–69. https://doi.org/10.1021/cg070296l..

[66] Das, Sonali, Javier Pérez-Ramírez, Jinlong Gong, Nikita Dewangan, Kus Hidajat, Bruce C. Gates, and Sibudjing Kawi. "Core-Shell Structured Catalysts for Thermocatalytic, Photocatalytic, and Electrocatalytic Conversion of CO_2." *Chemical Society Reviews* 49, no. 10 (2020): 2937–3004. https://doi.org/10.1039/c9cs00713j.

[67] Wang, Xuewen, Gang Liu, Gao Qing Lu, and Hui-Ming Cheng. "Stable Photocatalytic Hydrogen Evolution from Water over ZnO-CdS Core-Shell Nanorods." *International Journal of Hydrogen Energy* 35, no. 15 (2010): 8199–8205. https://doi.org/10.1016/j.ijhydene.2009.12.091.

[68] Zeng, Haibo, Zhigang Li, Weiping Cai, Bingqiang Cao, Peisheng Liu, and Shikuan Yang. "Microstructure Control of Zn/ZnO Core/Shell Nanoparticles and Their Temperature-Dependent Blue Emissions." *The Journal of Physical Chemistry B* 111, no. 51 (2007): 14311–17. https://doi.org/10.1021/jp0770413.

[69] Guidelli, E. J., O. Baffa, and D. R. Clarke. "Enhanced UV Emission From Silver/ZnO And Gold/ZnO Core-Shell Nanoparticles: Photoluminescence, Radioluminescence, And Optically Stimulated Luminescence." *Scientific Reports* 5, no. 1 (2015). https://doi.org/10.1038/srep14004.

[70] Haldar, Krishna Kanta, Tapasi Sen, and Amitava Patra. "Au@ZnO Core–Shell Nanoparticles Are Efficient Energy Acceptors with Organic Dye Donors." *The Journal of Physical Chemistry C* 112, no. 31 (2008): 11650–56. https://doi.org/10.1021/jp8031308.

[71] Kumar, Santosh, Arabinda Baruah, Surendar Tonda, Bharat Kumar, Vishnu Shanker, and B. Sreedhar. "Cost-Effective and Eco-Friendly Synthesis of Novel and Stable N-Doped ZnO/g-C_3N_4 Core-Shell Nanoplates with Excellent Visible-Light Responsive Photocatalysis." *Nanoscale* 6, no. 9 (2014): 4830. https://doi.org/10.1039/c3nr05271k.

[72] Zhai, Jing, Xia Tao, Yuan Pu, Xiao-Fei Zeng, and Jian-Feng Chen. "Core/Shell Structured ZnO/SiO_2 Nanoparticles: Preparation, Characterization and Photocatalytic Property." *Applied Surface Science* 257, no. 2 (2010): 393–97. https://doi.org/10.1016/j.apsusc.2010.06.091.

[73] Sadollahkhani, Azar, Iraj Kazeminezhad, Jun Lu, Omer Nur, Lars Hultman, and Magnus Willander. "Synthesis, Structural Characterization and Photocatalytic Application of ZnO@ZnS Core-Shell Nanoparticles." *RSC Advances* 4, no. 70 (2014): 36940–50. https://doi.org/10.1039/c4ra05247a.

[74] Güy, Nuray, and Mahmut Özacar. "The Influence of Noble Metals on Photocatalytic Activity of ZnO for Congo Red Degradation." *International Journal of Hydrogen Energy* 41, 44 (2016): 20100–112. https://doi.org/10.1016/j.ijhydene.2016.07.063.

[75] Liu, H. R., G. X. Shao, J. F. Zhao, Z. X. Zhang, Y. Zhang, J. Liang, X. G. Liu, H. S. Jia, and B. S. Xu. "Worm-Like Ag/ZnO Core-Shell Heterostructural Composites: Fabrication, Characterization, and Photocatalysis." *The Journal of Physical Chemistry C* 116, no. 30 (2012): 16182–90. https://doi.org/10.1021/jp2115143.

[76] Zhang, Yunyan, and Jin Mu. "One-Pot Synthesis, Photoluminescence, and Photocatalysis of Ag/ZnO Composites." *Journal of Colloid and Interface Science* 309, no. 2 (2007): 478–84. https://doi.org/10.1016/j.jcis.2007.01.011.

[77] Yu, Changlin, Kai Yang, Yu Xie, Qizhe Fan, Jimmy C. Yu, Qing Shu, and Chunying Wang. "Novel Hollow Pt-ZnO Nanocomposite Microspheres with Hierarchical Structure and Enhanced Photocatalytic Activity and Stability." *Nanoscale* 5, no. 5 (2013): 2142. https://doi.org/10.1039/c2nr33595f.

[78] Wu, Jih-Jen, and Chan-Hao Tseng. "Photocatalytic Properties of Nc-Au/ZnO Nanorod Composites." *Applied Catalysis B: Environmental* 66, no. 1–2 (2006): 51–57. https://doi.org/10.1016/j.apcatb.2006.02.013.

[79] Kanwal, Mahwish, Saadia Rashid Tariq, and Ghayoor Abbas Chotana. "Photocatalytic Degradation of Imidacloprid by Ag-ZnO Composite." *Environmental Science and Pollution Research* 25, no. 27 (2018): 27307–20. https://doi.org/10.1007/s11356-018-2693-8.

[80] Huang, Mianli, Yu Yan, Wenhui Feng, Sunxian Weng, Zuyang Zheng, Xianzhi Fu, and Ping Liu. "Controllable Tuning Various Ratios of ZnO Polar Facets by Crystal Seed-Assisted Growth and Their Photocatalytic Activity." *Crystal Growth & Design* 14, no. 5 (2014): 2179–86. https://doi.org/10.1021/cg401676r.

[81] Vo, Nhu Thi Thu, Sheng-Jie You, Minh-Thuan Pham, and Viet Van Pham. "A Green Synthesis Approach of P-n CuO/ZnO Junctions for Multifunctional Photocatalysis towards the Degradation of Contaminants." *Environmental Technology & Innovation* 32 (2023): 103285. https://doi.org/10.1016/j.eti.2023.103285.

[82] Dhanalakshmi, Radhalayam, Juliano C. Denardin, T.V.M. Sreekanth, Maddaka Reddeppa, P. Rosaiah, Kisoo Yoo, and Jonghoon Kim. "Synergistic Magnetic Field-Photocatalysis for the Concurrent Removal of Inorganic and Organic Contaminants over PrFeO3/ZnO Heterostructures." *Inorganic Chemistry Communications* 156 (2023): 111189. https://doi.org/10.1016/j.inoche.2023.111189.

[83] Shen, Jyunhong, Antong Shi, Jiahui Lu, Xiangtao Lu, Hongyu Zhang, and Zhuwu Jiang. "Optimized Fabrication of Cu-Doped ZnO/Calcined CoFe–LDH Composite for Efficient Degradation of Bisphenol a through Synergistic Visible-Light Photocatalysis and Persulfate Activation: Performance and Mechanisms." *Environmental Pollution* 323 (2023): 121186. https://doi.org/10.1016/j.envpol.2023.121186.

[84] Lu, Siyu, Yuqin Ma, and Lang Zhao. "Production of ZnO-CoO$_x$-CeO$_2$ Nanocomposites and Their Dye Removal Performance from Wastewater by Adsorption-Photocatalysis." *Journal of Molecular Liquids* 364 (2022): 119924. https://doi.org/10.1016/j.molliq.2022.119924.

[85] Singh, Karanpal, Nancy, Harpreet Kaur, Pushpender Kumar Sharma, Gurjinder Singh, and Jagpreet Singh. "ZnO and Cobalt Decorated ZnO NPs: Synthesis, Photocatalysis and Antimicrobial Applications." *Chemosphere* 313 (2023): 137322. https://doi.org/10.1016/j.chemosphere.2022.137322.

[86] M, Mylarappa, N. Raghavendra, N.R. Bhumika, C.H. Chaithra, B.N. Nagalaxmi, and K.N. Shravana Kumara. "Study of ZnO Nanoparticle-Supported Clay Minerals for Electrochemical Sensors, Photocatalysis, and Antioxidant Applications." *ChemPhysMater*, 2023. https://doi.org/10.1016/j.chphma.2023.07.002.

[87] Busarello, Pamela, Samara de Quadros, Lizandra M. Zimmermann, and Eduardo G.C. Neiva. "Graphene Oxide/ZnO Nanocomposites Applied in Photocatalysis of Dyes: Tailoring Aqueous Stability of Quantum Dots." *Colloids and Surfaces A: Physicochemical and Engineering Aspects* 675 (2023): 132026. https://doi.org/10.1016/j.colsurfa.2023.132026.

[88] Kavitha, Thangavelu, Anantha Iyengar Gopalan, Kwang-Pill Lee, and Soo-Young Park. "Glucose Sensing, Photocatalytic and Antibacterial Properties of Graphene-ZnO Nanoparticle Hybrids." *Carbon* 50, no. 8 (2012): 2994–3000. https://doi.org/10.1016/j.carbon.2012.02.082.

[89] Malekshoar, Ghodsieh, Kanjakha Pal, Quan He, Aiping Yu, and Ajay K. Ray. "Enhanced Solar Photocatalytic Degradation of Phenol with Coupled Graphene-Based Titanium Dioxide and Zinc Oxide." *Industrial & Engineering Chemistry Research* 53, no. 49 (2014): 18824–32. https://doi.org/10.1021/ie501673v.

[90] Agarwal, Happy, S. Venkat Kumar, and S. Rajeshkumar. "A Review on Green Synthesis of Zinc Oxide Nanoparticles - An Eco-Friendly Approach." *Resource-Efficient Technologies* 3, no. 4 (2017): 406–13. https://doi.org/10.1016/j.reffit.2017.03.002.

[91] Osuntokun, Jejenija, Damian C. Onwudiwe, and Eno E. Ebenso. "Green Synthesis of ZnO Nanoparticles Using Aqueous *Brassica Oleracea* L. Var. *Italica* and the Photocatalytic Activity." *Green Chemistry Letters and Reviews* 12, no. 4 (2019): 444–57. https://doi.org/10.1080/17518253.2019.1687761.

[92] Aldeen, Thana Shuga, Hamza Elsayed Ahmed Mohamed, and Malik Maaza. "ZnO Nanoparticles Prepared via a Green Synthesis Approach: Physical Properties, Photocatalytic and Antibacterial Activity." *Journal of Physics and Chemistry of Solids* 160 (2022): 110313. https://doi.org/10.1016/j.jpcs.2021.110313.

[93] Sadiq, Hamad, Farooq Sher, Saba Sehar, Eder C. Lima, Shengfu Zhang, Hafiz M.N. Iqbal, Fatima Zafar, and Mirza Nuhanović. "Green Synthesis of ZnO Nanoparticles from *Syzygium Cumini* Leaves Extract with Robust Photocatalysis Applications." *Journal of Molecular Liquids* 335 (2021): 116567. https://doi.org/10.1016/j.molliq.2021.116567.

[94] Ameen, Fuad, Turki Dawoud, and Saleh AlNadhari. "Ecofriendly and Low-Cost Synthesis of ZnO Nanoparticles from *Acremonium Potronii* for the Photocatalytic Degradation of Azo Dyes." *Environmental Research* 202 (2021): 111700. https://doi.org/10.1016/j.envres.2021.111700.

[95] Akbari, Mohammad Zahir, Yifeng Xu, Chuanzhou Liang, Zhikun Lu, Siyuan Shen, and Lai Peng. "Synthesis of ZnO@VC for Enhancement of Synergic Photocatalytic Degradation of SMX: Toxicity Assessment, Kinetics and Transformation Pathway Determination." *Chemical Engineering and Processing - Process Intensification* 193 (2023): 109544. https://doi.org/10.1016/j.cep.2023.109544.

[96] Ranjith, Kugalur Shanmugam, Zehra Irem Yildiz, Mohammad Aref Khalily, Yun Suk Huh, Young-Kyu Han, and Tamer Uyar. "Membrane-Based Electrospun Poly-Cyclodextrin Nanofibers Coated with ZnO Nanograins by ALD: Ultrafiltration Blended Photocatalysis for Degradation of Organic Micropollutants." *Journal of Membrane Science* 686 (2023): 122002. https://doi.org/10.1016/j.memsci.2023.122002.

21 Prospective Study of Different Types of Valuable Biopolymers, Biosensors, and Biomarkers and Their Future Challenges

Nishant Shekhar, Ashlesha Kawale, Arti Srivastava, Manoj Kumar Bharty, and Pravat K. Swain

INTRODUCTION

Biopolymers, whether natural or synthetic, offer special attributes that render them inestimable across colorful diligence. This chapter examines several common or garden biopolymers, involving DNA and RNA, proteins, polysaccharides, and biodegradable plastics. Each of these biopolymers has distinct operations, from revolutionizing genetics through DNA and RNA to producing remedial proteins via recombinant DNA technology. Polysaccharides and biodegradable plastics are also arising as essential accoutrements with operations in colorful fields, involving drug and sustainability. Biosensors have the potential to detect cancer biomarkers, monitor cancer cells, and determine the effectiveness of chemotherapy agents, leading to early detection and improved treatment outcomes. Biosensor technology can provide fast and accurate cancer detection, reliable imaging of cancer cells, and monitoring of angiogenesis and cancer metastasis. Biosensors can also help eliminate ambiguity in common cancer screening methods, such as the prostate-specific antigen (PSA) test, by reducing false positives. Amperometric-based biosensors utilizing sequence-specific DNA as the recognition element can detect gene mutations associated with cancer, enabling early diagnosis. The use of biosensor technology in early cancer detection and more effective treatments can improve patients' quality of life and overall chance of survival, especially for cancers diagnosed at late stages [1]. Biopolymer-based soil treatment (BPST) has been implemented in geotechnical engineering for various applications such as dust control, soil strengthening, and erosion control. Biopolymer-Enhanced Soil Treatment (BPST) methods ensure engineering effectiveness while meeting environmental protection requirements. Biopolymers have shown promising results in ground improvement and earth stabilization practices in geotechnical and construction engineering. Biopolymers, such as dextran and starch, have been engineered and assessed in laboratory-scale studies for their effectiveness in soil behavior and erosion control. Dextran is a flexible biopolymer that forms coils with high density and low permeability in an aqueous medium. Starch, found in various plants, is another common natural biopolymer with different properties depending on the source [2]. Biopolymers, biosensors, and biomarkers are key components of contemporary biotechnology and healthcare. They play an essential function in plenty of programs, from drug development to sickness prognosis and environmental tracking. In this chapter, we are able to discover the extraordinary types of biopolymers, the use of biosensors, and the importance of biomarkers in current and destiny programs.

DOI: 10.1201/9781003479239-21

TYPES OF BIOPOLYMERS

Biopolymers are natural or synthetic biological or synthetic materials. They have unique qualities that make them incredibly valuable in different industries. Here we will discuss some common biopolymers.

DNA AND RNA

Deoxyribonucleic acid (DNA) and ribonucleic acid (RNA) are nucleic acid biopolymers that store genetic information. They have revolutionized fields like genetics, genomics, and molecular biology. DNA and RNA sequencing techniques, such as next-generation sequencing, have paved the way for personalized medicine, genetic disease diagnosis, and gene therapy [3].

PROTEINS

Proteins are essential biopolymers composed of amino acids. They have diverse functions: enzymes, structural components, and signaling molecules. In biotechnology, recombinant DNA technology is used to produce therapeutic proteins like insulin and antibodies, benefiting patients with diabetes, cancer, and autoimmune diseases [4].

POLYSACCHARIDES

Polysaccharides such as cellulose, starch, and chitin are primarily carbohydrate-based biopolymers. They find programs inside the meal industry, biodegradable substances, and drug delivery structures. For instance, cellulose-based nanomaterials are used in wound dressings and tissue engineering scaffolds.

BIODEGRADABLE PLASTICS

Biodegradable plastics, like polyhydroxyalkanoates (PHAs) and polylactic acid (PLA), are green options for standard plastics. They have programs in packaging, agriculture, and clinical gadgets. PHAs, specifically, preserve promise in lowering plastic pollutants.

BIOSENSORS

Biosensors represent sophisticated analytical devices that combine organic elements like enzymes, antibodies, or cells with physicochemical detectors to identify specific molecules or biological activities. Their applications span diverse fields, from healthcare to environmental monitoring. Biosensors are groundbreaking tools that possess a remarkable capability: the conversion of biological entities into electrical signals, opening up immense potential, particularly in the domain of cancer detection and monitoring.

This innovative technology offers a multitude of advantages, effectively addressing significant challenges in the field of oncology. Among the most promising aspects of biosensors is their capacity to detect emerging cancer biomarkers. With our ever-deepening understanding of cancer, new biomarkers continually emerge, each providing unique insights into the disease's progression. Biosensors can swiftly and accurately detect these biomarkers, even in minuscule quantities, providing valuable insights into the intricate landscape of cancer biology.

Moreover, biosensors play a pivotal role in evaluating the efficacy of cancer drugs. Traditional methods often involve time-consuming and resource-intensive processes, whereas biosensors deliver rapid and dependable results. This speed and precision have the potential to revolutionize the tailoring of treatment plans for individual patients, optimizing their chances of recovery.

In the realm of cancer imaging, biosensors introduce a new level of reliability. Their ability to target specific markers on cancer cells enables more precise and clearer imaging, facilitating the early detection of tumors and metastasis. This can significantly impact patient outcomes, as early intervention is often the key to successful cancer treatments. For instance, in the case of prostate cancer, a prevalent and often complex disease, biosensors have the potential to transform cancer research. Traditional methods, such as PSA testing, can sometimes yield ambiguous results. In contrast, biosensors can delve deeper into cancer genetics by identifying specific DNA sequences unique to cancer cells. These advancements are particularly significant in identifying BRCA1 and BRCA2 gene mutations linked to hereditary breast cancer, even predating the introduction of this technology [1,2,5]. The importance of biosensors in early cancer detection cannot be overstated. Many cancers are diagnosed at advanced stages when treatment options are limited and less effective. Biosensors offer a ray of hope in such cases, as they can provide a swift and accurate diagnosis, substantially enhancing patients' quality of life and overall well-being.

Biosensors stand as a beacon of innovation in the intricate landscape of cancer treatments. Their ability to translate biological signals into tangible data equips healthcare professionals with the tools they need to combat cancer more efficiently and expeditiously. As research continues to advance and biosensor technology evolves, we can anticipate more significant breakthroughs in cancer diagnoses and treatments, offering hope to patients and their families worldwide. Here are several examples of biosensor applications.

MEDICAL DIAGNOSTICS

Biosensors are used for diagnosing diseases and monitoring biomarkers. For instance, glucose biosensors have transformed the management of diabetes. Advanced biosensors can detect cancer biomarkers, infectious agents, and cardiac markers, enabling early disease detection.

ENVIRONMENTAL MONITORING

Biosensors play a vital role in monitoring environmental pollutants, such as heavy metals, pesticides, and toxins. They provide real-time data for assessing water quality, air pollution, and soil contamination, contributing to environmental conservation efforts.

FOOD SAFETY

Foodborne pathogen detection is crucial for ensuring food safety. Biosensors can rapidly identify pathogens such as *E. coli* and *Salmonella*, reducing the risk of foodborne illnesses. They also help in monitoring food quality and freshness.

BIOMARKERS

Biomarkers are measurable indicators of biological processes or disease states. They have become indispensable in modern medicine, enabling early diagnosis, prognosis, and treatment monitoring. A wearable tear bioelectronics platform for non-invasive monitoring of tear biomarkers, including alcohol and glucose, has been developed using a microfluidic electrochemical detector integrated into eyeglass's nose-bridge pad. The alcohol biosensor flow detector on the wearable eyeglasses-based platform showed a strong correlation between tear alcohol content and blood alcohol levels (BAC). Tear glucose monitoring can also be performed using the same wearable platform, with a waiting time of 15 minutes after a meal for accurate measurement. The integration of biosensors into wearable devices, such as eyeglasses, offers a convenient and portable solution for tear analysis. The ability to monitor tear biomarkers opens up possibilities for non-invasive monitoring of alcohol intake, glucose levels, and other nutrients. The development of wearable tear biosensors paves the way for future applications in healthcare, personalized medicine, and continuous monitoring of biomarkers [6].

THE ROLE OF BIOMARKERS IN BIOMARKER DETECTION

In the realm of biomarker detection, biomarkers assume a pivotal role, serving as specific molecules or indicators utilized to discern the presence or progression of various diseases or medical conditions. These biomarkers find their applications in chemo/biosensing and bioimaging techniques, contributing significantly to the enhancement of early disease diagnosis and subsequent treatment. Particularly, ratiometric fluorescence (FL) probes stand out as valuable tools in this domain. These probes offer inherent self-calibration mechanisms that can correct for various target-independent factors, making them extensively employed in chemo/biosensing and bioimaging for biomarker detection [7]. A wide array of biomarkers, ranging from nucleic acids to cell-surface biomarkers and specific molecules such as HSO_3A, can be effectively detected using ratiometric FL probes. This approach to biomarker detection using ratiometric FL probes brings about precise, quantitative, visual, and real-time analysis capabilities, all of which are instrumental in ensuring accurate disease diagnosis and ongoing disease monitoring. In summary, biomarkers play a pivotal role in the context of biomarker detection through the application of ratiometric FL probes, ultimately enabling early disease diagnosis and the precise monitoring of medical conditions [8]. Hereafter, we delve into some present and prospective applications of biomarkers in the following sections.

CANCER BIOMARKERS

Cancer is the second leading cause of death in the United States, with over 570,000 expected deaths in 2010. It can take over 200 distinct forms, and both environmental and genetic factors are associated with an increased risk of developing cancer. Prostate and breast cancers are the most common types in men and women, respectively. Cancer is not only deadly but also one of the most expensive diseases in the world. The average 5-year survival rate for all cancers has increased to 66%, thanks to technological advances in treatment and early diagnosis. However, certain cancers like liver, pancreatic, and lung still have very low survival rates [1]. The use of emerging biosensor technology could be instrumental in early cancer detection and more effective treatments, particularly for late-stage cancers that respond poorly to treatment. Cancer biomarkers like PSA and CA-125 are currently used for cancer screening and monitoring (Figure 21.1).

The stage at the time of the diagnosis seriously affects the survival rate of ovarian cancer patients. IVD of biomarkers related to ovarian cancer has become a breakthrough in the diagnosis of ovarian cancer. Emerging biomarkers (TEX, CTC, LPA, LSR, metabolites, miRNA, and ctDNA) and classic biomarkers (CA125 and HE4) show considerable potential. Various excellent detection technologies (such as microfluidic chips, optical biosensors, and electrochemical biosensors) combined with nanomaterials (polymer materials, carbon nanomaterials, quantum dots (QDs), and metal nanoparticles) improve the detection performance of these biomarkers (Source: Yang et al. [9]).

Future applications may involve personalized cancer therapy based on genomic biomarkers and thus tailor treatments to a patient's genetic profile. Ovarian cancer is a significant cause of cancer-related death in females worldwide, with a high mortality rate that has remained unchanged for the past half-century. Current imaging methods for ovarian cancer diagnosis lack sensitivity and resolution, leading to difficulty in the detection of mid- to late-stage ovarian cancer. In vitro diagnosis has the potential to enable early detection of ovarian cancer through non-invasive and dynamic analysis of biomarkers. However, conventional marker-based in vitro diagnostic devices have limitations in terms of false positives and low efficiency in early ovarian cancer diagnosis. Table 21.1 outlines some common cancer biomarkers and many other biomarkers associated with different types of cancers.

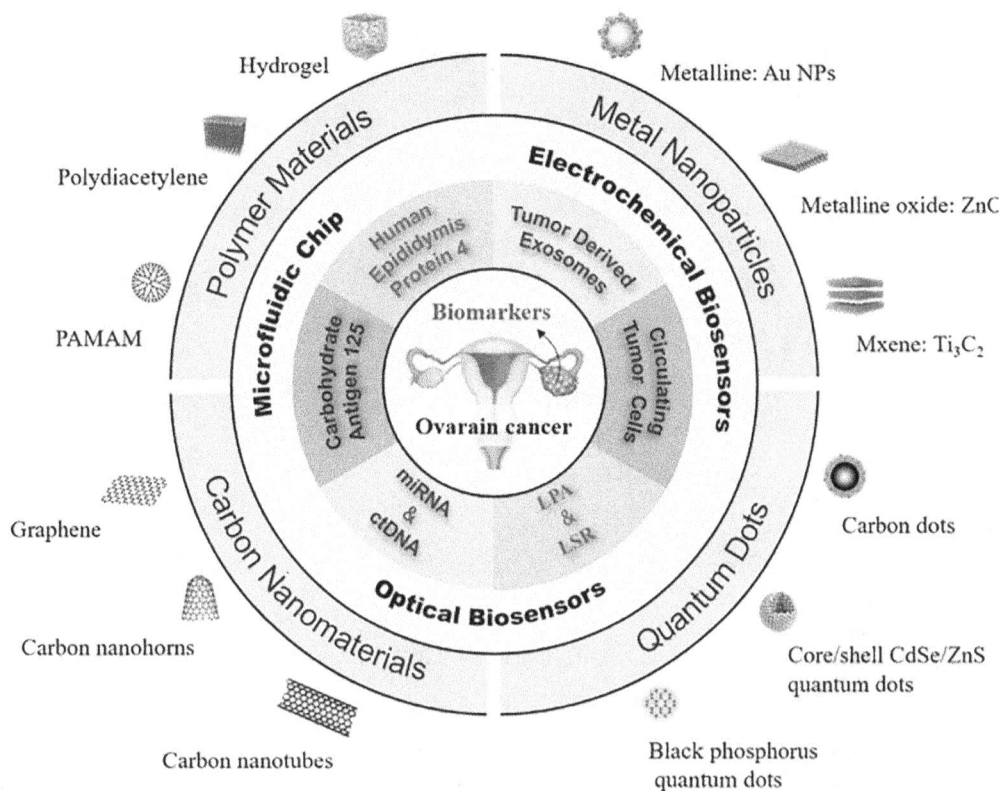

FIGURE 21.1 Ovarian cancer is one of the most common cancers in women all over the world.

TABLE 21.1
Different Types of Cancer and Their Clinical Significance

Biomarker	Cancer Type	Clinical Significance
CA-125	Ovarian cancer	Monitoring disease progression
PSA (Prostate-Specific Antigen)	Prostate cancer	Screening and monitoring
HER2/neu (Human Epidermal Growth Factor Receptor 2)	Breast cancer	Target for therapy
CEA (Carcinoembryonic Antigen)	Colorectal cancer	Monitoring and post-treatment surveillance
EGFR (Epidermal Growth Factor Receptor)	Lung cancer	Target for therapy
BRCA1 and BRCA2	Breast and ovarian cancers	Identifying high-risk individuals
AFP (Alpha-Fetoprotein)	Liver cancer	Screening and monitoring
CA 19-9	Pancreatic cancer	Monitoring treatment response
KRAS	Various cancers	Target for therapy and mutation detection

The discoveries of novel ovarian cancer biomarkers and advancements in nanomaterials have opened up possibilities for improving the sensitivity and specificity of in vitro diagnosis for early ovarian cancer diagnosis. Nanomaterial-based biosensors offer many advantages to in vitro diagnosis of ovarian cancer such as convenience, speed, selectivity, and sensitivity. These biosensors utilize nanomaterials with increased surface area for efficient affinity capture of ovarian cancer biomarkers. Integration of multiple detection technologies using nanomaterials in in vitro diagnosis devices is a promising approach for overcoming the challenges in early ovarian cancer diagnosis [9].

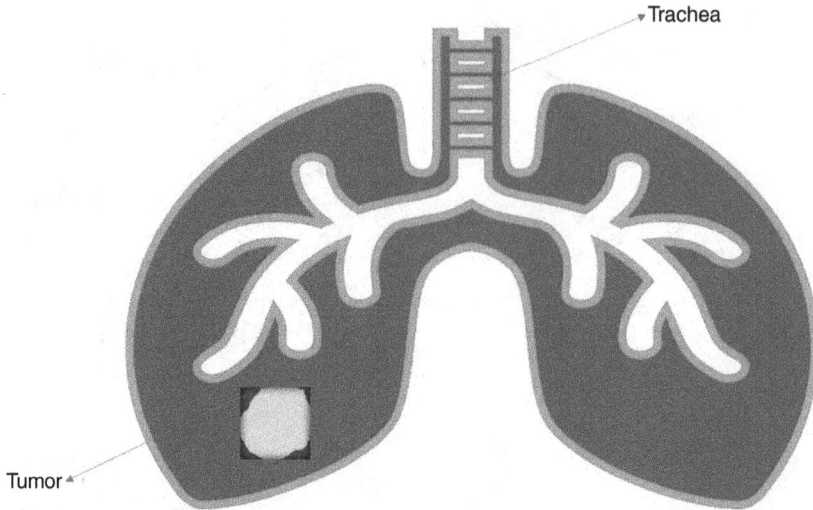

FIGURE 21.2 Lung cancer.

Lung cancers are a major worldwide health subject with excessive mortality rates, emphasizing the want for early detection and diagnosis (Figure 21.2).

Tumor markers, including neuron-precise enolase (NSE), cytokeratin 19 fragments (CYFRA 21-1), squamous mobile carcinoma antigen (SCCA), and tissue polypeptide antigen (TPA), have proven promising in detection of lung cancers. Biosensors offer a touchy, handy, and cost-effective technique for detecting lung cancer biomarkers. Electrochemical and optical biosensors were developed for the detection of lung cancer biomarkers, including SCCA and TPA, using nanomaterials. SCCA, a serine protease inhibitor, is a valuable biomarker for squamous cell carcinomas of the lung, cervix uteri, head and neck areas, and esophagus. Combining SCCA with CYFRA21-1 can be beneficial for monitoring non-small-cell lung cancer (NSCLC). Overall, biosensors offer a promising avenue for the detection of lung cancer biomarkers, such as SCCA and TPA, which may aid in early analysis and remedy-making plans [10].

NEUROLOGICAL BIOMARKERS

Biomarkers for neurodegenerative sicknesses like Alzheimer's and Parkinson's are under intensive research. These biomarkers ought to result in earlier disorder prognosis and the improvement of ailment-modifying therapies [11].

LIQUID BIOPSIES

Liquid biopsies are emerging as a non-invasive method to come across most cancers and different diseases via the evaluation of biomarkers in blood, urine, or other physical fluids. They offer promise for early detection and treatment tracking of most cancers.

INFECTIOUS DISEASE BIOMARKERS

Biomarkers are important in diagnosing and tracking infectious diseases such as HIV, hepatitis, and COVID-19. Future programs can also include the rapid detection of rising infectious marketers and the improvement of more effective vaccines. In the end, biopolymers, biosensors, and biomarkers are transformative elements in biotechnology and healthcare. Their numerous applications are

TABLE 21.2

Different Types of Aspects and Their Discussions

Aspect	Description
Definition	BPST is a soil improvement technique that uses biopolymers, natural or synthetic, to enhance the engineering properties of soils.
Types of Biopolymers	Natural biopolymers (e.g., guar gum, xanthan gum) and synthetic biopolymers (e.g., polyacrylamides).
Purpose	Improve soil stability and bearing capacity. Reduce soil erosion and sediment runoff. Enhance soil water retention and permeability.
Application Methods	Injection into the soil as a liquid solution. Mixing biopolymers with soil in a dry or slurry form.
Benefits	Environmentally friendly and biodegradable. Reduces soil erosion, improving sustainability. Enhances soil's mechanical properties.
Soil Types	Suitable for various soil types, including sandy, silty, and clayey soils.
Geotechnical Applications	Road construction and stabilization. Erosion control in slopes and embankments. Foundation improvement. Land reclamation.
Effect on Permeability	Biopolymers can improve or reduce soil permeability depending on the specific application and type of biopolymer used.
Long-term Stability	BPST can provide long-lasting soil improvements when used appropriately and maintained properly.
Environmental Considerations	Biopolymers are typically considered safe for the environment. Biodegradability reduces long-term impact. Proper disposal practices should be followed.

continually increasing, with promising tendencies on the horizon. As technological know-how and technology strengthen, these gears will keep improving our capacity to diagnose illnesses, reveal our surroundings, and broaden progressive answers for a sustainable future.

Biopolymer-based soil treatment (BPST) in geotechnical engineering: BPST has been carried out in recent years for various programs in geotechnical engineering, which include dust control, soil strengthening, and erosion management.

The efficacy of BPST in engineering has been established, particularly in meeting assembly environmental protection requirements, but similarly validation is wanted in terms of website applicability, durability, and economic feasibility. Biopolymers were engineered and assessed in laboratory scales, and their outcomes on soil conduct were reviewed in phrases of geotechnical engineering application and exercise, including soil consistency limits, energy parameters, hydraulic conductivity, soil–water traits, and erosion control. The economic feasibility and sustainability of BPST utility in floor development and earth stabilization practices have been mentioned. Biopolymers are taken into consideration as new, promising, environmentally friendly floor improvement materials for geotechnical and production engineering practices. Overall, incorporating statistics about BPST and its packages in geotechnical engineering can provide insights into the usage of biopolymers in soil remedy, which may be applicable to the fields of biosensors and biomarkers inside the context of environmental engineering and sustainability [2,12] (Table 21.2).

These points summarize key aspects of BPST, including its definition, types of biopolymers, purposes, application methods, benefits, suitability for different soil types, geotechnical applications, its impact on permeability, long-term stability, and environmental considerations.

BIOSENSORS AND BIOMARKERS FOR HEALTHCARE APPLICATIONS

Graphene functionalized with metallic oxide nanostructures has proven promising in the improvement of sensitive sensors and biosensors for healthcare packages. The combination of graphene and metallic oxide nanostructures has brought about enhanced catalytic activity, improved detection

TABLE 21.3

Organized Summary of Different Biosensors, Biomarkers They Detect, and Their Respective Healthcare Applications

Biosensor	Biomarker	Healthcare Application
Glucose Biosensor	Blood glucose levels	Diabetes management
Electrocardiogram (ECG)	Heart rate, rhythm	Cardiac monitoring
Pulse Oximeter	Blood oxygen saturation	Respiratory and cardiovascular care
DNA Biosensor	Genetic sequences	Genomic research and diagnostics
CRP (C-Reactive Protein)	Inflammation	Infection and disease monitoring
PSA (Prostate-Specific Antigen)	Prostate cancer	Prostate cancer screening
Troponin	Troponin levels	Cardiac injury detection
B-Type Natriuretic Peptide (BNP)	BNP levels	Heart failure diagnosis and management
D-Dimer	Blood clot presence	Deep vein thrombosis detection
pH Sensor	pH levels	Acid–base balance monitoring

limits, sensitivity, and selectivity in biosensing. Various types of biosensors were evolved using graphene and functionalized graphene, which includes protein sensors, electrochemical immune sensors, phytochrome sensors, cholesterol biosensors, glucose sensors, hydrogen peroxide sensors, and nicotinamide adenine dinucleotide detection sensors. Graphene-based totally glucose sensors were fabricated through immobilizing glucose oxidase enzyme on a graphene electrode floor, overcoming the limitations of electron transfer and improving sensitivity. Primarily graphene-based enzymatic biosensors have also been developed for the detection of hydrogen peroxide, which is considerable in various industries and medical applications [13–15] (Table 21.3).

BIOSENSORS AND BIOMARKERS IN CANCER RESEARCH

Electrochemical biosensors have shown great potential in cancer liquid biopsy, drug sensitivity monitoring, cancer on a chip, and in vivo detection. Different electrochemical measurement methods, such as differential pulse voltammetry (DPV) and square wave voltammetry (SWV), provide electrochemical biosensors that offer several advantages for cancer studies, which include extreme sensitivity, excessive selectivity, low price, short readout, and simplicity. Biosensors encompass a bioreceptor, biosurface/biointerface architecture, a transducer, and related electronics. Electrochemical biosensors have shown first-rate potential in liquid biopsy of most cancers, drug sensitivity tracking, most cancers-on-a-chip, and in vivo detection. Different electrochemical dimension methods, which include differential pulse voltammetry (DPV) and rectangular wave voltammetry (SWV), provide sensitive and multi-analyte analysis. Electrochemical biosensors were used for the detection of most cancer biomarkers at the molecular level (DNA, RNA, and proteins), organelle degree (exosomes), and cellular level (mobile counting, phenotypic and metabolism analysis, drug sensitivity tracking). Matrix metalloproteinases (MMPs) are crucial biomarkers associated with cancer development and metastasis. Electrochemical biosensors have been used for the detection of MMPs, together with MMP-7, and the usage of peptide cleavage techniques [15–18].

BIOSENSORS FOR DETECTION OF *MYCOBACTERIUM TUBERCULOSIS* AND TUBERCULOSIS BIOMARKERS

Electrochemical-based biosensors are an applicable topic to add to the fields of biosensors and biomarkers. These biosensors have advantages such as low-fee operation, rapid processing, simultaneous multi-analyte evaluation, and the ability to function with turbid samples. They are considered pioneering tools for factor-of-care diagnostics in communities with limited accessibility to reference laboratories. The use of nanomaterials, which includes gold nanoparticles, steel oxides, carbon nanotubes, magnetic nanoparticles, fullerenes, and graphene derivatives, has ended up a crucial part of electrochemical biosensing designs. These nanomaterials decorate the resulting signals and improve detection limits. Traditional techniques for tuberculosis prognosis are costly and require specialized facilities and trained technicians. Biosensors, alternatively, offer benefits together with reproducibility, continuous monitoring of analytes, and stringent detection limits even in attomolar scales. They have gained vast interest in the recent years. Immobilization of carbon nanotubes on biosensor surfaces may be hard. However, coating the nanotubes with conductive polymers can cope with this trouble and decrease nonspecific interactions with nucleic acid probes [19–21].

BIOSENSORS AND BIOMARKERS IN WASTEWATER-BASED EPIDEMIOLOGY

Biosensors serve as efficient and cost-effective intelligent sensing systems capable of rapidly and sensitively detecting viruses, even in wastewater samples. They offer an all-in-one solution encompassing sample preparation and the ability to identify specific targets using a unified platform, including advanced technologies such as CRISPR-powered systems.

Another valuable technology in the domain of viral detection within wastewater is lab-on-a-chip (LOC). These microfluidic devices consist of interconnected networks of microchannels equipped with various components like valves, mixers, pumps, and detectors. LOC systems are adept at executing intricate diagnostic tasks with a minimal need for human intervention, and they are conducive to cost-effective, large-scale production. Notably, LOC-based approaches have found extensive utility in clinical applications, including the detection of diseases such as Ebola virus and HIV.

In the context of wastewater-based epidemiology, it is pertinent to target specific biomarkers, particularly those associated with the inflammatory response, for the detection of COVID-19 outbreaks. These biomarkers can furnish invaluable insights into the presence and severity of the virus in wastewater samples. Consequently, ongoing research efforts are focused on the identification of biomolecules that can effectively target COVID-19 outbreaks based on the presence of inflammatory response biomarkers.

Additionally, the exploration of immunoglobulins, such as IgM and IgG antibodies, as potential biomarkers for the early detection of COVID-19 outbreaks in wastewater-based epidemiology is also underway. These antibodies have the potential to indicate current or past infections with the virus [22].

CARBON NANOTUBE BIOSENSORS FOR BIOMARKER DETECTION

Carbon nanotubes (CNTs) possess a unique set of properties that render them highly suitable for a range of biosensing applications. Notably, they can serve as scaffolds for immobilizing biomolecules, and their exceptional physical, chemical, electrical, and optical characteristics further enhance their utility. Functionalized CNTs, in particular, exhibit distinct behavior and improved stability compared to their pristine counterparts. Surface functionalization can enhance biocompatibility while mitigating toxicity, rendering them safer for cellular and in vivo applications.

CNT-based biosensors have been at the forefront of detecting various biomarkers, including those associated with cancer. These biosensors offer exceptional sensitivity and can perform multiplexed detection of multiple biomarkers simultaneously. The integration of CNTs into biosensing devices has paved the way for the development of highly sensitive electrochemical biosensors capable of early-stage biomarker detection. CNTs hold immense promise in advancing biosensor technology for biomarker detection, opening avenues for heightened sensitivity and lower detection limits [23–25].

BIOSENSORS AND BIOMARKERS IN AFFECTIVE SYSTEMS

i. Skin Conductivity (Sweat): Skin conductivity, regularly measured as sweat, is a common biomarker used for detecting feelings and pressure in affective systems [25].

ii. Antistress Hormones and Cortisol Metabolites: Antistress hormones and cortisol metabolites were diagnosed as primary strain biomarkers that would doubtlessly be used in destiny wearable sensors designed for affective systems.

iii. Oxytocin: Oxytocin, labeled as an antistress hormone, has proved promising as a pressure biomarker. However, further research is needed to completely apprehend its potential and application in strain detection.

iv. Physiological Changes: Emotional responses can trigger numerous physiological modifications, together with alterations in coronary heart charge, pores and skin conductivity, and oxygen saturation. These modifications can produce precious biomarkers to function in affective systems.

v. Multimodal Approaches: Utilizing multiple modalities, such as facial expression analysis, enhances the benefits of affective structures, picture processing, and speech evaluation, to hit upon and apprehend emotional states. This multimodal approach enhances the accuracy and reliability of emotion detection.

It's clear that a mixture of physiological measurements and multimodal approaches can provide valuable insights into emotional states and pressure degrees in affective systems. Further studies and improvement in this subject have the capability to result in advanced wearable sensors and systems for emotion and pressure tracking. Skin conductivity (sweat) is a generally used biomarker for emotion and strain detection in affective structures. Antistress hormones and cortisol metabolites have been recognized as the number one strain biomarkers that can be utilized in destiny wearable sensors for effective machines. Oxytocin, labeled as an antistress hormone, has been detected in biofluids and might have the capacity to serve as a strain biomarker, but further investigations are wanted. Physiological changes including heart rate, skin conductivity, and oxygen saturation also occur as emotional responses and can be used as biomarkers in affective structures. Facial expression, photo processing, and speech analysis have been used to hit upon emotional states in affective systems, indicating the benefit of multimodal procedures [26–32].

RESULTS AND DISCUSSION

Biopolymers encompass DNA, RNA, proteins, polysaccharides, and biodegradable plastics, each serving unique purposes. DNA and RNA revolutionize genetics and personalized medicine through sequencing. Proteins produced via recombinant DNA technology benefit patients with various conditions. Polysaccharides and biodegradable plastics find applications in food, materials, and drug delivery systems. Biosensors play pivotal roles across diverse domains. They will enable swift and accurate cancer biomarker detection, aiding in early diagnosis, precise imaging, and effective treatment planning. Biosensors mitigate false positives in common cancer screening methods. Sequence-specific DNA recognition elements are utilized to identify cancer-related gene mutations, facilitating early diagnosis and improved patient outcomes. They are also used to monitor environmental pollutants in real time and enhance efforts in water quality, air pollution, and soil contamination assessment. They also rapidly detect foodborne pathogens, thus reducing the risk of foodborne illnesses and contributing to food quality and freshness monitoring (Table 21.4).

SIGNIFICANT IMPLICATIONS OF BIOMARKERS FOR HEALTHCARE

Wearable tear bioelectronic platforms offer non-invasive monitoring of tear biomarkers, presenting opportunities for continuous healthcare and personalized medicine. Biosensors play a crucial role in detecting cancer biomarkers, potentially leading to earlier diagnosis and tailored treatments.

TABLE 21.4
Points That Further Illustrate the Wide-Ranging Applications and Significance of DNA, RNA, Proteins, Polysaccharides, and Biodegradable Plastics in Various Fields

Biopolymers	Applications
DNA and RNA	Enable the study of genetic variations and their implications for disease
Proteins	Facilitate the development of targeted therapies
	Include therapeutic antibodies used in immunology
	Serve as critical components in vaccines and biopharmaceuticals
Polysaccharides and Biodegradable Plastics	Improve food texture and shelf life
	Contribute to sustainable packaging solutions
	Reduce environmental impact of plastic waste

TABLE 21.5
Diverse Applications of Biosensors in Healthcare, Environmental Monitoring, and Food Safety

Biosensors	Applications
Swift and accurate cancer biomarker detection	Early diagnosis
	Precise imaging of cancer cells
	Effective treatment planning
Mitigation of false positives	Improve cancer screening methods
Sequence-specific DNA recognition	Identify cancer-related gene mutations
	Facilitate early diagnosis
Monitoring of environmental pollutants in real time	Real-time monitoring of pollutants in water, air, and soil
Rapid detection of foodborne pathogens	Reduced risk of foodborne illnesses
	Food quality and freshness monitoring

Notable examples include ovarian and lung cancer biomarker detection. Biomarkers for neurodegenerative diseases hold promise for early diagnosis and potential disease-modifying therapies.

Liquid biopsies provide non-invasive cancer detection and treatment monitoring, especially valuable for early diagnosis. Biomarkers are vital for diagnosing and monitoring infectious diseases, including HIV, hepatitis, and COVID-19. They may aid in the rapid detection of emerging infectious agents and improve vaccine development [27–30] (Table 21.5).

FUTURE PROSPECTS OF BIOPOLYMERS, BIOSENSORS, AND BIOMARKERS

1. Supercharged Biopolymers: We can expect new types of biodegradable materials with custom-made properties for things like drug delivery and eco-friendly products in future.
2. Tailored Medicine: Techniques are heading toward more precise medical treatments. Advanced biosensors will help doctors match treatments to individual patients, making medicine more effective and safer.
3. Environmental Guardians: Biopolymers and biosensors will help us closely monitor our environment. They'll spot pollution and other problems faster, helping us protect our planet better.
4. Early Warning for Diseases: In cancer research, we'll keep finding better ways to catch the disease early. Thus more lives can be saved, and there will be less suffering from treatments.

5. Tiny Tech: We can expect even smaller biosensors using materials like tiny carbon tubes and graphene in future. These will be super sensitive and useful for both health checks and environmental monitoring.

6. Wearables for Health: Wearable biosensors will become popular. You'll wear them to keep an eye on your health in real time.

7. Smart Computers: Computers will get better at understanding the data from biosensors. They'll help doctors make more accurate diagnoses and treatment plans.

8. Easy Testing: Getting tested for diseases will become easier. You might even do it yourself at home, especially in places where healthcare isn't easy to access.

9. Green Farming: Biopolymers will help farming become more eco-friendly. They'll improve soil quality and help grow more food sustainably.

10. Rules and Safety: As these technologies grow, there will be rules to make sure they're safe and work well in medicine and the environment.

In a nutshell, the future looks bright for these technologies. They'll make healthcare better, protect the environment, and improve our lives in many ways. Scientists, engineers, and doctors will lead the way in making these ideas a reality.

CONCLUSION

Biopolymers, biosensors, and biomarkers are transformative factors in biotechnology and healthcare, shaping numerous applications with promising developments on the horizon. As science and generation continue to develop, this equipment will in addition beautify our potential to diagnose illnesses, display our environment, and create progressive answers for a sustainable destiny. In the end, this chapter has provided a complete assessment of biopolymers, biosensors, and biomarkers, showcasing their immense significance and flexibility across numerous domain names. Biopolymers, encompassing DNA, RNA, proteins, polysaccharides, and biodegradable plastics, exhibit precise attributes that make them helpful in diverse industries, starting from genetics to sustainability.

Biosensors have emerged as modern analytical gadgets, harnessing the power of biological additives and physicochemical detectors to detect unique molecules or biological activities. Their capability applications are substantial, with a particular consciousness on detection and monitoring of most cancers. Biosensors offer speedy and accurate detection of cancer biomarkers, permitting early prognosis, specific imaging of cancer cells, and evaluation of treatment effectiveness. They also hold the promise for reducing false positives in screening methods of most cancers, revolutionizing the sector of oncology.

Furthermore, biopolymers play an essential role in geotechnical engineering, in which biopolymer-based soil remedy (BPST) techniques have proven effective in enhancing soil balance, controlling dust, and coping with erosion, all while meeting stringent environmental requirements. These sustainable techniques provide a promising direction ahead for floor development and earth stabilization practices. In the world of current biotechnology and healthcare, biopolymers, biosensors, and biomarkers are necessary components, serving critical roles in drug development, sickness prognosis, and environmental monitoring. Biopolymers are flexible substances with packages in various industries, and at the same time, as biosensors, they empower healthcare professionals with the equipment needed for speedy and correct diagnostics. Biomarkers have grown to be important signs in current medication, facilitating early detection, analysis, and remedy monitoring of disorders. As technology and research progress, these mechanisms will increasingly assume a crucial role in our ability to detect and diagnose diseases, display our surroundings, and develop progressive solutions for a sustainable future. The applications of biosensors and biomarkers in fields which include clinical diagnostics, environmental tracking, and food safety are usually expanding, promising a brighter and healthier destiny for individuals and the planet. In essence, biopolymers, biosensors, and biomarkers are catalysts for innovation and development, providing hope to patients, researchers, and communities worldwide. The transformative capabilities of these tools underscore their lasting significance in influencing the future of healthcare, technology, and sustainability.

ACKNOWLEDGMENT

The authors thank Guru Ghasidas Vishwavidyalaya Central University, Bilaspur, Chhattisgarh-495009, India. The authors also thank Dr. J N College, P.O., Rasalpur, Balasore-756021, Odisha, India; and Berhampur Degree College, Raj Berhampur, Balasore-756058, Odisha, India for providing the necessary facilities. The authors gratefully acknowledges the infrastructural support from the Department of Chemistry, Institute of Science, Banaras Hindu University, Varanasi-221005, U.P., India.

REFERENCES

[1] Bohunicky, Brian, and Shaker A. Mousa. "Biosensors: The new wave in cancer diagnosis." In Dr Kattesh V Katti (Ed.) *Nanotechnology, Science and Applications* (2010): pp. 1–10. Department of Radiology, University of Missouri, United States, Taylor & Francis. https://doi.org/10.2147/NSA.S13465

[2] Chang, Ilhan, Minhyeong Lee, An Thi Phuong Tran, Sojeong Lee, Yeong-Man Kwon, Jooyoung Im, and Gye-Chun Cho. "Review on biopolymer-based soil treatment (BPST) technology in geotechnical engineering practices." *Transportation Geotechnics* 24 (2020): 100385.

[3] Huertas, Cesar S., Olalla Calvo-Lozano, Arnan Mitchell, and Laura M. Lechuga. "Advanced evanescent-wave optical biosensors for the detection of nucleic acids: An analytic perspective." *Frontiers in Chemistry* 7 (2019): 724. https://doi.org/10.3389/fchem.2019.00724

[4] Soler, Maria, Cesar S. Huertas, and Laura M. Lechuga. "Label-free plasmonic biosensors for point-of-care diagnostics: A review." *Expert Review of Molecular Diagnostics* 19, no. 1 (2019): 71–81. https://doi.org/10.1080/14737159.2019.1554435

[5] Dai, Yifan, Rodrigo A. Somoza, Liu Wang, Jean F. Welter, Yan Li, Arnold I. Caplan, and Chung Chiun Liu. "Exploring the trans-cleavage activity of CRISPR-Cas12a(cpf1) for the development of a universal electrochemical biosensor." *Angewandte Chemie International Edition* 58, no. 48 (2019): 17399–17405. https://doi.org/10.1002/anie.201910772

[6] Sempionatto, Juliane R., Laís Canniatti Brazaca, Laura García-Carmona, Gulcin Bolat, Alan S. Campbell, Aida Martin, Guangda Tang et al. "Eyeglasses-based tear biosensing system: Non-invasive detection of alcohol, vitamins and glucose." *Biosensors and Bioelectronics* 137 (2019): 161–170. https://doi.org/10.1016/j.bios.2019.04.058

[7] Fothergill, Sarah Madeline, Caoimhe Joyce, and Fang Xie. "Metal enhanced fluorescence biosensing: From ultra-violet towards second near-infrared window." *Nanoscale* 10, no. 45 (2018): 20914–20929. https://doi.org/10.1039/C8NR06156D.

[8] Gui, Rijun, Hui Jin, Xiangning Bu, Yongxin Fu, Zonghua Wang, and Qingyun Liu. "Recent advances in dual-emission ratiometric fluorescence probes for chemo/biosensing and bioimaging of biomarkers." *Coordination Chemistry Reviews* 383 (2019): 82–103. https://doi.org/10.1016/j.ccr.2019.01.004

[9] Yang, Yuqi, Qiong Huang, Zuoxiu Xiao, Min Liu, Yan Zhu, Qiaohui Chen, Yumei Li, and Kelong Ai. "Nanomaterial-based biosensor developing as a route toward in vitro diagnosis of early ovarian cancer." *Materials Today Bio* 13 (2022): 100218. https://doi.org/10.1016/j.mtbio.2022.100218

[10] Gao, Xiaoshan, Shuyan Niu, Junjun Ge, Qingyu Luan, and Guifen Jie. "3D DNA nanosphere-based photoelectrochemical biosensor combined with multiple enzyme-free amplification for ultrasensitive detection of cancer biomarkers." *Biosensors and Bioelectronics* 147 (2020): 111778. https://doi.org/10.1016/j.bios.2019.111778

[11] Carneiro, Pedro, Simone Morais, and Maria Carmo Pereira. "Nanomaterials towards biosensing of Alzheimer's disease biomarkers." *Nanomaterials* 9, no. 12 (2019): 1663. https://www.mdpi.com/2079-4991/9/12/1663.

[12] Gavrilescu, Maria, Kateřina Demnerová, Jens Aamand, Spiros Agathos, and Fabio Fava. "Emerging pollutants in the environment: Present and future challenges in biomonitoring, ecological risks and bioremediation." *New Biotechnology* 32, no. 1 (2015): 147–156. https://doi.org/10.1016/j.nbt.2014.01.001.

[13] Kumar, Sudesh, Shikandar D. Bukkitgar, Supriya Singh, Pratibha, Vanshika Singh, Kakarla Raghava Reddy, Nagaraj P. Shetti, Ch Venkata Reddy, Veera Sadhu, and S. Naveen. "Electrochemical sensors and biosensors based on graphene functionalized with metal oxide nanostructures for healthcare applications." *ChemistrySelect* 4, no. 18 (2019): 5322–5337. https://doi.org/10.1002/slct.201803871

[14] Arshad, MK Md, Subash CB Gopinath, W. M. W. Norhaimi, and M. F. M. Fathil. "Current and future envision on developing biosensors aided by 2D molybdenum disulfide (MoS$_2$) productions." *Biosensors and Bioelectronics* 132 (2019): 248–264. https://doi.org/10.1016/j.bios.2019.03.005.

[15] Bollella, Paolo, Giovanni Fusco, Cristina Tortolini, Gabriella Sanzò, Gabriele Favero, Lo Gorton, and Riccarda Antiochia. "Beyond graphene: Electrochemical sensors and biosensors for biomarkers detection." *Biosensors and Bioelectronics* 89 (2017): 152–166. https://doi.org/10.1016/j.bios.2016.03.068

[16] Banerjee, Hirendra N., and Mukesh Verma. "Application of nanotechnology in cancer." *Technology in Cancer Research Treatment* 7, no. 2 (2008): 149–154. https://doi.org/10.1177/153303460800700208

[17] Ratajczak, Katarzyna, and Magdalena Stobiecka. "High-performance modified cellulose paper-based biosensors for medical diagnostics and early cancer screening: A concise review." *Carbohydrate Polymers* 229 (2020): 115463. https://doi.org/10.1016/j.carbpol.2019.115463

[18] Huang, Xiaolin, Yijing Liu, Bryant Yung, Yonghua Xiong, and Xiaoyuan Chen. "Nanotechnology-enhanced no-wash biosensors for in vitro diagnostics of cancer." *ACS Nano* 11, no. 6 (2017): 5238–5292. https://doi.org/10.1021/acsnano.7b02618.

[19] Golichenari, Behrouz, Rahim Nosrati, Aref Farokhi-Fard, Mahdi Faal Maleki, Seyed Mohammad Gheibi Hayat, Kiarash Ghazvini, Farzam Vaziri, and Javad Behravan. "Electrochemical-based biosensors for detection of *Mycobacterium tuberculosis* and tuberculosis biomarkers." *Critical Reviews in Biotechnology* 39, no. 8 (2019): 1056–1077. https://doi.org/10.1080/07388551.2019.1668348

[20] Chang, Jiafu, Xin Wang, Jiao Wang, Haiyin Li, and Feng Li. "Nucleic acid-functionalized metal-organic framework-based homogeneous electrochemical biosensor for simultaneous detection of multiple tumor biomarkers." *Analytical Chemistry* 91, no. 5 (2019): 3604–3610. https://doi.org/10.1021/acs.analchem.8b05599

[21] Huang, Rongrong, Nongyue He, and Zhiyang Li. "Recent progresses in DNA nanostructure-based biosensors for detection of tumor markers." *Biosensors and Bioelectronics* 109 (2018): 27–34. https://doi.org/10.1016/j.bios.2018.02.053.

[22] Barceló i Cullerés, Damià. "Wastewater-based epidemiology to monitor COVID-19 outbreak: Present and future diagnostic methods to be in your radar." *Case Studies in Chemical and Environmental Engineering* 2 (2020): 100042. https://doi.org/10.3389/fchem.2015.00059

[23] Tîlmaciu, C.-M., and Morris, M. C. "Carbon nanotube biosensors." *Frontiers in Chemistry* 3 (2015). https://doi.org/10.3389/fchem.2015.00059

[24] Sireesha, Merum, Veluru Jagadeesh Babu, A. Sandeep Kranthi Kiran, and Seeram Ramakrishna. "A review on carbon nanotubes in biosensor devices and their applications in medicine." *Nanocomposites* 4, no. 2 (2018): 36–57. https://doi.org/10.1080/20550324.2018.1478765.

[25] Dervisevic, Muamer, Maria Alba, Beatriz Prieto-Simon, and Nicolas H. Voelcker. "Skin in the diagnostics game: Wearable biosensor nano-and microsystems for medical diagnostics." *Nano Today* 30 (2020): 100828. https://doi.org/10.1016/j.nantod.2019.100828.

[26] Zamkah, Abdulaziz, Terence Hui, Simon Andrews, Nilanjan Dey, Fuqian Shi, and R. Simon Sherratt. "Identification of suitable biomarkers for stress and emotion detection for future personal affective wearable sensors." *Biosensors* 10, no. 4 (2020): 40. https://doi.org/10.3390/bios10040040

[27] Wang, Guixiang, Rui Han, Qun Li, Yinfeng Han, and Xiliang Luo. "Electrochemical biosensors capable of detecting biomarkers in human serum with unique long-term antifouling abilities based on designed multifunctional peptides." *Analytical Chemistry* 92, no. 10 (2020): 7186–7193. https://doi.org/10.1021/acs.analchem.0c00738

[28] Yang, Gaojian, Ziqi Xiao, Congli Tang, Yan Deng, Hao Huang, and Ziyu He. "Recent advances in biosensor for detection of lung cancer biomarkers." *Biosensors and Bioelectronics* 141 (2019): 111416. https://doi.org/10.1016/j.bios.2019.111416

[29] Jin, Xiaofeng, Conghui Liu, Tailin Xu, Lei Su, and Xueji Zhang. "Artificial intelligence biosensors: Challenges and prospects." *Biosensors and Bioelectronics* 165 (2020): 112412. https://doi.org/10.1016/j.bios.2020.112412

[30] Banerjee, Hirendra N., and Mukesh Verma. "Application of nanotechnology in cancer." *Technology in Cancer Research Treatment* 7, no. 2 (2008): 149–154. doi:10.1177/153303460800700208

[31] Ratajczak, Katarzyna, and Magdalena Stobiecka. "High-performance modified cellulose paper-based biosensors for medical diagnostics and early cancer screening: A concise review." *Carbohydrate Polymers* 229 (2020): 115463. https://doi.org/10.1016/j.carbpol.2019.115463

[32] Zhang, Junyu, Xiaojing Zhang, Xinwei Wei, Yingying Xue, Hao Wan, and Ping Wang. "Recent advances in acoustic wave biosensors for the detection of disease-related biomarkers: A review." *Analytica Chimica Acta* 1164 (2021): 338321. https://doi.org/10.1016/j.aca.2021.338321.

22 Sustainable Approach in the Treatment of Industrial Dyes Based on Inorganic Nanomaterials

N.G.R.H.R. Senevirathne, M. Dilshad Begum Golgeri,
Soumya V. Menon, Asha Kademane, Yashwanth Narayan,
and K.M. Nikhileshwar

INTRODUCTION

Rapid urbanization followed by industrialization is needed for fulfilling the demands of current scenario. Industries are our major sources of foreign exchange that forms a fundamental activity in the economy of any country (Jain et al. 2021; Golgeri et al. 2022). Most of the industrial effluents released after usage will be discharged into the environment which are rich in various toxic materials like organic compounds, harmful pollutants, heavy metals, synthetic dyes and many more such non-biodegradable toxicants. Among them, dye contamination is a serious concern as the usage of synthetic dyes has increased greatly in various industries ranging from paint industries, textile industries, leather industry, food, paper and pulp industries to cosmetics and pharmaceuticals (Mehta et al. 2021). Most of the dyes are known to be carcinogens that disturb the natural ecological balance (Robinson et al. 2001; Vaghela et al. 2005; John et al. 2020). Among various dyes, azo dyes represent up to 70% of dyestuffs used in textile and other industries. Dye-containing wastewater adversely affects the environment and human health (Shanker et al. 2017). There are several approaches available like physical, chemical and biological treatments. These techniques have their own limitations. However, updated and effective eco-friendly approaches of controlling dye pollutants is the primary goal of sustainable industrial treatment of toxic dyes are the promising ones. This can be achieved by adopting nanotechnology which seems to attract an increased attention from the researchers globally (Mumtaz et al. 2023; Shanker et al. 2017; Jain et al. 2021). Discovery of fullerenes and graphene has opened up doors of huge hopes in the development of nanotechnology. Usage of nanomaterials due to its large surface area, high reactivity and strong mechanical properties has revolutionized the efficacy of key techniques adopted in wastewater treatment and has provided innovative solutions for water purification and thus dye degradation (Cheriyamundath and Vavilala 2021; Golgeri et al. 2022).

DEFINITION OF NANOMATERIALS

Nanomaterials are the substances with a size range from 1 to ~100 nm. Use of nanomaterials has revolutionized in the last 15 years as the materials at nanosize have shown broad applications in broad areas ranging from health sector to food industries and sewage treatments (Armarego 2022). The physio-chemical properties of atoms or molecules differ in their nanosize from the corresponding materials in bulk size. These changes in the size have higher surface area and even reveal quantum effects due to its very small size. This property has widely accepted to incorporate in a

DOI: 10.1201/9781003479239-22

wide range of applications in treating wastewater for its decontamination process (Mumtaz et al. 2023). Nanomaterials are designed in such a considerable novelty and have applications in better redesigning the existing conventional methods of wastewater treatments, especially membrane and adsorption treatment processes (Karthigadevi et al. 2021). Advanced material science research has helped in development of many nanocomposites like polymeric, mineral, magnetic, carbon, graphene and cellulose/fibre based with potential applications in membrane-based wastewater treatment methods (Chaturvedi et al. 2020). Similarly, many nanoadsorbents/dendrimers, like metallic and mixed oxide based, have been developed with applications in adsorption-based wastewater treatment methods.

TYPES OF NANOMATERIALS

Nanomaterials can be classified into four types, namely, inorganic-based nanomaterials, organic-based nanomaterials, carbon-based nanomaterials, and composite-based nanomaterials (Figure 22.1) (Majhi, and Yadav 2021).

1. **Inorganic-Based Nanomaterials:**
   Inorganic-based nanomaterials include a variety of metals and metal oxides as the basis for nanomaterial synthesis. The most widely used metals include silver (Ag), gold (Au), copper (Cu), aluminium (Al), cadmium (Cd), zinc (Zn), lead (Pb), and iron (Fe). Metal oxides–based inorganic nanomaterials include copper oxide (CuO), iron oxide (Fe_3O_4), magnesium aluminium oxide ($MgAl_2O_4$), zinc oxide (ZnO), titanium dioxide (TiO_2), iron oxide (Fe_2O_3), cerium oxide (CeO_2), and silica (SiO_2).
2. **Organic-Based Nanomaterials:**
   These nanomaterials basically originate from organic materials that use noncovalent interactions for self-assembling and molecular designing which can transform into stable structures like micelles, ferritins, liposomes, cyclodextrins, and dendrimers (Majhi and Yadav, 2021). These types are from natural sources and hence easily degradable biologically, and they remain nontoxic and eco-friendly in the environment.

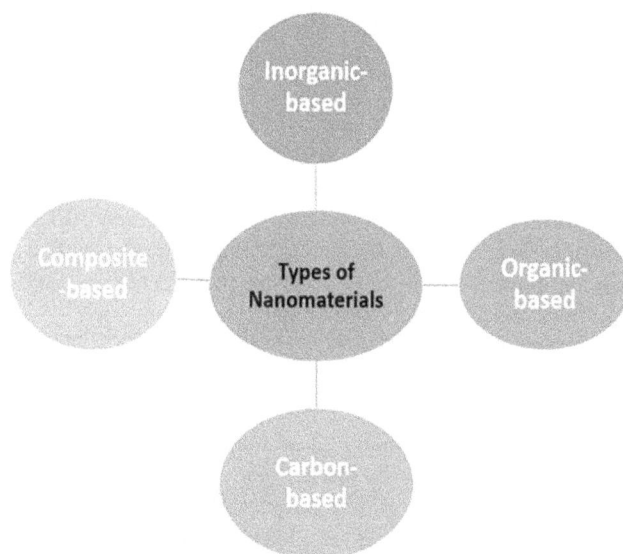

FIGURE 22.1 Different types of nanomaterials.

3. **Carbon-Based Nanomaterials:**

The sole property of catenation of carbon element that can form covalent bond with other carbons in various hybridization structures has gained advantages to use carbon sources in the synthesis of nanomaterial structures (Jain et al. 2021). These are morphologically found as ellipsoids or hollow tubes or even spheres. They include graphene, carbon nanotubes, fullerenes, and carbon nanofibers.

4. **Composite-Based Nanomaterials:**

The composite-based nanomaterials are framed by mutual combination of two or more components of nanoscale which exhibit different physical and functional properties. Such interfaces enhance the intrinsic performance and significantly show a variety of novel features that have been shown great applications in various toxic component treatments from wastewater bodies (Luo et al. 2014).

A broad spectrum of substances including elemental metals, metal oxides, and metal salts are used in inorganic nanoparticle synthesis. Inorganic nanomaterials are widely used in various sectors like chemical, electronics, photonics, energy, and medical industries (Karthigadevi et al. 2021; Mumtaz et al. 2023). Using inorganic nanomaterials as the principal component of nanotechnology used in various wastewater treatments has been gaining remarkable attention in the research community since the last decade (Majhi and Yadav 2021). Nanometals designed from gold and silver metals have been used in desalination processes for removal of toxic heavy metals such as arsenic, lead, and chromium from wastewater bodies (Mumtaz et al. 2023).

COMMONLY AVAILABLE MATERIALS

With the rapid expansion of population and its demands, urbanization, industrialization, and fossil fuel combustion have become major producers of pollutants into the natural water bodies. These industrial setups involve in the mass production of leather, plastic, paper, food, pharmaceutical, and textile products and thus make use of an array of different agents such as leaching agents, finishing chemicals, thickening agents, surface-active chemicals, metallic salts, and dyestuffs which are harmful to the flora and fauna. Most dyes cause skin discoloration, eye irritations, damaged vision, dizziness, sweating, nausea, severe headache, and other health issues. The dyes are water-soluble organic species, and hence their degraded metabolites cause deleterious effects to humans and other living organisms. High water solubility of these dyes makes it difficult to remove from water bodies. The use of dyes by human civilization runs back to the time around 3,500 B.C., and the sources from which the dyes are being derived have shifted from natural to synthetic origin. Once being released into the environment, these synthetic compounds cause major issues to life as a whole. The dye waste stream discharged from different industries shows high colour intensity, pH, suspended solids, metallic species, their salts and high chemical oxygen demand (COD), biological oxygen demand (BOD), and temperature (Joshi et al. 2021). These components are highly mutagenic and carcinogenic in nature and hence have become a serious hazard to the biotic and abiotic components. W.H. Perkins, in 1856, discovered the use of synthetic colour.

TYPES OF DYES

Traditionally dyes are categorized into natural and synthetic dyes based on their sources. The natural dyes are mainly derived from minerals, insects, and plants, while synthetic dyes are manufactured for specific purposes using various chromophores. Additionally, based on their chemical nature, synthetic dyes are classified into sulphur, phthalocyanine, anthraquinone, triarylmethane, and azo dyes. Furthermore, according to their mode of application, they may be pf direct dispersed, basic, reactive, and vat types (Regmi et al. 2018; Joshi et al. 2020).

Based on conjugated chromophoric side chains, they may be categorized into triphenyl methyl, nitrated, phthalein, nitro, azo indigo, and anthraquinone dyes (Hassaan et al. 2017). Wastewater treatment is defined as "the application of known available technologies to treat wastewater to such an extent that the quality of the treated water meets the specifications of governmental environmental regulatory agencies" (Mendes-Felipe et al. 2022; Lade 2021).

Apart from these methods (Figure 22.1), membrane filtration, oxidation, chemical precipitation, and chemical oxidation and materials like minerals, clays, and metal oxides conventional adsorbents such as commercial activated carbon, chitosan, and natural waste are extensively used. The efficacy and cost effectivity of these methods could be altered by the change of concentration, presence, and percentage of other pollutants, optimum amount required for efficient removal of each pollutant, etc. (Chong et al. 2010; Chen et al. 2019). The latest trend in the industrial dry removal from effluents is the use of organic and inorganic nanomaterials like inorganic metal nanoparticles, carbon nanotubes, etc. for effective removal of the dyes from water bodies, which helps remove the dye materials by adsorption-initiated photocatalytic degradation (Ajormal et al. 2020). Inorganic metal-based nanoparticles resemble excellent photocatalytic features (Figure 22.2), compared to their macromolecular counterparts (Melinte et al. 2019; Herrmann 2010).

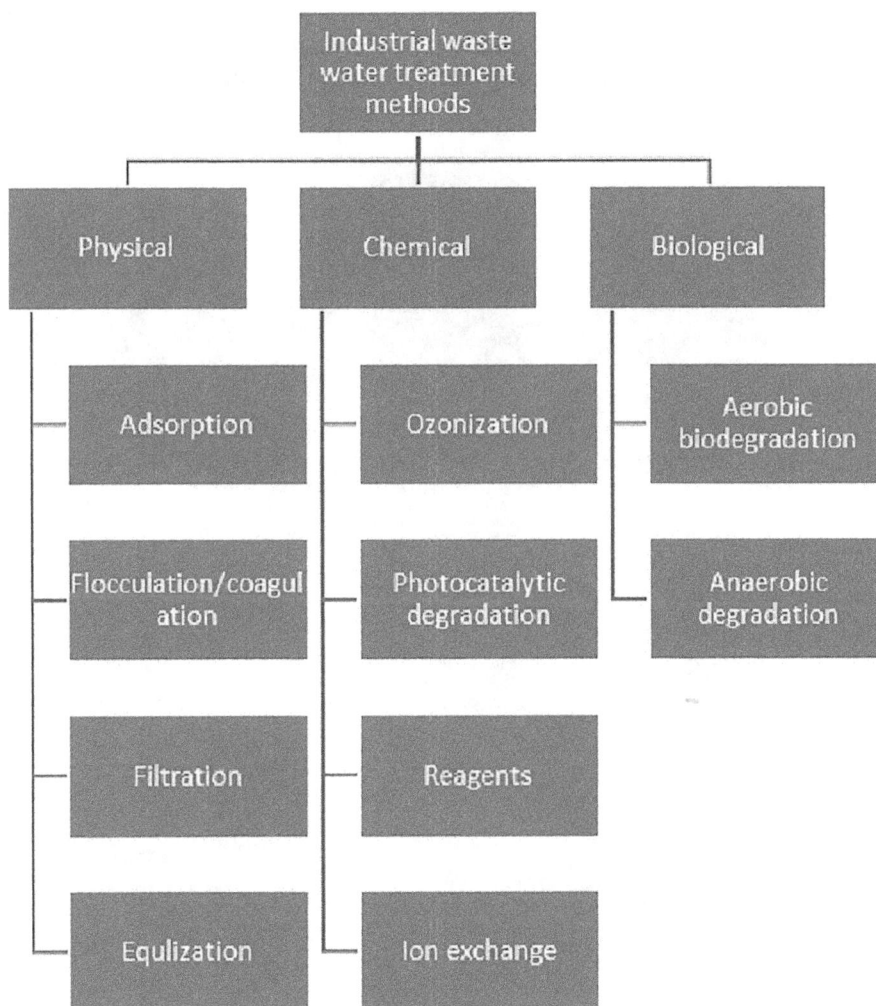

FIGURE 22.2 Conventional methods for industrial waste removal.

Hence, nanotechnology is an area that consistently grows in popularity in research due to the ability to manipulate its physical and chemical characteristics with high specificity, which allows extensive designing and synthesis of materials to achieve the desired purposes (Shrivastava et al. 2009).

FEATURES OF AN EFFECTIVE PHOTOCATALYTIC MATERIAL

The inorganic metal oxide nanoparticles have received considerable attention due to their advanced physical and chemical characteristics, influenced by their shape and size, and the material used in synthesis. Before getting popularity in large scale in the field of bio-remediation, metal oxide nanoparticles were initially debuted in various sectors like medicine, pharmaceuticals and biological sciences, material science, electronics and information technology, catalysis, and optical and sensor technology. The magnetic properties of metal oxide nanoparticles are usually based on their shape and size and the specific type of materials used in their synthesis process (Joshi et al. 2021) (Figure 22.3).

Designing of the nanoparticles with semiconductor properties such as metal and non-metal composites, and heterojunctions aims to enhance their photocatalytic abilities.

Therefore, the development of inexpensive and eco-friendly strategies to eliminate these hazardous materials from the environment and underground water has become a major challenge (Akbari et al. 2020; Ren et al. 2018; Chakhoum et al. 2018). The rapid growth rate of industry is resulted

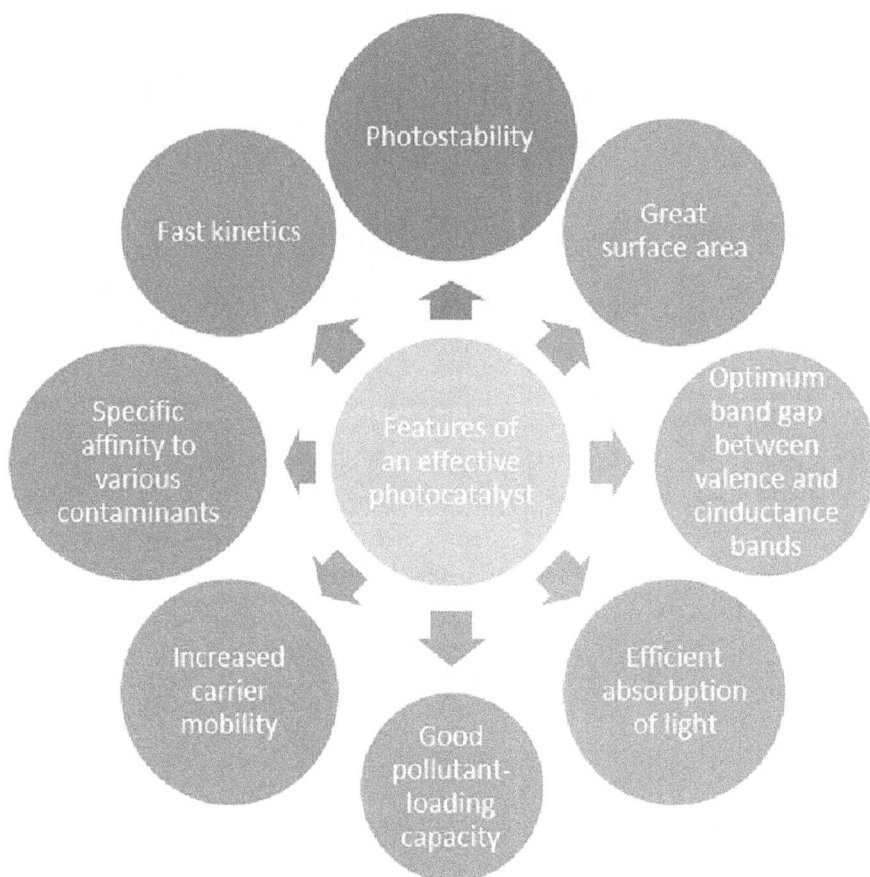

FIGURE 22.3 Features of an effective photocatalytic material (Tomar et al. 2020).

in several consequences such as environmental pollution, reduced ground water, and the rising temperature of earth sphere (Najafian et al. 2019). Most of the outstanding features of photocatalyst wastewater purification procedures include the rapid photocatalyst oxidation reaction at low concentrations, lack of byproducts formation in a multi-cycle implementation, and production of certain harmless products (Rafique et al. 2012; Motahari et al. 2014; Li et al. 2019a). The unique features of semiconductor photocatalysis method are the utilization of cheaper, nontoxic metal-based nanoparticles, possessing high chemical stability and reusable catalysts (Torki et al. 2018; Massaro et al. 2019), and also accommodating the probability of radiation that provides more efficacy in wastewater treatment (Abo-Farha 2010).

INORGANIC NANOMATERIALS USED IN DYE DEGRADATION

ZINC OXIDE NANOPARTICLES (ZnO_2)

Macromolecular zinc oxide (ZnO), a white colour powder with semiconductive properties, is a multifunctional material with unique physical and chemical properties (Kołodziejczak-Radzimska et al. 2014). It is a widely used material in manufacturing glassware, ceramics, cement, galvanized rubber, paint, adhesives, batteries, cameras, gas censors, etc. The use of nanoparticular ZnO in photocatalysis, cosmetics, sol-gel preparations having antimicrobial activity, textiles, and environmental remediation process has revolutionized industrial growth. ZnO NPs are also widely incorporated in solar cells and light-emitting diodes due to its photocatalytic properties (Joshi et al. 2021). ZnO acts as an active photocatalyst with a band gap of 3.37 eV. It is an active photocatalyst, and its wutzite form acts as n-type semiconductor with a direct band gap of 3.37 eV and a high exciton binding energy of 60 meV (Tomar et al. 2020; Sharma et al. 2018). ZnO, in particular nanosized ZnO, is a very promising semiconducting photocatalyst due to its redox potential, cost-effectiveness, and low toxicity. ZnO nanoparticles were observed to degrade 97% of Methylene blue under UV irradiation for 90 minutes (Bibi et al. 2020) and 85.91% of Remazol brilliant blue under UV irradiation for 60 minutes. The nanoparticles were tested on the dyes RB19 and RB21, and they were observed to give 100% and 91% degradation after 6 hours of UV irradiation. The effects of concentration of dye, length:diameter ratio of the particle granules, and the catalyst size for the degradation speed were depicted.

High photocatalytic efficiency, high response to UV light, and production of reactive oxygen species make these semiconductors stand out as nanoscale remediators. Doping the nanoparticles with other inorganic metals like Ni has been observed to drastically increase the antimicrobial activity (Chauhan et al. 2020a) and the photocatalytic properties (Gnanamozhi et al. 2020).

TiO_2 NANOPARTICLES FOR DYE DEGRADATION

TiO_2 nanoparticles play a role as effective photocatalysts in dye degradation, due to their electronic, optical, and mechanical properties, optimum gap between valence and conductance bands, and chemical stability. They are capable of degrading inorganic and organic pollutants. Also, they are highly efficient in integrating a wide range of organic and inorganic pollutants. They are generally used as constituents in solar cells, antibacterial agents, and wastewater treatments. According to Joshi et al. (2020), TiO_2 NPs were very efficient in degradation of Rhodamine B from the aqueous solutions. These are available rapidly, less toxic, and cost-effective (Sungur 2020; Haider et al. 2019). Improvement of photocatalytic efficacy could be achieved by integrating into organic polymers to form nanocomposites with advanced properties and nitrogen doping (Khade et al. 2015; Nabi et al. 2022). Polyvinylidene fluoride (PVDF)–TiO_2 mixed-matrix membranes via phase inversion have shown excellent removal efficiency due to increased hydrophilicity (Mendes-Felipe et al. 2022). Incorporation of carbon gels has shown enhanced photocatalytic properties (Ma et al. 2020; Yin et al. 2019).

GRAPHENE BIOMATERIAL, POLYMER-METAL OXIDE NANOCOMPOSITES FOR DYE DEGRADATION

Novel methods use inorganic metal oxides integrated into organic macromolecules to form nanocomposites (Joshi et al. 2021) using normal/reduced graphene (Ramanathan et al. 2019; Adly et al. 2019; Zhou et al. 2014), organic polymers (Gilja et al. 2017; Zhu et al. 2010; Wang et al. 2009), and biomaterials (Bahal et al. 2019; Adnan et al. 2020; Oliveira et al. 2020; Dehghani et al. 2020).

MnO$_2$ NANOPARTICLES FOR DYE DEGRADATION

Manganese dioxide (MnO$_2$)/pyrolusite is a material widely used in ceramic industries, glassmaking, batteries, etc. (Joshi et al. 2021). MnO$_2$ nanoparticles possess a unique 2D structure, which has enabled its excellent function generally in biomedicine, fluorescence sensing, magnetic resonance imaging, and cargo-loading functionality (Wu et al. 2019). The ability of MnO$_2$ nanoparticles to operate under visible light was able to successfully eliminate up to 97% of Crystal Violet dye (90 minutes) (Rahmat et al. 2019), 90.3% of Rhodamine B (120 minutes) (Chan et al. 2016), and 80% of Alizarin dye under UV irradiation (Hamza et al. 2020). Incorporation with activated carbon showed excellent efficacy with degradation of up to 98.53% of Congo red dye (5 minutes) (Khan et al. 2019).

MoS$_2$ NANOPARTICLES

Molybdenum sulphide nanoparticles function as photocatalysts in industrial dye removal. Direct redox reaction between the degraded organic dye products was done by surface adsorption, and the photogenerated holes migrated to the surface of the semiconductor catalyst, for example, Fe$_3$O$_4$@MoS$_2$/Ag$_3$PO$_4$_6% on degrading Rhodamine blue and Congo red (Guo et al. 2016). Later, the studies were able to uncover the existence of an indirect oxidation of hydroxyl radicals in the photocatalysis, and that these two reactions undergo mutual transformation with the change in reaction conditions (e.g., degradation of Methylene blue and Rhodamine blue; Zhao et al. 2017). MoS$_2$ nanomaterials are prepared using physical methods like exfoliation, sputtering, epitaxy, plasma, etc. or by chemical means such as hydrothermal, solvothermal, chemical vapour, etc. The materials of high purity is mandatory. Controlling of MoS$_2$ nanoparticles' structure and morphology and the composition of the catalyst, through (i) metallic and non-metallic modifications, (ii) conjunction with metal oxides, organic polymers and other compounds was studied to have a promising increase in dye-degrading efficacy of the nanomaterials. The rapid recombination of photogenerated electron–hole pairs resulting in low quantum efficiency poses a major drawback (Yuan et al. 2021).

SILVER AND GOLD NANOPARTICLES FOR DYE DEGRADATION

Metal nanoparticles such as silver, gold, and platinum along with the other metal-based nanoparticles possess a variety of properties, especially optical ones which contribute to their applications in the field of medicine, electronics, and agriculture (Rai et al. 2012).

Gold nanoparticles are synthesized by facilitating reduction in an aqueous medium (Freitas de Freitas et al. 2018), are low in toxicity to biological systems, and show extensive conformational flexibility, making it a risk-free candidate for safe operation in biological systems (Pan et al. 2020). Change in the size of the particles or the environmental conditions can also alter the optical properties exhibited (Soni et al. 2021). Silver nanoparticles, on the other hand, could be synthesized either through biological or chemical methods, but biological methods are preferred due its easy and rapid yield of materials with ideal size and morphology (Abdellatif et al. 2021). They possess similar mechanical, thermal, optical and electronic features as gold nanoparticles, and hence are utilized in medicine, food (Barkat et al. 2018), waste treatment operations like photocatalytic degradation

(Su et al. 2012), biosensors (Chen et al. 2007), and water filters (Jain et al. 2005). High photocatalytic efficiency, utilization of visible light, cost effectivity, and the eco-friendly (Kumar et al. 2012a, b) nature acquired through green synthesis (Chauhan et al. 2020b) are major recent advances in silver nanoparticles (Ganapathy Selvam et al. 2015).

IRON AND IRON OXIDE NANOPARTICLES FOR DYE DEGRADATION

Some of the types of iron oxide nanoparticles in use in agriculture, wastewater treatment, magnetic materials, and biomedicines are Fe_3O_4, γ-Fe_2O_3, and hematite (α-Fe_2O_3) nanoparticles (Kataria et al. 2019). As metal-based nanoparticles, they are exhibited with super para-magnetism, which is free from the difficulty in post-remedial detachment of the nanoparticles, coupled with the volume, larger surface area, etc. (Ali et al. 2016). Studies have depicted the efficacy in photocatalytic degradation of various dyes: 52% of Reactive red 4 (RR4) dye under UV irradiation (135 minutes) (Parhizkar et al. 2017), 95.61% of Trypan blue and 63.02% of Methylene blue under UV irradiation, after incorporation of Fe_3O_4 with ZnO nanoparticles (Długosz et al. 2021), >90% degradation of Methylene blue (Alagiri and Hamid 2014), and 40% of Rhodamine B and 59% of Atrazine dyes under UV irradiation (40 minutes) (Rincón Joya et al. 2019) were observed with the use of hematite (α-Fe_2O_3) (Joshi et al. 2021). Iron nanoparticles show better reactivity with oxygen (Huber 2005) owing to its larger surface area, availability in ultra-pure form, high thermal conductivity, and high magnetic properties compared to the macro-particulate iron (Roy et al. 2022).

SnO_2 NANOPARTICLES

SnO_2 nanoparticles are n-type semiconductors with efficient electrical and optical properties and a bandgap of 3.62 eV at ordinary temperature, contributing to its great photocatalytic performance and low resistivity. These nanoparticles are also used in various sensors (Shi et al. 2010; Wu et al. 2009).

Photocatalytic degradation of 90% of Methylene blue under visible light (120 minutes) using the hydrothermally synthesized SnO_2 nanoparticles (Viet et al. 2016), and rhodamine blue at 270 minutes under UV radiation (Li et al. 2018) were effective. Higher percentage of Methylene blue dye under UV irradiation (30 minutes) using smaller SnO_2 particles (Tammina et al. 2018) also was efficient. Titus et al. (2019) applied the green-synthesized SnO_2 NPs (using methanolic extract of *Arachis hypogaea*) to degrade Congo red under ultraviolet light; 89% degradation was recorded after 50 minutes. Green-synthesized, rod-shaped (21 nm) SnO_2 NPs (using methanolic extract of *Cyphomandra betacea*) were revealed to perform high photocatalytic degradation of Methylene blue dye (Elango et al. 2016).

CERIUM OXIDE (CeO_2) NANOPARTICLES

Cerium dioxide (CeO_2), or ceria, has the highest stability among the lanthanide series of elements, aided by its +4 oxidation. Nanoparticles derived from this also has applications in biomedicine, drug delivery, bio-scaffolding, and bioremediation (Xu et al. 2014). The highest rate of complete photodegradation of Methylene blue was achieved using 1.0 g/L CeO_2 (pH 11) under UV irradiation (125 minutes) (Pouretedal and Kadkhodaie 2010), in contrast to the degradation achieved under different parameters in a span of 175 minutes (Majumder et al. 2019). Moreover, complete degradation of trypan blue under UV irradiation was achieved within 135 minutes (Ravishankar et al. 2015).

COPPER AND COPPER OXIDE (CuO) NANOPARTICLES

Copper (II) oxide nanoparticles are generally brownish-black in colour, monoclinic in structure with increased thermal conductivity, stability, selectivity, photo-absorptivity, and antimicrobial activities (Yecheskel et al. 2013, Ishio et al. 2012, Wang et al. 1999). More than 96% degradation of Victoria blue and Direct red 81 (DR 81) was resulted in the use of CuO nanoparticles under UV irradiation (Chauhan et al. 2020a). A considerably higher rate was studied for degradation with green-synthesized CuO nanoparticles: 93% of Nile blue and 81% of Yellow 60 (120 minutes) with nanoparticles made from *Psidium guajava* leaves extract (Singh et al. 2019) and large percentage of Congo red degradation with nanoparticles made from *Drypetes sepiaria* leaves extract (Narasaiah et al. 2017). Copper nanoparticles are a more cost-effective substitute for gold and silver counterparts (Ingle et al. 2014) and therefore has a high demand in technology and agriculture (Mehata 2021).

NICKEL OXIDE NANOPARTICLES FOR DYE DEGRADATION

NiO is a semiconductor of high stability, zero toxicity, and increased photo-activity (Li et al. 2019b) which has a wide range of applications in comparison to other metal oxide nanoparticles and has superior ferromagnetic properties and chemical stability (Jaji et al. 2020). NO nanoparticles are in high demand due to wide-spread usage in synthesizing sensors (Ali 2010), memory storage devises (Vakili et al. 2014), cosmetics (Baruah et al. 2018), pharmaceuticals (Rajendrachari et al. 2021), in operations like drug delivery (Adane et al. 2021, Fakruddin et al. 2012), magnetic resonance imaging (Baig et al. 2021, Yaqoob et al. 2020), catalysis (Verma et al. 2021), biomedicine (Rafique et al. 2019), space technology (Madiha et al. 2018), optoelectronics and electronics (Marchiol et al. 2020), magnetism (Baruah et al. 2019), and photocatalysis (Rasheed et al. 2020, Mani et al. 2020), environmental science and energy (Saratale et al. 2018), and recently in treating cancer (Thirumurugan et al. 2017). NiO nanoparticles usually contain an energy band of about 3.4 eV (Jaji et al. 2020) which contributes to high photo-catalysed degradation potential of dyes. Electrons' excitation through irradiation of NiO-NPs can help destroy a wide range of contaminants. Nickel oxides are commonly utilized as it is a competent photocatalyst material and environment friendly in degradation of azo dyes. According to Roy et al. (2022), some of the major factors were pH, light source, amount of photocatalyst, the initial concentration of dyes, mineral ion content, and solvent type.

NANOCOMPOSITES

Nanocomposites are hybrid nanomaterials of 1–100 nm scale made in association of organic nanomaterials like natural and synthetic polymers, graphene (reduced/non-reduced), and inorganic nanoparticles like metal oxides. They could be categorized into three main categories, namely, (i) inorganic, (ii) polymeric [organic], and (iii) inorganic–organic hybrid nanocomposites (Mendes-Felipe et al. 2022), and six sub-classes: (i) graphene oxide–metal oxide nanocomposites, (ii) polymer/metal oxide nanocomposites, (iii) metal oxide–biological material nanocomposites, (iv) metal oxide–metal oxide nanocomposites, (v) metal oxide–silica nanocomposites, and (vi) metal–metal oxide nanocomposites (Joshi et al. 2022). These materials have a vast array of applications such as remediation of inorganic and organic pollutants, gas sensing, electronic devices, biomedical fields, photocatalytic degradation of organic dyes, etc. due to its good mechanical strength, porosity, toughness, electrical and thermal conductivities, dispersibility, and high mobility (Jain et al. 2005; Kataria, et al. 2019; Parhizkar et al. 2017). Different types of nanocomposites, methods of synthesis, and the dyes degraded by them are listed in Table 22.1. Organic–inorganic hybrid materials, synthesized by integrating nanomaterials into solid organic polymers, pose a beneficial modification. It has been observed to improve the efficacy with excellent chemical and physical properties, and nullify the faults exhibited in the nanomaterials individually. The organic polymer materials contribute with great mechanical and thermostable properties, while the nanoparticles enhance the separation efficiency and anti-fouling properties and hydrophilicity (Mendes-Felipe et al. 2022).

TABLE 22.1

Inorganic Nanomaterials Used in Industrial Dye Remediation, Methods of Synthesis, and Types of Dyes Degraded by Each Nanomaterial

Inorganic Nanoparticle	Method of Synthesis	Dyes Degraded
ZnO_2	Sol-gel, Green	Methylene blue B, Reactive red 120
TiO_2	Colloidal, Green	Methylene blue
MnO_2	Microemulsion	Crystal violet, Rhodamine B, Congo red, Alizarin red
MoS_2	Physical methods	Rhodamine blue
	Mechanical exfoliation, sputtering, epitaxy, plasma method	
	Chemical methods	
	Chemical vapour deposition method, hypothermal method, solvothermal method	
Silver and gold nanoparticles	Ag reduction of silver nitrate to aqueous silver metal ions	Methyl orange, Methylene blue, Rhodamine blue
Fe_2O_3	Microwave-assisted	Rhodamine B
		Methyl orange
		Eosin yellowish, Methylene blue
		Indigo carmine
Fe_3O_4	Microwave-assisted, Solvothermal	Rhodamine B
		Methylene blue
Fe^o	Used in ultra-pure form	Methylene blue
ZrO_2	Laser ablation, sol-gel, sono-chemical, hydrothermal method, chemical co-precipitation, combustion	Methyl orange
		Methylene blue
SnO_2	Colloidal, template	Methylene blue, Rhodamine blue
NiO_2	Sol-gel method	Methyl orange, Rhodamine blue, and Methylene blue
CeO_2	Sono-chemical, combustion	Trypan blue, Methylene blue
CuO	Microwave-assisted, vapour deposition, thermal deposition, radiolysis reduction, chemical reduction of metal salts of copper, electrochemical reduction, hydrazine hydrate and starch (RT)	

Inorganic Nanocomposites

1. Metal/Metal Oxide

Ag/ZnO	Combustion, chemical precipitation, chemical co-precipitation	Congo red
Cu–TiO_2/CuO	Chemical co-precipitation, pyrolysis, electrolysis deposition, thermal decomposition	Methylene blue
Zn/SnO_2	Pyrolysis, hydrothermal method, flame synthesis method, sol-gel method	Methylene blue
Ag/TiO_2	Hydrothermal method, green synthesis	Methylene blue
Co/ZnO	Chemical co-precipitation, sol-gel method	Malachite green, Methyl orange
Mg/ZrO_2	Green synthesis, sol-gel method, chemical co-precipitation	Methyl violet
		Methyl blue
Ag/ZrO_2	Microwave-assisted method, Chemical co-precipitation	Rhodamine B

2. Metal Oxide/Silica

Fe_3O_4/SiO_2	Stöber method, reverse microemulsion method, chemical co-precipitation	Methylene blue
		Procion red MX-5B

(Continued)

TABLE 22.1 (*Continued*)

Inorganic Nanomaterials Used in Industrial Dye Remediation, Methods of Synthesis, and Types of Dyes Degraded by Each Nanomaterial

Inorganic Nanoparticle	Method of Synthesis	Dyes Degraded
ZnO/SiO_2	Chemical precipitation, sol-gel method, sonication-dispersion method	Methylene blue Rhodamine B
TiO_2/SiO_2	Sol-gel method	Malachite green
SnO_2/SiO_2	Sol-gel method, hydrothermal method	Orange II Rhodamine B Crystal violet
ZrO_2/SiO_2	Stöber method, sol-gel method	Rhodamine B Methylene blue
3. Metal Oxide/Metal Oxide		
CuO/ZnO	Thermal decomposition, sol-gel method, wet impregnation, chemical co-precipitation	Congo red
Cu_2O-CuO/TiO_2	Sol-gel method, one-step electrolyte decomposition, chemical thermal oxidation	Reactive blue 49 (RB 49)
CuO/TiO_2	Thermal decomposition, chemical co-precipitation	Rhodamine B Methylene blue
Fe_3O_4/ZrO_2	Microwave-assisted, sol-gel method, ultrasonic-assisted method	Methyl red Methyl orange
4. Metal Oxide/Inorganic Molecules		
Fe_3O_4/H_2O_2	Gas-phase flame synthesis, reverse co-precipitation	Azo dye Rhodamine B

METHODS OF SYNTHESIS

Different synthesis methods have been developed to control the grain size, shape, stability, and surface chemistry of nanomaterials as shown in Figure 22.1 (Akbari et al. 2020; Duan et al. 2012). The basic synthetic methods used to synthesize metal oxide nanoparticles are generally categorized into (i) liquid phase, (ii) vapour state-based, and (iii) biological/green methods (Table 22.2). The selection of synthetic methods determines many physical and chemical characteristics of metal oxide nanoparticles. These properties include particle size, crystal appearance, shape of particles, dispersity, and type of intrinsic or extrinsic defects (Figure 22.4).

FACTORS AFFECTING DYE DEGRADATION EFFICACY

Concentration of the dyes in the water body, type of dye being degraded, concentration of other contaminant materials, band gap between the conductance and valence band of photocatalysts (Joshi et al. 2022), pH, light source, amount of photocatalyst, initial concentration of dyes, light intensity, radiation time, presence of mineral ions, and type of solvent were observed in several photocatalytic activity of different types of nanoparticles, since these nanoparticles exhibit properties that can be controlled by changing size, doping, surface modification, or sensitization, through which their reactivity can be altered or enhanced. Moreover, doping a certain amount of metal ions as well as deposition of organic or inorganic materials on the surface of semiconductor nanoparticles can enhance their photocatalytic activity by interfacial charge transfer and electronic interaction between the surface attachment and the host semiconductor (Mendes-Felipe et al. 2022). The mechanism of action of dye degradation is depicted in Figure 22.2.

TABLE 22.2

Methods of Inorganic Nanoparticle Synthesis for Water Remediation (Joshi et al. 2021)

Method	Description
1. Liquid Phase-Based Synthetic Methods	
Colloidal methods	Solutions containing different ions are mixed under controlled parameters, and an insoluble precipitate of easily separable nanoparticles is formed. Frequently used in the synthesis of metal, metal oxide, organics, and other nanoparticles.
Sol-gel methods	Based on the interactions between sol (containing solid particles) and the gel (containing solid macromolecule). Leading to hydrolysis of precursors. Widely used and well-established methods utilized mostly when materials with novel and predefined properties are used.
Sono-chemical method	Ultrasonic vibrations are used in the breaking of chemical bonds between material molecules, causing compression and relaxation, followed by a large amount of formation and collapsing of acoustic cavitations, releasing a large amount of energy into the medium.
Solvothermal method	Precursors are dispersed into a suitable solvent and then applied with moderate temperature and pressure to synthesize different nano-based materials, including metal oxide-based materials.
Microemulsion method	Two immiscible phases (oil and water) containing metallic precursors are separated through surfactant molecules and are stirred under normal conditions to achieve a homogenous mixture.
Microwave-assisted method	Reaction usually suffices with low energy and temperature but could be sped up by applying high heating rates, nucleation, and highly dispersed particles which enhances the ionic conductivity and polarization.
2. Vapour-State Synthesis Methods	
Laser ablation method	Nanoparticles through the irradiation of colloidal solutions originating from the bulk materials in water or other solutions. The particles size could be manipulated by adjusting on the laser fluence and nature of the liquid medium. Used to synthesize stable nanoparticles in the absence of stabilizing agents.
Template synthesis	The template methods involve the control of crystal growth and nucleation. Comprises three steps: (i) preparation of the template, (ii) applicability of suitable synthetic approaches (sol-gel, precipitation, etc.), and (iii) removal of the template (natural/synthetic template).
Combustion method	Also called self-propagating high-temperature synthesis and commonly used to synthesize metal oxide nanoparticles. Highly efficient and energy-saving, and can take place through the solid-, liquid-, and gas-phase combustions, and the maximum synthesis temperatures are limited by the thermodynamics of the systems. Combustion methods can be proceeded in solid, liquid, and gas phases.
3. Biological or green methods	Used in synthesizing metal oxide nanoparticles. Have gained much attention due to their efficiency, low cost, and most importantly, for being environmentally friendly. The biological extracts containing different secondary metabolites generally obtained from plants, algae, fungi, etc. are used as reducing, capping, or stabilizing agents during the synthesis of different metal and metal oxide nanoparticles.

MECHANISM OF ACTION

Photocatalytic reaction occurs through light absorption by the photocatalyst and excitation of the semiconductor photocatalysts electron from the valence band to the conduction band and recombination of e^-/h^+ pair due to the low stability of the photogenerated electrons and holes, resulting in degraded dye products (Figures 22.4 and 22.5).

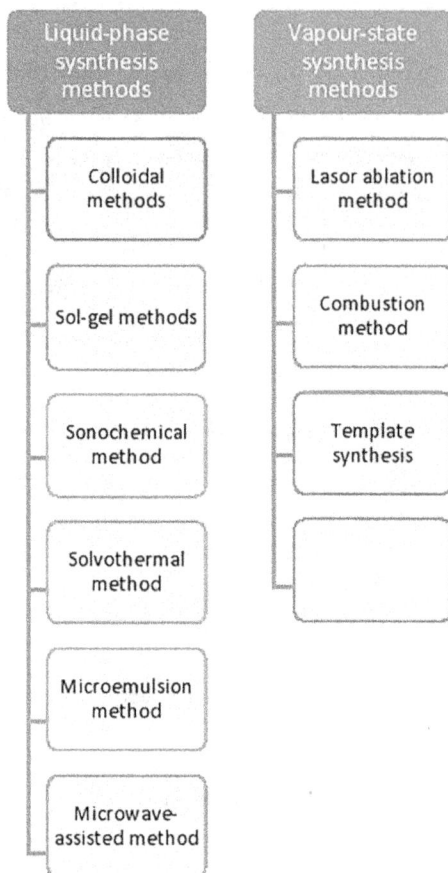

FIGURE 22.4 Different methods used in dye-degrading inorganic metal nanoparticle synthesis.

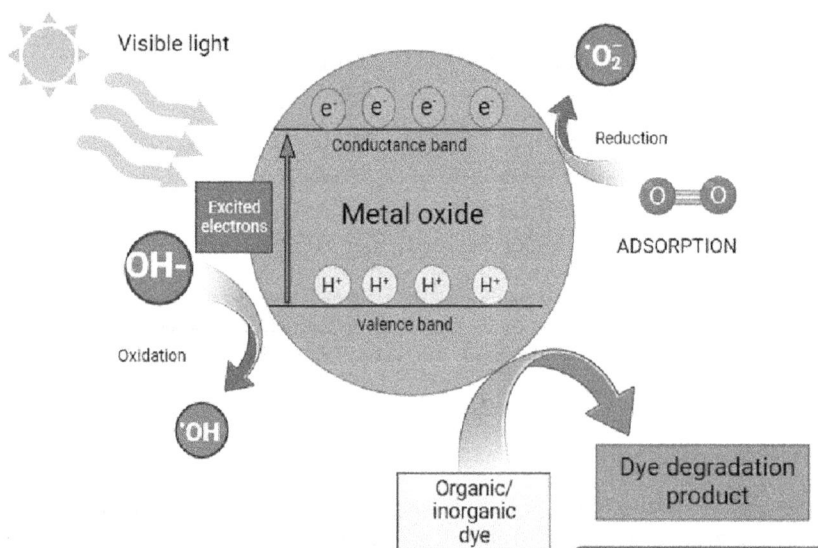

FIGURE 22.5 Photocatalytic dye degradation mechanism.

DYE DEGRADATION MECHANISMS BASED ON NANOADSORBENTS AND REDUCTANTS

Per the findings of a few research works, nanoparticles (NPs) possess substantial adsorption surface areas, effective catalytic capabilities, and conductivity, all of which contribute to their ability to facilitate the degradation of dyes.

The four primary categories of nanomaterials that can play a pivotal role in the degradation of dyes are nanophotocatalysts, nanoreductants, nanomembranes, and nanosorbents.

Nanoreductants: Reductive degradation is a fast and cheap method that is easy to apply for removing industrial-level dyes. Several types of particles, including nanoparticles, have been used to eliminate azo dyes. One of them is nanoscale zero-valent iron (nZVI), which forms necklace-like clusters due to its small size, as shown by experimental studies. However, nZVI systems are only highly reactive when the wastewater is acidic.

Another type of particles is bimetallic nZVI, which consist of zerovalent iron and a noble metal. In this case, the iron acts as an anode and gets oxidized to protect the noble metal, which acts as a cathode. The noble metal either transfers electrons directly to the pollutants on the surface of the bimetallic nZVI, or the hydrogen produced by the iron oxidation reacts with them. Different experiments have shown that Fe–Pd is more effective than other combinations.

A recent development in the field of metallic materials research is high-entropy alloys (HEAs), which have multiple elements in equal or near-equal proportions. HEAs have some advantages for dye degradation, such as fast decolourization, environmental friendliness, low cost, and low efficiency.

Other metallic materials that have been reported to have good decolourization properties for azo dyes are zerovalent iron, Mn–Al binary alloys, and Al-based metallic glass alloys.

Nanophotocatalysts: Photocatalytic reactions involving nanoparticles play a crucial role in eliminating various pollutants. These processes harness light energy and metallic nanoparticles, typically composed of semiconductor metals, to decompose numerous stubborn organic substances such as dyes, detergents, and volatile organic compounds. TiO_2 nanoparticles are among the most prevalent photocatalysts and have been proposed for dye degradation. Alternatively, silver nanoparticles, which can be generated through green synthesis, can be employed to treat dyes and other organic compounds. Gold nanoparticles (GNP) with a diameter of 43 nm have also been experimented for the degradation of Methylene blue, a dye frequently used as a representative pollutant. A sustainable method for synthesizing silver nanoparticles (AgNPs) using *Blumea lacera* leaf extract, which is a rich source of biomolecules that can act as reducing and stabilizing agents. This method is known as "green synthesis", and it is preferred over conventional methods because it is affordable, eco-friendly, and biocompatible. The silver salt used for this method is also cheaper than gold salt, which is another option for producing nanoparticles.

The AgNPs are then applied for the photocatalytic degradation of Methylene blue, an organic dye that contains harmful aromatic compounds. Removing these dyes from wastewater is a major challenge, and nanomaterials offer a promising solution. The study examines how the AgNPs enhance the degradation of Methylene blue in the presence of sodium borohydride ($NaBH_4$) as a catalyst. The study also measures the absorbance intensity of Methylene blue by $NaBH_4$ and the role and effectiveness of Ag as a catalyst.

Nanomembranes: Carbon nanomembranes (CNMs) are synthetic 2D carbon sheets with tailored physical or chemical properties. These depend on the structure, molecular composition, and surroundings on either side. Due to their molecular thickness, they can be regarded as interfaces without bulk separating regions of different gaseous, liquid, or solid components and controlling the materials exchange between them. Here, a universal scheme for the fabrication of 1 nm thick, mechanically stable, functional CNMs is presented. CNMs can be further modified, for example perforated by ion bombardment or chemically functionalized by the binding of other molecules onto the surfaces.

The underlying physical and chemical mechanisms are described, and examples are presented for the engineering of complex surface architectures, e.g., nanopatterns of proteins, fluorescent dyes, or polymer brushes. A simple transfer procedure allows CNMs to be placed on various support structures, which makes them available for diverse applications: supports for electron and X-ray microscopy, nanolithography, nanosieves, Janus nanomembranes, polymer carpets, complex layered structures, functionalization of graphene, and novel nanoelectronic and nanomechanical devices.

Nanosorbents: Carbon-based materials have demonstrated their superior adsorbent properties to remove organic and inorganic contaminants from water. Among carbon-based materials, carbon nanotubes (CNTs) have emerged as a possible alternative. CNTs represent a carbon allotrope in which a hexagonal lattice of carbon atoms in SP^2 hybridization (graphitic sheets) are rolled in a cylinder-shaped structure. CNTs are classified into two main categories depending on the number of graphene sheets rolled to form the tube: single-walled CNTs (SWCNTs) that consist on a single layer of carbon atoms, and multi-walled CNTs (MWCNTs) that consist up to dozens of carbon layers. Moreover, SWCNTs exhibit a diameter between 0 and 3 nm, while MWCNTs can reach up to 100 nm of diameter.

As efficient nanosorbents, biogenic nanoparticles have been developed among research groups worldwide for wastewater treatment from pharmaceutical industries. The formation of this kind of nanoparticles is given by the presence of bioactive molecules such as peptides, enzymes, vitamins, alkaloids, and phenolics, among others, obtained from different biosources (e.g., plant extracts, bacteria, fungi) that are combined with a metal salt solution under different conditions depending on the nature of precursor materials and the final intended use of NPs.

ADVANTAGES AND DISADVANTAGES

The use of photocatalytic wastewater treatment facilitates direct and complete degradation of pollutants completely, with no energy consumption, less toxicity, and high stability. Nanomaterials serve the benefit of larger surface area, high stability, and the luxury of designing from the physical and chemical characteristics to achieve the nanomaterials of desire (Joshi et al. 2022). Separation of photocatalyst compounds post-treatment from water bodies could lead to higher recovery cost and change in efficacy with the change in reaction conditions, which are major drawbacks observed. However, these drawbacks are eliminated with the emerging of magnetic metal oxide nanoparticle usage, immobilization of the photocatalysts onto a membrane support, and incorporation of organic materials and other inorganic nanoparticles in compensation of the characteristics required for undisturbed and highly efficient dye degradation amidst varying external factors (Mendes-Felipe et al. 2022). Despite being an equally sought material for bioremediation, care must be taken while adding nanoparticulate-inorganic fillers, since some materials like calcium carbonate, mica, and glass fibres have shown to mask some beneficial mechanical properties of the membranes (Kang and Cao 2014).

FUTURE PERSPECTIVES

Green synthesis, or biological synthesis, of nanoparticulate remediation materials is gaining attention amidst the rising concerns over environmental safety and the awareness towards the importance of green chemistry (Akbari et al. 2020). The membranes formed by metal–organic frameworks are one of the newest and most versatile developments in nanoparticle formation. The semiconductive nature of some metal–organic frameworks presented when exposed to light, makes them suitable for photocatalytic dye degradation mechanisms. Green synthesis needs to be highlighted in more detail, and bioavailability and cost-effectiveness of inorganic materials need to be explored in the future studies.

The textile and dyeing industry has grown rapidly in recent years, fuelled by the rising demand for coloured fabrics and clothing. However, this growth has also caused severe environmental problems due to the release of coloured wastewater into water bodies. This wastewater contains various organic dyes, which can harm aquatic life and human health. Therefore, researchers have been looking for innovative and sustainable ways to degrade dyes. One of the most promising solutions is the use of nanoparticles. In this essay, we will discuss the future prospects of nanoparticles in dye degradation, highlighting their advantages, challenges, and environmental benefits.

Nanoparticles are materials that have sizes at the nanoscale, usually between 1 and 100 nm. They have unique physical and chemical properties, which make them very effective in different applications, including environmental cleanup. For dye degradation, nanoparticles have several benefits:

Increased Catalytic Activity: Nanoparticles can act as powerful catalysts, speeding up the degradation of organic dyes through different mechanisms, such as photocatalysis and Fenton-like reactions. Materials like titanium dioxide (TiO_2), zinc oxide (ZnO), and iron-based nanoparticles have shown remarkable catalytic activity in degrading dye molecules.

Surface Modification: Nanoparticles can be modified with specific groups or coatings to improve their affinity for certain dye molecules. This adjustability allows for selective dye degradation, reducing the production of harmful byproducts.

Photocatalysis: Semiconductor nanoparticles like TiO_2 and ZnO have excellent photocatalytic properties when they are exposed to ultraviolet (UV) or visible light. This feature enables the degradation of dyes under sunlight, making the process energy-efficient and eco-friendly.

Large Surface Area: Nanoparticles have a large surface-to-volume ratio, which enables efficient adsorption of dye molecules and increases the likelihood of successful degradation reactions.

Green Synthesis of Nanoparticles: The green synthesis of nanoparticles is a new field of research. It uses natural extracts, microorganisms, or waste materials to produce nanoparticles in an eco-friendly way. These methods not only lower the carbon emissions but also offer low-cost and scalable options for large-scale nanoparticle production.

Nanoparticles and Advanced Oxidation Processes (AOPs): Nanoparticles can be combined with advanced oxidation processes, such as Fenton-like reactions and sono-chemical processes, to boost the degradation of dyes. By using nanoparticles with AOPs, the efficiency of dye removal can be greatly increased and the applicability of nanoparticle-based systems can be extended to a broader range of dye compounds.

Despite the great potential of nanoparticles in dye degradation, there are also some challenges that need to be overcome to fully utilize their potential:

Toxicity Issues: The possible toxicity of nanoparticles to aquatic organisms and humans raises concerns about their environmental impact. Research is ongoing to develop biodegradable nanoparticles or encapsulate them to prevent their release into the environment.

Scalability and Cost: The production of nanoparticles at a large scale can be costly and energy-consuming. Future research should focus on low-cost synthesis methods to make nanoparticles more available for industrial applications.

Stability and Reusability: Keeping the stability of nanoparticles during repeated use is essential for their practical application. Researchers are exploring ways to improve the durability and recyclability of nanoparticles in dye degradation processes.

Regulations and Safety: As nanoparticles become more widely used, regulations and safety guidelines must be developed to ensure their responsible use and disposal.

CONCLUSION

With the emerging population and the equally rising basic demands, water poses the most sought as well as the most difficult to recover from damage. The decrease of ground and underground water resources warns the need of saving, minimizing of waste, and the reusing of the water resources. In an era where humanity is challenged to suffice the remaining minute amount of fresh water available, usage and improvement of bioremediation techniques pose as the timeliest action. The constant advancement of high-end technologies like nanotechnology provides keen solutions for recovery of the depleting water resources, with high efficacy and less wastage in every manner.

REFERENCES

Abdellatif, Ahmed AH, Hamad NH Alturki, and Hesham M. Tawfeek. "Different cellulosic polymers for synthesizing silver nanoparticles with antioxidant and antibacterial activities." *Scientific Reports* 11, no.1 (2021): 84. https://doi.org/10.1038/s41598-020-79834-6

Abdullah Sani, Nor Syazwani, Wei Lun Ang, Abdul Wahab Mohammad, Alireza Nouri, and Ebrahim Mahmoudi. "Sustainable synthesis of graphene sand composite from waste cooking oil for dye removal." *Scientific Reports* 13, no. 1 (2023): 1931.

Abo-Farha, S. A. "Photocatalytic degradation of monoazo and diazo dyes in wastewater on nanometer-sized TiO$_2$." *Journal of American Science* 6, no. 11 (2010): 130–142.

Adane, Teshale, Amare Tiruneh Adugna, and Esayas Alemayehu. "Textile industry effluent treatment techniques." *Journal of Chemistry* 2021 (2021): 1–14. https://doi.org/10.1155/2021/5314404.

Adly, M. S., Sh M. El-Dafrawy, and S. A. El-Hakam. "Application of nanostructured graphene oxide/titanium dioxide composites for photocatalytic degradation of rhodamine B and acid green 25 dyes." *Journal of Materials Research and Technology* 8, no. 6 (2019): 5610–5622. https://doi.org/10.1016/j.jmrt.2019.09.029.

Adnan, Mohd Azam Mohd, Bao Lee PHOON, and Nurhidayatullaili Muhd Julkapli. "Mitigation of pollutants by chitosan/metallic oxide photocatalyst: A review." *Journal of Cleaner Production* 261 (2020): 121190. https://doi.org/10.1016/j.jclepro.2020.121190.

Ajormal, F., F. Moradnia, S. Taghavi Fardood, and Ali Ramazani. "Zinc ferrite nanoparticles in photo degradation of dye: Mini-review." *Journal of Chemical Reviews* 2 (2020): 90–102.

Akbari, Alireza, Zahra Sabouri, Hasan Ali Hosseini, Alireza Hashemzadeh, Mehrdad Khatami, and Majid Darroudi. "Effect of nickel oxide nanoparticles as a photocatalyst in dyes degradation and evaluation of effective parameters in their removal from aqueous environments." *Inorganic Chemistry Communications* 115 (2020): 107867.

Alagiri, M., and Sharifah Bee Abdul Hamid. "Green synthesis of α-Fe$_2$O$_3$ nanoparticles for photocatalytic application." *Journal of Materials Science: Materials in Electronics* 25 (2014): 3572–3577. https://doi.org/10.1007/s10854-014-2058-0.

Ali, Attarad, Hira Zafar, Muhammad Zia, Ihsan ul Haq, Abdul Rehman Phull, Joham Sarfraz Ali, and Altaf Hussain. "Synthesis, characterization, applications, and challenges of iron oxide nanoparticles." *Nanotechnology, Science and Applications* (2016): 49–67. https://doi.org/10.2147/NSA.S99986.

Ali, Hazrat. "Biodegradation of synthetic dyes-a review." *Water, Air, & Soil Pollution* 213 (2010): 251–273. https://doi.org/10.1007/s11270-010-0382-4.

Bahal, M., N. Kaur, Nidhi Sharotri, and Dhiraj Sud. "Investigations on amphoteric chitosan/TiO2 bionanocomposites for application in visible light induced photocatalytic degradation." *Advances in Polymer Technology* 2019 (2019). https://doi.org/10.1155/2019/2345631.

Baig, Nadeem, Irshad Kammakakam, and Wail Falath. "Nanomaterials: A review of synthesis methods, properties, recent progress, and challenges." *Materials Advances* 2, no. 6 (2021): 1821–1871. https://doi.org/10.1039/D0MA00807A.

Barkat, Md A., Sarwar Beg, Mohd Naim, Faheem H. Pottoo, Satya P. Singh, and Farhan J. Ahmad. "Current progress in synthesis, characterization and applications of silver nanoparticles: Precepts and prospects." *Recent Patents on Anti-Infective Drug Discovery* 13, no. 1 (2018): 53–69. https://doi.org/10.2174/1574891X12666171006102833.

Baruah, Debjani, Monmi Goswami, Raj Narayan Singh Yadav, Archana Yadav, and Archana Moni Das. "Biogenic synthesis of gold nanoparticles and their application in photocatalytic degradation of toxic dyes." *Journal of Photochemistry and Photobiology B: Biology* 186 (2018): 51–58. https://doi.org/10.1016/j.jphotobiol.2018.07.002.

Baruah, Debjani, Raj Narayan Singh Yadav, Archana Yadav, and Archana Moni Das. "*Alpinia nigra* fruits mediated synthesis of silver nanoparticles and their antimicrobial and photocatalytic activities." *Journal of Photochemistry and Photobiology B: Biology* 201 (2019): 111649. https://doi.org/10.1016/j.jphotobiol.2019.111649.

Bibi, I., S. Kamal, Z. Abbas, S. Atta, F. Majid, K. Jilani, A. I. Hussain, A. Kamal, S. Nouren, and A. Abbas. "A new approach of photocatalytic degradation of remazol brilliant blue by environment friendly fabricated zinc oxide nanoparticle." *International Journal of Environmental Science and Technology* 17 (2020): 1765–1772.

Chakhoum, M. A., A. Boukhachem, M. Ghamnia, N. Benameur, N. Mahdhi, K. Raouadi, and M. Amlouk. "An attempt to study (111) oriented NiO-like TCO thin films in terms of structural, optical properties and photocatalytic activities under strontium doping." *Spectrochimica Acta Part A: Molecular and Biomolecular Spectroscopy* 205 (2018): 649–660.

Chan, Yim-Leng, Swee-Yong Pung, Srimala Sreekantan, and Fei-Yee Yeoh. "Photocatalytic activity of β-MnO_2 nanotubes grown on PET fibre under visible light irradiation." *Journal of Experimental Nanoscience* 11, no. 8 (2016): 603–618. https://doi.org/10.1080/17458080.2015.1102342.

Chaturvedi, Garima, Amandeep Kaur, Ahmad Umar, M. Ajmal Khan, H. Algarni, and Sushil Kumar Kansal. "Removal of fluoroquinolone drug, levofloxacin, from aqueous phase over iron based MOFs, MIL-100(Fe)." *Journal of Solid State Chemistry* 281 (2020): 121029. https://doi.org/10.1016/j.jssc.2019.121029.

Chauhan, Ankush, Ritesh Verma, Swati Kumari, Anand Sharma, Pooja Shandilya, Xiangkai Li, Khalid Mujasam Batoo, Ahamad Imran, Saurabh Kulshrestha, and Rajesh Kumar. "Photocatalytic dye degradation and antimicrobial activities of Pure and Ag-doped ZnO using Cannabis sativa leaf extract." *Scientific Reports* 10 (2020a): 7881.

Chauhan, Moondeep, Navneet Kaur, Pratibha Bansal, Rajeev Kumar, Sesha Srinivasan, and Ganga Ram Chaudhary. "Proficient photocatalytic and sonocatalytic degradation of organic pollutants using CuO nanoparticles." *Journal of Nanomaterials* 2020 (2020b): 1–15. https://doi.org/10.1155/2020/6123178

Chen, Hao, Feng Gao, Rong He, and Daxiang Cui. "Chemiluminescence of luminol catalyzed by silver nanoparticles." *Journal of Colloid and Interface Science* 315, no. 1 (2007): 158–163.

Chen, Wensong, Jiahao Mo, Xing Du, Zhien Zhang, and Wenxiang Zhang. "Biomimetic dynamic membrane for aquatic dye removal." *Water Research* 151 (2019): 243–251.

Cheriyamundath, Sanith, and Sirisha L. Vavilala. "Nanotechnology-based wastewater treatment." *Water and Environment Journal* 35 (2021): 123–132. https://doi.org/10.1111/WEJ.12610.

Chong-Bang Zhang, Jiang Wang, Wen-Li Liu, Si-Xi Zhu, Han-Liang Ge, Scott X. Chang, Jie Chang, Ying Ge. "Effect of plant diversity on microbial biomass and community metabolic Profiles in a Full-Scale Constructed Wetland." *Journal of Ecological Engineering* 36 (2010): 62–68. https://doi.org/10.1016/j.ecoleng.2009.09.010.

Conceição, Pedro. *Human Development Report 2019: Beyond Income, Beyond Averages, Beyond Today: Inequalities in Human Development in the 21st Century* (2019). United Nations Development Programme, United Nation.

Dehghani, Mostafa, Humayun Nadeem, Vikram Singh Raghuwanshi, Hamidreza Mahdavi, Mark M. Banaszak Holl, and Warren Batchelor. "ZnO/cellulose nanofiber composites for sustainable sunlight-driven dye degradation." *ACS Applied Nano Materials* 3, no. 10 (2020): 10284–10295. https://doi.org/10.1021/acsanm.0c02199.

Długosz, O., K. Szostak, M. Krupiński, and M. Banach. "Synthesis of Fe_3O_4/ZnO nanoparticles and their application for the photodegradation of anionic and cationic dyes." *International Journal of Environmental Science and Technology* 18 (2021): 561–574. https://doi.org/10.1007/s13762-020-02852-4.

Duan, W. J., S. H. Lu, Z. L. Wu, and Y. S. Wang. "Size effects on properties of NiO nanoparticles grown in alkalisalts." *The Journal of Physical Chemistry C* 116 (2012): 26043–26051

Elango, Ganesh, and Selvaraj Mohana Roopan. "Efficacy of SnO_2 nanoparticles toward photocatalytic degradation of methylene blue dye." *Journal of Photochemistry and Photobiology B: Biology* 155 (2016): 34–38. https://doi.org/10.1016/j.jphotobiol.2015.12.010.

Elliott, A., W. E. Hanby, and B. R. Malcolm. "The near infra-red absorption spectra of natural and synthetic fibres." *Brazilian Journal of Applied Physics* 5 (1954): 377.

Fakruddin, Md, Zakir Hossain, and Hafsa Afroz. "Prospects and applications of nanobiotechnology: A medical perspective." *Journal of Nanobiotechnology* 10, no. 1 (2012): 1–8. https://doi.org/10.1186/1477-3155-10-31.

Freitasde Freitas, Lucas, Gustavo Henrique Costa Varca, Jorge Gabriel dos Santos Batista, and Ademar Benévolo Lugão. "An overview of the synthesis of gold nanoparticles using radiation technologies." *Nanomaterials* 8, no. 11 (2018): 939.

GanapathySelvam, G., and K. Sivakumar. "Phycosynthesis of silver nanoparticles and photocatalytic degradation of methyl orange dye using silver (Ag) nanoparticles synthesized from Hypnea musciformis (Wulfen) JV Lamouroux." *Applied Nanoscience* 5 (2015): 617–622.

Gilja, Vanja, Katarina Novaković, Jadranka Travas-Sejdic, Zlata Hrnjak-Murgić, Marijana Kraljić Roković, and Mark Žic. "Stability and synergistic effect of polyaniline/TiO₂ photocatalysts in degradation of azo dye in wastewater." *Nanomaterials* 7, no. 12 (2017): 412. https://doi.org/10.3390/nano7120412.

Gnanamozhi, P., Vengudusamy Renganathan, Shen-Ming Chen, V. Pandiyan, M. Antony Arockiaraj, Naiyf S. Alharbi, Shine Kadaikunnan, Jamal M. Khaled, and Khalid F. Alanzi. "Influence of Nickel concentration on the photocatalytic dye degradation (methylene blue and reactive red 120) and antibacterial activity of ZnO nanoparticles." *Ceramics International* 46 (2020): 18322–18330.

Golgeri, Dilshad Begum M., Syeda Ulfath Tazeen Kadri, Satish Kumar Murari, Dummi Mahadevan Gurumurthy, Muhammad Bilal, Ram Naresh Bharagava, Anyi Hu, Paul Olusegun Bankole, Luiz Fernando R. Ferreira, and Sikandar I. Mulla. "Physicochemical–biotechnological approaches for removal of contaminants from wastewater. In Vineet Kumar and Manish Kumar (Eds.) *Integrated Environmental Technologies for Wastewater Treatment and Sustainable Development* (2022): pp. 241–261. https://doi.org/10.1016/B978-0-323-91180-1.00010-7.

Guo, Na, Haiyan Li, Xingjian Xu, and Hongwen Yu. "Hierarchical Fe₃O₄@ MoS₂/Ag₃PO₄ magnetic nanocomposites: Enhanced and stable photocatalytic performance for water purification under visible light irradiation." *Applied Surface Science* 389 (2016): 227–239.

Haider, Adawiyah J., Zainab N. Jameel, and Imad HM Al-Hussaini. "Review on: Titanium dioxide applications." *Energy Procedia* 157 (2019): 17–29. https://doi.org/10.1016/j.egypro.2018.11.159.

Hamza, Muhammad, Ataf Ali Altaf, Samia Kausar, Shahzad Murtaza, Nasir Rasool, Rukhsana Gul, Amin Badshah, Muhammad Zaheer, Syed Adnan Ali Shah, and Zainul Amiruddin Zakaria. "Catalytic removal of alizarin red using chromium manganese oxide nanorods: Degradation and kinetic studies." *Catalysts* 10, no. 10 (2020): 1150. https://doi.org/10.3390/catal10101150

Hassaan, Mohamed A., Ahmed El Nemr, and A. Hassaan. "Health and environmental impacts of dyes: Mini review." *American Journal of Environmental Science Engineering* 1 (2017): 64–67.

Herrmann, Jean-Marie. "Photocatalysis fundamentals revisited to avoid several misconceptions." *Applied Catalysis B: Environmental* 99 (2010): 461–468.

Huber, Dale L. "Synthesis, properties, and applications of iron nanoparticles." *Small* 1, no. 5 (2005): 482–501. https://doi.org/10.1002/smll.200500006.

Ingle, Avinash P., Nelson Duran, and Mahendra Rai. "Bioactivity, mechanism of action, and cytotoxicity of copper-based nanoparticles: A review." *Applied Microbiology and Biotechnology* 98 (2014): 1001–1009. https://doi.org/10.1007/s00253-013-5422-8.

Ishio, S., T. Narisawa, S. Takahashi, Y. Kamata, S. Shibata, T. Hasegawa, Z. Yan et al. "L10 FePt thin films with [0 0 1] crystalline growth fabricated by SiO₂ addition-rapid thermal annealing and dot patterning of the films." *Journal of Magnetism and Magnetic Materials* 324, no. 3 (2012): 295–302. https://doi.org/10.1016/j.jmmm.2010.12.014.

Jain, Keerti, Anand S. Patel, Vishwas P. Pardhi, and Swaran Jeet Singh Flora. "Nanotechnology in wastewater management: A new paradigm towards wastewater treatment." *Molecules* 26, no. 6 (2021): 1797. https://doi.org/10.3390/molecules26061797.

Jain, Prashant, and T. Pradeep. "Potential of silver nanoparticle-coated polyurethane foam as an antibacterial water filter." *Biotechnology and Bioengineering* 90, no. 1 (2005): 59–63.

Jaji, Nuru-Deen, Hooi Ling Lee, Mohd Hazwan Hussin, Hazizan Md Akil, Muhammad Razlan Zakaria, and Muhammad Bisyrul Hafi Othman. "Advanced nickel nanoparticles technology: From synthesis to applications." *Nanotechnology Reviews* 9, no. 1 (2020): 1456–1480. https://doi.org/10.1515/ntrev2020-0109.

John, Jojy, Ramadoss Dineshram, Kaveripakam Raman Hemalatha, Magesh Peter Dhassiah, Dharani Gopal, and Amit Kumar. "Bio-decolorization of synthetic dyes by a halophilic bacterium *Salinivibrio* sp." *Frontiers in Microbiology* 11 (2020): 594011. https://doi.org/10.3389/fmicb.2020.594011

Joshi, Naveen Chandra, Ekta Joshi, and Ajay Singh. "Biological synthesis, characterisations and antimicrobial activities of manganese dioxide (MnO₂) nanoparticles." *Research Journal of Pharma Technology* 13 (2020b): 135–140. https://doi.org/10.5958/0974-360X.2020.00027.X.

Joshi, Naveen Chandra, Prateek Gururani, and Shiv Prasad Gairola. "Metal oxide nanoparticles and their nanocomposite-based materials as photocatalysts in the degradation of dyes." *Biointerface Research Applied Chemistry* 12 (2022): 6557–6579.

Joshi, Naveen Chandra, Shekhar Malik, and P. J. L. I. A. N. Gururani. "Utilisation of polypyrrole/ZnO nanocomposite in the adsorptive removal of Cu^{2+}, Pb^{2+} and Cd^{2+} ions from wastewater." Letters Applied Nano Bioscience, 10 (2021): 2339–2351. https://doi.org/10.33263/LIANBS103.23392351.

Kadri, Syeda Ulfath Tazeen, Satish Kumar Murari, Dummi Mahadevan Gurumurthy, Muhammad Bilal, Ram Naresh Bharagava, Anyi Hu, Paul Olusegun Bankole, Luiz Fernando R. Ferreira, and Sikandar I. Mulla. "Physicochemical-biotechnological approaches for removal of contaminants from wastewater." In Vineet Kumar and Manish Kumar (eds.) *Integrated Environmental Technologies for Wastewater Treatment and Sustainable Development* (2022): pp. 241–261. https://doi.org/10.1016/B978-0-323-91180-1.00010-7.

Kang, Guo-dong, and Yi-ming Cao. "Application and modification of poly(vinylidene fluoride) (PVDF) membranes-A review." *Journal of Membrane Science* 463 (2014): 145–165.

Kant, Rita. "Textile dyeing industry an environmental hazard." *Natural Science*, 4 (2012): 22–26.

Karthigadevi, Guruviah, Sivasubramanian Manikandan, Natchimuthu Karmegam, Ramasamy Subbaiya, Sivasankaran Chozhavendhan, Balasubramani Ravindran, Soon Woong Chang, and Mukesh Kumar Awasthi. "Chemico-nanotreatment methods for the removal of persistent organic pollutants and xenobiotics in water - A review." *Bioresource Technology* 324 (2021). https://doi.org/10.1016/j.biortech.2021.124678.

Kataria, Navish, and V. K. Garg. "Application of EDTA modified Fe_3O_4/sawdust carbon nanocomposites to ameliorate methylene blue and brilliant green dye laden water." *Environmental Research* 172 (2019): 43–54.

Khade, G.V.; Suwarnkar, M.B.; Gavade, N.L.; Garadkar, K.M. Green synthesis of TiO_2 and its photocatalytic activity. *Journal of Materials Science. Materials in Electronics* 26 (2015): 3309–3315. https://doi.org/10.1007/s10854-015-2832-7.

Khan, Idrees, Muhammad Sadiq, Ibrahim Khan, and Khalid Saeed. "Manganese dioxide nanoparticles/ activated carbon composite as efficient UV and visible-light photocatalyst." *Environmental Science and Pollution Research* 26 (2019a): 5140–5154. https://doi.org/10.1007/s11356-018-4055-y.

Kołodziejczak-Radzimska, Agnieszka, and Teofil Jesionowski. "Zinc oxide-from synthesis to application: A review." *Materials* 7, no. 4 (2014): 2833–2881.

Kumar, P., S. Senthamil Selvi, A. Lakshmi Prabha, K. Prem Kumar, R. S. Ganeshkumar, and Munisamy Govindaraju. "Synthesis of silver nanoparticles from *Sargassum tenerrimum* and screening phytochemicals for its antibacterial activity." *Nano Biomed Eng* 4, no. 1 (2012): 12–16.

Lade, Vikesh G. "Introduction of water remediation processes." In Bhanvase B, Sonawane S, Pandit A. (Eds.) *Handbook of Nanomaterials for Wastewater Treatment* (2021): pp. 741–777. Elsevier Publisher.

Lellis, Bruno, Cíntia Zani Fávaro-Polonio, João Alencar Pamphile, and Julio Cesar Polonio. "Effects of textile dyes on health and the environment and bioremediation potential of living organisms." *Biotechnology Research Innovation* 3 (2019): 275–290. https://doi.org/10.1016/j.biori.2019.09.001.

Li, Jiang, Xiaolu Zhu, Fengxian Qiu, Tao Zhang, Fengping Hu, and Xiaoming Peng. "Facile preparation of Ag/ Ag_2WO_4/g-C_3N_4 ternary plasmonic photocatalyst and its visible-light photocatalytic activity." *Applied Organometallic Chemistry* 33, no. 3 (2019a): e4683.

Li, Jin Feng, Esrat Jahan Rupa, Joon Hurh, Yue Huo, Ling Chen, Yaxi Han, Jong chan Ahn et al. "Cordyceps militaris fungus mediated zinc oxide nanoparticles for the photocatalytic degradation of methylene blue dye." *Optik* 183 (2019b): 691–697. https://doi.org/10.1016/j.ijleo.2019.02.081

Li, Yuanyuan, Qimei Yang, Zhongming Wang, Guoyu Wang, Bin Zhang, Qian Zhang, and Dingfeng Yang. Rapid fabrication of SnO_2 nanoparticle photocatalyst: computational understanding and photocatalytic degradation of organic dye." *Inorganic Chemistry Frontiers* 5, no. 12 (2018): 3005–3014. https://doi.org/10.1039/C8QI00688A.

Luo, Guangsheng, Le Du, Yunjun Wang, and Kai Wang. "Composite nanoparticles." In Li, Dongqing (ed.) *Encyclopedia of Microfluidics and Nanofluidics* (2014). Springer, Boston, MA. https://doi.org/10.1007/978-3-642-27758-0_243-3

Ma, Dongge, Jundan Li, Anan Liu, and Chuncheng Chen. "Carbon gels-modified TiO_2: Promising materials for photocatalysis applications." *Materials* 13 (2020): 1734.

Madiha Batool, Zahid Qureshi, Nida Mehboob, and Abdul Salam Shah. "Studie on malachite green dye degradation by biogenic metal nano cuo and cuo/zno nano composites" *Archives of Nanomedicine: Open Access Journal* 1, no. 4 (2018): 119. https://doi.org/10.32474/ANOAJ.2018.01.000119.

Majhi, Kartick Chandra, and Mahendra Yadav. "Synthesis of inorganic nanomaterials using carbohydrates." In Inamuddin, Rajender Boddula, Mohd Imran Ahamed, and Abdullah M. Asiri (eds.) *Green Sustainable Process for Chemical and Environmental Engineering and Science* (2021): pp. 109–135. https://doi.org/10.1016/B978-0-12-821887-7.00003-3.

Majumder, Deblina, Indranil Chakraborty, Kalyan Mandal, and Somenath Roy. "Facet-dependent photodegradation of methylene blue using pristine CeO$_2$ nanostructures." *ACS Omega*. 4, no. 2 (2019): 4243-4251. https://doi.org/10.1021/acsomega.8b03298.

Mani, Sujata, and Pankaj Chowdhary. "Dyes: Industrial applications and toxicity profile." In Chowdary P. and Raj A. (Eds.) *Contaminants and Clean Technologies* (2020): pp. 137–148. CRC Press, Taylor & Francis.

Marchiol, Luca, Michele Iafisco, Guido Fellet, and Alessio Adamiano. "Nanotechnology support the next agricultural revolution: Perspectives to enhancement of nutrient use efficiency." *Advances in Agronomy* 161 (2020): 27–116. https://doi.org/10.1016/bs.agron.2019.12.001.

Massaro, Marina, Carmelo G. Colletti, Bruno Fiore, Valeria La Parola, Giuseppe Lazzara, Susanna Guernelli, Nelsi Zaccheroni, and Serena Riela. "Gold nanoparticles stabilized by modified halloysite nanotubes for catalytic applications." *Applied Organometallic Chemistry* 33, no. 3 (2019): e4665.

Mehata, Mohan Singh. "Green synthesis of silver nanoparticles using Kalanchoe pinnata leaves (life plant) and their antibacterial and photocatalytic activities." *Chemical Physics Letters* 778 (2021): 138760. https://doi.org/10.1016/j.cplett.2021.138760.

Mehta, Malvika, Mahima Sharma, Kamni Pathania, Pabitra Kumar Jena, and Indu Bhushan. "Degradation of synthetic dyes using nanoparticles: A mini-review." *Environmental Science Pollution Research* 28 (2021): 49434–49446. https://doi.org/10.1007/s11356-021-15470-5

Melinte, Violeta, Lenuta Stroea, and Andreea L. Chibac-Scutaru. "Polymer nanocomposites for photocatalytic applications." *Catalysts* 9 (2019): 986.

Mendes-Felipe, Cristian, Antonio Veloso-Fernández, José Luis Vilas-Vilela, and Leire Ruiz-Rubio. "Hybrid organic-inorganic membranes for photocatalytic water remediation." *Catalysts* 12 (2022): 180.

Motahari, Fereshteh, Mohammad Reza Mozdianfard, Faezeh Soofivand, and Masoud Salavati-Niasari. "NiO nanostructures: Synthesis, characterization and photocatalyst application in dye wastewater treatment." *RSC Advances* 4, no. 53 (2014): 27654–27660.

Mumtaz, Zaroon Mehvish, Nazim Hussain, and Hafiz Muhammad Husnain Azam. "Applications of novel nanomaterials in water treatment." In Guillermo R. Castro, Ashok Kumar Nadda, Tuan Anh Nguyen, Swati Sharma, and Muhammad Bilal (eds.) *Nanomaterials for Bioreactors and Bioprocessing Applications* (2023): pp. 217–243. https://doi.org/10.1016/B978-0-323-91782-7.00002-3.

Nabi, Ghulam, Qurat-Ul Ain, M. Bilal Tahir, Khalid Nadeem Riaz, Tahir Iqbal, Muhammad Rafique, Sajad Hussain, Waseem Raza, Imran Aslam, and Muhammad Rizwan. Green synthesis of TiO$_2$ nanoparticles using lemon peel extract: Their optical and photocatalytic properties. *International Journal of Environmental Analytical Chemistry* 102, no. 2 (2022): 434–442. https://doi.org/10.1080/03067319.2020.1722816

Najafian, Hassan, Faranak Manteghi, Farshad Beshkar, and Masoud Salavati-Niasari. "Enhanced photocatalytic activity of a novel NiO/Bi$_2$O$_3$/Bi$_3$ClO$_4$ nanocomposite for the degradation of azo dye pollutants under visible light irradiation." *Separation and Purification Technology* 209 (2019): 6–17.

Narasaiah, Palajonna, Badal Kumar Mandal, and N. C. Sarada. "Biosynthesis of copper oxide nanoparticles from Drypetes sepiaria leaf extract and their catalytic activity to dye degradation." In *IOP Conference Series: Materials Science and Engineering* (2017): p. 022012. IOP Publishing. https://doi.org/10.1088/1757-899X/263/2/022012.

Oliveira, Larissa VF, Simona Bennici, Ludovic Josien, Lionel Limousy, Marcos A. Bizeto, and Fernanda F. Camilo. "Free-standing cellulose film containing manganese dioxide nanoparticles and its use in discoloration of indigo carmine dye." *Carbohydrate Polymers* 230 (2020): 115621. https://doi.org/10.1016/j.carbpol.2019.115621.

Pan, Mingfei, Jingying Yang, Kaixin Liu, Zongjia Yin, Tianyu Ma, Shengmiao Liu, Longhua Xu, and Shuo Wang. "Noble metal nanostructured materials for chemical and biosensing systems." *Nanomaterials* 10, no. 2 (2020): 209. https://doi.org/10.3390/nano10020209.

Parhizkar, Janan, and Mohammad Hossein Habibi. "Synthesis, characterization and photocatalytic properties of Iron oxide nanoparticles synthesized by sol-gel autocombustion with ultrasonic irradiation." *Nanochemistry Research* 2, no. 2 (2017): 166–171. https://doi.org/10.22036/NCR.2017.02.002.

Pouretedal, H. R., and A. Kadkhodaie. "Synthetic CeO$_2$ nanoparticle catalysis of methylene blue photodegradation: Kinetics and mechanism." *Chinese Journal of Catalysis* 31, no. 11–12 (2010): 1328–1334. https://doi.org/10.1016/S1872-2067(10)60121-0.

Rafique, Muhammad, Iqra Sadaf, M. Bilal Tahir, M. Shahid Rafique, Ghulam Nabi, Tahir Iqbal, and Kalsoom Sughra. "Novel and facile synthesis of silver nanoparticles using Albizia procera leaf extract for dye degradation and antibacterial applications." *Materials Science and Engineering: C* 99 (2019): 1313–1324. https://doi.org/10.1016/j.msec.2019.02.059.

Rafique, Uzaira, Anum Imtiaz, and Abida K. Khan. "Synthesis, characterization and application of nanomaterials for the removal of emerging pollutants from industrial waste water, kinetics and equilibrium model." *Journal of Water Sustainability* 2, no. 4 (2012): 233–244.

Rahmat, Muniba, Asma Rehman, Sufyan Rahmat, Haq Nawaz Bhatti, Munawar Iqbal, Waheed S. Khan, Sadia Zafar Bajwa, Rubina Rahmat, and Arif Nazir. "Highly efficient removal of crystal violet dye from water by MnO_2 based nanofibrous mesh/photocatalytic process." *Journal of Materials Research and Technology* 8, no. 6 (2019): 5149–5159. https://doi.org/10.1016/j.jmrt.2019.08.038.

Rai, Maharashtra K., S. D. Deshmukh, A. P. Ingle, and A. K. Gade. "Silver nanoparticles: The powerful nanoweapon against multidrug-resistant bacteria." *Journal of Applied Microbiology* 112, no. 5 (2012): 841–852.

Rajendrachari, Shashanka, Parham Taslimi, Abdullah Cahit Karaoglanli, Orhan Uzun, Emre Alp, and Gururaj Kudur Jayaprakash. "Photocatalytic degradation of Rhodamine B (RhB) dye in waste water and enzymatic inhibition study using cauliflower shaped ZnO 1926 JRM, 2022, vol.10, no.7 nanoparticles synthesized by a novel one-pot green synthesis method." *Arabian Journal of Chemistry* 14, no. 6 (2021): 103180. https://doi.org/10.1016/j.arabjc.2021.103180.

Ramanathan, S.; Selvin, S.P.; Obadiah, A.; Durairaj, A.; Santhoshkumar, P.; Lydia, S.; Ramasundaram, S.; Vasanthkumar, S. "Synthesis of reduced graphene oxide/ZnO nanocomposites using grape fruit extract and *Eichhornia crassipes* leaf extract and a comparative study of their photocatalytic property in degrading Rhodamine B dye." *Journal of Environmental Health Science and Engineering* 17 (2019): 195–207. https://doi.org/10.1007/s40201-019-00340-7.

Rasheed, Tahir, Sameera Shafi, Muhammad Bilal, Tariq Hussain, Farooq Sher, and Komal Rizwan. "Surfactants-based remediation as an effective approach for removal of environmental pollutants-A review." *Journal of Molecular Liquids* 318 (2020): 113960. https://doi.org/10.1016/j.molliq.2020.113960.

Ravishankar, Thammadihalli Nanjundaiah, Thippeswamy Ramakrishnappa, Ganganagappa Nagaraju, and Hanumanaika Rajanaika. "Synthesis and characterization of CeO_2 nanoparticles via solution combustion method for photocatalytic and antibacterial activity studies." *ChemistryOpen* 4, no. 2 (2015): 146–154. https://doi.org/10.1002/open.201402046.

Regmi, Chhabilal, Bhupendra Joshi, Schindra K. Ray, Gobinda Gyawali, and Ramesh P. Pandey. "Understanding mechanism of photocatalytic microbial decontamination of environmental wastewater." *Frontiers in Chemistry* 6 (2018): 33. https://doi.org/10.3389/fchem.2018.00033.

Ren, Xiaochen, Peng Gao, Xianglong Kong, Rui Jiang, Piaoping Yang, Yujin Chen, Qianqian Chi, and Benxia Li. "NiO/Ni/TiO_2 nanocables with Schottky/pn heterojunctions and the improved photocatalytic performance in water splitting under visible light." *Journal of colloid and interface science* 530 (2018): 1–8.

Rincón Joya, Miryam, José Barba Ortega, João Otávio Donizette Malafatti, and Elaine Cristina Paris. "Evaluation of photocatalytic activity in water pollutants and cytotoxic response of α-Fe_2O_3 nanoparticles." *ACS Omega* 4, no. 17 (2019): 17477–17486.

Robinson, Tim, Geoff McMullan, Roger Marchant, and Poonam Nigam. "Remediation of dyes in textile efuent: A critical review on current treatment technologies with a proposed alternative." *Bioresource Technology* 77 (2001): 247–255. https://doi.org/10.1016/s0960-8524(00)00080-8.

Roy, Arpita, HC Ananda Murthy, Hiwa M. Ahmed, Mohammad Nazmul Islam, and Ram Prasad. "Phytogenic synthesis of metal/metal oxide nanoparticles for degradation of dyes." *Journal of Renewable Materials* 10 (2022): 1911.

Saratale, Rijuta Ganesh, Ganesh Dattatraya Saratale, Han Seung Shin, Jaya Mary Jacob, Arivalagan Pugazhendhi, Mukesh Bhaisare, and Gopalakrishanan Kumar. "New insights on the green synthesis of metallic nanoparticles using plant and waste biomaterials: Current knowledge, their agricultural and environmental applications." *Environmental Science and Pollution Research* 25 (2018): 10164–10183. https://doi.org/10.1007/ s11356-017-9912-6.

Shanker, Uma, Manviri Rani, and Vidhisha Jassal. "Degradation of hazardous organic dyes in water by nanomaterials." *Environmental Chemistry Letters* 15 (2017): 623–642. https://doi.org/10.1007/ s10311-017-0650-2

Sharma, Mahima, Kannikka Behl, Subhasha Nigam, and Monika Joshi. "TiO_2-GO nanocomposite for photocatalysis and environmental applications: A green synthesis approach." *Vacuum* 156 (2018): 434–439.

Shi, Liang, and Hailin Lin. "Facile fabrication and optical property of hollow SnO_2 spheres and their application in water treatment." *Langmuir* 26, no. 24 (2010): 18718–18722. https://doi.org/10.1021/la103769d.

Shrivastava, Siddhartha, and Debabrata Dash. "Applying nanotechnology to human health: Revolution in biomedical sciences." *Journal of Nanotechnology* (2009). https://doi.org/10.1155/2009/184702.

Singh, Jagpreet, Vanish Kumar, Ki-Hyun Kim, and Mohit Rawat. "Biogenic synthesis of copper oxide nanoparticles using plant extract and its prodigious potential for photocatalytic degradation of dyes." *Environmental Research* 177 (2019): 108569. https://doi.org/10.1016/j.envres.2019.108569.

Soni, Vatika, Pankaj Raizada, Pardeep Singh, Hoang Ngoc Cuong, S. Rangabhashiyam, Adesh Saini, Reena V. Saini et al. "Sustainable and green trends in using plant extracts for the synthesis of biogenic metal nanoparticles toward environmental and pharmaceutical advances: A review." *Environmental Research* 202 (2021): 111622.

Su, Minhua, Chun He, Virender K. Sharma, Mudar Abou Asi, Dehua Xia, Xiang-zhong Li, Huiqi Deng, and Ya Xiong. "Mesoporous zinc ferrite: Synthesis, characterization, and photocatalytic activity with H_2O_2/ visible light." *Journal of Hazardous Materials* 211 (2012): 95–103.

Sungur, Ş. "Titanium dioxide nanoparticles." In *Handbook of Nanomaterials and Nanocomposites for Energy and Environmental Applications* (2020): pp. 1–18. Springer, Cham. https://doi.org/10.1007/978-3-030-11155-7_9-1.

Tammina, Sai Kumar, Badal Kumar Mandal, and Nalinee Kanth Kadiyala. "Photocatalytic degradation of methylene blue dye by nonconventional synthesized SnO_2 nanoparticles." *Environmental Nanotechnology, Monitoring & Management* 10 (2018): 339–350. https://doi.org/10.1016/j.enmm.2018.07.006.

Thirumurugan, A., E. Harshini, B. Deepika Marakathanandhini, S. Rajesh Kannan, and P. Muthukumaran. "Catalytic degradation of reactive red 120 by copper oxide nanoparticles synthesized by *Azadirachta indica*." In *Bioremediation and Sustainable Technologies for Cleaner Environment* (2017): pp. 95–102. https://doi.org/10.1007/978-3-319-48439-6_9

Titus, Deena, and E. James Jebaseelan Samuel "Photocatalytic degradation of azo dye using biogenic SnO_2 nanoparticles with antifungal property: RSM optimization and kinetic study." *Journal of Cluster Science* 30 (2019): 1335–1345. https://doi.org/10.1007/s10876-019-01585-w.

Tomar, Richa, Ahmed A. Abdala, R. G. Chaudhary, and N. B. Singh. "Photocatalytic degradation of dyes by nanomaterials." *Materials Today: Proceedings* 29 (2020): 967–973.

Torki, Firoozeh, and Hossein Faghihian. "Visible light degradation of naproxen by enhanced photocatalytic activity of NiO and NiS, scavenger study and focus on catalyst support and magnetization." *Photochemistry and Photobiology* 94, no. 3 (2018): 491–502.

Vaghela, Sanjay S., Ashok D. Jethva, Bhavesh B. Mehta, Sunil P. Dave, Subbarayappa Adimurthy, and Gadde Ramachandraiah. Laboratory studies of electrochemical treatment of industrial azo dye effluent. *Environmental Science and Technology* 39 (2005) 2848–2855. https://doi.org/10.1021/es035370c.

Vakili, Mohammadtaghi, Mohd Rafatullah, Babak Salamatinia, Ahmad Zuhairi Abdullah, Mahamad Hakimi Ibrahim, Kok Bing Tan, Zahra Gholami, and Parisa Amouzgar. "Application of chitosan and its derivatives as adsorbents for dye removal from water and wastewater: A review." *Carbohydrate Polymers* 113 (2014): 115–130. https://doi.org/10.1016/j.carbpol.2014.07.007.

Verma, Ayushi, Arpita Roy, and Navneeta Bharadvaja. "Remediation of heavy metals using nanophytoremediation." In *Advanced Oxidation Processes for Effluent Treatment Plants* (2021): pp. 273–296. Elsevier. https://doi.org/10.1016/B978-0-12-821011-6.00013-X.

Viet, Pham Van, Cao Minh Thi, and Le Van Hieu. "The high photocatalytic activity of SnO_2 nanoparticles synthesized by hydrothermal method." *Journal of Nanomaterials* 2016 (2016). https://doi.org/10.1155/2016/4231046.

Wang, Desong, Jie Zhang, Qingzhi Luo, Xueyan Li, Yandong Duan, and Jing An. "Characterization and photocatalytic activity of poly (3-hexylthiophene)-modified TiO_2 for degradation of methyl orange under visible light." *Journal of Hazardous Materials* 169, no. 1–3 (2009): 546–550. https://doi.org/10.1016/j.jhazmat.2009.03.135.

Wang, Xinwei, Xianfan Xu, and Stephen US Choi. "Thermal conductivity of nanoparticle-fluid mixture." *Journal of Thermophysics and Heat Transfer* 13, no. 4 (1999): 474–480. https://doi.org/10.2514/2.6486

Wu, Muyu, Pingfu Hou, Lina Dong, Lulu Cai, Zhudian Chen, Mingming Zhao, and Jingjing Li. "Manganese dioxide nanosheets: From preparation to biomedical applications." *International Journal of Nanomedicine* (2019): 4781–4800. https://doi.org/10.2147/IJN.S207666.

Wu, Shuisheng, Huaqiang Cao, Shuangfeng Yin, Xiangwen Liu, and Xinrong Zhang. "Amino acid-assisted hydrothermal synthesis and photocatalysis of SnO_2 nanocrystals." *The Journal of Physical Chemistry C* 113, no. 41 (2009): 17893–17898. https://doi.org/10.1021/jp9068762.

Xu, Can, and Xiaogang Qu. "Cerium oxide nanoparticle: A remarkably versatile rare earth nanomaterial for biological applications." *NPG Asia Materials* 6, no. 3 (2014): e90. https://doi.org/10.1038/am.2013.88.

Yaqoob, Asim Ali, Hilal Ahmad, Tabassum Parveen, Akil Ahmad, Mohammad Oves, Iqbal MI Ismail, Huda A. Qari, Khalid Umar, and Mohamad Nasir Mohamad Ibrahim. "Recent advances in metal decorated nanomaterials and their various biological applications: A review." *Frontiers in Chemistry* 8 (2020): 341. https://doi.org/10.3389/ fchem.2020.00341.

Yecheskel, Yinon, Ishai Dror, and Brian Berkowitz. "Catalytic degradation of brominated flame retardants by copper oxide nanoparticles." *Chemosphere* 93, no. 1 (2013): 172–177. https://doi.org/10.1016/j.chemosphere.2013.05.026.

Yin, Xu, Qianwen Liu, Yong Chen, Anlin Xu, Yi Wang, Yong Tu, and Weiqing Han. "Preparation, characterization and environmental application of the composite electrode TiO_2-NTs/SnO_2-Sb with carbon aerogels." *Journal of Chemical Technology & Biotechnology* 94, no. 10 (2019): 3124–3133.

Yuan, Ye, Rui-tang Guo, Long-fei Hong, Xiang-yin Ji, Zheng-sheng Li, Zhi-dong Lin, and Wei-guo Pan. "Recent advances and perspectives of MoS_2-based materials for photocatalytic dyes degradation: A review." *Colloids and Surfaces A: Physicochemical and Engineering Aspects* 611 (2021): 125836.

Zhao, Yongjie, Xiaowei Zhang, Chengzhi Wang, Yuzhen Zhao, Heping Zhou, Jingbo Li, and HaiBo Jin. "The synthesis of hierarchical nanostructured MoS_2/Graphene composites with enhanced visible-light photo-degradation property." *Applied Surface Science* 412 (2017): 207–213.

Zhou, Chunjiao, Wenjie Zhang, Huixian Wang, Huiyong Li, Jun Zhou, Shaohua Wang, Jinyan Liu, Jing Luo, Bingsuo Zou, and Jianda Zhou. "Preparation of Fe_3O_4-embedded graphene oxide for removal of methylene blue." *Arabian Journal for Science and Engineering* 39 (2014): 6679–6685. https://doi.org/10.1007/s13369-014-1183-7.

Zhu, Dandan, and Qixing Zhou. "Action and mechanism of semiconductor photocatalysis on degradation of organic pollutants in water treatment: A review." *Environmental Nanotechnology Monitoring and Management* 12 (2019): 100255. https://doi.org/10.1016/j.enmm.2019.100255.

Zhu, Yunfeng, Shoubin Xu, and Dan Yi. "Photocatalytic degradation of methyl orange using polythiophene/titanium dioxide composites." *Reactive and Functional Polymers* 70, no. 5 (2010): 282–287. https://doi.org/10.1016/j.reactfunctpolym.2010.01.007.

23 Employing Artificial Intelligence-Driven Computational Approaches to Devise Ground-Breaking Inorganic Nanomaterials

G. Subbulakshmi, G. Padma Priya, Soumya V. Menon, Vinutha Reddy, and Priyanshi Roshan

INTRODUCTION

Autonomous experimentation powered by artificial intelligence (AI) offers an intriguing prospect to revolutionise the discovery and development of inorganic materials. In this chapter, we examine the current advances in the design of self-driving laboratories, including the use of robots to automate the synthesis and characterisation of materials, as well as the use of AI to analyse experimental results and propose new experimental protocols. Connections are established in each example to relevant work in organic chemistry, where automation is more widespread. Characterisation approaches are presented mostly in the context of phase identification, as this activity is crucial for understanding what products have been produced during synthesis. Deep learning is used to analyse multivariate characterisation data and conduct phase identification.

In contrast to organic chemistry, the development of autonomous experimentation for inorganic materials is still in its early phases. Given the difficulties associated with handling solid powders, the scarcity of methods for reliably characterising bulk samples, and the lack of a rigorous theoretical framework describing the factors influencing synthesizability, the majority of existing work has demonstrated only partial automation of the experimental process. Within the thin-film community, for example, HT automation of synthesis and characterisation is commonly carried out to investigate the implications of composition and processing conditions on the characteristics of resultant samples [1].

The automation of synthesis and characterisation would enable high experimental throughput, allowing the researcher to analyse generated datasets and design future experiments [2].

Increased availability of synthesis data may also aid in the creation of AI that learns from experimental outcomes – not just to determine if a given synthesis effort succeeded or failed, but also to hypothesise why it succeeded or failed. Such predictions often need human researchers with a thorough grasp of potential response mechanisms. It is a daunting task to automate this process. However, we propose that a useful set of rules for understanding synthesis can be extracted from work being done in several related areas, such as theories on synthesizability, in situ characterisation of reaction pathways, and increasing availability of synthesis data. If advances in these areas are successful in enabling a self-driving synthesis laboratory, it will have far-reaching implications in the materials science community, allowing researchers to generate new compounds at an unprecedented rate while reducing the amount of time and labour spent.

DOI: 10.1201/9781003479239-23

Sample synthesis is the first important step in the automated optimisation of material properties and processes. Note that our preliminary discussion here is limited to the hardware requirements necessary to carry out the synthesis procedure with a given set of parameters, including the choice of starting materials and conditions. After the samples are prepared, appropriate characterisation methods must be used to reveal the objects of interest and obtain the experimental result [3].

SYNTHESIS AND CHARACTERISATION USING AI

These can be performed using robotic systems combined with real-time and online monitoring to ensure high accuracy and solve operational problems. However, the ease of automation varies depending on the synthesis method and form of the products. Accordingly, we divide our discussion into three main categories: batch or continuous solution-based synthesis, thin-film deposition, and solid-state synthesis from bulk powders. In addition, although the focus of this review is on inorganic materials, we often highlight organic chemistry platforms of related materials where automation is more common, learn from their success, and understand how similar methods can be extended to inorganic compounds.

One of the potential nanotechnology applications for effective drug delivery to the targeted region is the nanoparticle delivery system. However, the selection of a lead excipient, the prediction of miscibility/solubility parameters, drug loading capability, drug release rate, stability prediction, the transportation of nanoparticles through a complex network of blood vessels, and drug–target recognition and binding are all critical aspects of nanoparticle formulation development [4]. The applications of molecular computational models are, for example, computational fluid dynamics (CFD) simulations, dissipative particle dynamics (DPD) simulations, coarse-grained (CG) molecular dynamics (MD) modelling, quantum mechanical simulation methods, atomistic molecular dynamics, quantitative structure–activity relationships (QSAR), and discrete element modelling.

Understanding the complicated processes involved in the creation of nanoparticle formulation is aided by pharmacokinetic/pharmacodynamic modelling (PK/PD) and physiologically based pharmacokinetic (PBPK) modelling. The current study focuses on the computer simulation modelling methods used to produce nanoparticle formulations, as well as their use in the creation of various organic and inorganic nanoplatforms utilised in drug delivery [4].

The medication research and discovery process is critical in identifying a therapeutically active chemical for treating and preventing illness. In tackling the restrictions and issues associated with traditional drug research and discovery procedures, computational simulation has emerged as a new boon. A computational simulation is a near-realistic model that may be utilised to speed up drug research and discovery.

Almost every level of the drug development process has made substantial use of computational model tools. Target identification and characterisation, which entail establishing the function of a possible therapeutic target and its involvement in the disease, are the first steps in the drug development process. Following the discovery of the therapeutic target, the target's molecular mechanism is studied. Some of the computational approaches used to filter a large number of targets into a small set of identified active targets include pharmacophore mapping and inverse docking. A good target should be effective and safe, and should meet clinical and commercial requirements. A molecular target is involved in the illness process, and modulating the target is a necessary condition for having a therapeutic impact.

To optimise and validate the discovered targets, researchers employed advanced computational methods such as molecular docking, pharmacophore modelling, de novo design, virtual library design, quantitative structure–activity connection, and sequence-based approaches. Computational simulation has lately left its mark on the subsequent drug discovery and development stages, such as the preclinical trial stage, by playing an important role in the design and development of nanoformulations.

Engineered nanoparticles (ENPs) are widely used in many different applications, and the momentum is constantly growing. Evaluation of the biological effects of ENP is extremely important, and experimental and, more recently, computational methods have been proposed for this purpose. To computationally explore available data sets leading to practical applications, we developed and validated a QNAR-model (Quantitative Nanostructure-Activity Relationship), to predict cellular uptake of nanoparticles in pancreatic cancer cells. Our in silico workflow has been made available online via the Enalos In Silico Nano platform [5].

Nanotechnology-based detailing is a promising mechanical investigation plan with broader suggestions in biomedical applications and has been demonstrated to be an effective conveyance framework for treating the world's deadliest infections. Advancement of nanoformulation is challenging and complex because it includes numerous factors of preparation, exorbitant trials and mistakes in tests, and harmfulness concerns. To overcome such challenges, computational re-enactment can be utilised as a starting screening apparatus to conduct test trials, bring about noteworthy enhancement of preparation, and diminish the exploratory burden. Computational recreation is an up-and-coming breakthrough innovation in investigation of nanoparticles, and it guides the plan of modern nanoparticles and medicate conveyance frameworks such as Drug Delivery Systems (DDSs) with ideal medicate stacking, steadiness, and decreased harmfulness [6].

The innovation of medicate conveyance frameworks (DDSs) has illustrated an extraordinary execution and viability in generation of pharmaceuticals, because it is demonstrated by numerous FDA-approved nanomedicines that have an improved selectivity, reasonable medicate discharge energy, and synergistic restorative activities. Nevertheless, the methodical strategy and high-throughput development of nanomaterial-based DDSs for specific purposes remain far from an appointed time and are still in their early stages. This is mainly because of the difficulties scientists face in effectively obtaining, analysing, supervising, and comprehending intricate and constantly expanding sets of exploratory data, which makes the creation of DDSs with a desired set of functionalities essential [2]. At the same time, this errand is attainable for the data-driven approaches, high-throughput experimentation methods, automatisation of preparation, fake insights (AI) innovation, and machine learning (ML) approaches, which is alluded to as the fourth worldview of logical inquiry. Subsequently, an integration of these approaches into nanomedicine and nanotechnology can possibly quicken the judicious plan and high-throughput advancement of exceedingly proficient nanoformulated drugs and keen materials with pre-defined functionalities [7].

Nanotechnology, nanomedicine in specific, offers wide openings to specific designing of biomedical nanomaterials with desired properties and behaviours within the human body. This incorporates, but not constrained to, tunable half-life of pharmaceuticals, their aggregation in tissues and organs, selectivity, cellular take-up, discharge energy, and numerous other parameters. The parameters mentioned above are incomprehensible to control physically when managing with little drugs, since quantitative understanding of structure–activity connections (QSAR) is missing. At the same time, it is achievable for nanoformulated drugs in DDSs due to capacity of exact control over nanomaterial dimensionality, surface chemistry, surface charge, porosity, and other physicochemical properties.

Nanoformulated DDSs can (i) play the part of a detachment carrier to ensure particles from different natural components, such as corruption by proteins and resistant cells, (ii) serve a dynamic focusing on moieties to realise particular restorative activity of the medication in wanted tissue or organ, (iii) be a cytotoxic specialist to attain synergistic activity between the cage and cargo, (iv) act as an imaging methodology to precisely decide the localisation of pathology, (v) stimuli-responsive fabric to supply specific enactment of the medicate as it were in wanted tissue, (vi) bear immunomodulatory moiety when hoisted safe reaction is required and others, not to specify the gigantic assortment of these properties combinations which deliver rise to heaps of potential nanomedicines with completely unsurprising and multimodal organic behaviour. Ultimately, nanomedicines are broadly utilised in demonstrative, restorative, and theragnostic stages, subsequently covering nearly all the restorative subfields [6].

To date, the major deterrent for utilising nanomedicines broadly within the generation of drugs is their destitute clinical translation, which is due to the need of quantitative – and now and then indeed subjective – understanding how precise the nanomaterial properties are when associated with the organic components, leading to better results in the human body. Heuristic and pseudo-rational plan of nanomaterials was a common methodology for decades and, not shockingly, has not driven to a significant down-to-earth result within the frame of interpretation to the clinic since the fabric potential application was nearly continuously gone before by fabric union. Hence, in a general sense, a distinctive approach with required applications and properties has to be created to reveal the structure–activity connections of nanomedicines as well as its sound plan with completely controllable physicochemical and organic results [7].

Nanoscience in healthcare offers noteworthy progression within the demonstrative and restorative regions for imaging and biosensing, focused on medicate conveyance frameworks, etc. To amplify the applications in biomedical designing, counterfeit insights of AI and its innovation allows to analyse and decipher natural information, quicken sedate revelation and distinguish particular little particles or special compounds with prescient behaviour. Execution of such database frameworks for fast information investigation, treatment procedures, and novel speculations improvement, surprisingly makes strides the treatment results with the potential to quicken the high-throughput improvement and orderly plan of profoundly successful keen materials and nano formulations with pre-defined usefulness [1]. Particularly, optimising physicochemical parameters, compatibility, and drug-dose parameters with higher forecast proficiency (over 90%) is the region where AI holds the potential to actionably cognise the total nanotechnology potential. This chapter examines the investigation discoveries to quicken the clinical interpretation of nanoscience, offer the potential advancement of high-throughput experimentation-based, AI-assisted plan, and data-driven generation of nanosynthesised systems [2].

Stimuli-responsive or "intelligent" polymers have been broadly utilised for the plan of polymeric nanosystems such as polymeric micelles, polymerases, nanohydrogels, and crossover nanosystems comprising inorganic nanoparticles. Application of an inner or outside trigger such as pH, temperature, redox, attractive, light, or ultrasound causes these nanosystems to experience auxiliary changes which discharges the payload at the target site.

Some nanomaterials show significant promise for decreasing waste, cleaning up industrial contaminants, supplying drinkable water, and enhancing energy production and usage efficiency. The considerable potential to enhance environmental technologies (some of which date back to the Victorian period) is inextricably linked to engineered nanomaterials' tiny size, which results in markedly different characteristics than the corresponding bulk materials. Because of their small size, they have a high surface-to-volume ratio, which means they have more opportunities to interact with environmental toxins. Nanomaterials are, in some ways, "all surface." This is a very desirable feature for the treatment of water, wastewater, and hazardous waste [8].

CONCLUSION

Some nanomaterials have the potential to be superior adsorbents or catalysts, removing pollutants more efficiently and at a lower cost than conventional (material-intensive) techniques like ion-exchange resins and activated-carbon adsorption. Nanotechnology also has the potential to create multifunctional materials, for example, nano-architecture membranes for water treatment that combine chemically reactive nanoparticles to separate and degrade contaminants while simultaneously improving antifouling qualities. The good news is that many of our colleagues are making great strides in developing environmental nanotechnologies.

No matter the technology we are studying or interested in, the integration of AI into other technologies is unavoidable. Given the fact that AI has such a profound impact in the area of medicine these days, it is necessary to enhance AI, create a nano-based medication, and improve its delivery procedure to deal with medical domains such as cancer treatment, biomedicine, and nanobiology.

REFERENCES

[1] Kononova, Olga, Haoyan Huo, Tanjin He, Ziqin Rong, Tiago Botari, Wenhao Sun, Vahe Tshitoyan, and Gerbrand Ceder. "Text-mined dataset of inorganic materials synthesis recipes." *Scientific Data* 6, no. 1 (2019): 203. https://doi.org/10.1038/s41597-019-0224-1.

[2] Kasture, Kaustubh, and Pravin Shende. "Amalgamation of artificial intelligence with nanoscience for biomedical applications." In *Archives of Computational Methods in Engineering* (2023): pp. 1–19. https://doi.org/10.1007/s11831-023-09948-3.

[3] Nathan J. Szymanski, Yan Zeng, Haoyan Huo, Christopher J. Bartel, Haegyeom Kim, and Gerbrand Ceder. "Toward autonomous design and synthesis of novel inorganic materials." *Materials Horizons* 8 (2021): 2169–2198. https://doi.org/10.1039/D1MH00495F.

[4] Ashwini, T., Reema Narayan, Padmaja A. Shenoy, and Usha Y. Nayak. "Computational modeling for the design and development of nano based drug delivery systems." *Journal of Molecular Liquids* (2022): 120596. https://doi.org/10.1016/j.molliq.2022.120596.

[5] Melagraki, Georgia, and Antreas Afantitis. "Enalos InSilicoNano platform: An online decision support tool for the design and virtual screening of nanoparticles." *RSC Advances* 4, no. 92 (2014): 50713–50725. https://doi.org/10.1039/C4RA07756C.

[6] Serov, Nikita, and Vladimir Vinogradov. "Artificial intelligence to bring nanomedicine to life." *Advanced Drug Delivery Reviews* 184 (2022): 114194. https://doi.org/10.1016/j.addr.2022.114194.

[7] Zohuri, Bahman, and Farahnaz Behgounia. "Application of artificial intelligence driving nano-based drug delivery system." In *A Handbook of Artificial Intelligence in Drug Delivery* (2023): pp. 145–212. Academic Press. https://doi.org/10.1016/B978-0-323-89925-3.00007-1.

[8] Alvarez, Pedro J. "Nanotechnology in the environment-the good, the bad, and the ugly." *Journal of Environmental Engineering* 132, no. 10 (2006): 1233–1233. https://doi.org/10.1061/(ASCE)0733-9372 (2006)132:10(1233)

24 AI-Driven Innovation and Designing of Nanomaterials for a Better Tomorrow

G. Subbulakshmi, G. Padma Priya,
S. Thiyagaraj, and K.N. Bhoomika

INTRODUCTION

The International Union of Pure and Applied Chemistry (IUPAC) defines a nanoparticle as a particle with one or more characteristic dimensions between 1 and 100 nm. Numerous nanoparticles have unique electrical, optical, magnetic, or catalytic capabilities that differ from those of their macroscopic counterparts, despite having the same composition. These properties rely on the size, shape, and surface chemistry of the particles.

The self-assembly and self-organisation of atoms and molecules by chemical reactions, physical interactions, or the application of an external electric or magnetic field to these atoms and molecules are the subject of the new field of nanoscience, known as nanoarchitectonics [1]. By regulating each atom's reaction, it is possible to create new nanoparticles or improve the ones that already exist [2]. Its high surface area, semiconductor behaviour, biocompatibility, and thermal stability, among other characteristics, make it a significant advancement in the development of nanotechnology. They also have a wide range of uses in the fields of solar cell storage, fuel cells, sensors, medicines, and waste water treatment.

Smart polymers, a further application of nanoarchitectonics for biological objectives, have caught the interest of the scientific community [3]. Smart polymers are those types of polymeric materials that respond to physical or chemical stimuli by exhibiting unusual and reversible features. Changes in (i) electric/magnetic fields, (ii) temperature, (iii) pH, (iv) interaction with light, (v) contact with other chemical substances can all result in such stimulation [4]. Smart polymers can exhibit various surface peculiarities, shape alterations, induced sol-gel transition, swelling behaviour, and altered solubility when given the right stimulus [5]. In this regard, smart polymers, hydrogels, and nanoparticles have been studied for medication delivery to specific areas in acidosis, cancer therapy, and diabetic mellitus [6].

The exceptional features of nanomaterials, which are lacking at the macroscale, have led to extensive research on them and their use in a variety of sectors (such as medication delivery, electronics, heat transfer, structural composites, and pollution control). In order to understand the structure and characteristics of nanomaterials and discover new nanomaterials with ideal functionalities, conventional research techniques including laboratory investigations and molecular stimulation take a long time. Hence, as an artificial intelligence (AI) technique, machine learning (ML) can efficiently analyse an extensive amount of complex experimental data and identify important governing rules. By collecting chemical expertise from datasets, it helps accelerate chemical discovery. This method is especially helpful when it's unclear how the experiment's variables and results relate to one another. There are several popular ML models, such as the naïve Bayes (NB), support vector machine (SVM), decision tree (DT), random forest (RF), artificial neural network (ANN), logistic regression (LR), genetic algorithm (GA), k-nearest neighbour (KNN), and k-means models.

DOI: 10.1201/9781003479239-24

ARTIFICIAL INTELLIGENCE FOR NANOMATERIAL DESIGN

The innovative methods of nanoarchitectonics are rapidly accomplished with the aid of statistical techniques, experimental scheduling, and ML. AI is defined as a machine's or system's capacity to mimic human intelligence [7], and it has been shown to be extremely useful in the development of repeatable nanomaterials based on nanoarchitecture as well as in the estimation of physical, mechanical, and chemical properties [8].

AI systems try to emulate human behaviour and critical thinking [9]. AI can be broken down into subfields, each of which serves as a tool for designing nanoarchitectures. These technologies include (i) ML, (ii) natural language processing, (iii) computer vision, (iv) optimisation, (v) knowledge-based systems, (vi) automated planning and scheduling, and (vii) robotics [10] (Figure 24.1).

ML is a subfield of AI that can recognise patterns in data [11]. Statistical and mathematical techniques are employed in this learning process to create computer models that are then utilised to make predictions [12]. These models are created by splitting the dataset into train and test sets of data [13]. ML techniques can also be separated into supervised and unsupervised learning categories. Unlike unsupervised learning, which does not have the goal variable as a label in the training dataset, supervised learning includes the target variable in the training dataset [14].

On the other hand, deep learning (DL), a subfield of ML AI, uses ANNs to train models. An ANN mimics brain activity [15] and is thus defined as a collection of sophisticated computational and mathematical techniques that can solve problems involving regression, classification, and clustering while accurately mapping non-linear relationships within the studied phenomena [16]. The ANN typically consists of three layers: the input, the processing layers (also known as the hidden layers), and the output [17].

Simply defined, ANN connects inputs and outputs by learning, recognising patterns, or anticipating outcomes using microscopic building blocks (similar to neurons) to replicate the functioning of the human brain [18].

Regression, classification, and clustering are three possible subcategories of ML issues [19]. Classification seeks to predict ordinal variables and assign them to classes; clustering aims to divide the train set into clusters, that is, groups that demonstrate different features. Regression aims to predict a continuous variable as the target.

In order to lower costs associated with synthesis and the recommendation of optimised materials, AI appears to be a useful tool for designing nanomaterials [20]. In addition, employing AI during the design phase of hierarchical nanostructures makes their manufacture time-effective [21].

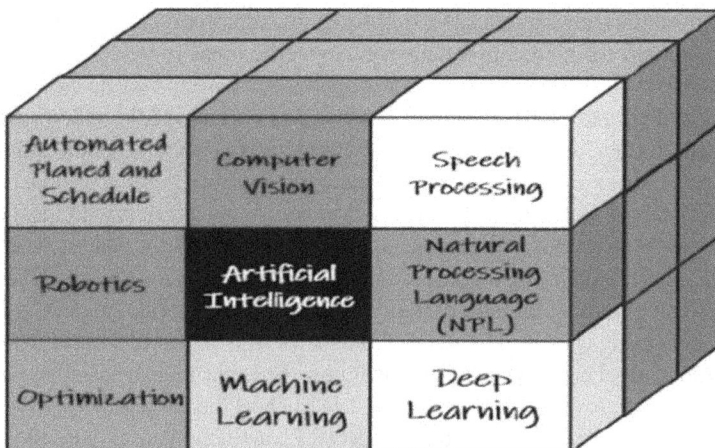

FIGURE 24.1 Main fields and subfields associated with artificial intelligence.

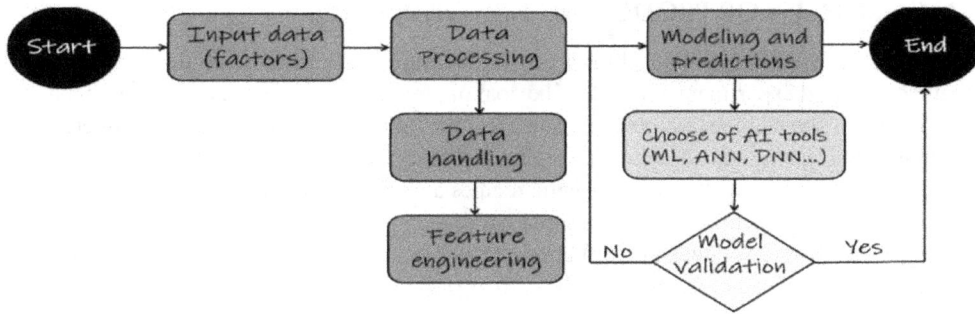

FIGURE 24.2 Flowchart of the main steps used in the implementation of AI for nanomaterial design.

STEPS INVOLVED IN SYNTHESIS OF NANOPARTICLES USING AI

The first step in the process is input data processing, where data are cleaned (chosen based on significance) and arranged for future model treatment. Following the modelling of the chosen data, general predictions may be drawn from them. The model is then validated. It is important to note that during the modelling stage, AI tools (such as ML, ANN, and DNN) are used to make predictions, and the algorithm has reached its conclusion if the model reviewed is trustworthy (Figure 24.2).

DISADVANTAGES OF AI IN SYNTHESIS OF NANOPARTICLES

In order to scale up the manufacturing of these unique hierarchical and adjustable nanostructures, the processing costs (the computer capacity to process some models and algorithms) and the amount of data needed for applying ML approaches need be overcome during the design of nanoarchitectures. For instance, a high-accuracy ML model may occasionally take a long time to run, necessitating more computing power than a model with a lesser accuracy that yet performs admirably (Figure 24.3)

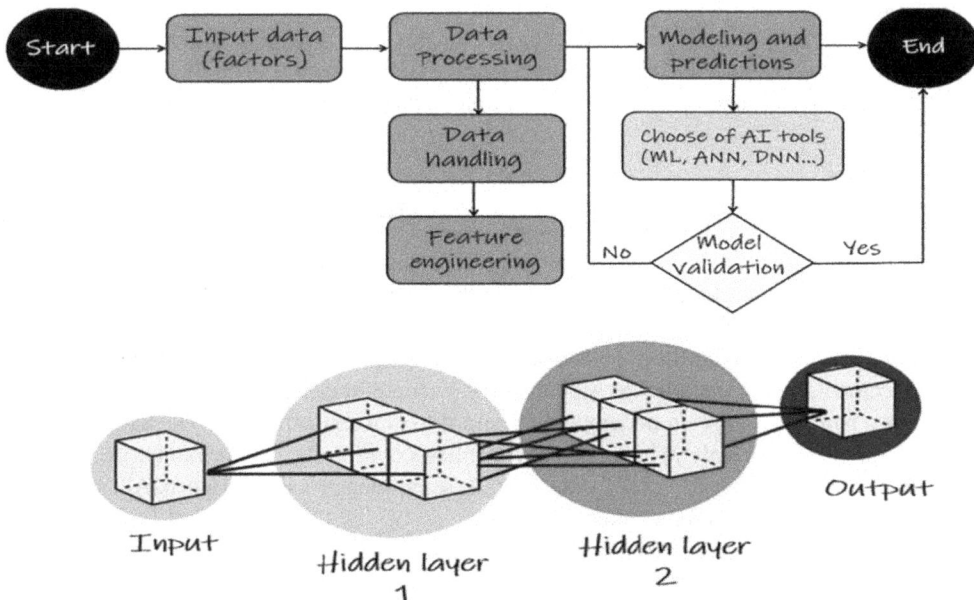

FIGURE 24.3 ANN and DNN structures.

ML ALGORITHMS FOR PRODUCING NANOPARTICLES

In nanoparticle synthesis, ML algorithms are typically employed for two tasks: the prediction of the synthetic output and experiment planning. The goal of prediction tasks is to develop a mathematical understanding of the relationship between an experimentally determined attribute of interest (such as nanoparticle size and shape or reaction yield) and a set of experimental input conditions. Because they need a dataset with examples of synthetic recipes and accompanying results from prior experiments to infer a relationship between the input parameters and the features of the output nanoparticles, supervised ML techniques for prediction are known as such. The training set is the portion of this dataset that is utilised to develop an ML model. Contrary to prediction algorithms, experiment planning algorithms seek efficient data collecting, frequently in the pursuit of an optimisation goal, rather than directly attempting to predict the future.

SYNTHESIS OF INORGANIC SEMICONDUCTOR NANOPARTICLES

The electrical and optical features of inorganic semiconductor nanoparticles, such as their wide absorption and constrained photoluminescence emission spectrum ranges, are beneficial. The continuous energy bands are constrained to discrete energy levels (quantum confinement effect) for nanoparticles with dimensions that are similar to or less than their Bohr exciton radius, which is generally 210 nm.

The name of these nanoparticles is quantum dots. Semiconductor nanoparticles' strong photostability, quantum yield, and capacity to precisely alter their emission wavelength make them advantageous for use in bioimaging, solar cells, optoelectronic devices, and quantum information technologies. The main classes of inorganic semiconductor nanoparticles include metal chalcogenide quantum dots (for instance, binary metal sulphide), selenide or telluride nanoparticles (for instance, CdS, ZnS, CdSe, and CdTe), ternary nanoparticles (for instance, $AgBiSe_2$ and $CuFeS_2$), and quaternary nanoparticles (for instance, $CuZnSnS_4$).

In general, hot injection and heat-up methods are used to synthesise these quantum dots. For a hot injection, a surfactant-containing hot solvent is quickly injected with the reagent. The heat-up approach involves heating the reaction mixture to the temperature at which crystallisation occurs in a batch procedure.

The type of precursor metal and chalcogenide reactants, the reaction, determines the optical and electrical properties of metal chalcogenide quantum dots, which rely on their size, shape, and composition ratio, reaction temperature, reaction time, ligands and reaction solvent.

CONCLUSION

Reducing the costs and time involved in the synthesis and use of nanomaterials requires the use of computational tools, such as AI-based techniques (ML and deep learning). AI can supply resources to create modifiable nanoparticles (nanomaterials with a focus on nanoarchitectonics) with increased surface area, improved mechanical and electrical characteristics, and special porosity. Yet there are a few drawbacks to achieving these objectives and overcome obstacles, for example the demand for greater computational capability to produce very accurate models. In spite of these difficulties, AI tools have shown promise in understanding and development of reliable mechanisms for the adsorption of pollutants, catalytic processes, the synthesis of reproducible nanomaterials, and even the scaling up of large-scale processes like membrane separation processes. Hence, AI can innovate and synthesise nanomaterials for a better tomorrow.

REFERENCES

[1] Ariga, Katsuhiko. "Nanoarchitectonics: A navigator from materials to life." *Materials Chemistry Frontiers* 1, no. 2 (2017): 208–211. https://doi.org/10.1039/C6QM00240D.

[2] Markovich, Gil, C. Patrick Collier, Sven E. Henrichs, Françoise Remacle, Raphael D. Levine, and James R. Heath. "Architectonic quantum dot solids." *Accounts of Chemical Research* 32, no. 5 (1999): 415–423. https://doi.org/10.1021/ar980039x.

[3] Ebara, M. "Smart polymers with nanoarchitectonics." In *Supra-Materials Nanoarchitectonics*, pp. 207–220. William Andrew Publishing, 2017. https://doi.org/10.1016/B978-0-323-37829-1.00009-2.

[4] Wells, Carlos M., Michael Harris, Landon Choi, Vishnu Priya Murali, Fernanda Delbuque Guerra, and J. Amber Jennings. "Stimuli-responsive drug release from smart polymers." *Journal of Functional Biomaterials* 10, no. 3 (2019): 34. https://doi.org/10.3390/jfb10030034.

[5] Yaqoob, Asim Ali, Muhammad Taqi-uddeen Safian, Mohd Rashid, Tabassum Parveen, Khalid Umar, and Mohamad Nasir Mohamad Ibrahim. "Introduction of smart polymer nanocomposites." In *Smart Polymer Nanocomposites*, pp. 1–25. Woodhead Publishing, 2021. https://doi.org/10.1016/B978-0-12-819961-9.00007-4.

[6] Wang, Zihao, Yageng Zhou, Teng Zhou, and Kai Sundmacher. "Identification of optimal metal-organic frameworks by machine learning: Structure decomposition, feature integration, and predictive modeling." *Computers & Chemical Engineering* 160 (2022): 107739. https://doi.org/10.1016/j.compchemeng.2022.107739.

[7] Jarrahi, Mohammad Hossein, David Askay, Ali Eshraghi, and Preston Smith. "Artificial intelligence and knowledge management: A partnership between human and AI." *Business Horizons* 66, no. 1 (2023): 87–99. https://doi.org/10.1016/j.bushor.2022.03.002.

[8] Abioye, Sofiat O., Lukumon O. Oyedele, Lukman Akanbi, Anuoluwapo Ajayi, Juan Manuel Davila Delgado, Muhammad Bilal, Olugbenga O. Akinade, and Ashraf Ahmed. "Artificial intelligence in the construction industry: A review of present status, opportunities and future challenges." *Journal of Building Engineering* 44 (2021): 103299. https://doi.org/10.1016/j.jobe.2021.103299.

[9] Sarker, Iqbal H. "Data science and analytics: An overview from data-driven smart computing, decision-making and applications perspective." *SN Computer Science* 2, no. 5 (2021): 377. https://doi.org/10.1007/s42979-021-00765-8.

[10] Vrigazova, Borislava. "The proportion for splitting data into training and test set for the bootstrap in classification problems." *Business Systems Research: International Journal of the Society for Advancing Innovation and Research in Economy* 12, no. 1 (2021): 228–242. https://doi.org/10.2478/bsrj-2021-0015.

[11] Yeturu, Kalidas. "Machine learning algorithms, applications, and practices in data science." In *Handbook of Statistics*, vol. 43, pp. 81–206. Elsevier, 2020. https://doi.org/10.1016/bs.host.2020.01.002.

[12] Malekian, Arash, and Nastaran Chitsaz. "Concepts, procedures, and applications of artificial neural network models in streamflow forecasting." In *Advances in Streamflow Forecasting*, pp. 115–147. Elsevier, 2021. https://doi.org/10.1016/B978-0-12-820673-7.00003-2.

[13] Jawad, Jasir, Alaa H. Hawari, and Syed Javaid Zaidi. "Artificial neural network modeling of wastewater treatment and desalination using membrane processes: A review." *Chemical Engineering Journal* 419 (2021): 129540. https://doi.org/10.1016/j.cej.2021.129540.

[14] Park, Y.-S., and S. Lek. "Artificial neural networks: Multilayer perceptron for ecological modeling." In *Developments in Environmental Modelling*, vol. 28, pp. 123–140. Elsevier, 2016. https://doi.org/10.1016/B978-0-444-63623-2.00007-4.

[15] Chauhan, Harsh, Jonathan Bernick, Dev Prasad, and Vijay Masand. "The role of artificial neural networks on target validation in drug discovery and development." In *Artificial Neural Network for Drug Design, Delivery and Disposition*, pp. 15–27. Academic Press, 2016. https://doi.org/10.1016/B978-0-12-801559-9.00002-8.

[16] Chaikittisilp, Watcharop, Yusuke Yamauchi, and Katsuhiko Ariga. "Material evolution with nanotechnology, nanoarchitectonics, and materials informatics: What will be the next paradigm shift in nanoporous materials?." *Advanced Materials* 34, no. 7 (2022): 2107212. https://doi.org/10.1002/adma.202107212.

[17] Li, Xianfeng, Jian Wang, Yangyang Guo, Tingyu Zhu, and Wenqing Xu. "Adsorption and desorption characteristics of hydrophobic hierarchical zeolites for the removal of volatile organic compounds." *Chemical Engineering Journal* 411 (2021): 128558. https://doi.org/10.1016/j.cej.2021.128558.

[18] Sugeno, Ayaka, Yasuyuki Ishikawa, Toshio Ohshima, and Rieko Muramatsu. "Simple methods for the lesion detection and severity grading of diabetic retinopathy by image processing and transfer learning." *Computers in Biology and Medicine* 137 (2021): 104795. https://doi.org/10.1016/j.compbiomed.2021.104795.

[19] Lyngdoh, Gideon A., and Sumanta Das. "Integrating multiscale numerical simulations with machine learning to predict the strain sensing efficiency of nano-engineered smart cementitious composites." *Materials & Design* 209 (2021): 109995. https://doi.org/10.1016/j.matdes.2021.109995.

[20] Schütt, Kristof T., Henning Glawe, Felix Brockherde, Antonio Sanna, Klaus-Robert Müller, and Eberhard KU Gross. "How to represent crystal structures for machine learning: Towards fast prediction of electronic properties." *Physical Review B* 89, no. 20 (2014): 205118. https://doi.org/10.1103/PhysRevB.89.205118.

[21] Tao, Huachen, Tianyi Wu, Matteo Aldeghi, Tony C. Wu, Alan Aspuru-Guzik, and Eugenia Kumacheva. "Nanoparticle synthesis assisted by machine learning." *Nature Reviews Materials* 6, no. 8 (2021): 701–716. https://doi.org/10.1038/s41578-021-00337-5.

25 Computational Techniques for Designing Novel Inorganic Nanomaterials

G. Subbulakshmi, G. Padma Priya, Soumya V. Menon,
Puli Mohith Vishnu Vardhan Reddy, and K.M. Nikhileshwar

INTRODUCTION

Nanoparticles are very small particles that can be found in nature or made by humans. They vary in their dimension from 1 to 100 nm. They have special material properties because of their tiny size. They have different uses than larger pieces of the same material because they have more surface area compared to their volume, and they interact with other substances in different ways. Nanoparticles can occur naturally in the environment or be made by humans. They are explored by many branches of science such as chemistry, physics, geology, and biology.

Nanoparticles are too small to be seen with normal optical microscopes, so they need electron microscopes or laser microscopes to be observed. Because of their small size, they can also be transparent when mixed with transparent media, unlike the other particles that disperse light that hits in return. They can pass from ordinary ceramic filter candles, so they need special nanofiltration methods to be separated from liquids. Since the size of a particle is 0.15–0.6 nm, most of its material is close to its surface. Therefore, the surface layer's properties may be more important than the others.

Nanosystems: Nanosystems are systems that work at a very small scale. Nanosystems can be grouped into different kinds based on how they look, what they do, or where they come from. Some kinds of nanosystems are as follows:
- Nanoparticles, which are small pieces of matter that have one dimension in the very small scale. Nanoparticles can have different forms, such as balls, sticks, or wires. Nanoparticles can also have different features, such as light, electricity, magnetism, or reaction. Nanoparticles can help with different things, such as delivering drugs, imaging, sensing, or cleaning the environment.
- Nanodevices, which are devices that do a specific thing at a very small scale. Nanodevices can be made of different materials, such as metals, semiconductors, or carbon. Nanodevices can also work in different ways, such as moving, electric, heat, or light. Nanodevices can help with different things, such as computing, communicating, storing, or changing energy.
- Nanomachines, which are machines that have many nanodevices that work together to do a complex thing. Nanomachines can be made of different materials, such as plastics, proteins, or DNA. Nanomachines can also work in different ways, such as putting themselves together, making copies of themselves, or fixing themselves. Nanomachines can help with different things, such as making things, moving things, or carrying things.
- Nanosystems, which are systems that have many nanomachines or other nanosystems that work together to do a better thing. Nanosystems can be made of different

DOI: 10.1201/9781003479239-25

materials, such as organic or inorganic. Nanosystems can also have different levels of difficulty, such as layered or modular. Nanosystems can help with different things, such as diagnosing diseases, treating diseases, or improving abilities. Nanosystems are copied from nature and made by technology. Nature has many examples that do amazing things with high speed and accuracy. For example, living cells are nanosystems that have various parts that do different jobs such as breaking down food, making copies of themselves, or talking to each other.

Molecules such as DNA and proteins are also nanosystems that keep information and speed up reactions. Technology has many tools and methods to make and change nanosystems with new features and functions. For example, scanning probe microscopy lets us see and touch single atoms and molecules. Making techniques such as chemical vapour deposition or sol-gel process let us make various nanoparticles and nanodevices. Putting techniques such as self-assembly or molecular recognition let us make complex nanomachines and nanosystems. Nanosystems have many benefits and problems for science and society. Nanosystems can change many areas such as medicine, engineering, energy, and environment by giving new answers and chances.

Nanotechnology: Nanotechnology is the science of manipulating materials at the molecular level to create functional systems. This field encompasses both current research and more advanced concepts. It involves the ability to construct things from the bottom up, using techniques and tools that are being developed today to produce complete, high-quality products. With nanotechnology, it may be possible to create lighter, stronger, and programmable materials that require less energy to produce, generate less waste, and offer greater fuel efficiency in land, sea, air, and space transportation. Nanocoating can make surfaces resistant to rust, scratches, and radiation. Electronic, magnetic, and mechanical devices and systems at the nanoscale may offer new levels of information processing. Carbon nanotubes are stronger than steel and more bendable than rubber. Nanotechnology could let us find and cure diseases at the molecule level; make new materials and devices with better performance; make and save clean and renewable energy; and clean and protect the environment from dirt and danger. These are some examples of what nanotechnology is and what it can do.

Nanotechnology is an interesting and hopeful field that has many possible benefits for people.

Inorganic Nanoparticles: Inorganic nanoparticles made of metals, such as silver, gold, platinum, or copper, can conduct electricity well and have different effects on light. For example, silver nanoparticles can show blue or green colour based on their size and shape whereas gold nanoparticles can switch colours from red to purple when they stick together. These nanoparticles aids in detection, processing images, catalysing reactions, enhance drug delivery systems, and developing nanoscale electronics. Inorganic nanoparticles made of metal oxides, such as iron oxide, zinc oxide, titanium dioxide, or cerium oxide, have different magnetic, optical, catalytic, and antibacterial properties [1].

For example, iron oxide nanoparticles can be pulled by a magnet and can heat up when exposed to a changing magnetic field. This makes them good for MRI and heat therapy. Zinc oxide nanoparticles can stop UV rays and kill bacteria. This makes them good for sunscreen and wound healing. Titanium dioxide nanoparticles can also stop UV rays and break down organic pollutants under sunlight. This makes them good for self-cleaning surfaces and water purification.

Cerium oxide nanoparticles can act as antioxidants and protect cells from oxidative stress. This makes them possible treatments for neurodegenerative diseases. There are different ways to make inorganic nanoparticles, such as using chemicals, gels, water, microwaves, or lasers. They can also be changed by covering them with organic materials or plastics to make them more stable, soluble, compatible, or functional. Inorganic nanoparticles have many benefits for science and technology. They can create new things and solve problems. But they also have some drawbacks and dangers for people and nature. So, it is important to learn how they affect living things and the environment. It is also important to make them safely and responsibly.

TYPES OF NANOPARTICLES

Carbon-Based Nanomaterials: These are nanomaterials that are mostly made of carbon atoms such as fullerenes, CNTs (carbon nanotubes), graphene, and CQDs (carbon quantum dots) [2]. They have outstanding mechanical, thermal, electrical, and optical properties. They are used for nanoelectronics, nanocomposites, sensors, drug delivery, and energy storage devices. Fullerenes and carbon nanotubes (CNTs) are two major types of carbon-based NPs. These are nanomaterials with a spherical hollow shape made of different isotopes of carbon. They have garnered attention because of their properties such as shape, size, tendency to accept electrons, strength, and versatility. These materials have carbon units arranged in pentagons and hexagons, where all the carbon atoms are sp2 hybridised. Few common fullerenes C60 with a diameter of 7.114 nm and C70 with a diameter of 7.648 nm look like graphite sheets that are rolled up. The sheets having one layer are called single-walled nanotubes.

Graphene, which is a carbon sheet that has excellent electrical and thermal conductivity, can be used for making transparent and flexible electronics. Carbon dots, which are carbon particles that glow in different colours depending on their characteristics, can be used for imaging and sensing biological systems [3].

Metal-Based Nanomaterials: These are nanomaterials made of metals such as gold, silver, iron, and metal oxides like titanium dioxide and zinc oxide. Localised surface plasmon resonance (LSPR), one of the properties of these nanoparticles, makes them unique [4]. They have optical, electrical, magnetic, and catalytic properties that vary with their shape, size, and composition. They are used for biomedical, environmental, and industrial applications. The controlled synthesis of metal nanoparticles with specific facets, sizes, and shapes is crucial for the development of advanced materials. Metal NPs have various uses in different medical fields due to their improved optical properties. For example, gold nanoparticle coating is commonly used in scanning electron microscopy (SEM) sample preparation to enhance the electron stream, resulting in higher quality SEM images [5].

Nanocomposites: These are composed of two or more different substances combined at the nanoscale, resulting in new materials with enhanced properties. These materials exhibit improved characteristics due to the synergistic effects of their constituent components. They can be organic–inorganic hybrids or mixtures of different nanoparticles. They are used for coatings, membranes, sensors, and nanofluids.

Dendrimers: These nanomaterials have a tree-like structure with a core and many layers of branches [6]. They have a precise size and shape, and can carry different functional groups on their surface. They are used for drug delivery, imaging, gene therapy, and catalysis.

APPLICATIONS

- **Photothermal Ablation:** This is a technique that uses nanoparticles to convert light energy into heat and destroy unwanted cells or tissues. For instance, gold nanoparticles can be used to selectively target and kill cancer cells by absorbing near-infrared light and generating heat that induces cell death [7].
- **Energy Storage Devices:** Nanoparticles have the potential to improve the performance of energy storage devices such as lithium-ion batteries, capacitors, and solar cells. By incorporating nanoparticles into these devices, their efficiency and capacity can be enhanced.

For instance, manganese dioxide nanoparticles can improve the capacitance, conductivity, and stability of supercapacitors. They can also act as catalysts for oxygen reduction reaction in fuel cells.

- **Nanotheranostics:** This is a field that combines nanotechnology and theranostics, which is an approach that integrates therapy and diagnosis of a disease on a single platform in health care. Nanotheranostics uses nanoparticles such as quantum dots, electrospun fibres, and biomacromolecule-based platforms to deliver drugs, image tissues, sense biomarkers, and monitor the therapeutic response. Nanotheranostics can be applied for various diseases such as cardiovascular diseases, cancer, Parkinson's disease, antimicrobial resistance, and orthopaedic diseases [8].

- **Sensors and Imaging:** Nanoparticles can be used as sensors and imaging agents for detecting and visualising various biological and chemical phenomena. For instance, gold nanoparticles can be used as colorimetric sensors for detecting analytes such as DNA, proteins, ions, and gases. Nanoparticles can also serve as contrast agents in optical imaging techniques such as dark-field microscopy, surface plasmon resonance, and surface-enhanced Raman scattering [2]. These techniques can be used to enhance the sensitivity and specificity of imaging, allowing for improved detection and analysis of biological and chemical samples [3].

ROLE OF NANOPARTICLES IN COMPUTATIONAL TECHNIQUES

Computational techniques are methods that use computers to solve problems or simulate systems. Computational techniques can be based on different theories or models, such as quantum mechanics, classical mechanics, statistical mechanics, or thermodynamics [6]. Computational techniques can also use different algorithms or software like molecular dynamics (MD), finite element method (FEM), density functional theory (DFT), and Monte Carlo (MC).

Nanomaterials play an important role in computational techniques for several reasons.

First, nanomaterials are often difficult or expensive to synthesise or characterise experimentally. Therefore, computational techniques can provide a cost-effective and efficient way to design, predict, or optimise the properties and functions of nanomaterials. Second, nanomaterials often exhibit novel phenomena or behaviours that are not well understood by existing theories or models. Therefore, computational techniques can provide a fundamental and comprehensive understanding of the underlying mechanisms and processes of nanomaterials [9].

Third, nanomaterials often have potential applications that require high performance or functionality. Therefore, computational techniques can provide a rational and systematic way to explore the possibilities and limitations of nanomaterials for various purposes:

- Some examples of how nanomaterials are used in computational techniques are as follows. Nanoparticles are discrete units of matter that have at least one dimension in the nanoscale. Nanoparticles can have different shapes, such as spheres, rods, or wires.
- Nanoparticles can also have different properties, such as optical, electrical, magnetic, or catalytic.
- Nanoparticles can be used for various purposes, such as drug delivery, imaging, sensing, or environmental remediation. Computational techniques can be used to study the structure, stability, reactivity, interaction, and transport of nanoparticles in different environments. For instance, DFT can be used to calculate the electronic structure and energy of nanoparticles and their adsorption or dissociation of molecules.
- MD can be used to simulate the dynamics and diffusion of nanoparticles and their interaction with solvents or biomolecules.
- MC can be used to model the aggregation or self-assembly of nanoparticles and their phase transitions.

- Nanowires are long structures that have two dimensions in the very small scale. Nanowires can have different materials, such as metals, semiconductors, or carbon. Nanowires can also have different features, such as electric flow or heat flow. Nanowires can help with different things, such as nanoelectronics, sensors, or energy change [10].
- Computational techniques can be used to study how electric flow and heat flow happen in nanowires in different situations. For example, DFT with NEGF can be used to find out the current-voltage features and the flow of nanowires and how they join with metal parts. MD can be used to see how heat moves, and how hot nanowires are, and how they touch with other things.
- FEM can be used to see how nanowires bend and stretch and how that affects their electric or heat features [10].
- Nanotubes are tube-like structures that have two dimensions in the very small scale. Nanotubes can have different materials, such as carbon, BN, or TMDs. Nanotubes can also have different features, such as strength or light. Nanotubes can help with different things, such as making things stronger, moving liquids, or making light.

Computational techniques can be used to study how nanotubes look, shake, flow, and shine in different cases. For example, DFT can be used to find out the electronic structure and energy of nanotubes and their mistakes or changes. MD can be used to see how nanotubes vibrate and make sounds and how they touch with liquids or gases. MC can be used to see how liquids flow and move through nanotubes and how that changes the liquid features.

FEM can be used to see how nanotubes make or take light and how they work with optical parts or paths. These are some ways that nanomaterials are used in computational techniques. Nanomaterials give a rich and varied place for computational techniques to look at and know the very small scale things and actions. Computational techniques also give a useful and strong way for nanomaterials to make and improve the very small scale features and functions. Nanomaterials and computational techniques are good for each other and work together to make nanoscience and nanotechnology better.

DESIGNING NANOTECHNOLOGY

To design nanoparticles, we need to know how to manage the making process, so that we can get the same results every time and also find out if we can guess the features of the nanostructures we want.

The following are ways to design inorganic nanoparticles:

- **Inorganic Nanoparticles**
  I. Ceria and ceria–zirconia are types of nanoparticles that can act as antioxidants and help treat diseases like ischaemic stroke. Using a method called CRAIT, researchers were able to send nanoparticles with drugs inside them to the core of breast tumours and shrink them without harming the rest of the body.
  II. Design nanomaterials for different energy devices, such as fuel cells and solar cells. The use of galvanic replacement reactions to make hollow nanocrystals of different mixed metal oxides, such as $Mn_3O_4/\gamma\text{-}Fe_2O_3$, from metal-oxide nanocrystals [10]. Carbon-based hybrid cellular nanosheets were used to make the cells with SnO_2 nanoparticles using a simple synthesis method.
  III. To design functional inorganic NPs into larger devices, they are arranged into superstructures by self-assembly. This can also be done with DNA-based and polymer-based NP assemblies that have controlled shapes and positions, and with quasicrystal line ordered structures. These superstructures have different properties than single NPs, such as optical chirality and dynamic structural change when stimulated [11].
  IV. Inorganic nanoparticles are widely used in biomedicine for diagnosis and therapy. To design them as water soluble, they are coated with surfactants and then with polymers.

The polymer-coated nanoparticles can be further modified with dyes, sensors, and biomolecules to make bio-functional nanoprobes.

V.   Some inorganic nanoparticles (NPs) are designed in such a way that they can maintain high potential phototoxicity, quick removal from blood, and stability acting as sonosensitisers in PDT (photodynamic therapy) and also as carriers for organic sonosensitisers, which helps overcome the limitations of organic small molecules for sonosensitisation.

Hence, it is relevant that development of new methods for achieving these nanomaterials with various shapes using different biotemplates along with the help of bioinformatics and AI is crucial breakthrough in the field of nanotechnology.

AI WITH NANOTECHNOLOGY

Nanoparticles are tiny particles with special properties that can boost AI in various ways. They can help make memory and storage devices smaller, faster, and more energy-efficient. They can also enable quantum computing, which can solve AI problems much faster than normal computers. Nanoparticles and AI are a powerful combination that can lead to new possibilities.

Nanoparticles can boost the processing power, sensing ability, and energy efficiency of AI hardware. Quantum dots can help create faster processors for complex AI tasks. Nanoparticles can also make sensors that detect various signals with high accuracy, improving the data quality for AI algorithms. Moreover, nanoparticles can reduce the power and heat of AI components, making them more durable and sustainable [2].

Nanoparticles can help create AI systems that mimic the human brain, using quantum dots to perform tasks such as pattern recognition and decision-making. Furthermore, nanoparticles can enhance data security and encryption, by using their physical and chemical features to store and protect AI-generated data. Nanoparticles are thus transforming the AI field in various ways, promising breakthroughs in the future [2].

ADVANTAGES OF AI AND NANOTECHNOLOGY

ATOMIC FORCE MICROSCOPY

AI and nanotech are often seen as mysterious and futuristic, but they are already used in everyday situations. One example is AFM, a method for imaging tiny objects. AFM produces more noise in the result data, which makes it difficult to see and specify details. AI can help filter out the noise by using neural and PCA, which are techniques for finding patterns in data. AI can do this faster and more accurately than humans. This makes AFM more efficient and reliable.

CHEMICAL MODELLING

Chemical modelling predicts how molecules interact [11]. It's used in bioscience and nanotech. Neural networks can help model nanotech materials more accurately by filtering out noise and measuring structural features. They can also learn from data of known systems and simulate complex phenomena. This can improve the production and innovation of nanotech.

NANOCOMPUTING

Nanocomputing can boost computing power by using novel materials and methods. AI can help us understand the quantum effects and physical systems involved in nanocomputing. This can lead to new architectures and designs that are not limited by transistors. Nanocomputing could also enable self-coding and self-building computers.

REFERENCES

[1] Zhang, Min, Hengrui Zhang, Jie Feng, Yunlong Zhou, and Bailiang Wang. "Synergistic chemotherapy, physiotherapy and photothermal therapy against bacterial and biofilms infections through construction of chiral glutamic acid functionalized gold nanobipyramids." *Chemical Engineering Journal* 393(2020): 124778. https://doi.org/10.1016/j.cej.2020.124778.

[2] Wang, Wei. "Imaging the chemical activity of single nanoparticles with optical microscopy." *Chemical Society Reviews* 47, no. 7 (2018): 2485–2508. https://doi.org/10.1039/C7CS00451F.

[3] Fan, Yuan, Shaobo Ou-yang, Dong Zhou, Junchao Wei, and Lan Liao. "Biological applications of chiral inorganic nanomaterials." *Chirality* 34, no. 5 (2022): 760–781. https://doi.org/10.1002/chir.23428.

[4] Haberfehlner, Georg, Franz-Philipp Schmidt, Gernot Schaffernak, Anton Hörl, Andreas Trügler, Andreas Hohenau, Ferdinand Hofer, Joachim R. Krenn, Ulrich Hohenester, and Gerald Kothleitner. "3D imaging of gap plasmons in vertically coupled nanoparticles by EELS tomography." *Nano Letters* 17, no. 11 (2017): 6773–6777. https://doi.org/10.1021/acs.nanolett.7b02979.

[5] Goris, Bart, Jan De Beenhouwer, Annick De Backer, Daniele Zanaga, K. Joost Batenburg, Ana Sánchez-Iglesias, Luis M. Liz-Marzán et al. "Measuring lattice strain in three dimensions through electron microscopy." *Nano Letters* 15, no. 10 (2015): 6996–7001. https://doi.org/10.1021/acs.nanolett.5b03008.

[6] Vecchio, Drew A., Samuel H. Mahler, Mark D. Hammig, and Nicholas A. Kotov. "Structural analysis of nanoscale network materials using graph theory." *ACS Nano* 15, no. 8 (2021): 12847–12859. https://doi.org/10.1021/acsnano.1c04711.

[7] Li, Yiwen, Ziwei Miao, Zhengwen Shang, Ying Cai, Jiaji Cheng, and Xiaoqian Xu. "A visible-and NIR-light responsive photothermal therapy agent by chirality-dependent MoO_{3-x} nanoparticles." *Advanced Functional Materials* 30, no. 4 (2020): 1906311. https://doi.org/10.1002/adfm.201906311.

[8] Chen, Daniel S., and Ira Mellman. "Oncology meets immunology: the cancer-immunity cycle." *Immunity* 39, no. 1 (2013): 1–10. https://doi.org/10.1016/j.immuni.2013.07.012.

[9] Winckelmans, Naomi, Thomas Altantzis, Marek Grzelczak, Ana Sánchez-Iglesias, Luis M. Liz-Marzán, and Sara Bals. "Multimode electron tomography as a tool to characterize the internal structure and morphology of gold nanoparticles." *The Journal of Physical Chemistry* 122(2018): 13522–13528. https://doi.org/10.1021/acs.jpcc.7b12379.

[10] Rao, C. N. R., S. R. C. Vivekchand, Kanishka Biswas, and A. Govindaraj. "Synthesis of inorganic nanomaterials." *Dalton Transactions* 34 (2007): 3728–3749. https://doi.org/10.1039/B708342D.

[11] Alkhazraji, Emad, A. Ghalib, K. Manzoor, and M. A. Alsunaidi. "Plasmonic nanostructured cellular automata." In *EPJ Web of Conferences*, vol. 139, p. 00001. EDP Sciences, 2017. https://doi.org/10.1051/epjconf/201713900001

26 Deep Diving into AI-Enhanced Innovative Approaches in Designing Inorganic Nanomaterials

Javed Akhtar, Trishna Kalita, Bitap Raj Thakuria,
Manash Jyoti Sarmah, and Himangshu Prabal Goswami

INTRODUCTION

In recent times, the field of material science has experienced a great transformation. This change has been facilitated by the two branches of computer science, namely, artificial intelligence (AI) and machine learning (ML), with the ability to develop systems of human intelligence and algorithms trained on data for predictions and decision-making. This synergy has opened up unprecedented avenues for innovation. In this captivating realm of AI-driven methods for exploring nanomaterials, we stand at the intersection of inquiry and technological advancement. The merging of fields such as physics, chemistry, software engineering, and data science holds the potential to transform how we utilise and employ nanomaterials [1].

Inorganic nanomaterials have garnered attention in the field of technology thanks to their structural and functional characteristics at the nano level. They provide a realm for exploration and progress in fields such as electronics, energy storage, healthcare, and environmental remediation. Nevertheless, the process of synthesising, characterising, and optimising these materials has always posed challenges [2].

The advent of AI has introduced a paradigm shift, enabling researchers to navigate this intricate landscape with newfound precision and efficiency. AI-driven methods and ML algorithms have demonstrated their value in expediting the exploration and creation of new inorganic nanomaterials. Through the analysis of datasets, these approaches enable the prediction of material properties and offer recommendations for ideal synthesis conditions. As a result, they are ushering in an era of innovation [3].

The potential of AI-driven nanomaterials knows no bounds. From creating cutting-edge sensors for early disease detection to designing nanoscale drug delivery systems, this deep dive explores the transformative impact of AI on the scientific frontier. It offers a glimpse into a future where nanotechnology and AI converge, reshaping our world.

The arrangement of this book chapter is as follows: We start with an exploration of data-driven approaches for nanomaterial design and their predictive modelling. Subsequently, we engage in a comprehensive examination of the requisite metrics essential for the development of ML models tailored to the domain of nanomaterials. Thereafter, we embark on an extensive investigation into the data-driven synthesis of nanomaterials. Lastly, we probe the augmentation of nanomaterial safety and efficacy through the integration of AI and ML, culminating in a concluding segment.

DOI: 10.1201/9781003479239-26

DATA-DRIVEN DESIGN OF MATERIALS FROM BULK TO NANO

In the past few years, there has been a growing interest in nanomaterials such as nanocrystals, nanorods, and nanoplates. This increased attention is largely due to advancements in material synthesis methods. These materials offer versatile physical and chemical tunability with enhanced system performance, ranging from inorganic semiconductors and metals to molecular crystals. As a result of their high surface-to-bulk ratio and strong environmental interactions, these are prone to defects and surface reconstruction [4,5]. To harness the full potential of these materials and stabilise their properties, researchers employ sophisticated techniques, including the use of surface-active agents and the creation of core-shell architectures, adding layers of complexity to their design and synthesis endeavours. Transitioning from the intricate synthesis techniques of these materials, we turn our attention to the equally complex realm of computational analysis. Traditional methods like periodic Density Functional Theory (DFT) work effectively for well-organised bulk materials that have a few hundred atoms per unit cell. However, when dealing with the complex structures and behaviours of nanomaterials, simulations need to account for larger numbers of atoms to incorporate hybrid structures, defects, and the impact of environmental factors such as pH, chemical potential, and solvents [1]. In the next section, we discuss data-driven methods for predicting the structures of nanomaterials.

STRUCTURE PREDICTION

Nanomaterials introduce a unique dimension to synthesis, encompassing parameters such as size, shape, and ligand chemistry, apart from conventional bulk synthesis considerations. Observations reveal that the stability of nanocrystals depends on their sizes, where metastable bulk phases can be stabilised due to more favourable surface energies compared to bulk ground states. This relationship underscores the significance of computational structure prediction in navigating the complex phase space of nanomaterials. Nanoclusters, a particular type of nanoparticle, stand out for their incredibly small size. They usually consist of up to 100–150 atoms and have dimensions below 1 nm. In contrast, nanoparticles have a wider range in terms of size, typically ranging from 1 to 100 nm. This distinction arises from the distinct characteristics of nanoclusters, including pronounced quantum confinement effects, off-lattice atomic arrangements, and variable bond lengths. Conversely, nanoparticles tend to exhibit atomic configurations resembling bulk crystal structures, resulting in more predictable properties. Identifying stable nanocluster geometries holds significant importance in computational materials research, offering validation for experimental measurements and guidance for synthesis.

While DFT, augmented with various corrections and models, makes computational analysis feasible for nanoclusters, larger entities such as nanoparticles or arrays of nanoclusters pose computational challenges due to their size and complexity. Researchers often turn to alternative, less computationally intensive methods, such as semi-empirical tight binding approaches and interatomic potentials, to estimate structure and energy properties. These approaches offer a balance between efficiency and accuracy [6]. Exploring the potential energy surface (PES) of nanostructures is vital but computationally challenging. Various techniques, including genetic algorithms and simulated annealing, help generate plausible structures [7,8]. Efficient structure generation often relies on leveraging prior knowledge and similarities between materials. For instance, Sokol et al. exemplified this approach by conducting a data mining study, demonstrating that low-energy structures of $(ZnO)_n$ clusters, when applied to other binary heteropolar materials like $(AlN)_n$, exhibit similar trends in energy. This tendency of specific elements to exhibit similar behaviours at the nanoscale can be quantified by comparing the relative stability rankings of structures involving each element [9]. An interatomic potential (IAP) is a practical tool that researchers use to mathematically describe the potential energy interactions between atoms in nanoparticles. It provides an empirical parameterisation of the potential energy surface specific to these small particles. In recent years, machine-learned potentials (MLPs) applied specifically to nanoparticle systems have advanced significantly and can now compete with traditional atomistic methods. ML potential regression methods offer distinct advantages over physics-based fundamental techniques, as these can be constructed without prior knowledge of the material, applied to a diverse array of input systems, and have demonstrated improvements in modelling the potential energy surface (PES) topology [10,11].

It is crucial to recognise that MLPs, like any parametric model, are inherently reliant on the quality of their training data [12]. Achieving high accuracy in IAP models typically necessitates a substantial dataset comprising approximately 50,000 first-principles calculations of nanoclusters, encompassing their structures, energies, and forces. This requirement can potentially be reduced to around 500 calculations by implementing active learning (AL) strategies that iteratively explore the energy landscape in regions where the model exhibits shortcomings, thereby enhancing the training dataset [13]. Notably, these advancements have been validated within a representative system of gold clusters. The active learning neural network (AL-NN) workflow efficiently trained a neural network model with around 500 total reference calculations. Extensive tests, involving over 1,100 configurations, showed that the AL-NN model accurately predicted DFT energies and forces for gold clusters of varying sizes with small deviations.

IAPs excel in modelling small and large structures but face challenges with medium-sized structures that fall between these two regimes. Bridging this gap is crucial for advancing nanoparticle structure prediction. Cluster expansions, including the Atomic Cluster Expansion (ACE) and equivalent MLPs, have demonstrated accuracy and performance in modelling nanoparticle surfaces and adsorbates [13–15].

Ensuring meticulous data reporting standards is vital, as high-quality data accelerates nanomaterial design and potential development. Databases such as OpenKIM and NIST offer valuable resources with fit potentials for various systems, facilitating further research and bridging the gap from bulk materials to nanostructures [16,17]. In the realm of materials science, the pursuit of ideal materials for specific applications is a multifaceted journey that often follows a systematic 'funnel-type investigation' approach. This approach starts with a broad selection of materials, taking into account fundamental properties, computational cost, and accuracy. As the process advances, it narrows down to the most promising candidates, incorporating specific criteria and computationally demanding properties. These criteria include stability under different conditions, band gap for optoelectronics, transport properties, defects, exciton binding energy, interfaces, surface adsorption, and surface reactivity. This is discussed briefly in the next section.

DESIGN METRICS FOR ML MODEL DEVELOPMENT

With a multitude of materials boasting diverse properties, the quest for the ideal material for a specific application can be a daunting and protracted process. Initially, a broad selection is considered, taking into account fundamental properties, accuracy, and computational cost. As discussed earlier in this discussion, we will explore several commonly used criteria employed in data-driven searches for nanosized materials tailored to specific applications.

Stability: Identifying and anticipating a material's stability under different conditions is one of the fundamental aspects of material design. High-throughput techniques are important in evaluating the stability of numerous materials. Metrics such as formation energy (ΔH_f), energy above the convex hull (E_{hull}), and decomposition enthalpy (ΔE_d) provide insights into a material's likelihood of undergoing phase transformations [18,19]. However, free energy is a more practical metric for real-world stability since it includes both energetics and entropy. Nanocrystals, characterised by their high surface-to-volume ratio, face specific stability challenges. Their stability is influenced by factors such as the chemical potential, solvent, entropy effects, size, configuration, and surface chemistry [20]. Nanocrystals have specific surface facets, each with unique stability and functionality implications.

Researchers can tailor nanocrystals for various applications by controlling the exposure of specific facets. Despite the gravity of nanomaterial stability, comprehensive high-throughput investigations in this area are sparse [21]. Developing phase diagrams considering the chemical environment and surface energy of bulk materials can be a valuable resource for assessing the stability of nanomaterials efficiently.

Band Gap: Materials used in optoelectronics and photocatalysis have a band gap as a vital metric. High-throughput computational investigations often begin with efficient but approximate methods like GGA-PBE functional [22]. To achieve higher accuracy in band gap calculations, more computationally demanding approaches such as hybrid functionals or GW theory are used [23]. However, implementing GW calculations can be challenging, and efforts have been made to improve their reliability. The work of Van Setten et al. and Rasmussen et al. underscores the critical role of establishing an accurate starting point, often referred to as the Kohn Sham energy, to enhance the reliability of band gap calculations [24,25]. ML plays a significant role in analysing vast amounts of band gap data and understanding structure–property relationships [26,27]. It can capture complex relationships between materials and their electronic structures by incorporating complex descriptors such as densities of states [28,29]. The degree of confinement and the dimensionality of the material directly influence the behaviour of band gaps in nanostructures. Different approaches are applied based on whether the confinement is weak, intermediate, or strong, which occurs as the size of the nanostructure approaches that of the exciton Bohr radius. These variations in confinement result in different relationships between the change in band gap (ΔEg) and the size of the nanostructure [30,31].

Transport Properties: Efficient carrier mobility is essential in semiconductor materials for optimal performance. Quantum mechanical principles govern carrier mobility, which is mathematically described through transport equations [32,33]. Effective mass, determined from electronic structure dispersion, plays a pivotal role in carrier mobility, and it is seen that smaller effective masses result in higher carrier mobility. In nanomaterials, carrier mobility is influenced by factors such as crystal size, orientation, impurities, surface passivation, and the presence of nanocrystalline domains or grains [34,35].

Defects: Defects are ubiquitous in materials and significantly impact material properties. Point defects induce charge transition levels inside the band structure of the material or within the band gap itself, and deep defects can act as recombination centres. Assessing defects in nanomaterials often requires long supercell calculations [36]. Recent advancements have brought automated defect calculations through packages such as PyCDT, PyLada, and PyDEF, enabling systematic defect analysis and the construction of databases cataloguing charged defect properties across diverse materials [37–39]. The Quantum Point Defect (QPOD) database focuses on intrinsic point defects in 2D semiconductors and insulators [40], whereas, due to their high surface-to-bulk ratio and non-stoichiometric composition nanocrystals exhibit unique defect behaviour [4,5,41]. Defects in nanocrystals are more concentrated on the surface, introducing localised states, and their formation in colloidal solutions presents additional challenges.

Exciton Binding Energy (E_b): Small exciton binding energy is preferred for solar cells, while larger binding energy is desirable in LEDs [42]. Exciton binding energy is generally computed by the computationally demanding Bethe–Salpeter equation. Dielectric constant (ϵ) is a material property that influences E_b. Challenges arise when transitioning from bulk to nanosized materials, as stronger electron-hole overlap increases binding energy. The dimensionality of the material also influences how the binding energy responds to size changes [43,44].

Interfaces: Interfacial alignment in nanomaterials significantly influences their performance in various applications, including sensors, catalysts, and electronic devices [45]. Proper alignment of energy levels and charge transport pathways at interfaces enhances conductivity and device efficiency. In electronic devices like solar cells, energy level alignment is critical for efficient charge transfer. DFT calculations generally used for band alignment in nanomaterials face difficulty due to limited knowledge of their chemical and structural features [46]. The DFT community lacks consensus on treating nanocrystalline thin-film interfaces because of inhomogeneity and anisotropy at the interface. For example, solid–liquid interfaces are complex due to surface dipoles affecting band positions. Further, higher order electron–phonon interactions and short-range orders do not allow accurate modelling of the interfacial dynamics. Recent computational frameworks address such issues but are case-specific and hence challenging for high-throughput workflows [47,48].

Surface Adsorption: Reactivity, catalytic activity, and surface reconstruction are generally understood through the study of surface adsorption. High-throughput workflows, like those developed by Montoya et al., employ software tools such as Automate and Pymatgen to efficiently calculate adsorption energies on solid surfaces [49–51]. These workflows involve slab generation, adsorption site identification, geometry optimisation, and error corrections to ensure accuracy. The application of high-throughput methods extends to inorganic semiconductors, where complex surface chemistries necessitate a detailed understanding of photocatalysis. High-throughput DFT workflows were developed by Andriuc et al. for computing the p-band centre of inorganic semiconductors essential for photocatalysis applications and by Rosen et al. to identify adsorption sites for heterogeneous catalysis applications in metal-organic frameworks (MOFs) [52–54].

Surface Reactivity: Surface reactivity is described using microkinetic modelling, which relies on thermodynamic properties and rate constants [55]. Reduction to low-dimensional descriptor space, which means simplifying the study of surface reactions by concentrating on one essential property (like adsorption energy), simplifies high-throughput screening [56,57]. Modelling nanomaterial surface reactivity is difficult as it generally includes various challenges such as particle size effects, active sites and their chemical environment, and the influence of various parameters such as pressure, temperature, and reactant concentrations. Data-driven design of materials with target reactivity requires well-structured and transferable databases with reliable metadata, reporting surface-related properties, chemical environments, and kinetics. Exploring nanomaterials using computational analysis and data-driven methods provides valuable insights that can greatly inform the experimental synthesis process.

Scientists can utilise AI and ML techniques for the synthesis of nanoscale materials by utilising the above-mentioned criteria and metrics. Moreover, AI and ML can assist in monitoring quality and detecting defects during the synthesis process. During the synthesis process, the aforementioned metrics are collected as data points. These data streams are continuously fed into ML algorithms. The algorithms recognise patterns associated with defects or deviations from the desired quality standards. This training involves using historical data that includes examples of both defect-free and defective nanomaterials obtainable from the metrics discussed above. As data flows in, ML models compare the current data to established patterns for desired quality. If any anomalies or deviations are detected, the models perform root cause analysis by examining the data to determine the likely factors contributing to the defects based on historical data and current conditions. This proactive approach allows for preventive actions to be taken before defects occur. ML makes it easier for researchers to understand and act upon the information since it provides quantitative measures of defects, such as defect density, size, or distribution, allowing for a detailed assessment of the quality of the nanomaterials. Such continuous monitoring of materials through metrics can identify any deviations from the desired properties, and ML models can offer guidance and adjustments to ensure that the final product meets the specified criteria. ML approaches and extensive datasets aim to streamline computational costs and enhance the exploration of diverse chemical spaces, particularly those relevant to nanomaterials. To sum up, incorporating AI and ML with the metrics discussed earlier will pave the pathway for experimental synthesis of nanomaterials with a more efficient and informed approach, enabling rapid screening of potential materials, optimisation of synthesis conditions, and real-time quality control, ultimately speeding up the development of customised nanomaterials for specific applications.

ADVANCING NANOMATERIAL SYNTHESIS THROUGH DATA-DRIVEN INSIGHTS

In the realm of nanomaterial synthesis, a myriad of parameters extends beyond bulk synthesis, including size, shape, and ligand chemistry. Experimental observations have revealed that nanocrystal phase stabilities are intricately tied to their sizes, whereas metastable bulk phases can achieve stability due to more favourable surface energies compared to their bulk ground-state counterparts [58,59].

The synthesis of nanomaterials introduces an added layer of complexity compared to bulk synthesis. Even minor modifications in synthesis conditions, such as temperature, mole ratios, or ligand variations, can yield profound changes in composition, dimensionality, morphology, and short-range order. These adjustments exert significant influence over the electronic and optical properties of nanomaterials, highlighting the intricate interplay of factors at this minuscule scale [60,61]. Nanoparticle databases have emerged as invaluable resources, offering access to vital data. The NOMAD (Novel Materials Discovery) Laboratory, for instance, serves as a curated repository, uniting experimental and computational data. It provides a versatile search engine, a Python Application Programming Interface (API), and a vast database with over 2 million clusters, supporting researchers in the field [62]. Additionally, the Open Catalysis Project is a computational database dedicated to catalyst structures, offering a repository with 1.3 million molecular relaxations, enhancing innovation and training [63]. For small-molecule organic and metal-organic structures, the Cambridge Structure Database (CSD) offers a treasure trove of data, boasting a collection exceeding 1 million structures [64]. Meanwhile, the Quantum Cluster Database explores pure element nanoclusters, featuring over 50,000 clusters calculated through DFT, providing valuable resources for research and innovation. A remarkable array of such databases is chronicled in detail in a review by Panneerselvam and Choi, collectively serving as a reservoir of knowledge and a catalyst for breakthroughs in nanomaterial synthesis [64].

The proliferation of experimental databases in biological and medical applications stands in stark contrast to the limited availability of comprehensive materials synthesis databases. This gap highlights the challenges unique to nanomaterials databases, stemming from data variability and the complex task of representing each structure with appropriate metadata [65].

A theoretically calculated database, for instance, may encompass a multitude of descriptors, ranging from atomic species and atomic coordinates to effective charge, implicit solvent models, calculation software, choice of functional and basis set, energy, forces, band gap, electronic structure, and beyond. Conversely, an entry for an experimentally observed nanoparticle in a database may need to incorporate details on atomic species, the presence of counterions, stabilising ligands, synthesis protocols, reagent concentrations, temperature, and percent yield. Such entries are also likely to feature data concerning the average particle size and the distribution of sizes. Efforts have been undertaken to establish universal descriptors that can harmonise the annotations of nanoparticle databases. Yan et al., for instance, have strived to standardise the annotations within nanoparticle databases, generating an impressive 2,142 nanodescriptors for each nanomaterial in their open database, Pubvinas [66]. These characteristics include bioactivities, such as protein adsorption and cytotoxicity, and physical characteristics. Notably, while Pubvinas has established fixed descriptors for their database, NOMAD has adopted a flexible, posterior data sorting approach, allowing for the continuous addition of descriptors as new data emerges [62].

The challenges associated with managing general material data, characterised by volume, variety, velocity, and veracity, necessitate efficient approaches for data exchange and utilisation. It is crucial to acknowledge that the data challenge becomes even more pronounced in nanomaterials, where data-driven research is in its infancy and the intricate landscape of property-synthesis-characterisation expands in complexity. In this context, the use of "FAIR" data principles—ensuring data is Findable, Accessible, Interoperable, and Repurposable—(a guideline for major databases) becomes imperative [67]. In the realm of experiments, the ultimate goal is to furnish a comprehensible set of guidelines that indicate how various tunable factors influence product formation. In the context of synthesis, paramount descriptors are the explicit reaction conditions that have yielded the product and the relative thermodynamic energies of potential structures. Employing phase diagrams and astute data visualisation techniques emerges as a potent strategy to achieve this goal. Within these intricate phase diagrams, variables such as partial pressure, chemical potentials, pH, and temperature serve as tuning knobs, offering the means to pinpoint promising regions within the high-dimensional parameter space, thereby facilitating the precise fabrication of a desired compound [68].

A promising avenue for further advancement lies in the development of easily interpretable reaction conditions on a larger scale. Taking cues from the strides made in bulk materials databases like the Materials Project, which have harnessed Phase Diagram and Pourbaix Diagram analyses to deliver synthesis-relevant data, there exists an analogous opportunity to devise similar analysis applications tailored for nanomaterials. Such tools would be highly coveted, yet their realisation hinges upon the availability of underlying data—a resource that is currently lacking [50,69].

We expect the convergence of computational insights, ML techniques, and experimental data to transform the area of nanomaterials synthesis research in the emerging era of data-driven research. By employing big data, researchers can find new directions in nanomaterials and accomplish previously unachievable objectives in sensing, material design, catalysis, and other areas.

ELEVATING NANOMATERIAL SAFETY AND EFFICACY WITH AI AND ML

Nanomaterials are revolutionising biomedical applications, but concerns regarding their potential toxicity persist. To address this, AI and ML are emerging as important tools. In this exploration, we focus on two crucial dimensions: predicting nanomaterial toxicity with ML algorithms and optimising nanomaterials' in vivo behaviour.

PREDICTING BIOLOGICAL TOXICITY OF NANOMATERIALS USING ML ALGORITHMS

The toxicity of nanomaterials remains a significant challenge for their application in biomedical sectors. The incorporation of AI in nanotechnology has emerged as a promising approach to address this bottleneck. The random forest (RF) technique was employed to develop a tissue-specific classification model for predicting neurotoxicity caused by nanoparticles within in vitro systems. The application of ML facilitated the classification and identification of hazardous nanoparticles in a cost-effective and timely manner. The developed classification model utilised RF and goodness-of-fit with additional robustness and predictability matrices to provide information on the exposure dose and duration, toxicological assay, cell type, and zeta-potential, enabling the prediction of neurotoxicity [70]. The cell shape index (CSI) and nuclear area factors (NAFs) were employed by Singh et al. as nanotoxicity descriptors of nanomaterials to establish standard operating procedures in nanotoxicology. The authors utilised an ML-based graph modelling and correlation-establishing approach to investigate the change in the phenotype of cells due to interactions with nanoparticles. The CSI and NAF, as nano descriptors, can serve as intuitive cell phenotypic parameters to define the safety of nanomaterials that are extensively used in consumer products and nanomedicine [71]. LDA, NB, MLogitR, SMO, AdaBoost, J48, and RF were some of the ML algorithms employed to predict the toxic effects of metal oxide nanoparticles (MONPs) on *Escherichia coli* [72]. The developed models offer a scientific basis for the design and preparation of safe nanomaterials.

A study employed an ML-based perturbation theory (PTML) quantitative structure toxicity relationship (QSTR) approach to predict the genotoxicity of MONPs under various experimental assay conditions. The authors developed a unified in silico model based on artificial neural networks that aimed to simultaneously predict the general toxicity profiles of nanomaterials. Using a dataset of 20 metal oxide nanomaterials, the model was trained, validated, and achieved 96.5% accuracy in predicting genotoxicity [73].

ML ALGORITHMS TO OPTIMISE IN VIVO FATE OF NANOMATERIALS

Within the human body, proteins swiftly adhere to the surface of nanomaterials, giving rise to protein coronas. These coronas play a vital role in shaping how cells recognise nanomaterials. The formation and composition of protein corona were predicted by combining a meta-analysis and a RF model. This predictive capability, in turn, aids in the judicious selection of nanoparticles that are both ideal and safe for a range of applications, including nanomedicine, biosensing, organ targeting, and various other applications [74].

Delivering drugs in precise amounts to specific targets inside the human body is crucial to avoid adverse effects and the onset of other medical conditions. The comprehension of the interaction between nanoparticles and the human body holds significant importance. A recent investigation employed all-atom molecular dynamics simulations (MDS) to determine the effect of silver nanoparticles (AgNPs) on the conformation and function of human serum albumin (HSA), a protein responsible for transport. The findings demonstrated that AgNPs interacted with HSA without significantly altering its tertiary and secondary structures, thereby preserving the protein's function. This discovery suggests that AgNPs hold promise as a tool for transporting conjugated drug molecules, particularly in biomedical applications where HSA plays a vital role in the circulatory system [75]. A computational strategy was carried out in the battle against COVID-19. A clustering algorithm and Schrödinger software were employed to develop a drug delivery system (DDS) capable of ensuring stable drug administration. This method yielded six potential carrier molecules that met predefined criteria based on their molecular structure and functional group positions. After conducting the computational analysis, glycyrrhizin emerged as a highly prospective contender for the purpose of drug delivery. Glycyrrhizin, nafamostatmesilate, underwent a transformation into micelle nanoparticles (NPs) to enhance drug stability and optimise the treatment of COVID-19. Validation of the spherical particle morphology (predicted by the computation) was accomplished through transmission electron microscopy (TEM). Dynamic light scattering (DLS) measurements and zeta-potential determination assessed particle size and stability, with UV spectrum analysis nafamostatmesilate loading efficiency above 90% [75].

CONCLUSION

The dynamic synergy between AI and nanoscience is reshaping the landscape of materials research, offering a plethora of reasons for optimism. The incorporation of AI-enhanced approaches for nano-sized inorganic materials, particularly in biomedical applications, represents a paradigm shift in our ability to design and synthesise nanomaterials to address pressing healthcare needs, heralding a promising future in scientific discovery and technological advancement. The exceptional capacity to forecast and optimise nanomaterial formations is the starting point for the voyage. While there are drawbacks to traditional approaches, data-driven options enabled by ML are expanding the realm of what is possible. Researchers are traversing the complex world of nanomaterials with accuracy and speed, exposing their distinctive features and unlocking their promise for practical applications by utilising massive datasets and advanced algorithms. AI-driven design metrics, considering factors such as stability, band gap, transport properties, and defects, are providing the blueprint for creating nanomaterials optimised for specific functions. The future holds potential for revolutionary discoveries, whether it's creating nanoscale medicine delivery systems or inventing cutting-edge sensors. In the realm of practical applications, AI-assisted nanotechnology is demonstrating its prowess by accurately predicting nanomaterial properties. This feature, which makes it possible to create nanomaterials that are suited to particular applications, is revolutionary. Numerous areas of study, including environmental research and medicine, could undergo a complete transformation as a result of these advancements, which range from extremely sensitive sensors to more effective drug delivery systems. Furthermore, evaluating the biological safety of nanomaterials is a critical task for AI. To ensure the development of safer nanomaterials for a wide variety of applications, ML algorithms are becoming important in predicting toxicity effects. As we embark on this thrilling journey at the confluence of technology and nanoscience, we may anticipate a world in which nanomaterials supercharged by AI push us towards a future full of limitless possibilities and beneficial developments.

REFERENCES

[1] Yang, Ruo Xi, Caitlin A. McCandler, Oxana Andriuc, Martin Siron, Rachel Woods-Robinson, Matthew K. Horton, and Kristin A. Persson. "Big data in a nano world: a review on computational, data-driven design of nanomaterials structures, properties, and synthesis." *ACS Nano* 16, no. 12 (2022): 19873–19891. https://doi.org/10.1021/acsnano.2c08411.

[2] Mikolajczyk, Alicja, Natalia Sizochenko, Ewa Mulkiewicz, Anna Malankowska, Bakhtiyor Rasulev, and Tomasz Puzyn. "A chemoinformatics approach for the characterization of hybrid nanomaterials: safer and efficient design perspective." *Nanoscale* 11, no. 24 (2019): 11808–11818. https://doi.org/10.1039/c9nr01162e.

[3] Barnard, A. S., and G. Opletal. "Predicting structure/property relationships in multi-dimensional nanoparticle data using t-distributed stochastic neighbour embedding and machine learning." *Nanoscale* 11, no. 48 (2019): 23165–23172. https://doi.org/10.1039/C9NR03940F.

[4] Goldzak, Tamar, Alexandra R. McIsaac, and Troy Van Voorhis. "Colloidal CdSe nanocrystals are inherently defective." *Nature Communications* 12, no. 1 (2021): 890. https://doi.org/10.1038/s41467-021-21153-z.

[5] Brinck, Ten, Stephanie, Francesco Zaccaria, and Ivan Infante. "Defects in lead halide perovskite nanocrystals: analogies and (many) differences with the bulk." *ACS Energy Letters* 4, no. 11 (2019): 2739–2747. https://doi.org/10.1021/acsenergylett.9b01945.

[6] Ghasemi, S. Alireza, Maximilian Amsler, Richard G. Hennig, Shantanu Roy, Stefan Goedecker, Thomas J. Lenosky, C. J. Umrigar, Luigi Genovese, Tetsuya Morishita, and Kengo Nishio. "Energy landscape of silicon systems and its description by force fields, tight binding schemes, density functional methods, and quantum Monte Carlo methods." *Physical Review B* 81, no. 21 (2010): 214107. https://doi.org/10.1103/PhysRevB.81.214107

[7] Johnston, Roy L. "Evolving better nanoparticles: genetic algorithms for optimising cluster geometries." *Dalton Transactions* 22 (2003): 4193–4207. https://doi.org/10.1039/B305686D.

[8] Wales, David J., and Jonathan PK Doye. "Global optimization by basin-hopping and the lowest energy structures of Lennard-Jones clusters containing up to 110 atoms." *The Journal of Physical Chemistry A* 101, no. 28 (1997): 5111–5116. https://doi.org/10.1021/jp970984n.

[9] Sokol, Alexey A., C. Richard A. Catlow, Martina Miskufova, Stephen A. Shevlin, Abdullah A. Al-Sunaidi, Aron Walsh, and Scott M. Woodley. "On the problem of cluster structure diversity and the value of data mining." *Physical Chemistry Chemical Physics* 12, no. 30 (2010): 8438–8445. https://doi.org/10.1039/C0CP00068J.

[10] Zuo, Yunxing, Chi Chen, Xiangguo Li, Zhi Deng, Yiming Chen, Jörg Behler, Gábor Csányi et al.. "Performance and cost assessment of machine learning interatomic potentials." *The Journal of Physical Chemistry A* 124, no. 4 (2020): 731–745. https://doi.org/10.1021/acs.jpca.9b08723.

[11] Weinreich, Jan, Anton Romer, Martín Leandro Paleico, and Jorg Behler. "Properties of α-brass nanoparticles. 1. Neural network potential energy surface." *The Journal of Physical Chemistry C* 124, no. 23 (2020): 12682–12695. https://api.semanticscholar.org/CorpusID:216384290.

[12] Lysogorskiy, Yury, Cas van der Oord, Anton Bochkarev, Sarath Menon, Matteo Rinaldi, Thomas Hammerschmidt, Matous Mrovec et al. "Performant implementation of the atomic cluster expansion (PACE) and application to copper and silicon." *npj Computational Materials* 7, no. 1 (2021): 97. https://doi.org/10.1038/s41524-021-00559-9.

[13] Loeffler, Troy D., Sukriti Manna, Tarak K. Patra, Henry Chan, Badri Narayanan, and Subramanian Sankaranarayanan. "Active learning a neural network model for gold clusters & bulk from sparse first principles training data." *ChemCatChem* 12, no. 19 (2020): 4796–4806. https://api.semanticscholar.org/CorpusID:219531431.

[14] Drautz, Ralf. "Atomic cluster expansion for accurate and transferable interatomic potentials." *Physical Review B* 99, no. 1 (2019): 014104. https://link.aps.org/doi/10.1103/PhysRevB.99.014104.

[15] Batzner, Simon, Albert Musaelian, Lixin Sun, Mario Geiger, Jonathan P. Mailoa, Mordechai Kornbluth, Nicola Molinari, Tess E. Smidt, and Boris Kozinsky. "E (3)-equivariant graph neural networks for data-efficient and accurate interatomic potentials." *Nature Communications* 13, no. 1 (2022): 2453. https://doi.org/10.1038/s41467-022-29939-5.

[16] Becker, Chandler A., Francesca Tavazza, Zachary T. Trautt, and Robert A. Buarque de Macedo. "Considerations for choosing and using force fields and interatomic potentials in materials science and engineering." *Current Opinion in Solid State and Materials Science* 17, no. 6 (2013): 277–283. https://doi.org/10.1016/j.cossms.2013.10.001.

[17] Hale, Lucas M., Zachary T. Trautt, and Chandler A. Becker. "Evaluating variability with atomistic simulations: the effect of potential and calculation methodology on the modeling of lattice and elastic constants." *Modelling and Simulation in Materials Science and Engineering* 26, no. 5 (2018): 055003. https://doi: 10.1088/1361-651X/aabc05.

[18] Yan, Qimin, Jie Yu, Santosh K. Suram, Lan Zhou, Aniketa Shinde, Paul F. Newhouse, Wei Chen et al. "Solar fuels photoanode materials discovery by integrating high-throughput theory and experiment." *Proceedings of the National Academy of Sciences* 114, no. 12 (2017): 3040–3043. https://www.pnas.org/doi/abs/10.1073/pnas.1619940114.

[19] Singh, Arunima K., Joseph H. Montoya, John M. Gregoire, and Kristin A. Persson. "Robust and synthesizable photocatalysts for CO2 reduction: a data-driven materials discovery." *Nature Communications* 10, no. 1 (2019): 443. https://doi.org/10.1038/s41467-019-08356-1.

[20] Wenhua Luo and Wangyu Hu. "Gibbs free energy, surface stress and melting point of nanoparticle." *Physical Review B* 425 (2013): 90–94. https://doi.org/10.1016/j.physb.2013.05.025.

[21] Shi, Hongqing, Amanda S. Barnard, and Ian K. Snook. "High throughput theory and simulation of nanomaterials: exploring the stability and electronic properties of nanographene." *Journal of Materials Chemistry* 22, no. 35 (2012): 18119–18123. https://doi.org/10.1039/C2JM32618C.

[22] Tran, Fabien, and Peter Blaha. "Accurate band gaps of semiconductors and insulators with a semilocal exchange-correlation potential." *Physical Review Letters* 102, no. 22 (2009): 226401. https://doi.org/10.1103/PhysRevLett.102.226401.

[23] Hinuma, Yoyo, Taisuke Hatakeyama, Yu Kumagai, Lee A. Burton, Hikaru Sato, Yoshinori Muraba, Soshi Iimura et al. "Discovery of earth-abundant nitride semiconductors by computational screening and high-pressure synthesis." *Nature Communications* 7, no. 1 (2016): 11962. https://doi.org/10.1038/ncomms11962.

[24] Van Setten, M. J., Matteo Giantomassi, Xavier Gonze, G-M. Rignanese, and Geoffroy Hautier. "Automation methodologies and large-scale validation for G W: towards high-throughput G W calculations." *Physical Review B* 96, no. 15 (2017): 155207. https://doi.org/10.1103/PhysRevB.96.155207.

[25] Rasmussen, Asbjørn, Thorsten Deilmann, and Kristian S. Thygesen. "Towards fully automated GW band structure calculations: what we can learn from 60.000 self-energy evaluations." *npj Computational Materials* 7, no. 1 (2021): 22. https://doi.org/10.1038/s41524-020-00480-7.

[26] Ward, Logan, Ankit Agrawal, Alok Choudhary, and Christopher Wolverton. "A general-purpose machine learning framework for predicting properties of inorganic materials." *npj Computational Materials* 2, no. 1 (2016): 1–7. https://doi.org/10.1038/npjcompumats.2016.28.

[27] Ward, Logan, Alexander Dunn, Alireza Faghaninia, Nils ER Zimmermann, Saurabh Bajaj, Qi Wang, Joseph Montoya et al. "Matminer: an open source toolkit for materials data mining." *Computational Materials Science* 152 (2018): 60–69. https://doi.org/10.1016/j.commatsci.2018.05.018.

[28] Wang, Vei, Gang Tang, Ren-Tao Wang, Ya-Chao Liu, Hiroshi Mizuseki, Yoshiyuki Kawazoe, Jun Nara, and Wen-Tong Geng. "High-throughput computational screening of two-dimensional semiconductors." *arXiv preprint arXiv:1806.04285* (2018). https://doi.org/10.1021/acs.jpclett.2c02972.

[29] Mahmoud, Chiheb Ben, Andrea Anelli, Gábor Csányi, and Michele Ceriotti. "Learning the electronic density of states in condensed matter." *Physical Review B* 102, no. 23 (2020): 235130. https://doi.org/10.1103/PhysRevB.102.235130.

[30] Brus, Louis. "Electronic wave functions in semiconductor clusters: experiment and theory." *The Journal of Physical Chemistry* 90, no. 12 (1986): 2555–2560. https://doi.org/10.1021/j100403a003.

[31] Smith, Andrew M., and Shuming Nie. "Semiconductor nanocrystals: structure, properties, and band gap engineering." *Accounts of Chemical Research* 43, no. 2 (2010): 190–200. https://doi.org/10.1021/ar9001069.

[32] Bardeen, J., and W. J. P. R. Shockley. "Deformation potentials and mobilities in non-polar crystals." *Physical Review* 80, no. 1 (1950): 72. https://doi.org/10.1103/PhysRev.80.72.

[33] Faghaninia, Alireza, Joel W. Ager III, and Cynthia S. Lo. "Ab initio electronic transport model with explicit solution to the linearized Boltzmann transport equation." *Physical Review B* 91, no. 23 (2015): 235123. https://doi.org/10.1103/PhysRevB.91.235123.

[34] Kolahi, Sanaz, Saber Farjami-Shayesteh, and Yashar Azizian-Kalandaragh. "Comparative studies on energy-dependence of reduced effective mass in quantum confined ZnS semiconductor nanocrystals prepared in polymer matrix." *Materials Science in Semiconductor Processing* 14, no. 3–4 (2011): 294–301. https://doi.org/10.1016/j.mssp.2011.07.002.

[35] Liu, Yao, Markelle Gibbs, James Puthussery, Steven Gaik, Rachelle Ihly, Hugh W. Hillhouse, and Matt Law. "Dependence of carrier mobility on nanocrystal size and ligand length in PbSe nanocrystal solids." *Nano Letters* 10, no. 5 (2010): 1960–1969. https://doi.org/10.1021/nl101284k.

[36] Kumagai, Yu, and Fumiyasu Oba. "Electrostatics-based finite-size corrections for first-principles point defect calculations." *Physical Review B* 89, no. 19 (2014): 195205. https://doi.org/10.1103/PhysRevB.89.195205.

[37] Goyal, Anuj, Prashun Gorai, Haowei Peng, Stephan Lany, and Vladan Stevanović. "A computational framework for automation of point defect calculations." *Computational Materials Science* 130 (2017): 1–9. https://doi.org/10.1016/j.commatsci.2016.12.040.

[38] Broberg, Danny, Bharat Medasani, Nils ER Zimmermann, Guodong Yu, Andrew Canning, Maciej Haranczyk, Mark Asta, and Geoffroy Hautier. "PyCDT: a Python toolkit for modeling point defects in semiconductors and insulators." *Computer Physics Communications* 226 (2018): 165–179. https://www.sciencedirect.com/science/article/pii/S0010465518300079.

[39] Stoliaroff, Adrien, Stéphane Jobic, and Camille Latouche. "PyDEF 2.0: an easy to use post-treatment software for publishable charts featuring a graphical user interface." *Journal of Computational Chemistry* 39, no. 26 (2018): 2251–2261. https://doi.org/10.1002/jcc.25543.

[40] Bertoldo, Fabian, Sajid Ali, Simone Manti, and Kristian S. Thygesen. "Quantum point defects in 2D materials-the QPOD database." *npj Computational Materials* 8, no. 1 (2022): 56. https://doi.org/10.1038/s41524-022-00730-w.

[41] Pannetier, Jean, J. Bassas-Alsina, Juan Rodriguez-Carvajal, and Vincent Caignaert. "Prediction of crystal structures from crystal chemistry rules by simulated annealing." *Nature* 346, no. 6282 (1990): 343–345. https://doi.org/10.1038/346343a0.

[42] Li, Feiming, Lan Yang, Zhixiong Cai, Ke Wei, Fangyuan Lin, Jie You, Tian Jiang, Yiru Wang, and Xi Chen. "Enhancing exciton binding energy and photoluminescence of formamidinium lead bromide by reducing its dimensions to 2D nanoplates for producing efficient light emitting diodes." *Nanoscale* 10, no. 44 (2018): 20611–20617. https://doi.org/10.1039/C8NR04986F.

[43] Efros, Al L., M. Rosen, Masaru Kuno, Manoj Nirmal, David J. Norris, and M. Bawendi. "Band-edge exciton in quantum dots of semiconductors with a degenerate valence band: dark and bright exciton states." *Physical Review B* 54, no. 7 (1996): 4843. https://doi.org/10.1103/PhysRevB.54.4843.

[44] María C. Geʹlvez-Rueda, Magnus B. Fridriksson, Rajeev K. Dubey, Wolter F. Jager, Ward van der Stam, and Ferdinand C. Grozema, Overcoming the exciton binding energy in two-dimensional perovskite nanoplatelets by attachment of conjugated organic chromophores, *Nature Communications* 11, no. 1 (2020): 1901. https://doi.org/10.1038/s41467-020-15869-7.

[45] Pan, Jinbo, and Qimin Yan. "Data-driven material discovery for photocatalysis: a short review." *Journal of Semiconductors* 39, no. 7 (2018): 071001. https://dx. doi.org/10.1088/1674-4926/39/7/071001.

[46] Sun, Wenhao, and Gerbrand Ceder. "Efficient creation and convergence of surface slabs." *Surface Science* 617 (2013): 53–59. https://www.sciencedirect.com/science/article/pii/S003960281300160X.

[47] Ping, Yuan, Ravishankar Sundararaman, and William A. Goddard III. "Solvation effects on the band edge positions of photocatalysts from first principles." *Physical Chemistry Chemical Physics* 17, no. 45 (2015): 30499–30509. https://doi.org/10.1039/C5CP05740J.

[48] Kharche, Neerav, James T. Muckerman, and Mark S. Hybertsen. "First-principles approach to calculating energy level alignment at aqueous semiconductor interfaces." *Physical Review Letters* 113, no. 17 (2014): 176802. https://doi.org/10.1103/PhysRevLett.113.176802.

[49] Montoya, Joseph H., and Kristin A. Persson. "A high-throughput framework for determining adsorption energies on solid surfaces." *npj Computational Materials* 3, no. 1 (2017): 14.https://doi.org/10.1038/s41524-017-0017-z.

[50] Jain, Anubhav, Shyue Ping Ong, Geoffroy Hautier, Wei Chen, William Davidson Richards, Stephen Dacek, Shreyas Cholia et al. "Commentary: the materials project: a materials genome approach to accelerating materials innovation." *APL Materials* 1, no. 1 (2013). https://doi.org/10.1063/1.4812323.

[51] Ong, Shyue Ping, William Davidson Richards, Anubhav Jain, Geoffroy Hautier, Michael Kocher, Shreyas Cholia, Dan Gunter, Vincent L. Chevrier, Kristin A. Persson, and Gerbrand Ceder. "Python Materials Genomics (pymatgen): a robust, open-source python library for materials analysis." *Computational Materials Science* 68 (2013): 314–319. https://www.sciencedirect.com/science/article/pii/S0927025612006295.

[52] Mamun, Osman, Kirsten T. Winther, Jacob R. Boes, and Thomas Bligaard. "High-throughput calculations of catalytic properties of bimetallic alloy surfaces." *Scientific Data* 6, no. 1 (2019): 76. https://doi.org/10.1038/s41597-019-0080-z.

[53] Andriuc, Oxana, Martin Siron, Joseph H. Montoya, Matthew Horton, and Kristin A. Persson. "Automated adsorption workflow for semiconductor surfaces and the application to zinc telluride." *Journal of Chemical Information and Modeling* 61, no. 8 (2021): 3908–3916. https://doi.org/10.1021/acs.jcim.1c00340.

[54] Rosen, Andrew S., Shaelyn M. Iyer, Debmalya Ray, Zhenpeng Yao, Alan Aspuru-Guzik, Laura Gagliardi, Justin M. Notestein, and Randall Q. Snurr. "Machine learning the quantum-chemical properties of metal-organic frameworks for accelerated materials discovery." *Matter* 4, no. 5 (2021): 1578–1597. https://www.sciencedirect.com/science/article/pii/S2590238521000709.

[55] Ishikawa, Atsushi, and Yoshitaka Tateyama. "A first-principles microkinetics for homogeneous-heterogeneous reactions: application to oxidative coupling of methane catalyzed by magnesium oxide." *ACS Catalysis* 11, no. 5 (2021): 2691–2700. https://doi.org/10.1021/acscatal.0c04104.

[56] Latimer, Allegra A., Ambarish R. Kulkarni, Hassan Aljama, Joseph H. Montoya, Jong Suk Yoo, Charlie Tsai, Frank Abild-Pedersen, Felix Studt, and Jens K. Nørskov. "Understanding trends in C-H bond activation in heterogeneous catalysis." *Nature Materials* 16, no. 2 (2017): 225–229. https://doi.org/10.1038/nmat4760.

[57] Medford, Andrew J., M. Ross Kunz, Sarah M. Ewing, Tammie Borders, and Rebecca Fushimi. "Extracting knowledge from data through catalysis informatics." *ACS Catalysis* 8, no. 8 (2018): 7403–7429. https://doi.org/10.1021/acscatal.8b01708.

[58] Yang, Ruo Xi, and Liang Z. Tan. "Understanding size dependence of phase stability and band gap in CsPbI3 perovskite nanocrystals." *The Journal of Chemical Physics* 152, no. 3 (2020). https://doi.org/10.1063/1.5128016.

[59] Scarabelli, Leonardo, Ana Sánchez-Iglesias, Jorge Pérez-Juste, and Luis M. Liz-Marzán. "A "tips and tricks" practical guide to the synthesis of gold nanorods." *The journal of physical chemistry letters* 6, no. 21 (2015): 4270–4279. https://doi.org/10.1021/acs.jpclett.5b02123.

[60] Shamsi, Javad, Alexander S. Urban, Muhammad Imran, Luca De Trizio, and Liberato Manna. "Metal halide perovskite nanocrystals: synthesis, post-synthesis modifications, and their optical properties." *Chemical Reviews* 119, no. 5 (2019): 3296–3348. https://doi.org/10.1021/acs.chemrev.8b00644.

[61] Draxl, Claudia, and Matthias Scheffler. "The NOMAD laboratory: from data sharing to artificial intelligence." *Journal of Physics: Materials* 2, no. 3 (2019): 036001. https://doi.org/10.1088/2515-7639/ab13bb.

[62] C. Lawrence Zitnick, LowikChanussot, Abhishek Das, Siddharth Goyal, Javier Heras-Domingo, Caleb Ho, Weihua Hu, Thibaut Lavril, Aini Palizhati, Morgane Riviere, Muhammed Shuaibi, Anuroop Sriram, Kevin Tran, Brandon Wood, Junwoong Yoon, Devi Parikh, and Zachary Ulissi. "An introduction to electrocatalyst design using machine learning for renewable energy storage." *arXiv preprint arXiv:2010.09435* (2020). https://doi.org/10.1021/acscatal.0c04525.

[63] Groom, Colin R., Ian J. Bruno, Matthew P. Lightfoot, and Suzanna C. Ward. "The Cambridge structural database." *Acta Crystallographica Section B: Structural Science, Crystal Engineering and Materials* 72, no. 2 (2016): 171–179. https://doi.org/10.1107/S2052520616003954.

[64] Panneerselvam, Suresh, and Sangdun Choi. "Nanoinformatics: emerging databases and available tools." *International Journal of Molecular Sciences* 15, no. 5 (2014): 7158–7182. https://doi.org/10.3390/ijms15057158.

[65] Yan, Xiliang, Alexander Sedykh, Wenyi Wang, Bing Yan, and Hao Zhu. "Construction of a web-based nanomaterial database by big data curation and modeling friendly nanostructure annotations." *Nature Communications* 11, no. 1 (2020): 2519. https://doi.org/10.1038/s41467-020-16413-3.

[66] Wilkinson, Mark D., Michel Dumontier, IJsbrand Jan Aalbersberg, Gabrielle Appleton, Myles Axton, Arie Baak, Niklas Blomberg et al. "The FAIR Guiding Principles for scientific data management and stewardship." *Scientific Data* 3, no. 1 (2016): 1–9. https://doi.org/10.1038/sdata.2016.18.

[67] Goldsmith, Bryan R., Jacob Florian, Jin-Xun Liu, Philipp Gruene, Jonathan T. Lyon, David M. Rayner, André Fielicke, Matthias Scheffler, and Luca M. Ghiringhelli. "Two-to-three dimensional transition in neutral gold clusters: the crucial role of van der Waals interactions and temperature." *Physical Review Materials* 3, no. 1 (2019): 016002. https://doi.org/10.1103/PhysRevMaterials.3.016002.

[68] Persson, Kristin A., Bryn Waldwick, Predrag Lazic, and Gerbrand Ceder. "Prediction of solid-aqueous equilibria: scheme to combine first-principles calculations of solids with experimental aqueous states." *Physical Review B* 85, no. 23 (2012): 235438. https://doi.org/10.1103/PhysRevB.85.235438.

[69] Furxhi, Irini, and Finbarr Murphy. "Predicting in vitro neurotoxicity induced by nanoparticles using machine learning." *International Journal of Molecular Sciences* 21, no. 15 (2020): 5280. https://www.mdpi.com/1422-0067/21/15/5280.

[70] Singh, Ajay Vikram, Romi-Singh Maharjan, Anurag Kanase, Katherina Siewert, Daniel Rosenkranz, Rishabh Singh, Peter Laux, and Andreas Luch. "Machine-learning-based approach to decode the influence of nanomaterial properties on their interaction with cells." *ACS Applied Materials & Interfaces* 13, no. 1 (2020): 1943–1955. https://doi.org/10.1021/acsami.0c18470

[71] Kar, Supratik, Kavitha Pathakoti, Paul B. Tchounwou, Danuta Leszczynska, and Jerzy Leszczynski. "Evaluating the cytotoxicity of a large pool of metal oxide nanoparticles to *Escherichia coli*: mechanistic understanding through In Vitro and In Silico studies." *Chemosphere* 264 (2021): 128428. https://www.sciencedirect.com/science/article/pii/S0045653520326230.

[72] Halder, Amit Kumar, André Melo, and M. Natália DS Cordeiro. "A unified in silico model based on perturbation theory for assessing the genotoxicity of metal oxide nanoparticles." *Chemosphere* 244 (2020): 125489. https://doi.org/10.1016/j.chemosphere.2019.125489.

[73] Findlay, Matthew R., Daniel N. Freitas, Maryam Mobed-Miremadi, and Korin E. Wheeler. "Machine learning provides predictive analysis into silver nanoparticle protein corona formation from physicochemical properties." *Environmental Science: Nano* 5, no. 1 (2018): 64–71. https://doi.org/10.1039/C7EN00466D.

[74] Hazarika, Zaved, and Anupam Nath Jha. "Computational analysis of the silver nanoparticle-human serum albumin complex." *ACS Omega* 5, no. 1 (2019): 170–178. https://doi.org/10.1021/acsomega.9b02340.

[75] Cho, Taeheum, Hyo-Sang Han, Junhyuk Jeong, Eun-Mi Park, and Kyu-Sik Shim. "A novel computational approach for the discovery of drug delivery system candidates for COVID-19." *International Journal of Molecular Sciences* 22, no. 6 (2021): 2815. https://doi.org/10.3390/ijms22062815.

27 Design of Medicinally Pertinent Multifunctional Inorganic Nanomaterials Using Artificial Intelligence

Manojna R. Nayak, Praveen K. Bayannavar, and Ravindra R. Kamble

INTRODUCTION

Computational approaches have grown tremendously over the past couple of years owing to their enhanced, simpler and improved skills in addressing academic, numerical, geographical, technological and statistical challenges with the assistance of machines. As a result, these approaches have become the epicentre of current research and will continue to dominate development and research in the future [1]. Machine learning (ML), artificial intelligence (AI), data mining and massive datasets, predicting and simulation, edge and cloud computing, and the Internet of Things are merely a few examples of computational tools geared towards juggling multiple tasks and are cross-disciplinary. AI is the most prevalent and contemporary of the aforementioned computing technologies [2].

AI is a sprawling phrase in the field of computation that encompasses techniques that include ML, deep learning (DL), computer vision (CV), and processing natural languages. These breakthroughs hold the prospective of enabling technology to replicate human intellect and carry out a variety of intricate tasks [3]. AI seeks to answer the following three basic inquiries: what information is necessary for any part of cognition, the manner in which that information ought to be utilised and how it needs to be expressed. Professor John McCarthy of Dartmouth initially used the phrase "Artificial Intelligence" in 1956 after noticing the ability of robots handling the issues that had been previously assumed to be unique to individuals, such as comprehending speech semantics and creating abstracts and notions. McCarthy pioneered the modern age of AI over 60 years ago when he and a team consisting of computer scientists and mathematicians showed that the computer programmes have the capacity of logical thinking *via* experimentation and mistakes [4]. On the other hand, Alan Turing, who created the Turing test in order to distinguish between people and machines, highlighted the potential that computers would adept to mimic human actions and potentially think beforehand [5]. Since that time, computing capacity has evolved substantially to the point where it is now possible to do computations instantly and to compare fresh data to data that has already been evaluated in the current moment. If we look into the present scenario, AI is advancing quickly, from SIRI to ALEXA to Google Assistant [6]. Although robots with anthropomorphic features are frequently depicted in science fiction, AI can refer to any aspect regarding e-commerce algorithms for forecasting to IBM's Watson machinery [7]. The surge in fascination with AI is eliciting a range of sentiments, including enthusiasm regarding how ML will supplement manual labour and concerns about their potential to eradicate employment [8]. An upsurge of excitement got generated since the debut of ChatGPT (Chat Generative Pre-trained Transformer, OpenAI Limited Partnership, San Francisco, USA) on 30 November 2022. Using cutting-edge natural language processing technologies, ChatGPT can produce conversations that sound realistic

DOI: 10.1201/9781003479239-27

and is capable of writing in a number of computer languages [9]. Folks spanning every aspect of acquaintance seem keen to try out this unique technology, and debates are raging, particularly in the realm of education sector. During the past few decades, the incorporation of technology based on AI in publications has matured, radically altering the manner in which scientific discourse operates. While the AI explosion affects our society, it may usher in an era of blissful utopia in which humans cohabit peacefully with robots, or it may upshot ramification in a nightmarish one riddle with cruelty, destitution, and misery.

Nanotechnology was initially proposed in 1959 in a discussion entitled "There's Plenty of Room at the Bottom" by Nobel Prize-winning physicist Richard Feynman. Feynman propagated utilising an array of conventionalised robotic arms to build a clone of their own which is one-tenth of the initial measurement, following which he employed the newly constructed pair of arms to build a substantially smaller set, and continuing this process to the point till anatomical resolution is attained [10]. The field of nanotechnology is a novel innovation with several legitimate uses. It pertains to an asset that's long-term prospects are entirely speculative. It belongs to a multidisciplinary area which has led to the emergence of numerous possibilities and advancements in domains that encompass applied neuroscience, engineering, mechanics, electrical engineering and many more in this present period. The use of soft computing approaches may facilitate networks having ideal inherent qualities such as development, self-repair and intricate systems, which might be utilised as a tool in nanotechnology-based autonomous system layout [11].

In accordance to the International Union of Pure and Applied Chemistry (IUPAC), nanomaterials are comprised of a particle with one or more distinctive measurements ranging from 1 to 100 nm [12]. In light of the essential features that enormous substances are devoid of, nanomaterials research has risen in prominence in the past few decades, touching practically every discipline and workplace. Shape, dimensions and crystal structure are three fundamental properties that regulate and impact the behaviour of nanomaterials. Countless investigative trials have already been done throughout the last 10 years to illustrate the efficacy of nanomaterials and its potential uses. Especially multifunctional inorganic nanomaterials are gaining its relevance in modern science owing to their programmable physical attributes that include melting point, moisture absorption, electrical as well as thermal insulation, catalytic behaviour, light penetration and diffusion, all of which culminate in improved efficacy over bulk states [13]. These distinctive qualities are connected with their dimensions: enormous surface area and remarkable surface actions contribute to their size-dependent traits owing to substantial variations in the surface area to total volume ratio. Inorganic nanomaterials of various designs and topologies have gained significance considering a variety of potential uses ranging from catalyst development to optical technology, although nanomaterial production and expansion are frequently strenuous as well as Edisonian procedures [14]. AI is transforming the manner in which things are designed and engineered [15]. Researchers and technologists are able to create inorganic nanomaterials featuring specified attributes that can be utilised using ML algorithms and other AI techniques. These materials may be used for everything from catalytic activity and dispersion of gases to conserving energy and healthcare.

Nanomedicine has achieved major milestones in the fields of curative and clinical research. For instance, inorganic nanomaterials modified therapeutic molecules and neuroimaging chemicals have significantly improved recuperation results as well as accuracy [16,17]. AI has lately witnessed the dawn of tremendous expansion spanning several areas, including medicine. AI has been used in research that extends several sectors of medicine to replicate clinicians' diagnosing traits. Innovative healthcare devices, AI-powered, are currently welcomed by laypeople, mainly since they permit a 4P version of healthcare (Predictive, Preventive, Personalised, Participatory) and thereby the autonomy of patients through means which were previously unthinkable [18]. AI-endowed healthcare innovations are fast turning into clinically beneficial interventions [19]. The idea of wisdom and expertise is critical to comprehending computational intelligence as well as its consequences in health. The greater the amount of information as well as data (information processing) we possess, the greater the likelihood we are able to formulate knowledge-driven judgements. Throughout the discipline of nanomedicine's inception, significant progress has been achieved in relation to leveraging heterogeneous chemistry to concurrently enhance nanomaterial frameworks employing numerous medicines. At this juncture, just a handful of clinical scenarios can be assisted by the use of AI, which includes the identification of irregular heartbeats, seizures related to epilepsy, and hypoglycaemia, or identifying an illness confirmed by histological investigation or imaging techniques [20–23].

The present chapter will address the comprehensive information about the use of AI in designing the various types of inorganic nanomaterials, with a particular emphasis on their significance in nanomedicine.

ARTIFICIAL INTELLIGENCE IN THE DESIGN OF INORGANIC NANOMATERIALS

Nanomaterials constitute the most refined substances available in the modern era. Their origin, structure, and categorisation are critical when it comes to their application and execution. Owing to their unique physicochemical traits as in contrast to conventional materials, multifunctional inorganic nanomaterials, in particular, have demonstrated promising potential in a variety of fields of science and research [24]. Inorganic nanomaterials are a subset of nanomaterials which comprise mainly of metals, oxides of metals, mesoporous, certain non-metals, and additional nanostructured materials with an inorganic core that determines their luminescent [25] conductive, electronic, and optical attributes. In contrast to organic materials, inorganic nanoparticles are environmentally friendly, hydrophilic in nature, biologically compatible, and extremely resistant to degradation [26]. The physical and chemical attributes of a substance in nanoform might change significantly from those of its bulk equivalent, contingent upon the substance's composition as well as its form, dimension, and modification. Following latent shifts, it is now feasible to better regulate their dimensions, form, and composition by choosing a compatible fabrication strategy and adjusting the process's circumstances. They may be synthesised utilising bottom-up or top-down techniques [27]. Nanomaterials may be formed of numerous components, which include metal oxides or composites, or solitary components, such as metals or carbon. Nanomaterials can be classified as nanoparticles, nanowires, or nanotubes based upon their dimensions.

Recently, spotlight has been drawn to the computational tools used in the synthesis and design of nanomaterials that possess the potential to speed up and shorten the synthesis process. AI is the endeavour of creating smart gadgets that are capable of emulating humanoid traits with little assistance from the outside world. ML, neural networks (NNs), DL, convolutional neural networks (CNNs), CV, and robotics are some of the components that make up AI (Figure 27.1).

MACHINE LEARNING (ML)

The capacity of a computer to adjust how it processes in response to freshly gathered knowledge is known as ML. The approach might be built around a straightforward if-then decision-making tree that leads to a conclusion, or it may utilise algorithms known as DL that simulate how the brain of an individual processes various sorts of input and develops correlations for application in neural systems for taking decisions [28]. The form, dimensions, and chemical composition of nanomaterials all influence an array-associated attributes. Nanomaterials must be synthesised with carefully regulated properties in order to be used in pharmaceutical monitoring, health care diagnosis, catalytic activity, solar power panels, and pharmaceuticals [29]. Considering the manufacture of nanomaterials, it frequently requires the incorporation of many chemicals and is carried out amid interrelated circumstances, and also it is a tedious, lengthy, and expensive operation [30]. The rapid creation of effective algorithms for the synthesis of nanomaterials and, possibly, the manufacture of novel varieties of nanomaterials is made possible by ML, which is an intriguing method in this regard.

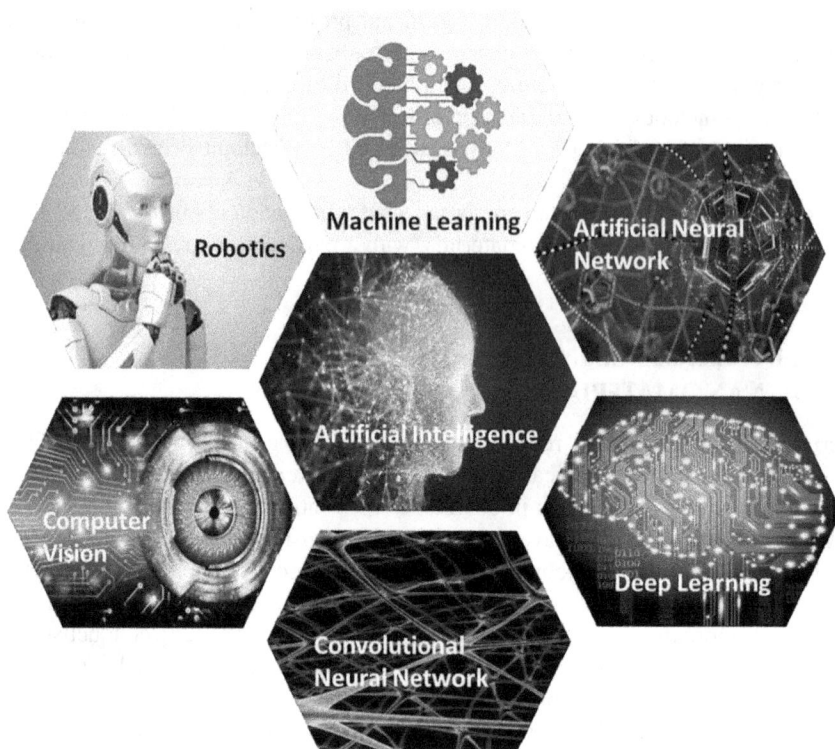

FIGURE 27.1   Components of artificial intelligence used to design inorganic nanomaterials.

Currently, ML approaches displayed significant benefits for foreseeing the characteristics of nanomaterials, avoiding several iterative investigations [31]. The main functions of ML techniques in generating nanomaterial include experimentation design and result assessment. The goal of forecasting assignments is to develop a conceptual comprehension of the connection among an experimentally determined attribute that is noteworthy (such as inorganic nanomaterial dimension and form or efficiency) and an assortment of simulated source circumstances. The most basic ML prognosis algorithm is linear regression, which establishes a linear relationship between simulated circumstances and nanomaterial attributes [32,33]. There are plenty of programmes accessible for setting up examinations. Scientific design, involving carefully analysing several experimental circumstances, is a component of specific algorithms. In comparison to conventional experimental planning, more effective heuristic and Bayesian optimisation algorithms use data derived through escalating the procedure in order to accomplish the intended objective (such as enhancing nanomaterials traits) [34]. As a result, fewer trials are needed to optimise nanoparticle attributes. By maximising factors including the amount of light, acceleration, spinning velocity, pH level, temperature, algorithms and neural systems, and different inorganic metal nanoparticles such as gold and silver may be synthesised in the appropriate size.

ARTIFICIAL NEURAL NETWORK (ANN)

ANN, a nonlinear statistical analytic technique, is notably suitable for the modelling of complex networks devoid of the requirement of precedent data. Indeed, through the use of an assortment of nonlinear operations, ANN provides a method of connecting the input data to the output data. Ultimately, ANN serves as a streamlined representation of natural networks that lacks complexity and is effective at expressing nonlinear interactions among input and output variables [35] Inorganic nanomaterials cannot be produced using traditional techniques because they necessitate lengthy preparatory times, expensive costs, and stringent synthesising ailments [36]. As a result, several attempts are being made to create a new environmentally friendly approach that will supersede the present methods of synthesising and enable for the quick, affordable, and toxic-free production of nanomaterials. By anticipating the outcomes of the empirical responses, NN analysis programmes have the potential to save expenditure and resources by helping to identify the ideal settings.

Within laboratory circumstances, the synthesis of nanomaterials is incredibly dependent upon a number of variables. As a result, it's crucial to fully establish the relationships between the factors at play and the attributes of nanomaterials in order to manage synthesis. An effective data processing technique based on replicating organic neurological structures is called an ANN. Mega-parallelism, substantial nonlinear behaviour, self-learning, endurance to dispersed input, and quick execution time are some of the benefits of ANN that have led to its widespread application in a variety of domains, including anticipating, categorisation, processing signals, computation of images, and chemical workflow management [37]. There are three different categories of computing nodes (Figure 27.2):

1. Input Layer
2. Hidden Layer
3. Output Layer

The input layer of an ANN is the top layer. The hidden layer is connected to each node in the input layer. In accordance with the collection of data or information getting utilised, the hidden layer may have one or more input components attached to it. An evident component pattern is not formed by the hidden layer. The depiction of datasets may be divided into two categories: supervised learning and unsupervised learning. The output layer is going to display the outcomes of the experiment. This is illustrated as a block design of an artificial neural networking in Figure 27.2 [38].

DEEP LEARNING (DL)

DL, a type of ML, is built by taking into account multiple datasets at the same time that are assessed and reconditioned for the following two separate evaluations and furthermore, till they

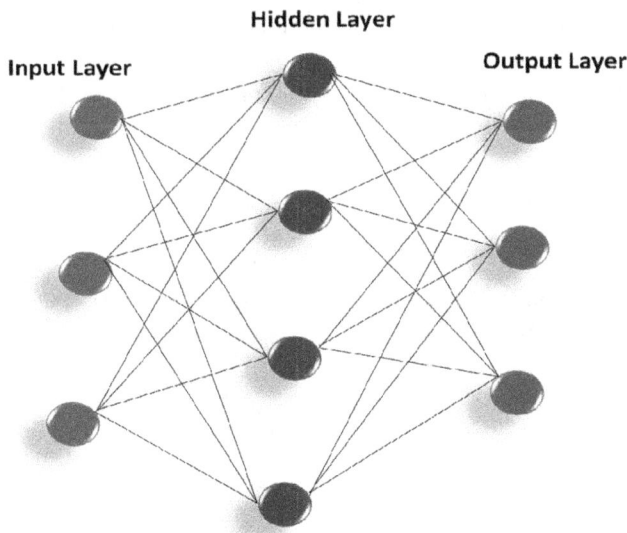

FIGURE 27.2   Three different categories of computing nodes of ANN.

acquire an output. The immensely redundant architecture of the human nervous system served as the inspiration for DL, which has its roots in Frank Rosenblatt's perceptual unit from the 1950s, which relied on McCulloch and Pitts' simplified model of a single neuron. Deep NNs with intricate topologies have gained broad interest and are employed in a variety of sectors as data volume and computational resources [39]. DL algorithms surpass superficial ML techniques and conventional data evaluation techniques for various purposes.

Deep NNs sporadically include multiple concealed layers that are arranged in a layered topology of neurons. In addition, they frequently include more complex circuits than ordinary ANNs. For example, contrary to employing a straightforward stimulation functionality, it might instead involve sophisticated processes (such as convolutions) or numerous impulses in a single neuron. Owing to combination of these attributes, deep NNs may be loaded with unprocessed information to autonomously find the representations required for a given nanomaterial [40,41].

Stacked neurons which develop additional connections to adjacent neurons, mimicking the organic architecture of the central nervous system, make up deep NNs. Each of these neurons transmits information to the remaining neurons that are linked, creating an intricate system that includes certain hidden levels. The rational foundation pieces of this include neurons, stimulation operation, input/output, weight, layers, and an optimisation algorithm (learning technique) [42]. Multiple neurons estimate a parameter known as the stimulus mechanism in each layer. Every interaction between a pair of neurons in different levels is also given a value. Initially, arbitrary weights are established, but they are perpetually adjusted in order to acquire the anticipated outcome of the precise output. The output is created as binary/multiple labels as the final outcome [43].

CONVOLUTIONAL NEURAL NETWORK (CNN)

CNNs, often known as convnets, have recently shown accomplishments that were initially believed to be exclusively at human level. A special kind of ANN called a CNN analyses data using numerous hidden layers and DL methods. Every single CNN has numerous hidden layers, while the interactions among the various CNNs are complicated (which is why the word convolutional). Algorithms can deal with and handle problems concerning spatially related databases wherein rows and columns cannot be convertible (such as picture data) [44]. Its design of networks consists of several kinds of steps which facilitate tiered feature acquisition of various properties in nanomaterials depending on the particular modelling objective [45].

CNNs concentrate upon the task of recognising images because the algorithms are capable of handling the input information represented as a series of matrices. The input layer, convolutional layer, pooling layer, and fully connected layers constitute only a few of the layers that make up a CNN (Figure 27.3).

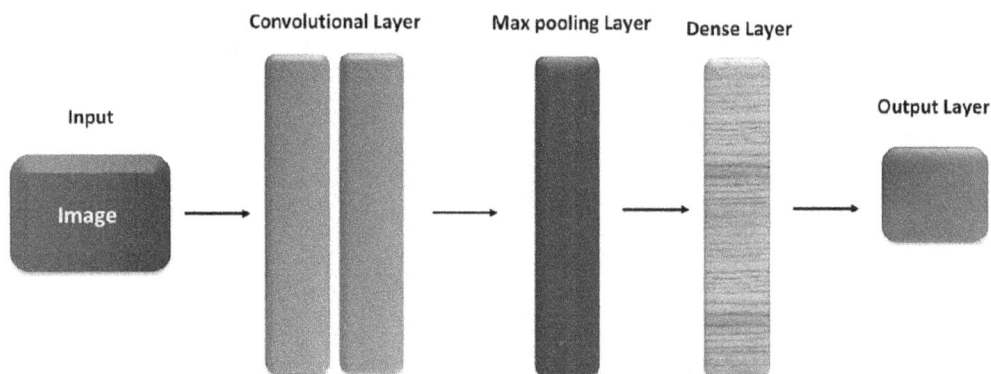

FIGURE 27.3   Schematic representation of layers that constitute CNN.

There are four main areas in which the instance of CNN's fundamental functioning may be categorised:

1. The input layer is going to retain the image's pixels, just like in various ANN variants.
2. By calculating the scalar sum among the masses of each of the neurons and the area related to its input volume, the convolutional layer's algorithm will recognise the final result of NNs whose particular portions of the data are their inputs.
3. After this, the pooling layer is going to merely downsample the input information across the plane of space, thereby lowering the total amount of variables activated.
4. Following that, the extremely dense layers are going to carry out identical functions as ordinary ANNs and make an effort to generate rating coefficients based upon impulses for purposes of classifying. ReLu (rectified liner unit) can be similarly employed within the aforementioned layers to enhance effectiveness.

CNNs have the capacity to alter the initial data source utilising convolutional and downsampling approaches employing this straightforward method of modification, producing class scores for evaluation and categorisation purposes in inorganic nanomaterials [46].

COMPUTER VISION (CV)

AI's field of CV permits systems and machines to extract beneficial data through digital photos, videos, and a variety of sight-based inputs and then to conduct operations or offer suggestions in response to the aforementioned data. While AI gives machines the ability to think, CV gives them the ability to perceive, sense, and comprehend. The primary objective of nanotechnology aims to examine and comprehend the elementary and structural characteristics of materials. This fresh and intriguing area of use for CV is evolving quickly, and significant studies are presently being made to create novel visual characterisation approaches to examine nanoscale pictures [47]. Exceptionally low signal-to-noise ratio (SNR), translucent samples, a restricted amount of pictures during perception, and intricate generation of images topology represent a few of the challenges associated with characterising nanomaterials.

The latest developments in the fields of CV and computational imaging, as well as the accessibility of reasonably priced, exceptionally well computing equipment, have made vision-based characterisation of nanomaterials, a significant area of study. SEM, TEM, and AFM are frequently employed as nanoscale scanning tools [48,49].

ROBOTICS

Robotics is the science of linking observation and motion intelligently. 'Intelligent' and the accompanying 'perception' are the important terms in the phrase. Modern intelligent machines often use sight, power, and sensory processing in addition to proprioceptive sense of the robot's internal state [50].

Present-day innovation in materials mainly focuses on the following important steps: (i) finding a novel composition and framework of concern, (ii) synthesising that material in a focused and sustainable manner and (iii) meticulously further processing the result to optimise its attributes [51]. This makes it vital to incorporate every step within a closed cycle in order to enable them operate quickly after one another and gain the most advantage of each other's input in order to speed up this process. Large-scale upfront models have helped with the early detection process, but synthesising and property optimisation still need incremental experimentation approaches, which may be challenging and laborious. Automated innovation made possible by self-driving labs, that seek to assist an individual scientist using robotic vehicles lead by ML, may provide a solution for such problems [52].

Self-driving laboratories (SDL), in which robotics is supplemented alongside AI techniques and innovations, require intricate remedies and unique systems of technology that enable the self-governing exploration of innovative substances and nanomaterials. SDL have the capacity to speed up and independently identify novel inorganic nanomaterials with innovative applications [53,54]. It entails the actions that are necessary to link the execution of the data-centred approach along with the traditional human-focused method of research in order to synthesise nanomaterials.

STEPS INVOLVED IN THE DESIGNING OF INORGANIC NANOMATERIALS USING AI

The technique of using AI to design inorganic nanoparticles entails multiple stages (Figure 27.4). These constitute the key steps that need to be followed.

1. **Data Source**:
   - Assessing the form, magnitude, surface chemistry, and sensitivity of the nanomaterials in order to achieve the necessary qualities and functions.
   - Compendium of data, including conceptual and practical data regarding the production and characteristics of inorganic nanomaterials.
2. **Data Processing**:
   - The transformation of raw data into immaculate accurate information.
3. **Model Selection**:
   - Choosing the right scheme for the development of inorganic nanomaterials.
4. **Training the Model**:
   - Using the dataset as input, the system is trained to optimise its variables.
   - The system is aware of the connections amongst the input attributes and the intended qualities of the nanomaterial.
5. **Model Validation**:
   - Assess, enhance, and tweak the model's efficiency for greater adequacy.
6. **Predictive Modelling**:
   - The model could potentially be utilised for predicting the characteristics of nanomaterials in accordance with the data supplied following verification and optimisation.
7. **Nanoparticle Design and Optimisation**:
   - Creation of potential designs in accordance with the suggested attributes by the chosen AI model.
   - Analysing the produced proposal designs and choosing the best ones.
8. **Experimental Validation**:
   - In-lab production of nanomaterials using an AI-generated blueprint.
   - Characterising the synthesised nanomaterials and evaluating the outcomes against assumptions made by AI.
9. **Feedback Loop**:
   - Enhance the procedure in accordance with the results of experimental validation to raise the preciseness and dependability of the AI model.

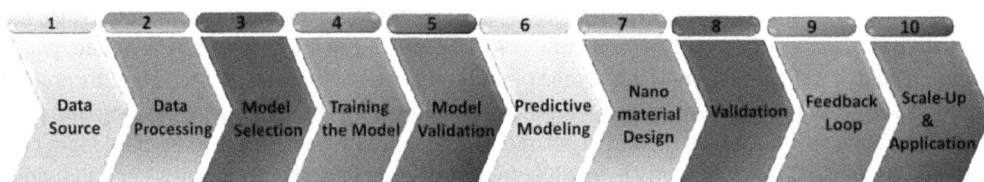

FIGURE 27.4 A flowchart illustrating each step performed while designing inorganic nanoparticles deploying artificial intelligence.

10. **Scale-Up and Application**:
   - Inorganic nanoparticles are designed and optimised for certain tasks at a bigger scale using a tried-and-true AI approach.

APPLICATIONS OF NANOMATERIALS IN MEDICINE

Nanomedicine is a moniker for the clinical use of nanotechnology in medicine, which uses nanoparticles for rejuvenation, tailored administration of drugs (nanotherapy), and diagnostics (nanodiagnosis) [55]. Through amazing results, the application of nanotechnology in healthcare is elevating the area to an entirely novel frontier. Prospects for the upcoming era of nanomedicine are significant since it offers personalised, effective treatment with fewer adverse reactions. Inorganic nanomaterials are benign, non-hazardous, sustainable, recyclable and extremely durable when juxtaposed with organic nanomaterials. According to their various shapes, dimensions, and physical and chemical features, they comprise nanoparticles, nanotubes, nanowires and quantum dots [56] (Figure 27.5)

NANOPARTICLES (NPS)

Considering all of NPs' measurements are within the nanoscale, they are known as zero-dimensional nanomaterials. Due to their structured porosity framework, ample surface area, simplicity of modification, and biological compatibility, inorganic nanoparticles (NPs) have attracted a great deal of curiosity for usage in healthcare fields [57].

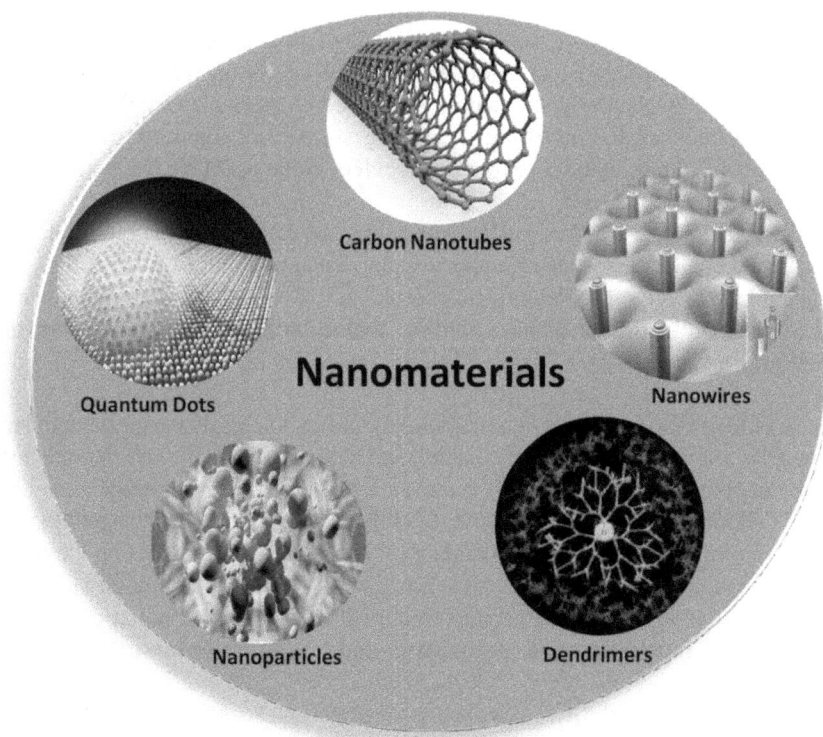

FIGURE 27.5 Different types of inorganic nanomaterials.

- Iron oxide NPs offer a broad spectrum of biological applicability because of their magnetism and extensive surface-to-volume ratio being well-balanced. The most glaring potential applications for these NPs include MRI scanning and magnetic hyperthermia treatment [58].
- The distinctive biological characteristics of gold nanoparticles make them extremely useful in the discipline of therapeutics. The Au NPs can penetrate inside of cells and aid in genome treatment as well as drug administration for medical purposes by using both DNA and RNA strands [59].
- To help with rapid diagnosis of illness, evaluation, and medication, gold nanoparticles are used for image processing, administration of medication, and cancer therapy [60].
- For research on cellular absorption of nanoparticle, targeting, and drug administration using fluorescence imaging, fluorescent mesoporous silica NPs have emerged as an effective tool [61].
- Multifunctional antimony and bismuth-based nanoparticles have drawn attention by enabling concurrent tumour detection and treatment [62].
- Silver-coated nanoparticles (NPs) are effective weapons for combating the risk of resistance from bacteria [63].
- Due to its distinctive characteristics, such as substantial area of specific surface, volume of pores, controlled particle dimension, and superior sustainability, mesoporous silica oxide NPs have been extensively explored potential applications in drug delivery, bioengineering and biosensors [64].

CARBON NANOTUBES (CNS)

In 1991, Iijima and colleagues made the discovery of CNs. These consist of cylinder-shaped carbon allotropes exhibiting nanometer-scale dimension and lengths that exceed millimetres [65,66]. CNs are typically created using graphite through a laser ablation process. CNs are divided into single-walled (SW) and multi-walled (MW) CNs on the basis of the number of layers within them. SWCNs are hexagonal close-packed clusters of a single graphene cylinder having a diameter between 0.4 and 2 nm. Featuring an exterior diameter of 2–100 nm and an interior diameter of 1–3 nm, coaxial graphene cylinders make up MWCNs, which can have two to many of them [67].

Owing to their distinctive features, such as substantial surface-to-volume ratios, elevated carrier adaptability, raised conductance and endurance, biological compatibility, adaptability, facile functionality, spectral characteristics, and optical transparency, CNs can be more dynamically applied in a variety of medical fields, such as drug delivery, molecular diagnostics, biosensors, tissue engineering, and carrying genes [68,69].

- Because of their distinctive physical traits and capacity to readily permeate cellular membranes, CNs are mostly used for medication and delivery of genes. CNs degrade naturally and are not obligated to be surgically removed following delivery of medication [70].
- For the evaluation and identification of a physiological analytic agent, such as a microbe, biomolecule, or organic structure, CNs are utilised as sensing carriers. With greater precision, glucose-oxidase biosensors with CNs managed blood sugar levels in patients with diabetes [71]. For various treatment tracking and evaluations, CNs-based dehydrogenase biosensors and peroxidase and catalase biosensors were likewise constructed [72].
- Their biological compatibility, resistance to degradation, stiffness, and ability to be customised with macromolecules to improve organs regrowth make them suitable for tissue engineering [73], wherein infectious or injured parts are swapped by biological stand-ins that can be recovered and preserve their original functions.

- In order to employ CNs as a middleman in radiotherapy and photothermal treatments, it is possible to make use of their distinctive qualities, such as their wide surface area, increased drug transport capacities, crosslinking and embedding proficiency, target-specific action, and strong near IR radiation absorption attributes [74]. CNs can infiltrate cancer cells because of their improved permeability and needle-like shape, which allows them to pierce cell membranes.
- CNs are also used as antioxidants and in the remedying of neurological disorders like Alzheimer's [70,75].

QUANTUM DOTS (QDS)

QDs are semiconductor NPs with several uses in therapeutics owing to their photoluminescent aspects. The most assessed QDs are combinations of elements from the II–VI groups (e.g., CdSe, CdTe), IV–VI groups (e.g., PbS, PbSe) and III–V groups (e.g., InP, GaP) from the periodic table [76].

- Graphene QDs are currently employed in the medical industry as antioxidant substances due to their high biocompatibility [77].
- Inorganic QDs are frequently used for biological fluorescence detection, and they offer an ideal environment for the administration of medications for treatment as well as the capacity to trace or evaluate the course of action instantaneously [78].
- Sugar-coated QDs are being employed in the management for tumours along with various disorders as a drug delivery mechanism [79].
- CdS QDs have been extensively investigated for how they could be utilised in non-Hodgkin lymphoma (NHL) cancer tumour in vitro diagnostics [80].
- CdSe/ZnS QDs have been employed in biomedical imaging to visualise several types of stem cells which include fatty tissue–based stem cells, stem cells from embryos, and bone marrow stem cells [81].

NANOWIRES (NWS)

NWs are single crystal nanomaterials possessing an exceptional contrast ratio and a diameter of 10 nm. Metallic substances or semiconductors can be used to make NWs using a variety of processes which include chemical vapour deposition, ablation by laser, thermal evaporation and alternating current electrodeposition [82].

- The identification of circulating tumour cells (CTC) is critical in carcinoma spread. Si NWs, also known as NanoVelcro chips, are utilised to selectively collect and discharge CTCs in bloodstreams [83,84].
- NWs were employed for recognising and measuring macromolecules via nanosensors and nanoprobes considering their characteristics like a considerable surface-to-volume ratio plus precise shape [85].
- Utilising Si NW-based sensors, several deadly viruses such as dengue [86], H1N1 [87], HIV [88] and influenza A H3N2 [89] were successfully identified.
- NWs are currently used for identification of biomarkers, infectious diseases and genetic material [85].
- Pt NWs and Fe/Pt composite NWs have been successfully used in focused chemotherapy techniques [90].
- Magnetic iron NWs are making significant contributions to nanomedicine due to their distinctive characteristics such as low toxicities and ease of manipulation with the magnetic field. Magnetic iron nanoparticles (NWs) are being employed in the fields of magnetic resonance imaging, magneto-mechanical, and photothermal therapies [91].

DENDRIMERS (DMS)

DMs are a distinct category of nanomaterials identified by Fritz Vogtle in 1978. They are nano-sized and have a clearly defined homogenous and monodisperse architecture made up of tree-like branches [92]. Dendrimers are often used in diagnosis and therapeutic applications considering to their features such as increased dissolution ability, accessibility, absorption, and target dissemination [93].

- Dendrimers play a role in composition of medication, administration, and drug transport as a result of the existence of cavities [94].
- Dendrimers are additionally implemented in image analysis, most notably in the form of structures composed of covalently bound chelators, like diethylene triamine pentaacetic acid (DTPA) capable of holding gadolinium salts [95].
- Usually when fused with fluorophores, phosphorus dendrimers are useful in combating cancer and are additionally employed in bioimaging [96].
- Aside from medication administration, dendrimers additionally feature antifungal, cytotoxic, and bactericidal properties [97].

FUTURE PERSPECTIVES

Nanomedicines have the potential to revolutionise contemporary healthcare by providing effective treatments and preliminary diagnosis. Nevertheless, there are several difficulties associated with their clinical implementation. Scaling-up, *in vivo* volatility and accessibility, possible toxicity, immunological response, administrative impediments, and clinical heterogeneity are a few obstacles that are currently impeding the clinical adaptation of nanomedicines. Through employing substantial, diverse information sources and converting it into accessible and enforceable information and decision-making instruments, AI-centred resources incorporated *via* an extensive variety of structure–internalisation interactions research have the capacity for enhancing every aspect of the nanomedicine design procedure while eschewing generations of trial-and-error in the creation of drugs. Innovations in the realm of computing studies which will depend on novel computer frameworks as well as data visualisations, hybrid innovations which employ life forms and nanotechnological gadgets, biological engineering, neurological science, along with a wide range of affiliated fields may be influenced by the merge of AI and nanotechnology.

CONCLUSION

Without a question, AI is a breakthrough discipline of computation that has poised to evolve as an essential aspect of a variety of future innovations. AI has progressed beyond fantasy to realism in just a matter of years. AI in nanomedicine, in particular, has evolved to replicate and potentially transcend human skills by giving real-time information, expediting processes, and reducing stress of physician. Despite the fact that AI has advanced significantly in health care, human supervision remains necessary and errors are still conceivable. Given the exponential growth of AI, an upheaval has occurred in every field, and individuals are concerned about losing their employment as AI advances. However, AI has facilitated more employment and possibilities for individuals across every area, and it is a prerequisite need of an individual to improve their inventiveness, analytical skills, and proficient decision-making talents in order to walk hand in hand with AI.

ACKNOWLEDGEMENT

One of the authors Manojna R. Nayak thank Karnatak University Dharwad for providing University Research Fellowship.

REFERENCES

[1] Raj, Dr Jennifer S. "A comprehensive survey on the computational intelligence techniques and its applications." *Journal of IoT in Social, Mobile, Analytics, and Cloud* 1, no. 3 (2019): 147–159. https://doi.org/10.36548/jismac.2019.3.002.

[2] Farris, Alton B., Juan Vizcarra, Mohamed Amgad, Lee AD Cooper, David Gutman, and Julien Hogan. "Artificial intelligence and algorithmic computational pathology: An introduction with renal allograft examples." *Histopathology* 78, no. 6 (2021): 791–804. https://doi.org/10.1111/his.14304.

[3] Das, Kaushik Pratim. "Nanoparticles and convergence of artificial intelligence for targeted drug delivery for cancer therapy: Current progress and challenges." *Frontiers in Medical Technology* 4 (2023): 1067144. https://doi.org/10.3389/fmedt.2022.1067144.

[4] Singh, Ajay Vikram, Daniel Rosenkranz, Mohammad Hasan Dad Ansari, Rishabh Singh, Anurag Kanase, Shubham Pratap Singh, Blair Johnston, Jutta Tentschert, Peter Laux, and Andreas Luch. "Artificial intelligence and machine learning empower advanced biomedical material design to toxicity prediction." *Advanced Intelligent Systems* 2, no. 12 (2020): 2000084. https://doi.org/10.1002/aisy.202000084.

[5] Mintz, Yoav, and Ronit Brodie. "Introduction to artificial intelligence in medicine." *Minimally Invasive Therapy & Allied Technologies* 28, no. 2 (2019): 73–81. https://doi.org/10.1080/13645706.2019.1575882.

[6] Brill, Thomas M., Laura Munoz, and Richard J. Miller. "Siri, Alexa, and other digital assistants: A study of customer satisfaction with artificial intelligence applications." *Journal of Marketing Management* 35, no. 15–16 (2019): 1401–1436. https://doi.org/10.4324/9781003307105

[7] Hasanzadeh, Akbar, Michael R. Hamblin, Jafar Kiani, Hamid Noori, Joseph M. Hardie, Mahdi Karimi, and Hadi Shafiee. "Could artificial intelligence revolutionize the development of nanovectors for gene therapy and mRNA vaccines?" *Nano Today* 47 (2022): 101665. https://doi.org/10.1016/j.nantod.2022.101665.

[8] Goralski, Margaret A., and Tay Keong Tan. "Artificial intelligence and sustainable development." *The International Journal of Management Education* 18, no. 1 (2020): 100330. https://doi.org/10.1016/j.ijme.2019.100330.

[9] Chen, Tzeng-Ji. "ChatGPT and other artificial intelligence applications speed up scientific writing." *Journal of the Chinese Medical Association* 86, no. 4 (2023): 351–353. https://doi.org/10.1097/JCMA.0000000000000900.

[10] Cavalcanti, Adriano. "Assembly automation with evolutionary nanorobots and sensor-based control applied to nanomedicine." *IEEE Transactions on Nanotechnology* 2, no. 2 (2003): 82–87. https://doi.org/10.1109/TNANO.2003.812590.

[11] Dhandapani, Kothandaraman, Krishnaveni Venugopal, and J. Vinoth Kumar. "Ecofriendly and green synthesis of carbon nanoparticles from rice bran: Characterization and identification using image processing technique." *International Journal of Plastics Technology* 23, no. 1 (2019): 56–66. https://doi.org/10.1007/s12588-019-09240-9.

[12] Vert, Michel, Yoshiharu Doi, Karl-Heinz Hellwich, Michael Hess, Philip Hodge, Przemyslaw Kubisa, Marguerite Rinaudo, and François Schué. "Terminology for biorelated polymers and applications (IUPAC Recommendations 2012)." *Pure and Applied Chemistry* 84, no. 2 (2012): 377–410. https://doi.org/10.1351/PAC-REC-10-12-04.

[13] Wang, Huilin, Xitong Liang, Jiutian Wang, Shengjian Jiao, and Dongfeng Xue. "Multifunctional inorganic nanomaterials for energy applications." *Nanoscale* 12, no. 1 (2020): 14–42. https://doi.org/10.1039/C9NR07008G.

[14] Li, Zhanfeng, Tingting Zhuang, Jun Dong, Lun Wang, Jianfei Xia, Huiqi Wang, Xuejun Cui, and Zonghua Wang. "Sonochemical fabrication of inorganic nanoparticles for applications in catalysis." *Ultrasonics Sonochemistry* 71 (2021): 105384. https://doi.org/10.1016/j.ultsonch.2020.105384.

[15] Moore, Julia A., and James CL Chow. "Recent progress and applications of gold nanotechnology in medical biophysics using artificial intelligence and mathematical modeling." *Nano Express* 2, no. 2 (2021): 022001. https://doi.org/10.1088/2632-959X/abddd3.

[16] Liu, Linbo, Mingcheng Bi, Yunhua Wang, Junfeng Liu, Xiwen Jiang, Zhongbin Xu, and Xingcai Zhang. "Artificial intelligence-powered microfluidics for nanomedicine and materials synthesis." *Nanoscale* 13, no. 46 (2021): 19352–19366. https://doi.org/10.1039/D1NR06195J.

[17] Ho, Dean, Peter Wang, and Theodore Kee. "Artificial intelligence in nanomedicine." *Nanoscale Horizons* 4, no. 2 (2019): 365–377. https://doi.org/10.1039/C8NH00233A.

[18] Chang, Anthony C. *Intelligence-Based Medicine: Artificial Intelligence and Human Cognition in Clinical Medicine and Healthcare.* Academic Press, Cambridge, MA, 2020.

[19] He, Jianxing, Sally L. Baxter, Jie Xu, Jiming Xu, Xingtao Zhou, Kang Zhang, The practical implementation of artificial intelligence technologies in medicine. *Nature Medicine* 25 (2019): 30–36. https://doi.org/10.1038/s41591-018-0307-0.

[20] Sung, Chih-Wei, Jiann-Shing Shieh, Wei-Tien Chang, Yi-Wei Lee, Ji-Huan Lyu, Hooi-Nee Ong, Wei-Ting Chen, Chien-Hua Huang, Wen-Jone Chen, and Fu-Shan Jaw. "Machine learning analysis of heart rate variability for the detection of seizures in comatose cardiac arrest survivors." *IEEE Access* 8 (2020): 160515–160525. https://doi.org/10.1109/ACCESS.2020.3020742.

[21] Wang, Yu-Chiang, Xiaobo Xu, Adrija Hajra, Samuel Apple, Amrin Kharawala, Gustavo Duarte, Wasla Liaqat et al. "Current advancement in diagnosing atrial fibrillation by utilizing wearable devices and artificial intelligence: A review study." *Diagnostics* 12, no. 3 (2022): 689. https://doi.org/10.3390/diagnostics12030689.

[22] An, Sora, Chaewon Kang, and Hyang Woon Lee. "Artificial intelligence and computational approaches for epilepsy." *Journal of Epilepsy Research* 10, no. 1 (2020): 8. https://doi.org/10.14581%2Fjer.20003.

[23] Porumb, Mihaela, Saverio Stranges, Antonio Pescapè, and Leandro Pecchia. "Precision medicine and artificial intelligence: A pilot study on deep learning for hypoglycemic events detection based on ECG." *Scientific Reports* 10, no. 1 (2020): 170. https://doi.org/10.1038/s41598-019-56927-5.

[24] Vaseghi, Zahra, and Ali Nematollahzadeh. "Nanomaterials: Types, synthesis, and characterization." *Green Synthesis of Nanomaterials for Bioenergy Applications* (2020): 23–82. https://doi.org/10.1002/9781119576785.ch2.

[25] Erdman, Aleksandra, Piotr Kulpinski, Tomasz Grzyb, and Stefan Lis. "Preparation of multicolor luminescent cellulose fibers containing lanthanide doped inorganic nanomaterials." *Journal of Luminescence* 169 (2016): 520–527. https://doi.org/10.1016/j.jlumin.2015.02.049.

[26] Soufi, Ghazaleh Jamalipour, and Siavash Iravani. "Eco-friendly and sustainable synthesis of biocompatible nanomaterials for diagnostic imaging: Current challenges and future perspectives." *Green Chemistry* 22, no. 9 (2020): 2662–2687. https://doi.org/10.1039/D0GC00734J.

[27] Yoo, Sung Chan, Junho Lee, and Soon Hyung Hong. "Synergistic outstanding strengthening behavior of graphene/copper nanocomposites." *Composites Part B: Engineering* 176 (2019): 107235. https://doi.org/10.1016/j.compositesb.2019.107235.

[28] Cai, Jiazhen, Xuan Chu, Kun Xu, Hongbo Li, and Jing Wei. "Machine learning-driven new material discovery." *Nanoscale Advances* 2, no. 8 (2020): 3115–3130. https://doi.org/10.1039/d0na00388c.

[29] Hiszpanski, Anna M., Brian Gallagher, Karthik Chellappan, Peggy Li, Shusen Liu, Hyojin Kim, Jinkyu Han, Bhavya Kailkhura, David J. Buttler, and Thomas Yong-Jin Han. "Nanomaterial synthesis insights from machine learning of scientific articles by extracting, structuring, and visualizing knowledge." *Journal of Chemical Information and Modeling* 60, no. 6 (2020): 2876–2887. https://doi.org/10.1021/acs.jcim.0c00199.

[30] Mekki-Berrada, Flore, Zekun Ren, Tan Huang, Wai Kuan Wong, Fang Zheng, Jiaxun Xie, Isaac Parker Siyu Tian et al. "Two-step machine learning enables optimized nanoparticle synthesis." *npj Computational Materials* 7, no. 1 (2021): 55. https://doi.org/10.1038/s41524-021-00520-w.

[31] Jin, Kun, Wentao Wang, Guangpei Qi, Xiaohong Peng, Haonan Gao, Hongjiang Zhu, Xin He et al. "An explainable machine-learning approach for revealing the complex synthesis path-property relationships of nanomaterials." *Nanoscale* (2023). https://doi.org/10.1039/D3NR02273K.

[32] Pellegrino, Francesco, Raluca Isopescu, Letizia Pellutiè, Fabrizio Sordello, Andrea M. Rossi, Erik Ortel, Gianmario Martra, Vasile-Dan Hodoroaba, and Valter Maurino. "Machine learning approach for elucidating and predicting the role of synthesis parameters on the shape and size of TiO2 nanoparticles." *Scientific Reports* 10, no. 1 (2020): 18910. https://doi.org/10.1038/s41598-020-75967-w.

[33] Huachen Tao, Tianyi Wu, Matteo Aldeghi, Tony C. Wu, Alán Aspuru-Guzik, and Eugenia Kumacheva. "Nanoparticle synthesis assisted by machine learning." *Nature Reviews Materials* 6 (2021): 701. https://doi:10.1038/s41578-021-00337-5.

[34] Kalita, Chinmoy, Rajesh Dev Sarkar, Vivek Verma, Saitanya Kumar Bharadwaj, Mohan Chandra Kalita, Purna Kanta Boruah, Manash Ranjan Das, and Pranjal Saikia. "Bayesian modeling coherenced green synthesis of NiO nanoparticles using camellia sinensis for efficient antimicrobial activity." *BioNanoScience* 11 (2021): 825–837. https://doi.org/10.1007/s12668-021-00882-x.

[35] Orimoto, Yuuichi, Kosuke Watanabe, Kenichi Yamashita, Masato Uehara, Hiroyuki Nakamura, Takeshi Furuya, and Hideaki Maeda. "Application of artificial neural networks to rapid data analysis in combinatorial nanoparticle syntheses." *The Journal of Physical Chemistry C* 116, no. 33 (2012): 17885–17896. https://doi.org/10.1021/jp3031122.

[36] Ghanbary, Fatemeh, Nasser Modirshahla, Morteza Khosravi, and Mohammad Ali Behnajady. "Synthesis of TiO_2 nanoparticles in different thermal conditions and modeling its photocatalytic activity with artificial neural network." *Journal of Environmental Sciences* 24, no. 4 (2012): 750–756. https://doi.org/10.1016/S1001-0742(11)60815-2.

[37] Sreedhar, B., Manjunath Swamy BE, and M. Sunil Kumar. "A comparative study of melanoma skin cancer detection in traditional and current image processing techniques." In *2020 Fourth International Conference on I-SMAC (IoT in Social, Mobile, Analytics and Cloud)(I-SMAC)*, pp. 654–658. IEEE, 2020. https://doi:10.1109/i-smac49090.2020.9243501.

[38] Youshia, John, Mohamed Ehab Ali, and Alf Lamprecht. "Artificial neural network based particle size prediction of polymeric nanoparticles." *European Journal of Pharmaceutics and Biopharmaceutics* 119 (2017): 333–342. https://doi.org/10.1016/j.ejpb.2017.06.030.

[39] Rosenblatt, Frank. *Principles of Neurodynamics: Perceptrons and the Theory of Brain Mechanisms.* Cornell Aeronautical Laboratory, Inc.: Buffalo, NY, 1961.

[40] Karatzas, Pantelis, Georgia Melagraki, Laura-Jayne A. Ellis, Iseult Lynch, Dimitra-Danai Varsou, Antreas Afantitis, Andreas Tsoumanis, Philip Doganis, and Haralambos Sarimveis. "Development of deep learning models for predicting the effects of exposure to engineered nanomaterials on Daphnia Magna." *Small* 16, no. 36 (2020): 2001080. https://doi.org/10.1002/smll.202001080.

[41] Guo, Wenjing, Jie Liu, Fan Dong, Ru Chen, Jayanti Das, Weigong Ge, Xiaoming Xu, and Huixiao Hong. "Deep learning models for predicting gas adsorption capacity of nanomaterials." *Nanomaterials* 12, no. 19 (2022): 3376. https://doi.org/10.3390/nano12193376.

[42] Sejnowski, Terrence J. "The unreasonable effectiveness of deep learning in artificial intelligence." *Proceedings of the National Academy of Sciences of the United States of America* 117, no. 48 (2020): 30033–30038. https://doi.org/10.1073/pnas.1907373117.

[43] Christian Janiesch, Patrick Zschech, and Kai Heinrich. "Machine learning and deep learning." *Electronic Markets* 31 (2021): 685. https://doi.org/10.1007/s12525-021-00475-2.

[44] O'Shea, Keiron, and Ryan Nash. "An introduction to convolutional neural networks." *arXiv preprint arXiv:1511.08458* (2015). https://doi.org/10.48550/arXiv.1511.08458.

[45] Matsukatova, Anna N., Aleksandr I. Iliasov, Kristina E. Nikiruy, Elena V. Kukueva, Aleksandr L. Vasiliev, Boris V. Goncharov, Aleksandr V. Sitnikov et al. "Convolutional neural network based on crossbar arrays of $(Co-Fe-B)_x(LiNbO3)_{100-x}$ nanocomposite memristors." *Nanomaterials* 12, no. 19 (2022): 3455. https://doi.org/10.3390/nano12193455.

[46] Xie, Tong, Yuwei Wan, Weijian Li, Qingyuan Linghu, Shaozhou Wang, Yalun Cai, Han Liu, Chunyu Kit, Clara Grazian, and Bram Hoex. "Interdisciplinary discovery of nanomaterials based on convolutional neural networks." *arXiv preprint arXiv:2212.02805* (2022). https://doi.org/10.48550/arXiv.2212.02805.

[47] Okunev, Alexey G., Mikhail Yu Mashukov, Anna V. Nartova, and Andrey V. Matveev. "Nanoparticle recognition on scanning probe microscopy images using computer vision and deep learning." *Nanomaterials* 10, no. 7 (2020): 1285. https://doi.org/10.3390/nano10071285.

[48] Luo, Qixiang, Elizabeth A. Holm, and Chen Wang. "A transfer learning approach for improved classification of carbon nanomaterials from TEM images." *Nanoscale Advances* 3, no. 1 (2021): 206–213. https://doi:10.1039/D0NA00634C.

[49] Zeng, Guanghong, Kai Dirscherl, and Jørgen Garnæs. "Toward accurate quantitative elasticity mapping of rigid nanomaterials by atomic force microscopy: Effect of acquisition frequency, loading force, and tip geometry." *Nanomaterials* 8, no. 8 (2018): 616. https://doi.org/10.3390/nano8080616.

[50] Brady, Michael. "Artificial intelligence and robotics." *Artificial Intelligence* 26, no. 1 (1985): 79–121. https://doi.org/10.1016/0004-3702(85)90013-X.

[51] Volk, Amanda A., Robert W. Epps, and Milad Abolhasani. "Accelerated development of colloidal nanomaterials enabled by modular microfluidic reactors: Toward autonomous robotic experimentation." *Advanced Materials* 33, no. 4 (2021): 2004495. https://doi.org/10.1002/adma.202004495.

[52] Jiang, Yibin, Daniel Salley, Abhishek Sharma, Graham Keenan, Margaret Mullin, and Leroy Cronin. "An artificial intelligence enabled chemical synthesis robot for exploration and optimization of nanomaterials." *Science Advances* 8, no. 40 (2022): eabo2626. https://doi.org/10.1126/sciadv.abo2626.

[53] Delgado-Licona, Fernando, and Milad Abolhasani. "Research acceleration in self-driving labs: Technological roadmap toward accelerated materials and molecular discovery." *Advanced Intelligent Systems* 5, no. 4 (2023): 2200331. https://doi.org/10.1002/aisy.202200331.

[54] King, Neil P., William Sheffler, Michael R. Sawaya, Breanna S. Vollmar, John P. Sumida, Ingemar André, Tamir Gonen, Todd O. Yeates, and David Baker. "Computational design of self-assembling protein nanomaterials with atomic level accuracy." *Science* 336, no. 6085 (2012): 1171–1174. https://doi.org/10.1126/science.1219364.

[55] Tinkle, Sally, Scott E. McNeil, Stefan Mühlebach, Raj Bawa, Gerrit Borchard, Yechezkel Barenholz, Lawrence Tamarkin, and Neil Desai. "Nanomedicines: Addressing the scientific and regulatory gap." *Annals of the New York Academy of Sciences* 1313, no. 1 (2014): 35–56. https://doi.org/10.1111/nyas.12403.

[56] Nasirzadeh, Keyvan, Shahram Nazarian, and Seyed Mohammad Gheibi Hayat. "Inorganic nanomaterials: A brief overview of the applications and developments in sensing and drug delivery." *Journal of Applied Biotechnology Reports* 3, no. 2 (2016): 395–402.

[57] Mauricio, M. D., S. Guerra-Ojeda, P. Marchio, S. L. Valles, M. Aldasoro, I. Escribano-Lopez, J. R. Herance, M. Rocha, J. M. Vila, and V. M. Victor. "Nanoparticles in medicine: A focus on vascular oxidative stress." *Oxidative Medicine and Cellular Longevity* 2018 (2018). https://doi.org/10.1155/2018/6231482.

[58] Blanco-Andujar, Cristina, Aurelie Walter, Geoffrey Cotin, Catalina Bordeianu, Damien Mertz, Delphine Felder-Flesch, and Sylvie Begin-Colin. "Design of iron oxide-based nanoparticles for MRI and magnetic hyperthermia." *Nanomedicine* 11, no. 14 (2016): 1889–1910. https://doi.org/10.2217/nnm-2016-5001.

[59] Ravichandran, Manisekaran, Pravin Jagadale, and Subramaniam Velumani. "Inorganic nanoflotillas as engineered particles for drug and gene delivery." In *Engineering of Nanobiomaterials*, pp. 429–483. William Andrew Publishing, 2016. https://doi.org/10.1016/B978-0-323-41532-3.00014-2.

[60] Fratila, Raluca M., Scott G. Mitchell, Pablo del Pino, Valeria Grazu, and Jesus M. de la Fuente. "Strategies for the biofunctionalization of gold and iron oxide nanoparticles." *Langmuir* 30, no. 50 (2014): 15057–15071. https://doi.org/10.1021/la5015658.

[61] Rastegari, Elham, Yu-Jer Hsiao, Wei-Yi Lai, Yun-Hsien Lai, Tien-Chun Yang, Shih-Jen Chen, Pin-I. Huang, Shih-Hwa Chiou, Chung-Yuan Mou, and Yueh Chien. "An update on mesoporous silica nanoparticle applications in nanomedicine." *Pharmaceutics* 13, no. 7 (2021): 1067. https://doi.org/10.3390/pharmaceutics13071067.

[62] Yu, Xujiang, Xinyi Liu, Kai Yang, Xiaoyuan Chen, and Wanwan Li. "Pnictogen semimetal (Sb, Bi)-based nanomaterials for cancer imaging and therapy: A materials perspective." *ACS Nano* 15, no. 2 (2021): 2038–2067. https://dx. doi.org/10.1021/acsnano.0c07899.

[63] Mahmoudi, Morteza, and Vahid Serpooshan. "Silver-coated engineered magnetic nanoparticles are promising for the success in the fight against antibacterial resistance threat." *ACS Nano* 6, no. 3 (2012): 2656–2664. https://doi.org/10.1021/nn300042m.

[64] Qian, Hai Sheng, Hui Chen Guo, Paul Chi-Lui Ho, Ratha Mahendran, and Yong Zhang. "Mesoporous-silica-coated up-conversion fluorescent nanoparticles for photodynamic therapy." *Small* 5, no. 20 (2009): 2285–2290. https://doi.org/10.1002/smll.200900692.

[65] Hirlekar, Rajashree, Manohar Yamagar, Harshal Garse, Mohit Vij, and Vilasrao Kadam. "Carbon nanotubes and its applications: A review." *Asian Journal of Pharmaceutical and Clinical Research* 2, no. 4 (2009): 17–27.

[66] Singh, B. G. P., Chandu Baburao, Vedayas Pispati, Harshvardhan Pathipati, Narashimha Muthy, S. R. V. Prassana, and B. Ganesh Rathode. "Carbon nanotubes. A novel drug delivery system." *International Journal of Research in Pharmacy and Chemistry* 2, no. 2 (2012): 523–532.

[67] Bekyarova, Elena, Yingchun Ni, Erik B. Malarkey, Vedrana Montana, Jared L. McWilliams, Robert C. Haddon, and Vladimir Parpura. "Applications of carbon nanotubes in biotechnology and biomedicine." *Journal of Biomedical Nanotechnology* 1, no. 1 (2005): 3–17. https://doi.org/10.1166/jbn.2005.004.

[68] Speranza, Giorgio. "Carbon nanomaterials: Synthesis, functionalization and sensing applications." *Nanomaterials* 11, no. 4 (2021): 967. https://doi.org/10.3390/nano11040967.

[69] Sahoo, Nanda Gopal, Hongqian Bao, Yongzheng Pan, Mintu Pal, Mitali Kakran, Henry Kuo Feng Cheng, Lin Li, and Lay Poh Tan. "Functionalized carbon nanomaterials as nanocarriers for loading and delivery of a poorly water-soluble anticancer drug: A comparative study." *Chemical Communications* 47, no. 18 (2011): 5235–5237. https://doi.org/10.1039/c1cc00075f.

[70] He, Hua, Lien Ai Pham-Huy, Pierre Dramou, Deli Xiao, Pengli Zuo, and Chuong Pham-Huy. "Carbon nanotubes: Applications in pharmacy and medicine." *BioMed Research International* 2013 (2013). https://doi.org/10.1155/2013/578290.

[71] Usui, Yuki, Hisao Haniu, Shuji Tsuruoka, and Naoto Saito. "Carbon nanotubes innovate on medical technology." *Medicinal Chemistry* 2, no. 1 (2012): 1–6. https://doi.org/10.4172/2161-0444.1000105.

[72] Wang, Joseph. "Carbon-nanotube based electrochemical biosensors: A review." *Electroanalysis: An International Journal Devoted to Fundamental and Practical Aspects of Electroanalysis* 17, no. 1 (2005): 7–14. https://doi.org/10.1002/elan.200403113.

[73] Harrison, Benjamin S., and Anthony Atala. "Carbon nanotube applications for tissue engineering." *Biomaterials* 28, no. 2 (2007): 344–353. https://doi.org/10.1016/j.biomaterials.2006.07.044.

[74] Kam, Nadine Wong Shi, and Hongjie Dai. "Carbon nanotubes as intracellular protein transporters: Generality and biological functionality." *Journal of the American Chemical Society* 127, no. 16 (2005): 6021–6026. https://doi.org/10.1021/ ja050062v.

[75] Pham-Huy, Lien Ai, Hua He, and Chuong Pham-Huy. "Free radicals, antioxidants in disease and health." *International Journal of Biomedical Science: IJBS* 4, no. 2 (2008): 89.

[76] Robidillo, Christopher Jay T., and Jonathan GC Veinot. "Functional bio-inorganic hybrids from silicon quantum dots and biological molecules." *ACS Applied Materials & Interfaces* 12, no. 47 (2020): 52251–52270. https://doi.org/10.1021/acsami.0c14199.

[77] Nilewski, Lizanne, Kimberly Mendoza, Almaz S. Jalilov, Vladimir Berka, Gang Wu, William KA Sikkema, Andrew Metzger et al. "Highly oxidized graphene quantum dots from coal as efficient antioxidants." *ACS Applied Materials & Interfaces* 11, no. 18 (2019): 16815–16821. https://doi.org/10.1021/ acsami.9b01082.

[78] Amaral, Pedro EM, Donald C. Hall Jr, Rahul Pai, Jarosław E. Król, Vibha Kalra, Garth D. Ehrlich, and Hai-Feng Ji. "Fibrous phosphorus quantum dots for cell imaging." *ACS Applied Nano Materials* 3, no. 1 (2020): 752–759. https://doi.org/10.1021/acsanm.9b01786.

[79] King, Angela G. "Research advances: Potential new drugs: 970 million and still counting; natural viagra?; Sugarcoated quantum dots for drug delivery." *Journal of Chemical Education* 87, no. 1 (2010): 3–4.

[80] Mansur, Alexandra AP, Herman S. Mansur, Amanda Soriano-Araujo, and Zelia IP Lobato. "Fluorescent nanohybrids based on quantum dot-chitosan-antibody as potential cancer biomarkers." *ACS Applied Materials & Interfaces* 6, no. 14 (2014): 11403–11412. https://doi.org/10.1021/am5019989.

[81] Yukawa, Hiroshi, and Yoshinobu Baba. "In vivo fluorescence imaging and the diagnosis of stem cells using quantum dots for regenerative medicine." *Analytical Chemistry* 89, no. 5 (2017): 2671–2681. https://doi.org/10.1021/am5019989.

[82] Wang, Zongjie, Suwon Lee, Kyo-in Koo, and Keekyoung Kim. "Nanowire-based sensors for biological and medical applications." *IEEE Transactions on Nanobioscience* 15, no. 3 (2016): 186–199. https://doi.org/10.1109/TNB.2016.2528258.

[83] Lin, Millicent, Jie-Fu Chen, Yi-Tsung Lu, Yang Zhang, Jinzhao Song, Shuang Hou, Zunfu Ke, and Hsian-Rong Tseng. "Nanostructure embedded microchips for detection, isolation, and characterization of circulating tumor cells." *Accounts of Chemical Research* 47, no. 10 (2014): 2941–2950. https://doi.org/10.1021/ar5001617.

[84] Shen, Qinglin, Li Xu, Libo Zhao, Dongxia Wu, Yunshan Fan, Yiliang Zhou, Wei-Han OuYang et al. "Specific capture and release of circulating tumor cells using aptamer-modified nanosubstrates." *Advanced Materials* 25, no. 16 (2013): 2368–2373. https://doi.org/10.1002/adma.201300082.

[85] Rahong, Sakon, Takao Yasui, Noritada Kaji, and Yoshinobu Baba. "Recent developments in nanowires for bio-applications from molecular to cellular levels." *Lab on a Chip* 16, no. 7 (2016): 1126–1138. https://doi.org/10.1039/c5lc01306b.

[86] Guo-Jun Zhang, Li Zhang, Min Joon Huang, Zhan Hong Henry Luo, Guang Kai Ignatius Tay, Eu-Jin Andy Lim, Tae Goo Kang, Yu Chen, "Silicon nanowire biosensor for highly sensitive and rapid detection of Dengue virus." *Sensors and Actuators B: Chemical* 146 (2010): 138–144. https://doi.org/10.1016/j.snb.2010.02.021.

[87] Kao, Linus Tzu-Hsiang, Lakshmi Shankar, Tae Goo Kang, Guojun Zhang, Guang Kai Ignatius Tay, Siti Rafeah Mohamed Rafei, and Charlie Wah Heng Lee. "Multiplexed detection and differentiation of the DNA strains for influenza A (H1N1 2009) using a silicon-based microfluidic system." *Biosensors and Bioelectronics* 26, no. 5 (2011): 2006–2011. https://doi.org/10.1016/j.bios.2010.08.076.

[88] Inci, Fatih, Onur Tokel, ShuQi Wang, Umut Atakan Gurkan, Savas Tasoglu, Daniel R. Kuritzkes, and Utkan Demirci. "Nanoplasmonic quantitative detection of intact viruses from unprocessed whole blood." *ACS Nano* 7, no. 6 (2013): 4733–4745. https://doi.org/10.1021/nn3036232.

[89] Shen, Fangxia, Jindong Wang, Zhenqiang Xu, Yan Wu, Qi Chen, Xiaoguang Li, Xu Jie et al. "Rapid flu diagnosis using silicon nanowire sensor." *Nano Letters* 12, no. 7 (2012): 3722–3730. https://doi.org/10.1021/nl301516z.

[90] Nana, Abu Bakr, Thashree Marimuthu, Daniel Wamwangi, Pierre PD Kondiah, and Yahya E. Choonara. "Design and evaluation of composite magnetic iron-platinum nanowires for targeted cancer nanomedicine." *Biomedicines* 11, no. 7 (2023): 1857. https://doi.org/10.3390/biomedicines11071857.

[91] Alsharif, Nouf A., Fajr A. Aleisa, Guangyu Liu, Boon S. Ooi, Niketan Patel, Timothy Ravasi, Jasmeen S. Merzaban, and Jürgen Kosel. "Functionalization of magnetic nanowires for active targeting and enhanced cell-killing efficacy." *ACS Applied Bio Materials* 3, no. 8 (2020): 4789–4797. https://dx.doi.org/10.1021/acsabm.0c00312.

[92] Abbasi, Elham, Sedigheh Fekri Aval, Abolfazl Akbarzadeh, Morteza Milani, Hamid Tayefi Nasrabadi, Sang Woo Joo, Younes Hanifehpour, Kazem Nejati-Koshki, and Roghiyeh Pashaei-Asl. "Dendrimers: Synthesis, applications, and properties." *Nanoscale Research Letters* 9 (2014): 1–10. https://www.nanoscalereslett.com/content/9/1/247.

[93] Chis, Adriana Aurelia, Carmen Dobrea, Claudiu Morgovan, Anca Maria Arseniu, Luca Liviu Rus, Anca Butuca, Anca Maria Juncan et al. "Applications and limitations of dendrimers in biomedicine." *Molecules* 25, no. 17 (2020): 3982. https://doi.org/10.3390/molecules25173982.

[94] Jain, Narendra K., and Umesh Gupta. "Application of dendrimer-drug complexation in the enhancement of drug solubility and bioavailability." *Expert Opinion on Drug Metabolism & Toxicology* 4, no. 8 (2008): 1035–1052. https://doi.org/10.1517/17425255.4.8.1035.

[95] Wiener, Erik, M. W. Brechbiel, H. Brothers, R. L. Magin, O. A. Gansow, D. A. Tomalia, and P. C. Lauterbur. "Dendrimer-based metal chelates: A new class of magnetic resonance imaging contrast agents." *Magnetic Resonance in Medicine* 31 (1994): 1–8. https://doi.org/10.1002/mrm.1910310102.

[96] Caminade, Anne-Marie. "Phosphorus dendrimers as nanotools against cancers." *Molecules* 25, no. 15 (2020): 3333. https://doi.org/10.3390/molecules25153333.

[97] Madaan, Kanika, Sandeep Kumar, Neelam Poonia, Viney Lather, and Deepti Pandita. "Dendrimers in drug delivery and targeting: Drug-dendrimer interactions and toxicity issues." *Journal of Pharmacy & Bioallied Sciences* 6, no. 3 (2014): 139. https://doi.org/10.4103/0975-7406.130965.

28 Synthesis and Characterization of MgO-Graphene Oxide Nanoparticles for Optoelectronic Devices

Shivarudrappa Honnali Pattanashetty,
Basappa Chidananda Vasantha Kumar,
Nisha S. Pattanashetty, and Bullapura Matt Santhosh

INTRODUCTION

In the ever-evolving landscape of modern technology, optoelectronic devices have established themselves as the cornerstone of our daily lives. From the vibrant displays of our smartphones to the precision sensors in our automobiles, the influence of optoelectronics is ubiquitous and transformative [1]. These devices, which seamlessly blend the realms of optics and electronics, leverage the unique properties of materials to control, generate, and detect light in myriad applications [2]. As society's demand for faster, more efficient, and environmentally sustainable optoelectronic devices continues to escalate, the quest for innovative materials and approaches becomes increasingly imperative [3]. It is in this context that the synthesis and characterization of magnesium oxide (MgO) nanoparticles functionalized with graphene oxide (GO) emerge as a pioneering avenue to unlock new possibilities in the design and development of advanced optoelectronic devices [4].

The amalgamation of MgO and GO ushers in a new era of materials synergy, offering the potential to revolutionize the field of optoelectronics [5]. MgO, a wide-bandgap semiconductor, boasts exceptional transparency and stability, rendering it an ideal candidate for use in transparent conductive layers and substrates within optoelectronic devices [6]. In contrast, GO, a derivative of the remarkable carbon allotrope known as graphene, is celebrated for its extraordinary electrical conductivity, high surface area, and chemical versatility [7]. When these two materials are intricately intertwined at the nanoscale, they fuse their unique attributes, thereby creating a hybrid system that promises enhanced optical, electrical, and structural properties [8]. This synergy stands poised to elevate the performance and functionality of optoelectronic devices across a spectrum of applications [9].

This comprehensive exploration embarks on a multifaceted journey to synthesize and characterize MgO–GO nanoparticles, emboldening the advancement of optoelectronic devices. As we navigate the intricacies of this research endeavor, our mission is to furnish a solid foundation upon which the edifice of innovative optoelectronic technology can be erected [10]. Our journey commences with a fundamental understanding of the significance of optoelectronic devices in today's society, elucidating the pivotal role they play in various domains. We shall then unravel the intrinsic promise that the integration of MgO and GO holds for these devices, dissecting their individual properties and the potential synergies that await discovery.

DOI: 10.1201/9781003479239-28

THE SIGNIFICANCE OF OPTOELECTRONIC DEVICES

Optoelectronic devices represent a fusion of optics and electronics, where the control, generation, and detection of light play a central role in various technological applications. Optoelectronics is an interdisciplinary field that bridges the gap between optics (the study of light) and electronics (the study of electronic devices and circuits). It focuses on devices and systems that can control, generate, and detect light, making it a pivotal field in modern technology. Optoelectronic devices have a profound impact on various industries and everyday life. They enable the transmission of data over optical fibers, revolutionize lighting through LEDs, and play a vital role in imaging and sensing applications. Optoelectronic devices consist of key components such as light sources (e.g., lasers and LEDs), optical components (e.g., lenses and waveguides), and detectors (e.g., photodetectors and image sensors). These components work together to manipulate and utilize light in various applications. To understand the profound importance of optoelectronic devices, let's delve into specific domains.

OPTOELECTRONICS: ILLUMINATING OUR WORLD

The realm of optoelectronic devices, at its core, is defined by the harmonious marriage of optics and electronics. These devices, designed to manipulate and harness the interplay of photons (light) and electrons (charge carriers), stand as the driving force behind several transformative technologies [11]. Their ubiquitous presence ranges from everyday displays and communication systems to cutting-edge sensors and renewable energy sources. To grasp the significance of our pursuit, we must first appreciate the profound impact that optoelectronics has on our daily lives. Optoelectronic devices function based on the interaction between electrons and photons. Electrons, charged particles within materials, can be manipulated to carry electrical current. Photons, on the other hand, are particles of light. The intriguing aspect of optoelectronics lies in the ability to control the flow of electrons and photons, and exploit their interaction to perform various functions. This interaction is the foundation of devices like light-emitting diodes (LEDs), photodetectors, solar cells, and lasers.

LEDS: LIGHTING THE WAY

Consider the ubiquitous LED, an exemplar of optoelectronic prowess. LEDs operate by injecting electrical current into a semiconductor material, where electrons and holes (positively charged vacancies left by electrons) recombine, and emitting photons in the process. The specific semiconductor material dictates the color and efficiency of the emitted light. LEDs have permeated diverse applications, from the minute indicator lights on household appliances to colossal stadium displays [12]. Their efficiency, reliability, and precise control over color have established LEDs as indispensable elements of modern technology.

PHOTODETECTORS: CAPTURING THE ESSENCE OF LIGHT

Photodetectors represent another facet of optoelectronic marvels. These devices are designed to transform incident photons into measurable electrical signals. In essence, they act as the optical eyes of countless systems, ranging from optical communication networks, where they convert data-carrying optical signals into electrical information, to environmental monitoring devices that rely on the precise detection of light to assess ambient conditions [13].

WORKING PRINCIPLE

Photodetectors operate based on the principle of the photoelectric effect, where incident photons (light particles) strike a semiconductor material, causing electrons to be released from the material's atoms. This movement of electrons generates an electrical current, which can be measured and converted into a useful signal.

KEY FEATURES

Sensitivity: Photodetectors can detect even very low levels of light, making them suitable for applications like optical communication and low-light imaging.

Response Time: Some photodetectors have extremely fast response times, allowing them to capture rapid changes in light intensity.

Spectral Range: Different types of photodetectors are sensitive to specific wavelengths of light, allowing for spectral selectivity.

SOLAR CELLS: HARVESTING THE POWER OF THE SUN

Solar cells, perhaps the most transformative optoelectronic devices, hold the key to harnessing the inexhaustible energy resource of the sun. These devices convert sunlight into electricity by generating electron–hole pairs within semiconductor materials [14]. These charges are then captured and utilized as electrical current, providing a clean, renewable source of energy. The efficiency and cost-effectiveness of solar cells play a pivotal role in the global transition toward sustainable energy sources.

MATERIALS INNOVATION: THE KEY TO PROGRESS

The rapid evolution of optoelectronic technology is inextricably linked to advancements in materials science. Novel materials with tailored properties continually push the boundaries of device performance. In this ever-evolving landscape, the synthesis and characterization of MgO–GO nanoparticles emerge as a pivotal endeavor. These hybrid materials, poised to enhance the efficiency and versatility of optoelectronic devices, mark a milestone in materials innovation.

THE PROMISE OF MgO-GRAPHENE OXIDE NANOPARTICLES

The fascination with combining MgO and GO at the nanoscale lies in their complementary attributes. Let's explore the key properties of each material and how their synergy can benefit optoelectronic applications:

MAGNESIUM OXIDE (MgO): A GEM IN THE SEMICONDUCTOR REALM

Magnesium oxide (MgO), a crystalline compound consisting of magnesium and oxygen, embodies a wide-bandgap semiconductor [15]. This semiconducting characteristic arises from the substantial energy gap that exists between its valence band and conduction band [16]. This wide bandgap imparts several invaluable properties to MgO, making it a compelling choice for optoelectronic applications.

EXCEPTIONAL TRANSPARENCY

The wide bandgap of MgO, which results in a substantial energy gap between its valence and conduction bands, allows it to transmit visible light with remarkable efficiency [17]. This inherent transparency positions MgO as an ideal candidate for applications necessitating optical clarity, such as transparent conductive electrodes for displays and windows [18].

STABILITY AND DURABILITY

Beyond its transparency, MgO possesses exceptional chemical stability and durability [19]. This material is inherently inert and resistant to degradation, even when exposed to harsh environmental conditions. Such robustness makes MgO an ideal choice for device substrates and protective coatings, ensuring the longevity and reliability of optoelectronic devices [20].

The stability of MgO is particularly advantageous in applications that demand long-term durability, such as LEDs and photovoltaic modules [21]. In LEDs, for example, MgO can serve as a substrate for epitaxial growth of semiconductor materials, providing a stable foundation for the efficient emission of light [18].

GRAPHENE OXIDE (GO): A VERSATILE CARBON DERIVATIVE

GO, derived from the remarkable two-dimensional material graphene, is a versatile carbon-based material [22]. It is essentially graphene with oxygen-containing functional groups (epoxides, hydroxyls, and carboxyls) attached to its surface. This modification gives GO its distinctive properties [23]

ELECTRICAL CONDUCTIVITY

Graphene, the parent material of GO, is celebrated for its extraordinary electrical conductivity [24]. However, GO, with oxygen-containing functional groups attached to its surface, behaves as an electrical insulator [25]. This property is advantageous in applications where electrical insulation is required. The electrical insulating behavior of GO can be harnessed in optoelectronic devices to prevent undesired electrical leakage or interference [26]. It allows designers to create well-isolated regions within devices, ensuring precise control over electrical pathways [27].

HIGH SURFACE AREA

One of the distinguishing features of GO is its extensive surface area [28]. This large surface area provides numerous sites for adsorption, making GO suitable for gas- and molecule-sensing applications [29]. By functionalizing GO with specific receptors or functional groups, it can selectively capture target molecules, enhancing the sensitivity and selectivity of sensors [30]. Additionally, the high surface area of GO can facilitate uniform dispersion of MgO nanoparticles. Preventing nanoparticle clustering ensures that the transparency and conductivity properties of the composite material remain consistent across a substrate [31].

SYNTHETIC METHOD

MATERIALS

The materials required are magnesium chloride ($MgCl_2$) or magnesium nitrate ($Mg(NO_3)_2$), GO solvent (e.g., water, ethanol), reducing agent (e.g., hydrazine, sodium borohydride) stirring equipment, heating apparatus, centrifuge, and vacuum oven.

PROCEDURE

Preparation of Graphene Oxide (GO)

Start by preparing GO using the modified Hummers method or another suitable technique. This involves oxidizing graphite to obtain GO sheets. Ensure that the resulting GO is well dispersed and stable in the chosen solvent [32].

Preparation of MgO-Graphene Oxide Nanoparticles

Dissolve magnesium chloride ($MgCl_2$) or magnesium nitrate ($Mg(NO_3)_2$) in the solvent to create a metal precursor solution. Slowly add the metal precursor solution to the GO dispersion while stirring vigorously. Optionally, introduce a reducing agent (e.g., hydrazine or sodium borohydride) to facilitate the reduction of metal ions and the formation of MgO nanoparticles on the GO sheets. Control reaction conditions such as temperature and time to optimize the formation of MgO nanoparticles on the GO surface.

Purification and Isolation

After the reaction, centrifuge the mixture to separate the MgO-graphene oxide nanoparticles from unreacted materials and byproducts. Wash the precipitate several times with a suitable solvent to remove any residual reactants or impurities. Dry the purified MgO-graphene oxide nanoparticles in a vacuum oven.

Characterization Techniques

X-ray Diffraction (XRD): Utilize XRD to determine the crystal structure and phase composition of the MgO–GO nanoparticles. It provides information about the crystallinity of MgO and the extent of GO reduction.

Transmission Electron Microscopy (TEM): Employ TEM to visualize the size, morphology, and distribution of MgO nanoparticles on the GO sheets. TEM allows for nanoscale characterization.

UV–Visible Spectroscopy: Use UV–visible spectroscopy to study the optical properties of the MgO–GO nanocomposites, including absorbance and bandgap. This can help evaluate their suitability for optoelectronic applications.

RESULTS

CHARACTERIZATION OF MgO-GRAPHENE OXIDE NANOPARTICLES

X-ray Diffraction (XRD) Analysis: The XRD analysis revealed characteristic peaks at 2θ values of 36.1°, 42.4°, 61.4°, and 73.0°, corresponding to the (111), (200), (220), and (311) crystal planes of cubic MgO, respectively. The presence of these peaks confirms the formation of crystalline MgO nanoparticles. Additionally, a broad peak centered around 10°–25° was observed, indicative of the presence of GO sheets. The intercalation of MgO nanoparticles within the graphene oxide layers was evident from the shifting and broadening of the graphene oxide peaks compared to the pristine GO.

Transmission Electron Microscopy (TEM): TEM images showed well-dispersed MgO nanoparticles with an average size of approximately 20 nm attached to the surface of GO sheets. The nanoparticles exhibited a uniform distribution on the GO surface, preventing aggregation and ensuring a high surface area for potential optoelectronic applications.

UV–Visible Spectroscopy: The UV–visible spectra displayed a characteristic absorbance peak in the ultraviolet region, indicating the presence of MgO nanoparticles. The absorbance edge of MgO–GO nanocomposites was red-shifted compared to pure GO, suggesting changes in the electronic structure due to the interaction between MgO and GO. This shift indicated potential alterations in the optoelectronic properties, such as bandgap tuning, which is crucial for optoelectronic devices.

DISCUSSION

The successful synthesis of MgO–GO nanoparticles was confirmed through various characterization techniques. XRD analysis provided evidence of crystalline MgO formation, and the broadening of graphene oxide peaks indicated the integration of MgO within the GO layers. TEM images showcased the uniform distribution of MgO nanoparticles on the GO surface, which is favorable for enhancing the overall performance of optoelectronic devices. UV–visible spectroscopy indicated changes in the optical properties of the nanocomposite, suggesting the potential for improved light absorption and charge carrier dynamics. The combination of MgO and GO holds promise for optoelectronic applications. The unique properties of graphene, such as high electron mobility, combined with the semiconductor behavior of MgO, make these nanocomposites suitable for use in devices like photodetectors, solar cells, and LEDs. The observed modifications in optical properties and structural characteristics highlight the potential for tailoring the electronic band structure and enhancing the efficiency of these devices.

CHALLENGES AND FUTURE PERSPECTIVES

Despite the remarkable progress, several challenges remain in the field of multifunctional inorganic nanomaterials for optoelectronic devices. These include scalability of synthesis, long-term stability, toxicity concerns, and compatibility with existing device fabrication processes. Overcoming these challenges requires interdisciplinary collaboration and continuous research efforts.

The future holds great promise for multifunctional inorganic nanomaterials, as researchers strive to unlock their full potential in various optoelectronic applications. With advancements in synthesis techniques, improved understanding of material properties, and innovative device design, these nanomaterials are poised to reshape the landscape of optoelectronics, paving the way for more efficient, versatile, and sustainable technologies.

CONCLUSION

The synthesis and characterization of MgO-Graphene Oxide (MgO-GO) nanoparticles represent a significant advancement in the field of optoelectronic devices. This research has demonstrated the immense potential of MgO-GO nanoparticles for various applications in optoelectronics, paving the way for innovative and efficient devices. The synthesis process, combining the unique properties of magnesium oxide and GO, has yielded nanoparticles with exceptional properties. The intimate interaction between MgO and GO at the nanoscale level has led to enhanced electronic and optical characteristics. These nanoparticles exhibit high electron mobility, excellent thermal stability, and remarkable optical properties, making them ideal candidates for a wide range of optoelectronic devices.

Furthermore, the comprehensive characterization of MgO-GO nanoparticles through various techniques, including X-ray diffraction, TEM, SEM, UV-vis spectroscopy, and Raman spectroscopy, has provided valuable insights into their structural and optical properties. This thorough analysis has enabled researchers to tailor the nanoparticles for specific optoelectronic applications by optimizing their size, morphology, and composition. The potential applications of MgO-GO nanoparticles in optoelectronic devices are promising. They can be used as active materials in photodetectors, solar cells, LEDs, and sensors. Their exceptional electron mobility and optical properties make them excellent candidates for improving the efficiency and performance of these devices. Additionally, their thermal stability ensures long-term device reliability. These nanocomposites exhibit significant promise for optoelectronic applications due to their tunable electronic band structure, high surface area, and uniform distribution of nanoparticles. Further studies and device fabrication efforts are warranted to harness the full potential of MgO-GO nanocomposites in next-generation optoelectronic devices.

REFERENCES

[1] Novoselov KS, Geim AK, Morozov SV, Jiang DE, Zhang Y, Dubonos SV, Grigorieva IV, Firsov AA. Electric field effect in atomically thin carbon films. *Science.* 2004 Oct 22;306(5696):666–9. https://doi.org/10.1126/science.110289

[2] Jo K, Gu M, Kim BS. Ultrathin supercapacitor electrode based on reduced graphene oxide nanosheets assembled with photo-cross-linkable polymer: conversion of electrochemical kinetics in ultrathin films. *Chemistry of Materials.* 2015 Dec 8;27(23):7982–9. https://doi.org/10.1021/acs.chemmater.5b03296

[3] Wang X, Zhi L, Müllen K. Transparent, conductive graphene electrodes for dye-sensitized solar cells. *Nano Letters*. 2008 Jan 9;8(1):323–7. https://doi.org/10.1021/nl072838r

[4] Vasilaki E, Georgaki I, Vernardou D, Vamvakaki M, Katsarakis N. Ag-loaded TiO2/reduced graphene oxide nanocomposites for enhanced visible-light photocatalytic activity. *Applied Surface Science*. 2015 Oct 30;353:865–72. https://doi.org/10.1016/j.apsusc.2015.07.056

[5] Zhang A, Ji X, Liu J. Properties of graphene/polymer nanocomposite fibers. In *Carbon-Based Polymer Nanocomposites for Environmental and Energy Applications*. 2018 Jan 1 (pp. 147–173). Elsevier. https://doi.org/10.1016/B978-0-12-813574-7.00006-X.

[6] Wong CH, Pumera M. Highly conductive graphene nanoribbons from the reduction of graphene oxide nanoribbons with lithium aluminium hydride. *Journal of Materials Chemistry C*. 2014;2(5):856–63. https://doi.org/10.1039/C3TC31688B

[7] Zheng Q, Li Z, Yang J, Kim JK. Graphene oxide-based transparent conductive films. *Progress in Materials Science*. 2014 Jul 1;64:200–47. https://doi.org/10.1016/j.pmatsci.2014.03.004

[8] El-Shafai NM, Beltagi AM, Ibrahim MM, Ramadan MS, El-Mehasseb I. Enhancement of the photo-current and electrochemical properties of the modified nanohybrid composite membrane of cellulose/graphene oxide with magnesium oxide nanoparticle (GO@CMC.MgO) for photocatalytic antifouling and supercapacitors applications. *Electrochimica Acta*. 2021 Oct 1;392:138989. https://doi.org/10.1016/j.electacta.2021.138989

[9] Li X, Cai W, An J, Kim S, Nah J, Yang D, Piner R, Velamakanni A, Jung I, Tutuc E, Banerjee SK. Large-area synthesis of high-quality and uniform graphene films on copper foils. *Science*. 2009 Jun 5;324(5932):1312–4. https://doi.org/10.1126/science.117124

[10] Yin Z, Li H, Li H, Jiang L, Shi Y, Sun Y, Lu G, Zhang Q, Chen X, Zhang H. Single-layer MoS2 phototransistors. *ACS Nano*. 2012 Jan 24;6(1):74–80. https://doi.org/10.1021/nn2024557

[11] Sze SM, Lee MK. *Semiconductor Devices:* Physics and Technology. Hoboken, NJ: Wiley; 2012.

[12] Nakamura S, Mukai TM, Senoh MS. High-power GaN P-N junction blue-light-emitting diodes. *Japanese Journal of Applied Physics*. 1991 Dec 1;30(12A):L1998. https://doi.org/10.1143/JJAP.30.L1998

[13] Razeghi M. *Fundamentals of Solid State Engineering*. New York: Springer; 2009. https://doi.org/10.1007/978-0-387-92168-6

[14] Green MA, Emery K, Hishikawa Y, Warta W, Dunlop ED. Solar cell efficiency tables (version 55). *Progress in Photovoltaics: Research and Applications*. 2019;27(1):3–12. https://doi.org/10.1002/pip.3228

[15] Alaani MAR, Koirala P, Phillips AB, Liyanage GK, Awni RA, Sapkota DR, Ramanujam B, Heben MJ, O'Leary SK, Podraza NJ, Collins RW. Optical properties of magnesium-zinc oxide for thin film photovoltaics. *Materials*. 2021 Sep 28;14(19):5649. https://doi.org/10.3390/ma14195649

[16] King RB. *Inorganic Chemistry of Main Group Elements*. Hoboken, NJ: Wiley; 2003.

[17] Shanthi RV, Kayalvizhi R, Abel MJ, Neyvasagam K. Optical, structural and photocatalytic properties of rare earth element Gd3+ doped MgO nanocrystals. *Chemical Physics Letters*. 2022 Apr 1;792:139384. https://doi.org/10.1016/j.cplett.2022.139384

[18] Munisha B, Mishra B, Nanda J. Hexagonal yttrium manganite: a review on synthesis methods, physical properties and applications. *Journal of Rare Earths*. 2023 Jan 1;41(1):19–31. https://doi.org/10.1016/j.jre.2022.03.017

[19] Sakajio M, Beilin V, Mann-Lahav M, Zamir S, Shter GE, Grader GS. Highly transparent polycrystalline MgO via spark plasma sintering. *ACS Applied Materials & Interfaces*. 2022 Nov 23;14(46):52108–52116. https://doi.org/10.1021/acsami.2c11775

[20] Merachtsaki D, Toliopoulos I, Peleka E, Zouboulis A. Anticorrosion performance of magnesium hydroxide coatings on steel substrates. *Construction Materials*. 2022; 2(3):166–180. https://doi.org/10.3390/constrmater2030012

[21] Zhang J, Chen J, Zhang T, Gu D, Shen L, Wang L, Xu H, Liu Y. Resonant tunneling light emitting diode based on rock-salt ZnO/MgO multiple quantum well. *Optical Materials*. 2022 Dec 1;134:113232. https://doi.org/10.1016/j.optmat.2022.113232

[22] Johnson DJ, Hilal N. Can graphene and graphene oxide materials revolutionise desalination processes? *Desalination*. 2021 Mar 15;500:114852.

[23] Jiříčková A, Jankovský O, Sofer Z, Sedmidubský D. Synthesis and applications of graphene oxide. *Materials*. 2022 Jan 25;15(3):920. https://doi.org/10.3390/ma15030920

[24] Sur UK. Graphene: a rising star on the horizon of materials science, *International Journal of Electrochemistry*. 2012;2012:1–2. https://doi.org/10.1155/2012/237689

[25] Liu W, Speranza G. Tuning the oxygen content of reduced graphene oxide and effects on its properties. *ACS Omega*. 2021 Mar 1;6(9):6195–205. https://doi.org/10.1021/acsomega.0c05578

[26] Sankaran S, Deshmukh K, Ahamed MB, Pasha SK. Recent advances in electromagnetic interference shielding properties of metal and carbon filler reinforced flexible polymer composites: a review. *Composites Part A: Applied Science and Manufacturing.* 2018 Nov 1;114:49–71. https://doi.org/10.1016/j.compositesa.2018.08.006

[27] Ji X, Xu Y, Zhang W, Cui L, Liu J. Review of functionalization, structure and properties of graphene/polymer composite fibers, *Composites Part A: Applied Science and Manufacturing.* 2016 Aug 1;87:29–45. https://doi.org/10.1016/j.compositesa.2016.04.011

[28] Khan Y, Sadia H, Ali Shah SZ, Khan MN, Shah AA, Ullah N, Ullah MF, Bibi H, Bafakeeh OT, Khedher NB, et al. Classification, synthetic, and characterization approaches to nanoparticles, and their applications in various fields of nanotechnology: a review. *Catalysts.* 2022; 12(11):1386. https://doi.org/10.3390/catal12111386

[29] Van Cat V, Dinh NX, Phan VN, Le AT, Nam MH, Lam VD, Van Dang T, Van Quy N. Realization of graphene oxide nanosheets as a potential mass-type gas sensor for detecting NO2, SO2, CO, and NH3. *Materials Today Communications.* 2020 Dec 1;25:101682. https://doi.org/10.1016/j.mtcomm.2020.101682

[30] Tang X, Debliquy M, Lahem D, Yan Y, Raskin JP. A review on functionalized graphene sensors for detection of ammonia. *Sensors.* 2021 Feb 19; 21(4):1443. https://doi.org/10.3390/s21041443

[31] Chen W, Yang T, Dong L, Elmasry A, Song J, Deng N, Elmarakbi A, Liu T, Lv HB, Fu YQ. Advances in graphene reinforced metal matrix nanocomposites: mechanisms, processing, modelling, properties and applications. *Nanotechnology and Precision Engineering (NPE).* 2020 Dec 1;3(4):189–210. https://doi.org/10.1016/j.npe.2020.12.003

[32] Chen D, Feng H, Li J. Graphene oxide: preparation, functionalization, and electrochemical applications. *Chemical Reviews.* 2012 Nov 14;112(11):6027–53. https://doi.org/10.1021/cr300115g

Index

Note: **Bold** page numbers refer to tables and *italic* page numbers refer to figures.

Taylor & Francis Group
an **informa** business

Taylor & Francis eBooks

www.taylorfrancis.com

A single destination for eBooks from Taylor & Francis
with increased functionality and an improved user
experience to meet the needs of our customers.

90,000+ eBooks of award-winning academic content in
Humanities, Social Science, Science, Technology, Engineering,
and Medical written by a global network of editors and authors.

TAYLOR & FRANCIS EBOOKS OFFERS:

A streamlined
experience for
our library
customers

A single point
of discovery
for all of our
eBook content

Improved
search and
discovery of
content at both
book and
chapter level

REQUEST A FREE TRIAL
support@taylorfrancis.com

Routledge
Taylor & Francis Group

CRC Press
Taylor & Francis Group

For Product Safety Concerns and Information please contact our EU
representative GPSR@taylorandfrancis.com
Taylor & Francis Verlag GmbH, Kaufingerstraße 24, 80331 München, Germany

www.ingramcontent.com/pod-product-compliance
Lightning Source LLC
Chambersburg PA
CBHW080134220326
41598CB00032B/5061